Pollution
Causes, Effects and Control
5th Edition

Edited by

R M Harrison
University of Birmingham, UK
Email: r.m.harrison@bham.ac.uk

RSCPublishing

ISBN: 978-1-84973-648-0

A catalogue record for this book is available from the British Library

© The Royal Society of Chemistry 2014*

Published by The Royal Society of Chemistry,
Thomas Graham House, Science Park, Milton Road,
Cambridge CB4 0WF, UK

Registered Charity Number 207890

Visit our website at www.rsc.org/books

*Please note that, where indicated, copyright is shared.

Preface

The subject of pollution remains high in the public consciousness and has been a significant factor on the political agenda of both developed and developing countries for a number of years. The subject is now seen as a priority area for research and for technological developments. It is therefore a fast-moving field and one where books require updating on a rather frequent basis. The First Edition of this book was published in 1983 and arose from the collation of course notes from a Residential School held at Lancaster University in 1982, supplemented with additional chapters to give a fuller overview of the field. Subsequent editions have expanded the coverage so as to provide a fairly full overview of the field of chemical and radioactive pollution. The level of treatment remains much the same, being essentially introductory, although covering some more advanced aspects. The very high sales achieved by the book suggest that this has been very popular and *Pollution* is used both as a teaching text and a reference book by practitioners requiring broad knowledge of the field.

In a fast-moving field it is necessary to scrutinise contents carefully and to ensure thorough updating. The Fourth Edition of *Pollution* contained one wholly new chapter on *Clean Technologies and Industrial Ecology* reflecting the growing importance of pollution prevention as opposed to end-of-pipe controls. Whilst authorship of the majority of the other chapters remained in the same hands, a large proportion of the chapters were thoroughly revised to reflect new developments in the field, and extended to improve coverage. In the Fifth Edition, overall coverage is similar to the Fourth Edition, but greatly updated to include major new developments such as nanomaterials, as well as new scientific insights and legislative changes. A wholly new chapter on *Climate Change* is included, reflecting the high societal importance of this issue and the need for an authoritative view on the science.

Once again, the chapter authors have been selected on the basis of their established reputation in the field and their ability to write with clarity of presentation. I am delighted that a high proportion of those who wrote for the Fourth Edition have updated their contributions for the Fifth Edition. A number of those in fast-moving areas have completely re-written their contributions. Inevitably, given the length of time between the editions, some authors have changed, and it is a pleasure to welcome distinguished newcomers to the team.

Comparing the Fifth with the First Edition of this book, I am struck by the explosion in knowledge in this vital area. Environmental pollution is now a very major area of research, consultancy and technological development, and I hope that this book goes some way towards providing an authoritative knowledge base for those working within the field.

Roy M. Harrison
Birmingham

Pollution: Causes, Effects and Control, 5th Edition
Edited by R M Harrison
© The Royal Society of Chemistry 2014
Published by the Royal Society of Chemistry, www.rsc.org

Contents

Pollution: Causes, Effects and Control, 5th Edition
Edited by R M Harrison
© The Royal Society of Chemistry 2014
Published by the Royal Society of Chemistry, www.rsc.org

Chapter 3   Drinking Water Quality and Health   60
John K. Fawell

Chapter 4   Water Pollution Biology   80
William M. Mayes

List of Contributors

Mohamed Abou-Elwafa Abdallah, *Department of Analytical Chemistry, Faculty of Pharmacy, Assiut University, 71526 Assiut, Egypt*

Mike Ashmore, *Stockholm Environment Institute, University of York, York YO10 5DD, UK*

Jon Ayres, *Institute of Occupational Health, School of Health and Population Sciences, University of Birmingham, Edgbaston, Birmingham B15 2TT, UK*

Martin G. Bigg, *Professor of Environmental Technology and Director, Environmental Technologies iNet, University of the West of England, Frenchay Campus, Coldharbour Lane, Bristol BS16 1QY, UK*

Elise Cartmell, *Cranfield Water Science Institute, Cranfield University, Cranfield, Bedfordshire, MK43 0AL, UK*

Roland Clift, *Centre for Environmental Strategy, University of Surrey, Guildford, Surrey GU2 7XH, UK*

Chris D. Collins, *Soil Research Centre, Department of Geography and Environmental Science, University of Reading, Reading, RG6 6DW, UK*

Juana Maria Delgado-Saborit, *Division of Environmental Health and Risk Management, School of Geography, Earth and Environmental Sciences, University of Birmingham, Edgbaston, Birmingham B15 2TT, UK*

Gev Eduljee, *SITA UK, SITA House, Grenfell Road, Maidenhead SL6 1ES, UK*

John K. Fawell, *(visiting professor), Cranfield Water Science Institute, Cranfield University, Cranfield, Bedfordshire, MK43 0AL, UK*

Rachel L. Gomes, *Department of Chemical and Environmental Engineering, Faculty of Engineering, University of Nottingham, University Park, Nottingham, NG7 2RD, UK*

Stuart Harrad, *Division of Environmental Health & Risk Management, University of Birmingham, Edgbaston, Birmingham B15 2TT, UK*

Roy M. Harrison, *School of Geography, Earth and Environmental Sciences, University of Birmingham, Edgbaston, Birmingham, B15 2TT, UK*

C. Nicholas Hewitt, *Lancaster Environment Centre, Lancaster University, Lancaster LA1 4YQ, UK*

Claire Holman, *Brook Cottage Consultants, Elberton, South Gloucestershire, BS35 4AQ, UK*

William Howarth, *Kent Law School, University of Kent, Canterbury, Kent, CT2 7NS, UK*

Oliver A. H. Jones, *School of Applied Sciences, RMIT University, GPO Box 2476, Melbourne, Victoria 3001, Australia*

John N. Lester, *Cranfield Water Science Institute, Cranfield University, Cranfield, Bedfordshire, MK43 0AL, UK*

Pollution: Causes, Effects and Control, 5th Edition
Edited by R M Harrison
© The Royal Society of Chemistry 2014
Published by the Royal Society of Chemistry, www.rsc.org

A. Robert MacKenzie, *School of Geography, Earth & Environmental Sciences, Division of Environmental Health and Risk Management, University of Birmingham, Edgbaston, Birmingham B15 2TT, UK*

Richard Macrory, *Centre for Law and the Environment, Faculty of Laws, University College London, Endsleigh Gardens, London WC1H OEG*

William M. Mayes, *Centre for Environmental and Marine Sciences, University of Hull, Scarborough, YO11 3AZ, UK*

Robert L Maynard, *School of Geography, Earth & Environmental Sciences, University of Birmingham, Edgbaston, Birmingham B15 2TT, UK*

Simon J. T. Pollard, *Department of Environmental Science and Technology, School of Applied Sciences, Cranfield University, MK43 0AL, UK*

Francis D. Pope, *School of Geography, Earth & Environmental Sciences, Division of Environmental Health and Risk Management, University of Birmingham, Edgbaston, Birmingham B15 2TT, UK*

Martin R. Preston, *School of Environmental Sciences, University of Liverpool, 4 Brownlow Street, Liverpool L69 3GP, UK*

Keith P. Shine, *Department of Meteorology, University of Reading, Earley Gate, Reading RG6 6BB, UK*

Stuart T. Wagland, *Department of Environmental Science and Technology, School of Applied Sciences, Cranfield University, MK43 0AL, UK*

Martin L. Williams, *Science Policy, Environmental Research Group, King's College London, Franklin-Wilkins Building, 150 Stamford Street, London, SE1 9NH, UK*

CHAPTER 1

Chemical Pollution of the Aquatic Environment by Priority Pollutants and its Control[†]

OLIVER A.H. JONES*[a] AND RACHEL L. GOMES[b]

[a] School of Applied Sciences, RMIT University, GPO Box 2476, Melbourne, Victoria 3001, Australia; [b] Department of Chemical and Environmental Engineering, Faculty of Engineering, University of Nottingham, University Park, Nottingham, NG7 2RD, UK
*Email: oliver.jones@rmit.edu.au

1.1 INTRODUCTION

It is difficult to imagine the modern 21st century lifestyle without the mobile phones, tablet PCs and social media that the majority of the general public have become accustomed to. Such technology is heavily reliant on chemicals and chemical technology. For example, solvents are widely used in electronics as solders and for cleansing, stripping, and degreasing operations and encapsulations. Solvents are also the cause of a significant portion of workplace hazards and exposure problems in not only the electronics industry but many others as well, for example agriculture, biotechnology and pharmaceuticals. The chemical industry is an important pillar of the modern world economy and the chemical industry affects nearly every part of our daily life.

Biological and physico-chemical processes operating in aquatic systems can remove pollutants from circulation, fix them more or less indefinitely, or degrade them to less harmful compounds. The self-purification capacity of many aquatic systems has led to their use for the indiscriminate disposal of society's waste in the past. While the pollutants themselves are often invisible to the naked eye, their impact on water resources and aquatic life is often quite conspicuous. Pollutant discharges may cause fish kill events, noxious smells or even change the colour of the water, all of which are easily perceptible to the casual observer. However, there are also many chemical pollutants that may cause harm to the health of a watercourse while not affecting its outward appearance. It can be both difficult and expensive to remediate water pollution, and the future use of the water may be affected by the presence of chemicals.

[†]This chapter is based upon an earlier contribution by B. Crathorne, Y. J. Rees and S. France which is gratefully acknowledged.

Pollution: Causes, Effects and Control, 5th Edition
Edited by R M Harrison
Published by the Royal Society of Chemistry, www.rsc.org

Awareness of the issues involved with the presence of chemicals in the environment has been high since Rachel Carson drew attention to the negative effects of the indiscriminate use of pesticides in the early 1960's.[1] Since that time, a growing environmental movement and the wide-ranging impact of social and digital media means that the public is often bombarded with sensational headlines and stories about pollution by both the scientific and mainstream press: from the greenhouse gas emissions generated by shipping food around the globe through to heavy metals from waste electrical and electronic equipment, and, most recently, nanoparticles and the other emerging environmental contaminants such as disinfection by-products, pharmaceuticals and hydraulic fracturing or fracking substances (see section 1.5.4).[2] In many such articles, terms such as "contamination" and "pollution" are often used somewhat interchangeably. It is important however, to make the distinction between them.

Contamination is simply the presence of a substance in a given sample where there is no evidence of harm. *Pollution* is contamination that results in, or can result in, adverse biological effects to individuals or communities.[3] All pollutants are therefore contaminants but not all contaminants are pollutants. This means that differentiating pollution from contamination cannot be done on the basis of chemical analyses alone because such analyses provide no information on factors such as bioavailability or toxicity which influence whether a chemical presence actually causes harm.[3] In addition, not all contaminants or pollutants are chemical in origin. Many different forms of pollution in the aquatic environment exist. These can be summarised as:

Chemical

- Toxicity: acute or chronic toxicity causing severe damage (including death) to aquatic or human life.
- Sub-lethal toxicity: such as endocrine disruption, physical impairment, reduction of immunological/biochemical function or changes in biodiversity.[4]
- Deoxygenation: lack of oxygen in the water reducing biodiversity.

Biological

- Spread of non-native and or invasive species to new systems.
- Eutrophication: excess nutrients giving rise to excessive growths of some organisms.

Physical

- Temperature: usually heat, for example from power station cooling systems.
- pH level changes; changes in H^+ levels in a water body may affect both chemical and biological processes; *e.g.* acid rain linked to reduced shell formation ability in molluscs.
- Aesthetic: visual nuisance caused, *e.g.* litter, algal blooms, discoloration and smells.
- Noise: seismic surveying, shipping, boat traffic, pile driving and navy sonars are all sources of marine noise pollution that can affect the health of marine mammals.[5,6]
- Light: increasing intentional and unintentional illumination of the coastal zone and near-shore (and increasing the deep sea) can interfere with the feeding, reproductive and migratory behaviour of some species.[7,8]

It is important in pollution regulation to remember the "Source–Pathway–Receptor" model. Even the most potent toxin is harmless as long as it is isolated or contained and a compound designed to target a specific receptor is unlikely to have an effect in an organism that lacks such a receptor. It is also helpful to keep in mind one of the underlying principles of toxicology; namely that the dose makes the poison. All chemicals - even water and oxygen - can be toxic in certain amounts (although not all organisms respond the same way to chemicals at all stages of their life cycles). For example excessive heat will kill many species, either directly or by reducing the amount of O_2 than can be dissolved in the water body concerned. Many serious pollution incidents are caused by spills of seemingly harmless substance such as milk or sugar. These substances are not directly toxic in themselves; in fact they have the opposite effect. Their high organic content increases bacterial growth, which causes a concomitant decrease in dissolved oxygen levels.[9] In some cases the milk itself could also be contaminated, for instance by radiation following a radiological accident[10] such as the Chernobyl disaster which contaminated fields and animals across Europe in the late 1980 s[11] and the more recent Fukashima incident in Japan in March 2011; the impact and fallout of which was felt (albeit weakly) as far away as Western Europe.[12,13] Such incidents are of course thankfully, very rare.

So, in the 21st century, society cannot function in the way in which it has become accustomed without producing pollution but left unchecked, such pollution will eventually undermine the functioning of said society. Consumers are currently encouraged to do their part, for example to reduce food miles by shopping locally, and to offset their carbon footprint by funding an equivalent carbon dioxide saving elsewhere (for example by investing in renewable energy projects). However, these are small savings. To ensure chemical pollution does not cause serious and irreparable damage to the environment there must be checks and balances in place to minimise the release of certain pollutants and the harm they could potentially cause. Such techniques may be economic and/or legal instruments. However, not everything can be regulated and it would not be economically or physically viable to do so. Thus, despite the fact that almost anything can be a pollutant, certain chemicals have been identified in regulations at a national, or increasingly international level, as being priority chemicals for control. Such pollutants generally meet one or more of the following criteria:[14]

- They are frequently detected by environmental monitoring programs.
- They are toxic at low concentrations.
- They bioaccumulate.
- They are persistent.
- They are carcinogens.

For many of these substances the precautionary principle has been applied. Here the target is for no contamination to occur but there are different approaches and philosophies as to how to achieve the best environmental results.

1.2 POLLUTION CONTROL PHILOSOPHY

The public tend to think of pollution control in terms of mandatory regulations and there is no doubt that these are very important for environmental protection but they are only part of the solution. Other tools such as, environmental education, economic instruments, market forces and stricter enforcements all have roles to play in pollution control. Given the range of control measures available for environmental protection, preventing and controlling the release of priority chemicals to the aquatic environment can still be complex and challenging.

Controlling pollution to an environment has tended to rely on standards or objectives that are in some way measurable. The types of standards may be broadly divided into standards set by reference either to the *target* being protected, or the *source* of the pollution. The latter being further divided into standards covering emissions, process, product and use (see Table 1.1).

Standards may also be considered *precise* where there is a defined quantifiable minimum or maximum value for a particular or range of pollutants, or *imprecise* (classifying the health of a river or lake as fit to support fish for example), requiring the use of Best Available Techniques (BAT) or Best Practicable Environmental Option (BPEO). Together these provide an integrated framework where the use of one is not to the exclusion of another.

Table 1.1 Examples of standards utilised in pollution control. (Adapted from Ref. 15).

Type of standard	Description
Environmental quality standards	Concerned with the effect on a particular target. The degree of concentration in surface water for certain pollutants, substances or groups of substances identified as priority on account of the significant risk they pose to or *via* the aquatic environment. *EU Directive in Environmental Quality Standards (Directive 2008/105/EC); Annex IX (Dangerous Substances Directive and associated Daughter Directives); Annex X (Water Framework Directive Priority List Substances).*
Emission standards	Concerned with setting specific limits regarding the nature and volume of a pollutant present in a liquid discharged from a point source to a sewer or "controlled water". *Environmental Permitting (England and Wales) Regulations 2010* require environmental permits setting the maximum content of a pollutant in a liquid discharge from a point source to sewer or controlled water.
Process standards	Details the process to be carried out or sets performance requirements that a whole process or part of a process must achieve. Can include further stipulations about the technology or operational factors. The *Urban Waste Water Treatment (England and Wales) Regulations 1994* requirement for secondary or equivalent treatment.
Product standards	Controls the characteristics or concentration of an item being produced in order to protect against damage that a product may cause during its life cycle. Encourages recovery and recycling at the point of disposal.
Use standards	Controls the marketing or use of the product, measuring any risks associated with the consequences of the use of a product. The *REACH Enforcement Regulations 2008 for the Registration, Evaluation, Authorisation and Restriction of Chemicals (REACH).*

For environmental protection, the traditional *command and control* (CAC) practices are also complemented and supplemented by market-based economic instruments (EIs). EIs function through their impact on market signals, utilising prices or economic incentives/deterrents to achieve environmental objectives. There are five broad categories of EIs covering: charges, subsidies, deposit or refund schemes, the creation of a market in pollution credits, and enforcement incentives.[16]

EIs offer the incentive and power for an industry or consumer to realign their rights and responsibilities and act in a more environmentally responsible manner.[17] Some EIs are self-standing, whilst others work within the regulatory framework linking costs to the prevention, reduction or clean-up of pollution. For example, in England, the Environment Agency (EA) in carrying out any works, operations or investigations to prevent or remediate water pollution is entitled to recover expenses reasonably incurred from any responsible person under the Water Resources Act 1991 and 2009 Amendment.[18]

These market-based EIs, along with other alternative control procedures, such as voluntary schemes and information systems, have been developed in response to growing awareness from governments of the need to increase the range of tools for controlling chemicals and encouraging environmentally responsible behaviour. Given the range of control measures available for environmental protection, preventing and controlling the release of priority chemicals to the aquatic environment can still be difficult.

Chemicals have the potential to gain entry to the aquatic environment at any stage in their life cycle (see Figure 1.1) and entry may be through a variety of avenues.

The routes through which priority chemicals may enter the aquatic environment can be broadly categorised into *point* and *non-point* (diffuse) sources.[19] A *point* source release is from a discrete

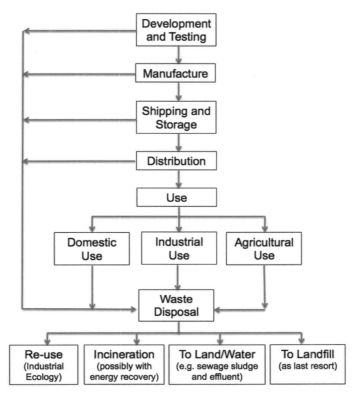

Figure 1.1   Life-cycle of chemicals in products. (Adapted from ref. 14).

location, be that a pipe or some other single identifiable localised source, *e.g.* effluent from a sewage treatment plant or an oil spill. *Non-point*, or diffuse pollution, sources are emitted indirectly from multiple discharge points and tend to be intermittent, occurring less frequently or in less quantity to point sources, *e.g.* unconfined runoff from agricultural or urban areas into a water body.[19]

Pollution control has traditionally focused on point sources due to the comparative ease in identifying and regulating a single pollution locale entering a water body. Strict requirements have been introduced to tackle the largest point sources on discharges to water and sewer, *e.g. The Urban Waste Water Treatment (England and Wales) Regulations 1994* and the *2003 Amendment*.[20] This has encouraged industry to develop technologies able to reduce or remove chemical pollutants in the effluent to meet these regulations, which has led to substantial improvements in the quality of the receiving water body over the past years. However, sewage can still act as a conduit for pollutants to affect water quality due to sewer overflow, pipe failure, or where control measures have failed.[20]

Similarly, the improvements in water quality brought about through the control measures imposed for point sources have also led to the realisation of, and additional focus on, the relative contribution of non-point sources to water pollution. Attention and control measures have therefore come to focus on and incorporate these non-point sources to facilitate further improvements in water quality. Environmental policy and practice in the last 20 years or so increasingly highlighted the need to develop a more holistic approach to environmental control and this has, and is influencing the philosophy of pollution control (see Table 1.2).

However, in some cases, the fines water companies face for polluting the environment are relatively small. For instance, although a sustained reduction in pollutant discharges from waste-water treatment works has been achieved in England and Wales through regulation over the past 20 years, in 2011 industrial sites caused 39% (240) of all serious pollution incidents. This is more than

Table 1.2 Moving towards a holistic approach in pollution control philosophy.

Holistic approach	*Traditional approach*	*Example control measures*
1 Integrated control measures	Fragmented reactive regulation covering a single environmental media or industry	– *Environmental Permitting Regulations (England and Wales) 2007 and Amendments.*
2 Trans-boundary considerations	Country-specific issue and control approach	– Multilateral environmental agreements including the Stockholm Convention Protecting human health and the environment from persistent organic pollutants
3 Complementary and supplementary control measures	Limited range of control measures	– Economic instruments – Voluntary schemes – Information systems
4 Life-cycle considerations	End-of-pipe control	– Strategic Approach to International Chemicals Management
5 Considering the impacts of chemical mixtures	Substance-by-substance approach	– *Regulation (EC) No. 1107/2009 Plant Protection Products*

in 2010 (172 or 27%).[21] This increase was due to more incidents from water company owned assets and waste management facilities. Indeed, in 2011, water company owned assets caused 120 serious pollution incidents in the UK (half of the incidents from sites regulated by the Environment Agency). This is almost double the number of incidents in 2010 (65 incidents) and the same as recorded by the EA in 2000. Of the 120 incidents, 101 were within the sewer or water network, and 19 were from permitted sites such as wastewater treatment works.[21] Some of these spills were of quite toxic substances. For instance, virtually all releases of tributyltin (TBT) to water in 2011 came from water companies.[21]

Environmental prosecutions in England and Wales in 2011 also make sober reading. In total 178 separate companies were fined for environmental offences in 2011, compared with 179 in 2010 and 317 in 2005. Total fines for the whole sector came to just over £3.8 million. This is lower than the total of £4.8 million in 2010 but this may be due to the large fines levied on the companies responsible for the 2005 Buncefield explosion.[21] These costs are very much lower than the investment required for even minor treatment plant upgrades and thus whilst acknowledging the efforts of water companies to reduce environmental contamination, it seems highly unlikely that purely punitive legal instruments are able to prevent aquatic pollution in this way unless the law on environmental pollution in the UK is changed substantially.

It is also probable that removing all possible pollutants from wastewater is likely to be not only physically almost impossible and economically undesirable; it also may not be the best approach for the protection of the environment. Aside from the high energy usage and associated increases in CO_2 and other greenhouse gas emissions, improved effluent quality also increases the amount of sludge produced, which requires environmentally sound disposal. Balancing desired improvements in the quality of effluent discharges, with the desire to reduce energy consumption and sludge production during treatment, poses a considerable challenge to the water industry. In 2007 Jones *et al.* went as far as to suggest that it may be time to address a paradigm of wastewater treatment, which has previously been unchallenged; namely that increasing effluent quality can only be environmentally beneficial. In fact, when subjected to life-cycle analysis, large-scale investment into increasingly energy intensive treatments is seen to be environmentally unsustainable. This is because the benefits of improved effluent quality are often outweighed by the negative effects on the wider environment when process construction and operation are looked at as a whole.[22]

The question then becomes one of diminishing returns and how much extra water utilities, and their customers, are willing to pay to remove an extra nanogram of a compound from wastewater effluent, even if a health effect is unlikely. It is also of note that even removing all pollutants and contaminants from sewage effluent would have no effect on the contributions of these compounds to the environment from other sources, such as agriculture and landfill leachates. Thus, market-based EIs, along with other alternative control procedures such as voluntary schemes and information systems, have been developed in response to the growing awareness from governments of the need to increase the range of tools for controlling chemicals and encouraging environmentally responsible behaviour.

1.2.1 Integrated Control Measures

Controlling pollution is neither a single environmental compartment nor single industry issue. However, in the past, regulatory controls were developed in response to a particular environmental issue. This led to a range of legislation and regulatory bodies responsible for individual sectors but without due consideration of the consequences from imposing control of one sector in relation to others.[15]

More recently, there has been a move to unify concepts of environmental protection. The Integrated Pollution Control (IPC) under Part 1 of the *Environment Protection Act 1990* played an important role in introducing a more holistic control philosophy to environmental management. The IPC was superseded by Pollution Prevention and Control (PPC), which implemented the *EU Directive on Integrated Pollution Prevention and Control (IPPC) (2008/1/EC)* and is still adhered to *via* the *Environmental Permitting Regulations (England and Wales) 2007 and Amendments*. These regulations take a wide range of environmental impacts into account and apply to a diverse range of industries (termed "installations" which may have multiple processes) that now include landfill sites, intensive agriculture, large pig and poultry units, and food and drink manufacturers. Each installation is required to have a permit containing emission limit values and more wide ranging criteria in the consideration of BAT. Permit conditions also have to address energy efficiency, waste minimisation, prevention of accidental emissions and site restoration.

1.2.2 Trans-boundary Considerations

Both pollutants and water bodies fail to recognise national borders. Priority pollutants tend to persist in the environment for long periods and can therefore be transported long distances. Transport can occur between environmental compartments *e.g.* deposition from the air to a water body some distance away, or within one environmental compartment *e.g.* as the water body meanders across national borders and/or acts as a border.

The global scope for environmental pollution has resulted in growing international co-operation and action including multilateral environmental agreements (*e.g.* the Basel, Rotterdam and Stockholm Conventions), with one of the key objectives of European Union environmental policy, as set out by the Lisbon Treaty, being to promote international measures to deal with regional or worldwide environmental problems. Regulation such as the *Water Framework Directive* also encourages coordination of all aspects of the WFD implementation across borders, from setting objectives to developing programmes of measures. There are also numerous initiatives specifically to protect the aquatic environment from priority pollutants, particularly for marine waters. The Oslo and Paris Commission (OSPAR) is one example. OSPAR is the mechanism by which 15 governments (Belgium, Denmark, Finland, France, Germany, Iceland, Ireland, Luxembourg, The Netherlands, Norway, Portugal, Spain, Sweden, Switzerland and United Kingdom) of the western coasts and catchments of Europe, together with the European Community (EC), cooperate to protect the marine environment of the North-East Atlantic. The Commission was started in 1972 with the Oslo Convention against dumping. It was broadened to cover land-based sources and the offshore industry by the Paris Convention of 1974. These two conventions were unified, up-dated and extended in the 1992 OSPAR Convention. A new annex on biodiversity and ecosystems was also adopted in 1998 to cover non-polluting human activities that can adversely affect the sea. Finland is also involved in this convention even though it is not on the western coasts of Europe. This is because several Finish rivers flow to the Barents Sea and Finland has been involved in the efforts to control the dumping of hazardous waste in the Atlantic and the North Sea for many years. Other external partners are Luxembourg and Switzerland, which are counted as contracting parties due to their location within the catchments of the Rhine. OSPAR itself aims to reduce discharges, emission and losses of specific hazardous substances continually, with the ultimate aim of reducing concentrations in the marine environment to near background values for naturally occurring substances and close to zero for anthropogenic compounds.

1.2.3   Complementary and Supplementary Control Measures

Traditionally the instruments for controlling chemicals have been regulations setting limits for discharges to water or banning a chemical from specific or all uses. However, regulation alone has long been recognised as a rather blunt instrument with which to achieve continuous improvement, particularly for diffuse sources.[23] In relation to priority pollutant control, there may be difficulties with encouraging replacements of priority chemicals by other, more benign substitutes. This has led to the use of alternative control procedures that complement and supplement regulatory control measures. These approaches include economic instruments, voluntary schemes and information systems, which work with the normal market forces, supply and demand, consumer choice *etc.*, to control chemicals and encourage environmentally responsible behaviour.

It is worth highlighting here that, depending on the application, identifying suitable chemical alternatives can be problematic. For instance, pharmaceuticals are necessary for human health and may need to be recalcitrant in order to work. For example, a major component of the contraceptive pill is ethinylestradiol (EE2). EE2 possesses an ethinyl group on the 17 carbon making it more resistant to degradation than estradiol, which is quickly inactivated by the liver. The synthesis of EE2 paved the way for oral contraceptives but EE2 has also been identified as contributing to the "feminisation of the male fish" present in water bodies which receive sewage effluent.[24] Were EE2 to be substituted for another chemical, the replacement would still need to possess a similar activity and be recalcitrant for use as an oral contraceptive but not to the extent that it may survive the sewage treatment plant (see section 1.5.2).

1.2.4   Life-cycle Considerations

Consideration of the life cycle of a chemical – from production to disposal – has led to recognition of a hierarchy of approaches to priority pollutant control. *The Waste (England and Wales) Regulations 2011* support the revised *Waste Framework Directive (2008/98/EC)* to ensure the recovery of waste or its disposal without endangering human health and the environment. These regulations introduced the concept of the waste hierarchy with emphasis placed on the prevention, reduction, re-use and recovery of waste:

  – Replace: use another, more environmentally friendly chemical.
  – Reduce: use as little of the priority pollutants as possible.
  – Manage: use in a carefully managed way to minimise accidental or adventitious loss and waste.

This is supported through the Environmental Permits System. To obtain a permit for an activity, applicants must demonstrate that the BAT will be used to "prevent, minimise or render harmless polluting releases". Policy frameworks also exist to promote the sound management of chemicals throughout their life cycle. For example the Strategic Approach to International Chemicals Management (SAICM) was established by the International Conference on Chemicals Management (ICCM) in February 2006 with the main objective to ensure that, by the year 2020, chemicals are produced and used in ways that minimise significant adverse impacts on the environment and human health.[25]

1.2.5 The Impacts of Chemical Mixtures

An issue currently receiving much consideration is how best to deal with mixtures of chemicals. Environmental risks from chemicals have traditionally been assessed as single chemicals or on a "substance-by-substance" approach, neglecting the impact of chemical mixtures. However, chemicals may enter as a:

- Waste or by-product into the aquatic environment, and as such are mixed with other chemicals.
- Single substance into the aquatic environment, which may already contain other chemicals to form a mixture of chemicals.

These chemicals may interact when they are mixed in an additive, synergistic or antagonistic way. They may produce breakdown products, by-products or react to form new substances in the waste stream or in the receiving aquatic environment. Whilst controls on individual chemicals can be effective, they face difficulties in determining the risk from, and regulation of the impact of chemical mixtures to the environment. Newer control measures are beginning to regulate under realistic conditions, the EC regulations on the authorisation of plant protection products containing certain active substances in Chapter II state 'they shall not have any harmful effects on human health, including that of vulnerable groups, or animal health, taking into account known cumulative and synergistic effects where the scientific methods accepted by the Authority to assess such effects are available' (see section 1.5.3).[26]

In addition, to provide a more integrated view of the state of the environment, ecological monitoring is essential. Given that the ultimate aim of chemical control is protection of the environment it is only through measuring ecological quality that the effectiveness of pollution control measures can be determined. Ecological quality is highlighted in the *Water Framework Directive* and this media-orientated form of legislation may provide valuable options for improving the protection of the environment from the risks of chemical mixtures.[27]

In summary, pollution control has evolved to utilise a combination of measures from regulation through to market-based economic instruments, voluntary agreements and information systems. This toolbox has enabled governments, regulatory agencies and industrial initiatives to adopt a holistic approach to better respond to both point and non-point sources of pollutants and replace, reduce or manage chemicals and their entry to the aquatic environment.

1.3 REGULATION OF DIRECT DISCHARGE SOURCES

1.3.1 The Water Framework Directive

The regulation of direct discharges of chemical pollutants to watercourses is generally the responsibility of individual nations. However, in many cases this is complicated by the fact that catchment areas cross international boundaries. Therefore both water and pollution control policies are often trans-national issues, making regulation a complex and time-consuming process. To take a salient example, 60% of the territory of the European Union (EU) as a whole lies in trans-boundary river basins and policy is decided in the European parliament and then implemented at Member State level. Individual states vary greatly in their economy and government, and compliance with EU directives can (and often does) vary significantly.

Regulation of direct discharges in Europe is moving very strongly towards control from the EU parliament. European Union legislation provides for measures against chemical pollution of surface waters. There are two components: the selection and regulation of substances of European Union (EU)-wide concern (the priority substances) and the selection by Member States of substances of national or local concern (river basin specific pollutants) for control at the relevant level.

Early European water legislation began, in a "first wave", with standards for rivers and lakes used for drinking water abstraction in 1975, and culminated in 1980 in setting binding quality targets for drinking water. It also included quality objective legislation on fish waters, shellfish waters, bathing waters and groundwaters. Its main emission control element was the *Dangerous Substances Directive*. A number of directives applying to members states have had an impact on water over the years including the following:[15]

- *The Birds Directive (79/409/EEC)*.
- *The Drinking Water Directive (80/778/EEC)* as amended by *Directive 98/83/EC*.
- *The Environmental Impact Assessment Directive (85/337/EEC)*.
- *The Sewage Sludge Directive (86/278/EEC)*.
- *The Urban Waste Water Treatment Directive (91/271/EEC)*.
- *The Plant Protection Products Directive (91/414/EEC)*.
- *The Nitrates Directive (91/676/EEC)*.
- *The Habitats Directive (92/43/EEC)*.
- *The Integrated Pollution Prevention Control Directive (96/61/EC)*.
- *The Major Accidents (Seveso) Directive (96/82/EC)*.

Historically, there has been a dichotomy in approach to pollution control at European level – with some controls concentrating on what is achievable at source, through the application of technology, and some dealing with the needs of the receiving environment in the form of quality objectives. Each approach has potential flaws. Source controls alone can allow a cumulative pollution load, which may be exceedingly detrimental to the environment, where there is a concentration of pollution sources. Quality standards can underestimate the effect of a particular substance on the ecosystem, due to the limitations in scientific knowledge regarding dose–response relationships and the mechanics of transport within the environment. There are a number of measures taken at Community level to tackle particular pollution problems. Key examples are the *Urban Waste Water Treatment Directive* and the *Nitrates Directive*, which together tackle the problem of eutrophication as well as health effects such as microbial pollution in bathing water areas and nitrates in drinking water, and the *Integrated Pollution Prevention and Control Directive* (IPPC) which deals with chemical pollution.

As a result of the increasing complexity of European water policy, the system underwent a thorough restructuring process in the late 1990s. This resulted in new, far reaching legislation in regard to regulation of the contamination of the aquatic environment in the form of *Directive 2000/60/EC* of the European Parliament and of the Council establishing a framework for the Community action in the field of water policy, or, in short the *EU Water Framework Directive* (or even shorter, the WFD). This directive is a combined approach using both emission limit values and quality standards. It was finally adopted in the year 2000 on the 23rd of October as *Directive 2000/60/EC* (as amended by *European Parliament and Council Decision No 2455/2001/E*).

Transposition into national law in the UK occurred through the following regulations: *The Water Environment (Water Framework Directive) (England and Wales) Regulations 2003 (Statutory Instrument 2003 No. 3242) for England and Wales; The Water Environment and Water Services (Scotland) Act 2003 (WEWS Act) and The Water Environment (Water Framework Directive) Regulations (Northern Ireland) 2003 (Statutory Rule 2003 No. 544) for Northern Ireland.*

Compared to previous water legislation, the *Water Framework Directive* (WFD) takes a more holistic view of the pressures and pollution on the water environment. Reducing chemical pollution from diffuse sources is considered under several articles. For example Article 10 establishes a combined approach for point and diffuse sources requiring emission controls, permits and/or best environmental practices to reduce 33 priority substances/substance groups and 14 priority hazardous substances (see Table 1.3). Whilst Article 16(6) states 'it shall identify the appropriate cost-effective and proportionate level and combination of product and process controls for both point and diffuse sources and take account of Community-wide uniform emission limit values for process controls'.

The WFD took an innovative approach to water pollution control. It is based not on national administrative or political boundaries, but on natural, geographical and hydrological formations: river basins. Member States were required to identify all the river basins lying within their national territory and to assign them to individual river basin districts. For each river basin district a "river basin management plan" (RBMP) was required to be established and updated every six years. River basins covering the territory of more than one Member State were assigned to an international river basin district (IRBD). Representatives from each state in an IRBD must work together for the management of the basin. Such areas require the cooperation and joint objective-setting across Member State borders, or in some cases, such as the Rhine, beyond the EU territory. Through the development of RBMPs, regulatory bodies and stakeholders can identify any necessary actions to address the pressures on the water environment.

The RBMPs aim to:

– Prevent deterioration, enhance and restore bodies of surface water, achieve good chemical and ecological status of such water by 2015 at the latest and to reduce pollution from discharges and emissions of hazardous substances.
– Protect, enhance and restore the status of all bodies of groundwater, prevent the pollution and deterioration of groundwater, and ensure a balance between groundwater abstraction and replenishment.
– Preserve protected areas.

The ultimate objective is to achieve "good ecological and chemical status" for all Community waters by 2015.

Good chemical status is defined in terms of compliance with all the quality standards established for chemical substances at the European level. The Directive also provides a mechanism for renewing these standards and establishing new ones by means of a prioritisation mechanism for hazardous chemicals. This system aims to ensure at least a minimum chemical quality, particularly in relation to very toxic substances, everywhere in the community.

Good ecological status is defined in *Annex V* of the WFD, in terms of the quality of the biological community (*e.g.* fish, benthic invertebrates, aquatic flora), as well as the hydrological (river bank structure, river continuity or substrate of the river bed), physico-chemical characteristics (*e.g.* temperature, BOD, pH and nutrient conditions) and chemical characteristics (the latter refer to refers to environmental quality standards for river basin specific pollutants; see Table 1.3). These standards specify maximum concentrations for specific water pollutant which, if exceeded, deprive that water body of "good ecological status". There are of course complications; due to ecological variability across the EU no absolute standards for biological quality can be set which apply in all states. Therefore the controls are specified so as to allow a slight departure from the biological community, which would be expected in conditions of minimal anthropogenic impact. A set of procedures for identifying that point for a given body of water, and establishing particular chemical or hydro-morphological standards to achieve it, is provided as part of the legislation, together with a system for ensuring that each Member State interprets the procedure in a consistent way. If this sounds somewhat complicated, it is, but this is perhaps not so surprising for such a complex piece of legislation, covering so many countries.

There are a number of objectives in respect of which the quality of water is protected. The key ones at European level are general protection of the aquatic ecology, specific protection of unique and valuable habitats, protection of drinking water resources, and protection of bathing water. All these objectives must be integrated for each river basin. It is clear that the last three – special habitats, drinking water areas and bathing water – apply only to specific bodies of water (those supporting special wetlands; those identified for drinking water abstraction; those generally used as bathing areas). In contrast, ecological protection should apply to all waters: the central requirement is that the environment be protected to a high level in its entirety.

However the objectives for which water is protected apply only in specific areas and not everywhere. Therefore, the obvious way to incorporate them is to designate specific protection zones within the river basin which must meet these different objectives. The overall plan of objectives for the river basin will then require ecological and chemical protection everywhere as a minimum, but where more stringent requirements are needed for particular uses, zones will be established and higher objectives set within them.

An important point to note about the WFD is that all pressures must be managed. This includes: diffuse source pollution (*e.g.* pesticide and nutrient run-off from agriculture); point source pollution (*e.g.* wastewater discharges); abstractions (*e.g.* for domestic or industrial use); and physical modifications, such as dams, land drainage and so forth. It is also interesting that temporary deterioration of bodies of water is not in breach of the requirements of the WFD if it is the result of circumstances which are exceptional or could not reasonably have been foreseen and which are due to an accident, natural causes or force majeure (a circumstance beyond the control of the parties, such as a war or riot).

For both ecological and chemical status, the WFD establishes the list of priority substances. A total of 33 substances or groups of substances are on the list of priority substances for which environmental quality standards were set in 2008, including selected existing chemicals, plant protection products, biocides, metals and other groups, such as polyaromatic hydrocarbons (PAH) that are mainly combustion by-products and polybrominated diphenylethers (PBDE) that are used as flame retardants. The complete list of these substances can be seen in Table 1.3.

A slightly different approach is taken to groundwater in the EU. The presumption is that, broadly speaking, groundwater should not be polluted at all. For this reason, setting chemical quality standards is not the best approach, as it gives the impression of an allowed level of pollution, which can be taken up. As such, very few such standards have been established at European level for particular issues (nitrates, pesticides and biocides), and those that have been set must always be adhered to. For general protection however, another approach has been taken. It is essentially a precautionary one and comprises a prohibition on direct discharges to groundwater, and (to cover indirect discharges) a requirement to monitor groundwater bodies so as to detect changes in chemical composition and to reverse any anthropogenically induced upward pollution trend.

Implementation of the WFD has been a long and complicated process since it came into force in 2000, involving extensive data gathering and mapping of river basin districts; the establishment of appropriate water management authorities and monitoring bodies; and economic and environmental analysis. Unfortunately the implementation of the WFD is a long way behind schedule. At the time of writing (January 2013) the latest estimates from the European Commission indicate that only 53% of EU water bodies will be in good condition by 2015. Similarly, the critical deadline for creation of the RBMP across all Member States was 2010. In early 2013, however, only twenty-three of the twenty-seven Member States have adopted and reported these. Four states (Belgium, Spain, Portugal and Greece) are behind schedule. Belgium and Spain have not yet adopted and

Table 1.3　Priority Substances and Certain Other Pollutants according to *Annex II of Directive 2008/105/EC.*

Number	CAS Number	Name of substance[a]	Identified as a priority hazardous substance
1	15972-60-8	Alachlor	
2	120-12-7	Anthracene	Y
3	1912-24-9	Atrazine	
4	71-43-2	Benzene	
5	n/a	Brominated diphenylether[b]	Y
	32534-81-9	Pentabromodiphenylether (congener numbers 28, 47, 99, 100, 153 and 154)	
6	7440-43-9	Cadmium and its compounds	Y
7	85535-84-8	Chloroalkanes, C10-13 v	Y
8	470-90-6	Chlorfenvinphos	
9	2921-88-2	Chlorpyrifos (Chlorpyrifos-ethyl)	
10	107-06-2	1,2-Dichloroethane	
11	75-09-2	Dichloromethane	
12	117-81-7	Di(2-ethylhexyl)phthalate (DEHP)	
13	330-54-1	Diuron	
14	115-29-7	Endosulfan	Y
15	206-44-0	Fluoranthene[c]	
16	118-74-1	Hexachlorobenzene	Y
17	87-68-3	Hexachlorobutadiene	Y
18	608-73-1	Hexachlorocyclohexane	Y
19	34123-59-6	Isoproturon	
20	7439-92-1	Lead and its compounds	
21	7439-97-6	Mercury and its compounds	Y
22	91-20-3	Naphthalene	
23	7440-02-0	Nickel and its compounds	
24	7440-02-0	Nonylphenols	Y
	104-40-5	(4-nonylphenol)	Y
25	1806-26-4	Octylphenols	
	140-66-9	(4-(1,1′,3,3′-tetramethylbutyl)-phenol)	
26	608-93-5	Pentachlorobenzene	Y
27	87-86-5	Pentachlorophenol	
28	not applicable	Polyaromatic	Y
	50-32-8	hydrocarbons	Y
	50-32-8	(Benzo[*a*]pyrene)	Y
	191-24-2	(Benzo[b]fluoranthene)	Y
	207-08-9	(Benzo[*g,h,i*]perylene)	Y
	193-39-5	(Benzo[*k*]fluoranthene)	Y
		(Indeno[1,2,3-*cd*]pyrene)	
29	122-34-9	Simazine	
30	not applicable	Tributyltin compounds	Y
	36643-28-4	(Tributyltin-cation)	Y
31	12002-48-1	Trichlorobenzenes	
32	67-66-3	Trichloromethane (chloroform)	
33	1582-09-8	Trifluralin	

The additional eight pollutants, below fall under the scope of *Directive 86/280/EEC* and are included in *List I of the Annex to Directive 76/464/EEC*, are not in the priority substances list. However, environmental quality standards for these substances are included in the *Environmental Quality Standards Directive 2008/105/EC.*

(6a)	56-23-5	Carbon tetrachloride	
(9a)	309-00-2	Aldrin	
	60-57-1	Dieldrin	
	72-20-8	Endrin	
	465-73-6	Isodrin	

Table 1.3   (*Continued*)

Number	CAS Number	Name of substance[a]	Identified as a priority hazardous substance
(9b)	n/a	DDT total[d]	
	50-29-3	*p*-,*p'*-DDT	
(29a)	127-18-4	Tetrachloroethylene	
(29b)	79-01-6	Trichloroethylene (1)	

[a]Where groups of substances have been selected, typical individual representatives are listed as indicative parameters (in brackets and without number). For these groups of substances, the indicative parameter must be defined through the analytical method.
[b]These groups of substances normally include a considerable number of individual compounds. At present, appropriate indicative parameters cannot be given.
[c]Fluoranthene is on the list as an indicator of other, more dangerous polyaromatic hydrocarbons.
[d]DDT total comprises the sum of the isomers 1,1,1-trichloro-2,2 bis(*p*-chlorophenyl) ethane (CAS number 50-29-3; EU number 200-024-3); 1,1,1-trichloro-2 (*o*-chlorophenyl)-2-(*p*-chlorophenyl) ethane (CAS number 789-02-6; EU Number 212-332-5); 1,1-dichloro-2,2 bis (*p*-chlorophenyl) ethylene (CAS number 72-55-9; EU Number 200-784-6); and 1,1-dichloro-2,2 bis (*p*-chlorophenyl) ethane (CAS number 72-54-8; EU Number 200-783-0).

reported all of their plans while Portugal and Greece have not adopted or reported any plans, possibly due to on-going effects of the global financial crisis in those countries. In contrast the UK has identified 16 river basin districts: 11 in England and Wales, 3 in Scotland and 4 in Northern Ireland (including 3 international RBDs). In the Severn Trent Basin District in England, 29% and 37% of surface waters are classified as "good" or better in terms of ecological and biological status, respectively.[28]

The reasons for failure for achieve a higher percentage of "good" status is due to:

– Diffuse source pollution from agriculture.
– Point source discharges from wastewater treatment plants.
– Physical modifications, including urbanisation, land drainage and flood protection.

In part to try and rectify these problems the European Commission launched a major new strategy, called *The Blueprint for Safeguarding Europe's Waters in 2013*. This document assesses the state of EU waters and makes a number of recommendations on how to improve them. EU environment ministers adopted *Conclusions on the Blueprint* in December 2012. However, it is only a strategy document, so legislation will flow from it only where required, as part of the usual EU legislative process. It remains to be seen how the implementation of the WFD will proceed given the continuing financial issues in the Eurozone. However, the WFD is not the only important piece of chemicals legislation in the EU.

1.3.2   REACH Regulations

The EU chemicals industry provides a significant contribution to EU net exports.[29] It is one of the European Union's most international, competitive and successful industries. Even after the global financial crisis, the European chemicals industry (including the European Union and the rest of Europe) is still in a strong position, posting results of €642 billion in 2011, 23.4% of world chemicals sales in value terms.[30] Societal concern over human health and the environment has prompted significant change in EU regulations on chemical substances in recent years. One of the most important of these is the *Registration, Evaluation, Authorisation and Restriction of Chemical substances (REACH) regulations on chemicals and their safe use (European Parliament and Council*

Regulation No 1907/2006 (*Corrigendum* 29 May 2007) and *Directive 2006/121/EC* (*Corrigendum* 29 May 2007).

The aim of REACH is to improve the protection of human health and the environment through the better and earlier identification of the intrinsic properties of chemical substances. Entering into force on 1st June 2007 the legislation is interesting from a regulatory perspective both for its scope (REACH is perhaps one of the most ambitious pieces of environmental legislation ever attempted) and for its "prevention is better than cure" philosophy (*i.e.* that upstream measures should be seen as preferable to downstream, cleaning up, solutions). The regulation also calls for the progressive substitution of the most dangerous chemicals when suitable alternatives have been identified.[31]

Under *Annex 17* of REACH, certain chemical substances may be restricted with limits on their marketing or use. In accordance with the precautionary principle, substances of very high concern will be subject to careful attention with the aim to have replacement by suitable alternative substances or technologies. Chemical pollution *via* diffuse sources is considered within REACH:[31]

– Evaluation (Chapter 1) states that 'Priority shall be given to registrations of substances which have or may have Persistence, Bioavailability, Toxicity (PBT), very persistent and very bioaccumulative (vPvB), sensitising and/or carcinogenic, mutagenic or toxic for reproduction (CMR) properties, or substances classified as dangerous according to *Directive 67/548/EEC* above 100 tonnes per year with uses resulting in widespread and diffuse exposure'.
– Granting of authorisations (Chapter 2) specifies that 'When granting the authorisation, and in any conditions imposed therein, the Commission shall take into account all discharges, emissions and losses, including risks arising from diffuse or dispersive uses, known at the time of the decision.'

REACH aims to enhance innovation and competitiveness of the EU chemicals industry by forcing innovation in how chemicals are assessed. Although the stated aims of the Directive were to streamline and improve the former legislative framework on chemicals of the European Union (EU), reading of the documentation might at first appear to imply the opposite since REACH makes industry responsible for assessing and managing the risks posed by chemicals and providing appropriate safety information to their users. REACH is, unsurprisingly, also one of the most difficult sets of regulations for industry to engage with.

REACH places greater responsibility on industry to manage the risks from chemicals and to provide safety information on the substances. It requires the evaluation not just of new chemicals but of large numbers of existing chemicals, even those in place for many years (*e.g.* for around 30 000 existing untested chemicals that are in common use) to safeguard the environment. Manufacturers and importers are required to gather information on the properties of their chemical substances, which will allow their safe handling, and to register the information in a central database run by The European Chemicals Agency (ECHA) in Helsinki. The ECHA acts as the central point in the REACH system: it manages the databases necessary to operate the system; coordinates the in-depth evaluation of suspicious chemicals; and is building up a public database in which consumers and professionals can find hazard information.

Under these rules it is the responsibility of industry to generate the necessary evidence of safety/toxicity, to explain how the risks are managed in practice and to submit the documentation to regulators. To evaluate so many chemicals using traditional animal tests would take decades, use millions of animals and cost billions of euros. Thus an intended side effect of REACH is to encourage the use of alternative methods for toxicity testing such as Quantitative Structure–Activity Relationship (QSAR) models, indeed animal testing is seen as a last resort. In this way the regulations explicitly encourage innovation in testing and thus represent a huge change in EU regulatory toxicology.

REACH requires that companies which use the same chemical:

- Share their existing data.
- Collaborate on further evaluation and registration.
- Use alternative methods where possible.

REACH also requires industry to submit evaluation dossiers that document safe use of the chemicals concerned. This data will include physicochemical, toxicological and ecotoxicological measurements. There is little choice but to use some form of *in silico* testing for this because they make use of existing experimental data, reduce or replace further animal testing and are much faster at lower costs. However it is a huge challenge to ensure reliability as well as speed, since there are so many chemicals to assess in a short space of time. Many in the chemical industry are also worried that regulators will not accept their safety assessments if *in silico* methods are used.

The ECHA has responded to this by trying to reassure manufacturers that while QSARS and *in silico* testing must be fit for purpose and not a one size fits all approach, there is a place for them in risk assessment in Europe. It is felt that the best way to use them is for additional evidence, in the weight of evidence approach including *in vitro*, *in silico* and existing datasets.[32] *In silico* methods are also only acceptable if they are well documented so that the authority can perform a transparent, independent assessment of the data. The use of *in silico* models by regulators may sound strange but these methods have been in use for calculating environmental fate and physical chemical parameters for many years and the weight of evidence approach is a logical extension which can reduce uncertainty, protect the environment and reduce animal testing. The predictive ability of *in silico* methods also enables a proactive approach to toxicity within product development. Toxicity evaluation can be brought "upstream" in the product development and decision making processes, so that chemicals and products are developed to be less toxic or non-toxic.

Both sides are in the acceptance phase of *in silico* models and developers and users need to understand how these models are used in regulatory decision-making. This will make it easier for the authorities and industry to apply and accept such tests.

1.4 REGULATION OF DIFFUSE SOURCES

Historically, regulation has focused on point sources of pollution leading to substantial improvements in water quality. With control measures in place to tackle the reduction and management of point source pollution, the contribution from diffuse sources to water pollution has become more apparent. Understanding and measuring diffuse sources of pollution to a water body is complex for several reasons:

- The term "diffuse pollution" is an umbrella term, covering multiple and diverse sources.
- The hydrologic cycle can enable pollutants to enter a water body *via* several diffuse pathways including precipitation, run-off and infiltration/exfiltration.
- Water bodies are individual in terms of their heath and characteristics and thus can be more or less sensitive to pollution.
- Recognising diffuse pollution to a water body and identifying the source(s) and contribution(s) has proved challenging.

The Environment Agency as the regulatory body of England and Wales (Scotland, Ireland and Northern Ireland each have their own equivalents and Wales split from England in 2013) spent over £140 million in 2008–2009 on its water quality work in England, including an estimated £8 million directly on diffuse pollution.[33] Risk assessments have identified that the percentage of water bodies at risk of failing WFD objectives from diffuse pollution are 87%, 50% and 68% for rivers, lakes and groundwater, respectively.[34] Though this considers on a risk basis and not impact, it is clear that diffuse pollution is considered a greater risk than point source pollution for many water bodies.

Sources of diffuse pollution can be broadly categorised into rural (agricultural) and urban (city) pollution. Rural pollution from agricultural activities is considered a significant contributor to diffuse pollution. Changes in agricultural practices over the last 50–60 years, including the increasing reliance on chemical use have led to nutrient, pesticide and soil washout into water bodies. This is confirmed by the economic costs arising from diffuse agricultural pollution. For example, between 2004–2005 and 2008–2009, water companies in England spent some £189 million removing nitrates and £92 million removing pesticides from their water supplies in order to meet water supply quality requirements.[35]

Urban pollution covers run-off from impermeable surfaces in the urban environment *e.g.* roads, railways and pavements or poorly plumbed drainage systems. Containing chemical pollutants including metals, pesticides and oils, diffuse urban pollution has received less attention compared with agricultural diffuse sources. However recent UK government White Papers prepared by government have set out their strategy for tackling diffuse water pollution arising from urban sources:

- *The Natural Choice (June 2011 Natural Environment White Paper)* committed the government to developing a strategy to identify and address the most significant sources of pollution from urban sources.[36]
- *Water for Life (December 2011 Water White Paper)* committed the government to consulting on a national strategy on urban diffuse pollution in 2012.[37]

A third category of diffuse pollution may also be included arising from abandoned mines. In England and Wales, they are considered to be responsible for 9% of rivers at risk of failing to meet the WFD targets of "good" chemical and ecological status.[38] With thousands of mines having been abandoned, heavy metal pollution can diffusely enter water bodies *via* mine water discharge, waste heaps and the re-suspension of contaminated river sediments found kilometres downstream. Abandoned mines highlight both the longevity and transport of chemical pollution and deficiencies in past management and regulatory practices.[39]

Controlling diffuse pollution is considered more challenging than point pollution, as only the latter responds well to investments in treatment technology and the setting of discharge permits to mitigate chemical pollutants entering water bodies. Whilst permitting is a well-practised mechanism for establishing clear standards, the diverse nature of diffuse pollution sources makes it difficult for regulatory agencies to monitor and enforce.

Significant changes in management practices and land use activity are therefore required to control diffuse pollution. The methods or "the desirable practices that should be adopted" for managing potential pollutants and reduce their diffuse pollution to the aquatic environment rely on:[40]

- Source management: reducing source of input levels of potential pollutants.
- Transport management: reducing the transfer of pollutants to sensitive areas.

For example, some pesticides in the UK are only approved for use on farms provided they are not applied within a specified buffer zone width from the top of the bank of a watercourse.[23] "Encouraging" the adoption of these methods can be achieved through a combination of basic and supplementary measures. Basic measures are compulsory and supported by legislation, whilst supplementary measures include economic incentives, voluntary initiatives, educational programmes *etc.* Legislation to identify the significant sources of pollution and progressing plans to improve the water quality is not new *e.g.* the *Bathing Water Directive* (76/160/EEC).[41]

It is perhaps not surprising that the *Water Framework Directive* also covers diffuse pollution sources. However, concern has arisen over whether effective control of diffuse pollution can be achieved within the desired timeframe imposed by the WFD.[42]

Chemical classification systems exist for both groundwater and surface water and it is recognised under the WFD that these two water environments interact with, and impact on one another. For chemical classification, the WFD and other existing EU legislation such as *Annex IX* of the *Dangerous Substances Directive* (and associated *Daughter Directives*) and *Annex X of the WFD* are of significance. Environmental standards for chemicals help define the classification status by informing us of the amount of chemical that is safely allowable, causing no harm to the ecology of that water environment. There exists a high level of confidence for environmental standards developed from existing standards, classification systems and regulatory regimes. However limited monitoring data may exist for other chemical substances, thus requiring an assessment of risk (*e.g.* Predicted No Effect Concentrations) before the development and implementation of new environmental standards. The more recent *Directive of Environmental Quality Standards* states that:[43]

'Chemical pollution of surface water presents a threat to the aquatic environment with effects such as acute and chronic toxicity to aquatic organisms, accumulation in the ecosystem and losses of habitats and biodiversity, as well as a threat to human health. As a matter of priority, causes of pollution should be identified and emissions should be dealt with at source, in the most economically and environmentally effective manner'.

The framework within the WFD also recognises that environmental standards alone may underestimate the impact of a particular chemical due to transport within the water body and/or mixture effects. It is also recognised that agriculture must make a contribution in the first cycle improvements and that this will involve a combination of incentive, advisory and regulatory measures.

The Common Agricultural Policy (CAP) is one such measure that can be utilised to control diffuse agricultural pollution. Initiated in 1962, CAP is the agricultural policy of the European Union and a form of protectionism that ensures farmers have a fair standard of living, and the provision of a safe, stable and affordable food supply. The latest proposals for the CAP include managing water resources and protecting biodiversity, which could greatly contribute to the WFD objectives.[44] The CAP offers potential to drive environmental change and minimise the risk of diffuse pollution to water. Farm subsidies are now linked to demonstrating observance of cross compliance conditions including a requirement to manage land in good agricultural and environmental conditions.

Other incentives to change behaviour and bring about improvements to control diffuse pollution and achieve WFD objectives include initiatives such as:

 – Catchment Sensitive Farming.
 – Water Protection Zones.
 – Sustainable Urban Drainage Systems.

The Department for Environment Food and Rural Affairs (DEFRA) launched the Catchment Sensitive Farming (CSF) initiative in 2006, delivered jointly with the Environment Agency and Natural England. The CSF initiative targets specific highly sensitive catchments within River Basin Districts and focuses on reducing pollution from farming into the water environment. Supported through advice and free training, farmers are encouraged to take voluntary action and implement changes in their farming practices to reduce diffuse pollution. The CSF Capital Grant Scheme is also available to support the installation and/or improvement of facilities (*e.g.* watercourse fencing). This proactive programme has a secondary aim to reduce the risk of regulation being applied to achieve required water quality standards.

If voluntary measures such as CSF are found to be ineffective, Water Protection Zones (WPZ) have been selected as a regulatory instrument. Amendments to the existing provisions of the *Water Resources Act 1991* widen the use of WPZs to prevent or mitigate diffuse pollution to water bodies in compliance of WFD objectives.[18] WPZs may be applied to a geographical area at any scale that requires extra protection to control a pollutant or polluting activity causing water pollution. One pilot and eight candidate WPZs have been designated though they are a relatively untried control measure. Several candidate sites exist in England and Wales as part of the first cycle of planning and action imposed by the WFD. However, to date, only one statutory WPZ exists – for the industrial area around the River Dee. The risk posed to the River Dee from pollution associated with the adjacent chemical and petrochemical industries have led to the Dee becoming one of the most regulated rivers in Europe. The Order prohibits, without the consent of the Environment Agency, the keeping or use within the catchment area of "controlled substances" above defined threshold amounts. Control measures for diffuse pollution have tended to focus on agricultural sources. Sustainable Urban Drainage Systems (SuDS) are a variety of management practices, control structures, and strategies designed to sustainably drain water whilst minimising pollution and managing the impact on water quality of the local water environment. However it is recognised that persistent pollutants may accumulate in SuDS sediments but that the contamination will be contained in a known area, and can therefore be better managed and treated than if it was in the wider aquatic environment.[45]

A "controlled substance" means any substance which is:

1. A dangerous substance.
2. A fuel, lubricant or industrial spirit or solvent which is a liquid under normal conditions or which is kept as a liquid within a site.
3. A medicinal product.
4. A food which is a liquid under normal conditions.
5. A feeding stuff which is a liquid under normal conditions.
6. An inorganic fertiliser.
7. A cosmetic product.
8. A substance identified by its manufacturer as being toxic, harmful, corrosive or irritant, but does not include:
   - Controlled waste that is kept, treated or disposed of under a Waste Management Licence.
   - Radioactive waste.
   - Any fuel, whether kept within a site and used exclusively for the production of heat or power.
   - Any substance contained in an exempt pipe-line.
   - Any substance at a site for a period of 24 hours or less.
   - Any substance which is a gas or vapour under normal conditions.

The minimum quantities subject to control are:

- In the case of food and feeding stuffs other than defined dangerous substances, an amount in excess of 500 litres.
- In other cases, an amount equal to or in excess of, 50 litres when the substance is present in a single container but otherwise 200 litres.

Control measures for diffuse pollution have tended to focus on agricultural sources. Sustainable Urban Drainage Systems (SuDS) are subject to a variety of management practices, control structures and strategies designed to sustainably drain water, whilst minimising pollution and managing the impact on water quality of the local water environment. It is recognised that persistent pollutants may accumulate in SuDS sediments but that the contamination will be contained in a known area and can therefore be better managed and treated than if it was in the wider aquatic environment.[3]

In addition to control measures that focus on the target of pollution, are land use practices and management. A major contribution to controlling diffuse sources is the use of "product controls". This type of approach shifts the emphasis from "end-of-pipe", to focus on the manufacture and use of the chemicals at a stage before they become wastes. The general aims of these controls are either to make the product more environmentally acceptable, or to restrict or prohibit the use of certain substances in product formulations.

A more recent example of product controls is the REACH regulations (previously discussed in section 1.3.2).

The need to control sources of diffuse pollution is recognised beyond Europe. In the United States, a number of measures have been taken to control diffuse sources from both agricultural and urban sources. The 1972 amendments of the *Clean Water Act* (CWA) was the first instance of when diffuse pollution was recognised. The 1987 amendments to CWA required each state to prepare a *State Nonpoint Source Assessment Report* and offered matched federal funds to encourage states to develop and implement management programmes. There are a plethora of federal statutory laws in addition to the CWA related to diffuse sources of pollution *e.g. Federal Environmental Pesticide Control Act, Toxic Substance Control Act* and *Surface Mining Control and Reclamation Act.*[46]

Pollution of the water environment has both environmental and economic costs and is a global issue. The contribution from diffuse sources has led to development of a range of control measures to mitigate diffuse chemical pollution to the water environment. Both point sources, as well as diffuse sources, need to be considered when targeting measures for the effective reduction in pollutant concentrations.[47] Though regulatory measures are necessary, control strategies should include educational and economic considerations to encourage buy-in from the regulated sectors. This issue is clearly important with regards to the *European Union Water Framework Directive*, eutrophication and river water quality.

1.5 CASE STUDIES

1.5.1 Disinfection By-Products (DBPs)

Disinfection is the inactivation or removal of those microorganisms in water with the potential to cause infection and therefore to harm people. Disinfection is achieved through the use of chemical or physical methods, applied under certain conditions. Disinfection by-products (DBPs) are the chemical compounds formed by the reaction of a water disinfectant with a precursor (*e.g.* humic acids) in the water undergoing disinfection. The type of DBP formed is dependent on the water disinfectant used (see Table 1.4) with potential risks from cancer, reproductive and developmental health effects leading to regulation of DBPs in drinking water.[48,49]

Whilst a necessary requirement for drinking water supply, disinfection may also be applied in wastewater treatment plants (WWTPs). If disinfection is practised, it will be as the final stage of WWTPs prior to discharge of the effluent to the aquatic environment. With the focus on the aquatic environment, it is WWTPs that may act as the conduit for DBPs.

In England and Wales, disinfection is not a requirement for all WWTP and depends on the aquatic environment that the effluent is being discharged into. Where effluent is discharged to user areas, bathing or shellfish waters, there are legislative drivers (*e.g. Bathing Water Directive, Shellfish Water Directive*) to meet mandatory requirements to protect public health and the environment from faecal pollution. Where a WWTP does not achieve this, disinfection techniques may be utilised.

Table 1.4   Common disinfection methods and associated disinfection by-products.

Disinfection method		*Example disinfection by-products*
Chemical	Chlorine	Trihalomethanes (THMs), haloacetic acids (HAA), chlorite,
	Chlorine dioxide	halonitromethanes, nitrosamines (*e.g.* nitrododimethylamine),
	Chloramine	iodo-THMs.
	Ozone	Aldehydes (*e.g.* formaldehyde), bromate.
Physical	Ultraviolet light	None of concern. May reduce the chlorination DBP formation potential.

It is the Environment Agency's policy to use disinfection techniques that do not involve chemical addition, preferring ultraviolet (UV) treatment or membrane filtration to achieve microbiological objectives beyond those expected from secondary treatment in WWTPs. Chlorination may only be considered as an interim disinfectant and for a limited period if water quality is poor and there are clear ecological or human benefits. However, chemical forms of disinfection for discharges directly into, or in close proximity to shellfish waters are never accepted.[50]

The control measures for disinfecting wastewater effluent in England and Wales mean that chlorine is rarely used. Consequently, these measures also indirectly control and mitigate the presence of chlorine-derived DBPs entering the aquatic environment. However chlorine-based disinfection methods within the WWTP context are practised elsewhere in the world. For example, the National Pollutant Discharge Elimination System (NPDES) permit programme (authorised under the *Clean Water Act*) controls water pollution by regulating point sources that discharge pollutants into United States waters. In controlling the transmission of waterborne diseases, wastewaters that pose a disease risk are disinfected prior to discharge from the WWTP.[51] The NPDES does provide a monitoring and enforcement policy for WWTPs disinfecting with chlorine *e.g.* for some trihalomethanes (THMs) in wastewater effluents. However, regulations do not yet exist on controlling and removing DBP precursors in WWTPs. This holds particular relevance with wastewater reuse, for instance wastewater effluent used for groundwater recharge that is subsequently abstracted for drinking water treatment and supply.[52]

The aquatic environment may also be the recipient of DBPs derived from ozonation. In recent years, ozone has been considered and trialled in WWTPs as an advanced/tertiary treatment technology to mitigate oestrogenic chemicals in wastewater effluent to reduce their impact on the aquatic organisms in the receiving water environment. Though not utilised as a disinfection method in this context, it is appreciated that ozonation may still produce by-products and that control measures are required, within the operational context to limit the formation of by-products in the wastewater effluent.[53]

1.5.2   Oestrogenic Chemicals

Chemicals which possess oestrogenic activity (*i.e.* are able to mimic the action of, or inhibit, hormones such as oestrogen) form an extremely wide and diverse group. Chemicals in this category include chlorinated pesticides such as DDT, detergents such as alkylphenol ethoxylates and their breakdown products, plasticisers such as phthalates, and dioxins. Oestrogenic chemicals come under the umbrella term of "endocrine disrupters", which display a greater range of biological activity and effects.[54]

The presence of these pollutants in aquatic environments is of global concern due to observations of severe reproductive abnormalities *e.g.* intersexuality in aquatic organisms often called "feminisation" of the male fish.[55] This issue is a relatively recent phenomenon dating back to the mid-1990s. Wastewater effluent was determined to be the point source for these oestrogenic pollutants entering the water environment, with steroid oestrogens as the main contributors and to a lesser extent alkylphenols and their ethoxylates.[56]

The synthetic steroid oestrogen, 17α-ethinylestradiol is the active ingredient in the contraceptive pill and therefore also considered a pharmaceutical. Natural steroid oestrogens, as their name indicates, are naturally produced within the human body and have no industrial use or application. Conversely, nonyl- and octylphenols are high production volume chemicals utilised in a wide range of industrial applications, and the building blocks (and breakdown products in the aquatic environment) for the alkylphenol ethoxylate group of non-ionic surfactants.

With growing evidence for endocrine disruption and increasing public concern, the European Commission has established a legislation-based strategy for endocrine disrupters. Short-, medium- and long-term actions are discussed from the establishment of a priority list of substances for further evaluation of their endocrine disrupting effects through to legislative actions including the *Water Framework Directive*, *Groundwater Directive*, *REACH* and *Placing of Plant Protective Products Directive*. Further control measures are discussed for those chemicals originally identified as oestrogenic pollutants of the aquatic environment namely:

- Natural and synthetic steroid oestrogens including: 17β-oestradiol (E2), oestrone (E1) and 17α-ethinyloestradiol (EE2).
- Alkylphenol and ethoxylates including: nonylphenol (NP), octylphenol (OP) and ethoxylates (NPE, OPE).

The Environment Agency has been working collaboratively with the Water Industry, the Water Services Regulation Authority (OFWAT) and the Department for Environment Food and Rural Affairs (DEFRA) to investigate the pollution of water by steroid oestrogens. As part of their Asset Management Plans (AMP), the single most important process for controlling, mitigating and reducing the impact of water company activities on the environment, the Endocrine Disrupters Demonstration Programme (EDDP) involved work programmes to deliver better water quality, a cleaner water environment and inform on the WFD. The EDDP evaluated the efficacy and cost effectiveness of existing and improved WWTPs for reducing steroid oestrogen emissions in order to inform future decisions on an appropriate regulatory strategy for oestrogenic domestic wastewater effluents. The next AMP commenced in 2010 for a further five years where steroid oestrogens are still of concern and assessed under the Chemical Investigation Programme (CIP).

In addition to these programmes, a national catchment-based risk assessment of steroid oestrogen emissions from WWTPs in England and Wales was undertaken. This looked to identify river reaches at risk of exceeding concentrations of steroid oestrogens which were indicative of adverse biological effects under both low and mean flow conditions The risk assessment also included a review of the risk class boundaries, and established that the currently proposed total steroid oestrogen predicted no effect concentration (PNEC) of 1 ng per liter E2 equivalent remained valid for distinguishing 'no risk' sites from 'at risk' sites.[57] More recently in January 2012, the European Commission proposed that the EU's Member States limit the annual average concentrations of EE2 in surface waters to no more than 0.035 ng per liter. The UK government estimates that wastewater effluent from around 1360 of the WWTPs would fail the proposed environmental standards for EE2.[58] The UK CIP culminated in 2012 with proposals for environmental quality standards (EQSs) for a number of additional Priority and Priority Hazardous Substances under the WFD and some revisions to existing EQS. The proposals included the addition of both oestradiol (E2) and EE2 as priority substances. The proposals are still being debated in Europe, and it remains to be seen as to whether and how they will go forward.[59]

In contrast to steroid hormones, nonylphenol, octylphenol and their ethoxylates may enter the water environment *via* point (*e.g.* WWTPs) and/or diffuse sources (*e.g.* agricultural application as a co-formulant in pesticides). For nonylphenol, octylphenol and their ethoxylates there exists a wide range of control mechanisms summarised as:

- Control emissions through discharge consents for point source discharges;
- Market control at source through restrictions on marketing and use of substance restrictions on use of nonylphenol imposed across Europe; and
- Voluntary initiatives with suppliers agreeing not to promote octylphenol as a substitute for nonylphenol and not to manufacture or import and reformulate or remove as a matter of urgency.

Nonylphenol and nonyphenol ethoxylates have been restricted in the European Union as a hazard to human and environmental safety. Both nonylphenol and ocytlphenol are classed are Priority/Priority Hazardous Substances in the field of water policy. In December 2012, nonylphenol and octylphenoletoxilate (which breaks down to octylphenol) were included as Substances of Very High Concern (SVHCs) to the REACH Candidate List. In April 2004 the chemical supply industry and secondary users produced a voluntary agreement for risk reduction for nonylphenol, octylphenol and their ethoxylates. Octylphenol is a high production volume chemical and was the most likely immediate replacement for nonylphenol, which was assessed for risks to the environment and human health under the *Existing Substances Regulation (ESR) 793/93/EEC* and since 1976 has been the subject of marketing and use restrictions under *Council Directive 76/769/EEC* (which placed restrictions on the marketing and use of certain dangerous substances and preparations).[25] In August 2010, the EPA released the *Nonylphenol and Nonylphenol Ethoxylates Action Plan* to address concerns over potential ecological and other effects from the manufacturing, processing, distribution in commerce, and uses of NP and NPEs.[25] However, the global nature of these chemicals and their effects are well recognised. Though a range of control measures exist for these surfactants in the UK, recent reports have identified that imported clothes can contribute a significant proportion of NP and ethoxylates detected in UK Rivers.

1.5.3 Pesticides

Pesticides are a class of biocide and are responsible for killing, repelling, or controlling any form of plant or animal life that is considered to be a pest.[60] They cover a wide range of products including insecticides, fungicides and weed killers for agricultural (plant protection products) and non-agricultural applications. Pesticides, by their very design and use, are common chemical pollutants in the aquatic environment. The multiple sources of pesticides, combined with pathways to a water body necessitate a range of control measures.

Twenty percent of rivers in England and Wales are considered at risk of failing the WFD standards from pesticides and a similar percentage of pesticide concentrations have been detected in groundwater.[2] Their persistent nature means that even with legislation to mitigate or ban use, pesticides pose a long term threat to water quality being present in the aquatic environment at levels that fail to comply with standards.[60] For instance, atrazine for non-agricultural uses was banned in 1993, with the European Commission excluding atrazine from a re-registration process in 2003 because the registrants did not supply sufficient water monitoring data. In England and Wales, simazine and atrazine were subject to restricted use in 2006 and were withdrawn completely in 2007. However, atrazine and simazine, amongst other pesticides, are still the most frequently detected compounds in European groundwater.[61] Additional concerns arise from:

- Pesticide substitution due to restricted use *e.g.* diuron in place of atrazine, which itself has subsequently been banned as a plant protection product; and

– Pesticide metabolites resulting from the breakdown of parent pesticide in the environment, which are often biologically active themselves and therefore subject to registration and control as well.

The control philosophy for pesticide pollution to the aquatic environment encompasses a wide range of measures. Restrictions on the marketing and use of certain pesticides can be covered by legislation *e.g.* the *Plant Protection Products Regulation* which also considers the impacts of chemical mixtures whereby plant protection products containing certain active substances in Chapter II state 'they shall not have any harmful effects on human health, including that of vulnerable groups, or animal health, taking into account known cumulative and synergistic effects where the scientific methods accepted by the Authority to assess such effects are available'.[62]

The *Directive on the Sustainable Use of Pesticides* has an overall objective to establish 'a framework to achieve a sustainable use of pesticides by reducing the risks and impacts of pesticide use on human health and the environment and promoting the use of Integrated Pest Management and of alternative approaches or techniques such as non-chemical alternatives to pesticides'. The production and use of certain pesticides can also be curtailed through multilateral environmental agreements such as the Stockholm Convention on Persistent Organic Pollutants (POPs) and the Aarhus Protocol on Persistent Organic Pollutants.

Legislation also sets quality standards for pesticides and their active substances in the water environment. Under the WFD, existing standards for pesticides remain and continuously evolve with newer legislation. Such legislation *e.g. Groundwater Directive* can include limits for the sum of individual pesticides present in that water source.[63] Translation of EU legislation to the national level can also be swayed by public pressure. For instance, the Netherlands, Denmark and Sweden have introduced national programmes of pesticide level reductions, and national targets in terms of the area under organic farming are found in half of the EU Member States.

Complementary and supplementary control measures are needed for both urban and rural (agricultural) diffuse sources of pesticide pollution. To deal with diffuse pollution from agriculture there is the Catchment Sensitive Farming scheme, the Voluntary Initiative for controlling pesticides, and reform of the European Union's Common Agricultural Policy to name a few. Changes in land management and actions by farmers to reduce diffuse water pollution from agriculture fall into four broad categories:

– Reducing the risk from inputs of pesticides to land and changing the pattern and/or timing of use.
– Maximising uptake of these products by the crops for which they are intended.
– Slowing the transportation of pesticides into watercourses.
– Changing the land use.

Education and information campaigns have been utilised to try to reduce diffuse urban pollution and direct pollution *via* disposal down the drain but there is still further work to be done in this area since pesticides continue to be a problem in the aquatic and terrestrial environment and in foods.[64]

1.5.4 Emerging Contaminants of Concern (ECC)

ECCs are chemicals that are not currently regulated because they are new chemicals (*e.g.* nanomaterials) or because we are only now becoming aware of their environmental or human health risks (*e.g.* nonylphenols). The term covers a variety of chemicals, including persistent organic pollutants (*e.g.* nonylphenols), personal care products (*e.g.* soaps and sun screens), nanomaterials and pharmaceuticals (*e.g.* painkillers, heart medications and anti-cancer drugs). These chemicals are finding their way into the environment *via* industrial and municipal waste and emission streams. Research is starting to find ECCs in soil and groundwater with the potential that new types of land contamination are being created,[65,66] for which there is little guidance with regard to assessing risks to receptors or approaching remediation. In addition, their continual input to the environment, which enables them to behave as persistent or pseudo-persistent compounds, even if they are not inherently stable.

In order to develop effective management strategies to minimise the risks of the release of compounds to the environment, it is necessary to understand fully the potential sources, and the subsequent fate and behaviour of the compounds in question. However, little information of this sort is available on many of the compounds in this class and governments are currently identifying and working out how to manage these chemicals.[65] For example, WorkSafe Australia recently called for carbon nanotubes to be labelled as a hazardous chemical.[67] In Europe, however, there are no nano-specific regulations in the EU and regulators are sponsoring research projects to evaluate risks associated with nanoparticles in recent years (*e.g.* NanoFATE; https://wiki.ceh.ac.uk/display/nanofate/Home). At the time of writing, nanoparticles in the EU are regulated under REACH. However, there is a lack of data on a number of issues concerning emerging contaminants, including:

(i) Types of ECCs likely to be found in surface or ground waters.
(ii) Sources of ECCs.
(iii) Fate and behaviour of these chemicals.
(iv) Environmental and human health risks of identified ECCs.
(v) Determination of high priority emerging chemicals of concern for specific geographical areas.

Future work will enable a better understanding of priority ECCs and risks to aquatic environments, which will in turn enable regulatory agencies to take a proactive role in managing ECCs as these chemicals are folded into the regulatory envelope.

One of the most recent issues concerning ECCs related to those used in hydraulic fracturing (or fracking). This is an industrial process used to extract fossil fuel reserves that lie deep underground. Wells are drilled and injected with large quantities of water mixed with specially selected chemicals at high pressures that allow petroleum or gas reserves to flow to the surface.[68] While the increased economic and energy output benefits are increasingly welcomed by governments, particularly that of the USA, the potential for water contamination as well as land destruction, air pollution, and geologic disruption has raised concerns about the merits of production activities used during extraction. Legislation at present is weak and disjointed and looks likely to be overcome by financial considerations for many years to come.

1.6 CONCLUSIONS

Environmental regulation and policy is often the source of controversy. Even 50 years after the publication of *Silent Spring* and the huge strides made in the regulation of pollution of the aquatic environment there is still work to be done. New challenges such as fracking chemicals and nanoparticles continue to confront regulators, while long standing substances such as pesticides and heavy metals are still with us. Regulators and policy makers must also keep in mind trade-offs between the beneficial effects of a chemical (*i.e.* increased food production as a result of pesticide application) and the risks of environmental/ecological degradation from its use. In some areas, there is a need for more research since the benefits or risks (in some cases both) are poorly defined. However, continued vigilance on the part of regulators and the public together with new and innovative research on the development of alternative sustainable technologies from scientists and engineers suggest that the environment of the future can be both productive and green.

REFERENCES

1. R. Carson, *Silent Spring*, Houghton Mifflin Harcourt, Boston, USA, 1962.
2. M. Stuart, D. Lapworth, E. Crane and A. Hart, *Sci. Total Environ.*, 2012, **416**, 1–21.
3. P. M. Chapman, *Environ. Int.*, 2007, **33**, 492–501.
4. E. L. Johnston and D. A. Roberts, *Environ. Pollut.*, 2009, **157**, 1745–1752.
5. H. Bailey, B. Senior, D. Simmons, J. Rusin, G. Picken and P. M. Thompson, *Marine Pollut. Bull.*, 2010, **60**, 888–897.
6. A. Codarin, L. E. Wysocki, F. Ladich and M. Picciulin, *Marine Pollut. Bull.*, 2009, **58**, 1880–1887.
7. M. H. Depledge, C. l. A. J. Godard-Codding and R. E. Bowen, *Marine Pollut. Bull.*, 2010, **60**, 1383–1385.
8. R. E. Kochevar, *Effects of Artificial Light on Deep Sea Organisms: Recommendations for Ongoing Use of Artificial Lights on Deep Sea Submersibles: Technical Report to the Monterey Bay National Marine Sanctuary Research Activity Panel*, Monterey Bay Aquarium, Monterey, USA, 1998.
9. E. Perraudin and S. Mudge, *Chem. Ecol.*, 2002, **18**, 213–221.
10. A. J. Elliott, B. T. Wilkins and P. Mansfield, *Marine Pollut. Bull.*, 2001, **42**, 927–934.
11. A. V. Nesterenko, V. B. Nesterenko and A. V. Yablokov, *Ann. N. Y. Acad. Sci.*, 2009, **1181**, 287–327.
12. N. A. Beresford, C. L. Barnett, B. J. Howard, D. C. Howard, C. Wells, A. N. Tyler, S. Bradley and D. Copplestone, *J. Environ. Radioact.*, 2012, **114**, 48–53.
13. A. Ioannidou, S. Manenti, G. Luigi and F. Groppi, *J. Environ. Radioact.*, 2012, **114**, 119–125.
14. B. Crathorne, Y. J. Rees and S. France, in *Pollution: Causes, Effects and Control*, ed. R. M. Harrison, Royal Society of Chemistry, London, UK, 4th edn, 2001, pp. 2–3.
15. S. Bell and D. McGillivray, in *Environmental Law*, ed. S. Bell and D. McGillivray, Oxford University Press, Oxford, 6th edn, 2008, pp. 222–252.
16. OECD, *Taxation and the Environment: Complementary Policies*, OECD Publishing, Paris, France, 1993.
17. UNEP, *The Use of Economic Instruments in Environmental Policy: Opportunities and Challenges*, United Nations Environment Programme, Geneva, 2004.
18. Parliament of the United Kingdom, *Statutory Instruments No. 3104, The Water Resources Act 1991 (Amendment England and Wales) Regulations*, London, UK, 2009.
19. A. Porteous, *Dictionary of Environmental Science and Technology*, 4th edn, Wiley, Chichester, UK, 2008.
20. DEFRA, *Sewage Treatment in the UK: UK Implementation of the EC Urban Waste water Treatment Directive*, 2002.
21. Environment Agency, *Sustainable Business Report 2011*, Bristol, UK, 2012.
22. O. A. H. Jones, P. G. Green, N. Voulvoulis and J. N. Lester, *Environ. Sci. Technol.*, 2007, **41**, 5085–5089.
23. B. J. D'Arcy, *Proceedings of the 2004 Water Institute of Southern Africa (WISA) Biennial Conference, Cape Town, South Africa*, 2004, **206**, 1552–1560.
24. C. Desbrow, E. J. Routledge, G. C. Brighty, J. P. Sumpter and M. Waldock, *Environ. Sci. Technol.*, 1998, **32**, 1549–1558.
25. UNEP, *SAICM Texts and Resolutions of the International Conference on Chemicals Management*, 2006, 123.

26. EU, *Off. J. Eur. Union*, 2009, **309**, 1–49.
27. D. J. Spurgeon, O. A. H. Jones, J.-L. C. M. Dorne, C. Svendsen, S. Swain and S. R. Stürzenbaum, *Sci. Total Environ.*, 2010, **408**, 3725–3734.
28. Environment Agency, *Water for Life and Livelihoods: River Basin Management Plan Severn Trent Basin District*, Environment Agency, Bristol, UK, 2009.
29. A. D. Chandler, *Shaping the Industrial Century: The Remarkable Story of the Evolution of the Modern Chemical and Pharmaceutical Industries*, Harvard University Press, Boston, USA, 2009.
30. European Chemical Industry Council (CEFIC), *The European Chemical Industry in Worldwide Perspective Facts and Figures 2012*, Auderghem, Belgium, 2013.
31. European Union, *Off. J. Eur. Union*, 2006, **396**, 1–849.
32. W. De Coen, Head of Evaluation, the European Chemicals Agency (ECHA), personal comm. to O. A. H. Jones, 2013.
33. N. A. Office, *Tackling Diffuse Water Pollution in England. Report by the Controller and Auditor General*, HC 188 Session 2010–2011, London, UK, 2010.
34. Environment Agency, *The Unseen Threat to Water Quality: Diffuse Water Pollution in England and Wales Report*, Environment Agency, Bristol, UK, 2007.
35. J. Howarth, *J. Environ. Law*, 2011, **23**, 129–141.
36. DEFRA, *The Natural Choice: Securing the Value of Nature (Government White Paper)*, Cm 8082, Norwich, UK, 2011.
37. DEFRA, *Water for Life (Government White Paper)*, CM 8230, Norwich, UK, 2011.
38. Environment Agency, *Abandoned Mines and the Water Environment*, SC030136/SR41, Bristol, UK, 2008.
39. P. L. Young, *Sci. Total Environ.*, 1997, **194–195**, 457–466.
40. P. Dampney, P. Mason, G. Goodlass, J. Hillmanm, *Methods and Measures to Minimise Diffuse Pollution of Water from Agriculture – A Critical Appraisal. Final report to DEFRA*, NT2507, 2002.
41. European Union, *Off. J. Eur. Union*, 1975, **31**, 1–7.
42. A. D. Reeves, E. A. Kirk and K. L. Blackstock, *Law, Science and Policy: An International Journal*, 2007, **3**, 123–156.
43. European Union, *Off. J. Eur. Union*, 2008, **348**, 84–97.
44. European Commission, *The Common Agricultural Policy: A Partnership between Europe and Farmers*, Brussels, Belgium, 2012.
45. C. Wilson, R. Clarke, B. J. D'Arcy, K. V. Heal and P. W. Wright, *Proceeding of the 7th IWA International Specialised Conference on Diffuse Pollution and Basin Management*, Dublin, Ireland, 2003.
46. K. Loague and D. L. Corwin, in *Encyclopedia of Hydrological Sciences: Water Quality and Biogeochemistry*, ed. M. G. Anderson, John Wiley & Sons, Ltd, Hoboken, USA, 2006, pp. 1427–1439.
47. J. J. Rothwell, N. B. Dise, K. G. Taylor, T. E. H. Allott, P. Scholefield, H. Davies and C. Neal, *Sci. Total Environ.*, 2010, **408**, 841–855.
48. L. C. Backer, D. L. Ashley, M. A. Bonin, F. L. Cardinali, S. M. Kieszak and J. V. Wooten, *J. Exposure Anal. Environ. Epidemiol.*, 2000, **10**, 321–326.
49. S. D. Richardson, M. J. Plewa, E. D. Wagner, R. Schoeny and D. M. DeMarini, *Mutat. Res./Rev. Mutat. Res.*, 2007, **636**, 178–242.
50. Environment Agency, *Water Discharge Permitting: Disinfection of Wastewater*, Bristol, United Kingdom, 2011.
51. P. B. Dorn and J. H. Rodgers, *Environ. Sci. Technol.*, 1989, **8**, 893–902.
52. L. B. Barber, S. H. Keefe, D. R. Leblanc, P. M. Bradley, F. H. Chapelle, M. T. Meyer, K. A. Loftin, D. W. Kolpin and F. Rubio, *Environ Sci. Technol.*, 2009, **43**, 4843–4850.

53. G.-G. Ying, R. S. Kookana and P. Dillon, *Water Res.*, 2003, **37**, 3785–3791.
54. R. L. Gomes, M. D. Scrimshaw and J. N. Lester, *Trends Anal. Chem.*, 2003, **22**, 697–707.
55. S. Jobling, M. Nolan, C. R. Tyler, G. Brighty and J. P. Sumpter, *Environ. Sci. Technol.*, 1998, **32**, 2498–2506.
56. S. Jobling, *Pure Appl. Chem.*, 1998, **70**, 1805–1827.
57. R. J. Williams, A. C. Johnson, V. Keller, C. Wells, M. G. R. Holmes and A. R. Young, *Using Science to Create a Better Place: Catchment Risk Assessment of Steroid Oestrogens from Sewage Treatment Works*, SC030275/SR3, Environment Agency, Bristol, UK, 2008.
58. N. Gilbert, *Nature*, 2012, **491**, 503–504.
59. Environment Agency, *England and Wales Fisheries Group Briefing Note, Item 9. Update on Steroid Oestrogens*, Bristol, UK, 2012.
60. J. Stenersen, *Chemical Pesticides: Mode of Action and Toxicology*, CRC Press, Boca Raton, USA, 2004.
61. R. Loos, G. Locoro, S. Comero, S. Contini, D. Schwesig, F. Werres, P. Balsaa, O. Gans, S. Weiss, L. Blaha, M. Bolchi and B. M. Gawlik, *Water Res.*, 2010, **44**, 4115–4126.
62. EU, *Off. J. Eur. Union*, 2009, 1–49.
63. EU, *Off. J. Eur. Union*, 2006, **37**, 19–31.
64. Pesticide Residues Committee, *Pesticide Residues Monitoring Report, Second Quarter Report 2005*, Pesticides Residues Committee, York, UK, 2006.
65. A. C. Johnson and B. Park, *Environ. Sci. Technol.*, 2012, **31**, 2582–2587.
66. D. J. Lapworth, N. Baran, M. E. Stuart and R. S. Ward, *Environ. Pollut.*, 2012, **163**, 287–303.
67. NICNAS, *National Industrial Chemical and Notification and Assessment Scheme for SafeWork Australia*, Canberra, 2012, p. 118.
68. H. Hatzenbuhler and T. Centner, *Water*, 2012, **4**, 983–994.

CHAPTER 2

Chemistry and Pollution of the Marine Environment

MARTIN R. PRESTON

School of Environmental Sciences, University of Liverpool, 4 Brownlow Street,
Liverpool L69 3GP, UK
Email: preston@liv.ac.uk

2.1 INTRODUCTION

The marine environment is a dominant feature of the Earth's surface and plays a key role in the evolution, maintenance and future of the planet's climate and ecosystems. Recent trends in marine research have been able to use the major increase in good quality data provided by advanced ship-borne instrumentation and remote sensors to move away from the relatively modest and spatially isolated data sets of the past, to a rather more holistic view of the oceans as part of the atmosphere–land–ocean–sediment continuum. Modern research is therefore able to focus more on process and mechanisms rather than 'snapshots' of isolated geographical positions. Nowhere is this more evident than in work endeavouring to unravel the role of the oceans in the global carbon cycle as part of studies of the influence of human activities on climate.[1,2]

The oceans are to a considerable extent an environmental corridor through which natural and anthropogenic material pass during their transport between original source and ultimate resting place (source ↔ sink). This is not however a passive or chemically 'conservative' transition but rather a dynamic biogeochemical processing system in which individual chemicals may undergo major changes (or even complete degradation) before achieving their ultimate fate. This applies equally to those materials that are of natural origin (such as the products of weathering processes, terrestrial biological activity and atmospheric photochemistry) and those that derive from human activity (industrial products, agricultural and domestic wastes). In the latter case it is sometimes useful to distinguish between Type 1 materials, which are essentially natural in origin and where problematic behaviour results from excessive releases as a result of human activity, and Type 2 materials, which are those which have no natural analogues and which are entirely the product of human activity.[3] Examples of Type 1 materials include nutrient chemicals from sewage treatment and agriculture, and trace metals from mining or refining processes, whereas Type 2 chemicals

Pollution: Causes, Effects and Control, 5th Edition
Edited by R M Harrison
© The Royal Society of Chemistry 2014
Published by the Royal Society of Chemistry, www.rsc.org

comprise of a wide variety of (primarily) synthetic organic chemicals used as, for example, pesticides, pharmaceuticals, food additives and cosmetics. Artificial radionuclides can fall into either category depending on whether they exist naturally as part of the Earth's natural radioactive decay series (uranium/thorium as well as products formed by cosmic radiation in the atmosphere), or whether they have no natural presence (*e.g.* technetium).

Both Type 1 and Type 2 materials may impinge on natural marine systems but the major difference is that Type 1 chemicals are often capable of being processed by pre-existing organisms which have had the opportunity to evolve alongside them. Type 2 chemicals often have no natural (*i.e.* biological) degradation mechanisms and (unless they are chemically unstable and decay inorganically) tend to be much more persistent and potentially troublesome. Examples of this type of behaviour include some organo-chlorine pesticides and industrial organo-halogens. A detailed understanding of the impact of anthropogenically derived material on marine systems therefore requires both knowledge of the natural process that can occur within the marine environment and a detailed understanding of the behaviour of synthetic chemicals that is superimposed on the natural background.

The present chapter is intended to provide an overview of the most important features of the interactions between potential pollutants and the marine system, so that the relationships between natural processes and pollutant-driven environmental changes can be evaluated. In the treatment adopted, some general features of the source ↔ sink flux pathways in the World Ocean are identified. Following this, pollutants are divided into a number of categories, each of which is treated individually.

2.2 GENERAL FEATURES OF THE OCEANIC ENVIRONMENT

The World Ocean, including coastal and marginal seas, covers almost three quarters of the Earth's surface ($361\ 110 \times 10^3\ km^2$), and has an average depth of ~ 3800 m.

Four principal parameters must be considered when attempts are made to describe the distribution of material in the oceans:

- (i) *Source terms* by which material is delivered to the ocean reservoir.
- (ii) *Circulation patterns* which govern the transport of material within the reservoir.
- (iii) *Biogeochemical processes* which govern the reactivity of material in the reservoir.
- (iv) *Sink terms* by which the material is removed from the reservoir.

Some of the complexities of the system as they apply to coastal regions are shown in Figure 2.1.

2.2.1 Sources of Chemicals to the Oceans

The ocean reservoir is continuously subjected to series of material fluxes which are delivered *via* a number of transport pathways. With respect to the input of pollutant material, river run-off, atmospheric deposition and a number of anthropogenic pathways are the dominant natural input mechanisms:

- (i) *River run-off* delivers both particulate and dissolved material to the surface ocean at the land/sea margins with discharge usually being *via* estuaries, *i.e.* regions where fresh and saline waters mix. Estuaries are regions of intense physical chemical and biological activity and the river flux undergoes considerable modification as it passes through them.
- (ii) *Atmospheric deposition* delivers both particulate material ('dry' deposition) and a combination of dissolved and particulate material ('wet' deposition, or 'precipitation scavenging') to the whole ocean surface, *i.e.* atmospheric inputs are not confined to the land/sea margins, and so need not pass through the estuarine environment. However, the strength of the *atmospheric signal* is strongest in coastal regions closest to the continental material sources.

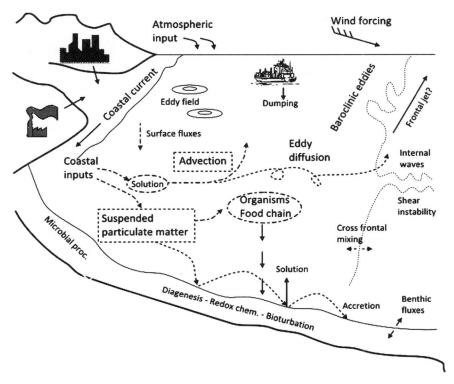

Figure 2.1   Some of the processes determining pollutant behaviour in the coastal zone. (Adapted from Simpson).[4]

(iii) *Anthropogenic pathways* by which pollutants are delivered to the oceans include:
   (a) Dumping, (*e.g.* for the disposal of sewage sludge, radioactive waste, dredge spoil, military hardware, off-shore structures).
   (b) Deliberate coastal discharges (*e.g.* coastal pipeline discharges from power stations, industrial plants, oil refineries, radioactive material reprocessing plants.
   (c) Off-shore operational discharges from tankers or other ships.
   (d) Accidents involving ships (including submarines), offshore exploration or production facilities or the wells associated with them.

2.2.2   Circulation Patterns

Once material is delivered to the oceans it is subjected to transport *via* the marine circulation systems. On a global-scale circulation in the surface ocean is mainly wind-driven, whereas in the deep-ocean it is gravity-driven. However, the strengths of both the river run-off and the atmospheric deposition signals bringing pollutant material to the oceans are strongest in coastal and marginal seas, and here the water circulation patterns are constrained by local influences.

2.2.3   Sea Water Reactivity – Biogeochemical Processes

Sea water is not simply a static reservoir in which the material supplied to it has accumulated over geological time. Thus, rather than being thought of as an 'accumulator', sea water should be regarded as a 'reactor' from which material is continually being removed on time scales (residence times) which vary considerably from one element to another.

The driving force behind the removal of elements from sea water to their marine sink, mainly sediments, is particulate ↔ dissolved reactivity. Vertical water column profiles of elements can provide data on the type of particulate ↔ dissolved reactivity that affects them in the oceans, and on this basis the elements in sea water can be divided into three principal types:

(i) *Conservative-type unreactive major elements.* These exhibit largely invariant concentration profiles down the water column, and have relatively large oceanic residence times (usually $> \sim 10^6$ yr). Conservative-type elements are mainly the sea salt-forming elements, such as sodium, potassium and chlorine.

(ii) *Scavenging-type reactive trace elements.* These are usually trace elements which are involved in passive particle scavenging throughout the water column and are not recycled, which results in a surface enrichment-subsurface depletion down-column concentration profile. Scavenging-type trace elements have oceanic residence times which can be as low as a few hundred years, and include Al, Mn and Pb.

(iii) *Nutrient-type reactive trace elements.* These are involved in active biological removal mechanisms. They exhibit a surface depletion-subsurface enrichment, and as a result of their involvement with biota in the major oceanic biological cycles they are involved in a major re-cycling stage similar to that exhibited by nutrients. Nutrient-type elements have residence times in the order of thousands of years.

Pollutants delivered to the oceans enter the circulation-driven transport and biogeochemically-driven removal processes, and are finally deposited in the sediment sink. Thus, pollution can affect the water, the biota and the sediment compartments of the ocean reservoir, and tends to have its greatest impact in coastal and marginal seas.

A wide range of pollutant, and potentially pollutant, substances can affect the marine environment and thus, the individual pollutants described in this chapter have been selected to cover examples of a variety of substances having different effects on the marine environment. The pollutants selected in this way are oil, sewage/nutrients, persistent organic compounds, trace metals and artificial radio-nuclides. Each type of pollutant is treated individually below.

2.3 SOURCES, MOVEMENT AND BEHAVIOUR OF INDIVIDUAL POLLUTANTS OR CLASSES OF POLLUTANT

2.3.1 Oil

The global demand for oil has increased rapidly from around 1.57×10^9 tonnes in 1965 to 3.95×10^9 tonnes in 2012.[5] A large proportion of this oil is transported around the world by sea and processed at refineries that are commonly in coastal regions. The global tanker fleet, which includes crude carriers, currently comprises more than 13000 vessels with a total DWT (deadweight tonnage) of 56 599 ktonnes.[6] Tanker routes run between producers (dominated by the Middle East) and consumers in the Americas, Europe and Far East. Global consumption patterns have changed over the past decade with China and India becoming much more significant as oil consumers as their economies have rapidly developed. Tankers follow economically optimal routes which tend to follow coastlines so operational problems deriving from adverse weather, equipment failure or navigational errors can result in threats to coastlines quite rapidly after the difficulties arise. However, the modern tanker fleet is better able to minimise oil spills through better construction techniques, notably the International Maritime Organisation requirement for double hulls in all vessels of 5000 dwt constructed after 3rd July 1993. Older ships must be scrapped or have double hulls fitted by an age of 25 years or by 2015 whichever is sooner.[7] The net result is that spills greater than 7 tonnes are estimated to occur at a rate of ~ 3 per 100×10^9 tonne miles.[8]

The public perception of the magnitude and significance of marine oil spills is greatly influenced by a relatively small number of high profile events. Examples of these include the wrecking of the oil tankers Torrey Canyon (1967), Amoco Cadiz (1978), Exxon Valdez (1989), Sea Empress (1996) and Prestige (2002), the deliberate destruction of onshore wells and coastal facilities at the end of the first Gulf War (1991) and the blowout at the BP Macondo Well in the Gulf of Mexico (2010) – all live on in common memory. However, such events are rare and, despite the publicity, the effects are generally relatively short lived in as much as the impacts are virtually indiscernible to the public within 2–5 years.[9] This is because most spills have so far occurred in temperate or tropical waters where rates of physical weathering and biological degradation are higher. The Exxon Valdez spill is the major exception to this generalisation because it occurred in Alaskan waters where the low temperatures (and consequent low degradation rates), restricted impact zone and poorly organised response all combined to cause major damage to an environment that, pre-spill, had been of great biological productivity and diversity. The ever increasing search for oil in ever more hostile deep-water and high latitude environments suggests that the risks of major long-term environmental damage may start to rise again after many years of decline.

2.3.1.1 The Composition of Crude Oil

Although oil is transported around the world in both crude and refined forms, marine transport is dominated by crude tankers so that will be the main focus in this section. Crude oil is a very complex mixture of many different chemicals. The exact composition varies from oil field to oil field, well to well and even within a single well as its reservoir is depleted. Added to this are the complexities of the environments affected by oil where the effects depend not only on the exact nature of the oil but also the quantity spilled, the prevailing weather conditions and the ecological characteristics of the affected region.[10] An indication of the physical-chemical properties of the major components of crude oil is shown in Table 2.1.

2.3.1.2 Fluxes of Oil to the Marine Environment

Estimating the quantities of oil reaching the marine environment from the myriad of potential sources has historically been very difficult. Estimates of some key parameters (notably the atmospheric inputs) have been subject to large uncertainties.[12] In the most recent attempt to produce a reliable set of data GESAMP have published the estimates shown in Table 2.2.[13]

Table 2.1 Typical physical-chemical properties for hydrocarbon groups. (Adapted from Doerffer).[11]

Hydrocarbon	Molecular weight (approx.)	Aqueous Solubility $(g\ m^{-3})$	Vapour pressure (Pa)	Density $(kg\ m^{-3})$	Oil–water partition coeff.
Lower alkanes (C_3–C_7)	72	40	70000	800	20000
Higher alkanes ($>C_8$)	120	0.8	2000	800	1000000
Benzenes	100	200	1500	800	4000
Naphthenes	160	20	5	800	40000
Higher polycyclics	200	0.1	0.003	800	8000000
Residues	–	0	0	800	∞

Table 2.2 Estimated average annual inputs of oil, in metric tonnes per year (tonnes yr⁻¹), into the sea from ships and other sea-based activities. Data for accidents covers the years 1988–97. Data are either calculated or measured. Large uncertainties of estimates are shown in brackets. (Adapted from GESAMP).[13]

		Ranges (t y⁻¹)	Est. or Avg. (t y⁻¹)
(1) Ships – Discharges			
Operational discharges	Machinery space bilge oil		1880
	Fuel oil sludge		186120
	Oily ballast (fuel tanks)		
Operational - cargo-related	Tank washing + oil in ballast	250–68000	19250
	VOC emissions-tankers		68000
Accidents	Tankers/barges	46000–256000	157900
	Non-tankers		5300
	– Sunken ships (casualties) [not possible to estimate]		
	– Dry docking		2900
	– Scrapping/recycling of ships		14830
	Subtotal		~457000
(2) Offshore E&P Operational			16350
	Accidents		600
	Pipelines		2800
	Subtotal		19750
(3) Coastal Facilities Coastal refineries		45000–180000	112500
	Accidents		2400
	Subtotal		114900
(4) Other Inputs	(4.1) Reception facilities no data		
	(4.2) Small craft activity OR using NRC (2003) methods 53,000	2.14–5.6×10⁶	3900000
			53000
	(4.3) Natural Seeps	0.02–2.0×10⁶	600000
	(4.4) Other sources (unknown)		200
	Subtotal		>653200
Totals – ships (1)			457000
– ships (1) plus offshore (2)			477000
– ships (1) plus offshore (2) plus coastal facilities (3)			592000
– small craft activity			53000
– natural oil seeps			600000
– GRAND TOTAL (minus oil seeps)			645000 t a⁻¹
– GRAND TOTAL (all inputs)			1245000 t a⁻¹

The figure of total discharges of 645 000 (excluding natural seeps) is considerably lower than those produced by the NAS (National Academy of Sciences) and NRC (National Research Council) in the 1980s and 1990s, which ranged up to 3.27×10^6 t a^{-1} though it does not take account of the $\sim 780\,000$ m^3 ($\sim 650\,000$–700 000 t) released during the major blowout at the BP Macondo well site in the Gulf of Mexico in 2010.

Of particular relevance to the Macondo spill is the natural seeps figure of 600 000 t shown in Table 2.1. Natural oil seeps are known to have existed for thousands of years, particularly in regions such as the Persian Gulf and the Gulf of Mexico. This makes crude oil a Type 1 pollutant with the result that a variety of marine organisms (particularly bacteria) have evolved which can feed on and degrade crude oil. Biodegradation of the Macondo spill oil has been the subject of a number of recent reports.[14–21]

The top 20 tanker accidents by size are shown in Table 2.3 from which it can readily be seen that the occurrence of major spills from this source has decreased markedly since the 1970s, with only one top 20 spill dating from the period since the year 2000.

However there are sufficient common characteristics to allow some general statements about the effects of marine spills

2.3.1.3 The Behaviour and Fate of Spilled Oil

Once crude oil is released onto the sea surface, a number of different processes immediately begin to act on it which influence its composition and environmental toxicity. These include: evaporation, dissolution and advection, dispersion, photochemical oxidation, emulsification, adsorption onto suspended particulate material, biodegradation and sedimentation (see Figure 2.2), each of which has different effects on the oil (see Table 2.3). In addition, the action of wind, surface waves and currents drive the oil slick away from its point of release. Much of the existing research on

Table 2.3 Top 20 tanker accidents by size (+ the Exxon Valdez for comparison). (Data from ITOPF).[8]

Ranking by spill size	Shipname	Year	Location	Spill Size (tonnes)
1	Atlantic Empress	1979	Off Tobago, West Indies	287 000
2	Abt Summer	1991	700 nautical miles off Angola	260 000
3	Castillo De Bellver	1983	Off Saldanha Bay, South Africa	252 000
4	Amoco Cadiz	1978	Off Brittany, France	223 000
5	Haven	1991	Genoa, Italy	144 000
6	Odyssey	1988	700 nautical miles off Nova Scotia, Canada	132 000
7	Torrey Canyon	1967	Scilly Isles, UK	119 000
8	Sea Star	1972	Gulf of Oman	115,000
9	Irenes Serenade	1980	Navarino Bay Greece	100,000
10	Urquiola	1976	La Coruna Spain	100,000
11	Hawaiian Patriot	1977	300 nautical miles off Honolulu	95 000
12	Independenta	1979	Bosphorus, Turkey	95 000
13	Jakob Maersk	1975	Oporto, Portugal	88 000
14	Braer	1993	Shetland Islands, UK	85 000
15	Khark 5	1989	120 nautical miles off Atlantic coast of Morocco	80 000
16	Aegean Sea	1992	La Coruna, Spain	74 000
17	Sea Empress	1996	Milford Haven, UK	72 000
18	Nova	1985	Off Kharg Island, Gulf of Iran	70 000
19	Katina P	1992	Off Maputo, Mozambique	66 700
20	Prestige	2002	Off Galicia, Spain	63 000
35	Exxon Valdez	1989	Prince William Sound, Alaska, USA	37 000

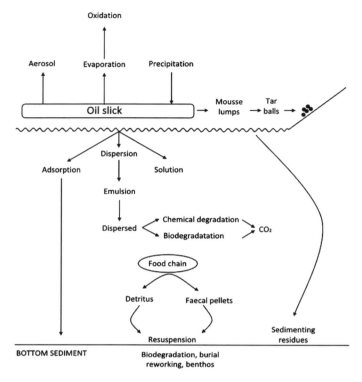

Figure 2.2   The fate of a marine oil slick.

movements and weathering relates to surface slicks, however the 2010 Macondo blowout also highlighted the need to understand what happens when the oil release is from the (deep) sea floor. The research currently underway on this topic is likely to produce some useful additions to our understanding of this topic which is likely to become more important in the future.

2.3.1.4   *The Environmental Impact of Marine Oil Spills*

The damage to coastlines is very much a function of the coastal ecosystem as well as the prevailing weather conditions and the degree of weathering of the oil by processes such as those identified in Table 2.4. The toxicity of crude oil decreases with weathering but its intractability to treatment increases. Decisions about appropriate interventions therefore need to take into account both the nature and condition of the oil but also the ecosystems at risk. To aid in this process various indices of coastal variability have been formulated. These indicate, for example, that high energy coastlines (such as exposed rocky headlands, or eroding wave-cut platforms) are amongst the least vulnerable, with sheltered tidal flats, salt marshes, mangrove stands and coral reefs being most sensitive.[22,23] The risks of major undersea oil releases such as that seen in the 2010 Macondo spill are not fully understood and are likely to present a significant point of dispute between the oil industries, scientists and local communities for some considerable time.

The most obvious problems of oil pollution are those associated with the aftermath of major events such as those identified in Section 2.3.1. These high profile events clearly demonstrate the major features of oil pollution damage to marine organisms and have been recently reviewed by Mearns *et al.*[24] The general effects are summarised in Table 2.5.

Table 2.4   The effects of weathering process on an oil slick.

Process	Rate	Effects	Controlling factors
Evaporation	Rapid.	Major influence in first 24–48 h removes volatiles (*e.g.* <C4 n-alkane) and reduces acute toxicity.	Oil composition, wind speed and temperature, water temperature and roughness.
Dissolution	Fairly rapid.	Removes more polar components from slick producing high sub-slick concentrations.	Oil composition, wind speed/ water roughness, water temperature.
Dispersion	Variable though more rapid in early stages.	Formation of oil in water emulsions. Dispersion enhances degradation rates.	Oil composition, wind speed/ water roughness.
Emulsification	More rapid in early stages.	Formation of water in oil emulsions ("chocolate mousse"). Oil in this state is untreatable by chemical techniques.	Oil composition, wind speed/ water roughness.
Photochemical oxidation	Dependent on light.	Oxidation produces aliphatic/ aromatic acids, alcohols, ethers, dialkyl peroxides which are more soluble (and possibly more toxic than parent molecules.	Oil composition, light intensity.
Adsorption	Dependent on suspended particle loading.	Removal of oil from surface slick to underlying sediment, potential impact on benthos.	Dependent on suspended particle loading.
Biodegradation	Initially fairly rapid then more slowly. High T dependency.	Rapid removal of *n*-alkanes followed by other susceptible species.	Oil composition, local bacterial/fungal populations, degree of oil dispersion, dissolved oxygen content and availability of nutrients.

Table 2.5   The types of damage caused to marine organisms by oil spills.

Affected organisms	Nature of damage
Plankton	Minor local damage, possible growth inhibition of phytoplankton by shading effect of slick.
Seaweeds	Major damage to slick-affected inter-tidal species. Recovery rapid but removal of grazers may cause excessive growth in future years.
Invertebrates	Large scale mortality in littoral communities though acute toxicity and smothering. Recovery of populations may take years.
Fish	Normally only minor casualties but pollution of spawning grounds/migration routes can cause greater damage. Exxon Valdez incident an exception.
Seabirds	Diving birds badly affected by oil. Death through drowning, hypothermia or toxic effects of ingested oil.
Marine mammals	Rarely affected but coastal populations (*e.g.* seal colonies) are vulnerable. Exxon Valdez incident a significant exception to this general rule.

Experience has shown that in some circumstances attempts to clean-up oil spills have caused more damage than if the oil had been left entirely alone; this was particularly the case in the Torrey Canyon incident.[25] Various guidelines have therefore been produced which indicate how to select the most appropriate treatment strategy from the various options available. Particularly clear (if a little dated) examples of these include two field guides published by CONCAWE on inland and coastal oil spill control and clean-up techniques.[23,26]

2.3.1.5   Control and Clean-up Techniques

2.3.1.5.1   Prevention.   Nearly all marine pollution incidents involving oil pollution are avoidable. Routine oil discharges from tank washing and accidents involving tankers, have historically been amongst the most important sources of oil to the marine environment and have led to the development of a number of strategies designed to reduce the levels or risks of pollution. For the most part, pollution prevention has received the most attention with techniques such as the Load on Top (LOT), Crude Oil Washing (COW) and Clean or Segregated Ballast Tanks (CBT or SBT) being introduced. SBT is the norm for new tankers and when combined with the oil reception facilities required under international conventions, such as *Annex 1* of the *International Convention for the Prevention of Pollution from Ships* (*MARPOL 73/78; 1992*), have dramatically reduced routine oil discharges to the oceans. More recent requirements for double hull construction to be incorporated into all new tankers have been contentious both on grounds of cost and effectiveness, and it yet remains to be seen how great the benefits from this strategy will be. In principle the double hulls should allow for better protection against spillage in the event of either grounding or collision.

2.3.1.5.2   Oil Spill Treatment Technologies.   Oil spill treatment technologies can be divided into a number of main types: chemical (including chemical enhancement of microbial processes), physical containment, and recovery systems, adsorption and burning (see Table 2.6). Useful reviews of this subject have been provided by Doerffer[11] and more recently by Fingas[27] and ITOPF.[28]

2.3.2   Sewage and Nutrients

In a significant number of countries, dumping of sewage sludge at sea was for many years a major route for the transmission of organic matter, nutrient elements and a variety of metals, synthetic organic chemicals and pathogenic microorganisms to the marine environment. In the UK a number of coastal sites in, for example, the Thames Estuary, the Firth of Clyde and Liverpool Bay were historically used for the disposal of sewage sludge arising from secondary treatment plants for around a century until the practice ceased in 1998. Nevertheless, problems still occasionally arise from raw sewage contamination of coastal waters, most often when storm water storage facilities are overwhelmed by surges in inputs. Such facilities provide buffering capacity for treatment plants

Table 2.6   A summary of oil spill treatment techniques.

Treatment process	*Advantages/disadvantages*
Chemical	Modern treatments effective on appropriate oil types. Dispersion enhances degradation. Only applicable to fresh oil, requires good logistical support, may enhance local toxicity whilst protecting coastlines. Other treatments (*e.g.* herding) applicable in confined areas.
Physical (containment)	Effective for small spills in enclosed or calm waters. Poor efficiency in rough weather.
Physical (recovery)	Effective for small spills in enclosed or calm waters. Poor efficiency in rough weather. Significant disposal problems with recovered oil.
Adsorption - sinking	Removes oil to sediment. Trades short term protection for long term benthic contamination.
Burning	May work on some fresh oils but not generally used. Leaves tarry residues. Possible use in ice covered regions in combination with herding agents.[29]

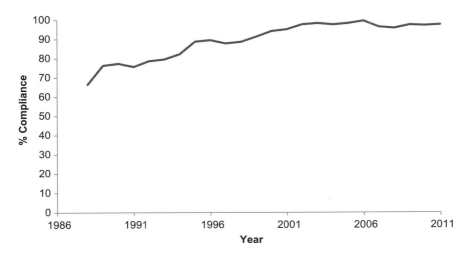

Figure 2.3   Percentage compliance of UK bathing beaches with EU Directives.[31]

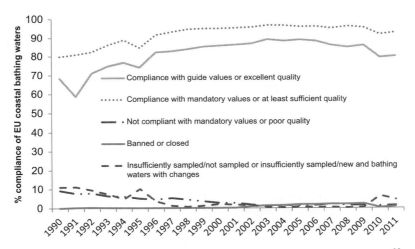

Figure 2.4   Summary percentage compliance of EU bathing waters with EU Directives.[32]

but, because the UK operates a combined rainwater/wastewater system, they are susceptible to exceed available capacity during extreme weather events.

With the advent of the *EU Bathing Water Quality Directives (Directive 76/160/EEC)*, which is in the process of being superseded by *Directive 2006/7/EC* (see text box for details), there has been a notable improvement in coastal water quality both in the UK and across the EU (see Figures 2.3 and 2.4), at least from the point of view of the occurrence of anthropogenically derived pathogenic organisms, though the *Good Beach Guide* regularly highlights problem areas.[30]

2.3.2.1   Problems associated with Biological Oxygen Demand (BOD)

Problems with pathogenic organisms are only one of a number of problems associated with wastewaters. A key environmental parameter is the Biological Oxygen Demand (BOD) which assesses the relative amount of dissolved oxygen required during the degradation of organic wastes. Whether or not a high BOD waste discharge to a natural water causes a serious environmental problem depends almost entirely on the characteristics of the receiving system. In essence, if the input of BOD is greater than the ability of the receiving water to supply new oxygen then there will be major problems of oxygen depletion and, in extreme cases, total anoxia. If the BOD input and the new oxygen supply are similar in magnitude then some oxygen depletion may be seen, and only if the renewal of oxygen is much greater than the supply of BOD will the discharge be innocuous (at least in this respect). Overall, therefore, the major problems arise from the discharge of untreated sewage to rivers, estuaries or enclosed coastal waters.

Numerous rivers and estuaries around the world have suffered from the effects of the inputs of high BOD wastes and although significant improvements have been made in some cases (*e.g.* the Rivers Thames and Mersey, UK) others have so far defied improvement. Pollution problems are somewhat compounded in tropical climates where the higher temperatures reduce the ability of natural waters to retain dissolved oxygen. Within rivers and estuarine systems the effects of severe oxygen depletion on organisms within the estuary are considerable. Many benthic animals, estuarine fish species, and those migratory species which pass from land to sea or *vice versa* as part of their natural life cycle, are effectively barred from transit through anoxic waters. Life in

EU DIRECTIVES RELATING TO WATER QUALITY IN MARINE WATERS

Old Standards – *Directive 76/160/EEC* of the European Parliament and of the Council
The I (*or Imperative*) standards, which should not be exceeded are:

- 10 000 total coliforms per 100 ml of water
- 2000 faecal coliforms per 100 ml of water

In order for a bathing water to comply with the Directive, 95% of samples (*i.e.* at least 19 of the 20 taken) must meet these standards, plus other criteria.
The higher G (*Guideline*) standards, which should be achieved where possible, are:

- no more than 500 total coliforms per 100 ml of water, and
- no more than 100 faecal coliforms per 100 ml of water in at least 80% of samples (*i.e.* 16 or more of the 20), and
  no more than 100 faecal streptococci per 100 ml of water in at least 90% of samples (*i.e.* 18 or more of the 20).

New Standards – *Directive 2006/7/EC* of the European Parliament and of the Council
For coastal waters and transitional waters

	A Parameter	B Excellent quality	C Good quality	D Sufficient	E Reference methods of analysis
1	Intestinal enterococci (cfu/100 ml)	100^a	200^a	185^b	ISO 7899-1 or ISO 7899-2
2	Escherichia coli (cfu/100 ml)	250^a	500^a	500^b	ISO 7308-3 or ISO 7308-1

[a]Based upon a 95-percentile evaluation (see *Annex II*).
[b]Based upon a 90-percentile evaluation (see *Annex II*).

underlying sediments is also considerably restricted. Only when oxygen levels can be maintained at a consistently high level over all tidal states can the full, natural range of organisms be sustained. The changing status of the River Mersey estuary as water quality has improved is an interesting case study in environmental recovery.[33]

Evidence for the recovery of ecosystems, even those as damaged as in the New York 12-mile dump site which was for many years the world's largest, suggests that some taxa staged major, and relatively rapid recoveries after the cessation of dumping in 1987.[34]

2.3.2.2 The Nature of Sewage and Nutrients

Sewage effluents contain large quantities of micronutrient elements such as nitrogen and phosphorus. Sewage treatment systems remove BOD but only increase the degree of mineralisation of the nutrients within the waste. So that, for example, a treated waste will contain primarily nitrate whereas an untreated one is dominated by ammonia and organic nitrogen species. Nutrient inputs to marine systems have been a matter for considerable concern because of the potential eutrophication effects that they may cause.[35] Climate change may also be implicated in this process with changes in ambient temperature and run-off influencing rates of biological processing and dilution factors. A simple view of eutrophication is that additional nutrients lead to the formation of excess biomass which, in turn, leads to an increase in BOD with subsequent oxygen depletion effects as the biomass decays.

It is not only an increase in nutrient concentrations that can cause deleterious effects. Anthropogenically induced alterations in nutrient ratios can also cause changes in prevailing phytoplankton species because of their differing physiological properties and requirements for nutrients. As a general rule marine phytoplankton assimilate nitrogen and phosphorus in a ratio of about 16 : 1 and under normal conditions in marine systems nitrogen is the growth limiting element. However, efforts to reduce nutrient inputs over recent years have primarily focused on phosphorus, with the result that nitrogen to phosphorus ratios have tended to increase. As a consequence, phosphorus has then become the growth limiting element in some regions (*e.g.* parts of the North Sea), so that phosphorus supplies are effectively exhausted during the spring/early summer phytoplankton bloom leaving a considerable excess of nitrogen unconsumed.[36] This excess nitrogen may be advected to other regions where it raises the total nitrogen levels, alters the prevailing N : P ratio, and has been implicated not only in changing dominant plankton species but also in stimulating toxin production in some species.

Within the North Eastern Atlantic region the Oslo and Paris Commissions (OSPARCOM) produce regular reports of the state of the region.[37] The most recent report identifies the principal pieces of legislation relating to nutrient releases in the EU (see Table 2.7).

It also identifies the reductions in nutrient inputs to problem areas for 2005 relative to 1985 (note that not all these areas are marine). Most countries in PASPARCOM regions II and II had met the desired 50% reduction for phosphorus but not for nitrogen (see Figure 2.5). This point has been noted by Grizzetti *et al.*[38] who have calculated that during the last 20 years, Europe has discharged 4.1–4.8 Tg yr^{-1} of nitrogen and 0.2–0.3 Tg yr^{-1} of phosphorus to its coastal waters.

2.3.2.3 Marine Sewage Contamination and Public Health

EU Directives with respect to sea food and public health include (in addition to the bathing water directives mentioned above): *Live Bivalve Molluscs Directive (91/492/EEC)*, *Fishery Products Directive (91/493/EEC)* and *Commission Decision on the Microbiological Criteria Applicable to the Production of Cooked Crustaceans and Molluscan Shellfish (93/51/EEC)*.

Table 2.7 EU legislation relating to nutrient releases.[37]

1 **EU Urban Waste Water Treatment Directive (991/271/EEC)**
    Connection of industry and households to waste water treatment
    Higher level treatment of waste water
    Designation of water areas sensitive to nutrient inputs

2 **EU Nitrates Directive (91/676/EEC)**
    Good agricultural practice
    Designation of water zones vulnerable to nitrogen losses

3 **EU Integrated Pollution Prevention and Control (IPPC) Directive (2008/1/EC)**
    Industrial and agricultural point sources
    Beast Available Techniques
    Emission and discharge limits

4 **EU Water Framework Directive (2000/60/EC)**
    Normative definitions describing good ecological status of a water body
    River basin management plans

5 **EU National Emissions Ceiling Directive (2001/81/EC)**
    Ceilings for air emissions of nitrogen

6 **MARPOL Annex VI**
    Emission control standards for ships
    Emission control sea areas with stricter ship standards

7 **UNCE Convention on Long-range Transboundary Air Pollution (Gothenburg Protocol)**
    Industrial and agricultural point sources
    Emission targets for nitrogen
    Transboundary air transport of nitrogen

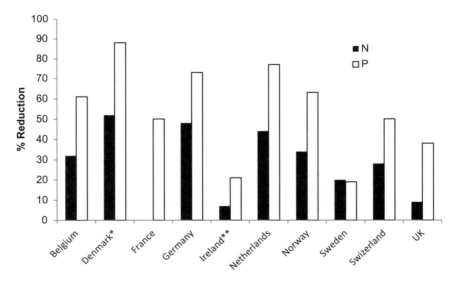

Figure 2.5 Percentage reduction in nitrogen and phosphorus releases by some EU countries between 1985 and 2005 (*1985–2003, **1995–2005).[37]

Sewage debris on beaches and in coastal waters is clearly an aesthetic problem, but it also poses a health risk. There is not only a direct risk of infection by sewage derived pathogenic organisms, but there are also considerable risks associated with the consumption of contaminated and improperly prepared sea food. Some examples of such infections are given in Table 2.8.

Table 2.8 Sewage derived pathogenic organisms and their effects on humans.[39]

Aetiological agent	Mode of transmission to humans[a]	Diseases/symptoms
Salmonella typhi	Fish or shellfish	Typhoid
S. paratyphi	- do -	Paratyphoid
S. typhimurium	- do -	Salmonellosis; gastroenteritis
S. enteritidis	- do -	- do -
Vibrio parahaemolyticus	- do -	Diarrhoea, abdominal pain
Clostridium botulinum	- do -	Botulism (high case fatality rate)
Staphylococcus aureus	- do -	Staphylococcal intoxication, nausea, vomiting abdominal pain, prostration
Clostridium perfringens	- do -	Diarrhoea, abdominal pain
Erysipelothrix insidiosa	Skin lesions	Erysipeloid - severe wound inflammation
Hepatitis virus	Shellfish	Infectious hepatitis
Heterophyes heterophyes	Fish or shellfish	Heterophyiasis – abdominal pain, mucous diarrhoea (eggs may be carried to brain, heart *etc.*)
Paragonimus westermani (P. ringer)	Crabs, crayfish or contaminated water	Flukes in lungs and other organs
Anisakis marina	Marine fish (notably herring)	Anisakiasis – eosinophilic enteritis
Angiostrongylus cantonensis	Shrimp or crabs	Eosinophilic meningitis

[a]Note that in most cases contamination of the organism is not transmitted to humans unless the food has been inadequately stored or cooked.

It should be noted that the lifetimes of most terrestrially derived bacteria in sea water are relatively short: typically of the order of 12–24 hours. This limited lifetime is due to the natural antibiotic properties of sea water which derive from a combination of the high salt concentration, low concentrations of inorganic and organic chemicals with antibiotic properties and exposure to natural UV radiation in surface waters. However, not all bacteria or viruses necessarily die. Some bacteria may, for example, enter a dormant phase (non-platable bacteria) which does not show up in routine test procedures, but which can become active again if ingested by bathers.

The *Seafood Directive (EEC/91/492)*, which became operative in January 1993, sets limits for the number of faecal coliforms, *Salmonella*, and toxins such as those which cause Paralytic Shellfish Poisoning (PSP) or Diarrhetic Shellfish Poisoning (DSP) (see Table 2.9). These toxins are produced by dinoflagellates, such as *Alexandrium* and *Protogonyaulax*, and are accumulated in shellfish, the consumption of which is capable of causing human deaths. The relationship between blooms of these dinoflagellates and pollution is not clear because such events do not take place exclusively in waters recognised as being polluted; however, such a link cannot be entirely discounted.[37]

2.3.3 Persistent Organic Compounds

Listings of potentially dangerous chemical such as those identified by *Annex II of Directive 2008/105/EC* (environmental quality standards in the field of water policy) are dominated by organic chemicals – see Tables 2.10(a) and (b). The criteria for inclusion of a chemical in that, and other priority pollutant lists, are that they have relatively high environmental toxicity (to humans and other organisms), they are persistent and liable to undergo significant biomagnification and they are produced in sufficient quantities to represent a potential threat.

Of the huge number of chemicals that might be considered to be potential pollutants only comparatively few have been studied in any detail in marine systems and many of these are halogenated species.[41–43]

Table 2.9 Microbiological standards for marine foodstuffs from *EU Directive EEC/91/492*.

Products	Microorganisms	Microbiological standard	Status/action
Live bivalve molluscs	*Salmonella* spp.	absent in 25 g	w
”	*Escherichia coli*	<230/100 g	w
”	Faecal coliforms	<300/100 g	w
”	Paralytic shellfish poison (PSP)	(80 mg/100 g)	w
”	Diarrhetic shellfish poison (DSP)	Negative in bioassay	a
Cooked crustaceans and molluscan shellfish	*Salmonella* spp.	Absent in 25 g, n = 5, c = 0	w, n, r
”	Other pathogens and toxins thereof	Not present in quantities such as to affect health	w, n, r
Whole products	Mesophilic aerobic bacteria (30 °C)	m = 10 000, M = 100 000, n = 5, c = 2	r
Shelled or shucked products	*Staphylococcus aureus*	m = 100, M = 1000, n = 5, c = 2	w, n, r
”	*Escherichia coli* (on solid medium)	m = 10, M = 100, n = 5, c = 1	n, r
”	Thermotolerant coliforms (44 °C on solid medium)	m = 10, M = 100, n = 5, c = 2	n, r
Shelled or shucked products except crabmeat	Mesophilic aerobic bacteria (30 °C)	m = 50 000, M = 500 000, n = 5, c = 2	r
Crabmeat	Mesophilic aerobic bacteria (30 °C)	m = 100 000, M = 1 000 000, n = 5, c = 2	r

Where in column 3, n = number of samples; c = number of "defect" samples; in column 4, w = withhold from market; n = notify competent authorities of findings and action taken; and r = review the methods and checking at CCPs (Critical Control Point).[40]

2.3.3.1 The Stockholm Convention

The concern about organic pollutants of the environment that arose in the 30 years or so after the publication of Rachel Carson's book *Silent Spring* culminated in the creation of the Stockholm Convention that became fully operative in 2004. By identifying key chemicals on the basis of (primarily) their persistence, toxicity and potential to biomagnify a 'dirty dozen'; see Table 2.11(a), chemicals were initially identified and moves were made to ban production and use. Subsequently other chemicals were added to the list as shown in Table 2.11(b), and others may be added at a later date because the convention operates a continuing review programme of potential problem chemicals.

The 'dirty dozen' have been fairly extensively investigated in marine systems but the more recent additions to the convention have so far received less attention. However, in both cases the focus is not on the likelihood of direct toxic effects but much more on sub-lethal effects, particularly those that influence reproductive success through *inter alia* disruption of endocrine systems.

2.3.3.2 Halogenated Compounds

2.3.3.2.1 *Chlorinated Pesticides (see also Chapter 4).* Dominant amongst the older formulations were the organochlorine pesticides such as DDT, dieldrin, aldrin and endrin (the "drins"), lindane (γ-hexachlorocyclohexane), hexachlorobenzene (HCB) and toxaphene. Despite bans on many persistent organochlorine pesticides in the industrialised world, usage of DDT at least is still

Table 2.10a Priority substances identified in *Annex 2* of *EU Directive 2008/105/EC*.

Number	CAS number[a]	EU number[b]	Name of priority substance[c]	Identified as priority hazardous substance
−1	15972-60-8	240-110-8	Alachlor	
−2	120-12-7	204-371-1	Anthracene	√
−3	1912-24-9	217-617-8	Atrazine	
−4	71-43-2	200-753-7	Benzene	
−5	not applicable	not applicable	Brominated diphenylether[d]	√
	32534-81-9	not applicable	Pentabromodiphenylether (congener numbers 28, 47, 99, 100, 153 and 154)	
−6	7440-43-9	231-152-8	Cadmium and its compounds	√
−7	85535-84-8	287-476-5	Chloroalkanes, C10-13[e]	√
−8	470-90-6	207-432-0	Chlorfenvinphos	
−9	2921-88-2	220-864-4	Chlorpyrifos (Chlorpyrifos-ethyl)	
−10	107-06-2	203-458-1	1,2-Dichloroethane	
−11	75-09-2	200-838-9	Dichloromethane	
−12	117-81-7	204-211-0	Di(2-ethylhexyl)phthalate (DEHP)	
−13	330-54-1	206-354-4	Diuron	
−14	115-29-7	204-079-4	Endosulfan	√
−15	206-44-0	205-912-4	Fluoranthene[f]	
−16	118-74-1	204-273-9	Hexachlorobenzene	√
−17	87-68-3	201-765-5	Hexachlorobutadiene	√
−18	608-73-1	210-158-9	Hexachlorocyclohexane	√
−19	34123-59-6	251-835-4	Isoproturon	
−20	7439-92-1	231-100-4	Lead and its compounds	
−21	7439-97-6	231-106-7	Mercury and its compounds	√
−22	91-20-3	202-049-5	Naphthalene	
−23	7440-02-0	231-111-4	Nickel and its compounds	
−24	25154-52-3	246-672-0	Nonylphenols	√
	104-40-5	203-199-4	(4-nonylphenol)	√
−25	1806-26-4	217-302-5	Octylphenols	
	140-66-9	not applicable	(4-(1,1′,3,3′-tetramethylbutyl)-phenol)	
−26	608-93-5	210-172-5	Pentachlorobenzene	√
−27	87-86-5	201-778-6	Pentachlorophenol	
−28	not applicable	not applicable	Polyaromatic hydrocarbons	√
	50-32-8	200-028-5	(Benzo[a]pyrene)	√
	205-99-2	205-911-9	(Benzo[b]fluoranthene)	√
	191-24-2	205-883-8	(Benzo[g,h,i]perylene)	√
	207-08-9	205-916-6	(Benzo[k]fluoranthene)	√
	193-39-5	205-893-2	(Indeno[1,2,3-cd]pyrene)	√
−29	122-34-9	204-535-2	Simazine	
−30	not applicable	not applicable	Tributyltin compounds	√
	36643-28-4	not applicable	(Tributyltin cation)	√
−31	12002-48-1	234-413-4	Trichlorobenzenes	
−32	67-66-3	200-663-8	Trichloromethane (chloroform)	
−33	1582-09-8	216-428-8	Trifluralin	

[a]CAS: Chemical Abstracts Service.
[b]EU number: European Inventory of Existing Commercial Substances (EINECS) or European List of Notified Chemical Substances (ELINCS).
[c]Where groups of substances have been selected, typical individual representatives are listed as indicative parameters (in brackets and without number). For these groups of substances, the indicative parameter must be defined through the analytical method.
[d]Only Pentabromobiphenylether (CAS number 32534 81 9).
[e]These groups of substances normally include a considerable number of individual compounds. At present, appropriate indicative parameters cannot be given.
[f]Fluoranthene is on the list as an indicator of other, more dangerous polyaromatic hydrocarbons.

Table 2.10b These eight pollutants, which fall under the scope of *Directive 86/280/EEC(1)* and which are included in List I of the *Annex to Directive 76/464/EEC*, are not in the priority substances list. However, environmental quality standards for these substances are included in the *Environmental Quality Standards Directive 2008/105/EC*.
Amended by EU Directives 88/347/EEC and 90/415/EEC.

	CAS number	*Name of other pollutant*
(6a)	56-23-5	Carbon-tetrachloride[a]
(9b)	not applicable	DDT total[a,b]
	50-29-3	p,p′-DDT[a]
(9a)		Cyclodiene pesticides
	309-00-2	Aldrin[a]
	60-57-1	Dieldrin[a]
	72-20-8	Endrin[a]
	465-73-6	Isodrin[a]
(29a)	127-18-4	Tetrachloroethylene[a]
(29b)	79-01-6	Trichloroethylene[a]

[a]This substance is not a priority substance but one of the other pollutants for which the EQS are identical to those laid down in the legislation that applied prior to 13 January 2009.
[b]DDT total comprises the sum of the isomers 1,1,1-trichloro-2,2 bis(*p*-chlorophenyl) ethane (CAS number 50-29-3; EU number 200-024-3); 1,1,1-trichloro-2-(*o*-chlorophenyl)-2-(*p*-chlorophenyl) ethane (CAS number 789-02-6; EU Number 212-332-5); 1,1-dichloro-2,2 *bis*(*p*-chlorophenyl) ethylene (CAS number 72-55-9; EU Number 200-784-6); and 1,1-dichloro-2,2 *bis* (*p*-chlorophenyl) ethane (CAS number 72-54-8; EU Number 200-783-0).

Table 2.11a The 'Dirty Dozen' identified by the Stockholm Convention.

Aldrin	*Hexachlorobenzene*
Chlordane	Mirex
DDT	Polychlorinated biphenyls (PCBs)
Dieldrin	Polychlorinated dibenzodioxins
Endrin	Polychlorinated dibenzofurans
Heptachor	Toxaphene

Table 2.11b The nine new POPs added to the Convention.

α-Hexachorocyclohexane	Hexabromobiphenyl
β-Hexachlorocyclohexane	Lindane
Chlordecone	Pentachlorobenzene
Commercial octabromodiphenyl ether (hexabromodiphenyl ether and heptabromodiphenyl ether)	Perfluorooctane sulfonic acid (PFOS), its salts and perfluorooctane sulfonyl fluoride (PFOS-F)
Commercial pentabromodiphenyl ether (tetrabromodiphenyl ether and pentabromodiphenyl ether	

important on a global scale because of its cheapness and effectiveness which still make it an attractive option for developing countries (see Table 2.12).[44]

2.3.3.2.2 Other Chlorinated Chemicals. Other persistent chlorinated organic chemicals which have proved to be of some concern in marine systems include polychlorinated dibenzo-*p*-dioxins (PCDD) and polychlorinated dibenzofurans (PCDF products of PCB combustion; some 2378

Table 2.12 Average reported insecticide use for vector control by WHO region (2000–2009), in metric tons of active ingredient per year.[44]

	Residual spraying			
WHO Region[a]	OC	OP	C	PY
Africa	805	19	19	24
Americas	0	97	4	164
Eastern Mediterranean	0	26	5	15
Europe	0	2	0	1
Western Pacific	0	1	0	39
All	4429	627	30	282

[a]Canada and the United States (Americas region) and Australia and Japan (Western Pacific region) were not targeted, whereas in the European region, only Armenia, Azerbaijan, Georgia, Kyrgyzstan, Tajikistan, Turkey, Turkmenistan, and Uzbekistan were targeted.
Abbreviations: C, carbamates; OC, organochlorines (DDT only); OP, organophosphates; PY, pyrethroids.

PCDD/F isomers exist); including 2,4-D (2,4-dichlorophenoxy acetic acid), 2,4,5-T (2,4,5-trichlorophenoxy acetic acid), and MCPA (2-methyl-4,6,-dichlorophenoxy acetic acid).[45] 2,4-D and 2,4,5-T became notorious through their use as defoliants in the Vietnam War (Agent Orange) and because of the contamination of the product with the dioxin, 2,3,7,8-TCDD (2,3,7,8-tetrachloro dibenzo-*p*-dioxin). The 'dioxins' and the PCDF have become widespread contaminants of marine systems, although their links with any deleterious effects in the oceans remain slight. Dioxin-like toxicity behaviour has also been reported for a group of compounds related to PBDEs (polybrominated diphenyl ethers) in fish.[46]

2.3.3.2.3 PCBs (see also Chapter 4). PCBs are widely distributed amongst all marine systems and are of particular concern in northern high latitude regions, where it has been estimated that 97% of the use of PCBs occurred.[47,48] In addition, some other associated compounds such as polychlorinated naphthalenes are also common in Arctic regions.[49]

The toxicity of PCBs to marine organisms varies considerably, but the co-planar PCBs exhibit the greatest mammalian toxicity. However, such acute toxicity is unknown in marine organisms which are more likely to suffer sub-lethal effects from the biomagnification of PCBs through the food web (see Table 2.13).

2.3.3.2.4 The Origins of Environmental Hazards. The environmental problems associated with the chlorinated pesticides and PCBs derive from similarities in their physical, chemical and biological properties. These may be summarised as persistence, widespread distribution amongst most environmental compartments, propensity to undergo biomagnification and high toxicity (including non-lethal effects; note however, that DDT has a low mammalian toxicity); in other words, nearly all of the properties that make a pollutant a high risk. Some organochlorines (OCs) also exhibit endocrine disrupting properties[52] which is of general concern but is highlighted in recent studies of human dietary exposure.[53,54]

The organochlorines are relatively water insoluble and involatile. They also have high octanol–water partition coefficients (*e.g.* log K_{ow} = 4.46–8.18 for different PCB congeners) and are therefore lipophilic. The low apparent volatility is, however, misleading. Under certain conditions (notably high temperatures and humidity) evaporative losses of applied pesticides can be high, and this accounts for their widespread occurrence through subsequent transport as vapour or condensates on atmospheric aerosol particles (see Table 2.14).

The lipophilic nature of the organochlorines leads not only to their rapid uptake and storage in fatty tissues, but also to their slow elimination because the fat reserves are only called upon at

Table 2.13 The distribution of PCBs between different environmental compartments.[50] (See also congener specific inventory update by Jonsson *et al.*)[51]

Environment	PCB Load (t)	Percentage of PCB Load
Terrestrial and coastal		
Air	500	0.13
River and Lake Water	500	0.94
Sea water	2400	0.64
Soil	2400	0.64
Sediment	130 000	35
Biota	4300	1.1
Total (A)	*143 000*	*39*
Open Ocean		
Air	790	0.21
Sea water	230 000	61
Sediment	100	0.03
Biota	270	0.07
Total (B)	*231 000*	*61*
Total Load (A + B)	**374 000**	**100**

Table 2.14 Estimated fluxes of PCBs to the ocean surface.[55]

	Flux ng m^{-2} a^{-1}		
	Arochlor 1242	Arochlor 1254	Total
Particles			
Dry	16	41	57
Wet	250	650	900
Gas Phase	22090	1709.5	3919
Total	2375	2301	4672
Total flux to Oceans ($\times 10^6$ g a^{-1})	**8.6**	**8.3**	**16.9**

stages in the life cycle when energy demands are high or food supplies low. It is this feature which leads to their biomagnification. The extent of biomagnification is large. Concentrations of organochlorines in sea water are typically of the order of pg l^{-1} to low ng l^{-1}, rising to 10 s of ng g^{-1} in marine invertebrates, low mg g^{-1} values in mussels and up to 10–100 of mg l^{-1} in fatty tissue of top predators such as seals, pelicans and terrestrial hawk species.

2.3.3.2.5   *The Effects of Marine Organochlorine Pollution.* At the planktonic level, primary production rates are reduced at DDT or PCB concentrations above ~ 1 mg l^{-1}. This is a low concentration but still very much higher than those likely to be found in sea water. Amongst marine invertebrates and fish 96-h LC_{50} values generally fall within the range 1–100 mg l^{-1}. However, bivalve molluscs are very resistant with 96-h LC_{50} values > 10000 mg l^{-1}.

Amongst the top predators the main deleterious effects have been eggshell thinning in birds (DDE, PCBs and BFRs)[56–58] and interference with the reproductive and immune system in mammals (PCBs).[52,59,60] A considerable number of top predator birds showed major declines in abundance during the period when DDT usage was at its greatest and since the banning of these chemicals in the industrialised world, these populations have mostly demonstrated a major recovery. Where populations have not recovered other factors, such as decline in suitable habitats, are probably responsible.

A number of seal populations, most notably in the Dutch Wadden Sea and the Baltic Sea, have exhibited reproductive abnormalities attributed to PCBs which have had a significant impact on populations.[56] The main symptoms of PCB poisoning are changes in the uterus, implantation or abortion/premature pupping. There have also been suggestions that PCBs may cause carcinogenic, teratogenic and immunological effects. An outbreak of phocine distemper virus in the common seal populations in much of the North Sea in 1988 has sometimes been attributed to depression of immune systems by PCBs, but the evidential basis for this is weak and such epidemics have also been reported in seal populations in other, less contaminated, regions.

2.3.3.3   Other Persistent Organic Compounds

As indicated above, the organochlorines are by no means the only organic compounds which are of concern as marine contaminants. There are many other chemicals which are sufficiently common, toxic and persistent to represent potential threats. These include, for example other pesticides, polycyclic aromatic compounds (PAH), plasticizers (*e.g.* phthalate esters), Bisphenol A, detergent residues, organic solvents *etc*. It is not possible within this chapter to review the behaviour and effects of all of these compounds. However, one aspect of organic pollution which has become particularly important recently is the relationship between the reported number of abnormalities in sexual and reproductive development in wildlife and humans coinciding with the introduction of so called 'oestrogenic' or 'endocrine disrupting' chemicals.[24,61–64] The list of compounds has continued to grow and includes(as mentioned above) several members of the DDT family of compounds, chlordecone (Kepone), brominated fire retardants, various sterols, various fluorinated compounds, plasticisers (phthalate esters) and numerous others. Of these, *p,p'*-DDE is amongst the most potent. Since the review by *Vos et al.*[65] research in this area has grown considerably with the number of implicated compounds similarly increasing. It is clear that the consequences of widespread disruption of fertility by organic contaminants are both subtle and potentially very serious, although a recent review of endocrine disruption for the European Commission by Kortenkamp *et al.*[66] has been criticised as being too anecdotal.[67]

2.3.4   Trace Metals

A large number of trace metals are transported to the oceans from natural sources. However, these natural sources are supplemented by releases from anthropogenic processes which, for some metals, can exceed natural inputs.

Trace metals are found in the water, biota and sediment compartments of the marine system, but potentially the most hazardous environmental effects to human health arise when they enter the food chain. The relationship between the total concentration of a trace metal in the environment and its ability to cause toxic effects in organisms is complex, and two important constraints must be considered; *i.e.* the speciation of the metals, and the condition of the organisms.

2.3.4.1   Metal Speciation

All organisms have a requirement for certain trace metals which must be present in their diet (or growth medium) to sustain healthy development. Such 'essential' metals include iron, copper, vanadium, cobalt and zinc. Other, non-essential, metals exert neither beneficial nor deleterious effects if present at sufficiently low concentrations, but often become increasingly harmful as the concentrations increase. However, it is the speciation of a trace metal, *i.e.* the way in which it is partitioned between host associations in the water and particulate phase, and not its total concentration, which constrains its effect on the environment. For example, not all forms of a trace metal are 'bioavailable', and numerous incidents have demonstrated that it is the organic forms of metals which tend to be of the most damaging to marine organisms. A useful review of metal speciation in the marine environments has been written by Hirose.[68] It has also recently been pointed out that speciation is a dynamic equilibrium, subject not only to local changes in chemical conditions but also to global scale influences such as increasing CO_2 concentrations.[69]

2.3.4.2 Condition of the Organism

Numerous factors relating to the condition of the organisms exposed to contaminants have a direct bearing on whether or no harmful effects are observable. These include such diverse aspects as the stage of development (egg, larva, juvenile, adult), size/age, sex, previous history of exposure (organisms exposed to toxins in the past may have selected for genetic features making them more robust to contaminant challenges or may simply have built up immunity), location (particularly relevant for inter-tidal organisms (even amongst the same species and population), availability and nature of food supplies (organisms short of food may be more vulnerable) and finally, the general environmental conditions such as pH, Eh, light intensity, temperature, salinity, dissolved oxygen *etc.*

The combination of all of the chemical, biological and environmental factors makes it very difficult to obtain sensible toxicity information even for acute toxicity measurements conducted under laboratory conditions. In 'real world' scenarios where acute toxicity caused by metals is very rare, establishing reliable cause–effect relationships is extremely challenging. Luoma and Rainbow[70] have provided a useful review of the manifestation of metal effects in nature.

In the current chapter, the hazardous effects of trace metals on the oceanic biogeochemical system are illustrated in the following sections with respect to mercury, lead and tin; all of which have been shown to be capable of generating stress in the marine environment.

2.3.4.3 Mercury (see also Chapter 4 and 20)

For many years mercury has been on lists of the highest priority environmental chemicals and probably represents the best known example of metal pollution in the marine environment through its role as the causative agent of 'Minimata' disease. Mercury production is around 2000 t a^{-1} and is declining from a peak of perhaps 6000 t a^{-1} through the discontinuation of many traditional uses. It must, however, be remembered that there are other sources of mercury to the environment including volcanic emissions and fossil fuel combustion. Nevertheless mercury still has important uses in *e.g.* batteries and some low-energy light bulbs.[71] A review of mercury in the environment has recently been published[72] in which there is a review of mercury in marine systems.[73]

Within natural waters the effects of mercury are dependent on its speciation and there are essentially three main chemical forms that are of potential significance: namely elemental mercury (0), divalent mercury (Hg^{2+}) and methyl mercury (CH_3Hg). Four factors strongly influence the environmental effects of mercury:

(i)   Mercury has a particularly high affinity for organic species, which results in its accumulation in marine biota.

(ii) Inorganic mercury can undergo bio-mediated transformation into 'alkylated' forms (methyl and dimethyl mercury) which are particularly toxic species of the element.[74] This occurs almost entirely under aerobic conditions in the marine environment. Methyl mercury concentrations in fish are related to trophic level with the highest concentrations in top predators.[75]

(iii) Mercury is non-conservative in estuaries where it tends to be accumulated in fine-grained near-shore sediments from which it can be remobilised (see ref. 76).

(iv) In the particulate form, mercury has a strong association with organic particles, *via* which it is readily transmitted to biota.

The residence time of mercury in the marine boundary layer has been estimated to be ~ 10 days[77] and in the oceans 350 years.[78] Methylation and photodecomposition rates have been estimated to be in the ranges of 11–47 and 8–40 pmol m^{-2} d^{-1}, respectively.[79] These figures indicate a high degree of biogeochemical activity and a rapid removal from the water column. Open-ocean concentrations of reactive mercury (which includes 'labile' organo-Hg associations, but not the more stable organo-Hg associations) are around ~ 2 pM[80] although concentrations can rise to >50 pmol l^{-1} in polluted coastal waters.[81] The WHO maximum tolerable limits for mercury consumption are 0.3 mg week, of which not more than 0.2 mg should be as methylmercury. These levels may be exceeded in societies where the diet is heavily fish-based, particularly where top-predators are consumed.[82]

The speciation of mercury is critically determined by redox conditions with mercury being rendered biologically unavailable in anoxic conditions where it forms the highly insoluble HgS. It is the methyl form that represents the greatest biohazard and this may derive (as in the Minimata incident) either from direct industrial releases or from bacterially mediated methylation reactions in surface sediments.[74]

Minimata disease arose from the consumption of contaminated seafood from Minimata Bay, Japan.[83] The first incidences were reported in the late 1950s and recent estimates put the total of affected people at over 2250 victims of whom nearly 1800 have died. Current risk analysis suggests that the populations that are at greatest risk of mercury exposure are subsistence fishing communities and pregnant mothers where the foetus may be particularly vulnerable to methyl mercury in the diet.[84]

2.3.4.4 Lead

The global cycle of lead has been strongly perturbed by anthropogenic effects, and the metal is an example of the input of a contaminant to the natural environment mainly through its release into the atmosphere. For example it has been estimated that in the late 1980s the global anthropogenic emission of Pb to the atmosphere was ($\sim 332 \times 10^6$ kg yr^{-1}), which exceeded the natural emissions ($\sim 12 \times 10^6$ kg yr^{-1}) by a factor of ~ 28.[85]

Lead has been widely used in a number of industrial applications for thousands of years. In the mid to late 20th century the main anthropogenic input to the environment was *via* the combustion of fossil fuels, especially from the use of lead alkyls (tetraethyl and tetramethyl lead) as an anti-knock additive in combustion engine fuel. However, the use of lead in fuels has decreased markedly over the past few years with the reduction of the anti-knock additive in Europe and North America and environmental concentrations have similarly declined.[86,87]

Whilst the anthropogenic origins of much environmental lead are well established, instances of acute lead toxicity are rare. One major incident on the Mersey Estuary UK, in which there was major mortality in bird populations, derived from emissions of alkyl lead products from a manufacturing plant.[88–90] Unlike mercury, the marine environment is not thought to be a significant source of lead to human populations.

2.3.4.5 Tin

Tin is now recognised as being potentially a very serious marine pollutant and is an example of a substance introduced into the marine environment for a specific purpose, rather than one which enters the system either as a by-product of anthropogenic activity, or as a result of accidental release. Many observers regard tributyl tin (TBT) as the most dangerous substance ever deliberately released into the marine environment.

Tin has many industrial uses, but it is the organotin compounds, notably TBT, which have given rise to the greatest concern. The principal uses of organotin compounds have been as stabilisers for

PVC and as biocides, especially anti-fouling paints used on ships hulls. Organotin compounds in the marine environment derive from two main sources:

(i) *Via* the bacterial methylation of inorganic tin.
(ii) By the sea water leaching of alkyl and aryl tin from some types of anti-fouling paints which may affect non-target species, especially in areas such as harbours, boat yards and marinas where there are large numbers of ships.

Concern over the environmental effects of organo-tin compounds have led to a number of restrictions on their use in marine systems notably the *EU Regulation No. 782/2003* which banned the use of anti-fouling paints containing TBT and also banned ships with TBT antifouling from visiting EU ports. Of global importance is the International Convention for the Control of Harmful Antifouling Systems on Ships (AFS Convention – which entered into force in 2008). These (and other) restrictions have led to a decline in organo-tin concentrations in marine systems.[91–93]

The principal concern over TBT relates to its role in the phenomenon of 'imposex' which occurs when exposure to an environmental chemical results in changes in the physical structures or the reproductive system. Gastropods are particularly susceptible and *Nucella lapillus* (dog whelks) demonstrate imposex through the development of a penis in females. This phenomenon can be induced at very low environmental concentrations (<50 ng TBT Sn L^{-1})[94] and was an unrecognised internationally widespread problem for some time before analytical techniques were sufficiently advanced to measure such small concentrations. Whilst observed instances of imposex have generally been in decline there are still problems with ships with TBT applied before the recent bans[95] and there are also concerns about the longer term legacy of TBT in sediments.[96]

2.3.5 Radioactivity

The release, or even the potential release, of radioactive substances to the environment is always a major issue. A recent review of radioactivity in the marine environment has recently been produced by the International Atomic Energy Agency (IAEA)[97] and a related bibliography has recently been compiled by the National Oceanic and Atmospheric Administration (NOAA) in response to the tsunami damage to the Japanese nuclear reactors at the Fukushima Daiichi power station.[98]

Seawater is naturally radioactive but over the years the background levels have been supplemented by anthropogenic sources. Principal amongst these latter sources have been global fallout from weapons testing programmes, operational releases from nuclear installations, accidental releases from nuclear installations (notably Chernobyl and Fukushima), dumping of nuclear wastes into the world's oceans, nuclear submarine accidents, contributions from weapons testing sites and satellite burnup.[97] Not all of these sources are of the same magnitude of significance so only a few specific examples will be examined in this chapter.

2.3.5.1 *The Natural Radioactivity of Sea Water*

Sea water and marine sediments are naturally radioactive (see Table 2.15). Sea water itself has a radioactivity of around 12.6 Bq L^{-1}, with that of marine sands and muds being around 200–400 Bq kg^{-1} and 700–1000 Bq kg^{-1}, respectively. Most of the radiation comes from the isotope potassium-40, but there are numerous members of the uranium and thorium decay series present as well. In addition, the creation of lighter radioisotopes through the interaction of cosmic rays and atmospheric gases, with the products subsequently transferred to the ocean surface, also makes a contribution (see Table 2.15). Chemically, radionuclides behave almost identically to their stable counterparts (where these exist). They are, therefore, partitioned between water, sediments and

Table 2.15 The concentrations of some radionuclides in sea water.[99]

Radionuclide	Concentration ($Bq\ kg^{-1}$)
Potassium-40	11.84
Tritium	0.022–0.11
Rubidium-87	1.07
Uranium-234	0.05
Uranium-238	0.04
Carbon-14	0.007
Radium-228	$(0.0037–0.37)\times10^{-2}$
Lead-210	$(0.037–0.25)\times10^{-2}$
Uranium-235	0.18×10^{-2}
Radium-226	$(0.15–0.17)\times10^{-2}$
Polonium-210	$(0.022–0.15)\times10^{-2}$
Radon-222	0.07×10^{-2}
Thorium-228	$(0.007–0.11)\times10^{-3}$
Thorium-230	$(0.022–0.05)\ \times10^{-4}$
Thorium-232	$(0.004–0.29)\times10^{-4}$

biota according to their behavioural properties. Examples of this chemically determined behaviour include:

(i) Cs-137, which is largely water soluble and behaves as a fairly conservative property of sea water moving with the prevailing currents.
(ii) Plutonium-239/240 isotopes, which are highly non-conservative and form strong associations with fine-grained sediments.

One of the more important consequences of the natural radioactivity of marine systems is that marine organisms have evolved in a comparatively radioactive environment, so that the additional radiation introduced as a result of human activities is to a considerable extent a Type I rather than a Type II pollutant problem.

2.3.5.2 Radiation Releases from Weapons Testing Programmes

Since the first atomic weapons test in 1945 there have been over 2000 tests instigated by a small number of countries namely: USA (1032), USSR/Russia (715), France (210), China (45), UK (45), India (3), Pakistan (2) and North Korea (1).[100] In general, only those tests conducted up to the introduction of the Partial Nuclear Test Ban Treaty in 1963 resulted in significant releases to the marine environment. In the years between 1963 and 1996, testing was confined to underground sites and post 1996 testing has been prohibited amongst those nations that have ratified the Comprehensive Nuclear-Test Ban Treaty. The majority of the early tests were conducted in the Northern hemisphere and the resultant fallout of the key radionuclides Sr-90 and Cs-137 also predominantly fell in this region (see Figure 2.6a and b).

2.3.5.3 Releases from Nuclear Power Plants

Under normal operating conditions releases of radioactive materials from operational nuclear power plants are small. Regular reports of releases and resultant exposures for UK establishments are published by a number of government departments.[101] Isotopes that are regularly monitored in marine systems and food include: Co-60, Sr-90, Zr-95, Nb-95, Ru-106, Sb-125, Cs-134, Cs-137,

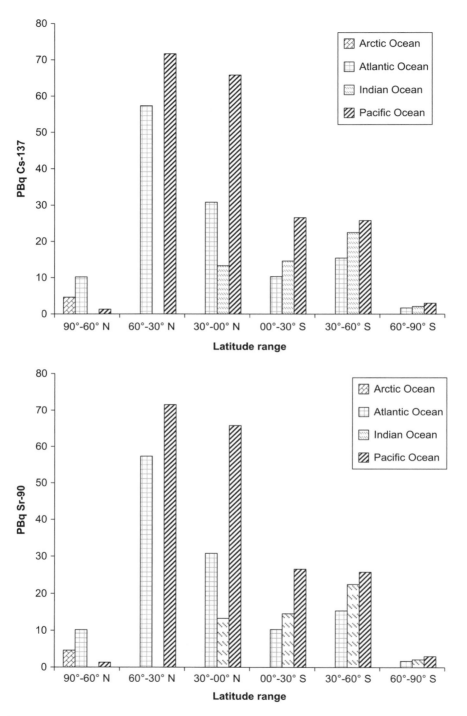

Figure 2.6 Fallout of (a) Sr-90 and (b) Cs as a function of latitude. (Data from ref. 97).

Table 2.16   Amounts of intermediate and low level radioactive waste dumped in the Northeast Atlantic.[99]

Year	α emitters (TBq)	β/γ emitters (TBq)	Tritium (TBq)
1974	15.5	40 700[a]	
1975	28.9	1130	1100
1976	32.6	1200	775
1980	70.3	3075	3630
1981	77.7	2930	2750
1982	51.8	1830	2860
Total 1948–82	**660**	**38 000**	**15 000**

[a]Combined total of β/γ emitters and tritium.

Ce-144, Am-241, Pu-239 and Pu-240. Calculated exposures to members of the public are very small relative to both natural background levels and international regulations.

Major incidents involving reactors have occurred most notably Three Mile Island, USA (1979), Chernobyl, Ukraine (1986) and Fukushima Daiichi, Japan (2011). The first two incidents occurred through operator error, whereas the most recent incident came about as a result of the inundation of the power station site by seawater following a tsunami. Both Chernobyl and Fukushima resulted in widespread dispersal or radioactivity although, whilst traces of iodine-131 from Japan were detected in UK air samples, only the Cs-137 from Chernobyl has had a lasting legacy in this country.[101] The debris field from the Japanese tsunami has been migrating across the Pacific and has begun to arrive on the coasts of North America and Canada. However, because the debris was washed into the ocean before the major problems at the reactors manifested themselves, it currently seems unlikely that radiation-contaminated material is involved. However, locally nuclear workers in particular were exposed[102] the Japanese coast has been contaminated and in the longer term isotopes such as Cs-137 may cross the Pacific.[103]

2.3.5.4   Direct Dumping of Low Level Waste

Disposal of low-level, packaged radioactive waste in the oceans was used by a number of countries until 1983 when the authorising body, the London Dumping Convention, (now the London Convention) took the political decision to suspend it. This decision was confirmed in 1995 when the members of the Convention voted to cease all marine waste disposal. The total radiation inventory dumped in the north-east Atlantic is given in Table 2.16.

2.3.5.5   Naval Sources of Radioactivity to the Oceans

Since 1945 at least 50 nuclear warheads and nine reactors have been introduced to the world's oceans mainly as a result of accidents to submarines (see ref. 104 for a useful overview). The incident of greatest concern appears to be the loss of a Soviet 'Mike Class' submarine, the *Komsomolets* containing the reactor, and two nuclear armed torpedoes, which sank some 270 miles north of Norway in April 1989, after a fire on board. The proximity of this wreck to land, and the damage sustained by the vessel prior to sinking, mean that radiation is leaking from the wreck. However, any kind of recovery operation is potentially very hazardous and the risks of retrieval are assessed to outweigh the likely benefits.[104,105]

2.3.6   The Effects of Artificial Radioactivity on the Marine Environment

Humans are the most sensitive of living organisms to the effects of radiation, and so historically, all radiological population measures have been designed to protect humans on the understanding that all other species will therefore automatically receive adequate protection.[106] More recently the view of the International Commission on Radiological Protection (ICRP) towards the effects of radiation on non-human organisms has been modified somewhat,[107,108] with the ICRP moving towards a more sophisticated assessment of wider effects. In its 2007 recommendations,[109] ICRP indicated its intentions to develop a clearer framework 'in order to assess the relationships between exposure and dose, and between dose and effect, and the consequences of such effects, for non-human species, on a common scientific basis'. How this approach will affect marine systems has yet to be determined.

The application of 'Critical Pathway Analysis' to nuclear discharges permits the identification of people or populations whose location or behaviour makes them high outliers in the radiation dose spectrum.[101] The general conclusions for the UK are that the highest (mid-1970s) Sellafield dose rates were about two orders of magnitude below those likely to cause observable effects at the population level. Since then discharges have declined by several orders of magnitude. Radiation from Chernobyl had an effect on the marketability of sheep from the high fells until June 2012 when the final restrictions were lifted. The effects of Chernobyl on the marine environment were much more transient. The Fukushima incident has currently resulted in restrictions on fishing and the sale of seafood from the coastal regions close to the accident site. It is not clear how long these restrictions will persist.

2.4 CONCLUSIONS

The influence of human activities on the marine environment can be detected in even the most remote regions. The main impacts are, not surprisingly, in coastal waters which are both closest to the sources of pollutants and the most physically, chemically and biologically active zones. In such coastal areas waste discharges combine with other pressures, such as coastal developments and fishing, to produce deleterious effects which may be environmentally significant. The situation in the more remote oceanic regions has not yet become as serious though high latitude regions are not immune (*e.g.* suggested link between hermaphroditism in polar bears and environmental chemicals),[110] and human activities are normally only detectable though the presence of artificially produced (Type II) chemicals. With the exception of lead, all Type I pollutants normally fall within the range of natural variability. Problems of chemical persistence are still evident, most obviously in the major 'garbage patch' in the Pacific Ocean where huge quantities of non-biodegradable plastics and other materials are accumulating.

In the decade that has elapsed since the previous edition of this book, the focus of scientific and social concern about the marine has to some extent changed from what might be termed 'acute' events (*e.g.* major oil spills which have declined in number) to more subtle effects such as the extent and influence of endocrine disrupting chemicals that appear to be widespread in the system. In general we have become better at not killing or damaging things outright but the long-term effects of human disturbances of natural systems have yet to be played out.

REFERENCES

1. R. G. Williams and M. J. Follows, *Ocean Dynamics and the Carbon Cycle*, Cambridge University Press, University Press, Cambridge, 2011.
2. S. R. Emerson and J. I. Hedges, *Chemical Oceanography and the Marine Carbon Cycle*, Cambridge University Press, Cambridge, 2009.
3. R. Chester and T. Jickells, *Marine Geochemistry*, 3rd edn., Wiley-Blackwell, 2012.
4. J. H. Simpson, in *Understanding the North Sea System*, H. Charnock, K. R. Dyer, J. Huthnance, P. S. Liss, J. H. Simpson and P. B. Tett, Chapman and Hall, London, 1993, pp. 1–4.
5. www.bp.com/statisticalreview. BP Statistical review of World Energy 2013 (Last accessed 1/9/2013).
6. http://www.lloydslistintelligence.com/llint/tankers/index.htm Lloyds List Intelligence, Tankers. (Last accessed 1/8/2013).
7. http://www.imo.org/ourwork/environment/pollutionprevention/oilpollution/pages/constructionrequirements.aspx.

8. http://www.itopf.com/information-services/data-and-statistics/statistics/ International Tanker Owners Pollution Federation, Information Services. (Last accessed 1/8/2013).
9. N. Wheeler, The Sea Empress Oil Spill, Cardiff, Terence Dalton Publishers, 1998.
10. GESAMP, *Impact of Oil and Related Chemicals on the Marine Environment*, GESAMP (IMO/FAO/UNESCO-IOC/UNIDO/WMO/IAEA/UN/UNEP Joint Group of Experts on the Scientific Aspects of Marine Environmental Protection), 1993.
11. J. W. Doerffer, *Oil Spill Response in the Marine Environment*, Pergamon Press, Oxford, 1992.
12. NRC, *Oil in the Sea III*, The National Academies Press, Washington, DC, 2002.
13. GESAMP, *Estimates of Oil Entering the Marine Environment from Sea-based Activities*. 75, GESAMP (IMO/FAO/UNESCO-IOC/UNIDO/WMO/IAEA/UN/UNEP Joint Group of Experts on the Scientific Aspects of Marine Environmental Protection), 2007.
14. J. Baelum, S. Borglin, R. Chakraborty, J. L. Fortney, R. Lamendella, O. U. Mason, M. Auer, M. Zemla, M. Bill, M. E. Conrad, S. A. Malfatti, S. G. Tringe, H. Y. Holman, T. C. Hazen and J. K. Jansson, *Environ. Microbiol.*, 2012, **14**, 2405–2416.
15. A. R. Diercks, R. C. Highsmith, V. L. Asper, D. J. Joung, Z. Z. Zhou, L. D. Guo, A. M. Shiller, S. B. Joye, A. P. Teske, N. Guinasso, T. L. Wade and S. E. Lohrenz, *Geophys. Res. Lett.*, 2010, **37**.
16. J. C. Dietrich, C. J. Trahan, M. T. Howard, J. G. Fleming, R. J. Weaver, S. Tanaka, L. Yu, R. A. Luettich, C. N. Dawson, J. J. Westerink, G. Wells, A. Lu, K. Vega, A. Kubach, K. M. Dresback, R. L. Kolar, C. Kaiser and R. R. Twilley, *Cont. Shelf Res.*, 2012, **41**, 17–47.
17. P. Dittrick, *Oil Gas J.*, 2010, **108**, 28–28.
18. S. K. Griffiths, *Environ. Sci. Technol.*, 2012, **46**, 5616–5622.
19. A. Horel, B. Mortazavi and P. A. Sobecky, *Environ. Toxicol. Chem.*, 2012, **31**, 1004–1011.
20. I. A. Mendelssohn, G. L. Andersen, D. M. Baltz, R. H. Caffey, K. R. Carman, J. W. Fleeger, S. B. Joye, Q. X. Lin, E. Maltby, E. B. Overton and L. P. Rozas, *Bioscience*, 2012, **62**, 562–574.
21. S. Mitra, D. G. Kimmel, J. Snyder, K. Scalise, B. D. McGlaughon, M. R. Roman, G. L. Jahn, J. J. Pierson, S. B. Brandt, J. P. Montoya, R. J. Rosenbauer, T. D. Lorenson, F. L. Wong and P. L. Campbell, *Geophys. Res. Lett.*, 2012, **39**.
22. E. R. Gundlach and M. O. Hayes, *Mar. Technol. Soc. J.*, 1978, **12**, 18–27.
23. Concawe, *A Field Guide to Coastal Oil Spill Control and Clean-up Techniques*, CONCAWE, 1981.
24. A. J. Mearns, D. J. Reish, P. S. Oshida, T. Ginn and M. A. Rempel-Hester, *Water Environ. Res.*, 2011, **83**, 1789–1852.
25. J. E. Smith, *'Torrey Canyon' Pollution and Marine Life*, Cambridge University Press, Cambridge, 1970.
26. H. Caspers, *Int. Rev. Gesamten Hydrobiol. Hydrograph*, 1983, **68**, 897–897.
27. M. Fingas, *Oil Spill Science and Technology*, Elsevier Science, 2010.
28. ITOPF, *Response to Marine Oil Spills*, Weather by Seamanship International, Livingston, UK, 2012.
29. I. Buist, S. Potter, T. Nedwed and J. Mullin, *Cold Reg. Sci. Technol.*, 2011, **67**, 3–23.
30. http://www.goodbeachguide.co.uk/ Marine Conservation Society. Good Beach Guide. (Last accessed 1/8/2013).
31. http://www.defra.gov.uk/statistics/environment/ Department for Food, Environment and Rural Affairs. Statistics at Defra. (Last accessed 1/8/2013).
32. European Environment Agency, 'Percentage of inland bathing waters in the European Union per compliance category'. http://www.eea.europa.eu/data-and-maps/figures/percentage-of-eu-coastal-and-inland-bathing-waters-meeting-the-non-mandatory-guide-levels-of-the-bathing-water-directive-1992-to-2006-for-eu-2. Last accessed 1/8/2013.
33. P. D. Jones, *Mar. Pollut. Bull.*, 2006, **53**, 144–154.
34. J. J. Vitaliano, S. A. Fromm, D. B. Packer, R. N. Reid and R. A. Pikanowski, *Mar. Ecol. Prog. Ser.*, 2007, **342**, 27–40.
35. J. M. O'Neil, T. W. Davis, M. A. Burford and C. J. Gobler, *Harmful Algae*, 2012, **14**, 313–334.
36. North Sea Task Force and International Council for the Exploration of the Sea, *North Sea Quality Status Report 1993*, Olsen & Olsen, 1993.
37. http://qsr2010.ospar.org/en/downloads.html OSPAR Commission Quality Status report 2010. (Last accessed 1/8/2013).

38. B. Grizzetti, F. Bouraoui and A. Aloe, *GCB*, 2012, **18**, 769–782.

39. GESAMP Principles for Developing Coastal Water Quality Criteria, GESAMP Reports and Studies No. 5, p. 36.

40. H. H. Huss, L. Ababouch and L. Gram, *Assessment and Management of Seafood Safety and Quality*, Food and Agriculture Organisation of the UN, Rome, 2004.

41. D. Barceló, *Emerging Organic Pollutants in Waste Waters and Sludge*, Springer–Verlag, Berlin Heidelberg New York, 2004.

42. D. Barceló, *Emerging Organic Pollutants in Waste Waters and Sludge*, Springer–Verlag, Berlin Heidelberg New York, 2005.

43. K. Kümmerer, *Pharmaceuticals in the Environment, Sources, Fate, Effects and Risks*, Springer, Berlin Heidelberg New York, 2nd edn, 2004.

44. H. van den Berg, M. Zaim, R. S. Yadav, A. Soares, B. Ameneshewa, A. Mnzava, J. Hii, A. P. Dash and M. Ejov, *Environ. Health Perspect.*, 2012, **120**, 577–582.

45. D. Broman, University of Stockholm, PhD Thesis, 1990.

46. X.–X. Sue and P. Zhan, *Toxicol. Environ. Chem.*, 2008, **90**, 829–835.

47. K. Breivik, A. Sweetman, J. M. Pacyna and K. C. Jones, *Sci. Total Environ.*, 2002, **290**, 181–198.

48. K. Breivik, A. Sweetman, J. M. Pacyna and K. C. Jones, *Sci. Total Environ.*, 2002, **290**, 199–224.

49. T. F. Bidleman, P. A. Helm, B. M. Braune and G. W. Gabrielsen, *Sci. Total Environ.*, 2010, **408**, 2919–2935.

50. S. Tanabe, *Environ. Pollut.*, 1988, **50**, 5–28.

51. A. Jonsson, O. Gustafsson, J. Axelman and H. Sundberg, *Environ. Sci. Technol.*, 2003, **37**, 245–255.

52. B. M. Braune, P. M. Outridge, A. T. Fisk, D. C. G. Muir, P. A. Helm, K. Hobbs, P. F. Hoekstra, Z. A. Kuzyk, M. Kwan, R. J. Letcher, W. L. Lockhart, R. J. Norstrom, G. A. Stern and I. Stirling, *Sci. Total Environ.*, 2005, **351**, 4–56.

53. E. C. Bonefeld-Jorgensen, *Rural Remote Health*, 2010, **10**.

54. J. L. Domingo and A. Bocio, *Environ. Int.*, 2007, **33**, 397–405.

55. E. Atlas, T. F. Bidleman and C. S. Giam, ed. J. S. Waid, *PCBs in the Environment*, CRC Press Inc, vol. 1 , 1987, p. 240.

56. C. Bredhult, B.–M. Backlin, A. Bignert and M. Olovsson, *Reprod. Toxicol.*, 2008, **25**, 247–255.

57. M. Breitholtz, C. Hill and B. E. Bengtsson, *Ambio*, 2001, **30**, 210–216.

58. C. Miljeteig, G. W. Gabrielsen, H. Strom, M. V. Gavrilo, E. Lie and B. M. Jenssen, *Sci. Total Environ.*, 2012, **431**, 92–99.

59. M. Houde, A. O. De Silva, D. C. G. Muir and R. J. Letcher, *Environ. Sci. Technol.*, 2011, **45**, 7962–7973.

60. R. J. Letcher, J. O. Bustnes, R. Dietz, B. M. Jenssen, E. H. Jorgensen, C. Sonne, J. Verreault, M. M. Vijayan and G. W. Gabrielsen, *Sci. Total Environ.*, 2010, **408**, 2995–3043.

61. A. O. Cheek, *Rev. Biol. Trop.*, 2006, **54**, 1–19.

62. M. H. Depledge and Z. Billinghurst, *Marine Pollut. Bull*, 1999, **39**, 32–38.

63. E. Oberdorster and A. O. Cheek, *Environ. Toxicol. Chem.*, 2001, **20**, 23–36.

64. C. Porte, G. Janer, L. C. Lorusso, M. Ortiz-Zarragoitia, M. P. Cajaraville, M. C. Fossi and L. Canesi, *Comp. Biochem. Physiol., C: Toxicol. Pharmacol*, 2006, **143**, 303–315.

65. J. G. Vos, E. Dybing, H. A. Greim, O. Ladefoged, C. Lambre, J. V. Tarazona, I. Brandt and A. D. Vethaak, *Criti. Rev. Toxicol.*, 2000, **30**, 71–133.

66. A. Kortenkamp, O. Martin, M. Faust, R. Evans, R. McKinlay, F. Orton and E. Rosivatz, *State of the Art Assessment of Endocrine Disrupters*, European Commission, 2012.

67. L. R. Rhomberg, J. E. Goodman, W. G. Foster, C. J. Borgert and G. Van Der Kraak, *Crit. Rev. Toxicol.*, 2012, **42**, 465–473.

68. K. Hirose, *Anal. Sci.*, 2006, **22**, 1055–1063.

69. F. J. Millero, R. Woosley, B. Ditrolio and J. Waters, *Oceanography*, 2009, **22**, 72–85.

70. S. N. Luoma and P. S. Rainbow, *Metal Contamination in Aquatic Environments – Science and Lateral Management*, Cambridge University Press, Cambridge, 2008.

71. http://minerals.usgs.gov/minerals/pubs/commodity/mercury/mcs-2012-mercu.pdf U.S. Geological Survey, Mineral Commodity Summaries, January 2012. Mercury. (Last accessed 1/8/2013).

72. F. S. Bank, ed., *Mercury in the Environment*, University of California Press, Berkeley, Los Angeles, 1st edn, 2012.
73. F. J. Black, C. H. Conaway and A. R. Flegal, in *Mercury in the Environment*, ed. *M. S. Bank*, University of California Press, Berkley and Los Angeles, 1st edn, 2012, pp. 167–222.
74. A. M. Graham, A. L. Bullock, A. C. Maizel, D. A. Elias and C. C. Gilmour, *Appl Environ. Microbiol.*, 2012, **78**, 7337–7346.
75. N. B. Al-Majed and M. R. Preston, *Marine Pollut. Bull*, 2000, **40**, 298–307.
76. F. J. G. Laurier, D. Cossa, J. L. Gonzalez, E. Breviere and G. Sarazin, *Geochim. Cosmochim. Acta*, 2003, **67**, 3329–3345.
77. I. M. Hedgecock and N. Pirrone, *Environ. Sci. Technol.*, 2004, **38**, 69–76.
78. G. A. Gill and W. F. Fitzgerald, *Geochim. Cosmochim. Acta*, 1988, **52**, 1719–1728.
79. C. R. Hammerschmidt and K. L. Bowman, *Marine Chem*, 2012, **132**, 77–82.
80. C. H. Lamborg, C. R. Hammerschmidt, G. A. Gill, R. P. Mason and S. Gichuki, *Limnol. Oceanogr. Methods*, 2012, **10**, 90–100.
81. N. A. M. BuTayban and M. R. Preston, *Marine Pollut. Bull*, 2004, **49**, 930–937.
82. N. B. Al-Majed and M. R. Preston, *Environ. Pollut.*, 2000, **109**, 239–250.
83. M. Harada, *Crit.Rev. Toxicol.*, 1995, **25**, 1–24.
84. WHO, *Evaluation of Certain Food Additives and Contaminants*, Technical Report Series 922, Geneva, 2004.
85. J. O. Nriagu, *Global Planetary Change*, 1990, **2**, 113–120.
86. Y. Lind, A. Bignert and T. Odsjo, *J. Environ. Monitor*, 2006, **8**, 824–834.
87. C. Barbante, C. Turetta, G. Capodaglio and G. Scarponi, *Int. J. Environ. Anal. Chem.*, 1997, **68**, 457–477.
88. K. R. Bull, W. J. Every, P. Freestone, J. R. Hall, D. Osborn, A. S. Cooke and T. Stowe, *Environ. Pollut., Ser. A*, 1983, **31**, 239–259.
89. D. Osborn, W. J. Every and K. R. Bull, *Environ. Pollut., Ser. A*, 1983, **31**, 261–275.
90. J. P. Riley and J. V. Towner, *Marine Pollut. Bull*, 1984, **15**, 15–158.
91. S. Diez, M. Abalos and J. M. Bayona, *Water Res*, 2002, **36**, 905–918.
92. S. Diez, J. Jimenez and J. M. Bayona, *Cien. Mar*, 2003, **29**, 547–560.
93. F. Zanon, N. Rado, E. Centanni, N. Zharova and B. Pavoni, *Environ. Monit. Assess.*, 2009, **152**, 35–45.
94. M. M. Santos, L. F. C. Castro, M. N. Vieira, J. Micael, R. Morabito, P. Massanisso and M. A. Reis-Henriques, *Comp. Biochem. Physiol., C: Toxicol. Pharmacol*, 2005, **141**, 101–109.
95. S. Galante-Oliveira, I. Oliveira, N. Ferreira, J. A. Santos, M. Pacheco and C. Barroso, *J. Environ. Monit.*, 2011, **13**, 304–312.
96. A. C. Birchenough, N. Barnes, S. M. Evans, H. Hinz, I. Kronke and C. Moss, *Mar. Pollut. Bull.*, 2002, **44**, 534–543.
97. IAEA, *Worldwide Marine Radioactivity Studies (WOMARS) – Radionuclide Levels in Oceans and Seas*, International Atomic Energy Agency, Austria, 2005.
98. NOAA, *Anthropogenic Radionuclides in the Marine Environment: A Selected Bibliography*, 2011.
99. R. B. Clark, *Marine Pollution*, Oxford University Press, Oxford, 5th edn, 2002.
100. SIPRI, *Armaments, Disarmament, and International Security*, Oxford University Press, Oxford, 2007.
101. CEFAS, *Radioactivity in Food and the Environment 2011*, 2012.
102. P. D. McLaughlin, B. Jones and M. M. Maher, *Br. J. Radiol.*, 2012, **85**, 1222–1225.
103. H. Wang, Z. Wang, X. Zhu, D. Wang and G. Liu, *Chin. Sci. Bull.*, 2012, **57**, 3518–3524.
104. O. M. Johannessen, *Radioactivity and Pollution in the Nordic Seas and Arctic: Observations, Modeling and Simulations*, Springer, New York, 1st edn, 2009.
105. S. Hoibraten, P. E. Thoresen and A. Haugan, *Sci. Total Environ.*, 1997, **202**, 67–78.
106. ICRP, Recommendations of the International Commission for Radiological Protection (Users edition) *Annals of the ICRP*, 2007, vol. 103.
107. J. Valentin, *Ann. ICRP*, 2003, **33**, 207–266.
108. A. D. Wrixon, *J. Radiol. Protect*, 2008, **28**, 161–168.
109. Anon, *Ann. ICRP*, 2007, **37**, 1–332.
110. L. E. Carmichael, P. Krizan, S. P. Blum and C. Strobeck, *J. Mammal.*, 2005, **86**, 160–169.

CHAPTER 3

Drinking Water Quality and Health

JOHN K. FAWELL[†‡]

Water Science Institute, Cranfield University, Cranfield, Bedfordshire, MK43 0AL, UK
Email: john.fawell@cranfield.ac.uk

3.1 INTRODUCTION

An adequate supply of safe drinking water is a major prerequisite for a healthy life. The importance of clean water and the link between contaminated or putrid water and illness was recognized in the distant past, even though the actual cause of disease was not properly understood until the latter half of the 19th century. Herodotus who lived in the 5th century BC records that Cyrus the Great, who lived a century before, carried boiled water stored in flagons of silver when he travelled. This was sound practice because boiling would have killed all of the pathogenic microorganisms and silver ions would have acted as a bacteriastat. Contaminated water was recognized by its foetid smell and unpleasant taste although we now know that sweet tasting water can still carry pathogens.

The Romans developed a sophisticated water supply system bringing clean water from uncontaminated sources and separating sewage disposal from drinking water supply. Most of this knowledge of water supply was lost after the fall of the Roman Empire, although some remnants can be detected in certain monasteries. As cities grew the problems associated with water supply also grew and in the 19th century outbreaks of typhus and cholera brought misery and fear to many of the great cities of Europe and North America. Abstracting water upstream of where sewage entered rivers was a great benefit as was the introduction of filtration through gravel and sand. Attempts were made to introduce disinfection using chlorine and ozone in the 19th century. John Snow, who identified the Broad Street pump as the source of cholera in the outbreak in that part of London in 1854, is reported to have attempted to treat the source with calcium hypochlorite. However, it was not until the 20th century that disinfection with chlorine was introduced on a wide scale with the first permanent chlorination plant at Middelkerke in Belgium in 1902. Chlorination

---

[†]Visiting Professor
[‡]This chapter is based upon an earlier contribution co-authored with Geoff Stanfield, whose contribution is gratefully acknowledged.

Pollution: Causes, Effects and Control, 5th Edition
Edited by R M Harrison
© J K Fawell and The Royal Society of Chemistry 2014
Published by the Royal Society of Chemistry, www.rsc.org

spread rapidly in the United States but it was not until the 1937 Croydon typhoid outbreak and the advent of the Second World War that universal chlorination was introduced in the UK.

The first drinking water standards related primarily to microbiological contaminants, although a number of inorganic contaminants were recognized. It was not until the development of advanced analytical techniques, such as gas chromatography-mass spectrometry in the 1970s, that organic contaminants in drinking water were recognized as an issue. This in turn resulted in a major change in the way that drinking water was regulated and provided the impetus for the development of new methods of water treatment. Unfortunately, waterborne disease is still a major cause of death in many parts of the world, particularly in children, and it is also a significant economic constraint in many subsistence economies. In developed countries there have been enormous changes in our knowledge of water contaminants and in the way that water is treated and supplied. The major risks associated with significant microbiological contamination have largely been overcome with the advent of both sewage treatment and the ability to treat large quantities of drinking water to a high level of microbiological and chemical purity. However, contamination can still occur and there is a constant need to maintain a high level of vigilance.

3.2 DRINKING WATER SOURCES

Drinking water is abstracted from three basic sources: lowland rivers and reservoirs, upland reservoirs and groundwater. All public water supplies in the UK receive treatment that reflects the quality of the raw water and the variability in that quality. Lowland surface water sources contain a high level of natural and anthropogenic inorganic and organic matter and are much more prone to pollution. They receive the highest level of treatment. Upland sources usually suffer less anthropogenic contamination but they usually contain high levels of naturally occurring organic matter from decaying vegetation. This results in the brown colouration seen in many upland streams. There can also, on occasion, be microbiological contamination from farm animals and wildlife. Upland waters are usually soft waters low in inorganic constituents. Groundwaters are usually low in organic matter and are less vulnerable to both microbiological and chemical contamination. They frequently contain high levels of inorganic substances from the rocks through which the water percolates.

Drinking water treatment is intended to remove microorganisms and chemical contaminants. However, the process itself can result in the formation of other contaminants, such as the trihalomethanes, from the reaction of chemical oxidants with naturally occurring organic matter. This requires a balance to be struck between the benefits of the chemical oxidants in destroying microorganisms and the potential risks from the by-products.

In many countries a significant proportion of the population does not receive water through the public water supply. Their water comes from small local supplies, which may be wells, springs, streams, or small reservoirs fed by streams or springs. The level of treatment received by these supplies is very variable and is often minimal, while source protection is often poor. This problem is compounded by limited resources to make improvements and so such supplies are potentially more at risk of contamination.

3.3 DRINKING WATER TREATMENT AND SUPPLY

Drinking water treatment, as applied to public water supplies, consists of a series of barriers in a treatment train which will vary according to the nature and vulnerability of the source and the requirements of the supply. Broadly for surface waters these will be coagulation and flocculation, filtration and oxidation.

Coagulation and flocculation using a coagulant, such as aluminium sulphate or ferrous sulphate will remove particulates, including microorganisms, and dissolved organic matter. Filtration includes slow sand filtration, which will remove particles, microorganisms and some organic matter; rapid gravity filtration to remove particles and associated microorganisms; and granular activated carbon (GAC) which removes many dissolved organic substances by adsorption. Oxidants are used both as part of the process for removal of chemicals and as disinfectants, which include chlorine, chlorine dioxide, ozone and ultraviolet irradiation. However, chlorine and chloramine are the main disinfectants used to provide a residual to help maintain hygienic conditions through the distribution system. Advanced treatment is now common for lowland surface waters, particularly membrane filtration, which provides a more comprehensive physical barrier, used for removal of microorganisms and chemical contaminants, depending on the pore size. Increasingly micro and nanofiltration using membranes are becoming the norm for sewage impacted surface waters and reverse osmosis is widely used for desalination and for treating wastewater to a standard where the effluent can be indirectly reused. The major targets of advanced treatment are to remove natural organic matter that is the precursor of disinfection by-products and also emerging contaminants which are discussed in section 3.7.2.7.

Water supplies in any one area can be quite complex with several different sources. Sources may also be remote from the area supplied, requiring long distribution systems, and they are sometimes changed over time in response to demand, water quality problems or operational difficulties. As demand for water has increased, particularly where availability might be limited, efforts are being made to make supplies even more flexible so that demand can be balanced over a wider area.

3.4 SOURCES OF CONTAMINATION

Water can be contaminated from both natural and anthropogenic sources. Most surface waters contain natural organic matter in the form of large complex humic and fulvic acids. All will contain inorganic substances dissolved from the minerals through which the water percolates and most will contain particles of clay and sand washed from the surrounding catchment, banks and bottom sediments. There are a number of possible sources of man-made contaminants, some of which are more important than others. Discharges from industrial premises and sewage treatment works are point sources and as such are more readily identifiable and controlled, in the UK, by the Environment Agency (see Chapter 21). However, diffuse sources, such as run-off from agricultural land and hard surfaces, such as roads, are not so easily controlled. Such sources can give rise to a significant variation in the contaminant load over time. Drinking water treatment can remove many contaminants, but it can also change the contaminants present, *e.g.* chlorine reacts with phenol to give chlorophenols, or generate other contaminants such as disinfection by-products, *e.g.* trihalomethanes (THMs). Contamination from distribution may arise from materials used in distribution such as iron which can corrode to release iron oxides, or from ingress into the distribution system. This latter phenomenon is a particular problem when the components of oil, spilt into the surrounding soil, diffuse through plastic pipes giving rise to taste and odour problems. Contamination can also take place in consumer's premises from materials used in plumbing or from back siphoning of liquids into the distribution system as a consequence of improper connections, in breach of regulations. Such contaminants can be either chemical or microbiological. It is recommended that water is not drunk from taps supplied from storage tanks, which may be open to the atmosphere or accessible by vermin because they can be easily contaminated, or from the hot water system, which may increase dissolution of metals from the plumbing.

Chemical contaminants may be present most of the time, in which case the issue for consumers is chronic exposure and chronic toxicity, or for short periods, such as in a spill that penetrates treatment, in which case the issue is likely to be acute toxicity following short-term exposure. Although the average water usage for a family of two adults and two children in the UK is about 475 l per day, only a very small proportion of this is used for drinking or cooking. An adult drinks about 1 to 2 l per day, a significant proportion of which is boiled. Children drink smaller quantities of water but this can be greater than adults when considered in relation to bodyweight. Normally, in carrying out risk assessments of chemical contaminants in drinking water, 1 l is allowed for a 10 kg child while a bottle-fed infant will ingest up to 150 ml kg^{-1} bodyweight, all of which will normally be boiled. Exposure to contaminants can also arise as a consequence of washing and showering. In general, exposure to volatile contaminants can be approximately doubled by skin exposure and inhalation.

3.5 DRINKING WATER GUIDELINES AND STANDARDS

The basis on which the quality of drinking water is judged in Europe is the *Directive on the Quality of Water for Human Consumption*,[1] which is transcribed into UK law in the *Drinking Water Regulations*.[2] The Directive lays down the minimum standards that must be met, but member states may extend the parameters covered or apply more stringent standards. The regulations cover a range of contaminants in public water supplies and specify sampling requirements for those contaminants. Similar regulations also apply to private supplies but enforcement and surveillance lies with local authorities.[3] Separate regulations apply to bottled waters, while water that is classified as natural mineral water is covered by a different directive.

Many of the parameters are based on *WHO Guidelines* and the third and fourth editions of the *WHO Guidelines for Drinking Water Quality* have also provided a stimulus for change with the introduction of Water Safety Plans which are discussed later in this chapter.[4] The Directive specifies three groups of parameters. These are microbiology (see Table 3.1), chemical contaminants (see Table 3.2) and finally indicator parameters (see Table 3.3), which are not directly related to health and which member states can apply in a more flexible way. There are only a limited number of contaminants covered by the drinking water directive because it is not possible, or desirable, to set legally binding standards for all possible contaminants. However, the directive does contain a catch-all clause that means that nothing else should be present in numbers or concentrations that could be of concern for health. The *WHO Guidelines* cover more contaminants than the standards and are an important source of information on what would be considered acceptable concentrations.[5]

One of the most frequently encountered questions regarding drinking water quality relates to the consequences of a standard being exceeded. To make an appropriate assessment of the implications

Table 3.1 Bacteriological quality standards and guidelines, comparison of *EC Drinking Water Directive* and *WHO Guidelines*.

Parameter	EC Directive	WHO Guidelines	Comments
E. coli	0 in 100 ml	0 in 100 ml	WHO – may use Thermo-tolerant coliforms
Enterococci	0 in 100 ml	Not included	
Clostridium perfringens and spores	0 in 100 ml	Not included	Indicator parameter for surface waters triggers investigation
Colony count 22 °C	No abnormal change	Not included	Indicator parameter for distribution
Total coliforms	0 in 100 ml	0 in 100 ml	Indicator parameter for distribution

Table 3.2 *EC Directive Standards* for chemical parameters and comparison with *WHO Guidelines*.

Parameter	EC Directive	WHO Guidelines	Comments
Acrylamide	0.1 µg l^{-1}	0.5 µg l^{-1}	Control through product spec[a]
Antimony	5 µg l^{-1}	20 µg l^{-1}	WHO revised guideline
Arsenic	10 µg l^{-1}	10 µg l^{-1}	WHO provisional (health data)
Benzene	1 µg l^{-1}	10 µg l^{-1}	WHO 10^{-5} cancer risk
Boron	1 mg l^{-1}	2.4 mg l^{-1}	WHO revised guideline
Bromate	10 µg l^{-1}	10 µg l^{-1}	WHO provisional (Toxicology)
Cadmium	5 µg l^{-1}	3 µg l^{-1}	
Chromium	50 µg l^{-1}	50 µg l^{-1}	WHO provisional (health data)
Copper	2 mg l^{-1}	2 mg l^{-1}	WHO acute value EC relates to plumbing
Cyanide	50 µg l^{-1}	none	WHO health-based value for emergencies
1,2-dichloroethane	3 µg l^{-1}	30 µg l^{-1}	WHO 10^{-5} cancer risk
Epichlorohydrin	0.1 µg l^{-1}	0.4 µg l^{-1}	Control through product spec
Fluoride	1.5 mg l^{-1}	1.5 mg l^{-1}	
Lead	10 µg l^{-1}	10 µg l^{-1}	EC average over week at tap. Applies 2013, 25 µg l^{-1} 2003 to 2013. Relates to plumbing
Mercury	1 µg l^{-1}	6 µg l^{-1}	
Nickel	20 µg l^{-1}	70 µg l^{-1}	WHO revised. EC relates to plumbing
Nitrate	50 mg l^{-1}	50 mg l^{-1}	Applied as sum of ratio of value against WHO GV for nitrate and nitrite. EC also 0.1 mg l^{-1} nitrite ex-treatment works
Nitrite	0.5 mg l^{-1}	3 mg l^{-1}	
Pesticides each	0.1 µg l^{-1}	Individual pesticides	EC is precautionary value not based on health, includes 'relevant' degradation products
Total	0.5 µg l^{-1}		
Polycyclic aromatic hydrocarbons	0.1 µg l^{-1}	Not included	EC four specified PAH based on early WHO GV. WHO state value for fluoranthene unnecessary
Benzo[a]pyrene	0.01 µg l^{-1}	0.7 µg l^{-1}	EC based on 1984 WHO GV
Selenium	10 µg l^{-1}	40 µg l^{-1}	
Tetrachloro- and trichloroethene	10 µg l^{-1}	40 µg l^{-1} & 70 µg l^{-1}	EC precautionary value
Trihalomethane total	100 µg l^{-1}	Individual THMs	WHO applied as sum of ratios of individual THMs
Vinyl chloride	0.5 µg l^{-1}	5 µg l^{-1}	WHO 10^{-5} cancer risk. Applied through product spec

[a]Analytical methods becoming available.

of a particular exceedance, it is important to know the basis on which the standard is set. For example, not all values in the standards are derived from data relating to health. Some, such as the standards for iron, manganese and aluminium, are based on preventing discolouration and dirty water problems. The current standard for polycyclic aromatic hydrocarbons (PAH) is what might be expected in an unpolluted supply. The pesticide parameter of 0.1 µg l^{-1} is based on a political, precautionary approach that reflected the limit of analytical detection when the original directive was formulated.

When a health-based standard for a chemical is exceeded, the key questions that need to be asked are: how long will exposure to the high concentration last and what is the margin of safety in the guideline value or standard? Most standards are set on the basis of long-term exposure and some averaging of exposure is acceptable, while most contain a significant margin of safety, which allows

Table 3.3 Comparison of *EC Directive* indicator parameters and *WHO Guidelines*.

Parameter	EC Directive	WHO	Comments (WHO only set GVs on basis of health)
Aluminium	200 µg l^{-1}	No GV	Prevents discolouration
Ammonium	0.5 mg l^{-1}	No GV	Indicator of possible bacterial, sewage and animal waste pollution
Chloride	250 mg l^{-1}	No GV	Corrosion and taste
Colour	Acceptable to consumers and no abnormal change	No GV	Indicator of pollution and operational problems
Conductivity	2.500 µS cm^{-1} at 20 °C		Corrosion
pH	≥ 5 and ≤9.5	No GV	Operational parameter
Iron	200 µg l^{-1}	No GV	Discolouration. 2 mg l^{-1} not hazard to health (WHO 1993)
Manganese	50 µg l^{-1}	No GV	Discolouration
Odour	Acceptable to consumers and no abnormal change	No GV	Indicator of pollution and operational problems
Oxidizability	5 mg l^{-1} O2	No GV	Operational parameter not needed if Total Organic Carbon (TOC) measured
Sulfate	250 mg l^{-1}	No GV	Corrosion, taste
Sodium	200 mg l^{-1}	No GV	Taste
Taste	Acceptable to consumers and no abnormal change	No GV	Indicator of pollution or operational problems
Total Organic Carbon (TOC)	No abnormal change	No GV	Indicator of pollution and disinfectant demand
Turbidity	Acceptable to consumers and no abnormal change	No GV	Indicator of pollution and operational problems

some leeway. It is important not just to take standards and guidelines at their face value but to examine the rationale behind them.

3.6 MICROBIOLOGICAL CONTAMINANTS

One of the great scourges of cities in Europe and North America in the 19th century was outbreaks of waterborne diseases such as cholera and typhoid. The development of sewerage systems and drinking water treatment were major factors in breaking the cycle of contamination of drinking water sources by infected faecal matter and continuing infection of the population. This was undoubtedly one of the great public health triumphs of the 19th century and its importance is still being demonstrated by outbreaks of these diseases in developing countries when drinking water treatment breaks down.

The problem is that faecal matter contains pathogenic organisms. However, since detection and enumeration of the pathogens is still extremely difficult, standards are stipulated in terms of microbiological indicators of faecal contamination (see Table 3.1). The indicators used are *Escherichia coli* and faecal streptococci, which must not be detectable in a 100 ml sample. The assumption is that if the indicators are detected, the pathogens, including viruses and parasites, could also be present and therefore appropriate action is required. One further indicator parameter is the spores of *Clostridium perfringens*, which are not as susceptible to disinfection and provide a measure of the efficacy of filtration. In particular a change in numbers of *C perfringens* spores can indicate a problem in the filtration process.[4]

Drinking water is not, however, sterile and bacteria can be found in the distribution system and at the tap. Most of these organisms are harmless, but some opportunist pathogens such as *Pseudomonas aeruginosa* and Aeromonas spp. may multiply in distribution or in the plumbing of buildings given suitable conditions.[6] Currently there is considerable debate as to whether these organisms are responsible for a significant level of waterborne, gastrointestinal disease in the community but they are important in hospitals where there is a vulnerable population many of whom have wounds. Research is continuing but there are difficulties, such as the reporting of minor gastrointestinal illness, confirmation that drinking water is the source of the infection and the potential for secondary spread of infection from person to person, masking the source. More recently quantitative microbial risk assessment has been used to estimate the risks from a number of pathogens and so the treatment requirements to remove a given load of the pathogen in the source water.[4]

A number of organisms are emerging as potential waterborne pathogens and some have emerged as significant pathogens that do give rise to detectable waterborne outbreaks of infection.[4] The most important of these is *Cryptosporidium*, a protozoan gastrointestinal parasite, which gives rise to severe, self-limiting diarrhoea and for which there is, currently, no specific treatment. *Cryptosporidium* is excreted as oocysts from infected animals, including man, which enables the organism to survive in the environment until ingested by a new host. There are many species of *Cryptosporidium* but not all are pathogenic to man. However, typing takes time and the starting assumption is for pathogenicity. Significant concentrations might be expected in farm slurry and in some wastewater discharges, which will contaminate surface water, and groundwater with close connection to surface water. This organism has given rise to a number of waterborne or water associated outbreaks in the UK and elsewhere, although the actual numbers of cases in any outbreak are usually relatively small. By contrast an outbreak of cryptosporidiosis in Milwaukee, in the United States, resulted in many thousands of cases, and probably a number of deaths among the immunocompromised members of the population. However, water is not the only source of infection as there is person to person spread following contact with faecal matter from infected animals and there have been outbreaks involving milk and swimming pools. The organism is resistant to the concentrations of chlorine used in water treatment and the primary barriers are normally coagulation, flocculation, and filtration but disinfection with ozone and UV are also effective.[4] It is detected by filtering samples, staining with an immunofluorescent tag and examining under a microscope. Unfortunately this method gives no indication whether the oocysts are viable or not and oocysts have been detected in water supplies without evidence of a detectable outbreak of clinical disease. This could be due to non-viable organisms or a high level of immunity in the exposed population. One of the great difficulties facing water suppliers and health professionals is what to do if small numbers of oocysts are detected in a supply in the absence of evidence of an outbreak of cryptosporidiosis in the community. Investigative action is an imperative, but the imposition of a 'boil water order' leads to a number of difficulties, not least of which is deciding the conditions under which the order will be lifted.

In the UK there is a treatment standard for cryptosporidium, which is the first occasion on which a standard has been stipulated in terms of a pathogen rather than an indicator organism. Additionally the standard requires continuous monitoring at treatment works which have been assessed as being at risk. The standard is not absolute in that it requires water to be filtered at a rate of 40 l per hour and that, on average, there must be less than one oocyst in 10 l.[2]

Cryptosporidium is an emergent pathogen, rather than a new pathogen since it was recognised in the latter part of the 19th century. However, its importance as a threat to water supplies has only been recognized through the development of improved diagnostic techniques and more sensitive epidemiological procedures. There has been no increased disease burden on drinking water consumers; indeed this continues to decline, but there is a better understanding of the causal agents. Whilst the use of chlorine and other oxidants has been of unquestionable value in the provision of safe drinking water, the emergence of cryptosporidium as a waterborne pathogen demonstrates the potential folly of over-reliance on such treatments. The provision of safe drinking water can only be achieved through the use of multiple barriers to infectious agents, since no single barrier can be expected to consistently provide adequate protection from the wide spectrum of pathogens that can be present. It is vital that water treatment continues to utilize the combined removal capacities of physical treatments, such as coagulation and filtration, and disinfection.

It is possible that as diagnostic techniques advance other pathogens will emerge and this will require the security of the treatment barriers to be reinforced. Any association with water is unlikely to come from direct monitoring of water supplies, but from surveillance data of diseases in the community, good laboratory diagnostics and epidemiological investigation.

Private supplies show evidence of a high frequency of microbiological contamination and this is clearly a significant concern for such supplies, particularly where treatment is minimal and there is a possibility of faecal contamination from livestock. These circumstances can give rise to contamination by some of the more virulent pathogenic organisms such as cryptosporidium or pathogenic *E. coli, e.g. E. coli* O157. This is of particular concern when groups of individuals who may be immunologically naive are exposed.

Although the common waterborne diseases of the 19th century are now almost unknown in developed countries, it is vital that vigilance is maintained at a high level because these diseases are still common in many parts of the world. The seventh cholera pandemic, which started in 1961, arrived in South America in 1991 and caused 4700 deaths in one year.[7] There are still an estimated 12.5 million cases of *Salmonella typhi* per year and waterborne disease is endemic in many developing countries. In this age of rapid global travel the potential for the reintroduction of waterborne pathogens still remains. In addition, as our knowledge of microbiological pathogens improves, we are able to identify organisms that were only suspected previously. These include viruses such as the Norwalk-like viruses, named after a major waterborne outbreak in North America, and a range of emerging pathogens including *Campylobacter*, also a major cause of food poisoning.

3.7   CHEMICAL CONTAMINANTS

3.7.1   Inorganic Contaminants

3.7.1.1   Lead

Lead in drinking water comes from plumbing materials and lead pipes in particular. Lead has been used in plumbing since Roman times and the word 'plumber' even comes from the Latin for lead. The Romans were aware of the possible dangers of lead water pipes and Charles Penney expressed his concern about the action of soft water from Loch Katrine on lead pipes in Glasgow, and its consequences, in the 19th century. However, in more recent times the problem of frank lead poisoning has been addressed and concerns centre around more subtle effects resulting from low doses.

The most widely recognized of these effects is the potential for lead to adversely affect the neurological development of young children resulting in learning deficits as measured by IQ.[8] These effects in children of four years and above show a probable reduction in IQ by 1–3 points per 10 μg dl^{-1} increase in blood lead up to 25 μg dl^{-1}. Above that the relationship changes and below about 10 μg dl^{-1} blood lead it is much more difficult to detect any change because of background 'noise' from other factors such as nutrition, parental IQ, socio-economic status and the type of environment the child lives in. The Joint Expert Committee on Food Additives and Contaminants (JECFA) has re-examined the data on lead and has withdrawn the provisional tolerable weekly intake (PTWI) on which the *WHO Guideline* value was based.[8] The findings at blood lead levels below 5 μg dl^{-1} are controversial. WHO has considered the new evaluation and has decided to maintain the current guideline value, with the clear implication that lead concentrations should be as low as is reasonably practical but should be considered in the context of overall action to reduce lead in the environment.[4,5] Exposure to lead, of course, includes air, dust and food, as well as water. The efforts that have been made to reduce lead over the past thirty years have resulted in a significant reduction in the average blood lead in the population, especially children, for whom average blood lead levels are now well below 5 μg dl^{-1}.[9] Since lead levels in the environment have continued to decline it is expected that average blood lead levels will also have continued to decline. The groups most at risk from high lead levels are, therefore, the foetus, bottle-fed infants whose intake of water in relation to bodyweight is much greater than other groups, and children.

The concentrations of lead in drinking water have steadily fallen with time as a consequence of lead pipe removal and treatment to reduce lead levels. The majority of lead arises from contact with service connections to the mains and from lead plumbing, usually where the properties were built before 1960. Some may also arise from the use of lead-based solders, although this is no longer officially permitted, and from fittings containing a proportion of lead in the metal to improve milling properties. Where water is known to be "plumbosolvent" (it has the propensity to dissolve lead from pipework), it is treated before entering the supply, usually with small amounts of phosphate. This not only reduces the level of lead in the water but possibly reduces the bioavailability of the dissolved lead. In some circumstances, minute particles of lead can be found in the water but the extent that these are bioavailable is uncertain. However, they can give rise to high concentrations in water samples, which may give a misleading picture of actual lead exposure in the property.

The standard for lead will fall to 10 μg dl^{-1} from 25 μg dl^{-1} from December 2013 and will be measured in unflushed samples at consumers' taps. However, care is needed in assessing the significance of any exceedances of the lead standard in a single sample, particularly first draw samples when the water has been standing in the pipes. The difficulty in dealing with lead in drinking water is that high lead is associated with individual properties. It is, therefore possible to find widely differing lead concentrations in drinking water even from adjacent properties. Routine sampling can identify areas with the greatest problems, but solving the problem finally requires investigation at the household level. It is also important that when old lead pipe is replaced, or new copper pipe is installed, that old-style lead solder is not used. Although the use of lead solder in drinking water systems is not legal, there are still occasions when it is used and does give rise to elevated lead concentrations.

3.7.1.2 Nitrate

Nitrate is found in both surface water and groundwater, primarily as a consequence of agricultural activity, but also from sewage, and in many sources is close to the standard for drinking water. This is not simply due to the use of excessive amounts of artificial fertilizers but the presence of excess nitrate from any source when there is insufficient plant growth to take up that excess and when rainfall is present to cause leaching from the soil. Indeed one of the major sources of nitrate in some groundwater was the ploughing of ancient grassland following the Second World War. While the appearance of nitrate in surface water following a leaching event can be quite rapid, it may take many years for nitrate to reach some groundwater, which can mean that that, even when surface inputs stop, the levels in groundwater will continue to rise for some time after. In some cases this can be more than 10 years.

For most individuals, vegetables are the main source of exposure to nitrate, although significant levels of nitrate and nitrite may be found in some cured meats. Vegetables and fruit usually contain between 200 and 2500 mg kg^{-1} nitrate with vegetables such as beetroot, lettuce and spinach being particularly high in nitrate. The intake of nitrate from the diet varies from about 40 to 130 mg day^{-1} so water with nitrate levels above the drinking water standard of 50 mg l^{-1} can make a significant contribution to daily intake.

Nitrate, along with nitrite which has similar actions, is often cited as a contaminant of concern for health but, there remain significant uncertainties as to harmful effects. Nitrate is secreted in human saliva and there is now evidence, from studies of the metabolism and enterosalivary circulation of nitrate in mammals, that nitrate, converted into nitrite in the oral cavity, is a key part of an important resistance mechanism against infectious disease through the production of oxides of nitrogen.[5,10]

At one time it was considered that nitrate and nitrite were a major cause of methaemoglobinaemia, or 'blue-baby' syndrome, in bottle-fed infants. Even this condition appears to be significantly influenced by the simultaneous presence of microbiological contamination because diarrhoeal disease has also been shown to induce methaemoglobinaemia in the absence of external sources of nitrate.[5] However, cases are seen at drinking water nitrate concentrations in excess of 100 mg l^{-1}. Nitrate is reduced to nitrite in the body and this subsequently reduces haemoglobin to methaemoglobin with a consequential reduction in oxygen carrying capacity. When the proportion of methaemoglobin rises from normal levels of <1 to 3% to about 10%, clinical symptoms are observed. Bottle-fed infants under three months of age are particularly vulnerable to this condition. The condition is now unusual in developed countries and there has not been a case associated with a public supply recorded in the UK since 1972. However, in some countries nitrate is still a significant issue, particularly in relation to shallow wells in agricultural areas supplying small communities with limited resources.

There has been considerable interest in the possibility that nitrate, reduced to nitrite, can react with secondary amines in the stomach giving rise to the formation of *N*-nitroso compounds. Some of this family of compounds have been shown to be potent carcinogens in a range of laboratory animals and so concern was raised that high nitrate levels could be a contributory cause of cancer. However, in spite of many studies, WHO concluded that there is no convincing evidence of a relationship between nitrates in drinking water and cancer in exposed populations.[4,5] This conclusion is supported by the majority of subsequent epidemiological studies.

There have been a number of additional theories as to the possible contribution of nitrate in drinking water to other human diseases and conditions. One of the recent suggestions has been that there is an association with childhood diabetes mellitus. However, there were flaws in the initial studies, particularly in the measurement of exposure, and subsequent studies in the UK and the Netherlands have failed to observe any significant association.[11,12] One condition that requires further research is the potential to interfere with iodine uptake causing thyroid disease.[5] While this is theoretically possible and is similar to other anions such as chlorate and perchlorate, it remains uncertain whether concentrations in drinking water are sufficiently high to be of significance.

3.7.1.3 Arsenic

Arsenic occurs in drinking water naturally as a consequence of dissolution from arsenic bearing rocks into groundwater. Although problems in the UK are limited, arsenic is a major concern in many parts of the world including the Indian sub-continent, South America, the Far East and the United States. Arsenic is the only contaminant that has been shown to be the cause of human cancers as a consequence of exposure through drinking water. It gives rise to a range of adverse effects including hyperkeratosis and cancer of the skin, peripheral vascular disease and a number of other cancers and adverse effects. In some countries, arsenic contamination ranks second only to microbiological contamination in its importance. However, determining a suitable safe level is difficult as many local factors are important, including nutritional status and exposure assessment needs to take into account uptake by crops, particularly rice, and uptake in cooking. JECFA[13] has carried out an assessment in which the existing PTWI was withdrawn. WHO has retained the guideline value of 10 µg l^{-1} which is the basis of the European standard, on the grounds that the risks from concentrations below this level are very uncertain and because arsenic removal at these concentrations is difficult it is necessary to achieve some sort of compromise. An additional complication is that many of the areas affected consist of small supplies with limited resources, which is the case in parts of Europe.

3.7.1.4   Fluoride

Fluoride is found naturally in water sources, particularly groundwater, in many parts of the world, sometimes in concentrations well in excess of 10 mg l^{-1}. It is also added to drinking water in small quantities in a number of countries, including some supplies in the UK, to supplement natural fluoride, which has been shown to be beneficial in preventing dental caries at a concentration of about 1 mg l^{-1}. However, above this concentration there is an increasing risk of adverse cosmetic effects in teeth (dental fluorosis) and, at high concentrations, of skeletal fluorosis. This latter condition arises from increasing density of the bones and distortion of the joints, which can give rise to severe skeletal deformation. There are many factors that appear to influence the risk of such adverse effects, including the intake of fluoride from other sources, volume of water ingested and nutritional status, particularly calcium intake. In general, significant effects are not seen below about 4 mg l^{-1}, which is the health based maximum contaminant level goal set for drinking water by the United States Environmental Protection Agency.[30] However, water supplies with high fluoride concentrations are still a major problem in many countries, particularly in parts of Africa (the Rift Valley), China and the Indian sub-continent.[14,15]

3.7.1.5   Aluminium

Aluminium has been widely used in drinking water treatment as a coagulant and flocculating agent to remove particulate matter, including microorganisms, and soluble, natural organic matter (NOM), which is the precursor to chlorination by-products. It is also found in raw waters although the chemical form will vary and will on occasion be found as inert aluminosilicates in clay particles. In the 1970s it was noted that some patients receiving kidney dialysis developed a severe form of dementia associated with high aluminium levels in the fluid used in the dialysis procedure. When the levels of aluminium were less than about 50 μg l^{-1}, dialysis dementia did not occur. At the same time it was shown that soluble aluminium salts injected into the brains of cats caused severe neurotoxicity. When aluminium was found in the lesions of patients, in parts of the Pacific, with Parkinsonism Dementia and Amyotrophic Lateral Sclerosis, concern was expressed that aluminium could also be associated with Alzheimer's Disease (AD), a major cause of pre-senile and senile dementia in Europe and North America. Aluminium was also found in the lesions characteristic of AD and a number of ecological type epidemiological studies reported an association between aluminium in drinking water and AD.[16]

However, exposure through drinking water is a very small part of daily exposure and the biological plausibility of the hypothesis was uncertain. A number of subsequent epidemiological studies, including a well conducted case control study, found no association between aluminium in drinking water and AD. However, studies in human volunteers using isotopic aluminium and advanced analytical techniques indicate that aluminium in drinking water is poorly absorbed.[17] In addition drinking water contributes only a very small proportion of daily aluminium intake compared to food. Since the first studies on aluminium and AD, there has been an increase in understanding of AD and there is extensive evidence that genetic factors play a major role in susceptibility to the formation and development of the neurological lesions characteristic of the condition. The majority view of scientists working in this field is that aluminium in drinking water is unlikely to play a primary causal role in the development of AD, particularly at concentrations below 100 μg l^{-1}. WHO concluded "There is no evidence to support a primary causal role of aluminium in AD, and aluminium does not induce AD pathology in any species, including humans."[5] JECFA re-evaluated aluminium in 2007 and proposed a PTWI of 1 mg kg^{-1} of body weight for aluminium from all sources. Using the WHO approach to determining a guideline would give a value of 0.9 mg l^{-1}.[18] However, aluminium polyhydroxides deposit in distribution and can cause significant dirty water problems. WHO suggests that well-run larger treatment works should be able to meet a residual value in drinking water of less than 100 µg l^{-1}, while smaller works should always be able to meet a residual value of 200 µg l^{-1}.[4]

3.7.1.6 Hardness

Hardness is a consequence of calcium, magnesium and other dissolved polyvalent metal ions. Hardness is responsible for scum formation with soap, scum formation in boiled water and deposition of scale in kettles and pipework in distribution and buildings. Although it is of no concern in relation to adverse health effects there has been a long-running debate as to possible beneficial effects in terms of cardiovascular disease. Magnesium, in particular, is known to impact on cardiovascular muscle and if individuals have a less than sufficient intake of these minerals then drinking water could provide sufficient extra intake to achieve sufficiency. WHO convened an expert group in 2011 to consider hardness. The conclusion was that "Although there is evidence from epidemiological studies for a protective effect of magnesium or hardness on cardiovascular mortality, the evidence is being debated and does not prove causality. Further studies are being conducted. There are insufficient data to suggest either minimum or maximum concentrations of minerals at this time, as adequate intake will depend on a range of other factors."[19,20]

3.7.1.7 Other Inorganic Contaminants

There are several inorganic contaminants and constituents in water that are of significance to drinking water. These include boron, copper, iron and manganese. Boron is found at high concentrations in sea water and was used in detergents, although much less so now. In desalination, boron has the potential to be a major problem because it is not well removed by most reverse osmosis membranes and the cost of removal can be prohibitive. The standard in Europe is 1 mg l^{-1} but WHO has re-evaluated the data on boron and has revised its guideline value to 2.4 mg l^{-1}, which is of great importance for desalination projects in many parts of the world.[4,21]

Copper is primarily found in drinking water as a consequence of copper pipes in buildings. The primary concern with copper relates to high concentrations that build up in copper plumbing after a period of standing. The WHO guideline of 2 mg l^{-1} is based on acute studies in humans and reflects the fact that the irritancy of copper to the gastrointestinal tract is a concentration dependent phenomenon rather than dose dependent. At above about 1 mg l^{-1} copper can stain sanitary ware.[4,5]

Iron occurs naturally in raw water and can also be present as a result of the corrosion of cast iron pipes. The natural status of iron in oxygenated waters is as low solubility ferric oxides which can form deposits in distribution that give rise to discoloured water when disturbed. This is exacerbated by the simultaneous presence of aluminium floc. Iron in drinking water is not considered to be an issue for public health but can cause severe discolouration and staining of laundry at concentrations above about 0.3 mg l^{-1}.[4]

Manganese also occurs naturally in raw waters and also causes severe discolouration when it deposits in distribution as insoluble oxides. Standards to prevent discolouration and staining are usually between 0.05 and 0.1 mg l^{-1}. Manganese does stay in solution to a greater extent than normal in some acid waters but this would normally be corrected in public water supplies in Europe. There have been studies that suggest that manganese in water at quite low concentrations, below the standards, might have an adverse effect on IQ in young children.[22] However, these studies all have flaws and there is a need for them to be repeated before any conclusions can be drawn.

3.7.2 Organic Contaminants

3.7.2.1   Introduction

There are many organic substances that can reach drinking water sources, either as a consequence of discharges of wastewater or run-off from urban or agricultural areas.[24] Most are present in extremely small concentrations of no more than a few micrograms per litre and usually much less, but occasionally much higher concentrations can be encountered as a consequence of spills. Organic contaminants are found in the greatest number in surface water, particularly if impacted by sewage/wastewater discharges, but some can reach groundwater. It is not possible or productive to consider most of the potential organic contaminants but WHO has reviewed the occurrence and effects of a substantial number.[4]

3.7.2.2   Disinfection By-products

The use of chemical disinfectants, such as chlorine and ozone, in water treatment has significantly contributed to the microbiological safety of drinking water all over the world. For example, eight out of ten waterborne outbreaks, between 1937 and 1986, involving public water supplies were associated with defective chlorination. All of the 13 outbreaks involving private supplies over the same period were also a result of problems with the disinfection process.[23] The notorious outbreak in Walkerton in Ontario in 2000 that resulted in 7 deaths from *E. coli* O157 was also primarily due to a failure of chlorination.[24] However, chlorine and ozone are oxidants and will react with organic material such as humic and fulvic acids and with inorganic constituents such as bromide. These reactions result in the formation of unwanted disinfection by-products (DBPs) but it was not until 1974 that Rook[25] demonstrated the presence of chloroform in the drinking water of Rotterdam. Since then intensive research has resulted in the identification of many more by-products of chlorination and ozonation, some of which are present at concentrations of only a few nanograms per litre. The discovery of chloroform and other trihalo-methanes in drinking water coincided with the discovery that chloroform increased liver and kidney tumours in laboratory animals. A weak association between chlorinated drinking water and cancer of the rectum, colon and bladder was found in epidemiological studies.[26] The only association which remains is with bladder cancer but again the association is weak and currently cannot be shown to be causal. As more DBPs, have been identified at very low concentrations it is the finding of a number of nitrogenous compounds that may provide a plausible mechanism for this weak association. All of the epidemiological studies have some limitations in design and/or methodology. One major problem in this respect is the problem of measuring exposure when these cancers relate to long-term exposure and considerable efforts have been made to reduce the concentrations of DBPs in drinking water so current levels of exposure may give a misleading picture of exposure in the past.[26]

   More recently concern has centred on the possibility of adverse birth outcomes being associated with drinking chlorinated water. In theory the problem of measuring exposure is smaller since the possible effects are unlikely to be related to paternal exposure and the critical exposure period will be up to nine months. The outcomes of interest which have emerged so far are low birth weight, preterm delivery, spontaneous abortions, stillbirth and various birth defects including defects of the central nervous system, heart, oral cleft, respiratory system and neural tube. The strongest evidence was for an association between low birth weight and total THMs but a large study carried out in the UK to examine THM concentrations in relation to still birth and low birth weight did not find any significant association.[27] In addition, the toxicological data do not, so far, provide evidence for a biologically plausible mechanism. The chlorination by-products that are regulated are the THMs and haloacetic acids or HAAs, which are the two groups present in the greatest concentrations, although only THMs are currently regulated in Europe. Both of these groups, and most other DBPs, can be reduced by optimising the removal of precursor natural organic matter in drinking water treatment.

Another regulated DBP is bromate, which is formed by the reaction of ozone with bromide ions. It is also formed in the electrolytic generation of chlorine if the brine is high in bromide. Bromate was shown to be genotoxic *in vitro* and carcinogenic in some laboratory animals.[4,5] As a consequence a drinking water guideline was proposed by WHO and was adopted widely around the world. Since the development of the WHO guideline more research has shown that the dose–response curve is non-linear at low doses because bromate is effectively reduced at several stages before it can reach intracellular DNA.[28] The standard for bromate is, therefore, probably much more conservative than it needs to be.

Chlorate is formed in aging and poorly stored hypochlorite solutions. It is not strictly a DBP but a contaminant. However, although there is no standard in Europe, it is of concern for small supplies because it has the capacity to interfere with iodine uptake. Whether this is of significance in humans remains uncertain but WHO has developed a guideline value and has recommended that hypochlorite storage should be managed to minimise chlorate formation, because it can reach significant concentrations that also result in reduced availability of free chlorine and greater amounts of hypochlorite solution being required to ensure adequate disinfection.[4]

Investigation of chloramination to provide a residual disinfectant in distribution revealed the potential for the formation of nitrogenous by-products, particularly *N*-nitrosodimethylamine (NDMA) which is a known and potent carcinogen in laboratory animals. WHO developed a guideline value of 100 ng l^{-1} but studies in the UK showed that concentrations can readily be maintained well below 10 ng l^{-1} in drinking water supplies and in most cases NDMA was not detected.[4,29]

In spite of the concerns regarding DBPs it is important to keep these in perspective when balanced against the measurable risks from waterborne pathogens, particularly in developing countries and small supplies. WHO has emphasised that disinfection should not be compromised in meeting guidelines or standards for DBPs.

3.7.2.3 Pesticides

One of the most common groups of contaminants cited by the media, and various pressure groups, as a problem in drinking water is pesticides. This concern arises from exceedances of the European drinking water standard of 0.1 μg l^{-1} for any pesticide and 0.5 μg l^{-1} for total pesticides.[1] This standard is a political rather than a scientifically based standard and, therefore, its relevance to public health is limited. Pesticides are a very diverse group of substances with widely differing chemical characteristics and toxicity. WHO[4,5] and the USEPA[31] are two organizations that have promulgated health-based guidelines or standards for pesticides in drinking water and these values have been rarely exceeded in drinking water in the UK. Advanced water treatment processes have been installed in part to combat this problem with great success. This has been supported by restrictions on some of the major polluters, particularly atrazine. However, there remain problems with a number of herbicides, particularly as a consequence of wash out after heavy rainfall and a problem with metaldehyde, a molluscicide used in slug pellets. This substance was thought to break down rapidly in soil but once present in water is extremely persistent and very difficult to remove. Although concentrations are well below health-based values it occurs above the standard and is, therefore, a challenge for drinking water suppliers in regions where it is extensively used to combat slugs and snails in wet periods.

3.7.2.4 Endocrine Disrupters (see also Chapters 1 and 20)

Endocrine disrupting chemicals have become a general concern in the light of possible reductions in sperm quality over time and the increase in a number of cancers which are potentially influenced by endocrine activity, such as testicular cancer and breast cancer. Much of the evidence for endocrine disruption comes from aquatic wildlife. Tributyltin oxide (TBTO) causes imposex in dog whelks and some other molluscs at very low concentrations; pulp and paper mill effluent has been shown to cause masculinisation in fish. However, it was the widespread finding of ovarian tissue in male fish exposed to treated sewage effluent and the induction of vitellogenin, an egg yolk protein under the control of oestrogen, in the blood of male trout exposed to treated effluent. A number of chemicals have been shown to bind or block oestrogen and androgen receptors in yeast and cell culture assays. Some have also been shown to possess activity in whole animals but they are very much less active than natural hormones. In some cases, such as the phthalates, there is evidence for species differences in metabolism, which imply that they will not possess the same activity in primates as in rodents. Studies by the Environment Agency on fractionating sewage effluent showed that the primary cause of the activity in fish was the natural hormones, oestrone and 17β-oestradiol. These hormones are excreted as the water soluble glucuronides and sulfates but are broken down by microbiological activity in sewage treatment to yield the parent hormone, some of which will also be adsorbed or broken down. There is also a contribution from the synthetic steroid ethinyl oestradiol and, in one instance, alkyl phenols. In this case, on the River Aire in Yorkshire, there were a significant number of wool scourers which discharged to the sewage treatment works and which were using alkylphenol ethoxylate based industrial detergents. These were breaking down in treatment to give smaller molecules with shorter ethoxylate chains and the parent alkylphenols. When most of these companies switched to alternative detergents the vitellogenin response in trout was also reduced.

Most of these substances are hydrophobic and are readily removed in water treatment. Laboratory studies have also shown that water treatment techniques remove or break down these substances fairly readily. These findings have been confirmed by specific chemical analysis at very low limits of detection.[31,32] Epoxy resins used in relining cast iron water mains have also been examined for the presence of bisphenol A and F as contaminants since these have been proposed as possible endocrine disrupters, but, so far, none has been found in water as it is supplied.[33] An additional pointer for drinking water which is reassuring is that two of the countries that reported the greatest apparent effects (increased testicular cancer and decreased sperm quality) in man, Scotland and Denmark, do not practise re-use of waste-water and so no sewage effluent is in the water used for water supply. In addition Scotland uses primarily upland surface water, while Denmark uses clean groundwater, so there is no possibility of waterborne exposure.

3.7.2.5   *Polycyclic Aromatic Hydrocarbons (PAH)*

Polycyclic aromatic hydrocarbons (PAH) are a substantial family of substances found widely in the environment as a consequence of incomplete combustion. In 1776 Percival Potts studied chimney sweeps in London who at that time were exposed to soot all over the body, particularly as boys were sent up chimneys to sweep them. Soot is high in many PAH, including benzo[*a*]-pyrene. The sweeps showed a significant level of scrotal cancer, which is very rare in other circumstances. Subsequently some of the PAH, including benzo[*a*]pyrene, were shown to induce skin cancer in mice. Up to the 1950s, cast iron water mains were often lined with coal tar to protect them against corrosion. Most PAH are of very low water solubility but some do leach from coal tar linings, *e.g.* fluoranthene. Fluoranthene is the most frequently observed PAH in drinking water unless coal tar particles are present. WHO has examined the toxicity of fluoranthene and the evidence indicates that although it is mutagenic in bacteria it is not carcinogenic in animals. WHO concluded there was no need to propose a guideline value for fluoranthene, which is not included in the current standard in the European directive.[4] Where problems arise with PAH in drinking water, the cause is usually the disturbance of sediment in water mains. In circumstances where there was a significant concentration of the PAH that are included in the *European Directive*, the water would be discoloured and, possibly, would also give rise to taste and odour problems. Exposure would, therefore, be very short and the problem rapidly identified and dealt with.

3.7.2.6 Tri and Tetrachloroethene

These two substances which were widely used as chlorinated solvents have caused significant problems in some groundwater. They are not found at significant concentrations in surface waters because they are rapidly lost to the atmosphere. However, they are poorly adsorbed to soil and are degraded extremely slowly so when spilt on the ground they can move down to groundwater where they are very persistent. Improved controls on the way these substances are used and increased awareness by users has significantly reduced new contamination and the problem is now largely as a consequence of historical pollution. Both of these substances have shown evidence of carcinogenicity in laboratory animals, although they are, at worst, only weakly genotoxic. The International Agency for Research on Cancer (IARC)[34] re-evaluated the data on the epidemiology of renal cancer associated with exposure to both tri- and tetrachloroethene and concluded that there was some evidence to support a causal association. However, other groups would not agree with this conclusion and consider that the totality of the epidemiological data does not support a causal association. WHO has proposed guideline values of 30 µg l^{-1} and 70 µg l^{-1} for tri- and tetrachloroethene, respectively, on the basis of cancer in laboratory animal studies but caused by a non-genotoxic mechanism.[4] However, the EU has introduced a more stringent precautionary standard of 10 µg l^{-1} for the two substances in total.[1]

3.7.2.7 Emerging Contaminants

As analytical capability has increased over the years so has the ability to measure a much wider range of substances in water at ever lower concentrations, often the low nanogram per litre range. Some of these substances have entered the environment relatively recently, while some may have been there for many years but at concentrations below the prevailing detection limits. Most of these emerging substances are present as a consequence of their use in domestic and industrial settings and are discharged to sewer, while others are present in the aquatic environment from industrial sources. Like the endocrine disrupters, some of these substances are known to be biologically active and there are questions as to whether they can reach drinking water and what effects there might be. They include pharmaceuticals, personal care products (PCPs) and perfluorinated substances.[35]

The perfluorinated compounds were used in the manufacture of non-stick and dirt resistant coatings for cookware and fabrics. They were also used in detergents that were widely used in, amongst other things, fire-fighting foams. These foams were used extensively at airports both fighting real fires in aircraft and in practice. The various manufactured materials and foams break down to release the parent compounds such as perfluorooctanoic acid (PFOA) and perfluorooctane sulfonate (PFOS). These substances are both persistent and water soluble and are often found in groundwater below manufacturing facilities and airports. There are concerns about their potential toxicity to humans but further research is needed to determine the risks more accurately. This is particularly important as removal from drinking water is not easy and it will be important to strike a balance between caution over health concerns and the practical considerations of removal from water.[36]

Pharmaceuticals have been identified in drinking water for more than 30 years but the identification and measurement of a large number of pharmaceuticals in wastewater, river water and, to a much lesser extent, drinking water has been a much more recent phenomenon. Pharmaceuticals are used by the human population and are excreted in urine and faeces as metabolites or as the parent compound. Some are used on skin and are washed off in bathing but both routes lead to discharge to sewer. Some are removed or partially removed in wastewater (sewage) treatment but some will reach surface water where there may be further removal or breakdown. Some pharmaceuticals used in veterinary medicine may also reach surface water more directly in run-off from agricultural land. A small number will be detectable in drinking water having survived drinking water treatment. The concentrations are very low; WHO has reviewed the data and concluded that "available studies have reported that concentrations of pharmaceuticals in surface water, groundwater and partially treated water are typically less that $0.1 \ \mu g \ l^{-1}$ and concentrations in treated drinking water are generally below $0.05 \ \mu g \ l^{-1}$.[37] However, the majority of pharmaceuticals have not been found in drinking water in developed countries where most of the research has been focused, in spite of their having been detected in source waters. It is not only surface water that could be impacted but any groundwater heavily influenced by surface water or rapidly responsive to rainfall could be vulnerable. Although no standards have so far been set for any pharmaceutical in drinking water, the conclusions of the WHO expert group was that "The substantial margins of safety for individual compounds suggest that appreciable adverse impacts on human health are very unlikely at current levels of exposure in drinking-water."

Personal care products (PCPs) are widely used in domestic, and in some cases, industrial settings and comprise substances used for cleaning and laundry, toiletries and cosmetics. This is a highly diverse group and although there are limited data on the concentrations of a small number in water, the data are very limited.

3.8   WATER SAFETY PLANS (WSPs)

As indicated above WHO has changed the emphasis on assuring drinking water safety from end-of-pipe monitoring to a proactive and preventive approach based on an extension of Hazard Assessment and Critical Control Points (HACCP) as used in the food industry for many years.[4] This approach requires a full and detailed understanding of a water supply from source to tap with identification of the hazards at all stages, including in source water, assessment of the risks to consumers and establishment of suitable barriers and procedures to mitigate these risks and to ensure that the barriers continue to work effectively. The role of drinking water standards and guidelines is to assist in assessing the risks, to provide a benchmark for designing barriers and for verification that the risks have been mitigated. WSPs are rapidly becoming accepted as best practice for assuring the safety of drinking water supplies in most parts of the world.

3.9   CONCLUSIONS

Clean and wholesome drinking water is an essential requirement for a healthy life. There are many potential sources of contamination with both pathogens and chemicals. The control and treatment of drinking water has evolved to combat these threats. The primary requirement for clean drinking water is that of microbiological safety. Our knowledge about waterborne pathogens has increased significantly and we now face pathogens that are new or newly recognized, such as *Cryptosporidium*, which is resistant to chlorine unlike the bacterial faecal indicators. Consequently changes in the approaches to monitoring drinking water and in drinking water treatment have been required, particularly for those supplies derived from surface waters. They have also resulted in a new, more holistic, approach to drinking water standards and guidelines and the management of drinking water supplies.

In general the concentrations of chemical contaminants are very small but there appears to be a higher perception of chemical contaminants by consumers than of microbiological contaminants. There are many possible contaminants, both inorganic and organic, that can arise naturally or from man's activities. However, in developed countries, the contaminants that have received greatest attention have been those arising from domestic plumbing, particularly lead, those arising as a consequence of the use of chemical disinfection processes, such as chlorination and pesticides and nitrate from agriculture. In the developing world there are other substances of concern, in particular, arsenic and fluoride. However there are sometimes local issues that are of concern and WHO has developed guidance on assessing which chemicals might be of concern for a particular supply.[38] There are now rigorous standards for the most important of these contaminants and guidance on other contaminants from WHO and there is continued consideration of emerging chemical and microbiological contaminants in the *Guidelines for Drinking Water Quality* and the associated documents covering specific topics. These guidelines have been, and will continue to be, the primary basis for drinking water standards in Europe and most of the world. However, as the pressure on water resources continues to increase there will be a need to consider new sources of water, such as desalinated water and reuse of treated wastewater to augment drinking water supplies, which will bring new challenges. It will, therefore, be important to take another look at the way we manage water and the way in which we develop standards and guidelines. WHO has begun this process with the framework for safe water in the third and fourth editions of the *Guidelines for Drinking Water Quality*. The pressure to use new sources of water will also require that more holistic approaches to governing the managed water cycle from wastewater through to drinking water are adopted. For example the primary source of emerging chemical contaminants is treated wastewater. Because there are concerns about the impact on aquatic life as well as drinking water, it makes sense to develop more energy efficient removal processes for wastewater treatment that can remove these and possible future contaminants, along with new types of chemical products to ensure that they are less polluting and easier to remove. Such an approach requires a change in thinking to the long-term proactive approach proposed by WHO in the Framework for Safe Drinking Water introduced in the third edition of the *Guidelines* and reinforced in the fourth edition.

REFERENCES

1. *Council Directive 98/83/EC on the Quality of Water intended for Human Consumption, Off. J. Eur. Comm.*, 05/12/91998; http://eurlex.europa.eu/LexUriServ/LexUriServ.do?uri = OJ:L:1998:330:0032:0054:EN:PDF (accessed 10/06/2013).
2. *The Water Supply (Water Quality) Regulations 2000*, No. 3184, Regulations as amended by S.I. 2001/2885, S.I. 2002/2469, S.I.2005/2035, S.I. 2007/2734 and S.I. 2010/991, Applying from 20 April 2010 (unofficial consolidated version); http://dwi.defra.gov.uk/stakeholders/legislation/ws_wqregs2000_cons2010.pdf (accessed 10/06/2013).
3. The *Private Water Supplies Regulations 2009*, No. 3101; http://dwi.defra.gov.uk/stakeholders/legislation/pwsregs2009.pdf (accessed 10/06/2013).
4. World Health Organization, *Guidelines for Drinking-water Quality*, 4th edn., WHO, Geneva, 2011; http://www.who.int/water_sanitation_health/publications/2011/dwq_guidelines/en/index.html (accessed 10/06/2013).
5. *Background Documents on Chemical parameters in the WHO Guidelines for Drinking-water Quality*; http://www.who.int/water_sanitation_health/dwq/chemicals/en/index.html (accessed 10/06/2013).
6. J. Bartram J. Cotruvo, M. Exner, C. Fricker and A. Glasmacher, *Heterotrophic Plate Counts and Drinking Water Safety. The Significance of HPCs for Water Quality and Human Health*, IWA publishing on behalf of WHO, 2003.
7. E. Salazar-Lindo, M. Alegre, M. Rodriguez, P. Carrion and N. Razzeto in *Safety of Water Disinfection: Balancing Chemical and Microbial Risks*, ed. G. F. Craun, ILSI Press, Washington, DC, 1993, 401.
8. FAO/WHO, *Lead in Evaluation of Certain Contaminants in Food*, 73rd Report of the Joint FAO/WHO Expert Committee on Food Additives. Geneva, World Health Organization, WHO Technical Report Series, No 960, 2011, 162.

9. Institute for Environment and Health, *Recent UK Blood Lead Surveys*, Report R9, IEH, Cranfield, 1998; http://www.cranfield.ac.uk/health/researchareas/environmenthealth/ieh/ieh%20publications/r9.pdf (accessed 10/06/2013).

10. N. S. Bryan and J. Loscalzo, *Nitrite and Nitrate in Human Health and Disease*, Humana Press, New York, 2011.

11. J. M. Van Maanen, J. Albering, S. G. Van Breda, D. M. Curfs, A. W. Amberg, B. H. Wolffenbuttel, J. C. Kleinens and H. M. Reeser, *Diabetes Care*, 2000, **23**, 1750.

12. A. Casu, M. Carlini, A. Contu, G. F. Bottazzo and M. Songini, *Diabetes Care*, 2000, **23**, 1043.

13. FAO/WHO, *Arsenic in Evaluation of Certain Contaminants in Food, 73rd report of the Joint FAO/WHO Expert Committee on Food Additives, Geneva, World Health Organization*, WHO Technical Report Series, No. 959, 2011, 21.

14. J. Fawell, K. Bailey, J. Chilton, E. Dahi, L. Fewtrell and Y. Magara, *Fluoride in Drinking-water*, WHO Drinking-water Quality Series, IWA Publishing, London, 2006.

15. IPCS, *Fluorides*, World Health Organization, International Programme on Chemical Safety, Geneva, Environmental Health Criteria, 227, 2002.

16. IPCS, *Aluminium*, World Health Organization, International Programme on Chemical Safety, Geneva, Environmental Health Criteria, 194, 1997.

17. N. D. Priest, R. J. Talbot, D. Newton, J. P. Day, S. J. King and L. K. Fifield, *Hum. Exp. Toxicol.*, 1998, **17**, 296.

18. FAO/WHO, *Evaluation of Certain Contaminants in Food*, 67th Report of the Joint FAO/WHO Expert Committee on Food Additives, Geneva, World Health Organization, WHO Technical Report Series, No. 940, 2007, 33.

19. WHO, *Calcium and Magnesium in Drinking-water. Public Health Significance*, World Health Organization, Geneva, 2009; http://apps.who.int/iris/bitstream/10665/43836/1/9789241563550_eng.pdf (accessed 10/06/2013).

20. WHO, *WHO Meeting of Experts on the Possible Protective Effects of Hard Water against Cardiovascular Disease*, World Health Organization, Geneva, 2006.

21. J. Fawell, M. Y. Abdulraheem, J. Cotruvo, F. Al-Awadh, Y. Magara and C. A Ong, in *Desalination Technology*, ed. J. Cotruvo, N. Voutchkov, J. Fawell, P. Payment, D. Cunliffe and S. Lattemann, CRC Press and IWA Publishing, London, 2010, 91.

22. M. F. Bouchard, S. Sauve, B. Barbeau, M. Legrand, M. F. Brodeur, T. Bouffard, E. Limoges, D. C. Bellinger and D. Mergler, *Environ. Health. Perspect.*, 2011, **119**(1), 138.

23. N. J. Galbraith and R. Stanwell-Smith, *JIWEM*, 1987, **1**, 7.

24. S. E. Hrudey and E. J. Hrudey., *Can. J. Public Health*, 2002, **93**(5), 332.

25. J. J. Rook, *Water Treat. Exam.*, 1974, **23**, 234.

26. IPCS, *Disinfectants and Disinfection By-products*, World Health Organization, International Programme on Chemical Safety, Geneva, Environmental Health Criteria, 216, 2000.

27. M. J. Nieuwenhuijsen, M. B. Toledano, J. Bennet, N. Best, P. Hambly, C. de Hoogh, D. Wellesley, P. A. Boyd, L. Abramsky, N. Dattani, J. Fawell, D. Briggs, L. Jarup and P. Elliott, *Environ. Health Perspect.*, **116**, 216.

28. J. K. Chipman, J. L. Parsons and E. J. Beddowes, *Toxicology*, 2006, **17**, 187.

29. G. Dillon, S. Blake, P. Rumsby, L. Rockett, T. Hall, P. Jackson and A. Rawlinson, *NDMA – Concentrations in Drinking Water and Factors affecting its Formation*, DWI report DEFRA 7438, Drinking Water Inspectorate, London, 2008; http://dwi.defra.gov.uk/research/completed-research/reports/DWI70_2_210.pdf (accessed 10/06/2013).

30. United States Environmental Protection Agency, 2011 Edition of The Drinking Water Standards and Health Advisories, 2011; http://water.epa.gov/action/advisories/drinking/upload/dwstandards2011.pdf (accessed 10/06/2013).

31. J. K. Fawell, D. Sheahan, H. A. James, M. Hurst and S. Scott, *Water Res.*, 2001, **35**, 1240.

32. A. Wenzel, J. Muller and T. Ternes, *Study on Endocrine Disrupters in Drinking Water*, Final Report ENV.D.1/ETU/2000/0083, 2003.
33. J. K. Fawell and J. K. Chipman, *JCIWEM*, 2001, **15**(2), 92.
34. International Agency for Research on Cancer, *IARC Monograph*, Vol. 63, IARC, Lyon, 1995.
35. J. K. Fawell and C. N. Ong, *Int. J. Water Resour. Dev.*, 2012, **28**, 247.
36. Drinking Water Inspectorate, *Guidance on the Water Supply (Water Quality) Regulations 2001 specific to PFOS (Perfluorooctane Sulphonate) and PFOA (Perfluorooctanoic Acid) Concentrations in Drinking Water*, Information letter October 2009; http://dwi.defra.gov.uk/stakeholders/information-letters/2009/10_2009annex.pdf (accessed 10/06/2013).
37. WHO, *Pharmaceuticals in Drinking-water*, World Health Organization, Geneva, 2011; http://www.who.int/water_sanitation_health/publications/2011/pharmaceuticals/en/index.html (accessed 10/06/2013).
38. T. Thompson, J. Fawell, S. Kunikane, D. Jackson, S. Appleyard, P. Callan, J. Bartram and P. Kingston, *Chemical Safety of Drinking-water: Assessing Priorities for Risk Management*, World Health Organization, Geneva, 2007.

CHAPTER 4

Water Pollution Biology

WILLIAM M. MAYES[†]

Centre for Environmental and Marine Sciences, University of Hull, Scarborough, YO11 3AZ, UK
Email: w.mayes@hull.ac.uk

4.1 INTRODUCTION

4.1.1 The Role of Biology in Understanding Water Pollution

Pollution can be defined as "the introduction into the environment of substances or energy liable to cause hazards to human health, harm to living resources and ecological systems, damage to structure or amenity, or interference with legitimate uses of the environment".[1] Pollutants are therefore chemical, physical or biological in nature and can be measured with varying degrees of accuracy and precision in waters. The measured quantities can then be compared with standards of allowable concentrations, be they for aquatic life or potable water in some cases. Why then do we need to undertake quantitative studies of organisms and biological communities when we know that they can be defined with much less precision than chemical or physical parameters? There are a number of reasons why biological studies are important.

Firstly, the definition of pollution given above includes the adverse effects on living resources and ecological systems therefore these impacts need quantifying. People, of course, by drinking water, consuming aquatic resources, and using freshwaters as recreation areas, are also linked to the aquatic environment. Indeed, in many parts of the world, the river is often central to the life of the community. The effects of pollutants on aquatic life recorded by monitoring programmes can act as an early warning of potential harm to ourselves.

Secondly, animal and plant communities respond to intermittent pollution which may be missed in a chemical surveillance programme. Ambient water quality monitoring typically consists of samples taken once a month at designated sample stations. Such monitoring is unlikely therefore to characterise the full range of physico-chemical conditions dynamic stream systems exhibit. These could include short-term pollution associated with the flushing of deicing salts into rivers in a winter urban storm event, an acid flushing event in a headwater stream leading to short-term increases in bioavailable metal concentrations, or even a disreputable farmer or factory-owner

---

[†]This chapter is based upon an earlier contribution from Dr Chris Mason which is gratefully acknowledged.

Pollution: Causes, Effects and Control, 5th Edition
Edited by R M Harrison
© Mayes and The Royal Society of Chemistry 2014
Published by the Royal Society of Chemistry, www.rsc.org

discharging polluting effluent between sample visits. The pollution will impact on the most sensitive members of the aquatic community, acting as indicators of pollution. The amount of change in the community will be related to the severity of the incident. Because the community can only be restored to its former diversity by reproduction and immigration, its recovery is likely to be slow. If the intermittent pollution occurs with some frequency, the community will remain impoverished. The biologist will be able to detect such damage and suggest a more detailed surveillance programme, both biological and chemical, to find the source and frequency of pollution.

Thirdly, biological communities may respond to unsuspected or new pollutants in the environment. There are over 70 million registered inorganic and organic compounds of which 8400 chemicals have been comprehensively tested for toxicity.[2] Thousands are discharged to freshwaters, while only some 30 determinands will routinely be tested for given financial constraints in ambient water quality monitoring programmes. If there is a change in the biological community, however, then a wider screening for pollutants can be initiated. A famous example comes from male rainbow trout (*Oncorhynchus mykiss*), caged in the River Lea, north of London, below the point of discharge of effluent from a sewage treatment works, which exhibited marked increases in vitellogenin levels.[3] This compound is normally produced in the liver of female fish in response to the hormone oestradiol and is incorporated into the yolk of developing eggs. Clearly, some chemical in the effluent was behaving like a female hormone and the male trout provided an early warning of a potential problem requiring urgent investigation. Nonylphenols (widely used as anti-oxidants) were suspected of stimulating vitellogenin production in the trout. The effect has since been demonstrated in fish from a number of other rivers. Laboratory experiments have identified three such active compounds causing oestrogenic activity in treated sewage effluent, two naturally occurring hormones and an active ingredient of birth control pills.[3,4] There are many other substances, including DDT and some PCBs, which exert similar effects. For example, abnormalities in the eggs and hatchlings of snapping turtles (*Chelydra serpentina*) from the Great Lakes region of North America were considered to be caused by polychlorinated aromatic hydrocarbons.[5] Increases in testicular cancer and falling sperm counts in the human male may be related to oestrogen-mimics released into the environment,[6] and these endocrine disruptors have long since been of concern and research effort.

Finally, some chemicals are accumulated in the tissues of certain organisms, the concentrations reflecting environmental pollution levels over time. In any particular sample of water, the concentration of a pollutant may be too low to detect using routine methods, but nevertheless will be gradually accumulated within the ecosystem to levels of considerable concern in some species. Metals, organochlorine pesticides, and PCBs have caused particular problems in aquatic habitats and are potential threats to human health. Eels (*Anguilla anguilla*), for example, have historically been used to detect mercury below discharges of sewage effluent.[7]

4.1.2 Pollution Types and Interactions

Pollutants having biological impacts may derive from point sources, often discrete discharges known to the authorities, which have traditionally been the focus for regulatory monitoring. However, the previous example of endocrine disruptors emphasizes that we are not always aware of the ecological effects of compounds coming from apparently well-regulated sewage effluent discharges. This may also be the case with many emerging contaminants such as engineered nanoparticles and personal health care products which are considered later in the chapter. Alternatively, sources of pollution may be diffuse, entering watercourses from land drainage and run-off, such as fertilizers and pesticides applied to agricultural land. Management of our river systems has gained a greater appreciation of the importance of such diffuse sources in recent years with regulation focussing increasingly on catchment-scale processes as opposed to discrete point discharges. In

Europe, the *Water Framework Directive (WFD, 2000/60/EC)* has facilitated this major change in how we evaluate and manage our water environment. Biological monitoring is a crucial tool, given the aquatic communities present in our freshwaters represent a function of the broader catchment morphology, land use and other anthropogenic impacts acting over a range of timescales.

Much pollution is chronic, the watercourse receiving discharges continuously or regularly, and such pollution can generally be reduced to acceptable levels given the right regulatory framework. A greater problem is episodic or accidental pollution, which is unpredictable in space or time. Table 4.1 offers a range of examples of major accidental water pollution events and their biological effects. Such events typically receive much public and scientific attention, which can drastically improve our knowledge of the risks associated with, and biological effects of the pollutants involved. They can also have a major bearing on developing regulation to minimise the risks of such events reoccurring. For example, the Aznalcóllar and Baia Mare events were responsible for a major piece of legislation in Europe (the *Mining Waste Directive: 2006/21/EC*) intended to reduced the risk of subsequent mine and processing waste disasters in Europe. However, the Ajka red mud spill in Hungary highlights that even with appropriate regulatory frameworks, avoidable accidents do still happen (see Table 4.1). One of the recurring themes apparent in many of these acute pollution events is the problem of determining what the specific causal agents are for observed biological impacts amongst a mixture of contaminants. Most polluting effluents are usually a cocktail of potential toxicants and these will interact with each other and naturally occurring substances in a variety of ways. As such, as we aim to understand the effects of pollutants on biological communities we must be aware of interactions between chemical species and how these affect bioavailability and subsequent toxicity to organisms. These are often summarised into three main types of interaction:

(i) *Additive effects*: where the action of combined pollutants is equal to the sum of their individual effects, for example mixtures of zinc and copper, common in many of the mine waste disasters (see Table 4.1), act together additively.

(ii) *Synergistic effects*: where the effect of combined pollutants is greater than the sum of their individual effects. For example, synergistic effects of multiple organophosphate compounds were cited as important in the deaths of grayling (*Thymallus thymallus*) and brown trout (*Salmo trutta*) in the Rhine after the Sandoz chemical spill as individual concentrations of the relevant contaminants were several orders of magnitude below their individual lethal concentrations (see Table 4.1).

(iii) *Antagonistic effects*: where the effects of combined pollutants is less than the sum of their individual effects. For example calcium antagonizes the toxic effect of lead and aluminium, while abundant particulate iron oxides are likely to have been important in limiting the toxicity of chromium and arsenic after the Ajka red mud spill (see Table 4.1).

This review will describe the effects of major types of pollutants on aquatic life and will consider some of the key methods used for assessing the biological impacts of water pollution.

4.2   ORGANIC POLLUTION

Organic pollution results when large quantities of organic matter are discharged into a watercourse to be broken down by microorganisms which utilize oxygen to the detriment of the stream biota. The most important source of organic pollution globally is the discharges associated with untreated or treated sewage. Over 2.5 billion people live without improved sanitation and 780 million with unsafe drinking water sources[16] which poses an extreme risk to human health through parasitic infections and waterborne diseases such as dysentery and cholera; a health burden which falls disproportionately on children.

In countries with established water management infrastructure, sewage treatment works (STWs) can still be enduring sources of instream organic pollution, particularly where infrastructure predates any rapid population growth in the towns they serve. In Europe, the water industry does, however, have to meet increasingly stringent legislation on the quality of discharged effluents (*e.g.* the *Urban Waste Water Treatment Directive* and the demands of the *Water Framework Directive*) which has seen demonstrable improvements in levels of organic pollution in many streams over recent decades. As such, in many catchments the chief sources of organic pollution are more likely to be related to:

(i) Short-term storm pollution events in urban areas that see Combined Sewer Overflows discharge untreated sewage directly into streams and rivers.
(ii) Farm effluents, especially in areas of intensive pastoral farming, which are likely to be diffuse in origin.
(iii) Discharges from other industries such as dairies, breweries and food processing plants.

Biochemical Oxygen Demand is a simple measure of the potential of organic matter for deoxygenating water which is determined in the laboratory by incubating a sample of water for five days at 20 °C and determining the oxygen used. In most of Europe, BOD targets are typically set by the oxygen conditions associated with macro-invertebrates, which are generally most sensitive to organic pollution, with different values for upland and lowland rivers (with values <4 mg l^{-1} or <5 mg l^{-1} respectively for a 'Good' status for upland and lowland streams). However, more stringent standards are enforced for lowland rivers if they support salmonids.[17]

In recent surveys of pollution pressures in catchments across England and Wales, 7.9% of water bodies (with an associated stream length >4500 km) were deemed at risk of not meeting European standards for Biochemical Oxygen Demand (BOD) while 10.5% (>6100 km of river) were deemed at risk of not meeting ammonia standards (which is a decay product of the breakdown of animal and vegetable wastes and common to most organic pollution sources).[18] The distribution of these impacted catchments around major urban centres and some intensive areas of agriculture predominantly in eastern England suggest a range of sources. These data support general trends of chronic, gross organic pollution diminishing greatly in the last three decades, but there is still widespread mild organic pollution in many settings, which can cause enduring impacts to aquatic biota.[19]

Figure 4.1 outlines the general effects of an organic effluent on a receiving stream.[1] At the point of entry of the discharge there is a sharp decline in the concentration of oxygen in the water, known as the oxygen sag curve. At the same time there is a large increase in BOD as the microorganisms added to the stream in the effluent and those already present utilize the oxygen as they break down the organic matter. As the organic matter is depleted, the microbial populations and BOD decline, while the oxygen concentration increases, a process known as 'self-purification', assisted by turbulence within the stream and by the photosynthesis of algae and higher plants. The effluent will also contain large amounts of suspended solids which cut out the light immediately below the discharge, thus eliminating photosynthetic organisms. Suspended solids settle on the stream bed, altering the nature of the substratum and smothering many organisms living within it.

Under conditions of fairly heavy pollution, sewage fungus develops. This is an attached, macroscopic growth containing a whole community of micro-organisms, dominated by *Sphaerotilus natans*, which consists of unbranched filaments of cells enclosed in sheaths of mucilage, and by zoogloea bacteria. Sewage fungus may form a white or light brown slime over the surface of the substratum, or it may exist as a fluffy, fungus-like growth with long streamers trailing into the water.

Protozoans are chiefly predators of bacteria and respond to changes in bacterial numbers. Attached algae are eliminated immediately below the outfall due to the diminished penetration of

Table 4.1 Examples of major short-term aquatic pollution events and their biological impacts.

Event	Pollution issue	Biological effects	Comment
Sandoz Agrochemical Spill, Schweizerhalle, Switzerland, November 1986	• 5–8 tons of pesticides released into Rhine	• Most aquatic life in immediate downstream reaches killed. • Eels (*Anguilla anguilla* L.) were killed over 560 km downstream.[8]	• Highlights complexity of response and difficulty in identifying causal agents for biological effects in cocktails of pollutants. • Many individual contaminants were below LC_{50} values so synergistic effects may have been important.[8]
Cantara Loop herbicide/pesticide spill, Sacramento River, USA, July 1991	• 70 000 litres of metam sodium, a soil fumigant spilt after train derailment. • Breakdown products include volatile methyl isocyanate.	• 41 mile reach of river affected. One million fish (including 200 000 rainbow trout, *Onchorhynchus mykiss*), tens of thousands of amphibians and crayfish killed. Widespread damage to riparian trees. Estimated recovery time from <1 year (for periphyton) to 20 years (molluscs).[9]	• Impacts not just confined to instream organisms given the cloud of toxic gas released as the herbicide reacted with stream water. • Disparities in recovery rates between sessile and mobile organisms.
Aznalcóllar mine waste disaster, Spain, April, 1998.	• 6 000 000 m³ of acidic wastes (rich in As, Cd, Cu, Pb and Zn) released after dam collapse into the Agrio-Guadiamar river system. • Contains the internationally important Doñana wetland site.	• 30 000 tons of fish killed. • Near lethal metal concentrations reported in Greylag Geese (*Anser anser*). • Increased incidence of DNA damage in nestling white storks (*Ciconia ciconia*) and black kites (*Milvus migrans*) at least in part explained by As and metal contamination.[10] • 1241 tonnes of fish killed.[11]	• Long-term studies highlight difficulties in attributing biological effects to disaster in mobile organisms.

Event	Description	Effects	Significance
Baia Mare mine waste disaster, Romania, January and March 2000.	• Failure of two dams released 10 000 m³ of cyanide rich gold mine tailings.		• Underlying chronic pollution in the system made assessments of long-term recovery to baseline conditions difficult. • Major cross border disaster
Kingston fly ash spill, Emory River, Tenessee, USA, December, 2008	• 3.8 million cubic meters of fly ash released after storage pond dyke failure. • As, Se, Ba, Hg and V rich. • Fine grained material, physical smothering.[9]	• Physical smothering of streambed retarded recovery of macro-invertebrate communities. • Risk of methylmercury contamination of fish.[12]	• Highlights that physical impacts on aquatic biota can be as significant as chemical changes.
Ajka bauxite processing residue (red mud) spill, Hungary, October 2010	• 1 million m³ red mud released into tributaries of Danube after pond dyke failure. • High salinity, extreme pH (>13), elevated trace metals/ metalloids (*e.g.* Al, As, Cr, V), P-rich.[13]	• Major fish kills in 90 km reach of the Marcal River. • Numerous bioassays showed adverse response in contact with red mud.[14] • Short-term decline in planktonic rotifer diversity and abundance 250 km downstream in the Danube.[15]	• Multiple stressors which may have differing long-term significance: salinity (short term), metal/metalloid availability (medium term) and nutrient enrichment (long term) all likely to be important.

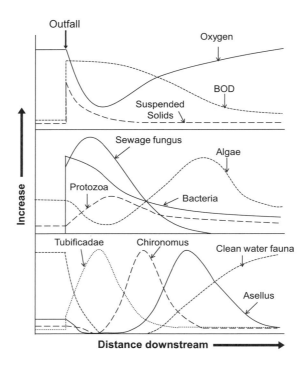

Figure 4.1 Changes in water quality and populations of organisms in a river below a discharge of an organic effluent.

light, but they gradually reappear below the zone of gross pollution, *Stigeoclonium tenue* being the initial colonizer. With the decomposition of organic matter large quantities of nitrates and phosphates are released, stimulating algal growth and resulting in dense blankets of the filamentous green *Cladophora* smothering the stream bed. Similarly, higher aquatic plants (macrophytes) may respond to the increased nutrient concentration, though only *Potamogeton pectinatus* is very tolerant of organic pollution.

Heavy organic pollution affects whole taxonomic groups of macroinvertebrates, rather than individual sensitive species and it is only in conditions of mild pollution that the tolerances of individual species within a group assume significance. The groups most affected are those which thrive in waters of high oxygen content and those which live on eroding substrata, the most sensitive being the stoneflies (*Plecoptera*) and mayflies (*Ephemeroptera*). The differences in tolerances of groups of macroinvertebrates form the basis of methods for monitoring, as will be described later in this chapter.

In the most severe pollution, the tubificid worms, *Limnodrilus hoffmeisteri* and/or *Tubifex tubifex*, are the only macroinvertebrates to survive. The organic effluent provides an ideal medium for burrowing and feeding, while in the absence of predation and competition, the worms build up dense populations, often approaching one million individuals per square metre of stream bed.

These worms contain the pigment haemoglobin, which is involved in oxygen transport, and they can survive anoxic conditions for up to four weeks. As conditions improve slightly downstream, the larvae of the midge *Chironomus riparius*, which also contains haemoglobin, thrives in dense populations and, as the water self-purifies, other species of this large family of flies appear, the proportion of *Chironomus riparius* gradually declining. Below the chironomid zone, the isopod crustacean *Asellus aquaticus* becomes numerous, especially where large growths of *Cladophora* occur. At this point, molluscs, leeches and the predatory alder fly (*Sialis lutaria*) may also be present in some numbers. As self-purification progresses downstream the invertebrate community diversifies, though some stonefly and mayfly species, which are sensitive even to the mildest organic pollution, may not recolonize the stream.

Fish are the most mobile members of the aquatic community and they can swim to avoid some pollution incidents. In conditions of chronic organic pollution they are absent below the discharge, reappearing in the *Cladophora/Asellus* zone, the tolerant three-spined stickleback (*Gasterosteus aculeatus*) being the first to take advantage of the abundant invertebrate food supply. Organic pollution is usually most severe in the downstream reaches of rivers and may prevent sensitive migratory species, such as Atlantic salmon (*Salmo salar*) and sea trout (*S. trutta*), from reaching their pollution-free breeding grounds in the headwaters.

4.3 EUTROPHICATION

There has been at least a doubling of reactive nitrogen at the Earth's surface since the widespread adoption of the Haber–Bosch process for fixing atmospheric nitrogen into ammonia.[20] Processing of ammonia for chemical fertilizers and their subsequent application on agricultural land has contributed to one of the major anthropogenic impacts on freshwater systems since the second half of the 20th century. Alongside this dramatic increase in nitrogen in freshwater systems, the global phosphorus cycle has been amplified by around 400% by human activities.[21]

It has already been described how the release of nutrients during the breakdown of organic matter stimulates the growth of aquatic plants. This addition of nutrients to a waterbody is known as eutrophication. Nitrogen and phosphorus are the two nutrients most implicated in eutrophication and, because growth is normally limited by phosphorus rather than nitrogen, it is the increase in phosphorus which stimulates excessive plant production in freshwaters. Beyond agricultural fertilizers, the main sources of excess nutrients to surface waters include phosphorus from STW, the washing of manure from intensive farming units into water, the burning of fossil fuels which increases the nitrogen content of rain and the felling of forests which causes increasing erosion and run-off.

4.3.1 Nutrient Pollution in Lakes

The concentrations of nitrate and phosphate in the water of Ardleigh Reservoir, a eutrophic waterbody in East Anglia, are shown in Figure 4.2.[22] Note that the concentration of nitrate increases during the late winter when fertilizer is applied to growing crops and is washed into streams feeding the reservoir in large amounts. By contrast, the concentration of phosphate peaks in late summer, when low flows in the feeder streams consist largely of treated sewage effluent while, at this time of year, much phosphate is released from the reservoir sediments into the water.[22]

Table 4.2 lists the guidelines for assessing the trophic status of a waterbody. Peak phosphorus concentrations in Ardleigh Reservoir were some 250 times the minimum concentration for assigning a waterbody as eutrophic, while peak concentrations of nitrogen were 10 times the minimum. One of the major biological effects of eutrophication is the stimulation of algal growth. As eutrophication progresses, there is a decline in the species diversity of the phytoplankton and a change in species dominance as overall populations and biomass increase.

Figure 4.3 illustrates the seasonal changes in biomass of the dominant groups of algae in Ardleigh Reservoir.[22] Typical of temperate lakes is the early peak of diatoms (*Bacillariophyta*), followed by a late spring peak of green algae (*Chlorophyta*). Eutrophic lakes are characterized by enormous summer growths of cyanobacteria, in the case of Ardleigh Reservoir mainly *Microcystis aeruginosa*.

The algal blooms associated with excessive amounts of nutrients have other consequences for the aquatic ecosystem. The macrophyte swards of many lakes have been eliminated as the light is

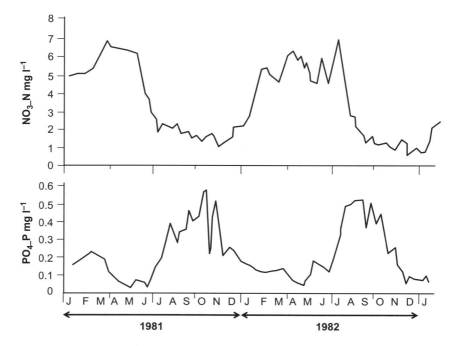

Figure 4.2 Concentrations (mg l^{-1}) of nitrate and phosphate in Ardleigh Reservoir, eastern England, over two years.[22]

Table 4.2 Eutrophication survey guidelines for lakes and reservoirs.

	Oligotrophic	*Mesotrophic*	*Eutrophic*
Total phosphorus (μg l^{-1})	<10	10–20	>20
Total nitrogen (μg l^{-1})	<200	200–500	>500
Secchi depth (m)	>3.7	3.7–2.0	<2.0
Hypolimnetic dissolved oxygen (% saturation)	>80	10–80	<10
Chlorophyll a (μg l^{-1})	<4	4–10	>10
Phytoplankton production (g C m^{-2} d^{-1})	7–25	75–250	350–700

reduced on the lake bed, preventing photosynthesis of germinating plants. Growths of epiphytic algae on the leaf surface may also restrict light uptake by aquatic plants so that they become scarce in the lake. Zooplankton use macrophyte canopies as refuges from fish. Without macrophytes, they are very vulnerable and this itself can accelerate the eutrophication process as the grazing pressure on phytoplankton is reduced in the absence of zooplankton, allowing denser algal blooms to develop.

The Norfolk Broads of eastern England are a series of shallow lakes, formed during medieval times when peat workings became flooded, and famous for their rich flora of Charophytes (stoneworts) and aquatic angiosperms, supporting a diverse assemblage of invertebrates. During the 1960s a rapid deterioration set in as nutrient concentrations increased and a survey of 28 of the main broads in 1972–73 revealed that eleven were devoid of macrophytes and only six had a well developed aquatic flora. The invertebrate fauna was similarly impoverished. Since then the situation has deteriorated further with additional losses of macrophytes and the Norfolk Broads now have some of the highest total phosphorus concentrations of world freshwater lakes.[23]

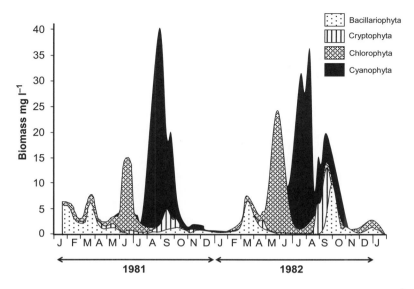

Figure 4.3 Seasonal variation in phytoplankton composition and total biomass (mg l^{-1} wet weight) in Ardleigh Reservoir, eastern England, over two years.[22]

It appears that shallow lakes have two alternative stable states over a range of nutrient concentrations.[24] They may either have clear water, dominated by aquatic vegetation, or turbid water with high algal biomass. The clear water, macrophyte community may be stabilized by luxury uptake of nutrients, making nutrients unavailable to plankton, by secreting chemicals to prevent plankton growth, and by sheltering large populations of grazing zooplankton which eat planktonic algae. Fish predation of zooplankton is reduced in the structured macrophyte sward.[25] The turbid water phytoplankton community may be stabilized by an early growing season, shading the later germination of macrophytes, and by producing large, inedible algae, which also acquire carbon dioxide more easily, especially at high pH. An absence of macrophytes increases the vulnerability of herbivores to predation by fish in the unstructured environment, while the predatory pike (*Esox lucius*) declines without cover for ambushing its prey, allowing even larger populations of planktivorous fish to survive.[26]

Fish communities also change as oligotrophic lakes become more eutrophic. Cold water fish with high oxygen requirements, such as salmonids and white-fish, are replaced by less demanding cyprinids, the commercial value of the fishery declining. Some algae at high densities produce toxins which kill fish. In the Norfolk Broads *Prymnesium parvum* has caused several large fish kills over the last four decades. The sudden collapse of an algal bloom may result in rapid deoxygenation of the water and large kills of fish. Cyanobacteria may also produce potent poisons which can induce rapid and fatal liver damage at low concentrations. Livestock, dogs and wildlife have been killed. Toxins do not always occur in blooms and can be highly variable with time, making them difficult to predict, detect and monitor.[27]

4.3.2 Nutrient Enrichment in Rivers and Groundwaters

The problems of eutrophication are not restricted to standing waters. Around 10 000 km of streams in England and Wales are deemed at risk of failing WFD guidelines for nitrate, the vast majority situated in areas of intensive lowland agriculture in eastern England.[18] Over 25 000 km of stream (which equates to 43% of all water bodies) were deemed at risk of phosphorus pollution which highlights the extent of the nutrient pollution problem.[18] The geographic distribution of streams impacted by excess phosphorus covers the same agricultural areas as those impacted by nitrate pollution, but also most major urban centres, reflecting a range of point and diffuse urban sources.[18]

Nitrate-rich drinking water presents potential health problems. In particular, babies under six months of age who are bottle fed may develop methaemoglobinaemia (blue baby syndrome), the nitrate in their milk being reduced to nitrite in stomach acids and then oxidizing ferrous ions in the haemoglobin of their blood, so lowering its oxygen carrying capacity. The disease only occurs when bacteriologically impure water, containing nitrate levels approaching 100 mg l^{-1}, is supplied, and is usually only an issue with water supply from private wells. It is worth cautioning however that a number of authors are questioning the simple association between nitrate and infant methaemoglobinaemia, and considering nitrate as a co-factor in one of several causes of the disease.[28] The *EU Drinking Water Directive (80/778/EEC)* standard for drinking water recommends that nitrates should not exceed 50 mg NO_3 l^{-1}, but concentrations in some lowland areas of Britain frequently exceed 100 mg NO_3 $^{-1}$. Recent studies have also warned of a potential "timebomb" with nitrate concentrations in the UK still rising in many areas despite efforts to curb nutrient pollution (notably the *EU Nitrates Directive* and WFD). Areas of eastern England with intensive agriculture, chalk lithology with a thick unsaturated zone and thin or absent superficial deposits are particularly susceptible.[29] This may lead to some strategically important aquifers becoming unsuitable for potable water supply unless they are treated. Furthermore, where river systems are predominantly fed by groundwater in these intensive agricultural areas, we may see similarly slow recoveries in nutrient status of the surface waters.[30]

4.3.3 Managing Nutrient Pollution

It is possible to reverse eutrophication although, in many cases, expensive remedial measures have proved much less successful than was hoped.[31] Methods include 'bottom-up' (nutrient control) and 'top-down' (biomanipulation) approaches. Bottom-up approaches involve controlling inputs of phosphorus because this element is normally limiting to plant growth and much of it in freshwaters is derived from point sources (*e.g.* sewage treatment works), whereas nitrogen enters the aquatic ecosystem diffusely, *via* land drainage. Phosphorus releases from STW have reduced in recent years with legislative pressure and increased investment in treatment technologies, while recent attempts to control diffuse agricultural releases in Europe has seen the setting of Nitrate Vulnerable Zones (NVZs) in areas with sensitive geology. This involves setting controls on the quantities and timings of fertilizer additions, and regulating the storage of silage to minimise risk of runoff. These source minimisation measures have been coupled with a range of control mechanisms to manage releases from land. Modifications to agricultural practice, such as the planting of buffer strips of grassland, woodland or wetland at field margins can help minimise nutrient loss from fields. Similarly, sediment detention ponds, the reinstatement of riparian wetlands (which may also serve dual benefits for flood control) and the installation of sub-surface gravel drainage from areas of farm hardstandings have also been recommended. Many researchers are also investigating novel, low-cost nutrient stripping technologies such as waste ferric oxides from coal mine drainage systems, which effectively minimise phosphorus release, with the pelletised ferric oxide suitable for deployment as a slow-release fertilizer when filter systems are routinely maintained.[32]

In some lentic systems (a non-flowing or standing body of freshwater such as a lake or pond) affected by eutrophication, biomanipulation has seen the management (or complete removal) of planktivorous fish, allowing the re-establishment of populations of the larger herbivorous zooplankton, which graze algae.[33] Fish removal has led to increases in water clarity, encouraging the growth of macrophytes. Once the macrophyte stable state has developed it should be possible, indeed it is desirable, to reintroduce fish, along with the predatory pike. Biomanipulation technology, however, is still being refined and the clear water phase may only be temporary, several lakes showing increased turbidity after the first few years of management. Some lakes have not responded as predicted and it appears that an increase in herbivorous wildfowl prevent the development of macrophytes.[34] The control of phosphorus is in most cases a necessary prerequisite to biomanipulation, which is likely to be most effective in shallow waters where macrophytes are a major component of the ecosystem. This can incorporate removal of phosphorus-rich surface sediments or iron addition to increase the phosphorus sorption capacity of the sediment.[35]

4.4 ACIDIFICATION

4.4.1 Acidification from Above: Sulfur and Nitrogen Oxides

The threats posed by acid rain and associated acidification of freshwaters have been documented from direct monitoring as early as the 1920s in Scandinavia, while by reconstructing past environments through analysing diatom communities in lake sediment cores these trends can be detected even earlier. For example studies of the diatom remains in cores of sediment from lochs in south-west Scotland have indicated that progressive acidification began around 1850, with an increase in *Tabellaria binalis* and *T. quadriseptata*, both species characteristic of acid waters.[36]

Acidification also represents an excellent example of a major global environmental pollution problem which, through a combination of legislation, technological improvements and reactive management, is now on trajectory that is at least tentatively positive in many parts of the world. Coal-fired power stations and metal smelters produce much of the polluting oxides of sulfur and nitrogen responsible for the problem, but domestic and other industrial sources are also significant, as are the exhausts of vehicles, particularly for nitrogen oxides (NO_x) (see also Chapters 7 and 8). The acids either fall directly into waterbodies as wet or dry deposition or are washed in from vegetation and soils within the catchment. The effects acidification has are largely a function of the bedrock geology of the receiving catchments, with three broad categories of water which differ in acidity:

(i) Those which are permanently acid, with a pH less than 5.6, low electrical conductivity, and an alkalinity close to zero. Such conditions occur in the headwaters of streams and in lakes, where the soils are strongly acid, or in the outflows of peat bogs.

(ii) Those which are occasionally acid, where pH is normally above 5.6 but because they have low alkalinity (usually less than 5.0 mg l^{-1} $CaCO_3$) the pH may drop below 5.6 periodically. These include streams and lakes in upstream areas of low conductivity on rocks unable to neutralize acid quickly. Such waters may show episodes of extreme acidity, for instance during snowmelt or following storms. These may be very damaging to aquatic life but the infrequency of acid events makes the problem difficult to detect.

(iii) Those which are never acid, the pH never dropping below 5.6 and the alkalinity always above 5 mg l^{-1} $CaCO_3$.

Much of northern and western Britain has a solid geology consisting of granites and acid igneous rocks; there is little or no buffering capacity. The situation is exacerbated in those catchments which have been extensively planted with conifer forests. The sulfate ion is very mobile and transfers acidity very efficiently from soils to surface waters. The nitrate ion behaves similarly but is normally quickly taken up by plant roots.

The effects of acidity on aquatic organisms are both direct (*e.g.* direct changes to ion regulation across the gill epithelium) and indirect, for example the increased availability of metal ions, notably aluminium which increases in solubility as pH falls. Table 4.3 provides a generalized summary of

Table 4.3 Sensitivities of aquatic organisms to lowered pH.

pH	Effects
6.0	Molluscs and most crustaceans disappear. White moss increases.
5.8	Salmon, char, trout and roach die; sensitive insects, phytoplankton and zooplankton die.
5.5	Whitefish, grayling die; crayfish lose calcium from exoskeleton
5.0	Perch, pike die, fish eggs do not hatch.
4.5	Eels, brook trout die.
4.0	Mayflies and frogs begin to die.

the sensitivity of aquatic organisms to lowered pH based on studies in Scandinavian and North American lakes. Changes in the community begin at pH 6.5 and most species have disappeared below pH 5.0 leaving just a few species of tolerant insects and some species of phyto and zooplankton. Considerable research has been directed towards the effects of acidification on fish because of their economic and recreational importance. The biological impacts caused by acidification became fully acknowledged after the extent of damage to salmonid populations in Scandinavia was revealed. For example, by 1975 it was recognised that over half a total of 2850 lakes in southern Norway had lost their brown trout (*Salmo trutta*) populations.[37]

The effect of acidity on fish is mediated *via* the gills. The blood plasma of fish contains high levels of sodium and chloride ions and those ions which are lost in the urine or from the gills must be replaced by active transport, against a large concentration gradient, across the gills. When calcium is present in the water it reduces the egress of sodium and chloride ions and the ingress of hydrogen ions. The main cause of mortality in acid waters is the excessive loss of ions such as sodium which cannot be replaced quickly enough by active transport. When the concentrations of sodium and chloride ions in the blood plasma fall by about a third, the body cells swell and extracellular fluids become more concentrated. To compensate for these changes, potassium may be lost from the cells, but, if this is not eliminated quickly from the body, depolarization of nerve and muscle cells occurs, resulting in uncontrolled twitching of the fish prior to death.

Aluminium has been shown to be toxic to fish in the pH range 5.0–5.5 and, during episodes of acidity, aluminium ions are frequently present in high concentrations. This can be particularly pronounced where water is draining conifer plantations or areas with extensive clay minerals in superficial deposits. Aluminium ions apparently interfere with the regulation by calcium of gill permeability so enhancing the loss of sodium in the critical pH range. They also cause clogging of the gills with mucus and interfere with respiration. The early developmental stages of fish are particularly sensitive to acidification[38,39] and it is thought that aluminium may interfere with the calcification of the skeleton of fish fry resulting in a failure of normal growth. This failure in recruitment results in a gradual decline of the fish population to extinction. Many invertebrate species, such as crustaceans, molluscs and caddis are absent from acidified streams even though their food supply may be present. Physiological stress is the most likely cause. Aquatic invertebrates need to actively take up sodium, chloride, potassium and calcium ions for survival. Uptake is dependent on external concentrations. In acid waters, ion concentrations may be too low while hydrogen and aluminium ions become dominant in the water. These ions are small and mobile and may be transported in instead of essential ions, upsetting the normal ionic balance, leading to death. Aluminium in acidic conditions has been shown to damage the ion-regulatory organs of caddis larvae, disrupting osmoregulation and leading to increased mortality. However the sensitivity to aluminium has been shown to differ between species.[40]

The simplification of the aquatic ecosystem due to acidification can also cascade to higher levels of the food chain, such as birds and mammals. The decline of the dipper (*Cinclus cinclus*), which feeds mainly on aquatic invertebrates was linked with acidification in Welsh upland streams.[41] A decline in pH of 1.7 units on one river over the period 1960–1984 resulted in a 70–80% decline in the dipper population.[41] If fish populations are eliminated from headwaters of streams by acidification, then the piscivorous otter (*Lutra lutra*) may not use them. In general, however, acidification results in a reduction in the carrying capacity for otters rather than a decrease in distribution.[42]

4.4.2 Recovery from Acidification

Amendments to the *US Clean Air Act* in the 1970s and 1980s and the *UNECE Convention on Long-range Trans-boundary Air Pollution* in 1979 mandated many industrial nations to radically cut their acidifying emissions. *Source control*, *via* the use of lime based scrubbers (which can generate useful gypsum products after stripping sulfur out of flue gases), the use of coal with a lower sulfur content and more efficient coal burning power stations has seen an observed fall in sulfur emission and deposition in many affected areas. SO_2 emissions in the UK have fallen by 85% from peak values in the 1970s and 1980s with an associated decline in sulfur deposition rates by at least 50%.[43] In the USA, emissions have fallen by around 80% relative to 1980 values. In addition to such pollution prevention, *reactive management* for acidification is based around direct liming of catchments and is practiced in many catchments in Scandinavia and North America.

Reduced atmospheric sulfate loadings have been accompanied by recovery in stream chemistry in most monitored areas, albeit falls in instream sulfate loadings have typically been less than the falls in deposition rate, suggesting some buffering or lag in response due to previously deposited sulfates in soils.[44] An important resource for assessing the longer term environmental recovery comes from long-term monitoring encompassing biological as well as chemical analysis tools. In the UK, the Acid Waters Monitoring Network (now the Upland Waters Monitoring Network) have maintained monitoring of deposition rates, stream chemistry as well as a suite of biological monitoring tools including diatoms, macro-invertebrates and macrophytes for over twenty five years at a broad geographic range of sites. The response such biological data show is more complex than that apparent in chemical parameters. Typical patterns include community shifts in diatom assemblages incorporating decline of acid indicating diatoms and an increase in those more indicative of less acidic conditions. An increasing number of acid-sensitive macro-invertebrates and a proportional rise in predatory species such as predatory caddisfly (notably *Cyrnus* sp. or *Polycentropus* sp.) and stoneflies (*Siphonoperla torrentium* and *Isoperla grammatica*) were also noted.[45] The more frequent presence of aquatic macrophytes sensitive to acid conditions (*e.g. Hyocomium armoricum, Chara virgata* and *Callitriche hamulata*) were observed at the majority of UWMN rivers and lakes where alkalinity had increased suggesting strong chemical drivers, while the appearance of juvenile brown trout has been documented in some systems once acid neutralising capacity approached or exceeded positive values.[45] However the biological recovery is often gradual, highly variable between catchments, and there are obvious lags in biological recovery after improvements in water quality. Biological recovery is always likely to be more complex than chemical recovery given limitations to the dispersal of acid-sensitive species and the fact that there may be certain chemical thresholds that elicit a biological response (*e.g.* some macrophytes utilise dissolved organic carbon from the water column in photosynthesis and only occur when this is present in less acid conditions) as opposed to a steady linear response.[45] Furthermore, many affected catchments are still subject to acid flushing events which can limit the recovery of more sensitive taxa, while recovery may also be impacted by land use. For example, there is evidence that chemical and ecological recovery is slower in areas of upland coniferous forestry in Wales.[43] One related side-effect of the decrease in mineral acidity in upland settings is the rise in organic acidity associated with increased release of dissolved organic carbon to surface waters. This has been a consistent phenomenon in recovering upland settings with organic-rich (peat) soils and may have implications for downstream biota, notably through increased water colour and reduced UV-B penetration through the water column.

Despite the marked reduction in acidifying emissions, there remain enduring challenges associated with acidification. The slow biological recovery and problems of short-term acid flush events require continued monitoring efforts, while the effects of NO_x emissions require particular attention given legislation has failed to reduce vehicle derived NO_x emissions as quickly as the SO_2 from industrial sources. Furthermore, most of the studies described above are from the North America and north western Europe. The world's largest coal burning nation, China, is currently experiencing rising SO_2 and NO_x emissions. While the former are anticipated to fall in coming years, recent modelling studies suggest that any fall in SO_2 deposition will be negated by rapidly increasing NO_x around China's main industrial cities.[46]

4.4.3   Acidification from Below: Acid Mine Drainage

The other major mechanism for the release of acids to freshwater systems arises from the problems of mine drainage. The process of mining exposes hitherto confined sulfide minerals, such as pyrite (FeS_2) or galena (PbS_2), to oxygen, water and chemosynthetic bacteria. These bacteria assist in the oxidation of sulfide minerals which both releases metals into solution and forms sulfuric acid. As such, mine drainage can have pH as low as 2 or 3 and can contain a cocktail of toxic metals and metalloids, most commonly arsenic, cadmium, copper, lead and zinc as well as iron which is less directly toxic. If the surrounding bedrock is not carbonate-rich to buffer the pH, such discharges can result in the extermination of much of the biota in the watercourse into which it flows. The problem usually arises after mine closure when pumping operations cease and water levels begin to rise and react with sulfide minerals in the mine voids. Mine drainage affects 19 000 km of streams and 73 000 ha of lakes in the USA alone,[47] around 10% of all rivers in England and Wales and is responsible for around half of all known emissions of cadmium, zinc and lead to the water environment of England and Wales.[48] The location of many former orefields in mountainous areas of otherwise high amenity value and quality compounds the issues. Furthermore, facilities storing fine mine wastes (tailings), despite legislative progress, can still cause localised devastation to aquatic systems if there are failures in retaining walls (see Table 4.1).

It is worth noting that while mining activity can lead to conditions that either facilitate or cause direct toxicity to aquatic organisms, in many cases it is the physical effects of streambed smothering that are most prominent. For example the impacts of amorphous iron oxyhydroxides which readily precipitate from iron-rich coal mine drainage prevent oxygen circulation through benthic sediments and have been cited as the causal agent (rather than direct toxicity) for impoverished invertebrate communities in some mine-impacted streams.[49] Similar physical smothering mechanisms may also be apparent at rivers impacted by siltation (*e.g.* from unregulated mining or quarrying) or sites affected by alkali pollution (*e.g.* chlor-alkali wastes or steel slag drainage).

4.5   TOXIC CHEMICALS

4.5.1   Modes of Action of Toxic Chemicals

Some aspects of toxic pollution have already been mentioned but it is now appropriate to describe the modes of action of toxic chemicals. Some of the major types of toxic compounds, which are not mutually exclusive, are:

   (i) Metals, such as zinc, copper, mercury, cadmium.
  (ii) Organic compounds, such as pesticides, herbicides, polychlorinated biphenyls (PCBs), phenols; organometals such as methylmercury.
 (iii) Gases, such as chlorine, ammonia.
  (iv) Anions, such as cyanide, sulfate, sulfite.
   (v) Acids and alkalis (*e.g.* caustic soda).

There are a number of terms in regular use in the study of toxic effects:

   (i) *Acute*: causing an effect (usually death) within a short period.
  (ii) *Chronic*: causing an effect (lethal or sub-lethal) over a prolonged period of time.
 (iii) *Lethal*: causing death by direct poisoning.
 (iv) *Sub-lethal*: below the level which causes death but which may affect growth, reproduction or behaviour so that the population may eventually be reduced.
  (v) *Cumulative*: the effect is increased by successive doses.

A schematic of a typical toxicity curve is given in Figure 4.4. The median periods for survival are plotted against a range of concentrations. The lethal concentration (LC) is used where death is the criterion of toxicity and has traditionally been the focus of toxicity studies. The number indicates the percentage of animals killed at that concentration and it is also usual to indicate the time of exposure. Thus 96-hour LC_{50} is the concentration of toxic material which kills fifty per cent of the test organism in ninety six hours. The incipient level is usually taken as the concentration at which fifty per cent of the population can live for an indefinite period of time. Where effects other than death are being sought, for example respiratory stress or behavioural changes, the term used is the effective concentration (EC) which is expressed in a similar way, *e.g.*, 96-hour EC_{50}. Other common terms used include the no observed effects concentration (NOEC), which assumes there is a threshold concentration below which no adverse effect is expected (*i.e.* response at lower concentrations is not significantly different from unexposed control treatments). Statistical testing of exposure concentrations against control concentrations (*i.e.* unexposed treatments), with the null hypothesis that there is no significant difference in average concentration will yield the NOEC along with the lowest observed effect concentration (LOEC). The latter is the lowest average concentration where response differs significantly from control treatments (see Figure 4.4).[50] The arithmetic mean of these two values gives the maximum allowable toxicant concentration (MATC).

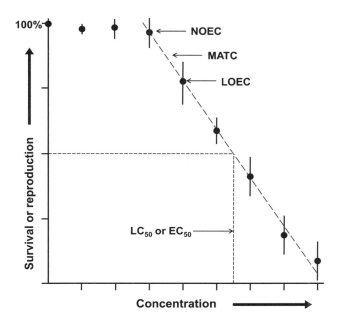

Figure 4.4   A schematic diagram of a typical toxicity curve. (Adapted from ref. 51).

While there is much discussion about the statistical validity of such measures, they are widely used by regulators in many settings.[50]

There has been a large amount of data collected on the acute toxicity of chemicals, especially to fish and invertebrates,[52] and this has undoubtedly been of great value in elucidating the mechanisms of toxicity. However the value of these data for practical river management is more questionable. Incidents resulting in large mortalities of fish and other organisms are usually accidents over which the river manager has no control. He or she can merely assess the damage and perhaps restock when conditions improve. In addition, detailed information on toxicity of a range of compounds is available for only a few test organisms, such as rainbow trout or *Daphnia*, and it is well known that even closely related species may show very different responses to particular pollutants. It is the sub-lethal effects of pollutants which are of particular concern in many field situations, for low levels of pollutants may result in the gradual loss of populations, without any overt signs of a problem.

Experiments on sub-lethal effects are more difficult to carry out because they invariably take longer and individuals under test may respond very differently to low levels of pollution. Furthermore, the reaction to pollutants may vary over the lifetime of an organism, early developmental stages often being more susceptible. It is therefore necessary to study the entire lifecycle of an organism under experimental conditions to find the weak link in its response to pollution and such long-term experiments, possibly over several generations, are essential to discover any carcinogenic, teratogenic or mutagenic effects of pollutants. Sub-lethal effects may be manifested at the biochemical, physiological, behavioural or life cycle level.[1] For example, novel experiments using electrical recordings from the nose (electro-olfactograms) and videos of the behavioural response of Pacific salmon (*Oncorhynchus kisutch*), demonstrated how the detection of salmonid olfactory cues was impaired due to copper exposure well below aquatic life standards[53] which could have a bearing on predator avoidance. Although it is possible in the laboratory to demonstrate small effects at very low levels of pollution, for example in biochemistry or on growth, it is essential to show that these are likely to reduce the fitness of an organism in its natural environment and are not merely within the organism's range of adaptation. Nevertheless these sub-lethal effects may be quite subtle and can be measured long before any outward toxic effects are manifested. They can be used as *biomarkers*, to show that an organism has been exposed to contaminants at levels which exceed the normal detoxification and repair capabilities.[54,55] A *biomarker* can be defined as a xenobiotically-induced variation in cellular or biochemical components or processes, structure, or functions that is measurable in a biological system or sample. Biomarkers can be used to predict what concentrations of a pollutant are likely to cause damage, rather than merely to measure concentrations when damage has been noted.

The liver enzymes (mixed-function oxidases) which metabolize xenobiotics are often used as biomarkers. In the Belgian River Meuse a once common fish, the barbel (*Barbus barbus*), is now scarce. The activities of three liver enzymes were strongly correlated with concentrations of PCBs in the liver. At high PCB levels there were marked alterations in the liver ultrastructure, with a change in mitochondrial membranes and an excessive growth of rough endoplasmic reticulum.[56] High PCB concentrations also reduced reproductive success in barbels,[57] which would explain the decline of fish populations in the Meuse. Organisms which are regularly subjected to toxic pollutants may develop tolerance to them. This may be achieved either by functioning normally at high concentrations of pollutants or by metabolizing and detoxifying pollutants.

Algae living in streams receiving mine drainage are highly tolerant to metals and this adaptation has been shown to be genetically determined. Metal tolerance has also been observed in invertebrates from metal contaminated streams, and in fish. Exposure stimulates the production of metallothioneins, low molecular weight proteins containing sulfur-rich amino acids which bind and detoxify some metals.

With the rapid advances in molecular biology and bioinformatics in recent years, new tools and approaches for toxicity screening are being generated. These focus on identifying the cellular response pathways, that when sufficiently perturbed by toxicants at environmentally-realistic exposure levels, are likely to result in adverse health effects.[58] Such adverse cellular responses have been identified for metal pollution (using metallothionein response) and hydrocarbon pollution (based on cytochrome P4501A, a protein which is involved in the detoxification of hydrocarbons)[59] amongst other groups of toxicants. The application of technologies developed in the pharmaceutical industry for drug development allows large numbers of chemicals to be screened each day to assess their molecular, biochemical or cellular effects. The large sample sizes permitted (compared to traditional approaches) also allows for the screening of many more chemicals, tested to a range of endpoints across different lifestages, species and genetic diversity within species. As such, these tools are going to reveal much about the effects of pollutants on aquatic organisms and will underpin regulatory tools for setting environmental thresholds in the 21st century.

4.5.2 Bioaccumulation and Biomagnification

Of particular concern to environmental toxicologists are those compounds which accumulate in tissues, especially some metals and Persistent Organic Pollutants (POPs) such as organochlorines (pesticides and PCBs). A chemical is said to *bioaccumulate* if it builds up in living organisms at concentrations higher than those in the surrounding environment. From often undetectable concentrations in water, organisms may accumulate levels of biological significance if the chemicals are difficult to break down and are absorbed at a rate greater than they are excreted from an organism. Furthermore, if these chemicals appear in progressively higher concentrations through the food chain they are said to have undergone *biomagnification*.

There have been various notorious examples of where such biomagnification in aquatic systems has even led to major public health issues. For example, the release of mercury, used as a catalyst in the production of acetaldehyde, from the Chisso Corporation chemical works into the enclosed Minamata Bay in south-west Japan led to methyl-mercury accumulation in shellfish and fish and the subsequent poisoning of at tens of thousands of people in the mid-20th century.[60] Methylmercury is readily absorbed across the intestine and into the bloodstream unlike elemental mercury and in this case caused severe neurological damage. It can also pass across the placenta and can severely affect the development of unborn children.[60] Prior to the Minamata disaster, Japan had already been subject to a major industrial pollution disaster in the early half of the 20th century, with the release of cadmium from metal mining into rivers in Toyoma Prefecture in north-east Japan. This cadmium accumulated in stream and floodplain sediments of the Jinzū River, areas widely used for rice cultivation. Cadmium is readily taken up by some cultivars of Asian rice (*Oryza sativa*)[61] and led to a major public health disaster known as "itai itai" disease; metal poisoning characterised by renal damage and weakening of bones. In both cases, a population with a narrow dietary range who were over-reliant on a particular local food source which became contaminated were severely affected.

Perhaps the most documented biomagnification example that spans numerous sites and trophic levels comes from monitoring the effects of PCBs on wild populations of vertebrates in the Great Lakes region of North America. In Green Bay, Lake Michigan, Forster's Terns (*Sterna forsteri*) exhibit impaired reproduction. The incubation is extended, few eggs hatch, the chicks have lower body weight and their livers are larger than normal. They show a high incidence of congenital deformities. The parents are inattentive at nesting and this further reduces their reproductive success. It was concluded, following a detailed toxicological analysis, that those PCB congeners which induce the enzyme aryl hydrocarbon hydroxylase (AHH) were the only contaminants present in sufficient amounts to cause the observed effects on eggs and chicks. More than 90% of the effect could be explained by two pentachlorobiphenyls. The behavioural abnormalities in adults were caused by total PCBs.[62] In addition to the Forster's Tern, six other species of fish-eating bird from the Great Lakes have exhibited growth deformities and physiological defects.[63] These symptoms have only been observed in the past three decades and it is considered that, in earlier years, symptoms were masked by the effects of DDE, which thinned eggs to such an extent that they did not survive long enough for abnormalities to be expressed. The biomagnification factor of PCBs from water to top carnivores may be as high as twenty-five million times.[64] Studies on humans also give cause for concern. Behavioural and neurological disorders have been found in the children of mothers who eat modest amounts of fish caught in the Great Lakes.[65] Parallel studies on rats have shown that these behavioural changes are still measurable two generations after exposure.[65]

There has been considerable, detailed research on PCBs in the Great Lakes but the problem, of course, is not confined to that region. For example, evidence suggests that PCBs have been largely responsible for the decline of the otter over large areas of Western Europe, in some places to extinction. PCBs are known to affect the reproductive success of mammals, being powerful endocrine disruptors. There is a strong relationship between the mean amount of PCBs in tissues of otters and the extent of the population decline, relationships which do not hold for other contaminants.[66] Using vitamin A in liver as a biomarker, a strong negative correlation was found between concentrations of this vitamin and PCB concentrations, which coincided with the incidence of disease,[67] the PCBs presumably affecting the immune system. A relationship between PCB concentrations and the incidence of disease has also been reported in harbour porpoises (*Phocoena phocoena*).[68]

With the restrictions placed on the manufacture and use of PCBs, concentrations have declined in tissues of otters in Britain to average levels in the early 1990s which are unlikely to have adverse health effects (see Figure 4.5).[69] The species has shown a considerable increase in range in rivers in lowland England from where it had been absent for four decades, and otters are now beginning to penetrate into the city of London. Similar studies in Sweden (see Figure 4.5) have also shown positive trends in otter populations coinciding with decreases in environmental and otter PCB concentrations.[70]

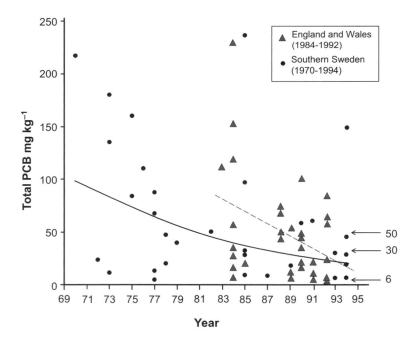

Figure 4.5   PCB concentrations (mg kg^{-1} lipid) in tissues of otters collected in England and Wales between 1984 and 1992[69] and Southern Sweden between 1971 and 1994.[70] (Trendlines taken from original sources). The arrow at 50 mg kg^{-1} illustrates the concentration of PCBs causing reproductive failure in mink (*Mustela vison*), that at 30 mg kg^{-1} is a more stringent standard based on the precautionary principle. The more stringent standard of 6 mg kg^{-1} is based on dose effects levels on vitamin A deficiency.

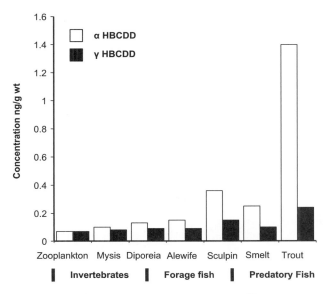

Figure 4.6 Biomagnification of HBCDD in Lake Ontario foodchain.[71] Data show two isomers of HBCDD, highlighting persistence of αHBCDD.

However, PCBs represent just one example of a group of chemicals, which through careful study, we now have an understanding of their behaviour across entire aquatic and terrestrial food chains. International agreement to manage such persistent, bioaccumulative and toxic substances comes *via* the Stockholm Convention on POPs, to which new substances are added on a regular basis (see also Chapter 18). A recent example includes the flame retardant HBCDD (hexabromocyclododecane), widely used in electronics, furniture and polystyrene tiles in construction materials. Studies have shown the biomagnification of HBCDD at levels similar to some PCBs (see Figure 4.6).[71] Like many POPs, HBCDD is *lipophilic*, so accumulates in fatty tissue, with concentrations in higher predators far exceeding those at lower trophic levels. Despite recommendations for a worldwide ban, HBCDD is still widely used and manufactured in some countries and is generally showing rising trends in the environment in media ranging from lake sediments to human breast milk.[72,73] As with many other POPs, given they accumulate over long timescales in higher trophic levels, the need for long-term monitoring of ecosystem effects will remain an area of great concern in the future.

4.6 THERMAL POLLUTION

4.6.1 Anthropogenic Impacts on Thermal Regime

Cooling water discharges from electricity generating stations are the main sources of pollution by heat. Such effluents also contain a range of chemical contaminants which, though small in relation to the volume of cooling water, may in fact have a greater impact on the ecology of the receiving stream.[74] Other anthropogenic activity can also impact on the thermal regime of streams, notably forestry operations, with some studies showing increases of over 7 °C in mean monthly maximum water temperatures in experimental catchments after felling of riparian trees (trees located near rivers, streams or lakes).[75] The location of much plantation forestry in headwater areas with low flow and therefore low thermal capacity can exacerbate such problems.

An increase in temperature alters the physical environment, in terms of both a reduction in the density of the water and its oxygen concentration, while the metabolism of organisms increases. Cold water species, especially of fish, are very sensitive to changes in temperature and they will disappear if heated effluents are discharged to the headwaters of streams. As the temperature increases, the oxygen consumption and heart rate of a fish will increase to obtain oxygen for increased metabolic processes but, at the same time, the oxygen concentration of the water is decreased. For example, at 18 °C, carp (*Cyprinus carpio*) can survive in an oxygen concentration as low as 0.5 mg l^{-1}, whereas at 35.8 °C the water must contain 1.5 mg l^{-1}. The swimming speeds of some species declines at higher temperatures, *e.g.* trout at 19 °C, making them less efficient predators. Resistance to disease may also change. The bacterium *Chondrococcus columnaris* is innocuous to fish below 10 °C but it invades wounds between 10 and 21 °C while it can invade healthy tissues above 21 °C. On the other hand, some organisms have the ability to adapt to altered thermal regime and there may even be some benefits in terms of increased productivity in some areas receiving heated effluents. Change in thermal regime can also lead to invasive or non-native species colonising areas and either out-competing the native biota or forming large colonies. In some cases this has caused major management problems, such as with zebra mussels (*Dreissena polymorpha*) clogging water discharge or intake pipes.

4.6.2 Climate Change

Much contemporary research is assessing the current and future impacts of climate change on the river environment. These impacts encompass direct changes to thermal regime with rising air temperatures in addition to the predicted increased incidence of extreme hydrological events, be they drought or flood. Flood events play a crucial role in the dynamic equilibrium of lotic systems (flowing freshwater systems such as rivers or streams) and connectivity of instream habitats with floodplain environments. However, frequent extreme events can reduce the habitat complexity of riparian environments while they may also be associated with chemical conditions challenging to instream biota (*e.g.* acid flushing events in upland settings). Of potentially greater ecological significance are the pressures associated with low flow periods which are likely to provide more frequent challenges to water managers in the future. Encouragingly, there are already examples of effective management strategies to deal with such circumstances. For example, the River Tyne in northern England, has seen numbers of Atlantic Salmon (*Salmo salar*) and Sea Trout (*Salmo trutta*) return to pre-industrial levels in recent decades. However, during recent drought years (1995, 2003, 2005) the river has been subject to major fish kills in summer months when returning salmon aggregate in estuarine reaches awaiting suitable flow conditions to progress to headwater spawning grounds.[76] Low flow conditions impart stress on various aspects of the aquatic ecosystem through:

  (i)   Reducing concentrations of dissolved oxygen.
  (ii) Increasing temperature.
  (iii) Reducing dilution of perennial pollutant sources (*e.g.* STW point discharges).
  (iv) Reducing the volume of water which minimises refuge areas from predators.

An integrated management system in the Tyne catchment sees automated monitoring of dissolved oxygen and temperature in estuarine reaches of the catchment inform environment managers as to the timing of critical poor water quality conditions. This prompts the controlled release of water from a major headwater reservoir. Fish count stations above the tidal limit demonstrate the success of this scheme with large numbers of salmonids migrating upstream after the release, akin to natural flood conditions.[76]

The challenges posed by a changing climate should therefore be viewed alongside other interacting pressures on freshwater systems such as increased abstraction and the discharge of polluted effluents.

Indeed a recent study in upland waters contaminated by mine pollution (notably Zn) in the Rocky Mountains showed a greater sensitivity of benthic invertebrate communities to increased UV-B exposure (another potential stress associated with more frequent low flow periods), which was suggested to be a consequence of the costs associated with tolerance to Zn pollution.[77] Such complex interactions are likely to be a common feature of the response of aquatic biota to changing climate.

4.7 RADIOACTIVITY

Chemically, radionuclides behave in the same way as their non-radioactive isotopes but, if they accumulate up the food chain the radioactive isotopes have much greater significance. Radionuclides come mainly from fall-out from weapons testing and the effluent from nuclear power stations. Because ionizing radiation is highly persistent in the environment, causing cancer and genetic disorders in humans, it has always attracted special concern and the release of radionuclides is strictly monitored and controlled.

Most of our information on the effects of radionuclide release has arisen out of research in the aftermath of the Chernobyl nuclear power station disaster in the Ukraine in April, 1986 (see also Chapter 19). The subsequent spread and deposition of radionuclides, (notably caesium (^{137}Cs) and strontium (^{90}Sr) which have half lives in the region of thirty years), over large areas of western Europe emphasized how potentially damaging and long-lived such pollution can be.

Research has highlighted the pathways by which aquatic organisms and potentially humans can be exposed to ionising radiation after fallout. Caesium behaves similarly to potassium and is soluble in typical surface waters (especially ion-poor upland waters) although once deposited on land, transfer to surface waters can take many years. Studies in the aftermath of the Tohoku earthquake/Fukushima Dai-ichi nuclear disaster in Japan in March 2011 suggested less than 1% of the deposited radiocaesium (^{134}Cs and ^{137}Cs) was transported out of affected catchments in the nine months following the disaster.[78] This highlights the need for long-term monitoring of the fate of ^{137}Cs through aquatic food webs. In studies on Finnish lakes in the months immediately after the Chernobyl disaster, planktivorous fishes such as whitefish (*Coregonus lavaretus*) had the highest concentrations suggesting rapid bioconcentration of ^{137}Cs in plankton and fish with the shortest food chain. Over time however and after 1987, activity concentrations suggested biomagnification of ^{137}Cs across trophic levels, with the piscivorous pike (*Esox lucius*) having activity concentrations 1.5–3.2 times higher than perch (*Perca fluviatilis*), which feeds primarily on zooplankton. These concentrations were in turn 1–2 times higher than planktivorous whitefish.[79] These studies also highlight that activity concentrations in perch and pike exceed human consumption guidelines (600 Bq kg^{-1}) twenty years after the Chernobyl disaster.

Because wild caught freshwater fish feature little in the diet of the vast majority of the local human population it was considered that the Chernobyl accident presented no cause for concern to human health away from the vicinity of the reactor. After Fukushima however, there have been concerns raised about a potential exposure pathway for humans from rice, given extensive use of river water for irrigating rice paddy fields in catchments affected by the fallout and the fact that Cs is readily taken up from soil solutions by many plants.

However, other species within the aquatic ecosystem are more exclusively piscivorous. The impact of fallout after Chernobyl has been demonstrated in studies of otter (*Lutra lutra*) from the UK. In south west Scotland in particular, an area of soft waters and with high elevation terrain heavily affected by Chernobyl fall-out, average radioactivity in otter faeces (spraints) were more than six times that in less affected areas of Wales (up to a maximum of 79 500 Bq kg^{-1} in one sample).[80] Unfortunately, there was no pre-Chernobyl sample but levels were still high in the following January, as has been found for other biological materials from this area.[80]

In highly contaminated areas around Chernobyl there has been extensive documentation of mutagenic effects of ionising radiation on aquatic life. These include increased incidence of aneuploidy (a chromosome in excess or missing) in channel catfish (*Ictalurus punctatus*), Crucian carp (*Carassius carassius*), carp (*Cyprinus carpio*) and tench (*Tinca tinca*) and increased DNA damage in channel catfish.[81] However, at lower dose rates further afield it is more challenging to disentangle the signal of ionising radiation from other mortality factors and environmental pressures; a common theme in ecotoxicology. New approaches to screening molecular level response to ionising radiation may be able to assist in partitioning such DNA damage in the future as has recently been demonstrated in marine mussels (*Mytilus edulis*).[82]

4.8 OIL

Compared to the marine situation (see Chapter 2), comparatively little work has been done on the effects of oil in freshwater ecosystems. Nevertheless the chronic pollution of freshwaters with hydrocarbons is widespread. Much of it derives from diffuse urban sources such as petrol and oil washed from roads together with the illegal discharge of engine oil. In England and Wales, the Environment Agency report on average 3000 pollution incidents involving oils and fuels each year,[83] and there have been various initiatives in recent years to minimise such incidents. These include educational initiatives aimed at better handling and storage of oils through to Sustainable Urban Drainage Systems (SuDS) to manage diffuse urban sources and prevent discharge of contaminated runoff into surface waters.

The water soluble components of crude oils and refined products may prove toxic to freshwater animals though the prediction of toxic effects is rather difficult owing to the complex chemical nature of discharges. American studies have shown eggs and young stages of organisms are especially vulnerable. In a study using the tree frog *Hyla cinerea*, hatching success of eggs was not influenced by the presence of oil but growth rates and metamorphosis were.[84] A study of pearl dace (*Margariscus margarita*) in pools contaminated with diesel fuel demonstrated severe pathological changes and a failure in reproduction which would eventually lead to the extermination of this fish population.[85]

In general terms, the aliphatic compounds of oils are relatively innocuous while the aromatic hydrocarbons, such as benzene, xylene and toluene are highly toxic. There are also marked species differences in susceptibility to these compounds, further adding to the difficulties of making predictions about toxicity. Some components of oils, such as PCBs and lead, will accumulate in tissues. Emulsifiers and dispersants, used to clean up spillages, are themselves often highly toxic as has been demonstrated in marine settings after the Deepwater Horizon spill.[86] The surface active agents which they contain make cell membranes more permeable and increase the penetration of toxic compounds into organisms. In this way mixtures of oils and dispersants are often more toxic than either applied separately.

The physical properties of floating oil are a special threat to higher vertebrates, especially aquatic birds because contamination reduces buoyancy through damage to the waterproofing of plumage, while oil may be ingested during subsequent preening. A further problem is the tainting of flesh, especially of fish, which is detectable to the human palate at very low levels of contamination and renders fish inedible. The major sources of taint are light oils and the middle boiling range of crude oil distillates but there are a number of other sources, such as exhaust from outboard motors, waste from petrochemical factories, refinery wastes, and all crude oils.

4.9 EMERGING CONTAMINANTS

In recent years there has been increasing concern about the impacts of so called 'emerging contaminants' on the environment. These encompass a broad range of synthetic compounds being increasingly used in everyday life and released into the environment generally without prior systematic risk assessments of their potential long-term effects on the terrestrial or aquatic environment. While the presence of these novel materials in the environment does not necessarily indicate negative effects on aquatic organisms, there is an urgent need to improve our knowledge base to accurately assess such risks. The key emerging contaminants include human and veterinary pharmaceuticals (and their metabolites), personal health care products, flame retardants (the example of HBCDD we considered previously) and engineered nanoparticles. Some of these have been released to the environment (notably medicines) for many decades, but it is only in recent years that improved analytical techniques have permitted direct measurement of them in a range of environmental matrices.

4.9.1 Nanomaterials

Nanotechnology concerns the development and exploitation of technologies of the size order 0.1–100 nm in diameter, and promises to be an area of scientific innovation and economic growth over coming decades. While we are aware of the very harmful effects of fine combustion-derived particulates (ultrafines) on pulmonary systems of mammals and humans, due to the novelty and rapid growth in nanotechnology applications, we have yet to gain a detailed understanding of potential hazards associated with the presence of engineered nanoparticles (ENPs) in the aquatic environment. Key sources and pathways for engineered nanoparticles to reach freshwater systems are summarised in Table 4.4. The most commonly cited ENP aquatic pollution problem concerns silver nanoparticles, which have been widely adopted in recent decades as antibacterial/antifungal agents in medicinal applications, biotechnology and textiles. The latter represents a key potential pathway into aquatic systems, where antibacterial textiles, such as the burgeoning market for antibacterial leisurewear, release silver ENPs during washing cycles into wastewater treatment systems. There they could be of potential significance to microbial communities essential for waste water purification. The presence of silver nanoparticles has been established in a range of wastewater treatment sites in various studies with the potential for a portion to pass through these systems highlighted. The key pathways by which aquatic biota could be adversely affected by ENP include direct ingestion, transfer across the body wall or *via* epithelial boundaries such as gills or olfactory organs.[87] The ingestion pathway may be exacerbated given ENPs aggregate to microalgae

Table 4.4 Examples of engineered nanoparticle sources and pathways to the aquatic environment.[89,90]

Nanoparticle	Application	Important pathways to the aquatic environment
Silver (n-Ag)	Antibacterial/ antifungal agent in textiles; medicinal uses.	Domestic washing of n-Ag containing textiles releases n-Ag into wastewater; sewage sludge application to land.
Nano titanium dioxide (n-TiO$_2$)	Paints; whitening agent; cosmetics and sunscreens.	Personal Health Care Products → wastewater effluents; industrial effluents; localised high concentrations in bathing waters.
Nano zinc oxide (n-ZnO)	UV filter in cosmetics.	Personal health care products → wastewater.
Fullerenes (C60)	Cosmetics, bactericides, drug delivery systems.	Personal health care products → wastewater.
Single walled carbon nanotubes (SWCNT)	Superconductors, optical and storage devices, fuel cells, catalysts.	Industrial effluents; poorly regulated disposal (*e.g.* landfill leachate).

and could cause bioconcentration of ENPs from the water column for zooplanktonic grazers such as copepods and cladocerans.[88]

Numerous workers have undertaken laboratory toxicity tests of ENPs on various common test species such as *Daphnia magna* and *Onchorynchus mykiss* which typically show a low acute toxicity.[90] However, silver nanoparticles have been demonstrated to accumulate in bacterial cell membranes and cause cell death,[90] while significant inhibition of growth of the commonly used aquatic plant bioassay, *Lemna minor* (Common Duckweed), has been observed at exposure rates of the same order of magnitude as those modelled in wastewater treatment facilities (but higher than those observed thus far in environmental settings).[91] Titanium oxides have been shown to inhibit photosynthesis in algae and fullerenes have been implicated in causing oxidative stress in the brains of juvenile Largemouth Bass (*Micropterus salmoides*) at low concentrations.[90,92]

However, most toxicological screenings of engineered nanoparticles are accompanied by various precautionary notes on findings.[88,90] Traditional standard toxicity tests are designed for ionic compounds in solution, whereas the behaviour of ENPs in the environment is governed by complex interactions with other chemical species, notably humic substances, which bind to particle surfaces and potentially limit bioavailability, as well as agglomeration and sedimentation which is difficult to replicate in laboratory tests. As such, there are widespread calls for new systematic approaches to exposure and effects studies of ENPs using a range of test species, appropriate environmental matrices and ENP concentrations, and tests on a range of organisms proceeding to various end-points to identify sub-lethal as well as lethal concentrations. Such work will play a key role in assessing the long-term risks of ENPs to the aquatic environment.

4.9.2 Human and Veterinary Medicines

We have already considered the example of oestrogenic pollution on rainbow trout which highlighted the importance of biomonitoring in identifying potential pollution impacts that would not be recognised by routine chemical monitoring alone. In recent years considerable research effort has been invested in assessing the effects of a range of emerging human and veterinary drugs on the water environment. As with oestrogenic compounds, it is the use, adsorption, metabolism and excretion of pharmaceuticals, be they prescribed or illicit, into sewage treatment systems, that provide the principal pathway for environmental exposure. Some drugs and metabolites will at least in part pass through treatment systems and into surface waters, while the application of sewage sludge to land as an agricultural amendment provides a major secondary pathway whereby compounds initially partitioned onto solid phases in sewage treatment works can be leached into the aquatic environment. Other exposure pathways can include release during manufacture and during disposal of unused medicines.[90] Pharmaceuticals such as hormones, steroids, antibacterials and illicit psychoactive drugs pose particular problems given they are designed to interact with human or animal receptors and can therefore trigger a biological response (and likely behavioural response) at incredibly low concentrations in the environment. Impacts have been documented for some common pharmaceuticals, for example, ibuprofen, widely used as an anti-inflammatory, has been shown to negatively impact the growth of aquatic plants such as *Lemna minor*, while stimulating cyanobacterial growth.[93] However, for many emerging substances such as illicit drugs, studies are establishing their presence in surface waters,[94] but we do not know yet what the long-term impacts on wildlife and humans are likely to be. An additional challenge posed by human and veterinary pharmaceuticals are the numerous transformation products or degradates that arise from parent chemicals either in the body or in the environment, some of which can be more toxic than the parent compound.[90] Therefore research is needed not only to test the impacts of an individual compound, but also to identify and assess the toxicity of degradates, which some recent legislation explicitly demands (*e.g.* the *EU Pesticides Directive: 91/414/EEC*). As with ENPs, traditional short-term toxicity tests may not be particularly appropriate for pharmaceuticals given exposure to medicinal products and degradates is likely to chronic, over very long periods of time, albeit at low concentrations.

4.10 BIOLOGICAL MONITORING OF POLLUTION IN FRESHWATERS

4.10.1 Laboratory Monitoring Techniques

The biological assessment of pollutants includes both laboratory and field techniques.[1] In the laboratory, the most widely used methods are toxicity tests for the preliminary screening of chemicals, for monitoring effluents to determine the extent of risk to aquatic organisms, and, for those effluents which are toxic, to determine which component is causing death so that it can receive special treatment. The simplest type of test is the static test in which an organism is placed in a standard tank in the water under investigation for 48–96 h. There are normally a series of tanks with test water of different dilutions. More sophisticated techniques involve the periodic replacement of test water or indeed continuous flow systems. Fish have traditionally been used as test organisms. In the United States, the main test species have been fathead minnows (*Pimephales promelas*) and bluegill sunfish (*Lepomis macrochirus*). Much toxicity work in Britain has been with rainbow trout but the tropical harlequin (*Rasbora heteromorpha*) has become increasingly popular because it is smaller and has a similar sensitivity to pollutants. Fish require large volumes of clean water for maintenance and, because tests need replicating, much space is needed. Furthermore there are obvious ethical objections to using vertebrates for routine toxicological assessments. There has therefore been much research effort in developing other test organisms across a range of trophic levels in whole effluent toxicity tests. The planktonic crustaceans *Daphnia magna* and *D. pulex* are widely used for they are easily cultured, have a high reproductive rate and are sensitive to a range of pollutants. Other invertebrates that have been used include the resilient brine shrimp *Artemia salina,* tubificid worms, zebra mussels *Dreissena polymorpha*, whose water filtration rates are measured, and the flagellate *Euglena gracilis*, in which movement is the measure of toxicity.[95–97] Bacterial tests are also available, of which one, the Microtox® test, is commercially available. It utilizes the bioluminescence of the marine bacterium *Allivibrio fischeri*, the reduction in light output being the measure of toxicity. The test is sensitive, precise and reproducible. In many applications a range of test organisms are used not only to reflect impact on different trophic levels, but potentially different environmental compartments, such as sediments, soils and waters.[14]

The revolution in molecular testing and bioinformatics highlighted previously will see major changes in the way we approach toxicity testing in the future. These new approaches should address the increasingly complex issues faced such as the differing exposure scenarios required for some emerging contaminants, testing mixtures of contaminants, assessing susceptibility at different lifestages and understanding the mechanisms of toxicity.

4.10.2 Field Monitoring Techniques

Fish and macroinvertebrate communities have been used as tools for monitoring the health of streams and rivers by monitoring agencies for over thirty years in some parts of Europe and North America. Macroinvertebrates have generally been the most favoured group for monitoring in the UK given they are relatively easy to study, are crucial components of aquatic food chains, sensitive to many pressures and their short lifespan (typically months to years) along with lack of mobility means that changes in community can be a used as a good indicator of impacts. Indeed community analysis in its simplest form can provide useful insight into invertebrate response to pollution on a site by site basis. For example, diversity indices take into account the number of species within the collection (species richness) and the relative abundance of species within the collection (evenness). It is argued that a community from an unstressed, *i.e.*, pollution free, environment will contain a large number of species (high richness), many at fairly low densities (high evenness), so that the calculated diversity index will be high. As pollution stress increases, species will gradually decline in number and disappear (low richness), while a few tolerant species will build up big populations in the absence of predation and competition (low evenness) resulting in a low diversity index. Diversity indices take no account of the tolerances of individual species to pollution however.

In recent years, the WFD has revolutionised monitoring of our water resources in Europe and now places a legal obligation on member states to incorporate biological monitoring, notably that of macroinvertebrates, fish, macrophytes and diatoms (termed 'phytobenthos') in measures of ecological quality of lakes and rivers. Given 'good ecological status' is a fundamental goal of the WFD, and is defined in terms of the quality of the structure and functioning of aquatic ecosystems, the role of biological monitoring has become fundamental to assessments of our water environment. This guidance has seen biological monitoring move beyond the traditional methods for monitoring specific pollutants and their effects to offer a broader assessment of ecosystem status. A range of established tools have been adapted to meet this need along with the development of new tools that seek to provide broader geographical coverage to define the ecological status of waterbodies. The use of a suite of biological indices has been shown in numerous cases to be most informative in determining more subtle responses to the varying pollutant pressures at a catchment scale[98] and provides a significant advance in being able to assess ecosystem effects of pollutants in the environment. The following section outlines protocols for biological monitoring in rivers in Great Britain. Similar monitoring protocols are in place for other European nations as well as for lakes and transitional and coastal water bodies.

4.10.2.1 *Macroinverterbrates*

The use of macroinvertebrates for routine monitoring of waters in Great Britain began in the 1970s. The Biological Monitoring Working Party (BMWP) devised a scoring system whereby individual organisms were given a score between 1 and 10 based on their perceived tolerance to pollution (see Table 4.5). Identification is required only to family level. Those species which are sensitive to

Table 4.5   List of benthic invertebrates and pressure sensitivity score.

Families	Score
Aphelocheiridae, Beraeidae, Brachycentridae, Capniidae, Chloroperlidae, Ephemerellidae, Ephemeridae, Goeridae, Heptageniidae, Lepidostomatidae, Leptoceridae, Leptophlebiidae, Leuctridae, Molannidae, Odontoceridae, Perlidae, Perlodidae, Phryganeidae, Potamanthidae, Sericostomatidae, Siphlonuridae, Taeniopterygidae	10
Aeshnidae, Astacidae, Calopterygidae, Cordulegastridae, Corduliidae, Gomphidae, Lestidae, Libellulidae, Philopotamidae, Psychomyiidae (incl. Ecnomidae)	8
Caenidae, Limnephilidae (incl. Apataniidae), Nemouridae, Polycentropodidae, Rhyacophilidae (incl. Glossosomatidae)	7
Ancylus group (Ancylidae, Acroloxidae, Ferrissia), Coenagrionidae, Corophiidae, Gammaridae (incl. Crangonyctidae & Niphargidae), Hydroptilidae, Neritidae, Platycnemididae, Unionidae, Viviparidae	6
Corixidae, Dendrocoelidae, Dryopidae, Dytiscidae (incl. Noteridae), Elmidae, Gerridae, Gyrinidae, Haliplidae, Hydrometridae, Hydrophilidae (incl. Hydraenidae, Helophoridae, Georissidae & Hydrochidae), Hydropsychidae, Hygrobiidae, Mesoveliidae, Naucoridae, Nepidae, Notonectidae, Planariidae (incl. Dugesiidae), Pleidae, Scirtidae, Simuliidae, Tipulidae	5
Baetidae, Piscicolidae, Sialidae	4
Asellidae, Erpobdellidae, Glossiphoniidae, Hirudinidae, Hydrobiidae (incl. Bithyniidae), Lymnaeidae, Physidae, Planorbidae (excl. Ancylidae), Sphaeriidae, Valvatidae	3
Chironomidae	2
Oligochaeta	1

pollution (such as stoneflies) are given a high score, tolerant species (such as tubificid worms) are given a low score. This tolerance was typically based around sewage/organic pollution. In early iterations, a total score for all taxa present was used as rough estimate of pollution at a site. However, an additional metric, Average Score Per Taxon (ASPT) was devised to overcome the signals of species richness and sampling effort could have on the initial score and was reported as the standard measure of quality along with the number of scoring families (Ntaxa).[99]

In the 1980s, a major step forward in biological assessment of rivers came with the River InVertebrate Prediction and Classification System (RIVPACS)[100] which adopted a reference condition approach. Such an approach has been fundamental to the bioassessment procedures adopted in the WFD and is founded on the desire to judge the impact of pollution or other disturbance on a biological community by comparing it to the fauna at a comparable site (in terms of morphological, geographical and chemical characteristics) that is not subject to anthropogenic stresses. This in itself provides its own challenge as true 'reference conditions' are difficult to attain given most water bodies will show some tangible human signal if it is looked for hard enough, for example from atmospheric deposition. However, in RIVPACS, through collation of a national invertebrate dataset using a standardised methodology, 268 suitable reference sites were identified across the UK. Using multivariate statistical analysis tools, these communities were clustered based on their similarity in composition which reflected the key physical and chemical characteristics at the sample site (*e.g.* high gradient acid upland, *versus* a slow moving stream overlying calcareous sedimentary bedrock). The physical variables that could best predict the community type included substrate type and alkalinity. RIVPACS then assesses observed community types and tests the probability of it belonging to one of the biological end groups identified in the cluster analysis based on presence of key taxa. This method has been demonstrated to be very robust in a range of settings and for a range of pollutant pressures beyond organic discharges, notably in metal contaminated streams.[101] A modified version of RIVPACS known as the River Invertebrate Classification Tools (RICT) are now applied for routine benthic invertebrate sampling as part of the WFD in the UK, with similar, albeit occasionally modified procedures in other member states (see Table 4.6). The comparison of the observed data with reference conditions produces an Ecological Quality Ratio (EQR). An EQR close to 1 indicates an invertebrate community close to its natural state with a value of zero showing a high level of pollution or disturbance. The EQR scores are then judged relative to boundary values that set the Biological Class of the water body with respect to invertebrate communities (see Table 4.6).[102]

4.10.2.2 *Macrophytes*

Macrophytes have been widely used as indicators of trophic status of both rivers and lakes as well as reliable indicators of acidification as we saw earlier. They have been widely used in the monitoring of metal pollution given the bioaccumulation of many metals in root and shoot tissue,[103] as well as in the actual phytoremediation of metal polluted waters.[104] Macrophytes are also good indicators of other key pressures on our surface waters beyond pollution, such as channel engineering and changes in flow regime (*e.g.* through impoundment or abstraction). The monitoring method applied in the UK for WFD, known as River LEAFPACS, requires summer macrophyte surveys of a 100 m reach of channel. Along with key physico-chemical variables, the presence of each macrophyte taxon and its percentage cover are estimated (see Table 4.6). Indices for the preference of each taxon with regard nutrient status and flow are applied to the observed data. The Nutrient Index (R) score ranges from low nutrient status values at 1 (*e.g.* bryophytes such as *Sphagnum* sp.) to 10 which reflects high nutrient status (*e.g.* Fennel-leaved pondweed, *Potamogeton pectinatus*, has an R of 9.6). The hydraulic index (H) again ranges between 1 and 10 with high scores reflecting low energy lowland rivers (*e.g.* Common Reed, *Phragmites australis,* has an H of 9.6) and

Table 4.6 Summary details for key biotic indices used in the WFD in England and Wales.

		Phytobenthos	*Macroinvertebrates*	*Macrophytes*	*Fish*
Sampling		Biofilm scraping from upper surfaces of cobbles or boulders. Where cobbles absent, collect from submerged portion of emergent macrophytes.	3 minute kick sample with 1 mm mesh size net. Additional manual search of 1 minute for invertebrates attached to submerged plants, stones or other solid surface.	100 m reach between 1st June and 30th September. Presence and percentage cover of macrophyte taxa recorded.	Electro-fishing or seine netting. Counts of fish present from single removal or catch-per-area sampling.
Measures		River Trophic Diatom Index – number of taxa and abundance with nutrient sensitivity score assigned for each taxon.	Number of taxa Average Score per Taxon (score relates to pressure sensitivity from 1 to 10).	Macrophyte nutrient index. Macrophyte hydraulic index. Number of macrophytes taxa. Number of functional groups. Percent of filamentous algae.	Prevalence of 23 species of fish indicating differing sensitivities to environmental disturbance.
EQR	High	0.93	0.85	0.8	0.81
	Good	0.78	0.71	0.6	0.58
	Moderate	0.52	0.57	0.4	0.4
	Poor	0.26	0.47	0.2	0.2
	Bad	<0.26	<0.47	<0.2	0

lower scores reflecting fast-flowing systems (*e.g.* Lax Notchwort, *Hygrobiella laxifolia* scores 2.1 and is common in rock faces in upland stream gorges; see Table 4.7). These scores are integrated with the total number of macrophytes, the percentage cover of filamentous algae (which is a strong eutrophication indicator) and the number of functional groups present in the observed species. The latter represents groups of organisms which exploit resources in a similar manner, which is a good measure of the microhabitat diversity in a stream reach. The scores for each of the individual characteristics (nutrient status, hydraulic index, number of macrophytes, functional groups and filamentous algae) are then compared with reference scores as with RIVPACS. These reference conditions are a function of various physical and chemical factors of the surveyed stream (*e.g.* altitude, latitude, channel slope, distance from source, alkalinity) and are obtained from statistical models based on extensive field datasets and model prediction. For each of the five characteristics a ratio is calculated comparing the observed macrophytes assemblage with the reference values. A combined overall ratio is then obtained, to give the EQR as with invertebrate monitoring to classify the biological status of the river with regard its macrophytes to be determined (see Table 4.7). A river therefore with a diverse flora or submerged and emergent plants would score highly, whereas a river with a low diversity of macrophytes and an abundance of filamentous algae would be classed as poor or bad.[105]

4.10.2.3 *Diatoms*

The development of diatom bioindicators in Europe was largely a result of the *Urban Wastewater Treatment Directive* in 1991 which produced metrics applied on restricted geographical scales alongside complementary monitoring data (notably chemical and macrophyte data).[106] Diatoms are sensitive indicators of nutrient enrichment and have been widely applied to other environmental pollutant pressures, notably acidification, but also acid mine drainage.[98] Table 4.6 summarises the

Table 4.7 Examples of nutrient and hydraulic indices for some common macrophytes found in river environments.

Taxon	Nutrient index (R)	Hydraulic index (H)
Azolla filiculoides (Water Fern)	9.71	8.98
Berula erecta (Water Parsnip)	8.24	8.17
Blue-green algal mats	5.10	5.20
Calliergon cuspidatum (Pointed Spear-moss)	3.49	3.72
Callitriche sp. (Water-starworts)	6.67	7.18
Caltha palustris (Marsh Marigold)	4.20	5.24
Carex riparia (Greater Pond Sedge)	9.06	8.89
Carex rostrata (Bottle Sedge)	2.64	4.71
Chara sp. (Stoneworts)	3.85	
Equisetum fluviatile (Water Horsetail)	3.92	6.01
Filamentous green algae	7.61	7.04
Glyceria maxima (Reed Sweetgrass)	9.64	8.96
Hygrobiella laxifolia (Lax Notchwort)	2.76	2.06
Juncus articulatus (Jointleaf Rush)	3.10	4.27
J. bulbosus (Bulbous Rush	1.89	4.35
Lemna minor (Common Duckweed)	8.80	8.59
Nardia sp. (Mat-grass)	1.40	3.40
Phragmites australis (Common Reed)	7.70	8.94
Potamogeton crispus (Curly-leaf Pondweed)	9.02	7.86
Sphagnum sp.	1.07	2.92
Stigeoclonium tenue	6.62	5.69
Typha latifolia (Greater Reedmace)	8.87	8.42
Zannichellia palustris (Horned Pondweed)	9.01	8.43

standardised monitoring protocols adopted in the UK under the WFD. Benthic diatom samples are collected by scraping biofilm off streambed rocks or the submerged shoots of tall emergent macrophytes such as *Phragmites australis, Typha latifolia* (Greater Reedmace) or *Sparganium erectum* (Branched Bur-reed).[107] The number of diatom species and abundance are then counted. A nutrient sensitivity score is assigned to each species ranging from 1 (low sensitivity) to 5 (high sensitivity). These nutrient sensitivity scores are used as multiplier for the abundance of each taxon. A river trophic diatom index is then calculated which adjusts the abundance and nutrient scores for the observed number of species in a sample. Reference values for computing an EQR as with the other biotic indices are based on sample season and stream alkalinity (see Table 4.6).[107]

4.10.2.4   Fish

Routine monitoring of fish populations for environmental assessments have been widely practiced over recent decades in North America and much of Europe. The WFD monitoring protocol for fish fauna in the UK builds on schemes devised in the 1990s for classifying the status of riverine fish populations.[108] The current methods follow these traditional approaches with improved habitat models to assess the presence and abundance of the 23 most prevalent fish species in England and Wales. These are classed as being of low (*e.g.* salmon), moderate (*e.g.* minnow) or high tolerance (*e.g.* eels) to environmental disturbance (see Table 4.8). Again, the reference site approach is used to compare observed communities with those predicted from reference data and statistical models. Electro-fishing or seine netting approaches are recommended to obtain catch-per area sample data. These data are then compared with reference values computed from geostatistical models that predict what fish community should be present based on geographic location and environmental variables. A comparison of the presence–absence predictions for the 23 species is then compared with the observed data to calculate an EQR (see Table 4.6).[109]

Table 4.8 List of fish indicator species classified according to tolerance to environmental disturbance.[109]

Tolerance	Species
Low	Salmon (*Salmo salar*), brown and sea trout (*S. trutta*), grayling (*Thymallus thymallus*), lamprey (*Lampetra planeri, L. fluviatilis, Petromyzon marinus*), bullhead (*Cottus gobio*)
Medium	Stone loach (*Barbatula barbatula*), barbel (*Barbus barbus*), Spined loach (*Cobitis taenia*), pike (*Esox lucius*), gudgeon (*Gobio gobio*), ruffe (*Gymnocephalus cernuus*), chub (*Leuciscus cephalus*), dace (*L. leuciscus*), minnow (*Phoxinus phoxinus*), rudd (*Scardinius erythrophthalmus*)
High	Bream (*Abramis brama*), bleak (*Alburnus alburnus*), Eel (*Anguilla anguilla*), common carp (*Cyprinus carpio*), 3-spined stickleback (*Gasterosteus aculeatus*), perch (*Perca fluviatilis*), roach (*Rutilus rutilus*), tench (*Tinca tinca*)

These biological surveys are then integrated with chemical and physico-chemical quality assessments and hydromorphological quality assessments to determine the overall ecological status of a water body. The former cover key water quality parameters, notably temperature, dissolved oxygen, pH and soluble reactive phosphorous concentration as well as other specific pollutants (*e.g.* As, Cr, Cu, Fe, Zn, phenol, toluene, ammonia, cyanide) routinely collected by environmental monitoring agencies. Hydromorphological quality assessments include hydraulic and geomorphological measures of the quality of channels such as channel substrate, connection with groundwater bodies, channel flow and riparian zone structure and connectivity. An overall status for the water body (designated as 'High', 'Good', 'Moderate', 'Poor' or 'Bad') is then assigned by taking the result of the different biological or physico-chemical quality element with the worst class, *i.e.* that which is most affected by human activity.[110] This monitoring procedure has been calibrated across different European nations and provides an ongoing means for assessing pressures, impacts and recovery of our water environment.

4.11 CONCLUSIONS

Freshwater systems are disproportionately diverse, occupying 0.8% of the Earth's surface area, yet containing 10% of all animals including over a quarter of known vertebrates. However, extinction rates for freshwater species have been estimated to be up to five times greater than terrestrial species given the multitude of pressures affecting our rivers, lakes and wetlands.[111] Humankind has left an indelible marker of its presence *via* impacts on freshwater systems through changing global cycles of major and trace elements, the release of synthetic compounds with long half-lives and the disruption of global water circulation *via* abstraction, river regulation and drainage of wetlands. These pressures will continue to present a range of research and management challenges through the 21st century, at a time when our changing climate is likely to impart additional stress on freshwater ecosystems.

Particular challenges concern the emerging contaminants given the fundamental improvements in our understanding of their environmental behaviour and effects required to underpin effective risk assessments and management. At the same time, many of the more 'traditional' river pollutants still pose considerable obstacles to progress. Ongoing investment in wastewater treatment infrastructure and deindustrialisation in Europe have been key factors in the over-riding improvements in chemical and ecological quality of rivers and lakes in recent years. However, such improvements have highlighted more challenging constraints to instream quality, for example from diffuse sources such as urban and agricultural runoff. Characterising these impacts and formulating management strategies will be crucial to improving the ecological status of our rivers and lakes. Furthermore, while there have been discernible improvements in water quality in some parts of the world, the global prognosis for pollution impacts on the hydrosphere still looks bleak.[112]

As such, the freshwater biologist has a crucial role to play as part of the interdisciplinary teams managing our water environment. This role stretches across scales that range from molecular (with novel tools for screening biomarkers as early warning systems for contamination) to the river basin scale; from effects on individuals to effects on entire ecosystems. The need for ever improving programmes of surveillance and monitoring to protect our water resources is stressed and biological monitoring plays a pivotal role in this. An expanding array of monitoring and analytical tools are available to assist in these endeavours, to help tease out sub-lethal effects of pollutants and to provide more effective early warning systems for the potential long-term impacts of pollutants on aquatic ecosystems.

REFERENCES

1. C. F. Mason, *Biology of Freshwater Pollution*, Pearson, Harlow, 4th edn, 2002.
2. USEPA, *ECOTOX databases,* 2013; www.epa.gov/med/Prods_Pubs/ecotox.htm (accessed 10/04/2013).
3. C. E. Purdom, P. A. Hardiman, V. J. Bye, N. C. Eno, C. R. Tyler and J. P. Sumpter, *Chem. Ecol.*, 1994, **8**, 275.
4. S. Jobling, R. Owen, Ethinyl oestradiol in the aquatic environment, in *Late Lessons from Early Warnings: Science, Precaution, Innovation*, European Environment Agency, 2013, p. 314.
5. C. A. Bishop, P. Ng, K. E. Pettit, S. W. Kennedy, J. J. Stegeman, R. J. Norstrom and R. J. Brooks, *Environ. Pollut.*, 1998, **101**, 143.
6. R. M. Sharpe and N. E. Skakkebaek, *Lancet*, 1993, **341**, 1392.
7. C. F. Mason and N. A.-E. Barak, *Chemosphere*, 1990, **21**, 695.
8. W. Giger, *Environ. Sci. Pollut. Res.*, 2009, **16**, S98.
9. Cantara Trustee Council, *Final Report on the Recovery of the Upper Sacramento River – subsequent to the 1991 Cantara Spill,* CTC, 2007.
10. R. Baos, R. Jovani, N. Pastor, J. L. Tella, B Jiminez, G. Gomex, M. J. Gonzalex and F. Hiraldo, *Environ. Toxicol. Chem.*, 2006, **25**, 2794.
11. I. Lázár and E. Kiss, in *Just Ecological Integrity: the Ethics of Maintaining Planetary Life*, ed. P. Miller and L. Westra, Rowman and Littlefield, MD, USA, 2002, ch. 14. p. 167.
12. L. Ruhl, A Vengosh, G. S. Dwyer, H. Hsu-Kim, A. Deonarine, M. Bergin and J. Kravchenko, *Environ. Sci. Technol.*, 2009, **43**, 6326.
13. W. M. Mayes, A. P. Jarvis, I. T. Burke, M. Walton, V. Feigl, O. Klebercz and K. Gruiz., *Environ. Sci. Technol.*, 2011, **45**, 5147.
14. O. Klebercz, W. M. Mayes, Á.D. Anton, V. Feigl, A. P. Jarvis and K. Gruiz, *J. Environ. Monit.*, 2012, **14**, 2063.
15. K. Schöll and G. Szövényi, *Bull. Environ. Contam. Toxicol.*, 2011, **87**, 124.
16. A. Prüss-Üstün, R. Bos, F. Gore and J. Bartram, *Safer Water, Better Health: Costs, Benefits and Sustainability of Interventions to Protect and Promote Health,* World Health Organization, Geneva, 2008.
17. WFD-UKTAG, *UK Environmental Standards and Conditions (Phase 1)*, UK Technical Advisory Group, 2008.
18. Environment Agency, *Combined (Point and Diffuse) Pressures on Rivers;* http://www.environment-agency.gov.uk/research/planning/100202.aspx (accessed 03/04/2013).
19. I. Durance and S. J. Ormerod, *Freshwater Biol.*, 2009, **55**, 388.
20. J. N. Galloway, J. D. Aber, J. W. Erisman, S. P. Seitzinger, R. W. Howarth, E. B. Cowling and B. J. Cosby, *Bioscience*, 2003, **53**, 341.

21. P. Falkowski, R. J. Scholes, E. Boyle, J. Canadell, D. Canfield and J. Elser, *Science*, 2000, **290**, 291.
22. M. M. Abdul-Hussein and C. F. Mason, *Hydrobiologia*, 1988, **169**, 265.
23. B. Moss, *Biol. Rev.*, 1983, **58**, 521.
24. M. Scheffer, S. H. Hosper, M.-L. Meijer, B. Moss and E. Jeppeson, *Trends Ecol. Evolut.*, 1993, **8**, 275.
25. P. Schriver, J. B. Egestrand, E. Jeppeson and M. Sùndergaard, *Freshwater Biol.*, 1995, **33**, 255.
26. M. Klinge, M. P. Grimm and S. H. Hosper, *Water Sci. Technol.*, 1995, **31**, 207.
27. L. A. Lawton and G. A. Codd, *J. IWEM*, 1991, **5**, 460.
28. L. Fewtrell., *Environ. Health Perspect.*, 2004, **112**, 1371.
29. L. Wang, M. E. Stuart, J. P. Bloomfield, A. S. Butcher, D. C. Goody, A. A. McKenzie, M. A. Lewis and A. T. Williams, *Hydrol. Process.*, 2012, **26**, 226.
30. M. G. Hutchins, A. Deflandre-Vlandas, P. E. Posen, H. N. Davies and C. Neal., *Environ. Model. Assess.*, 2010, **15**, 93.
31. P. Cullen and C. Forsberg, *Hydrobiologia*, 1988, **170**, 321.
32. K. V. Heal, P. L. Younger, K. A. Smith, S. Glendinning, P. F. Quinn and K. E. Dobbie, *Land Contamin. Reclamat*, 2003, **11**, 145.
33. B. Moss, J. Madgwick and G. Phillips, *A Guide to the Restoration of Nutrient-enriched Shallow Lakes*, Broads Authority, Norwich, 1996.
34. E. Bergman, L.-A. Hansson, A. Persson, J. Strand, P. Romare, M. Enell, W Graneli, J. M. Svensson, S. F. Hanrin, G. Cronberg, G. Andersson and E. Bergstrand, *Hydrobiologia*, 1999, **404**, 145.
35. M. Søndergaard, J. P. Jensen and E. Jeppesen, *Hydrobiologia*, 2003, **506**, 135.
36. R. W. Battarbee, *Hydrobiologia*, 1994, **274**, 1.
37. B. O. Rosseland, O. K. Skogheim and I. H. Sevaldrud, *Water Air Soil Pollut.*, 1986, **30**, 65.
38. M. Appelberg, E. Degerman and L. Norrgren, *Finnish Fish. Res.*, 1992, **13**, 77.
39. B. W. Stallsmith, J. P. Ebersole and W. G. Haggar, *Freshwater Biol.*, 1996, **36**, 731.
40. K.-M. Vuori, *Freshwater Biol.*, 1996, **35**, 179.
41. S. J. Tyler and S. J. Ormerod, *Environ. Pollut.*, 1992, **78**, 49.
42. C. F. Mason and S. M. Macdonald, *Water Air Soil Pollut.*, 1989, **43**, 365.
43. S. J. Ormerod and I. Durance, *J. Appl. Ecol.*, 2009, **46**, 164.
44. B. L. Skjelkvåle, J. L. Stoddard, D. S. Jeffries, K. Tørseth, T. Høgaåsen, J. Bowman, J. Mannio, D. T. Monteith, R. Mosello, M. Rogora, D. Rzychon, J. Vesely, J. Wieting, A. Wilander and A. Worsztynowicz, *Environ. Pollut.*, 2005, **137**, 165.
45. D. T. Monteith, A. G. Hildrew, R. J. Flower, P. J. Raven, W. R. B. Beaumont, P. Collen, A. M. Kreiser, E. M. Shilland and J. H. Winterbottom, *Environ. Pollut.*, 2005, **137**, 83.
46. Y. Zhao, L Duan, J. Xing, T. Larssen, C. P. Nielsen and J. Hao, *Environ. Sci. Technol.*, 2009, **43**, 8021.
47. R. L. P. Kleinmann and R. Hedin, in *Tailings and Effluent Management*, M. E. Chalkly, B. R. Conrad, V. I. Lakshmanan and K. G. Wheeland, Pergamon, New York, 1990, p. 140.
48. W. M. Mayes, H. A. B. Potter and A. P. Jarvis, *Sci. Total Environ.*, 2010, **408**, 3576.
49. A. P. Jarvis and P. L. Younger, *Chem. Ecol.*, 1997, **13**, 249.
50. C. J. van Leeuwen and T. G Vermiere, *Risk Assessment of Chemicals: An Introduction*, Springer, 2007.
51. P. C. Schulze, *Measures of Environmental Performance and Ecosystem Condition*, National Academies Press, Virginia. USA, 1999.
52. J. M. Hellawell, *Biological Indicators of Freshwater Pollution and Environmental Management*, Elsevier Applied Science, London, 1986.
53. J. F. Sandahl, D. H. Baldwin, J. J. Jenkins and N. L. Scholz, *Environ. Sci. Technol.*, 2007, **41**, 2988.

54. D. Peakall, *Animal Biomarkers as Pollution Indicators*, Chapman and Hall, London, 1992.
55. G. A. Fox, *J. Great Lakes Res.*, 1993, **19**, 722.
56. J. L. Hugla, J. C. Philippart, P. Kremers and J. P. Thome, *Netherlands J. Aquat. Ecol.*, 1995, **29**, 135.
57. J. L. Hugla, J. P. Thome and J. C. Philippart, *Cah. Etholog.*, 1993, **13**, 155.
58. USEPA, *The U.S. Environmental Protection Agency's Strategic Plan for Evaluating the Toxicity of Chemicals,*USEPA, Washington DC, USA, 2009.
59. P. R. Nogueira, J. Lourenco, S. Mendo and J. M. Rotchell, *Mar. Pollut. Bull.*, 2006, **52**, 1611.
60. T. Yorifuji, T. Tsuda and M. Harada, Minamata disease: a challenge for democracy and justice, in *Late Lessons from Early Warnings: Science, Precaution, Innovation*, European Environment Agency, 2013, p. 124.
61. J. Liu, K. Li, J. Xu, J. Liang, X. Lu, J. Yang and Q. Zhu, *Field Crops Res*, 2003, **83**, 271.
62. T. J. Kubiak, H. J. Harris, L. M. Smith, D. L. Starling, T. R. Schwartz, J. A. Trick, L. Sileo, D. E. Docherty and T. C. Erdman, *Arch. Environ. Contam. Toxicol.*, 1989, **18**, 706.
63. J. P. Giesy, J. P. Ludwig and D. E. Tillitt, *Environ. Sci. Technol.*, 1994, **28**, 128A.
64. USEPA, *Great Lakes Monitoring*, 2013; http://www.epa.gov/glindicators/fishtoxics/topfishb.html (accessed 03/03/2013).
65. T. Colborn, D. Dumanoski and J. P. Myers, *Our Stolen Future*, Dutton, New York, 1996.
66. S. M. Macdonald and C. F. Mason, *Status and Conservation Needs of the Otter (Lutra lutra) in the Western Palearctic*, Council of Europe, Nature and Environment, 67, 1994.
67. A. J. Murk, P. E. G. Leonards, B. van Hattum, R. Luit, M. E. J. van der Weiden and M. Smit, *Environ. Toxicol. Pharmacol.*, 1998, **6**, 91.
68. P. D. Jepson, P. M. Bennett, C. R. Allchin, R. J. Law, T. Kuiken, J. R. Baker, E. Rogan and J. K. Kirkwood, *Sci. Total Environ.*, 1999, **243–244**, 339.
69. C. F. Mason, *Chemosphere*, 1998, **36**, 1969.
70. A. Roos, E. Greyerz, M. Olsson and F. Sandegren, *Environ. Pollut.*, 2001, **111**, 457.
71. G. Tomy, W. Budakowski, T. Halldorson, M. Whittle, M. Keir and M. Alaee, *Environ. Sci. Technol.*, 2004, **38**, 2298.
72. M. Kohler, M. Zennegg, C. Bogdal, A. C. Gerecke, P. Schmid, N. V. Heeb, M. Sturm, H. Vonmont, H. P. Kohler and W. Giger, *Environ. Sci. Technol.*, 2008, **42**, 6378.
73. B. Fängström, I. Athanassiadis, T. Odsjö, K. Norén and Å. Bergman, *Mol. Nutr. Food Res.*, 2008, **52**, 187.
74. T. E. Langford, *Ecological Effects of Thermal Discharges*, Chapman and Hall, London, 1990.
75. D. Caissie, *Freshwater Biol.*, 2006, **51**, 1389.
76. D. Archer, P. Rippon, R. Inverarity and R. Merrix, *BHS 10th National Hydrology Symposium, Exeter*, 2008.
77. W. H. Clements, M. L. Brooks, D. R. Kashian and R. E. Zuellig, *Global Change Biol.*, 2008, **14**, 2201.
78. S. Ueda, H. Hasegawa, H. Kakiuchi, N. Akata, Y. Ohtsuka and S. Hisamatsu, *J. Environ. Radioact.*, 2013, **118**, 96.
79. M. Rask, R. Saxén, J. Ruuhijärvi, L. Arvola, M. Järvinen, U. Koskelainen, I. Outola and P. J. Vuorinen, *J. Environ. Radioact.*, 2012, **103**, 41.
80. C. F. Mason and S. M. Macdonald, *Water Air Soil Pollut.*, 1988, **37**, 131.
81. A. P. Møller and T. A. Mousseau., *Trends. Ecol. Evolut.*, 2006, **21**, 200.
82. O. D. Al-Amri, A. B. Cundy, Y. Di, A. N. Jha and J. M. Rotchell, *Environ. Pollut.*, 2012, **168**, 107.
83. Environment Agency, *What is Inland Oil Pollution?*, 2013; http://www.environment-agency.gov.uk/research/library/position/41233.aspx (accessed 28/03/2013).
84. P. A. Mahaney, *Environ. Toxicol. Chem.*, 1994, **13**, 259.
85. R. A. Khan, *Bull. Environ. Contamin. Toxicol.*, 1999, **62**, 638.

86. M. J. Hemmer, M. G. Barron and R. M. Greene, *Environ. Toxicol. Chem.*, 2011, **30**, 2244.
87. M. N. Moore, *Environ. Int.*, 2006, **32**, 967.
88. B. Campos, C. Rivetti, P. Rosenkranz, J. M. Navas and C. Barata, *Aquatic. Toxicol.*, 2013, **130-131**, 174.
89. F. Gottschalk, T. Sonderer, R. W. Scholz and B. Nowack, *Environ. Sci. Technol.*, 2009, **43**, 9216.
90. A. Boxall. *Ecology of Industrial Pollution*, ed. L. C. Batty and K. B. Hallberg, Cambridge University Press, Cambridge, UK, 2010, Ch. 5, pp. 101–125.
91. E. Gubbins, L. C. Batty and J. R. Lead, *Environ. Pollut.*, 2011, **159**, 1551.
92. E. Oberdörster, *Environ, Health. Perspect.*, 2004, **112**, 1058.
93. F. Pomati, A. G. Netting, D. Calamari and B. A. Neilan, *Aquat. Toxicol.*, 2004, **67**, 387.
94. E. Zuccato, S. Castiglioni, R. Bagnati, C. Chiabrando, P. Grassi and R. Fanelli, *Water. Res.*, 2008, **42**, 961.
95. M. Leynen, T. Van den Berckt, J. M. Aerts, B. Castelein, D. Berckmans and F. Ollevier, *Environ. Pollut.*, 1999, **105**, 151.
96. USEPA. *Whole Effluent Toxicity*, 2013; www.water.epa.gov/scitech/methods/cwa/wet/#methods (accessed 01/04/2013).
97. H. Tahedl and D. Häder, *Water Res.*, 1999, **33**, 426.
98. J. T. Zalack, N. J. Smucker and M. L. Vis, *Ecol. Indicators*, 2010, **10**, 287.
99. J. I. Jones, J. Davy-Bowker, J. F. Murphy, J. L. Pretty, *Ecology of Industrial Pollution*, ed. L. C. Batty and K. B. Hallberg, Cambridge University Press, Cambridge, UK, 2010, ch. 6, pp. 126–146.
100. J. F. Wright, D. Moss, P. D. Armitage and M. T. Furse, *Freshwater Biol.*, 1984, **14**, 221.
101. P. D. Armitage, M. J. Bowes and H. M. Vincent, *River Res. Applic*, 2007, **23**, 997.
102. WFD-UKTAG, *UKTAG River Assessment Methods: Benthic Invertebrate Fauna, River Invertebrate Classification Tool (RICT)*, SNIFFER, Edinburgh, UK, 2008.
103. A. J. M. Baker, W. H. O. Ernst, A. Ven der Ent, F. Malaisse and R. Ginocchio, *Ecology of Industrial Pollution*, L. C. Batty and K. B. Hallberg, Cambridge University Press, Cambridge, UK, 2010, Ch. 2, pp. 7–40.
104. W. M. Mayes, L. C. Batty, P. L. Younger, A. P. Jarvis, M. Kõiv, C. Vohla and Ü. Mander, *Sci. Tot. Environ.*, 2009, **407**, 3944.
105. N. J. Willby, J. Pitt, G. L. Phillips. The *ecological classification of UK rivers using aquatic macrophytes*, Environment Agency Science Report SC010080/SR. 2009.
106. M. G. Kelly, S. Juggins, R. Guthrie, S. Pritchard, B. J. Jamieson, B. Rippey, H. Hirst and M. L. Yallop, *Freshwater Biol.*, 2008, **53**, 403.
107. M. G. Kelly, S. Juggins, H. Bennion, A. Burgess, M. Yallop, H. Hirst, L. King, J. Jamieson, R. Guthrie and B. Rippey, *Use of Diatoms for Evaluating Ecological Status in UK Freshwaters*, Environment Agency Science Report SCO301030, 2007.
108. R. Wyatt, R. Sedgwick and H. Simcox, *River Fish Habitat Inventory Phase III: Multi-species Models*, Science Report: SC040028/SR, Environment Agency, 2007.
109. WFD-UKTAG, *UKTAG River Assessment Methods: Fish Fauna, Fisheries Classification Scheme 2 (FCS2)*, SNIFFER, Edinburgh, UK, 2008.
110. WFD-UKTAG. *Recommendations on Surface Water Classification Schemes for the Purposes of the Water Framework Directive*, 2007, UKTAG, pp. 61.
111. A. Ricciardi and J. B. Rasmussen, *Conserv. Biol.*, 1999, **13**, 1220.
112. C. J. Vörösmarty, P. B. McIntyre, M. O. Gessner, D. Dudgeon, A Prusevich, P. Green, S. Glidden, S. E. Bunn, C. A. Sullivan, C. R. Liermann and P. M. Davies, *Nature*, 2010, **467**, 555.

CHAPTER 5

Sewage and Sewage Sludge Treatment

ELISE CARTMELL[†]

Cranfield Water Science Institute, Cranfield University, Cranfield, Bedfordshire,
MK43 0AL, UK
Email: e.cartmell@cranfield.ac.uk

5.1   INTRODUCTION

It is estimated that the total volume of water supplied daily in the United Kingdom (UK) to homes, businesses and industry is 17 395 ML d^{-1} (Mega litres per day)[1] or approximately 275 L per capita per day.[2] Domestic use accounts for approximately 160 L d^{-1}. Nearly all of the water used is discharged to sewers yielding a total sewage flow of over 11 000 ML d^{-1} every day.[3]

The sewage from over 27 million properties in UK, which are connected to the public sewerage system, is treated by conventional sewage treatment processes. Most treated sewage is discharged to inland waters but some, from approximately six million people, is also discharged to sea. Approximately one to two million people are not connected to the sewerage system and are served by septic tanks, cesspits or package treatment plants.

To achieve this degree of sewage treatment requires over 9 000 sewage treatment works (STWs) serving populations which are distributed throughout the twelve water and sewage utilities in the UK (see Figure 5.1). The sewerage systems which carry the sewage to the site of treatment, or point of discharge, are of two types. Foul sewers carry only domestic and industrial effluent. In areas serviced in this way there are entirely separate systems for the collection of stormwater which is discharged directly to natural water courses. However, in older towns and cities considerable use has been made of combined foul and stormwater systems. The use of combined sewerage systems leads to very significant changes in the flow of sewage during storms. However, even in foul sewers significant changes in the flow occur due to variations in the pattern of domestic and industrial water usage which is essentially diurnal, and at its greatest during the day. Infiltration will also influence the flow in the sewage system. Although a properly laid sewer is watertight when constructed, ground movement and aging may allow water to enter the sewer if it is below the water table. The combined total of average daily flows to a sewage treatment works is called the dry

---

[†]This chapter is based upon an earlier contribution from John Lester and David Edge which is gratefully acknowledged.

Pollution: Causes, Effects and Control, 5th Edition
Edited by R M Harrison
© Cartmell and The Royal Society of Chemistry 2014
Published by the Royal Society of Chemistry, www.rsc.org

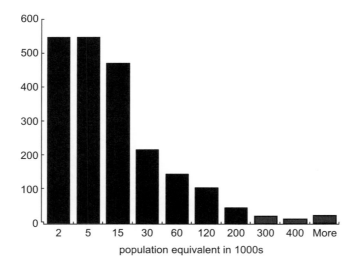

Figure 5.1   Profile of sewage treatment works in relation to population size served in the UK.

weather flow (DWF). The DWF is an important value in the design and operation of the sewage treatment works and other flows are expressed in terms of it. DWF is defined as the daily rate of flow of sewage (including both domestic and trade waste/industrial discharge), together with infiltration, if any, in a sewer in dry weather. This may be measured after a period of seven consecutive days during which the rainfall has not exceeded 0.25 mm.

The DWF may be calculated from the following equation:

$$DWF = PQ + I + E$$

Where:

P = population served
Q = average domestic water consumption (L d^{-1})
I = rate of infiltration (L d^{-1})
E = volume (L) of industrial effluent discharged to sewers in 24 hours.

5.1.1   Objectives of Sewage Treatment

Water pollution in the UK was already a serious problem by 1850. It is probable that the early endeavours to control water pollution were considerably stimulated by the state of the lower reaches of the River Thames which, at the point where it passed the Houses of Parliament, was grossly polluted. An early solution to these problems was sought through the construction of interceptor sewers. These collected all the sewage draining to the River Thames and carried it several miles down the river before discharging it to the estuary on the ebb tide. From this it moved towards the sea and in so doing received greater dilution. Despite these measures, and the passing in 1876 of the first Act of Parliament to control water pollution, the situation continued to deteriorate. The requirement for, and the objectives of, sewage treatment were first outlined by the Royal Commission on Sewage Disposal (1898–1915). The objectives of sewage treatment have developed significantly since this report; however, the standards described then are still applicable in many areas and this report provided the framework around which the UK wastewater industry has developed.

Originally, the objective of sewage treatment was to avoid pestilence and nuisance (disease and odour) and to protect the sources of potable supply. During sewage treatment disease-causing organisms may be destroyed or concentrated in the sludges produced; similarly, offensive materials may be concentrated in the sludges or biodegraded. As a consequence, the quantities of these agents present in the treated sewage effluent are much less than in the untreated sewage and their dilution in the receiving water far greater. The benefits of sewage treatment are not limited to greater dilution, however, since each receiving water has a certain capacity for 'self-purification'. Providing sewage treatment reduces the burden of polluting material to a value less than this capacity, then the ecosystem of the receiving water will complete the treatment of the residual materials present in the sewage effluent. Thus sewage treatment in conjunction with the selection of appropriate points for sewage effluent discharge has resulted in the elimination of waterborne disease in the UK and many other advanced countries. However, as the population has expanded and become urbanized with a concomitant development of water-consuming industries additional requirements have been placed upon sewage treatment. It is now the objective of sewage treatment in many parts of the UK to produce a sewage effluent which, after varying degrees of dilution and self-purification, is suitable for agricultural irrigation or abstraction for treatment to produce a potable supply. This indirect reuse for potable water supply affects over 30% of all water supplies in the UK. In addition, over two-thirds of irrigation water abstracted for agricultural raw food crop use is abstracted from rivers and streams with at least half of those sources within 4500 m of an upstream sewage treatment works outfall.[4]

5.1.2 The Importance of Wastewater Reuse

That the UK practises indirect reuse to a greater extent than most other countries may appear surprising given the annual rainfall. Indeed, that reuse should be important in global terms given the abundance of water on the earth's surface may also be considered improbable in all but the most arid regions. However, two important factors readily explain this situation: firstly a vast amount of the available water is too saline to be used as a potable supply (the salinity is too costly to remove in most cases), and secondly the non-uniform distribution of the population and the available water supply. The available water supply is determined by the rainfall, the ability of the environment to store water (essentially the size of lakes and rivers, which are small in the UK) and their location, *e.g.* Wales has an abundance of suitable water supplies, but limited population, whilst south east England has a large population with limited water resources.

It has been estimated that of the water falling on the United Kingdom, 50% is not available for use as a result of run-off to the sea. Of the remainder approximately 17% was utilized in the 1960s and this doubled by the 1990s, when consumption became stable. Thus, the potential reserves are very limited and will potentially become more so in certain areas as a result of climate change impacts. However, because demand and supply are not geographically proximate reuse is already essential. As a consequence the traditional concept of water supply employing single-purpose reservoirs impounding unused river water has been abandoned in favour of multi-purpose schemes designed to permit repeated use of the water before it reaches the sea. In these schemes sewage treatment plays a vital role, in addition to being an integral part of the hydrological cycle, and these water reuse schemes will continue to be a defining challenge for politicians, water managers and scientists over the coming years.

Water reclamation and reuse in urban environments is increasing in all parts of the world with both non-potable, indirect potable, and direct potable schemes all finding applications in increasingly water stressed regions. Source waters included reclaimed wastewater from domestic and/or commercial and industrial sources. Treatment of the reclaimed wastewater included the use of membranes and chemicals and final use of the treated wastewater included domestic toilet flushing through to fire-fighting supply and urban green space irrigation.

An example of a UK water reuse scheme is given in the following box.[5]

CASE STUDY: LONDON OLYMPIC PARK[5]

The London 2012 Olympic Park site incorporated a new water recycling plant, the Old Ford Water Recycling Plant, which provided sub-potable water for irrigating lawns and gardens as well as for toilet flushing and for the cooling towers of a combined cooling, heat, and power (CCHP) energy centre. The facility flanked the Olympic Park and took wastewater from the Northern Outfall Sewer to produce 570 000 litres a day of recycled water using membrane bioreactor (MBR) technology. The treatment chain consisted of 1 mm screens and then the MBR. Next, 0.04 μm ultrafiltration membranes were applied and a conventional granular activated carbon process further treats the water. The effluent is disinfected with sodium hypochlorite before being sent through the Olympic Park's nonpotable-water-distribution network. The whole operation was remotely operated from Thames Water's headquarters, 40 miles away in Reading. The scheme saves up to 83 Olympic-size swimming pools worth of drinking water every year.

As the largest water recycling facility built in the UK to date, the plant will serve as a benchmark for future water recycling plants in the increasingly water-scarce south of England.

5.1.3 Criteria for Sewage Treatment

Sewage is a complex mixture of suspended and dissolved materials; both categories constitute organic pollution. The strength of sewage and the quality of sewage effluent are traditionally described in terms of their suspended solids (SS) and biochemical oxygen demand (BOD); these two measures were either proposed or devised by the Royal Commission (1898–1915). The SS are determined by weighing after the filtration of a known volume of sample through a standard glass fibre filter paper, the results being expressed in mg L^{-1}. Dissolved pollutants are determined by the BOD they exert when incubated for five days at 20 °C (or BOD_5). Samples require appropriate dilution with oxygen-saturated water and suitable replication. The oxygen consumed is determined and the results again expressed in mg L^{-1}. The two standards for sewage effluent quality proposed by the Royal Commission were for no more than 30 mg L^{-1} of SS and 20 mg L^{-1} for BOD, the so called 30 : 20 standard. The Royal Commission envisaged that the effluent of this standard would be diluted 8 : 1 with clean river water having a BOD of 2 mg L^{-1} or less. This standard was considered to be the normal minimum requirement and was not enforced by statute because the character and use of rivers varied so greatly. It was intended that standards would be introduced locally as required. For example, a river to be used for abstraction of potable supplies would require a higher standard such as the 10 : 10 standard imposed by the Thames Conservancy. Whilst other countries which are members of the European Union have adopted 'uniform emission standards', that is, the same quality of effluent regardless of the state or use of the river, the UK has continued with its pragmatic approach whereby effluent standards are set depending on the 'water quality objectives' of the river, which in turn is determined by its function or use. In the 1970s the water industry's reliance solely on the 30 : 20 standard was abandoned and considerable importance was also placed upon the concentration of ammonia in the effluent with ammonia consents set around 5 mg N L^{-1}.[6] Overall, sewage treatment works now need to comply with the *Urban Wastewater Treatment Directive (UWWTD)*[7] (see Table 5.1) which sets secondary treatment as the normal standard, but requires tertiary treatment where qualifying discharges affect sensitive areas. An important aspect of the UWWTD is the protection of the water environment from nutrients, (specifically compounds of nitrogen and phosphorus) present in sewage where these substances have adverse impacts on the ecology of the water environment or abstraction source waters. The UWWTD is supported under the on-going *Water Framework Directive* (WFD).[8] The WFD

Table 5.1 General overview of the required final sewage effluent concentrations in the Urban Wastewater Treatment Directive.

Parameter	Guideline Consent Values	% Reduction
BOD$_5$	25 mg L^{-1}	70–90
COD	125 mg L^{-1}	75
TSS	35 mg L^{-1} ($>$10 k PE)	90
	60 mg L^{-1} (2 k–10 k PE)	70
P	2 mg L^{-1} (10 k–100 k PE)	80
	1 mg L^{-1} ($>$100 k PE)	
N	15 mg NL^{-1} (10 k–100 k PE)	70–80
	10 mg NL^{-1} ($>$100 k PE)	

Key: BOD$_5$ = Biochemical Oxygen Demand; COD = Chemical Oxygen Demand; TSS = Total suspended solids; P = total phosphorus; N = total nitrogen; PE = Population Equivalent.

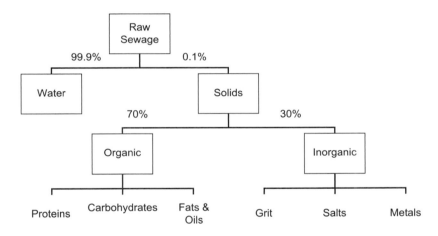

Figure 5.2 Main constituents of a typical crude sewage entering a sewage treatment works.

provides overarching European Union (EU) legislation in the field of water and the environment which has now extended the scope of pollution control measures further to protect surface waters.[9,10] In the future sewage treatment works may also need to comply with environmental quality standards (EQS) for hazardous chemicals contained in sewage effluents to further protect receiving waters and the flora and fauna they support. This is outlined in Chapter 1.

5.1.4 Composition of Sewage

Domestic sewage contains approximately 1000 mg L^{-1} of impurities of which about two-thirds are organic. Thus sewage is 99.9% water and 0.1% total solids upon evaporation (see Figure 5.2). When present in sewage, approximately 50% of this material is dissolved and 50% suspended. The main components are: nitrogenous compounds (proteins and urea); carbohydrates (sugars, starches and cellulose); fats (soap, cooking oil and greases). Inorganic components include chloride, metallic salts and road grit where combined sewerage is used. Thus sewage is a dilute, heterogeneous medium which tends to be rich in nitrogen.

5.2 SEWAGE TREATMENT PROCESSES

Conventional sewage treatment is typically a multi-stage process. After initial screening (preliminary treatment) much of the solid matter is settled out in primary treatment, then using bacteria organic material is broken down during secondary (biological) treatment. In some treatment works further tertiary and advanced treatment is required for additional sewage effluent polishing. These stages are presented schematically in Figure 5.3. In addition some form of sludge treatment facility is frequently employed, typically anaerobic digestion (see Section 5.3.4.1).

5.2.1 Preliminary Treatment

These treatment processes are intended to remove the larger floating and suspended materials. They do not make a significant contribution to reducing the polluting load, but render the sewage more amenable to treatment by removing large objects which could form blockages or damage equipment.

Floating or very large suspended objects are frequently removed by coarse bar screens. These consist of parallel rods with spaces between them which vary from 40 to 80 mm, through which the influent crude sewage must pass. Additional finer screens can also be utilised which are normally perforated plates or wires, varying in aperture size from medium (15–50 mm), fine (3–15 mm) to milli (0.25–3 mm). Material which accumulates on the screen may be removed manually at small works, but on larger works some form of automatic removal is used. The material removed from the screens contains a significant amount of putrescible organic matter which is objectionable in nature and may pose a disposal problem. Typically the screen material can be dewatered or washed on site and is transported to landfill for disposal. However, increasingly now reuse options are employed such as combustion with energy recovery.

If screens have been used to remove the largest suspended and virtually all the floating objects, then it only remains to remove the small stones and grit, which may otherwise damage pumps and valves, to complete the preliminary treatment. This is most frequently achieved by the use of constant velocity grit channels. The channels utilize differential settlement to remove only the heavier grit particles whilst leaving the lighter organic matter in suspension. A velocity of 0.3 m s^{-1} is sufficient to allow the grit to settle whilst maintaining the organic solids in suspension. If the grit channels are to function efficiently the velocity must remain constant regardless of the variation of the flow to the works (typically between 0.4 and 9 DWF). This is achieved by using channels with a parabolic cross section controlled by venturi flumes. The grit is removed from the bottom of the channel by a bucket scraper or suction, and organic matter adhering to the grit is removed by washing, with the wash water being returned to the sewage. Small sedimentation tanks from which

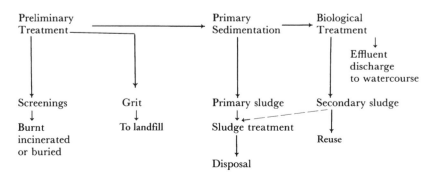

Figure 5.3 Flow diagram of a conventional sewage treatment works.

the sewage overflows at such a rate that only grit will settle out may also be used. These are compact and by the introduction of air on one side a rotary motion can be induced in the sewage which washes the grit *in situ*. However, these tanks do not cope with the variation in hydraulic load in such an elegant and effective manner as the grit channel.

To avoid the problems associated with the transportation of screenings, comminutors are frequently employed in place of screens. Unlike screens, which precede grit removal, the comminutors are placed downstream of the grit removal process. The comminutors shred the large solids in the flow without removing them. As a result they are reduced to a suitable size for removal during sedimentation. Comminutors consist of a slotted drum through which the sewage must pass. The drum slowly rotates carrying material which is too large to pass through the drum towards a cutting bar upon which it is shredded before it passes through the drum.

The total flow reaching the sewage treatment works is subjected to both these preliminary treatment processes. However, the works is only able to give full treatment up to a maximum design flow which is frequently three times the DWF. When the flow to the works exceeds this value the excess flows over a weir to the storm tanks which are normally empty. If the storm is short, no discharge occurs and the contents of the tanks are pumped back into the works when the flow falls below three DWF. If the storm is prolonged then these tanks may need to discharge to a nearby watercourse, inevitably causing some pollution. However, this excess flow has been subjected to sedimentation which removes some of the polluting material. Moreover, as a consequence of the storm, flow in the watercourse will be high, giving greater dilution.

5.2.2 Primary Sedimentation

The crude sewage (containing approximately 400 mg L^{-1} SS and 300 mg L^{-1} BOD) at a flow rate of approximately 3 DWF or less and with increased homogeneity as a result of the preliminary treatment process enters the first stage of treatment (primary sedimentation) to reduce the pollutant load. The aim of primary treatment is to reduce the BOD_5 of the incoming sewage by at least 20% and the total suspended solids by at least 50%. Circular (radial flow, diameter 3–40 m) or rectangular (horizontal flow, length 15–90 m, width 3–24 m) tanks equipped with mechanical sludge scraping devices are normally used (see Figure 5.4). However, on small works, hopper bottom tanks (vertical flow) are preferred; although more expensive to construct these costs are more than offset by savings made as a result of eliminating the requirement for scrapers (see Figure 5.4).

Removal of particles during sedimentation is controlled by the settling characteristics of the particles (their density typically 2500–2650 kg m^{-3}, size, and ability to flocculate), the retention time in the tank (hours), the surface loading (m^3 m^{-2} d^{-1}) and to a very limited degree the weir overflow rate (m^3 m^{-1} d^{-1}). Retention times are generally between 2 and 6 hours. However, the most important design criterion is the surface loading; typical values would be in the range 30 to 45 m^3 m^{-2} d^{-1}. The surface loading rate is obtained by dividing the volume of sewage entering the tank each day (m^3 d^{-1}) by the surface area of the tank (m^2). The retention time may be fixed independently of the surface loading by selection of the tank depth, typically 2 to 4 m, which increases the volume without influencing the surface area. Because they strongly influence the value for surface loading selected, the nature of the particles in the sewage is one of the most important factors in determining the design and efficiency of the sedimentation tank. Of the three factors mentioned before, flocculation is perhaps the most significant.

Four different types of settling can occur:

(i) *Class 1 Settling*: settlement of discrete particles in accordance with theory (Stokes' Law).
(ii) *Class 2 Settling*: settlement of flocculant particles exhibiting increased velocity during the process.

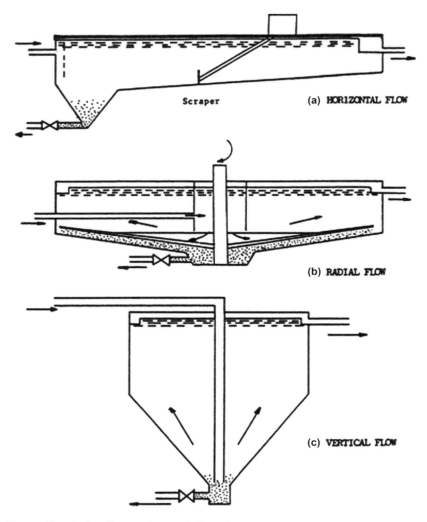

Figure 5.4   Types of typical sedimentation tank for primary sewage treatment.

(iii) *Zone Settling (Hindered Settlement)*: at certain concentrations of flocculant particles, the particles are close enough together for the interparticulate forces to hold the particles fixed relative to one another so that the suspension settles as a unit.
(iv) *Compressive Settling*: at high solids concentrations the particles are in contact and the weight of the particles is in part supported by the lower layer of solids.

During primary sedimentation, settlement is of the Class 1 or 2 types. However, in secondary sedimentation (see Section 5.2.4), zone or hindered settlement may occur. Compressive settlement only occurs in special sludge thickening tanks.

Primary sedimentation should remove approximately 55% of the suspended solids and, because some of these solids are biodegradable, the BOD_5 is also typically reduced by 20–35%. The floating scum is also removed and combined with the sludge. As a result the settled sewage from primary sedimentation has a SS of approximately 150 mg L^{-1} and a BOD_5 of approximately 200 mg L^{-1}. The solids are concentrated into the primary sludge which is typically removed at least once a day under the influence of hydrostatic pressure.

5.2.3 Secondary (Biological) Treatment

There are two principal types of biological sewage treatment:

(i) *Fixed Film Processes*: *e.g.* percolating filters (also referred to as a trickling or biological filters).
(ii) *Suspended Growth Processes*: *e.g.* activated sludge treatment.

Both types of treatment utilize two vessels, a reactor containing the microorganisms which oxidize/biodegrade the BOD_5, and a secondary sedimentation tank, which resembles the circular radial flow primary sedimentation tank, in which the microorganisms are separated from the final or secondary effluent.

The early development of biological sewage treatment is not well documented. However, it is established that the percolating filter was developed to overcome the problems associated with the treatment of sewage by land at 'sewage farms', where large areas of land were required for each unit volume of sewage treated. It was discovered that approximately 10 times the volume of sewage could be treated in a given area per unit by passing the sewage through a granular medium supported on underdrains designed to allow the access of air to the microbial film coating the granular bed.

The origins of the percolating filter are present in land treatment and its development was an example of evolution. The second and probably predominant form of biological sewage treatment, the activated sludge process, arose spontaneously and represents an entirely original approach. This process involves the aeration of freely suspended flocculant bacteria, 'the activated sludge floc' in conjunction with settled sewage, which together constitute the 'mixed liquor'. Activated sludge treatment continues the trend established by the change from land treatment to the percolating filter in that, at the expense of higher operating costs, it is possible to treat very much larger volumes of sewage in a smaller area.

The activated sludge process is probably the earliest example of a continuous bacterial (microbial) culture deliberately employed by man, and certainly the largest used to date. Development of the activated sludge process was announced by its originators Fowler, Ardern and Lockett in 1913, based upon their research at the Davyhulme Sewage Treatment Works, Manchester. These scientists very generously did not patent the process to facilitate its rapid and widescale application.

Development of these two forms of biological sewage treatment has been largely empirical and undertaken without the benefit of information about the fundamental principles of continuous bacterial growth, which began to be developed from the late 1940s when Monod published his work on continuous bacterial growth, although the relevance was not perceived until approximately 10 years later. This lack of microbiological knowledge is highlighted by the fact that the role of microorganisms in the activated sludge process was not fully accepted until after 1931; prior to this it was accepted by several workers that coagulation of the sewage colloids was the principal mechanism in the activated sludge process, although in the USA the role of bacteria in percolating filters was first recognized in 1889.

5.2.3.1 Percolating Filter

These units consist of circular or rectangular beds of broken rock, gravel, clinker or slag with a typical size in the range of 50–100 mm. In addition structured or loose plastic media can be utilised. The beds are between 1.5 and 2.0 m deep and of very variable diameter or size depending on the population to be served. The proportion of voids (empty spaces) in the assembled bed is normally in the range 45 to 55% for the non-plastic media and up to 97% for clean plastic media. The settled sewage trickles through interstices of the medium which constitutes a very large surface area on which a microbial film can develop. It is in this gelatinous film containing bacteria, fungi, protozoa and on the upper surface algae that the oxidation of the BOD_5 in the settled sewage takes place. The percolating filter is in fact a continuous mixed microbial film reactor. Settled sewage is fed onto the surface of the filter by some form of distributor mechanism. On circular filters a rotation system of radial sparge pipes is used which are usually reaction jet propelled although on larger beds they may be electrically driven. With rectangular beds electrically powered rope hauled arms are used.

The microorganisms which constitute the gelatinous film appear to be organized, at least near the surface of the filter where algae are present, into three layers (see Figure 5.5). The upper fungal layer is very thin (0.33 mm), beneath it the main algal layer is approximately 1.2 mm and both are anchored by a basal layer containing algae, fungi and bacteria of approximately 0.5 mm. However, algae do occur to some extent in all three layers. Beneath the surface, where sunlight is excluded and as a consequence the algae are absent, this structure is significantly modified, probably into a form of organization with only two layers. It has been calculated that photosynthesis by algae could provide only 5% or less of the oxygen requirements of the microorganisms in the filter. Furthermore, photosynthesis would only be an intermittent source of oxygen since it would not occur in the dark and algae are often present only in the summer months. Carbon dioxide generated by other organisms in the filter might, however, increase the rate of photosynthesis. It has been proposed that algae derive nitrogen and minerals from the sewage and that some may be facultative heterotrophs. The nitrogen fixing so-called 'blue-green algae', really bacteria, are frequently present in filters.

Whilst fungi are efficient in the oxidation of the BOD present in the settled sewage they are not desirable as dominant members of the microbial community. They generate more biomass than bacteria, per unit of BOD_5 consumed, thus increasing the sludge disposal problem. Moreover, an accumulation of predominantly fungal film quickly causes blockages of the interstices of the filter bed material, impeding both drainage and aeration. The latter may result in a reduction in the efficiency of treatment which is dependent upon the metabolic activity of aerobic microorganisms.

Protozoa and certain metazoans (macrofauna) play an important role in the successful performance of the biological filter, although the precise nature of this role is dependent on the extrapolation of observations made in the activated sludge process, which is more amenable to study. However, the similarity in the distribution of organisms within the two processes suggests strongly that their roles are the same in both. The protozoa in particular remove free-swimming bacteria, thus preventing turbid effluents since freely suspended bacteria are not settleable. Certain metazoans may also ingest free-swimming bacteria, but their most important function is to assist in breaking the microbial film which would otherwise block the filter. This film is 'sloughed off' with the treated settled sewage. Protozoa (principally ciliates and flagellates) tend to dominate in the

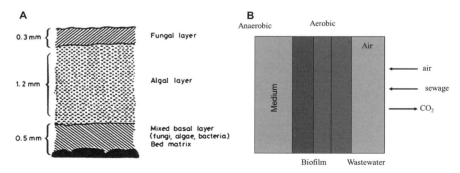

Figure 5.5 Cross-section of the surface layers of a percolating filter (A) and how those layers interact with the sewage and resulting redox conditions.

upper layers of the filter, whilst the macrofauna (nematodes, rotifers, annelids and insect larvae) dominate the lower layers.

If film is not removed satisfactorily, as the result of excessive fungal growth, the condition known as 'ponding' can develop. In this condition the surface of the filter is covered in settled sewage, air flow ceases, treatment stops and the bed becomes anaerobic. Ponding may also be caused by the growth of a sheet or felt of large filamentous algae, principally *Phormidian* sp., on the face of the filter. To minimize film production recirculation of treated effluent is often employed. This reduces film growth by dilution of the settled sewage, improves the flushing action for the removal of loose film, and promotes a more uniform distribution of the film with depth.

Treated sewage is subject to secondary sedimentation which is similar to primary sedimentation as a result of which the suspended sloughed off film is consolidated into humus sludge which then undergoes further treatment and the secondary effluent can then either be discharged to the receiving water or undergo further tertiary treatment.

5.2.3.2 Activated Sludge

In the activated sludge process the majority of biological solids removed in the secondary sedimentation tank are recycled (returned activated sludge or RAS) to the aerator. The feedback of most of the cell yield from the sedimentation tank encourages rapid adsorption of the pollutants in the incoming settled sewage, and also serves to stabilize the operation over a wide range of dilution rates and substrate concentrations imposed by the diurnal and other fluctuations in the flow and strength of the sewage. Stability is also provided by the continuous inoculation of the reactor with microorganisms in the sewage and airflows, which are ultimately derived from human and animal excreta, soil run-off, water and dust. The reactor of the activated sludge plant is usually in the form of long deep channels. Before entering these channels the returned activated sludge and settled sewage are mixed thereby forming the 'mixed liquor'. The retention time of the 'mixed liquor' in the aerator is typically three to six hours; during this period it moves down the length of the channel before passing over a weir, prior to secondary sedimentation. The sludge which is not returned to the aerator unit is known as surplus or waste activated sludge and has to undergo further treatment. In practice, the conditions in the aeration unit diverge from the completely mixed conditions commonly used for industrial fermentations and it may be best described as a continuous mixed microbial deep reactor with feed-back.

The design of the concrete tanks which form the reactor is strongly influenced by the type of aeration to be employed. Two types are available, compressed (diffused air) (see Figure 5.6) and mechanical (surface aeration) (see Figure 5.7). In the diffused air system much of the air supplied is required to create turbulence, to avoid sedimentation of the bacteria responsible for oxidation. Surface aeration systems introduce the turbulence mechanically and only provide sufficient air for bacterial oxidation. Both types of system aim to maintain a dissolved oxygen concentration of between 1 and 2 mg L^{-1}.

In the diffused air system the air is released through a porous sinter at the base of the tank and this system is characterized by long undivided channels which may be quite narrow. Mechanical aeration utilizes rotating paddles to agitate the surface thereby incorporating air and creating a rotating current which maintains the bacterial flocs in suspension. Each paddle is located in its own cell which has a hopper shaped bottom, this gives the plant the appearance of a square lattice (see Figure 5.7). However, beneath the face of the mixed liquor all the cells are connected, forming a channel. In both systems the channels are 2–3 m deep and 40–100 m long.

The success of the activated sludge process is dependent on the ability of the microorganisms to form aggregates (flocs) which are able to settle. It is generally accepted that flocculation can be explained by colloidal phenomena and that bacterial extracellular polymers play an important role, but the precise mechanism is not known. The significance of flocculation to the success of the

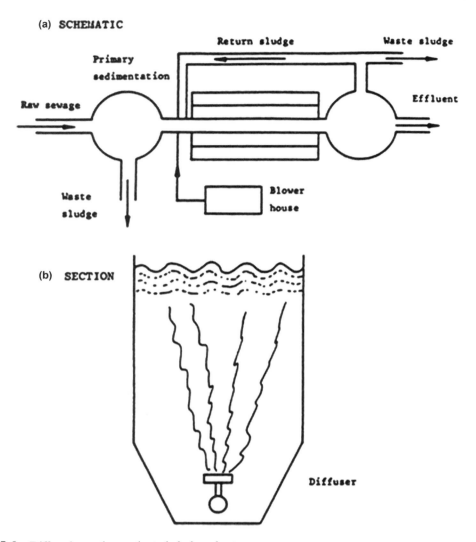

Figure 5.6 Diffused aeration activated sludge plant.

process is not the only characteristic to distinguish it from other industrial continuous cultures. There are four additional and very significant differences:

(i) The process utilizes a heterogeneous microbial population.
(ii) Growth occurs in a very dilute multi-substrate medium.
(iii) Many of the bacterial cells are not viable.
(iv) The objectives of the process, which are the complete mineralization of the substrates (principally carbon dioxide, water, ammonia and/or nitrate) with minimal production of both biomass and metabolites, are unique.

The heterogeneous population present in activated sludge includes bacteria, protozoa, rotifers, nematodes and fungi. The bacteria alone are responsible for the removal of the dissolved organic material, whilst the protozoa and rotifers 'graze', removing any 'free-swimming' and hence non-settleable bacteria; the protozoans and rotifers being large enough to settle during secondary

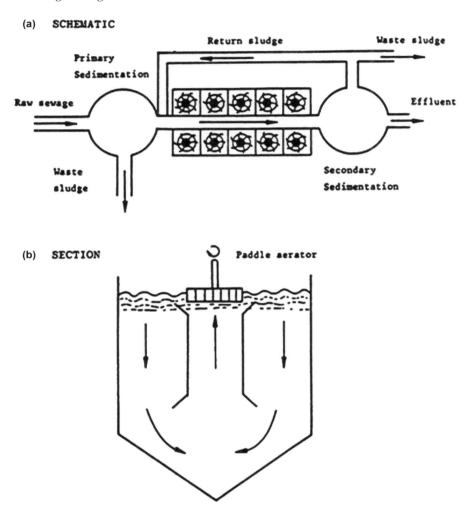

Figure 5.7 Mechanically aerated activated sludge plant.

Table 5.2 Importance of ciliated protozoa in determining effluent quality.

Effluent property	Ciliates absent	Ciliates present
Chemical Oxygen Demand (mg L^{-1})	198–254	124–142
Organic Nitrogen (mg L^{-1})	14–20	7–10
Suspended Solids (mg L^{-1})	86–118	26–34
Viable bacteria (10^7 mL^{-1})	29–42	9–12

sedimentation. The role of protozoa in activated sludge has been extensively studied; there are three groups involved: the ciliates, flagellates and amoebae. It is probably the ciliates (Ciliophora), which constitute the greatest number of species with the greatest number present in each species, that play the major role in the clarification process. The effect on effluent quality as a consequence of grazing by protozoa is summarized in Table 5.2. Not only do the protozoa remove free-swimming activated sludge bacteria but they play an important role in the reduction of pathogenic bacteria, including those which cause diphtheria, cholera, typhus and streptococcal infections. In the absence of protozoa approximately 50% of these types of organisms are removed, while in their presence removals rise to 95%. Nematodes have no significant role in the process, whilst the effects of the fungi are generally deleterious and contribute to, or cause, non-settleable sludge known as 'bulking'. Members of the following bacterial genera have been regularly isolated from activated sludge: *Pseudomonas*, *Acinetobacter*, *Comamonas*, *Lophomonas*, *Nitrosomonas*, *Zoogloea*, *Sphaerotilus*, *Azotobacter*, *Chromobacterium*, *Achromobacter*, *Flavobacterium*, *Alcaligenes*, *Micrococcu*s and *Bacillus*. Attributing the appropriate importance to each genus is a problem which confounds bacteriologists.

Of the principal groups of substrates listed in Section 5.1.4, only one single substrate (cellulose) was included. Each of the groups includes many substrates: for example the sugars identified in sewage include glucose, galactose, mannose, lactose, sucrose, maltose and arabinose, whilst the nitrogenous compounds include proteins, polypeptides, peptides, amino acids, urea, creatine and amino-sugars. Since bacteria normally only utilize a single carbon substrate or at the most two, this diversity of substrates in part explains the numerous genera of bacteria isolated from activated sludge because each substrate under most conditions will sustain one species of bacterium. Moreover, as a consequence of the large number of substrates present in the settled sewage the concentration of individual substrates is far less than the 200 mg L^{-1} of BOD_5 present, perhaps 20–40 mg L^{-1} for the most abundant and less than 10 mg L^{-1} for the less common ones. The concentration of each substrate is further reduced in the aeration tank by dilution with the returned activated sludge which is typically mixed 1 : 1 with settled sewage, resulting in a 50% reduction in substrate concentration.

The low substrate concentration means that the bacteria are in a starved condition. As a consequence many of them are 'senescent', *i.e.* in that phase between death, as expressed by the loss of viability, and breakdown of the osmotic regulatory system (the moribund state); thus the bacterium is a functioning biological entity incapable of multiplication. That bacteria could exist in this condition was established at an early stage in a series of inspired experiments by Wooldridge and Standfast who published their results in 1933. They determined the dissolved oxygen concentrations and bacterial numbers (by viable counts) in a series of BOD bottles containing diluted crude sewage on a daily basis. The viable count reached a maximum on the second day and thereafter fell rapidly. However, the consumption of oxygen increased by equal amounts until the fourth day and fell to a negligible value on the fifth day. There was no obvious relationship between viability and oxygen consumption. They tested experimentally the hypothesis that non-viable bacteria were apparently capable of oxygen uptake by destroying the capacity for division without significantly diminishing enzyme activity. Treatment of *Pseudomonas fluorescens* with a 0.5% formaldehyde solution prevented division but these bacteria exhibited vigorous oxygen uptake in both sewage and other media. Subsequently they were able to determine the presence of active oxidase and dehydrogenase enzymes in these non-viable bacteria. The effects of low substrate concentration on the viability of the bacteria are compounded by their specific growth rate. It is intended that biological wastewater treatment should result in the production of a final effluent containing negligible BOD. The biochemically oxidizable material in the effluent is composed of compounds originally present in the settled sewage, which have not been completely biodegraded, and bacterial products. Moreover this is to be achieved with the minimal production of biomass. These twin objectives are concomitant with the utilization of a bacterial population with a very low specific growth rate.

Unlike the percolating filter, bacterial growth in the activated sludge process is amenable to the type of description used by bacteriologists for conventional continuous cultures. However, although it is amenable to this type of treatment it inevitably appears to be very different from all other continuous cultures. The dilution rates (rate of inflow of settled sewage/aeration tank volume) used are invariably low by the standards of industrial fermentations, typically 0.25 h^{-1}, *i.e.* one quarter of the aeration tank volume is displaced every hour, therefore the hydraulic retention time is four hours. In the conventional single pass reactor the dilution rate and the specific growth rate (time required for a doubling of the population) are identical; that is the state in which the rate of production of cells through growth equals the rate of loss of cells through the overflow. However, in the activated sludge process, because of the recycling of the biomass, the specific growth rate is very much lower than the dilution rate, typically in the range $0.002–0.007 \ h^{-1}$. Since, under steady-state conditions, the bacteria are only able to grow at the same rate as they are lost from the system, recycling them dramatically lowers their specific growth rate and allows it to be controlled independently of the dilution rate. Under steady-state conditions the specific growth rate is equivalent to the specific rate of sludge wastage (mass of suspended solids lost by sludge wastage and discharged in the effluent in unit time as a proportion of the total mass in the plant) which is the reciprocal of the 'sludge age' or mean cell retention time which is typically 4–9 days. Thus, whilst the retention of the aqueous phase in the system is only four hours, the retention of the bacterial cells or sludge age is several days. The sludge age is a value which describes a great deal about the type of activated sludge plant: its purpose, quality of effluent and the bacteriological and bio-chemical states. The equations which can be used to calculate the sludge age or solids retention time (SRT) is outlined in the following equation:

$$\text{Sludge age}(\vartheta_x) = \frac{\text{Mass of solids undergoing aeration}}{(\text{Mass wasted} + \text{mass in effluent}) \ \text{per day}}$$

$$\vartheta = \frac{V \times X}{(Q_{was} \times X_{was}) + ((Q_i - Q_{was}) \times X_e)}$$

Where:

$X =$ Mixed Liquor Suspended Solids MLSS
$Q_{was} =$ Waste activated sludge flowrate
$X_{was} =$ Concentration of waste activated sludge
$X_e =$ Effluent suspended solids.

The activated sludge process may have up to four phases:

(i) Clarification, by flocculation of suspended and colloidal matter.
(ii) Oxidation of carbonaceous matter.
(iii) Oxidation of nitrogenous matter (see Section 5.2.3.5).
(iv) Auto-digestion of the activated sludge.

The occurrence of these four phases is directly dependent on increasing sludge age. Those processes which operate at low sludge ages give rapid removal of BOD_5 per unit time, but the effluent is of poor quality. Plants which have high sludge ages give good quality effluents but only a slow rate of removal. Low sludge ages result in actively growing bacteria, and consequently high sludge production, whilst bacteria grown at high sludge ages behave conversely. Figure 5.8 illustrates the relationship between the growth curve of the bacterial culture and the type of activated sludge plant. By operating continuously the activated sludge process functions only over a small region of the batch growth curve; this region is determined by the specific sludge wastage rate. The region selected determines the type of plant and its performance. These are summarized in Figure 5.8.

5.2.3.3 Dispersed Aeration

This type of process is rarely used and is not applicable to the treatment of municipal sewage but may be of use in the preliminary treatment of some industrial wastes. The bacteria are growing

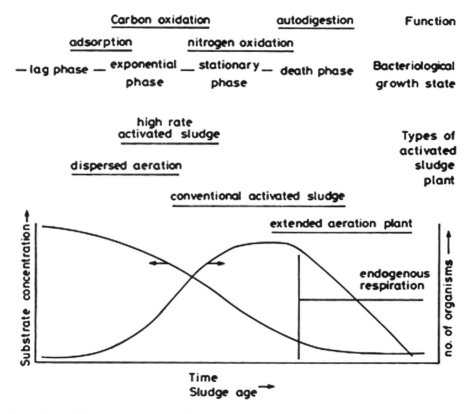

Figure 5.8   Relationship between batch culture and type of activated sludge plant.

rapidly (exponential phase), thus the process has the ability to remove a large quantity of BOD_5 per unit of biomass and as a consequence a small reactor may be used which is cheap to construct. However, because of their high rate of growth, the bacteria convert much of the BOD_5 into biomass, generating large volumes of sludge, flocculation is therefore limited and so additional treatment is essential to remove solids. Furthermore, although BOD_5 removal per unit biomass is high, the effluent BOD_5 is also high.

5.2.3.4   Conventional Activated Sludge

The previous process utilises actively growing bacteria in the exponential phase of growth. It achieves the oxidation of carbon compounds using an exclusively heterotrophic bacterial population. Conventional activated sludge plants operate in the stationary or declining growth phases utilising senescent bacteria. This very slow growth results in very low residual substrate concentrations and hence low values for effluent BOD_5. In addition, plants operating at sludge ages towards the upper end of this range contain autotrophic nitrifying bacteria. These organisms convert ammonia into nitrite and nitrate which further improves the quality of the effluent. In addition to maximizing effluent quality, conventional activated sludge plants limit the production of new cells. Bacteria which are growing slowly use much of the organic matter available in the maintenance of their cells rather than in the production of new cells. These features have made conventional activated sludge the most widely adopted biological sewage treatment process for medium and large communities. The rate of oxidation is highest at the inlet of the tank and it can be difficult to maintain aerobic conditions. Solutions to this problem have been adopted. With tapered aeration, rather than supplying air uniformly along the length of the tank, the air is concentrated at the beginning of the tank and progressively reduced along its length. The volume of air supplied remains unchanged but it is distributed according to demand. Alternatively stepped loading may be utilised. This aims to make the requirement for air uniform by adding the settled sewage at intervals along the tank, thus distributing the demand. Finally, the dissolved oxygen concentration can be controlled by linking the aeration requirements to the ammonia concentration in the settled sewage monitored using on-line ammonia probes.

5.2.3.5 *Nitrification*

The production of a final effluent with the minimum BOD_5 value is dependent upon the complete nitrification of the effluent, which involves the conversion of the ammonia present into nitrate. This is a two-stage process undertaken by autotrophic bacteria principally from the genera *Nitrosomonas* and *Nitrobacter*. Nitrification occurs in percolating filters and activated sludge plants operated in a suitable manner. The first stage involves the oxidation of ammonium ions to nitrite and follows the general equation:

$$NH_4^+ + 1.5O_2 \xrightarrow{Nitrosomonas} NO_2^- + 2H^+ + H_2O$$

In the second stage nitrite is oxidized to nitrate:

$$NO_2^- + 0.5O_2 \xrightarrow{Nitrobacter} NO_3^-$$

The overall nitrification process is described by the equation:

$$NH_4^+ + 2O_2 \rightarrow NO_3^- + 2H^+ + H_2O$$

Two important points are evident from this last equation. Firstly, nitrification requires a considerable quantity of oxygen. Secondly, hydrogen ions are formed and hence the pH of the wastewater will fall slightly during nitrification.

The settled sewage is effectively self-buffering but a fall of 0.2 of a pH unit is frequently observed at the onset of nitrification. In this autotrophic nitrification process, ammonia or nitrite provide the energy source, oxygen the electron acceptor, ammonia the nitrogen source and carbon dioxide the carbon source. The carbon dioxide is provided by the heterotrophic oxidation of carbonaceous nutrients, by reaction of the acid produced during nitrification with carbonate or bicarbonate present in the wastewater, or by carbon dioxide in the air. Whereas for carbonaceous removal the oxygen requirement is roughly weight for weight with the nutrients oxidized, in the case of ammonia removal by nitrification approximately seven times as much oxygen is required to achieve the removal of the same quantity of nutrient.

Nitrification significantly increases the cost of sewage treatment since more air is required. Furthermore, because these autotrophic organisms grow only slowly, longer retention periods are also required, resulting in higher capital costs. Nor does nitrification result in the production of an entirely acceptable sewage effluent under the UWWTD when effluent is discharged to sensitive areas. As a consequence denitrification is required. Here in this anoxic heterotrophic bacterial process, nitrite and nitrate replace oxygen in the respiratory mechanism and gaseous nitrogen compounds are formed (nitrogen gas, nitrous and nitric oxides).

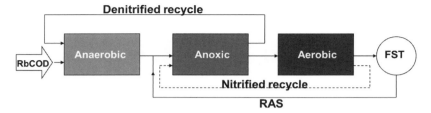

Figure 5.9   Typical configuration for combined nitrogen and phosphorus removal in an activated sludge plant. RbCOD = readily biodegradable COD; FST = final sedimentation tank; RAS = return activated sludge.

5.2.3.6   Enhanced Biological Phosphorus Removal

Now a number of activated sludge based processes have been developed for the combined removal of nitrogen and phosphorus. They employ combinations of anaerobic (no oxygen or nitrate), anoxic (nitrate but no oxygen) and aerobic (oxygen) compartments or zones (see Figure 5.9) and include the Bardenpho and the University of Cape Town (UCT) processes amongst others. In these combined biological nutrient removal processes, nitrification as described in Section 5.2.3.5 is occurring in the aerobic zone with solid retention times over six days. The nitrate containing effluent from the aerobic zone is recycled to an anoxic zone where denitrification takes place emitting nitrogen gas to the atmosphere and reducing the total nitrogen concentration of the wastewater. In the anaerobic zone, the volatile fatty acids in the wastewater are utilised by *Acinetobacter* to release stored phosphorus which is then taken up by phosphorus accumulating organisms in quantities excess to basic growth requirements in the aerobic zone. Wasting of the biomass containing the excess phosphorus thus reduced the phosphorus effluent concentrations.

5.2.4   Secondary Sedimentation

Both fixed and suspended growth types of biological treatment require sedimentation to remove suspended matter from the oxidized effluent. Tanks similar to those normally employed for primary sedimentation are generally employed, although at a higher loading of approximately $40 \, \mathrm{m^3 \, m^{-2} \, d^{-1}}$, at 3 DWF. Because of the lighter and more homogenous nature of secondary sludge, simple sludge scrapers are possible and scum removal is not necessary.

5.3   SLUDGE TREATMENT AND REUSE

Sludge treatment and reuse is a significant element of the cost of wastewater treatment, accounting for over 40% of all site costs. The collection, handling, processing and reuse of sludge are an operational chain of finite capacity which must be managed as a single process to ensure that reliable capacity and consistent products are provided. Often given insufficient attention in the past, it is the decision as to how the final product will be utilised that will influence design and process options for the entire treatment facility. Disposal options are restricted by the ban on sea dispersal under the UWWTD and controls on landfill. In the UK in 2010/11, 1465 thousand tonnes of dry solids required recycling. Approximately 80% of that treated sludge was recycled to agricultural land, 18% was sent for thermal destruction (including sludge that is sent for energy recovery or co-firing with other fuels) and only 0.6% of treated wastewater sludge was sent to landfill in 2010/11.

Sewage sludge is the combination of the product of primary sedimentation of sewage and the by-product of secondary and tertiary treatment processes. Primary sludges are odorous and liable to become putrescent with the potential to cause odour nuisance. Secondary and tertiary sludge consist largely of bacterial solids. When fresh it is much less offensive than primary sludges but is still liable to putresce. These sludges are usually thickened separately and then combined for sludge treatment, though co-settlement of surplus activated sludges in primary settlement tanks does occur and it can make the mixture less easy to settle and can increase process odour.

The purpose of sludge treatment is to secure the reliable, safe, sustainable and economic reuse of sludge. When recycled to agriculture, sludge must be treated to reduce pathogens and to stabilize the product to ensure minimum odour. When sludge is thermally destroyed treatment aims to achieve the maximum dry solids content to safeguard energy efficient combustion (*ca.* above 33% to ensure that the process is auto thermic).

5.3.1 Sources of Municipal Sludge

The production of sludge can be estimated from the contributing population and the type and extent of the treatment process. The higher the quality of effluent from a site, the greater the degree of treatment, the more sludge is produced. The primary settlement tanks will produce some 40 g capita^{-1} day^{-1} at about 5% dry solids content. Secondary treatment processes will add 20–50 g capita^{-1} day^{-1} of bacterial solids at 1–3% dry solids. Where chemical phosphorus treatment is practised then total sludge production will be increased by 10–25%. Settlement processes are managed so that the dry solids content of sludge is optimized to reduce volumes to be processed without being too thick for easy handling or so stale that it becomes septic and odourous. Sludge treatment is usually centralised into few processing centres. This requires that untreated sludges are transported from many smaller satellite treatment works.

As treatment processes will be designed with a maximum hydraulic capacity it is important to ensure reliable management of dry solids content. The volume of sludge at 2.5% dry solids is twice that at 5% dry solids and therefore would incur twice the transport or pumping costs and could overload processing capacity. The reliability of the treatment chain protects sewage treatment processes by ensuring that sludge is regularly removed from settlement tanks. In addition to managing volume, the quality of the sludge must be protected from damaging pollutants. Effective trade effluent control safeguards biological treatment processes, agricultural recycling, and thermal destruction.

5.3.2 Sludge Recycling Options

Treated sludge beneficially recycled in agriculture is often described as 'biosolids'. Agricultural recycling has for a long period of time been considered as the Best Practicable Environmental Option (BPEO). Biosolids contain useful amounts of nutrients, trace elements, and organic matter. Farmers value the product as a cost effective source of nutrient and soil conditioning. However, farmers are customers for the products and must be assured that the product delivers real benefits, that soil quality is protected, and that food grown on treated fields is safe. The increase in public concern about food safety, arising from several food related health incidents in the 1980s and 1990s has affected biosolids recycling. Untreated recycling to land used for food production was ended after 1999 as the result of an agreement (The Safe Sludge Matrix) between the UK water industry and the British Retail Consortium.[11]

The concerns and constraints surrounding agricultural recycling and the ending of disposal at sea have increased interest in energy recovery options. At times approximately half of the sludge produced has undergone thermal destruction. However, issues with costs, public perception, planning, sludge drying, legislation and insufficient energy recovery have resulted in a number of older incinerators being closed and a reduction in new plants being built. Nevertheless, thermal processes are liable to remain viable options for sludge in the future and interest in gasification/pyrolysis is increasing with larger scale plants planned.

5.3.3   Pre-treatment Handling

Sludge treatment processes cannot be considered in isolation. The process comprises the preparation and storage at satellite sites; transport, reception and processing at the treatment centre; and final reuse on farms for example. The capacity of the system to handle sludges is finite and bottlenecks in the system will cause delays in moving and processing sludge which can affect the efficient operation of the sewage treatment processes.

Thickening removes water from sludge to reduce volumes transported or processed. The process should produce a 6–8% dry solids product. Gravity thickening uses settlement to separate the water. A tank is equipped with valves at several levels in the tank to enable water layers to be identified and drained. Stirred settlement tanks, picket fence thickeners, avoid the problems of the development of layers of water within a tank by slow stirring. The thickened product is removed from the bottom of the tank and the water removed by overflow weirs at the top of the tank. Typical retention of a picket fence thickener is two days. However, for activated sludges and increasingly now primary sludges mechanical thickening is preferred. It can be achieved with centrifuges, filter belt presses and gravity belt presses. The thickening is achieved with the help of polyelectrolytes to aid floc formation. Polymer is added at a rate of 4–10 kg of active ingredient per tonne dry solids of sludge processed.

Dewatering removes water from sludge to achieve a cake of 20–35% dry solids. Dewatering may be placed after anaerobic digestion and before thermal destruction processes. It can be achieved with centrifuges, filter belt presses or plate presses. Polyelectrolytes are used to improve dewaterability of the sludge. The selection of equipment is usually based on pilot trials as the response of sludges can vary according to the process used. The amount of polymers used will vary from 4–8 kg active ingredient per tonne of dry solids. A key process parameter is the quality of returned liquors (centrate or filtrate); these can add significant loads to the treatment process and increasingly nutrient treatment processes such as Ammanox and struvite precipitation on sludge liquor are being installed.

5.3.4   Sludge Treatment Processes

Treatment for agricultural recycling requires that the product achieves pathogen reduction, is stable and, as far as is possible, is odour free. The critical control points for each process must be constantly monitored and recorded. The balance and availability of nutrients is different in different products. In general, drier and more processed product have lower agronomic values but are often more acceptable products to the customer.

5.3.4.1   Anaerobic Digestion

During digestion the organic matter present in the sewage sludge is biologically converted into a gas typically containing 65% methane and 35% carbon dioxide. The process is undertaken in a closed reactor usually equipped with a separate gas holder. The methane produced is used for maintaining the digester temperature, and for electricity production in combined heat and power (CHP) engines. The calorific value of biogas containing both methane and carbon dioxide is approximately 23.3 MJ m^{-3}. Increasingly though, in the UK, methane is also being upgraded by removing the carbon dioxide content so that the gas can be injected into the natural gas grid or used as a transport fuel with a calorific value then approaching upward of 37.5 MJ m^{-3}.

Methane production is more significant at elevated temperatures, when 0.35 L CH_4 can be produced for every g COD removed. Digesters are characterized by the temperature at which they operate: those in which gas production is optimum at 35 °C are described as 'mesophilic', whilst those at 55 °C are 'thermophilic'. These terms describe the temperature preferences of the bacteria undertaking the process. Heat exchangers are used to transfer heat from the treated sludge to the influent sludge. The additional heat is provided by the combustion of methane. For efficient operation the digester requires a robust mixing system which may be mechanical or utilize the gas produced in the process to provide turbulence. A conventional anaerobic digester is illustrated in Figure 5.10. The result of anaerobic digestion is to reduce the volatile solids present in the original sludge by 50% and the total solids by 30%. In addition, the unpleasant odour associated with the raw sludge is drastically reduced. During the 12 to 15 days required for sewage sludge digestion the sludge is stabilized and emerges with a slightly tarry odour.

Anaerobic digestion can be considered as a multi-stage process (see Figure 5.11). The initial step – HYDROLYSIS – involves the use of extracellular enzymes produced by hydrolytic bacteria which hydrolyse (break down and liquefy) complex polymers (such as lipids, polysaccharides and proteins) to simple soluble monomers (such as fatty acids, monosaccharides and amino acids). This makes the substrates more easily available for use by the acidogenic bacteria in the subsequent step. In general, proteins present in the waste are converted to amino acids, fats into long chain fatty

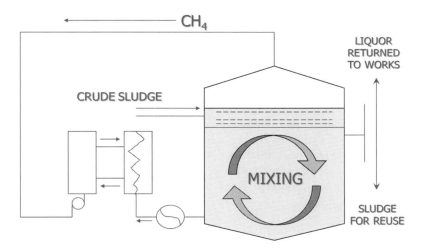

Figure 5.10 Simplified schematic diagram of an anaerobic digester.

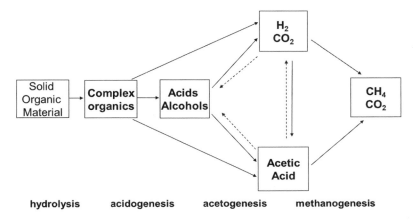

Figure 5.11 Typical biochemical steps involved in anaerobic digestion.

acids and carbohydrates into simple sugars. The liquefaction of complex compounds to more simple soluble substances is often the rate limiting step in digestion, since bacterial action at this stage proceeds more slowly than in the following steps. The rate at which hydrolysis takes place is governed by: substrate availability, bacterial population density, temperature and pH.

The second step – FERMENTATION – (also called ACIDOGENESIS) involves the fermentation of the soluble compounds produced during hydrolysis without the use of an additional electron acceptor or donor, and includes the degradation of soluble sugars and amino acids to a number of simpler products. For example, acetate, propionate and butyrate are produced from monosaccharides. Acetic acid is another major acidogenesis product produced from amino acids. Acidogenesis, as opposed to the following acetogenesis stage, can occur without an external electron acceptor and has higher free energy yields. Therefore the reactions can occur at higher hydrogen or formate concentrations and at higher biomass yields. Alternative fermentation products include lactate and ethanol. Lactate can often been seen during transient overload conditions in acidification reactors. Ethanol can be produced as an alternative to acetate at low pH conditions (*e.g.* <pH 5). The proportion of different by-products produced depends on the environmental conditions and especially on the particular bacterial species present.

Degradation of higher organic acids to acetate occurs during ACETOGENESIS. This is an oxidation reaction, but, unlike in acidogenesis an internal electron acceptor is not used. Instead an additional electron acceptor such as hydrogen ions or carbon dioxide is used to produce hydrogen gas or formate, respectively. These electron carriers must be maintained at a low concentration for the oxidation reaction to be thermodynamically possible with hydrogen and formate consumed by methanogenic organisms in the subsequent step.

The final step – METHANOGENESIS – produces methane from the breakdown products in the previous stage. This is achieved using obligate anaerobes, the growth rate of which is generally slower than the bacteria responsible for the preceding steps. Methane is produced from a number of different simple substrates: acetic acid, methanol, CO_2 and H_2. Of these, acetic acid and the closely related acetates are the most important substrate with approximately 70% of the methane produced during anaerobic digestion derived from degradation of these substrates according to the following equation:

$$CH_3COOH \longrightarrow CH_4 + CO_2 \text{ (aceticlastic methanogenesis)}$$

Methane forming bacteria can also use methanol or CO_2, or H_2 to produce approximately 30% of the methane according to the following equations:

$$CH_3OH + H_2 \longrightarrow CH_4 + H_2O$$

$$CO_2 + 4H_2 \longrightarrow CH_4 + 2H_2O.$$

Methanogens are pH sensitive, the required range being mildly acidic (pH 6.6–7.0) and problems are likely to be encountered if the pH falls.

Once in operation, with reasonable retention times and volatile solids loadings, the routine operation of digesters must include careful monitoring of certain parameters which are used to indicate whether the process is about to fail. Anaerobic digestion is quite sensitive to fairly low concentrations of toxic pollutants, such as heavy metals and chlorinated organics, and to variations in loading rates and other operational aspects. If the balance of the process is upset it is most likely that the methanogenic organisms become inhibited first. This results in a build-up of the intermediate compounds, volatile fatty acids, at the stage immediately prior to methane formation. These compounds include formic, acetic and butyric acids and can be monitored to determine the state of the process. The volatile acids are important because of their acidic nature. Normally digesters operate in the pH range of neutrality (pH 6.5–7.5). They also have some resistance to pH change. High concentrations of the volatile acids can cause a reduction in pH sufficient to inhibit bacterial activity to the extent where irreversible failure of the process occurs. Because of their capacity to resist changes in pH, volatile acid concentrations can build up to significant levels before pH change occurs. Therefore they can act as an early warning indicator of impending process failure. Normal levels of volatile acids in a sewage sludge digester are 250–1000 mg L^{-1}. If they exceed 5000 mg L^{-1} this could lead very quickly to failure. A good indication of process stability is therefore the volatile acid : alkalinity ratio. For example, volatile acid : alkalinity ratio of >0.8 indicates pH depression and methane production inhibition, whilst 0.3– 0.4 indicates some need for corrective action and a ratio of 0.1–0.2 indicates stable operation.

To ensure that mesophilic digestion processes achieve the necessary 99% (or 2 log) reduction in indicator pathogens, a secondary storage phase is normally needed. Batch storage as liquid or cake for 14 days is normally sufficient but each site must be evaluated to ensure process standards can be met. Where an enhanced treated product is required, mesophilic digestion may be preceded by a thermal treatment stage, pre-pasteurisation. Here batches of sludge are held at high temperature (>70 °C) for a period of time (>1 hour) to eliminate pathogens. Enhanced treatment process should achieve a 99.9999% (or 6 log) reduction in indicator pathogens.

Increasingly sludge pre-treatments are being applied prior to the methanogenic anaerobic digester to improve pathogen kill and/or improve methane production. Pre-treatments predominantly aim to improve hydrolysis which is considered to be a rate limiting step in digestion. In hydrolysis the breakdown of complex organics is slow even though the microorganisms that provide the catalytic enzymes grow quickly. Lignin can prevent access of the enzymes to the organics and often many of the organics are not readily degradable. Surplus activated sludge, in particular, is resistant to hydrolysis, with the digestibility of this secondary sludge often observed to be 50% that of primary sludge. Surplus sludge is resistant to digestion because the organic material is largely compartmentalised within microbial cells or sludge flocs. Cells maintain structural integrity during digestion because cross linkage of microbial glycan with peptide chains renders cell walls resistant to hydrolysis and biodegradation. Surplus activated sludge may contain up to 70% bacteria and the formation of flocs and micro – colonies enclosed by extracellular polymeric substances (EPS) also inhibit digestion.

A range of mechanical, thermal, chemical and biological pre-treatments have been investigated for their potential to improve the digestion. Pre-treatments are designed to disintegrate microbial cells and convert recalcitrant particulate organic matter in excess sludge into low molecular weight, biodegradable components. A typical sewage sludge pre-treatment applied in the UK is thermal hydrolysis which combines high temperatures (>135 °C) and pressure (>8 bar) to pasteurise and hydrolyse sludge before digestion. This can improve the efficiency of digestion and increase gas production. A biological pre-treatment process is enzymic hydrolysis. Here the biological steps to produce acetic acid are separated from the methanogenic steps through the use of additional reactors. There are up to six reactors in series operating at 42 °C with an overall retention time of 2 days to optimise hydrolysis and pathogen kill. Additional pathogen kill is achieved by operating the last two reactors at higher temperatures. The sludge is then digested in a single methanogenic reactor operating at 35 °C with a retention time of up to 20 days.

5.3.4.2 Composting

This is an aerobic stabilisation process utilising both thermophilic and mesophilic bacteria to break down organic matter in sludge. Sludge is generally composted in a ratio of 1 : 2 with straw or green waste to produce an enhanced product which is very acceptable to customers. However, because this aerobic process does not produce methane, which can be converted to electricity, attracting renewable energy financial incentives, composting is now less frequently applied to sewage sludge. Composting can be conducted in static piles, turned windrows or in contained systems. The optimum operating temperature is in the range 45–55 °C.

5.3.4.3 Combustion

Where agricultural recycling is not practical, energy recovery by thermal treatment is the usual alternative. Dried sludge has a similar calorific value to brown coal at 12–18 MJ kg^{-1}. The energy recovery options include combustion and gasification/pyrolysis. For the process to be energy positive then the recovered energy must exceed that used to dewater or dry the sludge prior to the thermal process. Combustion is the thermal conversion in air of the organic content of sludge producing heat and flue gas. The residual ash, which can comprise 30% of the sludge mass must be landfilled or can be incorporated into building materials. Gasification is a partial oxidation process in a low oxygen environment at temperatures of 900–1100 °C. The organic content of sludge is converted into syngas comprising CO, H_2 and CH_4, leaving a residual ash. Pyrolysis is the thermal degradation of the organic content in the absence of oxygen at 400–800 °C and usually precedes gasification. In addition to the gas the process produces an oil residue and a char. The gas from these processes can be burned to generate energy or used as a chemical feedstock.

REFERENCES

1. Water UK, *Water UK Sustainability Indicators 2010–2011*, 2012; http://www.water.org.uk/home/news/archive/sustainability/sutainability-indicators-10-11-31-01-2012 (accessed 19/02/2013).
2. UK Office for National Statistics, *Statistical Bulletin 2011 Census: Population Estimates for the United Kingdom*, 27 March 2011; http://www.ons.gov.uk/ons/dcp171778_292378.pdf (accessed 19/02/2013).
3. Environment Agency, *Transforming Wastewater Treatment to Reduce Carbon Emissions*, Report: SC070010/R2, 2009.
4. J. W. Knox, S. F. Tyrrel, A. Daccache and E. K. Weatherhead, A geospatial approach to assessing microbiological water quality risks associated with irrigation abstraction, *Water Environ. J.*, 2011, **25**(2), 282–289.
5. WEF Highlights London Strives To Make Olympics Sustainable Through Water-Reuse System, 26 July 2012; http://news.wef.org/london-strives-to-make-olympics-sustainable-through-water-reuse-system-2/ (accessed 02/09/2013).
6. Government of Great Britain, *Water Pollution Control Engineering*, HMSO, London, 1970.
7. European Communities, *Council Directive 91/271/EEC Concerning Urban Waste-water Treatment*, 1999.
8. EC European Commission, *Directive 2000/60/EC of the European Parliament and of the Council of 23 October 2000 establishing a framework for Community action in the field of water policy, 2000*; http://eur-lex.europa.eu/LexUriServ/LexUriServ.do?uri = CELEX:32000L0060:EN:NOT (accessed 23/05/2012).
9. EC European Commission, Priority Substances Daughter Directive – Directive 2008/105/EC of the European Parliament and of the Council of 16 December 2008 on Environmental Quality Standards in the Field of Water Policy, 2008; http://eur-lex.europa.eu/LexUriServ/LexUriServ.do?uri = OJ:L:2008:348:0084:0097:EN:PDF (accessed 23/05/2012).

10. EC European Commission, *Proposal for a Directive of the European Parliament and of the Council amending Directives 2000/60/EC and 2008/105/EC as regards Priority Substances in the Field of Water Policy*, Brussels, 31.1.2012 COM(2011) 876 final 2011/0429 (COD), 2012; http://ec.europa.eu/environment/water/water-dangersub/pdf/com_2011_876.pdf (accessed 12/06/2013).
11. British Retail Consortium (BRC), *Safe Sludge Matrix: Guidelines for the Application of Sewage Sludge to Agricultural Land*, 2001; http://www.adas.co.uk/LinkClick.aspx?fileticket = f_CX7x_v4nY%3D&tabid = 211&mid = 664 (accessed 27/02/2103).

CHAPTER 6

Treatment of Toxic Wastes

STUART T. WAGLAND* AND SIMON J. T. POLLARD

Department of Environmental Science and Technology, School of Applied Sciences, Cranfield University, MK43 0AL, UK
*Email: s.t.wagland@cranfield.ac.uk

6.1 INTRODUCTION

The exposure of the environment to materials of a toxic or hazardous nature is a subject of great concern, requiring careful management and handling to ensure that toxic or hazardous waste is dealt with in the most sensible manner possible. Whilst all waste materials (solid, liquid or gas) have the potential to cause harm to living organisms and the environment in the long term, waste of a hazardous nature has the potential to cause harm as soon as it is mismanaged.

In Europe vast quantities of waste are produced each year, with around 3.7% of all wastes produced by the EU-27 group of countries being classed as 'hazardous wastes'[†] (as of 2012). The proportion of hazardous wastes produced by the major waste producing countries across Europe range from 0.4–12.1%, which amounts to around 98 000 tonnes of waste classified as hazardous being produced each year.

Whilst the direct disposal of hazardous waste is a common route (landfill or incineration without energy recovery), the majority of these materials are treated and/or recycled/reused. Whilst the management of household waste is managed through the implementation of the waste hierarchy (see also Chapter 16), the handling of hazardous waste is also subject to this approach to a more sustainable way of dealing with waste produced through human activities. The waste hierarchy deems landfill disposal to be the least desirable waste management route:[1]

  (i) Reduce.
  (ii) Reuse.
 (iii) Recycle.
 (iv) Recovery (energy from waste).
  (v) Disposal (landfill).

[†]Data derived from: http://epp.eurostat.ec.europa.eu/statistics_explained/index.php/Waste_statistics

Pollution: Causes, Effects and Control, 5th Edition
Edited by R M Harrison
© Wagland and Pollard and The Royal Society of Chemistry 2014
Published by the Royal Society of Chemistry, www.rsc.org

The treatment options available broadly depend on the properties and physical state (solid or liquid) of the waste, and on the costs of the available options. The cost of landfill disposal of unwanted hazardous materials has increased dramatically in the past decade, largely due to the adoption of the 1999 *EU Landfill Directive 1999/31/EC*[2] and the increase in landfill taxes. The cost of landfilling hazardous materials was significantly greater than the disposal of non-hazardous materials even prior to the *Landfill Directive*, and this remains the case to date. As a result, producers of such waste will aim to treat and neutralise these materials to avoid the prohibitive costs of landfill. Treatment, reuse and recycling of hazardous waste allow for the production of inert or non-hazardous materials, which can be reused as a valuable product, or can be disposed of with significantly lower impact on the environment or to human health.

6.1.1 Definition of Toxic and Hazardous Wastes

The terms 'toxic' and 'hazardous' are interchangeable; under the *Waste Framework Directive 2008/98/EC*[1] 'toxic' is one of the categories of a 'hazardous' waste material. For the avoidance of confusion, this chapter will deal with waste materials for which either term is applied.

Generally, hazardous or toxic wastes are materials which have the potential to cause harm to the environment and/or to human health. A waste is broadly classified as hazardous if it exhibits any one of the following characteristics: (i) ignitable; (ii) corrosive; (iii) reactive, or (iv) toxic. These are broken down into 15 different categories (formerly 14 in the *1991 Hazardous Waste Directive 91/689/EEC*) in *Annex III* of the revised *2008 Waste Framework Directive 2008/98/EC*. These are shown in Table 6.1.

Table 6.1 Properties which classify a material as 'hazardous'.[1,3]

H1	'Explosive': substances and preparations which may explode under the effect of flame or which are more sensitive to shocks or friction than dinitrobenzene.
H2	'Oxidizing': substances and preparations which exhibit highly exothermic reactions when in contact with other substances, particularly flammable substances.
H3-A	'Highly flammable': <ul><li>liquid substances and preparations having a flash point below 21 °C (including extremely flammable liquids), or</li><li>substances and preparations which may become hot and finally catch fire in contact with air at ambient temperature without any application of energy, or</li><li>solid substances and preparations which may readily catch fire after brief contact with a source of ignition and which continue to burn or to be consumed after removal of the source of ignition, or</li><li>gaseous substances and preparations which are flammable in air at normal pressure, or</li><li>substances and preparations which, in contact with water or damp air, evolve highly flammable gases in dangerous quantities.</li></ul>
H3-B	'Flammable': liquid substances and preparations having a flash point equal to or greater than 21 °C and less than or equal to 55 °C.
H4	'Irritant': non-corrosive substances and preparations which, through immediate, prolonged or repeated contact with the skin or mucous membrane, can cause inflammation.
H5	'harmful': substances and preparations which, if they are inhaled or ingested or if they penetrate the skin, may involve limited health risks.
H6	'Toxic': substances and preparations (including very toxic substances and preparations) which, if they are inhaled or ingested or if they penetrate the skin, may involve serious, acute or chronic health risks and even death.
H7	'Carcinogenic': substances and preparations which, if they are inhaled or ingested or if they penetrate the skin, may induce cancer or increase its incidence.
H8	'Corrosive': substances and preparations which may destroy living tissue on contacts.
H9	'Infectious': substances containing viable micro-organisms or their toxins which are known or reliably believed to cause disease in man or other living organisms.

Table 6.1 (*Continued*)

H10	'Teratogenic': substances and preparations which, if they are inhaled or ingested or if they penetrate the skin, may induce non-hereditary congenital malformations or increase their incidence.
H11	'Mutagenic': substances and preparations which, if they are inhaled or ingested or if they penetrate the skin, may induce hereditary genetic defects or increase their incidence.
H12	Substances and preparations which release toxic or very toxic gases in contact with water, air or an acid.
H13	'Sensitizing': substances and preparations which, if they are inhaled or if they penetrate the skin, are capable of eliciting a reaction of hypersensitization such that on further exposure to the substance or preparation, characteristic adverse effects are produced.
H14	'Ecotoxic': substances and preparations which present or may present immediate or delayed risks for one or more sectors of the environment.
H15	Substances and preparations capable by any means, after disposal, of yielding another substance, *e.g.* a leachate, which possesses any of the characteristics listed above.

6.1.2 Sources

Hazardous waste materials arise from a wide range of activities, most of which are industrial processes such as oil refining and the manufacturing of chemical products. These can be distinctly split into liquid and solid wastes, which pose different treatment considerations. Common sources of hazardous wastes are summarised in Table 6.2, and additional detail is provided in the following sub-sections.

6.1.2.1 Liquid Wastes

Liquid wastes include general chemical wastes such as acids, bases and solvents, but also include large quantities of oil and oil–water mixtures. Acid wastes are generally produced through industrial manufacturing processes, and are often contaminated with metal compounds,[3] some of which are hazardous themselves (*i.e.* chromium VI).[4,5] Waste water and liquid wastes make up a significant proportion of total hazardous/toxics wastes produced each year. These materials typically include waste from water treatment, oil,[6] oil–water mixture and liquid manufacturing by-products.

6.1.2.2 Solid Wastes

Materials which can be classified as solid hazardous wastes include manufacturing wastes such as asbestos (now banned), however most arise as a post-consumer or post-treatment product. For instance, one common solid hazardous waste material is air pollution control (APC) residues (also referred to as 'fly ash') which are produced in the combustion or thermal treatment of waste materials,[7] which can, incidentally, be hazardous themselves. APC residues contain metals,[8] including heavy metals, which mean that they are toxic; also these materials have a high leaching potential and so can lead to the dispersion of the metals into the environment if care is not taken in the handling of them.

Contaminated soils removed from the site of origin are also classified as hazardous wastes. These soils may be contaminated with liquids such as oils[6] and toxic chemical waste, or otherwise could be contaminated with solid materials such as asbestos (see section 6.3.1).

6.1.2.3 Clinical Wastes

Refuse being produced from the healthcare industry can exhibit a wide range of properties as shown in Table 6.1. Wastes can include pharmaceutical and chemical wastes;[9] these materials can include substances which are toxic, carcinogenic and mutagenic (H6, H7 and H11, respectively).

Table 6.2 Common sources of hazardous wastes.

Source/Industry	Waste material
Mineral extraction	• Acidic wastes • Oil containing soils
Agriculture	• Chemical wastes (pesticides, fertilisers, herbicides)
Wood	• Preservatives • Paints
Chemical industry • Battery manufacture • Organic chemicals • Inorganic chemicals • Fireworks manufacture • Ammunition manufacture	• Acidic wastes/metal contaminated acids • Organic acids • Solvents • Oxidising agents • Flammable/explosive organic chemicals • Antibiotics and other pharmaceutical-related compounds • Asbestos (now banned) • Inorganic acids (H_2SO_4) • Bases • Activated carbon from chlorine production • Explosive wastes (ammunition and fireworks)
Metals	• Acidic wastes/metal contaminated wastes • Chromium containing wastes
Residual wastes (municipal and commercial)	• Batteries • Fluorescent tubes • Oils and fats
Vehicles (commercial garages and end-of-life vehicles)	• Air bags and fuel tanks (explosive) • Brake discs • Oil filters • Antifreeze
Oil refinery	• Oils • Oil sludges • Acids and bases, contaminated with oil
Waste management	• Ashes from incineration, such as bottom ashes and air pollution control (APC) residues

The specific 'clinical waste' category applies to wastes which are classed as 'infectious' (H9 in Table 6.1), and includes used sharps, bandages, gloves and any other material which is contaminated with human fluids and are deemed as potentially infectious.[10] Clinical wastes have been defined in the *Controlled Waste Regulations (1992)* as:[11]

"Any waste which consists wholly or partly of human or animal tissue, blood or other body fluids, excretions, drugs or other pharmaceutical products, swabs or dressings, or syringes, needles or other sharp instruments, being waste which unless rendered safe may prove hazardous to any person coming into contact with it; and any other waste arising from medical, nursing, dental, veterinary, pharmaceutical or similar practice, investigation, treatment, care, teaching or research, or the collection of blood for transfusion, being waste which may cause infection to any person coming into contact with it".[11]

6.1.3 Case Study: Detection of Hazardous Materials used in the Preservation of Wood

Chromated copper arsenate (CCA) is a water-based wood preservative containing copper oxide (CuO), arsenic pentoxide (As_2O_5) and chromic acid (CrO_3). The leaching of Cu, Cr and As into the environment, poses risks to environmental and human health.[12] Such risks have resulted in legislation globally, and a partial ban on the preservative in the UK through the *Controls on Dangerous Substances Regulations 2003*.[13]

Due to the regulations, the use of CCA has decreased, however due to the lifecycle of wood products, the percentage of CCA-treated wood waste is predicted to increase in the future.[14]

The detection of CCA-treated wood is a challenge to waste wood processors, and inadvertently CCA contaminated wood appears in recycled material such as wood mulch, which is used in landscaping applications.[15] This poses a significant overall environmental risk from the leaching into surface and ground water[16] and a significant human health risk through the ingestion of Cu, Cr or As by children from the mulch used in the playground.[15,17] As such, the detection of CCA in waste wood is essential in order to prevent environmental contamination. Methods such as X-ray fluorescence (XRF) are capable of determining the content of Cu, Cr or As, however the required equipment are expensive and are not commonly used in waste processing sites. Chemical methods of detection are documented, and the use of chelating agents such as Chromazural S (CS) and (1-(2-pyridylazo)-2-naphthol) (PAN) have been used successfully.[18] Both compounds can be used as a solution and sprayed onto waste wood, yielding a result in under one minute.[18] These compounds chelate with Cu, and an appropriate colour change indicates the presence of Cu, and therefore a Cu-containing compound such as CCA.

6.2   TREATMENT AND MANAGEMENT ROUTES

6.2.1   Introduction and Overview

There are a number of options when considering the treatment, management or disposal of toxic and hazardous wastes. As presented in section 1 of this chapter, more suitable methods of handling wastes are required in terms of minimising disposal and maximising the recovery of resources (including energy). As a result, out-right disposal is not a widespread or promoted practice (depending on the waste itself) and is avoided where additional value can be obtained from the waste. Nevertheless the destruction of certain materials (*e.g.* high temperature incineration) occurs where it is not economically or practically viable to recycle or recover resources.

This section presents some of the most common routes for handling toxic and hazardous waste materials.

In addition to liquid and solid wastes from chemical, household or industrial activities, animal by-products are materials which have not been covered previously in this chapter; a case study is provided to highlight the disposal considerations following an infectious disease outbreak.

6.2.2   Case Study: Animal Carcass Disposal following Disease Outbreak

An illustrative example of the multidimensional features of waste management is the disposal of animal carcases during or immediately following animal disease outbreaks.

When diseased animals die, are killed through natural disasters or to prevent suffering, or when they are culled to reduce the spread of infectious disease, arrangements have to be made for carcase disposal.[19] Historically for small numbers of animals, carcases were buried in clay-lined pits away from people and surface or groundwaters, with clay cover to prevent scavenging. For mass disposals, where thousands, tens of thousands and, in the extreme, millions of animals may be culled, governments have resorted to near-industrial means for the rapid disposal of these problematic wastes.[20] The scale of carcase disposals in these circumstances often overwhelms the waste management infrastructure for countries managing these events.[21] For example, since 1997 in Asia alone, over 200 million chickens and ducks have been killed by the highly pathogenic avian influenza (HPAI) virus H5N1, or culled in efforts to contain the disease. For individual outbreaks, the Canadian Ministry of Agriculture, Food and Fisheries incinerated and composted *ca.*

17 million culled birds following an outbreak of the H7N3 avian influenza virus in British Columbia (2004). Similar disposals have been required in the US, in Hong Kong, in Italy (1999–2001; H7N1 avian flu; *ca.* 13 million birds), in the Netherlands, Belgium and Germany, in Hungary, in the UK (2006; low pathogenic H7N3; *ca.* 49 000 poultry; incinerated and rendered; February 2007; highly pathogenic H5N1, *ca.* 159 000 turkeys rendered). The UK Foot and Mouth outbreaks in 1967, 2001 and 2007 resulted in the slaughter, for disease control purposes, of 442 000, >4 000 000 and 2160 animals, respectively.

The key risks from animal disposals are of onward infection, physical injury or chemical exposure to operational staff and the public in close proximity of operations;[19] of infection to other animals close by where disease agents may be persistent in the environment; and to the wider environment especially the long term pollution of groundwaters. Carcase disposal is best conceptualised as a multistage process requiring close and effective management because of the opportunities that exist for exposure to pathogens, chemicals and other hazardous agents along its path. Risk-based controls seek to reduce exposure at critical points as early and effectively as possible.[21] Modern carcase disposal is a multistage process initiated by the collection of fallen stock or the culling of live stock (*e.g.* birds, pigs, sheep, cattle, pigs), with the subsequent disinfection of carcases, their transport to localised collection points and potential off-site transport for waste processing (*e.g.* incineration, rendering, mass burial). Ensuring biosecurity and environmental protection is critical. Certain viruses are persistent in the environment, notably foot and mouth virus (FMDV; reportedly up to 28 days in soil in autumn) and carcases produce large volumes of blood, urine, faeces and milk that may contain disease agents and other pathogens. A typical disposal 'process chain' has five stages: (i) disinfection and collection; (ii) transport; (iii) pre-processing; (iv) processing; and (v) dealing with residues. A wide range of hazards are posed during disposal (see Table 6.3). Assessing the risk of harm from these requires information on the relative potency of the hazard; their likely exposure point concentration or the magnitude of exposure; the relative availability of exposure pathways; and the sensitivity and vulnerability of individual receptors.[22]

Table 6.3 provides a candidate list of potentially significant hazards posed by carcase disposal operations in the United Kingdom (UK), as they relate to four exotic animal diseases: avian influenza, Newcastle disease, FMDV and classical swine fever. The table represents hazardous agents believed to (i) pose potentially significant risks to human health, animal health or the environment; *and* (ii) that are capable of evading destruction; *and* (iii) that are also capable of presenting at sufficient doses to be of potential concern.

An assessment of the 2001 UK foot and mouth disease outbreak provides some detail on the extent to which potential hazards may be realised, as experienced during a national disease outbreak. The prevailing view has been that environmental impacts have been relatively minor.[23] This said, among the environmental (as opposed to economic, or sectoral) impacts reported part-way through the outbreak requiring effective management,[24] were:

   (i) A large number (*ca.* 200) of reported water pollution incidents from the surface run-off of blood and carcase fluids early on when slaughter rates outstripped disposal capacity; though few resulted in significant water pollution (*ca.* only three high category pollution incidents from slurry spill and disinfectant run-off).
  (ii) The generation of large quantities of ash from animal pyres (typically 15 tonne ash per 300 t pyre) required containment and disposal.
 (iii) The generation of high strength initial leachate (>75 000–100 000 mg L^{-1} chemical oxygen demand; >500–2000 mg L^{-1} total nitrogen; 400–3000 mg L^{-1} potassium) at landfills taking carcases.
 (iv) A high prevalence of odour complaints in the proximity of mass burial sites during disposal (300 for one large burial site).
  (v) Localised, short-term derogations in local air quality during pyre burning.[25,26]

Table 6.3 Significant hazards associated with carcase disposal.[22]

Biological agents	Chemical agents	Amenity, nuisance impacts
Foot and mouth disease	Ammonia, ammoniacal nitrogen and nitrates	Odour
Classical swine fever	Detergents (incl. phosphates, LAS-surfactants)	Noise
Newcastle disease virus	Disinfectants (formaldehyde, phenols, hypochlorite, peroxide, QAS-salts, FAM30, Virkon S)	Derived products (litters, slurry and manures)
Salmonella spp.	Biochemical oxygen demand (BOD)	Milk/treated milk (also BOD)
Bioaerosols (actinomycetes, *aspergillus* spp.)	Veterinary medicines (in carcase; *e.g.* pesticides, antibacterials, coccidiostats, barbiturates)	Smoke
Influenza (H5, H7)	Methane	Ash
Influenza (other A strains)	NO*x*	Waste treatment/wastewater
Campylobacter spp	Particulates	treatment residues (*e.g.*
Coxiella burnetii (Q-fever)	Polynuclear aromatic hydrocarbons (PAHs)	alkaline hydrolysis residues; wastewaters; sludges;
African swine fever	SO*x*	residual farm wastes)
Bovine spongiform encalopathy (BSE prion)	Kerosene and other accelerants (*e.g.* Feedol)	
Cryptosporidium spp.	Breakdown products (*e.g.* pesticide residues, metabolites)	
Scrapie	Heavy metals (Pb, As)	
Swine vesicular disease	Dissolved organic carbon, total organic carbon	
	Extremes of pH	
	Benzene, toluene, ethylbenzenes, xylenes	
	Wood resins and chemicals in wood preservatives (*e.g.* PCP, CCA)	
	Dioxins, furans, PCBs	Organic fertilizers and soil
	Insecticides (permethrins, organophosphates)	improvers

The selection of disposal options in these outbreaks has historically been based on the scale of the requirement, the technological capacity and the logistics of disposal, rather than on any prior analysis of exposures to public health, animal health and to the environment. On-farm burial, on-farm pyres, rendering, high temperature incineration, mass burial, landfill and composting have all been employed. In an attempt to bring a formality of approach to decisions on disposal, some jurisdictions have promoted a hierarchy of disposal strategies[23,27] that, whilst keeping a broad range of technological options open, seeks to set preferences for certain technologies over others. In brief, carcase destruction in licensed, engineered process facilities (incineration, rendering) is often preferred over controlled burial in landfill, pyre burning, or on-farm burial, where little process control or containment is assured. An assessment of opportunities for pathogenic and chemical exposure[21] during carcase disposal lends additional weight for a hierarchy of disposal[27] and highlights afresh the necessity for quick and effective action at the outbreak's premises, where the principal opportunities for exposure, and thus risk management exist.[28]

Decisions on carcase disposal should be made by reference to national biosecurity objectives and should address issues of cost and regulatory compliance. Governments should not be concerned solely with the cheapest or the safest disposal option at any cost; rather with developing a hierarchy of responses that maintains adaptive flexibility. The lessons from past events point to the need for well-organized and well-communicated contingency plans for exotic animal disease outbreaks,[29] specific attention to logistical issues during early stages,[30] and for long term monitoring programmes, especially where mass burials have been employed.[31]

6.2.3 Thermal Processes

Thermal processing of waste materials is not a new technology or management method, and has been common place for a number of years in the disposal of various types of wastes. These include municipal and commercial wastes, along with the topic of this chapter, hazardous wastes.

The incineration of waste materials, in particular with energy recovery, is seen as a sustainable and effective method of waste management where recycling is not practically possible.[32,33] However, despite the sustainable credentials of thermal treatment of hazardous and non-hazardous wastes, the incineration of waste is often an unpopular treatment option for a variety of reasons, some of which are based on historically poorly managed sites. [34,35]

In the waste hierarchy, thermal treatment can appear at the bottom as purely 'disposal'. This occurs when the process is simply designed to destroy the waste, and recover no energy from the material. However, if energy is recovered, then thermal treatment moves up one place in the hierarchy to 'recovery'. As such, the recovery of energy from the thermal treatment of wastes is regarded as a more sustainable and environmentally desirable option than landfilling or thermal treatment without energy recovery.

There are a wide range of thermal processes available in the treatment of waste, and in addition to the general types of process, there are a wide variety of configurations available. This section will introduce three commonly used combustion processes, including:

(i) Moving grate.
(ii) Rotary kiln.
(iii) Liquid injection.

Each process is suitable for different types (solid or liquid) of waste and quantities. In all processes a hot gas, which typically contains gaseous pollutants and particulates, is produced. Depending on the facility itself, the hot gases may pass through a heat exchanger, or boiler, to produce energy. In any case, the gases are then cleaned, typically by the use of liquid scrubbers, activated carbon and electrostatic precipitators.

6.2.3.1 Moving Grate

A feature of most modern waste incinerators built to manage residual wastes (municipal and commercial), moving grate technologies are common throughout the waste management industry. These technologies are capable of handling very large quantities of waste, and are typically most suitable for near-dry solid wastes. Hazardous and toxic wastes can be managed effectively in moving grate facilities, and energy is recovered through the heating of water to produce steam, which then passes through a turbine to produce electricity. In addition to electrical energy, the hot steam can be routed away from the facility to provide heat to nearby households or business premises. Such configuration is known as combined heat and power (CHP).

In moving grate, waste is loaded onto the moving grate *via* a feed hopper. The grate is made up of moving parts, which push the waste through the reactor. Oxidant (air or oxygen) is forced through the bottom of the grate to enhance combustion. An overview of a moving grate system is provided in Figure 6.1.

Fuel is either processed before being placed in the reactor, to remove potentially recyclable materials, or the ash is processed afterwards to remove metals. Due to the nature of hazardous wastes, these are generally placed directly into the reactor.

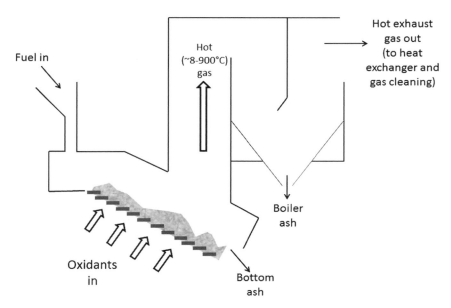

Figure 6.1   Schematic of a moving grate incinerator.

The advantages of a moving grate incinerator are:

(i)   Flexibility to handle a wide range of solid wastes.
(ii) Can be scaled up, or constructed in a modular fashion, to process very large quantities of waste.
(iii) High levels of operator control over the process.
(iv) Material recovery post-combustion including metals from the ash, and the reuse of the bottom ash as a construction aggregate.

The construction of the facility is expensive, and the public acceptance of waste incinerators, particularly large scale facilities such as those which use moving grate, is very poor. Maintenance is a consideration with all combustion processes, particularly those which use waste-derived materials due to the corrosive products formed, which can damage the reactor.[7,36]

6.2.3.2   Rotary Kiln

Rotary kiln incinerators, historically associated with the cement industry, are typically used to dispose of clinical wastes, however, they can also be used with sludge materials (sewage sludge or oily sludges from oil refinery wastes) and solid wastes.[37] Rotary-kiln incinerators are a common option in the thermal processing of a wide range of hazardous and toxic wastes due to their flexibility and level of control enabled over the process. Due to the design of these incinerators they are typically used on a relatively smaller scale than moving grate, which is covered in section 6.2.3.1; however rotary-kiln incinerators which can handle up to 10 tonnes per hour do exist on a commercial scale. The basic principle of a rotary-kiln, which makes this process unique, is that the cylindrical reaction chamber rotates to allow stirring and movement of the feedstock during combustion. A typical layout of a rotary-kiln incinerator is provided in Figure 6.2.

The waste material is loaded into the reactor by a hydraulic ram or screw-feeder; the reactor is constantly rotating as the waste moves down the reactor whilst being exposed to high temperatures of 1200–2000 °C where a high-temperature flame (fuelled by natural gas) is used as the ignition source.[38] The residence time can be controlled by the speed of rotation, to ensure complete

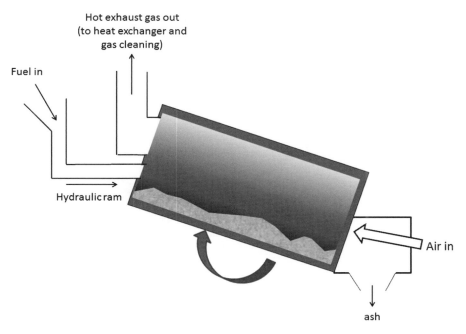

Fuel in

Hot exhaust gas out
(to heat exchanger and
gas cleaning)

Hydraulic ram

Air in

ash

Figure 6.2 A typical rotary-kiln reactor.

destruction of the feedstock. Ash is collected at the bottom of the reactor, and hot gases are exhausted at the top.

Advantages of rotary-kiln incinerators include:

(i) Flexibility to handle a wide range of fuels/feedstock.[37]
(ii) Very little, if any, pre-processing of the waste is required.[38]
(iii) Controlled residence time.
(iv) Air input and rotation allows thorough mixing, increasing the speed at which waste is destroyed.

The disadvantages include significant capital costs in the construction of rotary-kiln incinerators, along with the considerable cost of operation. As with moving grate, the maintenance is also an issue.

6.2.3.3 Liquid Injection

Large quantities of toxic wastes are liquids, as outlined previously, consisting of oils, acid wastes, wastes from the pharmaceutical industry and chemicals that are flammable. Chemical treatment is an option in some cases, but chemicals of certain characteristics, such as those which are flammable (H3-A and H3-B), toxic (H6) or carcinogenic (H7), chemical stabilisation may not be viable, and therefore thermal destruction becomes the most suitable option.

Liquid injection incineration is highly suitable for flammable chemical wastes, however the process can accommodate non-flammable materials. This is possible due to the ignition being fuelled by natural gas, and aqueous wastes then being injected into the burner as an aerosol.[39] A schematic of a liquid injection incinerator is shown in Figure 6.3.

The burner temperature is around 1600 °C, and liquid waste injected as an aerosol is rapidly broken down. The exhaust gases need to be managed in a manner described in the previously described thermal processes.

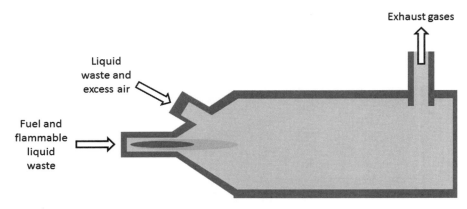

Figure 6.3 Liquid injection incineration (adapted from Oppelt, 1987).[39]

6.2.3.4 *Management of Thermal Residues*

There are two solid residues which arise from the combustion of wastes, bottom ash and boiler ash (fly ash).[7] Bottom ash is inorganic, assuming complete combustion, and can be reprocessed to be used as a road aggregate, or into construction materials. Care must be taken to ensure that the resulting products will not leach harmful substances (specifically metals) into the environment.[40] During combustion metals will be mobilised and will appear in the captured fly ash, either through filtration/ash capture or through electrostatic precipitation. Due to the elevated metal content, fly ashes are themselves categorised as hazardous.[41] Fly ashes are often treated through solidification/ stabilisation techniques, which are discussed later in section 6.2.4.3.

6.2.4 Chemical Processes

6.2.4.1 *Neutralisation and Precipitation*

The basic principles of neutralisation and precipitation are that when an acidic solution is neutralised by an alkali, H_2O and salts are formed.

$$H^+ + OH^- \rightarrow H_2O + salt$$

As a by-product of certain industrial processes, for example metal and chemical processing/ manufacturing, high quantities of strong acids are produced. These often contain significant quantities of metals due to the processes from which they are derived, and therefore require treatment prior to disposal. Acid neutralisation is a convenient process for such wastes, as it is relatively simple and significantly reduces the volume of waste for disposal. In terms of managing the metals contained within the liquid waste, the solubility of these metals decreases at higher pH. As a result, when the acidic waste is neutralised following the addition of an alkali, such as lime, the resulting salt will contain the previously dissolved metals. The salt can is then separated from the liquid effluent using a filter press, or by other techniques (*i.e.* centrifuging).

A filter press is a simple process, and works by pumping the liquid–solid mixture through semi-permeable screens, typically cloth. The filtered liquid can then be stored in a suitable tank or disposed of directly. This is dependent on site licence, and will be subject to thresholds of chemical components imposed by the licence. In the UK the water discharge licence will be issued under the relevant legislation.[11,42] For example, if the acid waste was primarily sulfuric acid then it is possible that the liquid effluent could contain a high concentration of sulfates. Sulfates are undesirable in the

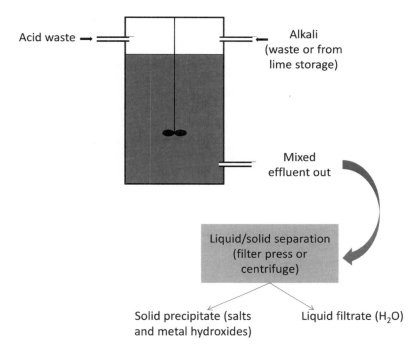

Figure 6.4   Schematic of the neutralisation and precipitation process.

sewage system as they cause the corrosion of the cement pipes. Other parameters which will need to be monitored include the chemical oxygen demand (COD, heavy metals and other metals such as zinc and copper.

A schematic overview of the process is provided in Figure 6.4.

The solid precipitate is deemed a hazardous waste material due to the high metal content, and so is landfilled in specifically designated hazardous waste landfills. Referring back to the waste hierarchy, these metals could be potentially recovered and reused.[3]

6.2.4.2   Solvent Recovery

Waste solvents are produced largely from the chemical industry, specifically including the pharmaceutical, paints and chemical production sectors. These solvents are contaminated and cannot be directly reused. Due to their flammable properties they can be disposed of in liquid injection incinerators. However with the waste hierarchy in mind, the recovery for reuse is a more sustainable and, in some cases, cost-effective treatment option.[43] Examples of solvents which can be extracted for reuse include dichloromethane (DCM) and toluene.[43,44] The recovery for reuse of solvents negates the need for the production of new solvents, most of which are fossil-derived.

The basic principles of solvent recovery are that solvents have a lower evaporation point than the contaminants which they contain, and so can be evaporated from a mixed waste solution. Two common approaches exist-thin-film evaporation or distillation; these approaches are effective in solvent removal, where the solvent is removed for potential reuse, and the contaminants remain. Solvent recovery/extraction is suitable for treating liquid wastes and contaminated soils.

6.2.4.3   *Solidification/Stabilisation*

The solidification of hazardous waste is a technique whereby the waste is encapsulated into a solid material. There are a wide range of processes associated with this technology; this section presents two of the more common approaches. These include the use of cement and pozzolan materials, with the latter also belonging to the lime-based technology.

The stabilisation step aims at immobilising contaminants by the conversion into less-soluble products;[45] this follows the same principles outlined in section 6.2.4.1, where the solubility of metals is reduced at higher pH levels. The intention of the overall process is the secure disposal of the hazardous waste; however other aims include improving the handling of the waste and converting the waste into less toxic products. These techniques are used as a final treatment stage prior to disposal in landfill.

Solidification/stabilisation (S/S) is either a cement-based or pozzolanic process, or a combination of the two.[46] Sludge materials and solid wastes of a hazardous nature are encapsulated in a secure and inert solid product; the use of only liquids in S/S processes is uncommon.[47] Cement is used is typically ordinary Portland cement (OPC),[48] and pozzolanic materials include residues from metal refinery and fly ash from thermal treatment processes.[46]

In cement processes the water content of the waste reacts chemically with the cement, producing hydrated silicates.[47] The water to cement ratio needs to be monitored, however once the correct ratios are achieved for the optimum workability of the cement a solid product suitable for transportation and disposal is produced.

A pozzolan material has no cementing properties itself, however will react with calcium hydroxide (*i.e.* lime) in the presence of water to form a hardened product.

Fly-ash is a hazardous material which is commonly used in S/S processes, and acts as both a bulking agent and a pozzolan.[47] Due to these favourable properties, the treatment of fly ash can comprise of OPC–fly-ash or lime–fly-ash mixtures. In both cases the OPC and lime act as setting agents.

Other waste materials which can be treated through the S/S route include hazardous sludges[49] such as 'red mud',[46] which is a product of aluminium production. Also, solid materials such as asbestos[50] and contaminated soils can also be treated through S/S.[47]

A common topic for academic research and environmental concern is the potential for the toxic compounds to leach out of the concrete over time,[48] especially when exposed to the harsh conditions of landfill. It is, however, widely accepted that the leaching capability of toxic compounds, metals and organic components is significantly reduced through S/S processes.

6.3   ENVIRONMENTAL AND HEALTH MANAGEMENT

6.3.1   Case Study: Severe Environmental Consequences of Poor Hazardous Waste Management (Spodden Valley, UK)

The environmental and health consequences of the poor management of hazardous materials and hazardous wastes can be severe in both the short and long term. There are numerous examples from around the world of environmental disasters due to hazardous materials, such as toxic waste dumping, oil spillages and gaseous emissions. The following case study presents an example from the UK where the poor control of a manufacturing process led to the exposure of workers, release of hazardous material to the surrounding area and the dumping of hazardous waste around the site. The results of this are a long history of illness of past workers and a heavily contaminated site in need of specialised and expensive remediation.

Severe environmental consequences of poor hazardous waste management: Spodden Valley, UK.
Asbestos is a well known carcinogen,[51] however it was heavily used from the late 1800s through until the late 1900s due to the highly favourable properties of the material. Asbestos is a

naturally occurring silicate contain material with valuable characteristics including heat and fire resistance.[52]

The management of asbestos-containing materials extracted from buildings requires careful management, subject to a number of regulations and health and safety protocols. However, the production of asbestos is also the source of many illnesses experienced by exposed workers. One such example is the Turner and Newall (prior to this, known as the Turner Brothers Asbestos Company) asbestos factory in Rochdale, UK. This factory was once a large employer in the local area, with over 4000 employees at one time.[53]

In the early 1900s company acknowledged the effects of asbestos on the health of its workers, and commissioned a research study into the particle abundance of asbestos fibres in the air around the site. These levels were found to be significantly higher than the safe levels, as considered at the time. The exposure of the workers to the asbestos fibres is an issue highlighted by poor working conditions and health and safety procedures, and these have had long-standing effects on those that once worked at the factory in the past. However it is the containment of residues and dust, and the management of waste materials from this site which has caused considerable damage to the environment. It was revealed by the company that around 6.8 tonnes per week of asbestos dust fibres were being removed from the ventilation filters. These fibres were then disposed of on the site in a disused coalmine shaft.[54] Around the same time local residents would observe dust fibres hanging from the trees near to the site.[54]

In 2004, after the asbestos factory was closed, the site was sold to a development company who planned on constructing 650 homes on the site. Prior to the submission of a planning application tree felling commenced, which caused a number of concerns by residents who were worried that asbestos dust would be unsettled. When the planning application was submitted it was noted that there was an absence of any asbestos contamination. This was disputed, and an independent study found considerable contamination in the soil.[54] The waste dumped in the disused mine shaft was discovered when trees were felled, exposing the extent of contamination.

The Greater Manchester Association of Trade Union Council demanded that on the grounds of safety the site be deemed permanently unsafe to develop, and that action should be taken to seal the site. One local councillor even described the site as 'Rochdale's Chernobyl'.[55]

In January 2011, the planning permission was refused on account of the site presenting a significant risk to human health due to asbestos contamination. To date the site remains undeveloped, however options are being considered in which to safely remediate the site to ensure that no further risks to human health is posed.

REFERENCES

1. Council of the European Union, *Directive 2008/98/EC on Waste and Repealing Certain Directives (Waste Framework Directive)*, *Off. J. Eur. Commun.*, 2008, *L 312/13*.
2. Council of the European Union, *Directive 1999/31/EC on the Landfill of Waste, Off. J. Eur. Commun.*, 1999, L 182, 1–19.
3. F. Rögener, M. Sartor, A. Bán, D. Buchloh and T. Reichardt, Metal recovery from spent stainless steel pickling solutions, *Resour. Conser. Recycling*, 2012, **60**, 72–77.
4. O. Krüger, U. Kalbe, W. Berger, F. G. Simon and S. L. Meza, Leaching experiments on the release of heavy metals and PAH from soil and waste materials, *J. Hazard. Mater.*, 2012, **207–208**, 51–55.
5. F. Macías, M. A. Caraballo and J. M. Nieto, Environmental assessment and management of metal-rich wastes generated in acid mine drainage passive remediation systems, *J. Hazard. Mater.*, 2012, **229–230**, 107–114.
6. L. J. Da Silva, F. C. Alves and F. P. De França, A review of the technological solutions for the treatment of oily sludges from petroleum refineries, *Waste Manage. Res.*, 2012, **30**(10), 1016–1030.

7. S. T. Wagland, P. Kilgallon, R. Coveney, A. Garg, R. Smith, P. J Longhurst, S. J. T. Pollard and N. Simms, Comparison of coal/solid recovered fuel (SRF) with coal/refuse derived fuel (RDF) in a fluidised bed reactor, *Waste Manage.*, 2011, **31**(6), 1176–1183.

8. L. S. Morf, R. Gloor, O. Haag, M. Haupt, S. Skutan, F. D. Lorenzo and D. Böni, Precious metals and rare earth elements in municipal solid waste – Sources and fate in a Swiss incineration plant, *Waste Manage*, in press.

9. D. Komilis, A. Fouki and D. Papadopoulos, Hazardous medical waste generation rates of different categories of health-care facilities, *Waste Manage.*, 2012, **32**(7), 1434–1441.

10. T. L. Tudor, W. K. Townend, C. R. Cheeseman and J. E. Edgar, An overview of arisings and large-scale treatment technologies for healthcare waste in the United Kingdom, *Waste Manage. Res.*, 2009, **27**(4), 374–383.

11. Her Majestry's Stationery Office (HMSO), *The Controlled Waste Regulations*; 1992.

12. T. G. Mercer and L. E. Frostick, Leaching characteristics of CCA-treated wood waste: A UK study, *Sci. Total Environ.*, 2012, **427–428**, 165–174.

13. Her Majestry's Stationery Office (HMSO), *The Environmental Protection (Controls on Dangerous Substances) Regulations*, 2003.

14. R. J. Murphy, P. Mcquillan, J. Jermer and R. D. Peek, *Preservative-treated wood as a component in the recovered wood stream in Europe- a quantitative and qualitative review, IRGWP 35th Annual Meeting, Slovenia, 2004*, Slovenia, 2004.

15. M. Guney, G. J. Zagury, N. Dogan and T. T. Onay, Exposure assessment and risk characterization from trace elements following soil ingestion by children exposed to playgrounds, parks and picnic areas, *J. Hazard. Mater.*, 2010, **182**(1–3), 656–664.

16. T. G. Townsend, H. Solo-Gabriele, T. Tolaymat and K. Stook, Impact of chromated copper arsenate (CCA) in wood mulch, *Sci. Total Environ.*, 2003, **309**(1–3), 173–185.

17. S. L Shalat, H. M. Solo-Gabriele, L. E. Fleming, B. T. Buckley, K. Black, M. Jimenez, T. Shibata, M. Durbin, J. Graygo, W. Stephan and G. Van De Bogart, A pilot study of children's exposure to CCA-treated wood from playground equipment, *Sci. Total Environ.*, 2006, **367**(1), 80–88.

18. G. Sawyer and M. Irle, Waste and Resource Action Programme (WRAP), *Development of Colour Indicator Techniques to Detect Chemical Contamination in Wood Waste for Recycling*, 2005.

19. J. Delgado, P. Longhurst, G. A. W. Hickman, D. M. Gauntlett, S. F Howson, P. Irving, A. Hart and S. J. T. Pollard, Intervention strategies for carcass disposal: Pareto analysis of exposures for exotic disease outbreaks, *Environ. Sci. Technol.* 2010, **44**(12), 4416–4425.

20. G. A. W. Hickman and N. Hughes, Carcase disposal: a major problem of the 2001 FMD outbreak, *State Vet. J.*, 2002, **12**, 27–32.

21. S. J. T. Pollard, G. A. W. Hickman, P. Irving, R. L. Hough, D. M. Gauntlett, S. F. Howson, A. Hart, P. Gayford and N. Gent, Exposure assessment of carcass disposal option in the event of a notifiable exotic animal disease: application to avian influenza virus, *Environ. Sci. Technol.*, 2008, **42**(9), 3145–3154.

22. S. J. Pollard, Disposal of diseased animals, in *Environmental Medicine*, ed. J. G. Ayres, R. M. Harrison, G. L. Nichols and R. L. Maynard, Edward Arnold Publishers Ltd, London, UK, 2010, pp. 574–581.

23. J. M. Scudamore, G. M. Trevelyan, M. V. Tas, E. M. Varley and G. A. W. Hickman, Carcass disposal: lessons from Great Britain following the foot and mouth disease outbreaks of 2001, *OIE Rev. Sci. Tech.*, 2002, **21**(3), 775–787.

24. Environment Agency, *The Environmental Impact of the Foot and Mouth Disease Outbreak: An Interim Assessment*, Environment Agency, Bristol, UK, 2001.

25. Department of Health, DETR, Food Standards Agency; Environment Agency and AEA Technology, *Foot and Mouth. Effects on health of emissions from pyres used for disposal of animals*, Department of Health, 2001.

26. I. Lowles, R. Hill, V. Auld, H. Stewart and C. Colhoun, Monitoring the pollution from a pyre used to destroy animal carcasses during the outbreak of Foot and Mouth Disease in Cumbria, United Kingdom, *Atmos. Environ.*, 2002, **36**(17), 2901–2905.

27. Department of Health, *A Rapid Qualitative Assessment of Possible Risks to Public Health from Current Foot and Mouth Disposal Options.* Main Report June 2001, Department of Health, London, UK, 2001.

28. World Organization for Animal Health (OIE), *Terrestrial Animal Health Code (2006)*, World Organization for Animal Health, Paris, France, 2006.

29. Department for Environment Food and Rural Affairs, *Defra's Framework Response Plan for Exotic Animal Diseases (for Foot and Mouth Disease, Avian Influenza, Newcastle Disease, Classical Swine Fever, African Swine Fever and Swine Vesicular Disease)*, presented to Parliament pursuant to s14A of the *Animal Health Act*, 2002, v1.0; Defra, London, UK, December 2006.

30. P. F. De Klerk, Carcass disposal: lessons from the Netherlands after the foot and mouth disease outbreak of 2001, *OIE Rev. Sci. Tech.*, 2002, **21**(3), 789–796.

31. D. H. Annells, G. H. Williams, R. Coulton, P. McKelvey and M. Tas, FMD burial site leachate disposal – a technical and commercial appraisal, *Proceedings Aqua Enviro Management of Wastewater Conference, Edinburgh, 15–17 April 2002*, Edinburgh, 2002; pp. 245–258.

32. S. Burnley, R. Phillips, T. Coleman and T. Rampling, Energy implications of the thermal recovery of biodegradable municipal waste materials in the United Kingdom, *Waste Manage.*, 2011, **31**(9–10), 1949–1959.

33. A. Porteous, Why energy from waste incineration is an essential component of environmentally responsible waste management, *Waste Manage.*, 2005, **25**(4 SPEC. ISS.), 451–459.

34. T. Jamasb and R. Nepal, Issues and options in waste management: a social cost–benefit analysis of waste-to-energy in the UK, *Resour. Conser. Recycling*, 2010, **54**(12), 1341–1352.

35. A. Porteous, Energy from waste: a wholly acceptable waste-management solution, *Appl. Energy*, 1997, **58**(4), 177–208.

36. C Velis, S. Wagland, P. Longhurst, B. Robson, K. Sinfield, S. Wise and S. Pollard, Solid recovered fuel: influence of waste stream composition and processing on chlorine content and fuel quality, *Environ. Sci. Technol.*, 2012, **46**(3), 1923–1931.

37. V. A. Cundy, T. W. Lester, C. Leger, G. Miller, A. N. Montestruc, S. Acharya, A. M. Sterling, D. W. Pershing, J. S. Lighty, G. D. Silcox and W. D. Owens, Rotary kiln incineration – combustion chamber dynamics, *J. Hazard. Mater.*, 1989, **22**(2), 195–219.

38. A. Paula Ottoboni, I. de Souza, G. JoséMenon and R. Joséda Silva, Efficiency of destruction of waste used in the co-incineration in the rotary kilns, *Energy Convers. Manage.*, 1998, **39**(16–18), 1899–1909.

39. E. T. Oppelt, Incineration of hazardous waste. A critical review, *J. Air Pollut. Control Assoc.*, 1987, **37**(5), 558–586.

40. S. Dugenest, J. Combrisson, H. Casabianca and M. F. Grenier-Loustalot, Municipal solid waste incineration bottom ash: characterization and kinetic studies of organic matter, *Environ. Sci. Technol.*, 1999, **33**(7), 1110–1115.

41. S. Sushil and V. S. Batra, Analysis of fly ash heavy metal content and disposal in three thermal power plants in India, *Fuel*, 2006, **85**(17–18), 2676–2679.

42. Her Majestry's Stationery Office (HMSO), *The Control of Pollution Act 1974 (Commencement No. 8) Order 1977*, HMSO, London, 1977.

43. G. V. D. Vorst, P. Swart, W. Aelterman, A. V. Brecht, E. Graauwmans, H. V. Langenhove and J. Dewulf, Resource consumption of pharmaceutical waste solvent valorization alternatives, *Resour. Conser. Recycling*, 2010, **54**(12), 1386–1392.

44. T. B. Hofstetter, C. Capello and K. Hungerbühler, Environmentally preferable treatment options for industrial waste solvent management. A case study of a toluene containing waste solvent, *Process Saf. Environ. Prot.*, 2003, **81**(3), 189–202.

45. S. Y. Hunce, D. Akgul, G. Demir and B. Mertoglu, Solidification/stabilization of landfill leachate concentrate using different aggregate materials, *Waste Manage.*, 2012, **32**(7), 1394–1400.
46. D. V. Ribeiro, J. A. Labrincha and M. R. Morelli, Potential use of natural red mud as pozzolan for Portland cement, *Mater. Res.*, 2011, **14**(1), 60–66.
47. J. R. Conner and S. L. Hoeffner, A critical review of stabilization/solidification technology, *Crit. Rev. Environ. Sci. Technol.*, 1998, **28**(4), 397–462.
48. J. Liu, X. Nie, X. Zeng and Z. Su, Long-term leaching behavior of phenol in cement/activated-carbon solidified/stabilized hazardous waste, *J. Environ. Manage.*, 2013, **115**, 265–269.
49. S. Bayar and I. Talinli, Solidification/stabilization of hazardous waste sludge obtained from a chemical industry, *Clean Technol. Environ. Policy*, 2013, **15**(1), 157–165.
50. Y. M. Chan, P. Agamuthu and R. Mahalingam, Solidification and stabilization of asbestos waste from an automobile brake manufacturing facility using cement, *J. Hazard. Mater.*, 2000, **77**(1–3), 209–226.
51. E. K. Park, K. Takahashi, Y. Jiang, M. Movahed and T. Kameda, Elimination of asbestos use and asbestos-related diseases: an unfinished story, *Cancer Sci.*, 2012, **103**(10), 1751–1755.
52. F. L. Pundsack, The properties of asbestos. I. The colloidal and surface chemistry of chrysotile, *J. Phys. Chem.*, 1955, **59**(9), 892–895.
53. M. C. Kidd, Homes to die for, *The Guardian*, 28 September, 2005.
54. D. Appleton, Call to probe 'every inch' of asbestos site, *Rochdale Observer*, 17 September, 2004.
55. Manchester Evening News, This could be our Chernobyl, *Manchester Evening News*, 13 August, 2007.

CHAPTER 7

Air Pollution: Sources, Concentrations and Measurements

ROY M. HARRISON

School of Geography, Earth and Environmental Sciences, University of Birmingham, Edgbaston, Birmingham, B15 2TT, UK
Email: r.m.harrison@bham.ac.uk

7.1 INTRODUCTION

Before commencing the description of individual air pollutants, it is useful to start with a consideration of the terminology. Air pollutants may exist in *gaseous* or *particulate* form. The former includes substances such as sulfur dioxide and ozone. Concentrations are commonly expressed either in mass per unit volume (μg m^{-3} of air) or as a volume mixing ratio (1 ppm $= 10^{-6}$ v/v; 1 ppb $= 10^{-9}$ v/v). Particulate air pollutants are highly diverse in chemical composition and size. They include both solid particles and liquid droplets and range in size from a few nanometres to hundreds of micrometres in diameter; concentrations are expressed in μg m^{-3}. Gaseous and particulate air pollutants may be separated operationally by use of a filter. Some substances are termed 'semi-volatile' and exist partitioned between particulate and vapour forms. An example is the polycyclic aromatic hydrocarbons.

Air pollutants emitted directly into the atmosphere from a source are termed *primary*. Thus, carbonaceous particles from diesel engine exhaust and sulfur dioxide from power stations are examples of primary pollutants. In contrast, *secondary* pollutants are not emitted as such, but are formed within the atmosphere itself. Thus, sulfuric acid and nitric acid, formed respectively from sulfur dioxide and nitrogen dioxide oxidation, are examples of secondary pollutants for which the atmospheric formation route far exceeds any primary emissions. The most commonly considered secondary pollutant is ozone, formed as a result of photolysis of molecular oxygen in the stratosphere, and of nitrogen dioxide in the troposphere (lower atmosphere).

This chapter will address the sources, concentrations and measurement methods for those pollutants considered most important in terms of human health effects and damage to crops and materials. Issues relating to the regional and global atmosphere will be considered in subsequent chapters.

Pollution: Causes, Effects and Control, 5th Edition
Edited by R M Harrison
© The Royal Society of Chemistry 2014
Published by the Royal Society of Chemistry, www.rsc.org

7.2 SPECIFIC AIR POLLUTANTS

7.2.1 Sulfur Dioxide

The major source of sulfur dioxide is the combustion of fossil fuels containing sulfur. These are predominantly coal and fuel oil since natural gas, petrol and diesel fuels have a relatively low sulfur content. Until recently, emissions of sulfur dioxide from diesel engines led to a small but perceptible increment in sulfur dioxide alongside busy roads, but recent years have seen a very substantial reduction in the sulfur content of both petrol (gasoline) and diesel fuels. The maximum permitted sulfur content of motor fuels in Europe is now 10 ppm. Figure 7.1 shows in diagrammatic form the sources of sulfur dioxide emissions by source category for the United Kingdom.[1] Combustion of coal in power stations is far the most major single source of SO_2 emissions.

Over past years, two source categories, both associated with coal burning have tended to dominate the UK situation with respect to sulfur dioxide. Urban ground-level concentration of SO_2 fell rapidly between 1970 and 1990 largely due to a decline in the burning of coal in domestic fireplaces for home heating. Airborne concentrations fell much faster than total emissions of sulfur dioxide because over that period the reduction in emissions from power stations was quite limited. Since around 1990, urban and rural concentrations of sulfur dioxide have become almost indistinguishable because they have a common source in power station plumes superimposed on a low background from diffuse sources. Total UK emissions declined by more than half between 1990 and 1996 due largely to a cut in emissions from power stations effected by installation of flue gas desulfurisation plants on some of the larger stations and a switch to electricity generation from combined cycle gas turbine plants burning natural gas. Emissions in 2010 were less than 10% of the 1990 levels. The main driver for this reduction has been international concern over acid rain problems (see Chapter 8) rather than domestic health concerns.

7.2.1.1 Measurements of Sulfur Dioxide

In the UK for many years, data were collected using very simple method developed originally for use in the *National Survey of Smoke and Sulfur Dioxide* which has now been discontinued. The method involves absorption of sulfur dioxide in hydrogen peroxide solution to form sulfuric acid. The resultant acid has traditionally been determined by acid–base titration which is subject to interference by other gaseous, acidic or basic compounds such as nitric acid or ammonia, respectively. As sulfur dioxide levels declined, so the reliability of measurements made by titration

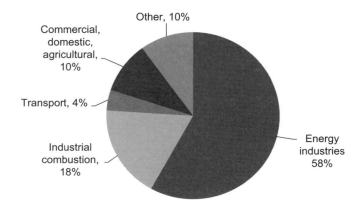

Figure 7.1   UK emissions of sulfur dioxide by sector, 2010.

Table 7.1 Summary of commonly employed methods for the measurement of air pollution.

Pollutant	Measurement technique	Sample collection period	Response time (continuous technique)[a]	Typical minimum detectable concentration
Sulfur Dioxide	Absorption in H_2O_2 and titration or sulfate analysis	24 hr		1 ppb
	Gas phase fluorescence		2 min	0.1 ppb
Oxides of Nitrogen	Chemiluminescent reaction with ozone		1 s	0.1 ppb
Total Hydrocarbons	Flame ionization analyser		0.5 s	10 ppb
Specific Hydrocarbons	Gas chromatography/flame ionization detector	[b]		<1 ppb
Carbon monoxide	Electrochemical cell		25 s	1 ppm
	Non-dispersive infrared		5 s	0.5 ppm
	Gas filter correlation		90 s	0.1 ppm
Ozone	UV absorption		30 s	1 ppb
Peroxyacetyl Nitrate	Gas chromatography/ electron capture detection	[c]		<1 ppb
Particulate Matter	High volume sampler	24 hr		5 $\mu g\ m^{-3}$
	TEOM	1 hr		4 $\mu g\ m^{-3}$

[a]Time taken for a 90% response to an instantaneous concentration change.
[b]Samples of air concentrated prior to analysis.
[c]Instantaneous concentrations measured on a cyclic basis by flushing the contents of a sample loop into the instrument.

reduced and many measurements were made by determination of sulfate by ion chromatography, which yields a result specific to sulfur dioxide.

The most commonly used instrumental technique for measurement of sulfur dioxide is based upon measurement of fluorescence excited by radiation in the region of 214 nm. Commercial instruments are available, capable of measurement of sulfur dioxide to less than 0.1 ppb, as well as source instruments with ranges into the thousands of ppm. The method is potentially subject to interferences from water vapour, which quenches the SO_2 fluorescence, and hydrocarbons capable of fluorescence at the same wavelength as SO_2. Commercial instruments are generally equipped with diffusion dryers and hydrocarbon scrubbers to overcome these problems. The commonly used techniques for analysis of SO_2 and other pollutants are summarized in Table 7.1 and are described in more detail elsewhere.[2,3]

7.2.2 Suspended Particulate Matter

Airborne particles are very diverse in character, including both organic and inorganic substances with diameters ranging from less than 10 nm to greater than 100 μm. Since very fine particles grow rapidly by coagulation and vapour condensation, and large particles sediment rapidly under gravitational influence, the major part (by mass) generally exists in the 0.1–10 μm range. A schematic representation of the typical size distribution for atmospheric particles appears in Figure 7.2. There are three peaks, or *modes*, in the distribution. The smallest one relates to the transient nuclei, which are very tiny particles formed by condensation of hot vapours, or gas to particle conversion processes. Thus, primary particles from motor vehicle exhaust and sulfuric acid formed from SO_2 oxidation are initially in the transient nuclei mode. Such particles, when emitted, are present in very high numbers and are subject to rather rapid coagulation both with other fine particles and also with coarser particles already in the atmosphere. Through this mechanism they enter the accumulation range of particles typically with diameters between about 100 nm and 2 μm. Such particles are also capable of growth through the condensation of low volatility materials. Removal of the

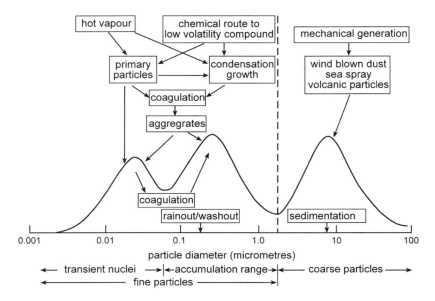

Figure 7.2 Schematic diagram of the size distribution and formation mechanisms for atmospheric aerosols (adapted from Ref. 4).

accumulation range particles by rainwater scavenging or dry deposition to surfaces is inefficient and such particles have a typical lifetime in the atmosphere of around one to two weeks. This renders them capable of very long range transport. The third mode, termed *coarse particles*, are mostly greater than 2 μm with sizes extending up to about 100 μm. This is comprised in the main of mechanically-generated particles such as wind-blown dust, sea spray and primary volcanic particles. These are formed by attrition of bulk materials and tend to be appreciably larger than transient nuclei or accumulation range particles.

The suspended particles in the atmosphere are typically referred to as the *atmospheric aerosol* and a number of other terms are used, which in the main are related to the method of collection or analysis.

In the UK, historically the largest set of measurements are for *black smoke* which (see later) is a measurement related to the blackness or soiling capacity of the particles. Other measures of suspended particulate matter depend upon a gravimetric determination of particle mass. The term *total suspended particulate matter* (TSP) has been used to describe the fraction of particles collected on a filter using the US high volume sampler. Since large particles, by virtue of their inertia, are not readily able to enter the inlet of such samplers, the measurement is dependent both on the orientation of the sampler with respect to the wind and the strength of wind, both of which influence the efficiency of particle aspiration.[2] To overcome this problem, the USEPA moved to use of a high volume sampler with a size selective inlet. This inlet has a 50% efficiency at 10 micrometres aerodynamic diameter and hence in simple terms may be considered to sample only those particles less than 10 micrometres in diameter, irrespective of orientation or wind speed. Measurements made using inlets meeting the EPA criterion are referred to as PM_{10}. A number of devices, other than the high volume sampler, are now available for PM_{10} measurement (see further on in this chapter). An inventory of UK emissions of primary particles as PM_{10} appears in Figure 7.3. This neglects secondary particles (which are formed in the atmosphere) and some diffuse sources such as resuspended soils and road dust which are hard to quantify. Inventories are also maintained for smaller particles such as $PM_{2.5}$ (*i.e.* particles below 2.5 μm diameter), $PM_{1.0}$ and even $PM_{0.1}$, termed the *ultrafine* particle fraction which has stimulated much interest as it is believed to have exceptional

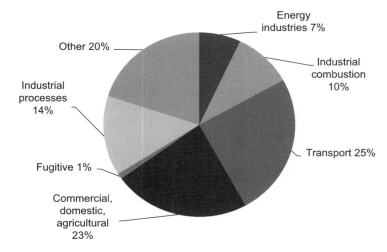

Figure 7.3 UK emissions of particulate matter (PM_{10}) by sector, 2010.

toxicity per unit mass compared to coarser particles. The road traffic contribution to the smaller size fractions is greater than to the PM_{10}, and is also larger in urban areas than in the UK as a whole.[4] Most developed countries have networks to monitor PM_{10} and $PM_{2.5}$, but not $PM_{0.1}$, although concentrations of the latter may be represented by counting the number of particles per unit volume of air.[5]

7.2.2.1 Black Smoke

Many older data relate to black smoke, although this is rarely used nowadays. Air is drawn through a cellulose filter upon which the particles are collected. At the end of the sampling period the ability of the filter surface to reflect light is determined quantitatively and the reflectance is related to a concentration of 'standard smoke' using a calibration graph constructed many years ago when urban particles were dominated by coal smoke. Research has shown that the method measures elemental carbon, and thus the dramatic reduction in the inefficient combustion of bituminous coal in domestic grates over recent years has led to a large drop in the emissions of particles containing black elemental carbon. Whilst inventories of emission of black smoke are no longer routinely compiled, the major source of black particles in UK cities is now emissions from diesel road vehicles. Black smoke concentrations are appreciably elevated alongside major roads and black smoke or elemental carbon can be a better indicator of vehicle emissions than gravimetrically determined particle concentrations.[6]

7.2.2.2 Gravimetrically Determined Particulate Matter: PM_{10} and $PM_{2.5}$

PM_{10} and $PM_{2.5}$ particulate matter are determined gravimetrically and therefore include both primary and secondary particles. In UK urban areas, two sources of $PM_{2.5}$ are dominant: one is road traffic emissions and the other is secondary particulate matter, mostly ammonium sulfate and ammonium nitrate particles. Whilst the former are black in colour, the latter are white and hence are not determined by the black smoke method. Other sources such as sea spray, abrasion and resuspension particles from traffic and wind-blown dust also contribute coarser particles to urban PM_{10}.[7,8] There are many other minor sources such as particles from building work and demolition, and biological particles such as pollens, spores and bacteria. The health effects of PM_{10} outlined in

Chapter 11 do not appear to be a strong function of chemical composition, although this remains a very active topic of research.

In addition to adverse effects on human health, airborne particles are responsible for the soiling of buildings and loss of visibility. Airborne particles both scatter and absorb light and thus cause deterioration in the quality of image transmission through the atmosphere which manifests itself as a loss of visibility. Although the efficiency of light scattering and absorption per unit mass of particles is dependent upon the size distribution and chemical composition of the particles, the aerosol mass loading is a fairly good predictor of visibility impairment.[9]

Typical UK urban background and roadside concentrations of PM_{10} together with those of other pollutants are shown in Table 7.2. These concentrations and those of black smoke are orders of magnitude below those associated with the major smog episodes of the 1950s and 60s. The smogs were caused primarily by low level emissions from coal combustion during periods of meteorology unsuitable for effective pollutant dispersal (low windspeeds and shallow mixing depth). The combination with fog (smog = smoke + fog) led to dramatic losses in visibility. The smog of December 1952 is believed to have caused some 4000 premature deaths. For many years after, UK pollution control policy focused on smoke and sulfur dioxide, and it is only since the 1980s that the major focus has transferred to motor traffic as the major source of urban air pollution.

Measurement of PM_{10}. The simplest method for determination of PM_{10} or $PM_{2.5}$ involves use of a high volume air sampler capable of drawing air through a filter at a rate of about 0.5–1 m^3 min^{-1} through a 10 μm or 2.5 μm size selective inlet. The sampler is run for 24 hours and the gain in mass of the filter, together with the volume of air passed, is used to calculate the airborne concentration over the 24 hour sampling period. Medium and low volume samplers operating at lower flow rates are also available.

Measurements in near-real time may be made using the Tapered Element Oscillating Micro-balance (TEOM). In this device, particles are collected on a filter which is attached is a vibrating element whose vibrational frequency changes with the accumulation of particles on its tip. Measurement of the vibrational frequency is used to estimate the mass of particles collected. An example of averaged diurnal data for various pollutants collected in London is shown in Figure 7.4 which clearly illustrates the influence of road traffic emissions on PM_{10}, NO_x and carbon monoxide concentrations. One draw-back of the TEOM is that the inlet is heated which causes loss of some semi-volatile constituents. Devices designed to correct for these losses are available.

Table 7.2 Annual mean and highest hourly concentrations of pollutants in North Kensington, London (urban background) and Marylebone Road, London (roadside) in 2012.

Pollutant	Annual period	Concentration ($\mu g\ m^{-3}$)	
		North Kensington	*Marylebone Road*
Carbon monoxide	annual mean	260	580
PM_{10}	annual mean	20	31
PM_{10}	highest hourly	106	133
Nitrogen dioxide	annual mean	37	96
Nitrogen dioxide	highest hourly	206	281
Ozone	annual mean	38	16
Ozone	highest hourly	186	110
Sulfur dioxide	annual mean	2	8
Sulfur dioxide	highest hourly	51	48
Benzene	annual mean	0.7 [a]	1.36

[a]value for London, Bloomsbury, 2011 as no data available for North Kensington.

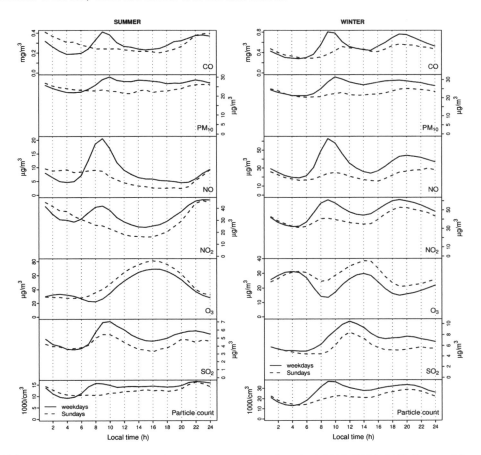

Figure 7.4   Diurnal pattern of mean hourly concentration for 1996–2008 (2001–2008 for particle count) in North Kensington, London.

Specific Components of Suspended Particulate Matter. Studies of airborne particles in the urban atmosphere of developed countries have shown that their composition can be described in terms of the following major components:

(i) Combustion particles comprising mainly fine (less than 2.5 μm diameter) particles of elemental and organic carbon derived predominantly from road vehicle traffic.
(ii) Secondary particles of mainly ammonium sulfate (or sulfuric acid), ammonium nitrate and organic matter predominantly within the fine (less than 2.5 μm diameter) size range.
(iii) Coarse particles arising largely from soil and road surface dust resuspended by traffic activity which are made up mainly of inorganic mineral components, abrasion particles (*e.g.* brake dust), and sea salt.

Further discussion of particle composition is given in Chapter 8. Whilst these three categories generally make up the major part of the particle mass, it is trace components associated with the various source categories which often stimulate the greatest interest and concern.

The components of airborne particles which give rise to most interest in the context of air pollution tend to be trace metals and trace organic compounds. A summary of concentrations of trace metals measured in Central London appears in Table 7.3. The table shows that most metals

Table 7.3 Trace metals in air in Central London in 1998 and 2011.

Metal	Concentration in 1998 (ng m^{-3})	Concentration in 2011 (ng m^{-3})
Arsenic	n/a	0.65
Cadmium	0.87	0.18
Chromium	2.9	3.4
Copper	17.7	16.2
Iron	832	419
Lead	38	9.4
Manganese	10.6	5.7
Nickel	8.1	1.55
Zinc	35	19.7
Vanadium	4.4	1.97

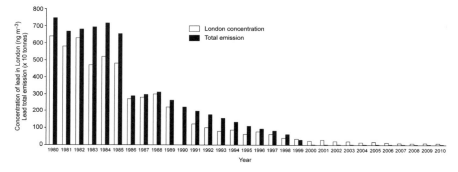

Figure 7.5 Time trend in annual emissions of lead from use in motor fuel, and concentrations of lead in air over the same period, 1980–2010.

declined substantially in concentration between 1998 and 2011, although some such as copper with a strong road vehicle source (brake wear) showed little change. Interest in trace metals relates primarily to their potential toxicity. Arsenic, chromium (vi) and nickel are believed to be genotoxic carcinogens and the World Health Organization has estimated unit risk factors for them.[10] Cadmium and mercury have also elicited a great deal of interest because of their high toxicity and there have been moves to control these elements throughout the developed world.

The trace metal which has stimulated far the greatest interest in relation to public health has been lead. The main source of airborne lead is its combustion in leaded petrol (gasoline) to which it is added as an octane improver in the form of tetraalkyllead compounds, notably tetramethyllead, $(CH_3)_4Pb$ and tetraethyllead, $(C_2H_5)_4Pb$. Upon combustion in the engine, these are emitted predominantly as an aerosol of fine particles of inorganic lead. Developed countries have without exception introduced regulations to limit the lead content of gasoline, and the United States has been essentially lead free for many years, with Europe becoming lead free at the end of 1999. The maximum permitted level of lead in UK gasoline stood in 1972 at 0.84 g L^{-1}, although such high concentrations were not generally used. By 1981, this had been reduced to 0.40 g L^{-1}, and in January 1986 fell sharply to 0.15 g L^{-1} in line with most other western European countries. The decline in the use of lead in petrol between 1980 and 2010 is documented in Figure 7.5, which also shows the decline of lead in air concentrations in central London. In general, it may be seen that the two have proceeded very much in parallel with one another, an example of a 'linear rollback' in which a primary pollutant with a single major source declines in proportion to the reduction in emissions. A small background concentration from other sources remains in the atmosphere.

The abrupt fall in the lead emissions from road traffic at the beginning of 1986 was exploited as the basis for an experiment to evaluate its impact upon population blood leads. In the event, blood leads fell slightly between 1985 and 1986, but little more, if at all, than between 1984 and 1985, and between 1986 and 1987 (when petrol lead remained almost constant) and only slightly more for groups heavily exposed to airborne lead than for less exposed groups.[11] This can be explained by lead from other sources, such as food and drink dominating lead intake at the time.

Concentrations of lead and most other metals in air are determined by use of air filtration to collect a sample. Non-destructive analytical methods are available, and these include X-ray fluorescence (both wavelength dispersive and energy dispersive) and instrumental neutron activation analysis. More commonly, destructive methods are used which involve dissolving the sample in oxidizing acids and analytical procedures such as atomic absorption spectrometry, inductively coupled plasma (ICP) emission spectrometry, or ICP-mass spectrometry.

The organic components of atmospheric particles of most interest as air pollutants are the semi-volatile groups of compounds, including the polynuclear aromatic hydrocarbons (PAH), polychlorinated biphenyls (PCB) and polychlorinated dibenzodioxins and dibenzofurans (PCDD and PCDF). The sampling, measurement and environmental behaviour of such compounds are discussed in Chapter 18.

7.2.3 Oxides of Nitrogen

The most abundant nitrogen oxide in the atmosphere is nitrous oxide, N_2O. This is chemically rather unreactive and is formed by natural microbiological processes in the soil. It is not normally considered as a pollutant, although it does have an effect upon stratospheric ozone concentrations (see Chapter 9) and there is much evidence that use of nitrogenous fertilizers is increasing atmospheric levels of nitrous oxide.

The nitrogen oxides of concern as local air pollutants are nitric oxide (NO) and nitrogen dioxide (NO_2). The major proportion of emitted NO_x (as the sum of the two compounds is known) is in the form of NO, although most of the atmospheric burden is usually in the form of NO_2. The major conversion mechanism is the very rapid reaction of NO with ambient ozone:

$$NO + O_3 \rightarrow NO_2 + O_2$$

The alternative third-order reaction with molecular oxygen is relatively slow at ambient air concentrations, although it can be of importance when NO concentrations reach about 1 ppm:

$$2NO + O_2 \rightarrow 2NO_2$$

This reaction can become important in severe winter pollution episodes when a catalytic cycle is established.[12]

The major source of NO_x is the high temperature combination of atmospheric nitrogen and oxygen, in combustion processes, there being also a lesser contribution from combustion of nitrogen contained in the fuel. An emission inventory for the UK appears in Figure 7.6.

Typical hourly average air concentrations of NO_x are normally in the range 10–100 µg m^{-3} in urban areas and less than 20 µg m^{-3} at rural sites, and higher close to heavy traffic. The proportion present as the more toxic pollutant, nitrogen dioxide, generally becomes greater the lower the total NO_x concentration. Time trends in annual mean NO_x emissions from road traffic and NO_2 concentrations in Central London appear in Figure 7.7. This can be contrasted with Figure 7.5 showing the linear rollback of lead in air concentrations as lead emissions from road traffic decreased. Over

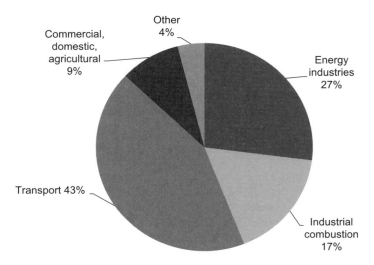

Figure 7.6   UK emissions of NO$_x$ by source sector, 2010.

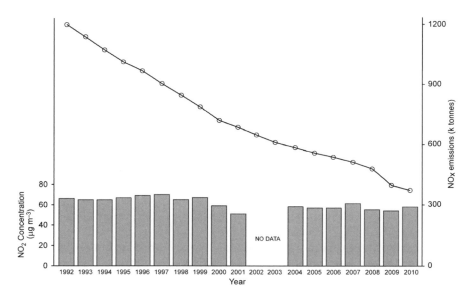

Figure 7.7   Time trend of UK emissions of NO$_x$ from road traffic (solid line, right axis) and concentrations of NO$_2$ in London (bars, left axis) over the same period, 1992–2010.

the period in question, road traffic will undoubtedly have been the main source of NO$_x$ in the atmosphere of London and the apparent lack of relationships between the NO$_x$ emissions and the NO$_2$ concentration can be largely explained by the atmospheric chemistry of nitrogen oxides. NO$_x$ as noted above is emitted predominantly in the form of NO depending upon reaction with ozone to convert it to NO$_2$. This reaction is limited mainly by the availability of ozone, and hence in an oxidant-limited situation, reductions in NO$_x$ emissions lead to a reduction in NO$_x$ concentrations, but an increase in the NO$_2$ to NO ratio. The typical relationship between hourly mean nitrogen dioxide and NO$_x$ appears in Figure 7.8 and it may be seen that over the predominant range of NO$_x$ concentrations (NO$_x$ between 40 ppb (about 76 µg m^{-3}) and 800 ppb), there is little change in NO$_2$

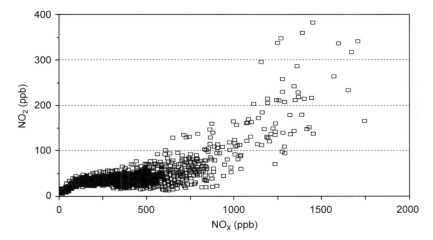

Figure 7.8 Relationship between hourly concentrations of NO_2 and NO_x (*i.e.* the sum of NO and NO_2) in Cromwell Road, London in winter months.

concentration. Thus, the annual mean concentration of NO_2, if it is made up from many hourly concentrations largely within this range is relatively insensitive to changes in NO_x. The implication is that major reductions in NO_x emissions are required to obtain modest reductions in NO_2 and hence to achieve the objectives the National Air Quality Strategy outlined later in this chapter. This problem has been exacerbated by an increased $NO_2 : NO_x$ ratio in the emissions from modern diesel vehicles.

7.2.3.1 Measurement of Oxides of Nitrogen

Analytical instrumentation for measuring oxides of nitrogen have been available for a good number of years. The currently favoured technique for the determination of oxides of nitrogen is based upon the chemiluminescent reaction of nitrogen oxide and ozone to give an electronically excited nitrogen dioxide which emits light in the 600–3000 nm region with a maximum intensity near 1200 nm:

$$NO + O_3 \rightarrow NO_2{}^* + O_2$$

$$NO_2{}^* \rightarrow NO_2 + h\nu$$

In the presence of excess ozone generated within the instrument, the light emission varies linearly with the concentration of nitrogen oxides from 1 ppb to 10^4 ppm. The apparatus is shown schematically in Figure 7.9.

The method is believed free of interference for measurement of NO and may be used to measure NO_x by prior conversion of NO_2 to NO in a heated stainless steel or molybdenum converter. Dependent upon which converter is used, some interference in the NO_x mode is likely from compounds such as peroxyacetyl nitrate and nitric acid. Some instruments incorporate two reaction chambers, one running permanently in the NO mode, the other analysing NO_x after NO_2 to NO conversion. Thus, pseudo-real-time NO_2 concentrations may be measured.

A widely used inexpensive technique for measuring nitrogen dioxide is based upon the use of diffusion tubes. These are straight, hollow tubes of length about 7 cm and diameter 1 cm, sealed at

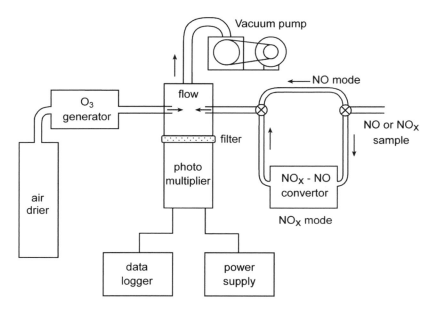

Figure 7.9   Schematic diagram of chemiluminescent analyser for oxides of nitrogen.

one end which is placed upwards, and open at the other end. At the sealed end a metal grid is coated with triethanolamine which acts as a perfect sink for nitrogen dioxide. Access of nitrogen dioxide from ambient air to the triethanolamine is by molecular diffusion along the tube, and analysis of nitrite collected by the triethanolamine reagent and application of Fick's Law, allows calculation of airborne concentrations of nitrogen dioxide. The tubes have a tendency to over-estimate nitrogen dioxide concentrations because wind-induced turbulence in the entry to the tube reduces the effective diffusion length, and due to chemical reactions converting NO to NO_2 within the tube. Nonetheless, diffusion tubes have been used very widely, and particularly in dense networks, to evaluate the spatial distribution of pollutants within an urban area.[13] Nitrogen dioxide has also been mapped across the entire United Kingdom on the basis of a network of diffusion tube samplers.

7.2.4   Carbon Monoxide

As exemplified by the inventories,[1] carbon monoxide is a pollutant very much associated with emissions from petrol vehicles. Within urban areas where concentrations tend to be highest, motor traffic is responsible for most of the emissions of carbon monoxide, and in the UK as a whole, road traffic accounted for 73% of total emissions in 1998, and 47% in 2010.

   The major sink process is conversion to CO_2 by reaction with the hydroxyl radical (see Chapter 8). This process is, however, rather slow and the reduction in CO level away from the source areas is almost entirely a function of atmospheric dilution processes. Carbon monoxide was in the past a problem in heavily trafficked areas especially in confined 'street canyons', but concentrations have reduced sharply since the introduction of catalytic converters.

7.2.4.1   Measurement of Carbon Monoxide

Non-dispersive infrared may be used to measure carbon monoxide in street air where levels encountered normally lie within the range 1–10 ppm. Using a long-path cell, the IR absorbance of polluted air at the wavelength corresponding to the C–O stretching vibration is continuously

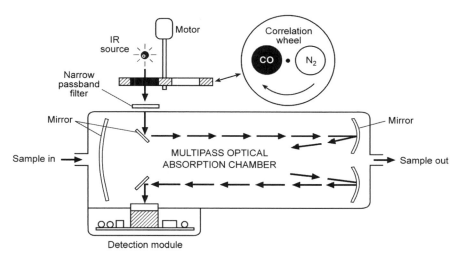

Figure 7.10 Schematic diagram of gas filter correlation analyser for carbon monoxide.

determined relative to that of reference air containing no CO. This is achieved without wavelength dispersion of the IR radiation by using cells containing CO at a reduced pressure as detectors for two beams which are chopped at a frequency of about 10 Hz and passed respectively through sample and reference cells. Absorption of radiation by the CO causes a differential pressure between the two detector cells which is sensed by a flexible diaphragm between them and used to generate an electrical signal. Because of partial overlap of absorption bands, carbon dioxide and water vapour interfere. The latter may be removed by passing the air sample through a drying agent, and the former interference by interposing a cell of carbon dioxide between the sample and reference cells and the detectors.

The other most common type of instrumental analyser is based upon gas filter correlation (see Figure 7.10). Infrared broad band radiation passes sequentially through gas cells containing carbon monoxide and molecular nitrogen contained within a spinning wheel, prior to passage of the radiation through a multi-pass optical cell through which ambient air is drawn. There are thus two beams separated in time but not space, one of which is absorbed by carbon monoxide in the ambient air passing through the sample cell; the other of which is already depleted in the wavelengths absorbed by carbon monoxide and is hence affected only by absorption by components other than carbon monoxide in the sample air. The difference in signal between the two beams is thus the result of absorption by carbon monoxide within the sample cell. Interferences from water vapour and carbon dioxide are claimed to be negligible. The instrument has a minimum detection limit of about 0.1 ppm, and reads up to 50 ppm.

7.2.5 Hydrocarbons

The major sources of volatile organic compounds (which are mainly but not exclusively hydrocarbons) in the UK atmosphere are shown in Figure 7.11. It may be seen that these are rather more diverse than for many of the pollutants. They include natural sources such as release from forest trees, which by convention are not included in the inventory. In urban areas, road transport is probably the major contributor, although use of solvents, for example in paints and adhesives, can be a very significant source. Emissions from road transport include both the evaporation of fuels and the emission of unburned and partially combusted hydrocarbons and their oxidation products from the vehicle exhaust.

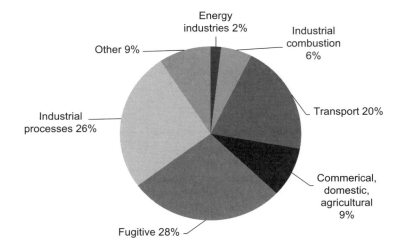

Figure 7.11   UK emissions of non-methane VOC by source sector, 2010.

Many sources emit a range of individual compounds and careful analytical work has shown measurable levels of in excess of 200 hydrocarbons in some ambient air samples. In the UK, the Hydrocarbon Network,[1] which makes automated hourly measurements of volatile organic compounds, reports data on some 25 individual hydrocarbons. Both benzene and 1,3-butadiene are subject to regulation through the UK National Air Quality Strategy (see section 7.4); typical concentrations of benzene appear in Table 7.2. Methane, which is not often measured, far exceeds the other hydrocarbons in concentration. The Northern Hemisphere background of this compound is approximately 1.8 ppm and rising, and elevated levels occur in urban areas as a result particularly of leakage of natural gas from the distribution system.

There are two major reasons for interest in the concentrations of hydrocarbons in the polluted atmosphere. The first is the direct toxicity of some compounds, particularly benzene and 1,3-butadiene, both of which are chemical carcinogens. The second cause of concern regarding hydrocarbons is due to their role as precursors of photochemical ozone (see Chapter 8). Compounds differ greatly in their potential to promote the production of ozone which has led to a system of classifying hydrocarbons according to their photochemical ozone creation potential (see also Chapter 8).

7.2.5.1   Measurement of Hydrocarbons

Determination of specific hydrocarbons in ambient air normally requires a pre-concentration stage in which air is drawn through an adsorbent such as a porous polymer or activated carbon, or a tube where freeze-out of the compounds by reduced temperature occurs, followed by injection into a gas chromatograph, which separates the compounds, which elute sequentially. Detection may be by flame ionization or mass spectrometer (GC–MS), the latter technique allowing a more definite identification of individual compounds. Automated systems are available for gas chromatographic measurements of hydrocarbon on a cyclic basis of about one hour to complete sampling and analysis. The UK Hydrocarbon Network was the first national network to adopt such instrumentation on a routine basis. It is also possible to measure 'total hydrocarbons' by passage of a full air sample to a flame ionization detector. Results are reported as ppb C (parts per billion carbon) since the response of the FID is related closely to the rate of introduction of organic carbon atoms into the flame. Non-methane hydrocarbons may be determined with such instruments by alternate selective removal of hydrocarbons other than methane from the air stream prior to analysis and determination by difference.

7.2.6 Secondary Pollutants: Ozone and Peroxyacetyl Nitrate

7.2.6.1 Ozone

Atmospheric reactions involving oxides of nitrogen and hydrocarbons cause the formation of a wide range of secondary products. The most important of these is ozone. In severe photochemical smogs, such as occur in southern California, levels of ozone have in the past exceeded 400 ppb.

In Europe the classic Los Angeles type of urban smog is not experienced. Nonetheless, the same chemical processes give rise to elevated concentrations of ground-level ozone, often in a regional phenomenon extending over hundreds of kilometres simultaneously. Thus, hydrocarbon and NO_x emissions over wide areas of Europe react in the presence of sunlight causing large scale pollution, which is further extended by atmospheric transport of the ozone.[14] The phenomenon is crucially dependent upon meteorological conditions and hence, in Britain, is observed on only perhaps 10–30 days in each year on average. Concentrations of ozone measured at ground-level can exceed 100 ppb during such 'episodes' and have on one severe occasion been observed to exceed 250 ppb in southern England. These levels may be compared with a background of ozone at ground-level arising from downward diffusion of stratospheric ozone and general tropospheric production of 20–50 ppb. This is seen in Figure 7.12 showing measurement data from southern Scotland.[15]

The UK air quality objective for ozone of 100 µg m^{-3} (about 50 ppb) measured as a rolling eight-hour average[16] is currently exceeded several times each year in most parts of the United Kingdom; the greatest number of exceedances occurring in rural areas of the south-eastern UK. The least frequent exceedances occur in the urban sites where high levels of nitric oxide emissions from road traffic suppress ozone (see Chapter 8). For the rural sites, there is a general gradient in ozone with highest levels in the south and east of Britain and lowest in the north and west. Damage to crop plants can occur at ozone concentrations as low as 40 ppb (this is elaborated on in Chapter 12).

Measurement of Ozone. The UV absorption of ozone at 254 nm may be used for its determination at levels down to 1 ppb. Interferences from other UV-absorbing air pollutants (such as mercury and hydrocarbons) may be minimized by taking two readings. The first reading is of the absorbance of an air sample after catalytic conversion of ozone to oxygen and the second is of an unchanged air sample, the difference in absorbance being due to the ozone content of the air. Instruments are available which perform this procedure automatically and give a read-out in digital form. Although truly continuous measurement of ozone levels is not possible, response is fast and readings may be taken at intervals of less than one minute. The instrument is shown diagrammatically in Figure 7.13.

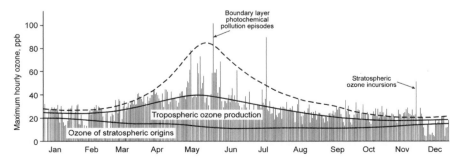

Figure 7.12 Contributions from the major ozone sources to the daily maximum hourly mean observed at ground-level at a rural site.[15]

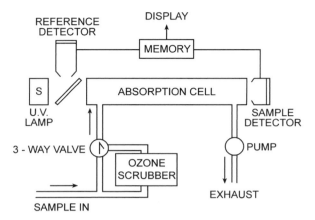

Figure 7.13   Schematic diagram of a UV-photometric analyser for ozone.

7.2.6.2   *Peroxyacetyl Nitrate (PAN)*

PAN is a product of atmospheric photochemical reactions and is a characteristic product of photochemical smog (see Chapter 8).

$$CH_3\text{-}C\text{-}O\text{-}O\text{-}NO_2 \quad PAN$$
$$\overset{\|}{O}$$

   Levels in southern California have been within the range 5–50 ppb on smoggy days. In Europe, the formation is far less favoured and concentrations are more usually well below 10 ppb (maximum hourly average), but few measurements are now made.

7.3   TEMPORAL PATTERNS OF AIRBORNE CONCENTRATION

Figure 7.14 shows the average annual cycle of concentrations of some of the major pollutants at the London North Kensington urban background sampling site.[17] The site is in a suburban area just west of the centre of London without any large local sources such as major roads. The annual cycle of the primary pollutants shows a maximum in the winter months and a minimum in the summer, most clearly visible in the case of CO and NO. These pollutants derive mainly from road traffic which shows little seasonal variation in emissions and the seasonal changes in concentration arise almost wholly from meteorological factors. The atmosphere is better mixed in the summer months of the year leading to better dispersion of pollutants and lower airborne concentrations of primary pollutants emitted at ground level. In the case of nitrogen dioxide, there is a similar but less marked seasonal cycle due to the complex influences of atmospheric chemistry on the formation and removal of nitrogen dioxide (see Chapter 8). Sulfur dioxide has diffuse ground-level sources but in London there is a significant source from power stations to the east of London and this probably explains the much reduced seasonality relative to the traffic-generated pollutants. The particle count represents primarily the ultrafine particles of less than 100 nm diameter which arise mainly from road traffic and have a substantial proportion of semi-volatile particles which evaporate, and do so most effectively at high temperatures; hence the seasonal pattern of particle count mirrors the annual temperature profile.

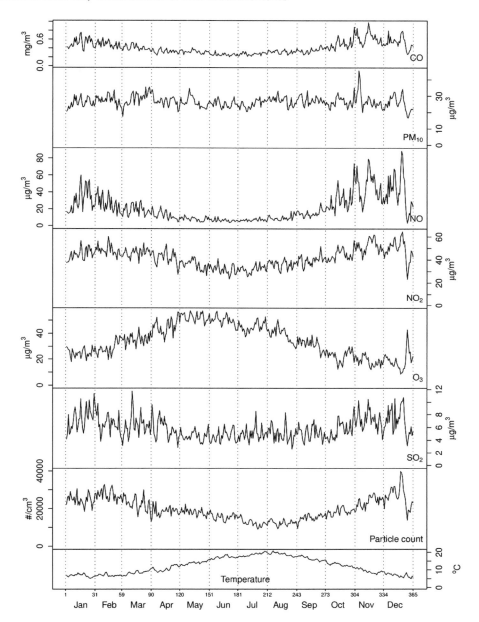

Figure 7.14 Average annual cycle of concentration for 1996–2008 time series of air pollutants (2001–2008 for particle count) in North Kensington, London.

The main source of ozone in London is in rural airmasses crossing the city which in the northern hemisphere typically show a spring maximum, as in the ozone profile at North Kensington. The situation is, however, a little more complex due to the interactions of oxides of nitrogen and the increased mixing depth in the summer months which dilutes the NO_x and brings more ozone into the boundary layer air (see also Chapters 8 and 10).

Figure 7.4 shows average diurnal patterns of the pollutants in summer and winter and for both weekdays (Monday to Saturday) and Sundays derived from the same measurement dataset at

London North Kensington. The traffic-generated pollutants, CO and NO, show a distinct morning rush-hour maximum when high traffic emissions coincide with a relatively poorly mixed atmosphere leading to elevated concentrations. The evening rush hour has a much lesser effect due to a better mixed atmosphere and concentrations tend to increase again overnight as the atmosphere becomes less well mixed (see Chapter 10). There is a secondary peak corresponding to the evening rush hour visible in the winter profiles of CO and NO which is not visible in summer.

Sulfur dioxide has a profile superficially similar to that of the traffic-generated pollutants but the morning peak, especially in winter, is later than that of the traffic-generated pollutants and is likely to be due to the emission plumes from high level sources (mainly power stations) being mixed down to ground level quite early in the day in summer but much later in winter due to poorer atmospheric mixing. The particle count behaves more similarly to the CO and NO but with less pronounced peaks.

Ozone shows an afternoon maximum in summer corresponding to a minimum in nitrogen dioxide which is primarily due to less ozone destruction in the airmasses moving from the rural areas across London. The behaviour of ozone is slightly more complex in winter but very closely anti-correlates with the behaviour of NO and NO_2 indicating that it is the traffic-generated emissions of NO_x which suppress ozone leading to the rush-hour minima observed. It is notable that winter concentrations of ozone are significantly lower than those in summer as was seen in Figure 7.14.

7.4   AIR QUALITY MANAGEMENT

In recent years national and local government authorities have been bringing quite sophisticated approaches to bear on the subject of air quality management. The basic elements of the approach are illustrated in Figure 7.15. The first stage of the process is to carry out an air quality assessment. This entails bringing together the available air quality data, and where necessary, making new measurements. These measurements are then assessed against an agreed set of air quality objectives. These air quality objectives are levels of air pollutant concentration which the strategy is aiming to achieve by a certain target date. If current concentrations exceed the air quality objectives, then model calculations are conducted to predict concentrations in the future target year taking account of measures already in place to reduce emissions within the intervening period. If concentrations in the target year are predicted to exceed the air quality objective, then an air quality management plan is devised and implemented in order to bring future concentrations into compliance with the objectives by the target year.

Air quality objectives are normally derived from air quality standards. An air quality standard is a health-based guideline which is a concentration, which, if achieved, will reduce the adverse effects of air pollution to a level which is zero or negligible at a population level. The latter refers to a concentration which may not be wholly safe for exquisitely sensitive individuals or which may imply a very small incidence of cancers as a result of breathing the pollutant. However, when viewed at the level of a large population, such effects are so small compared to the other risks of life as to be considered negligible. The air quality standards used in the UK which are derived from recommendations of the UK Expert Panel on Air Quality Standards (EPAQS) appear in Table 7.4. The UK National Air Quality Strategy[18] objectives as laid down in the year 2007 appear in Table 7.5a. The objectives are generally determined by what it is practicable to achieve on a given timescale and derive from EPAQS standards or from World Health Organization recommendations incorporated into European Union Limit Values. It will be seen from the tables that some objectives require full achievement of compliance with the air quality standard, whereas others lead only to a percentile compliance thus allowing some periods when the air quality standard is exceeded. Under UK legislation, for the majority of classical air pollutants excluding ozone (which

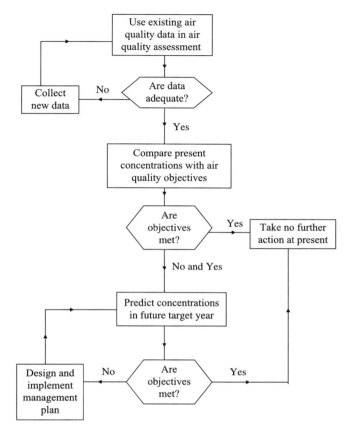

Figure 7.15 Diagram illustrating the Local Air Quality Management process.

Table 7.4 EPAQS recommended air quality standards.

Pollutant	Concentration[a]	Standard Measured as	EPAQS report
Benzene	5 ppb (16.25 µg m^{-3})	running annual mean	1994
1,3-Butadiene	1 ppb (2.25 µg m^{-3})	running annual mean	1994
Carbon monoxide	10 ppm (11.6 mg m^{-3})	running 8 h mean	1994
Lead	0.25 µg m^{-3}	annual mean	1998
Nitrogen dioxide	150 ppb (287 µg m^{-3})	1 h mean	1996
Ozone	50 ppb (100 µg m^{-3})	running 8 h mean	1994
Particles (PM$_{10}$)	50 µg m^{-3}	running 24 h mean	1995
Sulfur dioxide	100 ppb (266 µg m^{-3})	15 min mean	1995

[a]Conversion of ppb to µg m^{-3} and ppm to mg m^{-3} at 20 °C and 1013 mb.

is an international problem) responsibility is placed with local government to implement the objectives of the National Air Quality Strategy within their local boundaries.

Table 7.5b shows proposed new objectives for PM$_{2.5}$ which derive from the *EU Air Quality Framework Directive of 2008*. The latter contains an obligation for exposure reduction in addition to a Limit Value. The two are quite different concepts. The Limit Value is a maximum concentration which should not be exceeded at any of the specified site locations. It therefore forces the

Table 7.5a   Objectives to be included in the *Air Quality Regulations* for the purpose of local air quality management.[20]

Pollutant	Air Quality Objective Concentration ($\mu g\ m^{-3}$)	Measured as	Date to be achieved by
Benzene			
All authorities	16.25	running annual mean	31.12.2003
Authorities in England and Wales only	5.00	annual mean	31.12.2010
Authorities in Scotland and Northern Ireland only	3.25	running annual mean	31.12.2010
1,3 Butadiene	2.25	running annual mean	31.12.2003
Carbon monoxide			
Authorities in England, Wales and Northern Ireland only	10 000	maximum daily running 8 hour mean	31.12.2003
Authorities in Scotland only	10 000	running 8 hour mean	31.12.2003
Lead	0.5	annual mean	31.12.2004
	0.25	annual mean	31.12.2008
Nitrogen dioxide	200 (not to be exceeded > 18 times per year)	24 hour mean	31.12.2005
	40	annual mean	31.12.2005
Particles (PM$_{10}$) (gravimetric)[a]	50 (not to be exceeded > 35 times per year)	24-hour mean	31.12.2004
All Authorities	40	annual mean	31.12.2004
Authorities in Scotland only[b]	50 (not to be exceeded > 7 times per year)	24 hour mean	31.12.2010
	18	annual mean	31.12.2010
Sulfur dioxide	350 (not to be exceeded > 24 times per year)	1 hour mean	31.12.2004
	125 (not to be exceeded > 3 times per year)	24 hour mean	31.12.2004
	266 (not to be exceeded > 35 times per year)	15 minute mean	31.12.2005

[a]Measured using the European gravimetric transfer sampler or equivalent.
[b]These 2010 air quality objectives for PM$_{10}$ apply in Scotland only, as set out in the *Air Quality (Scotland) Amendment Regulations 2002*.

Table 7.5b   Proposed new PM$_{2.5}$ objectives (not included in Regulations).[20]

Region	Air Quality Objective Concentration	Measured as	Date to be achieved by
UK (except Scotland)[a]	25 $\mu g\ m^{-3}$	annual mean	2020
Scotland[a]	12 $\mu g\ m^{-3}$	annual mean	2020
UK urban areas	Target of 15% reduction in concentrations at urban background locations	3 year mean	Between 2010 and 2020

[a]The concentration cap is to be seen in conjunction with the 15% exposure reduction target.

improvement of air quality primarily at hotspot locations. The exposure reduction concept is that for a pollutant such as particulate matter with a linear relationship between exposure and adverse health consequences and no discernible no-effect level, the greatest benefit for public health is achieved by reducing exposure of the entire population rather than focussing on the hotspots with high concentrations. The EU requirement is for the averaging of concentrations across all urban background sampling stations, and a reduction in the three year mean concentration between measurements centred on 2010 and those centred on 2020. The exact level of reduction depends upon the 2010 concentration and is greatest for those countries with the highest concentrations.

7.5 INDOOR AIR QUALITY

Until rather recently, the emphasis in air quality evaluation has centred upon the outdoor environment. Recently, however, it has become clear that exposures to pollutants indoors may be very important (see also Chapter 20). The World Health Organization has published health-based guidelines for indoor air quality.[19]

Pollutants with indoor sources may build up to appreciable levels because of the slowness of air exchange. An example is oxides of nitrogen from gas cookers and flueless gas and kerosene heaters which can readily exceed outdoor concentrations. Kerosene heaters can also be an important source of carbon monoxide and sulfur dioxide. Building materials and furnishings can also release a wide range of pollutants, such as formaldehyde from chipboard and hydrocarbons from paints, cleaners, adhesives, timber and furnishings. The tendency towards lower ventilation levels (energy efficient houses) has tended to exacerbate this problem.

Pollutants with a predominantly outdoor source may be reduced to rather low levels indoors due to the high surface area : volume ratios indoors, leading to extremely efficient dry deposition of pollutants such as ozone and sulfur dioxide.

7.6 INTERNATIONAL PERSPECTIVE

The preceeding text has focussed very heavily upon the situation in the United Kingdom and upon regulations set by the European Union which apply across all EU countries. There are, however, benefits in taking an international perspective on air pollution issues. Figures 7.16 and 7.17 show concentrations of PM_{10} and sulfur dioxide, respectively, measured in cities across the world and demonstrate that, especially in less developed countries, very high air pollutant concentrations still exist. For many pollutants this is because the mitigation measures applied in developed countries are not being applied effectively in less developed countries, allowing excessive emissions of pollutants. An additional factor for particulate matter is that many such countries are in hot, dry regions of the world where wind blown soil and desert dust is a major source of particulate matter, and even in the absence of anthropogenic emissions, high concentrations of particulate matter may well arise. Another important international aspect is that air pollutants do not respect national boundaries and consequently pollutant levels in one country can be heavily affected by emissions in neighbouring countries and even by those in countries without a common boundary. The highest concentrations of ozone and particulate matter, for example, in the United Kingdom often arise in airmasses transported from the European mainland. This phenomenon of long-range transport emphasises the importance of regional and international agreements for the abatement of air pollution.

7.7 APPENDIX

7.7.1 Air Pollutant Concentration Units

Probably the most logical unit of air pollutant concentrations is mass per unit mass, *i.e.* μg kg^{-1} or mg kg^{-1}. This is, however, very rarely used. The commonest units are mass per unit volume (usually μg m^{-3}) or volume per unit volume, otherwise known as a 'volume mixing ratio' (ppm or ppb). For particulate pollutants the volume mixing ratio is inapplicable.

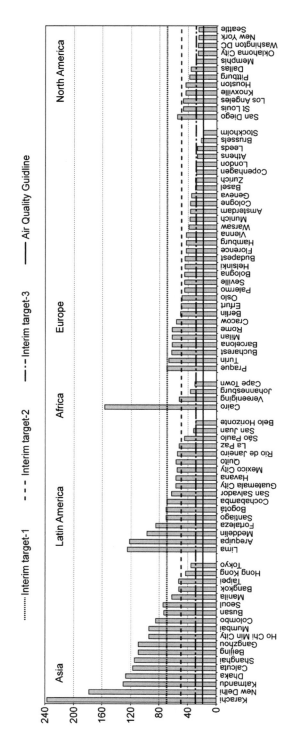

Figure 7.16 Annual average PM_{10} concentrations in selected cities worldwide ($\mu g\ m^{-3}$).

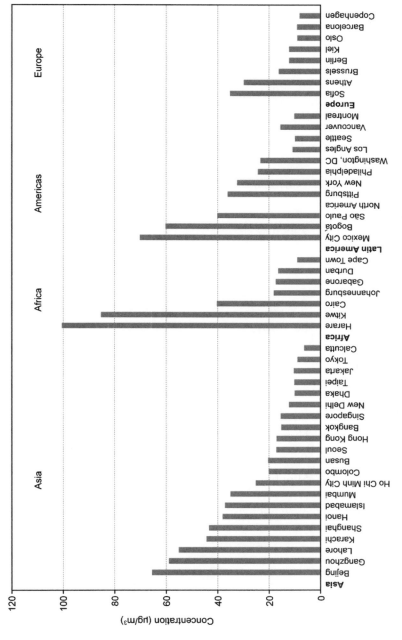

Figure 7.17 Annual average sulfur dioxide concentrations in selected cities worldwide (μg m^{-3}).

Much confusion arises in the interconversion of $\mu g\ m^{-3}$ and ppm. Whilst the volume mixing ratio is independent of temperature and pressure for an ideal gas (and air pollutant behaviour is close to ideal), the mass per unit volume unit is dependent on T and P conditions, and hence these will be taken into account.

7.7.1.1 Example 1

Convert 0.1 ppm nitrogen dioxide to $\mu g\ m^{-3}$ at 20 °C and 750 torr:

46 g NO_2 occupies 22.41 L at STP

46 g NO_2 occupies $22.41 \times \frac{293}{273} \times \frac{760}{750}$ L

$\quad = 24.37$ L at 20 °C and 750 torr

0.1 ppm NO_2 is 10^{-7} L NO_2 in 1 L, or

..... 10^{-4} L NO_2 in 1 m^3

10^{-4} L NO_2 at 20 °C and 750 torr contains $46 \times \frac{10^{-4}}{24.37} g$

$\quad = 189\ \mu g\ NO_2$

$\therefore NO_2$ concentration $= 189\ \mu g\ m^{-3}$

7.7.1.2 Example 2

Convert 100 $\mu g\ m^{-3}$ ozone at 25 °C and 765 torr to ppb:

48 g ozone occupies 22.41 L at STP

48 g ozone occupies $22.41 \times \frac{298}{273} \times \frac{760}{765}$ L

$\quad = 24.30$ l at 25 °C and 765 torr

100 μg ozone occupies $24.30 \times \frac{100 \times 10^{-6}}{48}$ L

$\quad = 50.6 \times 10^{-6}$ L at 25 °C and 765 torr

$\therefore$ Volume mixing ratio $= 50.6 \times 10^{-6}(L) \div 1000(L)$

$\quad\quad = 50.6 \times 10^{-9}$

$\quad\quad = 51$ ppb

(Note that for pressures quoted in millibars, the standard atmosphere, 760 torr $= 1013$ mb $= 1.013 \times 10^5$ Pa)

REFERENCES

1. Defra, UK-Air: Air Information Resource; http://uk-air.defra.gov.uk/ (accessed 13/06/2013).
2. R. M. Harrison and R. Perry, *Handbook of Air Pollution Analysis*, Chapman & Hall, London, 2nd edn, 1986.
3. J. P. Lodge, *Methods of Air Sampling and Analysis*, Lewis Publishers, Chelsea, Michigan, 3rd edn, 1989.
4. R. M. Harrison, H. ApSimon, A. G. Clarke, R. G. Derwent, B. Fisher, J. Hickman, D. Mark, T. Murrells, J. McAughey, F. Poolley, R. Richards, J. Stedman and Y. Vawda, *Source Apportionment of Airborne Particulate Matter in the United Kingdom*, The First Report of the

Airborne Particles Expert Group (APEG), Department of Environment, Transport and the Regions, London, 1999.

5. J. P. Shi, S. Xi, A. Khan, D. Mark, R. Kinnersley and J. Yin, *Philos. Trans. R. Soc. London, Ser. A*, 2000, **358**, 2567.

6. N. A. H. Janssen, G. Hoek, M. Simic-Lawson, P. Fischer, L. van Bree, H. ten Brink, M. Keuken, R. W. Atkinson, H. R. Anderson, B. Brunekreef and F. R. Cassee, *Environ. Health Perspect.*, 2011, **119**, 1691.

7. J. Yin and R. M. Harrison, *Atmos. Environ.*, 2008, **42**, 980.

8. J. Yin, R. M. Harrison, Q. Chen, A. Rutter and J. J. Schauer, *Atmos. Environ.*, 2010, **44**, 841.

9. R. J. Charlson, *J. Air Pollut. Control Assoc.*, 1968, **18**, 652.

10. World Health Organisation, Air Quality Guidelines for Europe, Second Edition, WHO Regional Publications, European Series, No. 91, 2000; http://www.euro.who.int/__data/assets/pdf_file/0005/74732/E71922.pdf (accessed 13/06/2013).

11. Department of Environment, *UK Blood Lead Monitoring Programme, 1984–1987. Results for 1987*, Pollution Report No. 28, HMSO, 1990.

12. R. M. Harrison, J. P. Shi and J. L. Grenfell, *Atmos. Environ.*, 1998, **32**, 2769–2774.

13. Quality of Urban Air Review Group, *Urban Air Quality in the United Kingdom*, QUARG, London, 1993.

14. Photochemical Oxidants Review Group, *Ozone in the United Kingdom*, 4th Report, DETR, London, 1997.

15. R. G. Derwent and P. J. A. Kaye, *Environ. Pollut.*, 1988, **55**, 191.

16. Expert Panel on Air Quality Standards, *Ozone*, HMSO, London, 1994.

17. A. Bigi and R. M. Harrison, *Atmos. Environ.*, 2010, **44**, 2004.

18. Department for Environment, Food and Rural Affairs, *The Air Quality Strategy for England, Scotland, Wales and Northern Ireland*, HMSO, 2007.

19. WHO Health Organization, Regional Office for Europe, *WHO Guidelines for Indoor Air Quality: Selected Pollutants*, Copenhagen, 2010.

20. Department for Environment, Food and Rural Affairs, *Local Air Quality Management*, Technical Guidance LAQM.TG(09), 2009; https://www.gov.uk/government/uploads/system/uploads/attachment_data/file/69334/pb13081-tech-guidance-laqm-tg-09-090218.pdf (accessed 13/06/2013).

CHAPTER 8

Chemistry of the Troposphere

ROY M. HARRISON

School of Geography, Earth and Environmental Sciences, University of Birmingham, Edgbaston, Birmingham B15 2TT, UK
Email: r.m.harrison@bham.ac.uk

8.1 INTRODUCTION

The atmosphere may conveniently be divided into a number of bands reflective of its temperature structure. These are illustrated by Figure 9.1 in Chapter 9. The lowest part, typically about 12 km in depth, is termed the *troposphere* and is characterized by a general diminution of temperature with height. The rate of temperature decrease, termed the *lapse rate*, is typically around 9.8 K km^{-1} close to ground level but may vary appreciably on a short-term basis. The troposphere may be considered in two small components: the part in contact with the earth's surface is termed the *boundary layer* and above it is the *free troposphere*. The boundary layer is normally bounded at its upper extreme by a temperature inversion (a horizontal band in which temperature increases with height) through which little exchange of air can occur with the free troposphere above. The depth of the boundary layer is typically around 100 m at night and 1000 m during the day, although these figures can vary greatly. The processes determining boundary layer height are introduced in Chapter 10. Pollutant emissions are generally into the boundary layer and are mostly constrained within it. Free tropospheric air contains the longer-lived atmospheric components, together with contributions from pollutants which have escaped the boundary layer, and from some downward mixing stratospheric air.

The average composition of the unpolluted atmosphere is given in Table 8.1. Some of the concentrations are uncertain and many are highly variable since:

(i) Analytical procedures for some components have only recently reached the stage where good data can be obtained.
(ii) Some components such as CH_4 and N_2O are known to be increasing in concentration at an appreciable rate.
(iii) It is questionable whether any parts of the atmosphere can be considered entirely free of pollutants.

Pollution: Causes, Effects and Control, 5th Edition
Edited by R M Harrison
© The Royal Society of Chemistry 2014
Published by the Royal Society of Chemistry, www.rsc.org

Table 8.1 Averaging composition of the dry unpolluted atmosphere. (Based upon Seinfeld and Pandis[1] and Brimblecombe).[2]

Gas	Average concentration (ppm)	Approx. residence time
N_2	780 820	10^6 years
O_2	209 450	5000 years
Ar	9340	⎫
Ne	18	⎬ not
Kr	1.1	⎪ cycled
Xe	0.09	⎭
CO_2	390	100 years
CO	0.12 (N. Hemisphere)	65 days
CH_4	1.85	15 years
H_2	0.58	10 years
N_2O	0.325	120 years
O_3	0.01–0.1	100 days
NO/NO_2	$10^{-6}-10^{-2}$	1 day
NH_3	$10^{-4}-10^{-3}$	5 days
SO_2	$10^{-3}-10^{-2}$	10 days
HNO_3	$10^{-5}-10^{-3}$	1 day

Table 8.1 also includes estimates of the lifetime of the various components. Those with a ground-level source and lifetimes of a few days or less are cycled mainly within the boundary layer. Components with longer lifetimes mix into the free troposphere more substantially and those with lifetimes of a year or more will penetrate the stratosphere to a significant degree. An indication of polluted air concentrations of some of these components is given in Chapter 7.

8.1.1 Pollutant Cycles

Pollutants are emitted from *sources* and are removed from the atmosphere by *sinks*. A typical cycle appears in Figure 8.1. Most pollutants have both natural and man-made sources; although the natural source is often of sizeable magnitude in global terms, on a local scale in populated areas pollutant sources are usually predominant.

Sink processes include both dry and wet mechanisms. Dry deposition involves the transfer and removal of gases and particles at land and sea surfaces without the intervention of rain or snow. For gases removed at the surface, dry deposition is driven by a concentration gradient caused by surface depletion; for particles this mechanism operates in parallel with gravitational settling of the large particles. The efficiency of dry deposition is described by the *deposition velocity*, V_g, defined as:

$$V_g\,(\mathrm{m\,s^{-1}}) = \frac{\text{Flux to surface }(\mu g\,m^{-2}s^{-1})}{\text{Atmospheric concentration }(\mu g\,m^{-3})}$$

Some typical values of deposition velocity are given in Table 8.2. For a gas, such as sulfur dioxide which has a fairly high V_g, dry deposition has little influence upon near-source concentrations, but may appreciably influence ambient levels at large downwind distances.

Wet deposition describes scavenging by precipitation (rain, snow, hail, *etc.*) and is made up of two components, *rainout* which describes incorporation within the cloud layer, and *washout* describing scavenging by falling raindrops. The overall efficiency is described by the

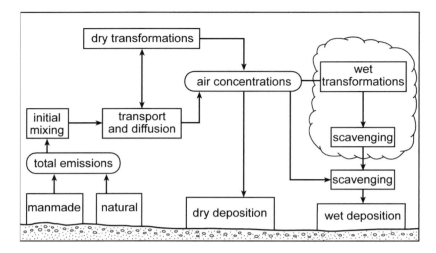

Figure 8.1 Typical atmospheric cycle of a pollutant (reprinted from *Environ. Sci. Technol.*, 1988, **22**, 241, with kind permission of the American Chemical Society).

Table 8.2 Some typical values of deposition velocity and corresponding lifetimes in boundary layers of 100 m and 1000 m in depth.

Pollutant	Surface	Deposition Velocity (cm s⁻¹)	Lifetime (H = 1000 m)	Lifetime (H = 100 m)
SO_2	grass	1.0	28 hours	2.8 hours
SO_2	ocean	0.5	56 hours	5.6 hours
SO_2	soil	0.7	40 hours	4.0 hours
SO_2	forest	2.0	14 hours	1.4 hours
O_3	dry grass	0.5	56 hours	5.6 hours
O_3	wet grass	0.2	5.8 days	13.9 hours
O_3	snow	0.1	11.6 days	27.8 hours
HNO_3	grass	2.0	14 hours	1.4 hours
CO	soil	0.05	23 days	26 hours
Aerosol (<2.5 µm)	grass	0.15	8 days	19 hours

scavenging ratio, W, often rather misleadingly referred to as the *Washout Factor*, and is defined as follows:

$$W = \frac{\text{Concentration in rainwater (mg kg}^{-1})}{\text{Concentration in air (mg kg}^{-1})}$$

Typical values of scavenging ratio are given in Table 8.3. A large value implies efficient scavenging, perhaps resulting from extensive vertical mixing into the cloud layer, where scavenging is most efficient. Alternatively, it may be due to a large particle size (*e.g.* Ca), efficiently collected by falling raindrops. A related deposition process termed *occult* deposition occurs when pollutants are deposited by fogwater deposition on surfaces. Pollutant concentrations in fogwater are typically much greater than in rainwater, hence the process may be significant despite the modest volumes of water deposited.

Another sink process involves chemical conversion of one pollutant to another (termed *dry transformations* as shown in Figure 8.1). Thus atmospheric oxidation to sulfuric acid is a sink for sulfur dioxide. For many pollutants, a major sink is atmospheric reaction with the hydroxyl radical (OH). Such reactions are described later in this chapter. Since pollutants are continually emitted

Table 8.3 Typical scavenging ratios.

Species	W
Cl^-	600
SO_4^{2-}	700
Na	560
K	620
Mg	850
Ca	1890
Cd	390
Pb	320
Zn	870

into and removed from the atmosphere, they have an associated atmospheric lifetime or *residence time* as defined in the following equation.

Many are chemically reactive and are transformed to other chemical species within the atmosphere. In some instances the products of such reactions, termed *secondary* pollutants, are more harmful than the *primary* pollutants from which they are formed. Thus an appreciation of atmospheric chemical processes is fundamental to any attempt to limit the adverse effects of air pollutant emissions.

Table 8.1 includes lifetimes (also termed residence times) of the atmospheric gases. In this context, lifetime, τ, is defined as:

$$\tau = \frac{A}{F}$$

where A = global atmospheric burden (Tg) (1 Tg = 10^{12} g)

F = global flux into and out of the atmosphere (Tg year^{-1})

This treatment assumes a steady-state between the input and removal fluxes. In the case of a pollutant remote from its source, the lifetime corresponds to the time taken to reduce its concentration to $^1/e$ (where e is the base of natural logarithms), or approximately one third of its initial level, if it is not replenished.

Sources of trace gases are not normally evenly spaced over the surface of the globe. Thus spatial variability in airborne concentrations occurs. For gases with an atmospheric lifetime comparable with the timescale of mixing of the entire troposphere (a year, or more), there is little spatial variation in concentration, as mixing processes outweigh the local variability in source strengths. For gases with short lifetimes, atmospheric mixing cannot prevent a substantially variable concentration. High spatial variability will also be associated with high temporal variability at one point as differing air mass sources and different mixing conditions will advect different concentrations of trace gas to the fixed receptor. An example of a rather well mixed gas of long lifetime is carbon dioxide which shows little variation in concentration over the globe and only small fluctuations at a given site (except very close to major combustion sources). At the other extreme, ammonia, which is chemically reactive and subject to efficient dry and wet deposition processes, is highly variable both spatially and temporally. The spatial variability may be described by the coefficient of variation (equal to the standard deviation of the mean concentration divided by the mean) which relates approximately to residence time as shown in Figure 8.2. More recent work on this concept has focussed on the temporal variability of concentrations, which can be explained by the relationship between the relative standard deviation of the natural logarithm of the concentration ($\sigma\ln(X)$) and the atmosphere lifetime, τ:

$$\sigma \ln(X) = A \tau^{-b}$$

where A and b are fitted parameters. Values of b typically range from 0 to 1, with $b = 0$ typifying a measurement point close to sources and $b = 1$ more typical of remote environments.[3,4]

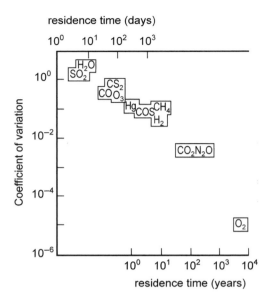

Figure 8.2   The relationship between coefficient of variation in spatially averaged concentration and residence time for atmospheric gases.

A process contributing to the flux of trace gases from the atmosphere is dry deposition. Atmospheric lifetimes with respect to this process are equal to H/V_g, where H is the depth of the mixed boundary layer, and hence lifetimes are shortest when the boundary layer is very shallow. Values of atmospheric lifetime for boundary layer depths of 100 metres and 1000 metres appear in Table 8.2. These values apply only to the lifetime in the boundary layer and therefore tell us nothing about the lifetimes of substances within the free troposphere above, which will be determined by chemical reactions and mixing processes.

Early books on tropospheric chemistry tended to consider the cycle of one substance in isolation from those of others. It is now well recognized that many of the important atmospheric chemical cycles are closely inter-linked and that a more integrated approach to study is appropriate. In this context, the hydroxyl radical has been recognised as having an immensely important role. It is responsible for the breakdown of many atmospheric pollutants, whilst its formation is dependent upon others. It is appropriate to commence a description of tropospheric chemistry with this short-lived free radical species.

8.2   ATMOSPHERIC CHEMICAL TRANSFORMATIONS

8.2.1   The Importance of the Hydroxyl Radical (OH)

The principal source of OH in the background troposphere is photolysis of ozone by light of short wavelengths ($\lambda < 315$ nm) to form singlet (excited state) atomic oxygen $O(^1D)$, which may either relax to the triplet (ground) state $O(^3P)$, or may react with water vapour to form OH:

$$O_3 + hv \rightarrow O(^1D) + O_2 \quad \lambda < 315\,\mathrm{nm} \tag{1}$$

$$O(^1D) + M \rightarrow O(^3P) + M \tag{2}$$

$$O(^1D) + H_2O \rightarrow 2OH \tag{3}$$

(M is an unreactive third molecule such as N_2)

Minor sources are available also through reactions of $O(^1D)$ with CH_4 and H_2:

$$CH_4 + O(^1D) \rightarrow CH_3 + OH \tag{4}$$

$$H_2 + O(^1D) \rightarrow H + OH \tag{5}$$

Photolysis of both HONO and H_2O_2 produces OH directly:

$$HONO + hv \rightarrow OH + NO \quad \lambda < 400\,nm \tag{6}$$

$$H_2O_2 + hv \rightarrow 2OH \qquad \lambda < 360\,nm \tag{7}$$

Formation from nitrous acid (HONO) may be of significance in polluted air. The route from hydrogen peroxide is not likely to represent a net source of OH since the main source of H_2O_2 is from the HO_2 (hydroperoxy) radical:

$$HO_2 + HO_2 \rightarrow H_2O_2 + O_2 \tag{8}$$

In polluted atmospheres, however, HO_2 is able to give rise to OH formation by a more direct route:

$$HO_2 + NO \rightarrow NO_2 + OH \tag{9}$$

A review of tropospheric concentrations of the hydroxyl radical found considerable variations in concentrations estimated by direct spectroscopic measurement, indirect measurement and modelling.[5] There are also genuine variations with latitude, season and the presence of atmospheric pollutants. The overall consensus was of tropospheric concentrations within the ranges $(0.5–5) \times 10^6\,cm^{-3}$ daytime mean and $(0.3–3) \times 10^6\,cm^{-3}$ 24 hr mean. A seasonal variation of about threefold is suggested by model studies.

As the following sections will demonstrate, the hydroxyl radical plays a central role, *via* the peroxy radicals with which it is intimately related, in the production of ozone and hydrogen peroxide. It also itself contributes directly to formation of sulfuric and nitric acids in the atmosphere, as well as indirectly contributing *via* ozone and hydrogen peroxide. Thus atmospheric processes leading to ozone formation will also tend to favour production of other secondary pollutants, including the strong acids HNO_3 and H_2SO_4.

8.3 ATMOSPHERIC OXIDANTS

8.3.1 Formation of Ozone

Mid-latitude northern hemisphere sites show background ozone concentrations typically within the range 20–50 ppb. These concentrations were for many years attributed solely to downward transport of stratospheric ozone and the seasonal fluctuation, with a pronounced spring maximum and broad winter minimum, relating to adjustments in the altitude of the *tropopause* (the boundary between the troposphere and the stratosphere).

In the late 1970s, Fishman and Crutzen[6,7] showed that ozone could be formed from oxidation of methane and carbon monoxide in the troposphere in processes involving the hydroxyl radical:

$$CH_4 + OH \rightarrow CH_3 + H_2O \tag{10}$$

$$CH_3 + O_2 + M \rightarrow CH_3O_2 + M \tag{11}$$

$$CO + OH \rightarrow CO_2 + H \tag{12}$$

$$H + O_2 + M \rightarrow HO_2 + M \tag{13}$$

In the presence of NO:

$$CH_3O_2 + NO \rightarrow CH_3O + NO_2 \tag{14}$$

$$CH_3O + O_2 \rightarrow CH_2O + HO_2 \tag{15}$$

$$HO_2 + NO \rightarrow OH + NO_2 \tag{9}$$

Thus *via* reactions of the peroxy radicals HO_2 and RO_2 (R = alkyl), NO is converted to NO_2. Then:

$$NO_2 + h\nu \rightarrow NO + O(^3P) \quad \lambda < 435 \, nm \tag{16}$$

$$O(^3P) + O_2 + M \rightarrow O_3 + M \tag{17}$$

The magnitude of this source of ozone is presently uncertain. However, there is much evidence to suggest that background northern hemisphere tropospheric ozone concentrations have approximately doubled since the turn of the century.[8] If this is indeed the case, the cause is almost certainly enhanced formation from the oxidation cycles of methane (whose concentration is known to be increasing), other less reactive hydrocarbons and carbon monoxide, involving anthropogenic nitrogen oxides. Such a source is also consistent with a spring maximum in ozone caused by reaction of hydrocarbons accumulated through the less reactive winter months. Figure 8.3 shows near-surface ozone concentrations measured in France and Italy in the nineteenth century compared with more recent data from Germany and Italy. These measurements provide strong evidence of an increase in ground-level ozone by a factor of approximately two over the past hundred years.

In polluted air there is abundant NO_2 whose photolysis leads to ozone formation. However, fresh emissions of NO lead to ozone removal and urban concentrations of ozone are usually lower than those in surrounding rural areas; the full cycle of reactions is:

$$NO_2 + h\nu \xrightarrow{J_1} NO + O(^3P) \quad \lambda < 435 \, nm \tag{16}$$

$$O(^3P) + O_2 + M \rightarrow O_3 + M \tag{17}$$

$$NO + O_3 \xrightarrow{k_3} NO_2 + O_2 \tag{18}$$

where J_1 and k_3 are the rate constants for the NO_2 photolysis and O_3 removal reactions, respectively.

All three reactions are rapid and an equilibrium is reached when the rate of ozone formation (equal to the rate of NO_2 photolysis if all $O(^3P)$ leads to O_3 formation) equals the rate of O_3 removal.

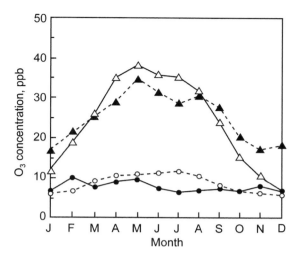

Figure 8.3 Average monthly concentrations of ozone at: ○ Moncalieri near Turin (Italy), 1869–1893; ● at Montsouris (France), 1876–1886; ▲ at Arkona (Germany), 1983; and △ Ispra (Italy), 1986–1989. (Adapted from ref. 8).

Then:

$$J_1[NO_2] = k_3[O_3][NO] \tag{19}$$

and

$$[O_3] = \frac{J_1\,[NO_2]}{k_3\,[NO]} \tag{20}$$

This is termed the *photostationary state*, and thus the ozone concentration is determined by the value of J_1, highest at peak sunlight intensity, and the ratio of NO_2 : NO. Hydrocarbons, and to a lesser extent CO, play a crucial role in producing HO_2 and RO_2 peroxy radicals which convert NO to NO_2 *without* consumption of O_3. For example, propene is attacked by hydroxyl:

$$CH_3CH = CH_2 + OH \rightarrow CH_3CH - CH_2OH \tag{21}$$

$$
CH_3CH - CH_2OH + O_2 \rightarrow CH_3CH - CH_2OH \\
\qquad\qquad\qquad\qquad\qquad\quad | \\
\qquad\qquad\qquad\qquad\qquad OO
\tag{22}
$$

$$
CH_3CH - CH_2OH + NO \rightarrow CH_3CH - CH_2OH + NO_2 \\
\quad\;\; | \qquad\qquad\qquad\qquad\qquad\qquad | \\
\quad\;\, OO \qquad\qquad\qquad\qquad\qquad\quad O
\tag{23}
$$

$$
CH_3CH - CH_2OH \rightarrow CH_2OH + CH_3CHO \\
\quad\;\; | \\
\quad\;\, O
\tag{24}
$$

$$CH_2OH + O_2 \rightarrow HCHO + HO_2 \tag{25}$$

In this case atmospheric photochemistry acts as a source of O_3 and aldehydes. It also leads to formation of HO_2, a source of hydrogen peroxide. None of the above processes is an effective free radical sink and hence, during hours of daylight, reactive free radicals such as OH and HO_2 are constantly recycled.

In remote atmospheres, photochemistry is typically a sink, rather than a source of ozone. The ozone is photolysed to generate hydroxyl radicals [as shown in reactions (1) to (3)], leading to subsequent formation of peroxy radical such as methylperoxy [see reactions (10) and (11)] or hydroperoxy [see reactions (12) and (13)]. Whereas in polluted atmospheres the peroxy radicals convert NO to NO_2 regenerating the OH radical [as shown in reactions (14), (15) and (9)], in clean atmospheres (less than about 20 ppt NO_x) there is insufficient NO and loss of peroxy radicals occurs through such processes as shown in reactions (8), or (26) which destroys ozone:

$$HO_2 + O_3 \rightarrow 2O_2 + OH \tag{26}$$

The apparent implication is that only massive cuts in NO_x emissions will reverse the increase in the northern hemisphere ozone background.

Polluted Atmospheres

In heavily polluted urban areas, subject to photochemical air pollution (*e.g.* Los Angeles), the concentrations of NO, NO_2 O_3, and other secondary pollutants tend to follow characteristic patterns, illustrated in Figure 8.4. Primary pollutants NO and hydrocarbons tend to peak with heavy traffic around 7–8 am. This is followed by a peak in NO_2 some time later when the atmosphere has developed sufficient oxidizing capability to oxidize the NO emissions of motor vehicles. The NO_2 : NO ratio is now high, favouring ozone production, and ozone peaks with peak sunlight intensity around noon, or a little later. In the United Kingdom, such diurnal profiles have been observed in urban areas during summer anticyclonic conditions, but peak ozone concentrations are generally rather later, around 3pm. The situation is more complex however, as in the UK much of the ozone is generated during long-range transport of air pollution, often from continental European sources.

Perhaps surprisingly, UK rural sites exhibit diurnal variations in ozone which are remarkably similar to those at urban sites. Figure 8.5 shows average variations at Stodday, near Lancaster, for various months in 1983. This diurnal change is not due to the same causes as that in urban areas, where fresh NO emissions destroy ozone in the night-time [see reaction (18)]. Measurement of vertical profiles of ozone and temperature (see Figure 8.6) show that nocturnal depletion of ozone is a surface phenomenon due to dry deposition, with little change in airborne concentrations at 50 m

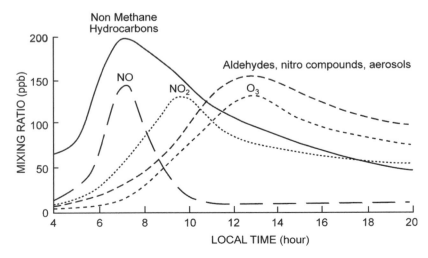

Figure 8.4 Typical mixing ratio profiles, as a function of time of day, in the photochemical smog cycle.

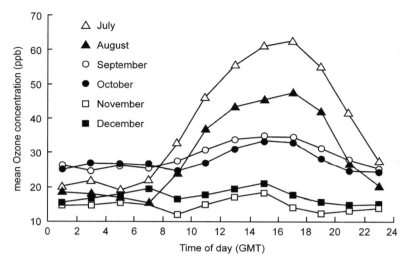

Figure 8.5 Mean diurnal variations of ground-level ozone, July–December, 1983, at Stodday, near Lancaster.[9]

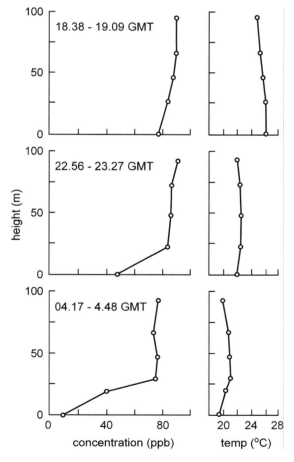

Figure 8.6 Vertical profiles of ozone and temperature on 11–12 July, 1983.[10]

Table 8.4 Percentage of total VOC emissions, POCP and POCP-weighted emissions for selected VOC compounds in northwestern Europe. (Data from ref. 11).

VOCs	Percent of emissions	POCP	Percent × POCP
All VOC	100		2640
Alkanes (total)	42.6		1265
Butane	8.9	31.4	280
Pentane	3.5	39.7	140
2-Methylbutane	3.3	33.7	109
2-Methylpentane	2.6	40.1	105
Heptane	2.0	34.9	70
Alkenes (total)	9.7		761
Ethene	2.9	100	287
Propene	1.2	117	145
Trichloroethene	2.0	28.5	56
Aromatics (total)	15.5		1074
Toluene	3.6	43.9	160
m-Xylene	2.3	85.6	174
1,2,4-Trimethylbenzene	1.4	110	157
3,5-Diethyltoluene	1.7	98	166
Oxygenates (total)	21.1		628
Ethanol	7.5	34.3	257
Formaldehyde	1.7	45.8	78

to 100 m altitude. The temperature profiles show the depletion occurs from a non-turbulent stable layer at the surface indicated by temperature rising with height (termed a *temperature inversion*). During the winter months, little diurnal change in ozone is seen (see Figure 8.5) since the atmospheric temperature structure shows a less marked change between night and day.

As mentioned in Chapter 7, many volatile organic compounds contribute to ozone formation through the chemical reactions exemplified above by methane and propene. Their individual contributions depend upon their concentrations and reactivity, generally with OH. The contributions of individual compounds per unit mass of emissions, termed their *photochemical ozone creation potential* (POCP), can be determined from numerical models of atmospheric photochemistry. Compounds contributing most to ozone formation in the UK atmosphere are listed in Table 8.4.

8.3.2 Formation of PAN

Peroxyacetyl nitrate (PAN) is of interest as a characteristic product of atmospheric photochemistry, as a reservoir of reactive nitrogen in remote atmospheres and because of its effects upon plants. The formation route is *via* acetyl radicals (CH_3CO) formed from a number of routes, most notably acetaldehyde oxidation:

$$CH_3CHO + OH \rightarrow CH_3CO + H_2O \tag{27}$$

$$CH_3CO + O_2 \rightarrow CH_3C(O)OO \tag{28}$$

$$CH_3C(O)OO + NO_2 \rightarrow CH_3C-OONO_2$$
$$\underset{\substack{\| \\ O \\ (PAN)}}{} \tag{29}$$

8.3.3 NO_y Compounds

The term NO_y refers to the sum of all odd nitrogen compounds in the atmosphere (*i.e.* not N_2O):

$$NO_y = \sum NO + NO_2 + HNO_3 + HONO + 2N_2O_5 + PAN$$

It thus contains not only the highly reactive species NO and NO_2 whose role in ozone formation is outlined above, but also products of NO_x oxidation (HNO_3 and N_2O_5; see Section 8.4.4) and reservoir species, HONO and PAN.

Nitrous acid (HONO) is unimportant in clean air, but can play an important role as source of hydroxyl radical through photolysis in polluted atmospheres:

$$HONO + h\nu \rightarrow OH + NO \tag{6}$$

The source of the HONO is currently not fully understood, but it appears to arise mainly from heterogeneous reactions of nitrogen dioxide and water on wet surfaces:

$$NO_2 + H_2O \rightarrow HNO_3 + HONO \tag{30}$$

Its photolysis [see reaction (6)] can be an important source of hydroxyl radical in the early morning within polluted atmospheres due to the build-up of HONO concentrations in the hours of darkness.

Peroxyacetyl nitrate is very important as a reservoir species of NO_x, as are its higher homologues such as peroxypropionyl nitrate (PPN). It is unstable with respect to dissociation [reverse of reaction (29)]:

$$\underset{\underset{O}{\|}}{CH_3C} - OONO_2 \leftrightarrow \underset{\underset{O}{\|}}{CH_3C} - OO + NO_2 \tag{31}$$

Cooler temperatures enhance the stability of PAN compounds and they can comprise a major component of NO_y in polar regions and in the mid and upper troposphere. When moved by the atmosphere to warmer locations, however, they can dissociate releasing NO_2 and hence acting as a reservoir for NO_x.

8.4 ATMOSPHERIC ACIDS

8.4.1 Weak Acids

Well known weak acids in the atmosphere which may contribute to corrosion processes and influence the pH of precipitation are carbon dioxide and sulfur dioxide. Atmospheric CO_2 at a concentration of 340 ppm leads to an equilibrium pH of 5.6 at 15 °C in otherwise unpolluted rainwater. This is normally taken as the boundary pH below which rain is considered acid. Sulfur dioxide is a stronger acid than CO_2 and at a concentration of only 5 ppb in air will, at equilibrium, cause a rainwater pH of 4.6 at 15 °C.[2] In many instances, this pH is not attained due to severe kinetic constraints upon achievement of equilibrium, as is the case with many atmospheric trace gases. Dissolved SO_2 after oxidation may contribute appreciably to total sulfate and acidity in urban rainwater.

Organic acids such as formic acid may be formed in the atmosphere from oxidation of aldehydes, in this case formaldehyde (methanal):

$$HO_2 + HCHO \leftrightarrow HOOCH_2O \leftrightarrow OOCH_2OH \tag{32}$$

$$OOCH_2OH + NO \rightarrow OCH_2OH + NO_2 \tag{33}$$

$$OCH_2OH + O_2 \rightarrow HCOOH + HO_2 \tag{34}$$

Carboxylic acids are believed to contribute significantly to rainwater acidity in remote areas, although their contribution is not at all well quantified. In polluted regions they are unlikely to be of much importance.

8.4.2 Strong Acids

Strong acids of major importance in the atmosphere are as follows:

(i) Sulfuric acid (H_2SO_4).
(ii) Nitric acid (HNO_3).
(iii) Hydrochloric acid (HCl).
(iv) Methanesulfonic acid (CH_3SO_3H).

8.4.3 Sulfuric Acid

Atmospheric oxidation of NO_2 proceeds *via* a range of mechanisms. Consequently, dependent upon the concentrations of the responsible oxidants, the oxidation rate is extremely variable with space and time but typically lies around 1% h^{-1}.

Several gas phase mechanisms have been investigated including photo-oxidation, reaction with hydroxyl radical, Criegee biradical, ground state atomic oxygen, $O(^3P)$, and peroxy radicals. Table 8.5, based upon the treatment of Finlayson-Pitts and Pitts,[12] summarizes the rate data and clearly indicates the overwhelming importance of the hydroxyl radical reaction in the gas phase.

The mechanism of the SO_2–OH reaction is as follows:

$$SO_2 + OH \rightarrow HOSO_2 \tag{35}$$

$$HOSO_2 + O_2 \rightarrow HO_2 + SO_3 \tag{36}$$

$$SO_3 + H_2O \rightarrow H_2SO_4 \tag{37}$$

Table 8.5 Homogeneous oxidation mechanisms for SO_2. (Based on ref. 12).

Oxidizing species	Concentration[a] (cm^{-3})	Rate constant ($cm^3\ molec^{-1}\ s^{-1}$)	Loss of SO_2 (% hr^{-1})
OH	5×10^6	9×10^{-13}	1.6
O_3	2.5×10^{12}	$<8\times10^{-24}$	$<7\times10^{-6}$
Criegee biradical	1×10^6	7×10^{-14}	3×10^{-2}
$O(^3P)$	8×10^4	6×10^{-14}	2×10^{-3}
HO_2	1×10^9	$<1\times10^{-18}$	$<4\times10^{-4}$
RO_2	3×10^9	$<1\times10^{-18}$	$<1\times10^{-3}$

[a]Assuming a moderately polluted atmosphere.

In the presence of water droplets, which can take the form of fogs, clouds, rain or hygroscopic aerosols, sulfur dioxide will dissolve opening the possibility of aqueous phase oxidation. Upon dissolution, the following equilibria operate:

$$SO_{2(g)} + H_2O \leftrightarrow SO_2.H_2O \tag{38}$$

$$SO_2.H_2O \leftrightarrow HSO_3^- + H^+ \tag{39}$$

$$HSO_3^- \leftrightarrow SO_3^{2-} + H^+ \tag{40}$$

These equilibria are sensitive to pH, and HSO_3^- is the predominant species over the range pH 2–7. The other consequence of these equilibria is that the more acidic the droplet, the greater the degree to which the equilibria move towards gaseous SO_2 and limit the concentrations of dissolved S(IV) species. Some rate constants are pH-dependent in addition.

The major proposed oxidation mechanisms for SO_2 in liquid droplets include the following:

(i) Uncatalysed oxidation by O_2.
(ii) Transition metal-catalysed oxidation by O_2.
(iii) Oxidation by dissolved oxides of nitrogen.
(iv) Oxidation by ozone.
(v) Oxidation by H_2O_2 and organic peroxides.

Relative rates for typical specified concentrations of reactive species are indicated in Table 8.6.

At high pH, all mechanisms in Table 8.6 are capable of oxidizing SO_2 at appreciable rates, generally far in excess of those observed in the atmosphere over a significant averaging period. Slower rates are observed, because of mass transfer limitations to the rate of introduction of SO_2 and oxidant to the water droplets and the previously noted effect of pH reduction (due to H_2SO_4 formation) on the solubility of SO_2. At pH 3, only the reaction with hydrogen peroxide is very fast, due to an increased rate constant at reduced pH compensating for the lower solubility of SO_2. Experimental studies indicate that this reaction can be very important in the atmosphere, but is limited by the availability of H_2O_2 which rapidly becomes depleted. The rate of formation of H_2SO_4 is therefore a function more of atmospheric mixing processes than of chemical kinetics in this case.

Several studies have emphasized the importance of SO_2 oxidation upon the surface of carbonaceous aerosols.[13,14] Other studies have shown much slower rates of oxidation,[15] and it remains unclear to what extent this mechanism contributes to the atmospheric oxidation of SO_2. It is likely to be limited rapidly by the aging of the soot particles surfaces.

Table 8.6 Aqueous phase oxidation mechanisms for SO_2. (Adapted from ref. 12).

		Oxidation rate (% hr^{-1})	
Oxidant	*Concentration*[a]	*pH 3*	*pH 6*
O_3	50 ppb (g)	3×10^{-2}	5×10^3
H_2O_2	1 ppb (g)	8×10^2	5×10^2
Fe-catalysed	3×10^{-7} M (l)	2×10^{-2}	5×10^1
Mn-catalysed	3×10^{-8} M (l)	3×10^{-2}	7
HNO_2	1 ppb (g)	5×10^{-4}	3

[a](g) and (l) denote gas and liquid phase concentrations, respectively. Conditions are 5 ppb gas phase SO_2 at 25 °C with no mass transfer limitations.

8.4.4 Nitric Acid

The main daytime route of nitric acid formation is from the reaction:

$$NO_2 + OH \rightarrow HNO_3 \tag{41}$$

The rate constant is 1.1×10^{-11} cm^3 molec^{-1} s^{-1} at 25 °C[12] implying a rate of NO_2 oxidation of 19.8% hr^{-1} at an OH concentration of 5×10^6 cm^{-3} (*cf.* Table 8.5). This is thus a much faster process than gas phase oxidation of SO_2. This process is not operative during hours of darkness due to near zero OH radical concentrations.

At night-time, reactions of the NO_3 radical become important which are not operative during daylight hours due to photolytic breakdown of NO_3. The radical itself is formed as follows:

$$NO_2 + O_3 \rightarrow NO_3 + O_2 \tag{42}$$

It is converted to HNO_3 by two routes. The first is hydrogen abstraction from hydrocarbons or aldehydes:

$$NO_3 + RH \rightarrow HNO_3 + R \tag{43}$$

Typical reaction rates imply formation of HNO_3 at about 0.3 ppb h^{-1} in a polluted urban atmosphere by this route.[12] This is modest compared to daytime formation from NO_2 and OH. The other night-time mechanism of HNO_3 formation is *via* the reaction sequence:

$$NO_3 + NO_2 \overset{M}{\leftrightarrow} N_2O_5 \tag{44}$$

$$N_2O_5 + H_2O \leftrightarrow 2HNO_3 \tag{45}$$

The reaction involving water is rate determining and may be fairly slow at low relative humidity, contributing HNO_3 at ~ 0.3 ppb hr^{-1}.[12] As humidity increases, and especially in the presence of liquid water, more rapid reactions may be observed. This is presently supported only by indirect evidence and requires further experimental investigation.

Aqueous phase oxidation of NO_2 is of little importance due primarily to low aqueous solubility of NO_2.

8.4.5 Hydrochloric Acid

Hydrochloric acid differs from sulfuric and nitric acids in that it is emitted into the atmosphere as a primary pollutant and is not dependent upon atmospheric chemistry for its formation. Coal combustion, if not fitted with flue gas scrubbing is an important source.

Another source which contributes HCl to the atmosphere arises because HCl is a more volatile acid than either H_2SO_4 or HNO_3 and thus may be displaced from aerosol chlorides, such as sea salt:

$$2NaCl + H_2SO_4 \rightarrow Na_2SO_4 + 2HCl \tag{46}$$

$$NaCl + HNO_3 \rightarrow NaNO_3 + HCl \tag{47}$$

This can be an appreciable source of HCl in areas influenced by maritime air masses.

8.4.6 Methanesulfonic Acid (MSA)

This strong acid is unlikely to be of importance in polluted regions but can contribute appreciably to acidity in remote areas. It is a major product of the oxidation of dimethylsulfide, $(CH_3)_2S$. A likely mechanism is:[12]

$$CH_3SCH_3 + OH \xrightarrow{M} \begin{array}{c} CH_3SCH_3 \\ | \\ OH \end{array} \tag{48}$$

$$\begin{array}{c} CH_3SCH_3 \\ | \\ OH \end{array} \rightarrow CH_3SOH + CH_3 \tag{49}$$

$$CH_3SOH + O_2 \xrightarrow{M} CH_3SO_3H \tag{50}$$

At night-time, the main breakdown mechanism of dimethylsulfide is reaction with NO_3 radicals, which can also lead to MSA formation. Dimethylsulfide is a product of biomethylation of sulfur in seawater and coastal marshes and its production shows a strong seasonal pattern, with peak emissions in spring and early summer.

Oxidation of dimethylsulfide can also lead to formation of sulfuric acid, both with and without the involvement of sulfur dioxide as an intermediate. Whether DMS oxidation gives MSA or sulfuric acid, the product is particulate and contributes to the number of cloud condensation nuclei. It has been postulated that biogenic dimethylsulfide may act as a climate regulator through this mechanism.[16]

8.5 ATMOSPHERIC BASES

Carbonate rocks such as calcite or chalk $(CaCO_3)$ or dolomite $(CaCO_3 . MgCO_3)$ exist in small concentrations in atmospheric particles and provide a small capacity for neutralizing atmospheric acidity. In western Europe, however, the major atmospheric base is ammonia. Although not emitted to any major extent by industry or motor vehicles, ammonia is arguably a man-made pollutant as it arises primarily from the decomposition of animal wastes; atmospheric concentrations relate closely to the density of farm animals in the locality. Release from chemical fertilizers can also be significant.[1]

In areas with a moderate or high ammonia source strength, ground-level atmospheric acidity is generally low. Sulfuric acid is present as highly neutralized $(NH_4)_2SO_4$, and HNO_3 and HCl predominantly as NH_4NO_3 and NH_4Cl. The latter two salts are appreciably volatile and may release their parent acids under conditions of low atmospheric ammonia, high temperature or reduced humidity.

$$NH_4NO_3 \leftrightarrow HNO_3 + NH_3 \tag{51}$$

$$NH_4Cl \leftrightarrow HCl + NH_3 \tag{52}$$

The relative neutrality of ground-level air at such locations may, however, not be reflective of far greater acidity at greater heights above the ground, as is shown from simultaneous sampling ground-level air and rainwater (see later). Additionally, once deposited in soils, ammonium salts are slowly oxidized to release strong acid in a process which may be represented as:

$$(NH_4)_2SO_4 + 4O_2 \rightarrow H_2SO_4 + 2HNO_3 + 2H_2O \tag{53}$$

Thus the neutralization process has only a temporary influence and ultimately causes additional acidification.

Deposition of ammonia and ammonium may also contribute directly to damage to vegetation, which has proved to be a particular problem in the Netherlands. Since the reaction of ammonia with the OH radical is slow, the main sinks lie in wet and dry deposition processes.

8.6 ATMOSPHERIC AEROSOLS AND RAINWATER

8.6.1 Atmospheric Particles

A substantial proportion of the atmospheric aerosol over populated areas is termed secondary as it is formed in the atmosphere from chemical reactions affecting primary pollutants. In the UK, water-soluble materials account typically for about 60% of the aerosol mass[17] and comprise nine or ten, major ionic components:

Anions: SO_4^{2-}, NO_3^-, Cl^-, (CO_3^{2-}).
Cations: Na^+, K^+; Mg^{2+}, Ca^{2+}, NH_4^+, H^+.

If concentrations are expressed in gram equivalents per cubic metre of air, some interesting relationships appear:

$$\text{Concentration (g equiv m}^{-3}) = \text{Concentration (g m}^{-3}) \times \frac{Z}{M}$$

where Z = Ionic charge and M = Molecular weight.

This is an expression of charge equivalents and hence if all major components are accounted for:

Σ anions = Σ cations

In marine aerosol:

$$Na^+ + Mg^{2+} \approx Cl^- \tag{54}$$

when expressed in gram equivalents.

This is a relationship very similar to that pertaining in seawater, suggesting the latter as the major source of these components. The aerosol over the oceans also contains sulfate in excess of that present in the sea (termed *non-sea* salt sulfate). This is present as finer particles than the coarse sea salt sulfate, and arises from pollutant inputs of sulfur dioxide together with sulfate formed from the oxidation of dimethyl sulfide (see section 8.4.6). Marine aerosol can penetrate considerable distances inland over continental interiors.

Over the continents, particles derived from the earth's crust (soil and rock fragments) comprise a major part of the aerosol.[18] Chemically these are composed largely of minerals such as clays (complex aluminosilicates), α-quartz and calcite. As well as these crustal particles, the UK atmosphere contains marine-derived particles, secondary aerosol and primary pollutant particles such as road vehicle emissions, comprising largely elemental carbon and organic compounds (see also section 7.2.2).

The inorganic secondary particles are comprised mainly of ammonium nitrate, and sulfate as either ammonium sulfate or sulfuric acid. Hence, in gram equivalents:

$$SO_4^{2-} + NO_3^- \approx NH_4^+ + H^+ \tag{55}$$

In UK air, the ratio $H^+ : NH_4^+$ is normally very low, although in other parts of the world it can be much higher. Close examination of a large data set[17] revealed that the NH_4^+, which is not accounted for by the above relationships, and Cl^-, not accounted for by association with Na^+ and Mg^{2+} in seawater, are approximately equal indicating the presence of NH_4Cl.

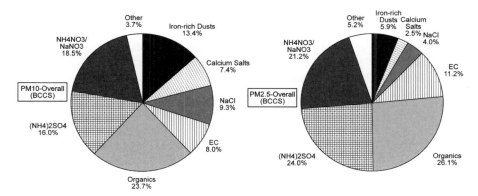

Figure 8.7 Comparison of mean composition of PM_{10} and $PM_{2.5}$ sampled in the central United Kingdom.[19]

K^+ and Ca^{2+} are mainly soil-derived, although some seawater contribution is likely, whilst CO_3^{2-}, not always observed, arises from carbonate minerals such as $CaCO_3$ in rocks and soils. K^+ in fine particles ($PM_{2.5}$) also arises from wood burning.

The major component composition of airborne particles sampled in the UK West Midlands appears in Figure 8.7.[19] It is presented as PM_{10} (left) and $PM_{2.5}$ (right). The composition of $PM_{2.5}$ is simpler, being comprised mainly of ammonium sulfate, ammonium nitrate, elemental carbon (EC) and organic compounds. PM_{10} is more diverse, as it contains a coarse fraction dominated by iron-rich dusts arising mainly from motor traffic (brake, tyre and road surface wear), calcium-rich dusts including soil and building materials generated during construction and demolition, and sea salt. Small amounts of these components appear in the fine fraction ($PM_{2.5}$), but far less than in the coarse particles contributing to PM_{10}.

The main chemical constituents of atmospheric particles are as follows:

(i) *Nitrates:* are the largest single component of airborne particles across much of western Europe and the western United States, and in the UK may represent an exceptionally large contribution to the mass of particles on high pollution days.[19] They form from the oxidation of NO_x to gas phase nitric acid (see section 8.4.4) which can either react with ammonia to form ammonium nitrate [see reaction (51)], or with sodium chloride to form coarser particles of sodium nitrate [see reaction (47)].

(ii) *Sulfates:* are formed from oxidation of sulfur dioxide (see above) and are in decline in most developed countries due to reductions in SO_2 emissions. The initially-formed sulfuric acid is neutralised by ammonia, first to ammonium bisulfate (NH_4HSO_4), and then to ammonium sulfate ((NH_4)$_2SO_4$) depending upon the availability of ammonia, and the formation rate of sulfuric acid. Unlike ammonium nitrate, ammonium sulfate is not appreciably volatile. Sulfates make a major contribution to particulate matter in the eastern United States, and in air masses originating in eastern Europe, where SO_2 abatement has lagged behind that in western Europe.

(iii) *Sodium chloride:* originates from sea salt, and often depleted in chloride, with nitrate combining with the sodium [see reaction (47)].

(iv) *Elemental carbon:* comprises graphitic particles from incomplete combustion, especially diesel engines, and in the past coal burning. Elemental carbon also know as *black carbon*, has major significance as an absorber of visible light, hence acting to warm the lower atmosphere.

(v) *Organic compounds:* are very diverse in composition and include both primary components, many of which accompany elemental carbon emitted from combustion processes, and secondary components arising from oxidation of volatile organic compounds. Hydrocarbons are typical of primary organic particles, while the secondary constituents include functional groups such as alcohols, aldehydes, ketones and carboxylic acids. Many compounds, both primary and secondary, are semi-volatile (see section 7.1) and can transfer between the condensed and vapour phases.

(vi) *Iron-rich dusts:* is a category which describes particles arising from non-exhaust emissions from road traffic. Much of the iron derives from the wear of vehicle brake components,[20] but this is accompanied by tyre and road surface particles, and possibly also dust resuspended from the road surface.

(vii) *Calcium-rich dusts:* have a number of sources, which include wind-blown soil, dust from construction and demolition sites, and possibly some road surface dust, although that is more likely to be included with the iron-rich dust.

Aerosol particles can play a substantial active role in atmospheric chemistry, rather than being simply an inert product of atmospheric processes. They can act as a surface for heterogeneous catalysis or, probably more importantly, can provide a liquid phase reaction medium. Soluble particles deliquesce at humidities well below saturation and many compounds, such as the abundant $(NH_4)_2SO_4$ probably exist predominantly in solution droplet form in humid climates such as that of the UK. As indicated above these droplets can form a medium for reaction of solutes. They can also provide a medium for reaction with gaseous components, for example, the dissolution of nitric acid vapour in sodium chloride aerosol, with release of hydrochloric acid vapour:

$$HNO_3 + NaCl \rightarrow NaNO_3 + HCl \tag{47}$$

One fascinating aspect of aerosol chemistry is the fact that at typical tropospheric temperatures both ammonium nitrate and ammonium chloride aerosols are close to dynamic equilibrium with their gaseous precursors:

$$NH_4NO_3(aerosol) \leftrightarrow NH_3(g) + HNO_3(g) \tag{51}$$

$$NH_4Cl(aerosol) \leftrightarrow NH_3(g) + HCl(g) \tag{52}$$

The position of these equilibria may be predicted from chemical thermodynamics, both for crystalline particles, and at higher humidities for solution droplets.[21] Deviations from equilibrium probably arise from kinetic constraints upon the achievement of equilibrium. The implication for the UK, where ammonia levels are fairly high, is that the ammonia concentration at a given atmospheric temperature and relative humidity controls the nitric acid concentration *via* the ammonium nitrate dissociation equilibrium. It appears also that very similar considerations apply to the concentration of HCl vapour in the air which is subject to control by the ammonium chloride dissociation equilibrium.

Atmospheric aerosols are responsible for the major part of the visibility reduction associated with polluted air. In the UK and other heavily industralized region, visibility reduction tends to correlate closely with the concentration of aerosol sulfate, a component tending to be present in sizes optimum for scattering of visible light. In very dry climates, visibility reduction is primarily associated with dust storms and is thus correlated with airborne soil.

8.6.2 Rainwater

Substances become incorporated in rainwater by a number of mechanisms. The major ones are as follows:

(i) Particles act as cloud condensation nuclei, *i.e.* they act as centres for water condensation in clouds and fogs when the relative humidity reaches saturation.

(ii) Particles are scavenged by cloudwater droplets or falling raindrops as a result of their relative motion.
(iii) Gases may dissolve in water droplets either within or below the cloud.

As mentioned earlier, in-cloud scavenging is referred to as *rainout*, whilst below-cloud processes are termed *washout*. Incorporation in rain is a very efficient means of cleansing the atmosphere as evidenced by the substantial improvements in visibility occasioned by passage of a front.

The dissolved components of rainwater are the same nine or, at high rainwater pH, ten major ions listed earlier for atmospheric aerosol. There are also insoluble materials, again similar to the insoluble components of the atmospheric aerosol. When the composition of rainwater collected in the Lancaster, UK, area is compared with that of aerosols collected over the same periods close to ground-level, the average composition as percentages of total anion or cation load is very similar for many components, but is markedly different for some ions in aerosol and rainwater.[22] The most obvious difference is in the $H^+ : NH_4^+$ ratio; rainwater has a far higher ratio and is thus much more acidic. This arises for two main reasons:

(i) The neutralizing agent, ammonia, has a ground-level source and is thus more abundant at ground-level where the aerosol is sampled than at cloud level where the major pollutant load is incorporated into the rain.
(ii) In-cloud oxidation processes of SO_2 (see above) may lead to appreciable acidification of the cloudwater.

8.6.3 Inter-relationships between Pollutants, Environmental Effects and Impacts

Figure 8.8 shows diagrammatically how the majority of pollutants, through their interactions, contribute to a number of effects. These linkages can be explained as follows:

(i) *Sulfur dioxide:* oxidises to sulfate aerosol in particulate matter which is a climate forcing agent (see Chapter 14); acid sulfates are acidic.
(ii) *Oxides of nitrogen*: contribute to formation of ground-level ozone and nitrate in particulate matter which are both climate forcing agents; oxidation to nitric acid leads to acidification; deposition of oxidised nitrogen is a cause of eutrophication.
(iii) *Carbon monoxide*: influences climate through formation of ozone in the global atmosphere and contributions to local ozone pollution.
(iv) *Ammonia*: is converted to ammonium in atmospheric particulate matter and oxidation of deposited ammonium salts leads to acidification. Deposition of both ammonia and ammonium leads to eutrophication.
(v) *Primary particulate matter*: contributes to climate forcing.
(vi) *Non-methane VOC*: are a contributor to formation of ground-level ozone and to secondary organic matter, both of which are climate forcing agents.
(vii) *Polycyclic aromatic hydrocarbons*: contribute directly to particulate matter.
(viii) *Methane*: is both a climate forcer in its own right and contributes to the formation of ground-level ozone.
(ix) *Heavy metals*: are a constituent of airborne particulate matter.

Climate forcers impact directly upon climate, but also indirectly upon ecosystems. Acidifying and eutrophying substances lead to effects upon ecosystems while ground-level ozone affects both ecosystems and human health. Airborne particulate matter also influences both ecosystems and human health.

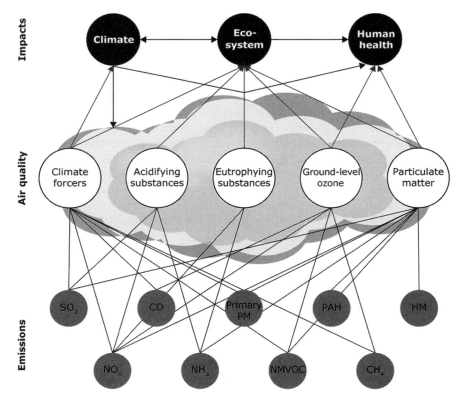

Figure 8.8   Relationship between air pollutants, environmental effects and impacts. From left to right the pollutants shown are as follows: sulfur dioxide (SO$_2$), nitrogen oxides (NO$_x$), carbon monoxide (CO), ammonia (NH$_3$) particulate matter (PM), non-methane volatile organic compounds (NMVOC), polycyclic aromatic hydrocarbons (PAH), methane (CH$_4$) and heavy metals (HM).

REFERENCES

1. J. H. Seinfeld and S. Pandis, *Atmospheric Chemistry and Physics*, Wiley, New York, 1998.
2. P. Brimblecombe, *Air Composition and Chemistry*, Cambridge University Press, Cambridge, 2nd edn., 1996.
3. J. N. Cape, M. Coyle and P. Dumitrean, *Atmos. Environ.*, 2012, **59**, 256–263.
4. T. Karl, P. J. Crutzen, M. Mandl, M. Staudinger, A. Guenther, A. Jordan, R. Fall and W. Lindinger, *Atmos. Environ.*, 2001, **35**, 5287–5300.
5. C. N. Hewitt and R. M. Harrison, *Atmos. Environ.*, 1985, **19**, 545.
6. J. Fishman and P. J. Crutzen, *J. Geophys. Res.*, 1977, **82**, 5897.
7. J. Fishman and P. J. Crutzen, *Nature (London)*, 1978, **274**, 855.
8. D. Anfossi, S. Sandroni and S. Viarengo, *J. Geophys. Res.*, 1991, **96D**, 17349.
9. I. Colbeck and R. M. Harrison, *Atmos. Environ.*, 1985, **19**, 1577.
10. I. Colbeck and R. M. Harrison, *Atmos. Environ.*, 1985, **19**, 1807.
11. J. G. Calvert, R. G. Derwent, J. J. Orlando, G. S. Tyndall and T. J. Wallington, *Mechanisms of Atmospheric Oxiation of the Alkanes*, Oxford University Press, 2008.
12. B. J. Finlayson-Pitts and J. N. Pitts Jr., *Atmospheric Chemistry*, Wiley, New York, 1986.
13. P. Middleton, C. S. Kiang and V. A. Mohnen, *Atmos. Environ.*, 1980, **14**, 463.
14. S. G. Chiang, R. Toosi and T. Novakov, *Atmos. Environ.*, 1981, **15**, 1287.

15. R. M. Harrison and C. A. Pio, *Atmos. Environ.*, 1983, **17**, 1261.
16. R. J. Charlson, J. E. Lovelock, M. O. Andreae and S. G. Warren, *Nature (London)*, 1987, **326**, 655.
17. R. M. Harrison and C. A. Pio, *Environ. Sci Technol.*, 1983, **17**, 169.
18. R. M. Harrison and R. E. van Grieken, *Atmospheric Particles*, Wiley, Chichester, 1998.
19. J. Yin and R. M. Harrison, *Atmos. Environ.*, 2008, **42**, 980–988.
20. J. K. Gietl, R. Lawrence, A. J. Thorpe and R. M. Harrison, *Atmos. Environ.*, 2010, **44**, 141–146.
21. A. G. Allen, R. M. Harrison and J. W. Erisman, *Atmos. Environ.*, 1989, **23**, 1591.
22. R. M. Harrison and C. A. Pio, *Atmos. Environ.*, 1983, **17**, 2539.

CHAPTER 9

Chemistry and Pollution of the Stratosphere

A. ROBERT MACKENZIE* AND FRANCIS D. POPE

School of Geography, Earth & Environmental Sciences, Division of Environmental Health and Risk Management, University of Birmingham, Edgbaston, Birmingham B15 2TT, UK
*Email: A.R.MacKenzie@bham.ac.uk

9.1 INTRODUCTION

The possibility of polluting the stratosphere was first raised in the 1970s. Compounds were identified that were capable of depleting the ozone layer, allowing more ultraviolet (UV) radiation to reach the ground. Because the chemical bonds in biological molecules – in DNA, for example – are broken down by UV radiation, increased UV radiation reaching the ground can affect human health as well as the health of aquatic and terrestrial ecosystems. The ozone-destroying compounds identified included nitrogen oxides, produced by the detonation of nuclear weapons or from the exhaust of supersonic aircraft, and chlorofluorocarbons (CFCs), which were used mainly as refrigerants and aerosol propellants.

Public and scientific concern over ozone escalated abruptly in 1985, when Farman, Gardiner and Shanklin[1] reported that the total ozone column over Antarctica in September and October had decreased by up to 40% over the past decade. No one had predicted such a dramatic change. The discovery of this 'ozone hole' over Antarctica precipitated a great deal of research on the chemistry and fluid dynamics of the stratosphere in a variety of experiments involving satellites, aircraft, balloons and ground stations. The research has shown, amongst other important results, clear evidence that ozone destruction has occurred within the lower stratosphere over the Arctic too.[2]

For reasons outlined below, Arctic ozone depletion is much more variable than Antarctic ozone depletion. Some northern winters have resulted in ozone depletion as severe – in terms of the amount of ozone destroyed relative to the unperturbed background concentrations – as Antarctic winters of the mid-1980s, whilst other northern winters have resulted in almost no depletion. Variability notwithstanding, the fragility of the ozone layer has been amply demonstrated.

Ozone, and in particular its absorption of solar UV, has a profound impact on the structure of the atmosphere. Figure 9.1 shows a typical mid-latitude temperature profile. Most ozone occurs in

Pollution: Causes, Effects and Control, 5th Edition
Edited by R M Harrison
© MacKenzie and Pope and The Royal Society of Chemistry 2014
Published by the Royal Society of Chemistry, www.rsc.org

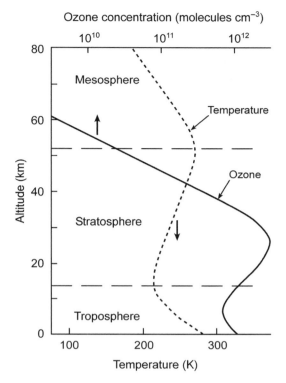

Figure 9.1 Typical ozone and temperature profiles in the mid-latitude atmosphere.

the stratosphere,[†] and the presence of ozone contributes to the characteristic temperature inversion and dynamic stability (*i.e.* hotter, lighter air above colder, denser air) of the stratosphere. The stratosphere extends from a lower limit of between 8 km and 16 km, depending on latitude, to an upper limit of between 45 km and 60 km, again depending on latitude.

A meridional (*i.e.* north–south) cross-section of the circulation in the wintertime stratosphere is shown schematically in Figure 9.2. Although most wind is westerly in winter and easterly in summer in the extra-tropical stratosphere, there is a slow overturning, caused by the interaction of planetary-scale internal waves on the mean stratospheric flow.

The meridional circulation is superimposed on rapid zonal (*i.e.* east–west) flow: winds are westerly in the wintertime extra-tropics, easterly in the summertime extra-tropics, and switch between easterly and westerly in the tropics roughly every two years. The slow overturning of the wintertime stratosphere brings air upwards, from the troposphere, in the tropics and downwards, from the mesosphere and upper stratosphere, at the pole. At a given altitude, therefore, air is 'older', in the sense that it has been in the upper atmosphere for longer, near the winter pole and air is 'younger' near the equator. Since CFCs are destroyed in the upper stratosphere and mesosphere, air near the winter pole is richer in the breakdown products of CFCs than is air near the equator. This gradient in the 'age' of stratospheric air is particularly steep in the subtropics and at the edge of the polar

---

[†]Three measures of local abundance are used in this chapter. Absolute abundances are reported as number densities, in molecules cm^{-3}, or as partial pressures, P_x, in nbar. These two concentration scales are related by the ideal gas law. Relative abundances are reported as volume mixing ratios (vmr), P_x/P_{air}, expressed as parts per million (ppmv, *i.e.* vmr$\times 10^6$), parts per billion (ppbv, *i.e.* vmr$\times 10^9$), or parts per trillion (pptv, *i.e.* vmr$\times 10^{12}$), as also used in Chapters 7 and 8. Because pressure decreases exponentially with height, the absolute concentration of chemicals in air parcels moving upwards or downwards will change, but the relative abundance will be conserved in the absence of chemical reactions.

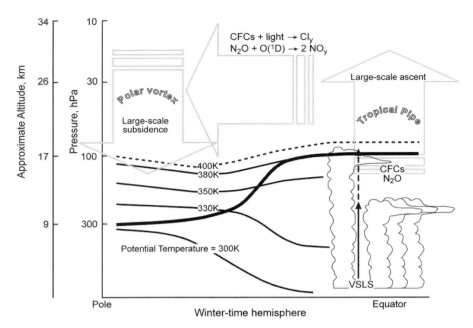

Figure 9.2 Schematic of the mean diabatic ("Brewer–Dobson") circulation in the wintertime hemisphere (broad arrows). Source gases enter the stratosphere in the tropics and are transported slowly to high altitudes where they are broken down by short wavelength light or the electronically excited $O(^1D)$ atom. Descent of air into the polar region brings down high mixing ratios of NO_y, Cl_y and H_2O (see main text for an explanation of the chemical formulae). Clouds in the polar vortex activate chlorine and so bring about ozone depletion. A strong westerly wind at the edge of the polar vortex prevents rapid latitudinal mixing. A similar barrier to mixing exists in the sub-tropics. The tropopause is shown by the thick line. Thin lines are isentropic surfaces. VSLS = very short-lived substances (see section 9.3, below). Vetical arrows denote the rapid upward transport of VSLS by tropical convection.

vortex, where barriers to horizontal mixing exist. The barrier at the polar vortex edge tends to contain the most dramatic ozone loss to within the polar lower stratosphere. Between the mixing barriers, in the mid-latitude 'surf zone', stirring and mixing, on isentropic surfaces,[‡] is rapid.

9.2   STRATOSPHERIC OZONE CHEMISTRY

Ozone occurs in trace amounts throughout the atmosphere but concentration peaks in the lower stratosphere between about 20 and 25 km altitude (see Figure 9.1). About 90% of all the ozone in the atmosphere resides in the stratosphere. Observations of the total amount of ozone in a vertical column have established the average column abundance and the general pattern of latitudinal, longitudinal, seasonal and meteorological variations.[§] Total or column ozone, per unit area, is

---

[‡] The stable stratification of the stratosphere is reflected in the fact that air tends to move on isentropic surfaces, *i.e.* surfaces of constant potential temperature. Potential temperature, θ, is defined as $\ln \theta = \ln T + (R/C_p)(p_0/p)$, where T is the temperature in Kelvin, p the atmospheric pressure (in hPa), at the altitude of interest, p_0 is a standard pressure surface (often taken as 1000 hPa), R is the specific gas constant of dry air and C_p is the specific heat capacity at constant pressure of dry air. θ is a measure of the entropy of air and so is conserved in adiabatic motion.

[§] It may appear strange, at first glance, to rely on an integrated measure of a chemical compound, but bear in mind that most of the ozone occurs in a layer, and that the UV-shielding effect of ozone relies on the total number of ozone molecules in the path sunlight takes to the Earth's surface. For both these reasons, integrated measures of ozone are meaningful and extremely useful.

recorded in Dobson Units (DU) or matm cm ('milli-atmo centimetres'), related to the thickness of an equivalent layer of pure ozone at standard pressure and temperature. If the ozone in a column of the atmosphere were concentrated into a thin shell surrounding the Earth at a standard pressure (typically 1000 hPa) it would be about 3 mm thick, which translates into an average total amount of ozone of 300 DU. Seasonal and latitudinal variations at sub-polar latitudes are about 20% of this value. The annual average total ozone is a minimum of approximately 260 DU at equatorial latitudes, and increases poleward in both hemispheres, to a maximum at sub-polar latitudes of about 400 DU. The high latitude maximum results from transport of ozone from the equatorial middle and upper stratosphere, the region of ozone production, to the polar lower stratosphere (see Figure 9.2).

9.2.1 Gas-Phase Chemistry

How is this ozone produced? Chapman[4] proposed a pure oxygen steady-state model that agreed well with the observations available at the time. The reactions were:

$$O_2 + h\upsilon \; (\lambda < 243 \, \text{nm}) \rightarrow O + O \tag{1}$$

$$O + O_2 + M \rightarrow O_3 + M \tag{2}$$

$$O_3 + h\upsilon \; (\lambda < 1180 \, \text{nm}) \rightarrow O + O_2 \tag{3}$$

$$O + O_3 \rightarrow O_2 + O_2 \tag{4}$$

(a fifth reaction, the self-reaction of oxygen atoms, is too slow to be important in the stratosphere). Reaction (2) becomes slower with increasing altitude, since the concentration of M which is any third body, *i.e.* pressure, is decreasing, while reactions (1) and (3) become faster, since the intensity of radiation, particularly short wavelength radiation, is increasing. Hence, O predominates at high altitudes and O_3 at lower altitudes. These two allotropes of oxygen (O and O_3) are converted one into the other very rapidly, so that it is convenient to consider them together using the term 'odd oxygen'. The rate of the loss of odd oxygen in reaction (4) was unknown at the time of Chapman's proposal but is now found to be too slow to destroy ozone at the rate it is produced globally. Therefore, for ozone to attain a steady state close to the observations, chemical processes must exist that provide more efficient routes for the loss of odd oxygen. The reaction sequence:

$$X + O_3 \rightarrow XO + O_2 \tag{5}$$

$$XO + O \rightarrow X + O_2 \tag{6}$$

$$\text{net} \; O + O_3 \rightarrow 2O_2$$

(*i.e.* two odd-oxygen species → two even-oxygen species)

achieves the same result as reaction (4). The species X is a catalyst – it is not consumed in the process – and can effectively destroy many ozone molecules before being removed by some other chemical process. Several chemical species have been identified with X, including H, OH, NO, Cl and Br. The catalytic cycles are very fast and are able to compete with the direct reaction even when concentrations of X and XO are 100 or 1000 times less than concentrations of the oxygen species. Because the interconversion of the various X/XO couples can be very rapid, it is often convenient to group the species together in chemical families. The various short-lived nitrogen compounds are grouped together as NO_x (*i.e.* $N + NO + NO_2 + NO_3$, where abundance is counted by number, *i.e.* moles, not mass), the various short-lived chlorine species are grouped together as ClO_x ($Cl + ClO + 2ClOOCl$), short-lived hydrogen species are grouped as HO_x ($H + HO + HO_2$), and so on. By extension, the total amount of potentially reactive nitrogen** is known collectively as NO_y (NO_x + HONO + $2N_2O_5$ + $HONO_2$ + HO_2NO_2 + $ClONO_2$), and the total amount of potentially reactive chlorine, the inorganic chlorine, is known collectively as Cl_y (ClO_x + HOCl + HCl + $ClONO_2$ + $2Cl_2$ + BrCl).

In addition to reactions (5) and (6), ozone may also be destroyed in cycles not involving atomic oxygen. Such ozone-specific cycles are important in the lower stratosphere because concentrations of atomic oxygen are low there. Variations on the general form are:

(i) Formation of an XO compound which itself reacts with ozone, for example:

$$OH + O_3 \rightarrow HO_2 + O_2 \tag{7}$$

$$HO_2 + O_3 \rightarrow OH + 2O_2 \tag{8}$$

$$net\ 2O_3 \rightarrow 3O_2$$

and:

$$NO + O_3 \rightarrow NO_2 + O_2 \tag{9}$$

$$NO_2 + O_3 \rightarrow NO_3 + O_2 \tag{10}$$

$$NO_3 + hv \rightarrow NO + O_2 \tag{11}$$

$$net\ 2O_3 + hv \rightarrow 3O_2$$

(ii) Formation of a compound from two XO species, leading to the recombination of oxygen *via*:
    (a) Elimination of O_2 in a thermal reaction:

$$Br + O_3 \rightarrow BrO + O_2 \tag{12}$$

$$OH + O_3 \rightarrow HO_2 + O_2 \tag{7}$$

$$HO_2 + BrO \rightarrow HOBr + O_2 \tag{13}$$

$$HOBr + hv \rightarrow OH + Br \tag{14}$$

$$net\ 2O_3 + hv \rightarrow 3O_2$$

    (b) Elimination of O_2 by photolysis of a different bond to the bond formed in the XO/YO combination:

$$Cl + O_3 \rightarrow ClO + O_2 \tag{15}$$

$$NO + O_3 \rightarrow NO_2 + O_2 \tag{9}$$

$$ClO + NO_2 + M \rightarrow ClONO_2 + M \tag{16}$$

$$ClONO_2 + M \rightarrow Cl + NO_3 \tag{17}$$

$$NO_3 + hv \rightarrow NO + O_2 \tag{11}$$

$$net\ 2O_3 + 2\,hv \rightarrow 3O_2$$

** In the definition of NO_y, $HONO_2$ is another way of writing HNO_3 (nitric acid) and is used here so that you can see the structural connection with pernitric acid (HO_2NO_2). In the definition of Cl_y, BrCl is the cross-halogen analogue of the standard elemental forms Cl_2 and Br_2.

A very important example of this last kind of ozone-specific cycle is the 'dimer cycle', which is responsible for the majority of the lower stratospheric ozone loss inside the polar vortices:

$$2(Cl + O_3 \rightarrow ClO + O_2) \tag{15}$$

$$ClO + ClO + M \rightarrow ClOOCl + M \tag{18}$$

$$ClOOCl + h\upsilon \rightarrow Cl + ClOO \tag{19}$$

$$ClOO + M \rightarrow Cl + O_2 \tag{20}$$

$$\text{net } 2O_3 + h\upsilon \rightarrow 3O_2$$

As well as these ozone-destroying catalytic cycles, X and XO species may be involved in null cycles that do not remove odd oxygen. For $X = NO$, we have:

$$NO + O_3 \rightarrow NO_2 + O_2 \tag{9}$$

$$NO_2 + h\upsilon \rightarrow NO + O \tag{21}$$

$$\text{net } O_3 + h\upsilon \rightarrow O_2 + O$$

(*i.e.* one odd-oxygen species → one even-oxygen species and one odd-oxygen species)

This cycle is in competition with the catalytic cycle [see reactions (9), (10) and (11)] and the NO_x tied up in this cycle is ineffective as a catalyst. Together with reaction (2), reactions (9) and (21) make up the cycle that defines the tropospheric photostationary state for ozone (see Chapter 8). The corresponding photostationary state for stratospheric ozone is more complicated, because of the X/XO cycles.

The termination of the catalytic and null cycles occurs by chemical conversion of radicals into more stable oxidation products. For example:

$$OH + NO_2 + M \rightarrow HNO_3 + M \tag{22}$$

The nitric acid (written chemically as HNO_3 or $HONO_2$) may be transported down into the troposphere and removed in rain or may be photolysed to regenerate OH and NO_2. This latter process is relatively slow, and HNO_3 is said to act as a *reservoir* of NO_x. Typically, in the lower stratosphere, more than 90% of the stratospheric load of NO_y is stored in this HNO_3 reservoir.[3] For ClO_x, about 40–80% of the stratospheric load is sequestered into hydrochloric acid (HCl),[4] *via*

$$Cl + CH_4 \rightarrow CH_3 + HCl \tag{23}$$

The chlorine tied up as stable HCl may be released by

$$OH + HCl \rightarrow H_2O + Cl \tag{24}$$

or by the heterogeneous reactions shown below. This ClO_x can then again participate in catalytic cycles leading to the removal of ozone. More temporary reservoir species such as $ClONO_2$, HOCl, N_2O_5 and HO_2NO_2, act to lessen the efficiency of ClO_x and NO_x species in destroying ozone in the lower stratosphere. These temporary reservoir species may be formed by the combination of two oxy-radicals (which requires the presence of a third body, M, to remove excess energy) or elimination of stable bimolecular oxygen in the reaction of two oxy-radicals, *i.e.* reaction (16) and:

$$NO_3 + NO_2 + M \rightarrow N_2O_5 + M \tag{25}$$

$$HO_2 + NO_2 + M \rightarrow HO_2NO_2 + M \tag{26}$$

$$HO_2 + ClO \rightarrow HOCl + O_2 \tag{27}$$

These reactions emphasize the coupling between the HO_x, NO_x and ClO_x families, and indicate that the effects of the families on ozone concentrations are not separable or additive. Members of one family can react with members of another to produce null cycles, chiefly by the production of NO_2, *e.g.*

$$NO + ClO \rightarrow NO_2 + Cl \tag{28}$$

$$NO + HO_2 \rightarrow NO_2 + OH \tag{29}$$

followed by reaction (21). The recycling of HO_x, NO_x and ClO_x from the reservoirs HNO_3, N_2O_5, HOCl, and $ClONO_2$ is by photolysis or heterogeneous reaction. Removal of odd hydrogen in the lower stratosphere occurs mainly by formation of water:

$$OH + HNO_3 \rightarrow H_2O + NO_3 \tag{30}$$

$$OH + HO_2NO_2 \rightarrow H_2O + NO_2 + O_2 \tag{31}$$

These reactions also release NO_x from its reservoirs. The water formed may be physically removed from the stratosphere, or converted back into HO_x by reaction with electronically excited O atoms.

Termination reactions become less efficient with increasing molecular weight in the halogen series, *i.e.* from Cl to I. The reaction of Cl with CH_4 [see reaction (23)], to produce HCl, limits the abundance of active chlorine in the stratosphere, but the analogous reaction of Br with CH_4 is endothermic (*i.e.* not energetically favourable) and can be neglected. The temporary reservoirs for bromine – HOBr and $BrONO_2$ – are very readily photolysed. As a result, BrO is the major form of Br_y in the stratosphere and so, on a molecule-for-molecule basis, bromine is about 50 times more efficient than chlorine at destroying O_3. Measurements of BrO show mixing ratios of around 5 pptv at 18 km. These are sufficient to account for about 25% of the ozone depletion in the Antarctic ozone hole and to make bromine cycles significant contributors to mid-latitude ozone loss.[5] Reactions involving iodine radicals have also been suggested, but current upper limits to the concentration of IO in the stratosphere, and measurements of the relevant chemical reaction rates, suggest that iodine plays a very minor role in stratospheric chemistry.[6] Figure 9.3 shows a model calculation of the ozone loss rate due to the various chemical families (HO_x, NO_x, ClO_x, BrO_x) as a function of altitude, for vernal equinox at 38° N.[7]

Hydrocarbon oxidation is closely related to all other reactive trace gas species and hence ozone photochemistry. Methane is the dominant stratospheric hydrocarbon and is a chemical source of water in the stratosphere, *via*:

$$OH + CH_4 \rightarrow CH_3 + H_2O \tag{32}$$

which is analogous to the Cl-loss process [see reaction (23)]. The CH_3 radical is further oxidized to CO_2 and another molecule of H_2O through intermediate products CH_3O_2, CH_3O, HCHO and CO. Below 35 km there is sufficient NO for CH_4 oxidation to be a net source of O_x ($O + O_3$) by the 'photochemical smog' reactions (see Chapter 8).

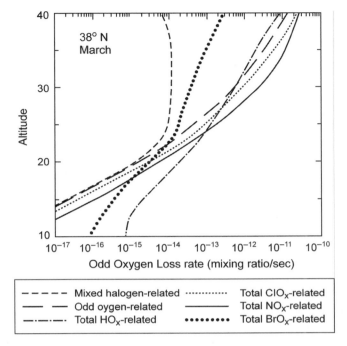

Figure 9.3 Calculated 24-hour average odd-oxygen loss rates from various chemical cycles for 38°N in March for non-volcanic aerosol loading. In these circumstances, the dominant odd-oxygen loss process below 22 km is due to reactions involving HO_x, with NO_x dominating between 23 km and 40 km. Under higher aerosol loading, coupled HO_x-halogen cycles become more important. Re-drawn from the two-dimensional model simulations of Garcia and Solomon.[7]

9.2.2 Heterogeneous Chemistry

Aerosol particles provide sites for reactions that would otherwise not occur in the stratosphere. There are many types of aerosol in the stratosphere, ranging from meteoritic mineral debris to water droplets.[8] Of primary importance in our current understanding of the chemistry of the stratosphere are sulfuric acid aerosol particles and polar stratospheric clouds (PSCs).

Sulfuric acid aerosol occurs globally, in a wide altitude band between about 15 and 30 km. The aerosol is formed primarily by oxidation of SO_2 injected by volcanic eruptions, and its formation is therefore highly intermittent. An additional small but constant source is the percolation of COS, emitted by microbes, up to the stratosphere, with subsequent oxidation to sulfuric acid. The eruption of Mt. Pinatubo in June 1991 increased the mass of aerosol in the stratosphere by more than an order of magnitude [from a total loading of around 2 Tg (*i.e.* 2 Mtonnes) to around 30 Tg]. The timescale for removal of the aerosol is of the order of one year,[9] so that these volcanic injections can have a significant global effect (see below). Above temperatures of about 215 K, at 50 hPa, the aerosol exists as concentrated sulfuric acid droplets. Below this temperature solid sulfuric acid hydrates are the thermo- dynamically favoured phase, but there is a considerable kinetic barrier to freezing which enables liquid particles to remain until the temperature approaches the frost point,[10] taking up water and nitric acid as they cool. The frost point for stratospheric air at 50 hPa is about 188 K (−85 °C).

Polar stratospheric clouds fall into two categories: type 1 PSCs are composed of nitric acid and water in volume ratios between 1 : 2 and 1 : 5, and form at temperatures about 2–5 K above the local frost point; type 2 PSCs are composed of water ice crystals and form, therefore, below the local frost point. PSCs evolve from the background sulfuric acid aerosol (at PSC temperatures, the sulfuric acid seed or 'core' makes a very small contribution to the overall composition of the PSC particle). The details of particle evolution in the lower stratosphere matter because the different particles have different reactivities. The thermodynamically favoured solid phase for sulfuric acid aerosol is sulfuric acid tetrahydrate (SAT), and that for nitric acid aerosol is nitric acid trihydrate (NAT).

Reactions involving particles are more complicated to describe than the simple molecular collisions occurring in the gas phase. Diffusion is important, and reactants partition between the gas and condensed phases according to their solubilities. In liquid particles, if reaction is slow, the reactants will have time to continuously adjust to their equilibrium partitioning. If the rate coefficient for reaction in/on the aerosol particles is faster than about 10^5 M^{-1} s^{-1}, then the rate of reaction becomes limited by transport of reactants into the particle and the effective volume available for reaction is reduced.[11] For very fast reactions the rate is dependent on the aerosol surface area rather than the volume. For solid particles, reaction rates depend linearly on the particle surface area (but then, of course, surface area is not as easily measured or calculated as for spherical liquid drops).

Table 9.1 lists some of the most important reactions occurring on sulfuric acid aerosol and PSC particles. The overall effect of heterogeneous reactions is to convert the chlorine and hydrogen in reservoir compounds into more active forms, whilst converting the NO_x in temporary reservoir compounds into nitric acid (the conversion of NO_x to less reactive forms of NO_y is often called *denoxification*). Many of the heterogeneous reactions are strongly temperature dependent, due to the variation of aerosol composition with temperature. For aerosol loadings typical of a volcanically quiescent period, heterogeneous reactions begin to activate chlorine at temperatures below about 200 K. This activation is relatively slow, however. When aerosol loadings are much increased following a volcanic eruption, the temperature at which chlorine activation begins is increased by up to 5 K and aerosol reactions can effectively compete with reactions on type 1 PSCs. One heterogeneous reaction that is not especially temperature dependent is the hydrolysis of N_2O_5.

In the polar regions, when temperatures fall below about 195 K, PSCs can form. These have a much larger total volume than the background aerosol, and so can convert chlorine reservoirs much more rapidly. At PSC temperatures, Cl_2 partitions into the gas phase, whereas HNO_3 remains in the particles. The conversion of chlorine reservoirs is known as *chlorine activation*. If the temperatures remain cold, the condensed phase can sediment out, leading to irreversible *denitrification* and *dehydration* of the air. Denitrification prevents the reformation of $ClONO_2$, and so increases the effective chain length of chlorine catalytic cycles. In sunlight, Cl_2 is photodissociated to release atoms that may then be converted into ClO. The small amount of NO_2, produced by photolysis of HNO_3, can convert the ClO back into $ClONO_2$. Hence we have

$$Cl_2 + h\nu \rightarrow Cl + Cl \qquad (33)$$

Table 9.1   Heterogeneous reactions of importance in stratospheric ozone chemistry.

Reactions	Consequences
$N_2O_5 + H_2O \rightarrow 2HNO_3$	Decreases NO_x mixing ratios throughout lower stratosphere.
$ClONO_2 + H_2O \rightarrow HOCl + HNO_3$	Denoxification. Chlorine activation at low temperatures (*NB*: HOCl is less easily photolysed than Cl_2).
$ClONO_2 + HCl \rightarrow Cl_2 + HNO_3$	Denoxification. Chlorine activation at low temperatures.
$HOCl + HCl \rightarrow Cl_2 + H_2O$	Chlorine activation at low temperatures.
$HOBr + HCl \rightarrow BrCl + H_2O$	Indirectly affects ClO_x and HO_x mixing ratios.
$BrONO_2 + H_2O \rightarrow HOBr + HNO_3$	Denoxification (minor route). Indirectly affects ClO_x and HO_x mixing ratios.

$$Cl + O_3 \rightarrow ClO + O_2 \tag{15}$$

$$ClO + NO_2 + M \rightarrow ClONO_2 + M \tag{16}$$

This chlorine nitrate is then available to react with any remaining HCl, to release more ClO_x until all the NO_x is removed. Hence within the polar vortices we would expect to find high concentrations of ClO (see below). Over the wintertime and springtime pole, methane concentrations are low as a result of downward transport from the upper stratosphere (see Figure 9.2) and NO_x concentrations can be low as a result of denitrification. Hence the usual chain-breaking steps of Cl reaction with CH_4, or ClO reaction with NO_2 are particularly slow. In the Antarctic, the reduced rates of termination reactions, together with continued cold temperatures and PSCs, mean that all the ozone destroying catalytic cycles are able to continue throughout September. In the Arctic, however, sporadic cold temperatures and denitrification mean that there is more continuous competition between radical production and radical termination.

9.3 NATURAL SOURCES OF TRACE GASES

Nitrous oxide (N_2O) is the dominant precursor of stratospheric NO_x. It is emitted predominantly by biological sources in soils and water.[12] Tropical forest soils are probably the single most important source of N_2O to the atmosphere. Tropical land-use changes and intensification of tropical agriculture indicate significant, growing sources of N_2O. N_2O mixing ratios in the stratosphere are growing at ~ 0.8 ppbv yr^{-1} and it has been estimated that human-induced N_2O emissions are the most significant ozone-depleting emission.[13]

Natural sources of Cl_y, which could conceivably have a marked influence on stratospheric ozone, are few. Methyl chloride, with an atmospheric abundance of approximately 550 pptv, is the dominant halogenated compound in the atmosphere. A production rate of around 4 Tg yr^{-1} is required to maintain this steady-state mixing ratio, about half of which is released from tropical land as a result of various biological processes and the rest comes from the oceans, from biomass burning, and from human activities.[12]

There are many naturally-occurring bromine compounds in the atmosphere, most of which have appreciable tropospheric reactivities. This reactivity makes it difficult to measure or calculate the input of each organo-bromine compound, or the inorganic bromine compounds derived from each organo-bromine compound, into the stratosphere. Halogenated compounds that can be broken down partially in the troposphere are called *very short-lived substances* (VSLS) in the stratospheric ozone literature and are a continuing focus of research because climate change may be altering the efficiency with which VSLS are transported into the stratosphere. The most abundant organo-bromine, methyl bromide (CH_3Br) is present in the troposphere at 7–8 pptv, and is declining as a result of controls on industrial emissions. There are relatively large two-way exchanges of methyl bromide into and out of soils and oceans, but overall they are considered to act as net sinks and a large source remains undetermined.[12] Other naturally-occurring organo-bromine compounds are dibromo- methane (CH_2Br_2), bromoform ($CHBr_3$), the mixed halogenated methane compounds (CH_2BrCl, $CHBr_2Cl$, $CHBrCl_2$), and ethylene dibromide ($C_2H_2Br_2$), all of which are produced by marine algae. They are broken down rather quickly in the troposphere, so that measurements at the tropical tropopause show much lower concentrations than at the surface. Nevertheless, reactive organo-bromine compounds can contribute about 10% of the total bromine entering the stratosphere. This contribution can be either by the transport of the residual amount of organo-bromine remaining after tropospheric degradation, or by transport of the inorganic bromine formed by that degradation.

9.4   ANTHROPOGENIC SOURCES OF TRACE GASES

CFCs, unlike most other gases, are not chemically broken down or removed in the troposphere but rather, because of their exceptionally stable chemical structure, persist and slowly migrate up to the stratosphere. Depending on their individual structure, different CFCs can remain intact for decades or centuries. Once in the stratosphere, photolysis of CFCs releases chlorine atoms that can take part in the catalytic destruction of ozone, for example:

$$CF_2Cl_2 + h\nu \rightarrow CF_2Cl + Cl \tag{34}$$

with the Cl released taking part in reaction (5). CFCs were used in a wide variety of industrial applications including aerosol propellants, refrigerants, solvents and foam blowing. The CFCs that have attracted most attention in ozone depletion are CFC-11 ($CFCl_3$) and CFC-12 (CF_2Cl_2); CFC-12 is the most abundant CFC, having a tropospheric mixing ratio close to that of methyl chloride (~ 540 pptv). Controls on their production and consumption (see section 9.8) have slowed the rate of accumulation of CFCs in the atmosphere and atmospheric concentrations of both CFC-11 and CFC-12 are decreasing by about 0.5 and 0.8 percent per year, respectively.[12] Total tropospheric chlorine in ozone-depleting substances peaked at 3.7 ppbv by 1994, with the stratospheric peak lagging by 3 to 6 years depending on latitude.[12]

Bromine is carried into the stratosphere in various forms such as halons and substituted hydrocarbons, of which methyl bromide is the predominant form. Stratospheric total bromine mixing ratios are a little over 20 pptv. As well as the biogenic sources, three major anthropogenic sources of methyl bromide have been identified: soil fumigation; biomass burning; and the exhaust of automobiles using leaded petrol. Bromine-containing halons – H-1211 ($CBrClF_2$) and H-1301 ($CBrF_3$) – have been widely used in fire protection systems. Their production has now ceased but emissions from existing fire protection systems are expected to continue for decades. Global background levels are about 3.5 pptv (H-1211) and 2.0 pptv (H-1301). H-1211 concentrations are decreasing but H-1301 concentrations are increasing.[12]

Carbon dioxide concentrations are increasing largely as a result of burning fossil fuels.[14] Although not directly involved in ozone chemistry, carbon dioxide is a 'greenhouse gas', and so can affect ozone concentrations indirectly. Greenhouse gases warm the troposphere by trapping infrared radiation released by the Earth's surface. Water vapour and carbon dioxide are the most important greenhouse gases, but others include methane, nitrous oxide, ozone, the CFCs and the HCFCs. Increased greenhouse warming in the troposphere leads to lower stratospheric temperatures, because less infrared radiation reaches the stratosphere. One consequence of stratospheric cooling is a slowing down of temperature-dependent (*e.g.* $O + O_3$, $NO + O$) ozone-destruction reactions. In the middle to upper stratosphere cooling would lead, therefore, to an increase in ozone. In the lower stratosphere, however, a decrease in temperature could increase the frequency and duration of PSCs, leading to more chlorine activation and hence more ozone depletion. In addition to its contribution as a greenhouse gas, methane has a direct chemical effect (see above). When methane is destroyed in the stratosphere, the chlorine and nitrogen ozone-destruction cycles are suppressed – *i.e.* reaction (23) and the 'smog' reactions of Chapter 8 compete more effectively with reactions (5), (6), (9), (10), and (11), *etc.* – thus increasing stratospheric ozone abundances. The destruction process also produces water vapour, which can indirectly destroy ozone through the production of OH. Current best estimates of the effects of increased greenhouse gases are that the recovery in polar ozone brought about by the Montreal Protocol will be delayed by 10–20 years relative to an atmosphere that is not experiencing climate change.

9.4.1   Direct Injection of Pollutants into the Stratosphere

NO_x emitted in the exhausts of supersonic aircraft (SST) can cause ozone depletion. Production of a new generation of SSTs is periodically mooted by the aircraft industry. Total ozone changes, for cruise altitudes of 16 and 20 km, are calculated by models to be a few percent for reasonable estimates of the size of the supersonic fleet. Emissions higher in the stratosphere lead to larger local ozone losses since the NO_x emitted remains in the stratosphere for longer. Although changes in NO_x have the largest impact on ozone, the effects of H_2O emissions contribute about 20% to the calculated ozone change.

Many subsonic flights pass through the lowermost parts of the stratosphere, especially at high latitudes (recall Figure 9.2). Subsonic aircraft flying in the North Atlantic flight corridor emit 44% of their exhaust emissions into the stratosphere. Models predict an ozone decrease in the lower stratosphere of less than 1%, but modelling the lower stratosphere is particularly difficult since a number of chemical processes are of comparable importance to each other, and to the transport processes. Industry projections are for a 5% per year increase in air traffic from 1995 and 2015, so the impact of aircraft on the atmosphere is likely to grow significantly.

The exhaust products of rockets contain many substances capable of destroying ozone. Attention has focused on the chlorine compounds produced from solid fuel rockets. Rockets that release relatively large amounts of chlorine per launch into the stratosphere include NASA's Space Shuttle (68 tons per launch) and Titan IV rockets (32 tons), and the European Space Agency's Ariane-5 (57 tons). Compared to the global release of CFCs, however, which is limited to 1×10^6 tons per year by the Montreal Protocol (and that is only 15% of the 1986 production), the chlorine source from rockets is tiny.

9.5 ANTARCTIC OZONE

The ozone hole continues to appear every austral spring, and was particularly severe in the 1990s. Figure 9.4 shows the monthly mean values of the ozone column over Halley Bay, Antarctica, for

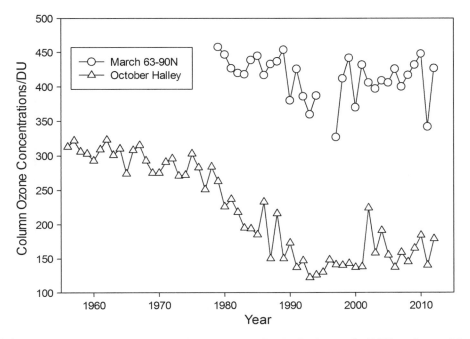

Figure 9.4 Mean October value of column ozone over Halley in the Antarctic (76°S) and mean March value of the Arctic region from 63–90°N. [We thank J. Shanklin (British Antarctic Survey, Cambridge) for the Halley data, and Greg Bodeker (Bodeker Scientific) for the Arctic data].

October. Prior to the mid-1970s the monthly mean column of ozone was approximately 300 DU. By 1985, the mean had fallen to below 200 DU and in the subsequent 5–6 years appeared to be leveling-off. However, the springs of 1992 and 1993 brought new record lows, with spot measurements below 100 DU for the first time, and the mean well below half that of the 1950s. The vertical profile of ozone above McMurdo station, Antarctica, is shown in Figure 9.5(a). In mid-August the profile was near normal but in less than two months 97% of the ozone between 14 and 18 km in altitude had been destroyed.

The major chemical causes for the severe ozone depletion over Antarctica have been discussed in section 9.2. The severity of polar depletion is caused by man-made chlorine pollution of the atmosphere in combination with unique wintertime meteorological conditions. The circulation of the winter stratosphere over the wintertime pole is dominated by the polar vortex (see Figure 9.2). This is a region of very cold air surrounded by strong westerly winds. Air within the vortex is largely sealed off from that at lower latitudes and the chemistry occurs in near isolation. In the lower stratosphere, temperatures within the vortex fall below 200 K; cold enough in the lower half of the stratosphere for heterogeneous chemistry on cold sulfuric acid aerosol, type 1 PSCs, and type 2 PSCs. Chlorine activation, often accompanied by denitrification and dehydration, results.

Figure 9.6 shows the evolution of ozone and related compounds on a single isentropic surface (the 465 K surface, about 19 km) as measured by the Microwave Limb Sounder (MLS) satellite instrument.[15] In late autumn (28 April) descent of air at the pole has brought down higher mixing ratios of ozone, nitric acid and water vapour from the middle and upper stratosphere. Other measurements have shown that higher mixing ratios of Cl_y are also present in this air. The formation of the polar vortex isolates the polar air from its surroundings. PSC formation has yet to begin and so ClO mixing ratios in the vortex are the same as outside the vortex, near zero on this scale.

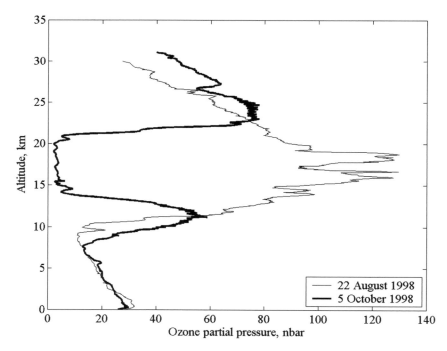

Figure 9.5a   Vertical ozone profiles above McMurdo, Antarctica, for 22 August 1998 and 5 October 1998. (Data courtesy of Terry Deshler, University of Wyoming).

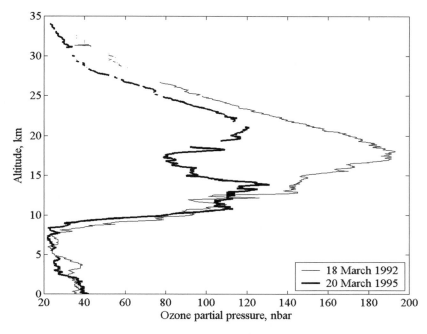

Figure 9.5b Vertical ozone profiles above Ny Ålesund, in the Arctic, for 18 March 1992 and 20 March 1995. (Data courtesy of the Alfred Wegener Institute). All profiles are taken inside the polar vortices.

By 2 June, temperatures have fallen below the threshold for type 1 PSC formation, as can be seen from the reduced mixing ratios of gas phase nitric acid in the MLS measurements. Chlorine activation takes place on the PSCs and the ClO formed is blown downwind, but there has not yet been enough exposure to sunlight for large-scale ozone depletion to be evident. By 17 August, temperatures are cold enough for type 1 and 2 PSCs to be present over large parts of the Antarctic continent, leading to reduced gas phase nitric acid and water vapour and to elevated ClO. Ozone depletion has begun on the sunlit outer rim of the vortex. By 1 November, temperatures are too high for PSC formation but nitric acid and water vapour are both reduced inside the vortex. This is due to denitrification and dehydration. ClO mixing ratios have returned to background levels almost everywhere since PSCs are no longer present and termination reactions have become effective. Ozone mixing ratios are now much reduced throughout the vortex, over an area larger than the Antarctic continent. During the 1990s, the area with ozone columns less than 220 DU approached 10% of the area of the Southern Hemisphere.

9.6 ARCTIC ZONE

The evolution of ozone and related compounds in the Arctic polar vortex during the winter of 1992/93 is shown in Figure 9.7, which can be compared to the evolution in the Antarctic vortex of 1992, shown in Figure 9.6. The initial descent into the vortex is less pronounced on 28 October in the Northern Hemisphere than on 28 April in the Southern Hemisphere, and the northern vortex is less fully developed than its southern counterpart. By 3 December the vortex has developed and descent, bringing down high mixing ratios of ozone, nitric acid and water vapour, has taken place. ClO mixing ratios are only elevated in a small region downwind of a patch of cold air over southern Finland. By 22 February there have been sufficient cold temperatures to activate chlorine throughout the vortex, but neither denitrification nor dehydration have taken place. Ozone mixing

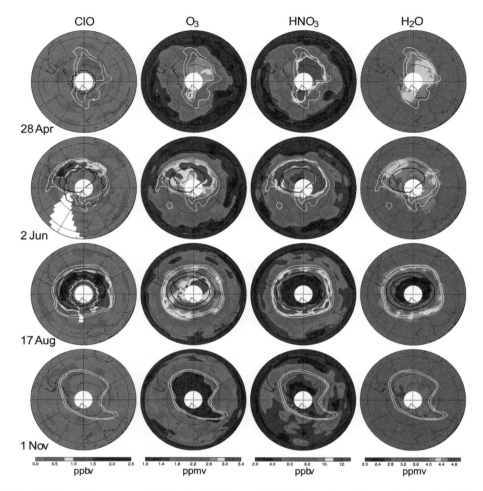

CIO   O₃   HNO₃   H₂O

Figure 9.6 MLS satellite maps of ClO, O_3, HNO_3, and H_2O for selected days during the 1992 southern winter. The satellite data are interpolated onto the 465 K isentropic surface, which is at about 19 km altitude. The maps are polar orthographic projections extending to the equator, with the Greenwich meridian at the top and the 30°S and 60°S latitude marked. No measurements were obtained poleward to 80°S and, for H_2O, equatorward of 30°S. The edge of the polar vortex is shown by the two black contours. The black contour near South America on 28 April surrounds air that has come from the polar vortex. Temperatures below which type 1 and type 2 PSCs can form are shown as white contours; the outer white contour is for type 1 PSCs. [Figure courtesy of Michelle Santee (Jet Propulsion Lab, California)].

ratios remain high. By 14 March, ClO mixing ratios have again returned to background levels, nitric acid and water vapour mixing ratios remain high, and no ozone hole has formed. However, that is not to say that chlorine-catalysed ozone destruction has not taken place: ozone mixing ratios in March are lower than those in February. Given the near-continuous descent of ozone-rich air onto this isentropic surface in the northern vortex during winter, this decreased ozone implies chemical loss. Further evidence of chemical loss in the northern polar vortex has been adduced from the careful matching up of ozonesondes that have sampled the same air parcel.[16] A statistically significant correlation is shown between ozone loss and hours of sunlight experienced between samplings.

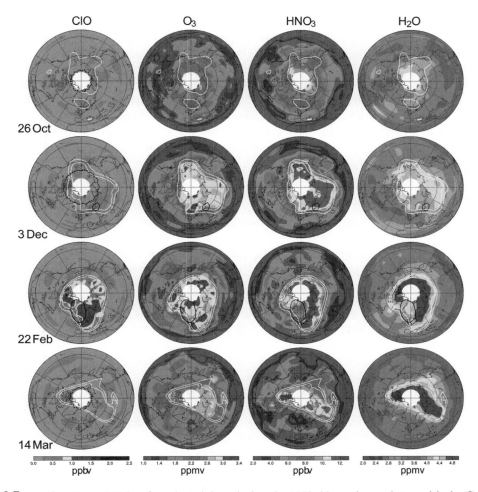

Figure 9.7 As for Figure 9.6, but for selected days during the 1992–93 northern winter, with the Greenwich meridian at the bottom of each map. [Figure courtesy of Michelle Santee (Jet Propulsion Lab, California)].

Figure 9.7 demonstrates that the Arctic has a warmer and weaker vortex than the Antarctic. The behaviour of the northern vortex can vary greatly from year to year. This variation can be seen in Figure 9.4, which shows the mean March column ozone between 63°N and 90°N. For winters which are cold early on, 1991–92 for example, large amounts of ClO can be formed, but it is formed when there is insufficient sunlight to cause severe ozone depletion.[2] Winters that are cold later, 1992–93 for example, produce more severe ozone depletion. The winter of 2010–11 produced particularly severe ozone depletion.[2] Figure 9.5(b) shows results from a pair of ozonesondes, launched from Ny Ålesund, into the polar vortex, at about the same time of year in 1992 and 1995. About half the ozone has been destroyed in the 18–20 km region in 1995 relative to 1992.

9.7 MID-LATITUDE OZONE

Ozone loss is not confined to polar regions. Trends for total ozone in the Northern and Southern Hemisphere middle latitudes are significantly negative in all seasons; trends in the equatorial region (20°S–20°N) are not statistically significant. Measurements of vertical profiles of ozone indicate that at altitudes of 40 km, over the period 1980 to 1996, ozone declined about 7% per decade at northern middle latitudes. Over the same period and region, depletion between 25 and 30 km was about 3% per decade, and about 7% per decade between 15 and 20 km. This 'high-low-high' signal in ozone depletion as a function of altitude is due to the effects of gas-phase chemistry at high altitudes and heterogeneous chemistry at low altitudes.

For the first two years after the Mt Pinatubo eruption, anomalously large downward ozone trends were observed in both hemispheres (see also Figure 9.4). Global total ozone values in 1992–93 were 1–2% more negative than those expected from the long-term trend. It seems likely that radiative, dynamical, and chemical perturbations, resulting from the injection of huge quantities of aerosol into the stratosphere, were responsible.

9.8   LEGISLATION

On September 16, 1987, a treaty was signed that has been hailed as 'the most significant international environmental agreement in history', 'a monumental achievement' and 'unparalleled as a global effort'.[17] It had been long believed that this particular agreement would be impossible to achieve because the issues were so complex and arcane and the initial positions of the negotiating parties so widely divergent.

The Montreal Protocol on Substances that Deplete the Ozone Layer was adopted in 1987 by 25 countries, and entered into force on 1 January 1989. The Protocol required each party's production and consumption of CFCs 11 ($CFCl_3$), 12 (CF_2Cl_2), 113 ($C_2F_3Cl_3$), 114 ($C_2F_4Cl_2$) and 115 (C_2F_5Cl), first to be frozen at 1986 levels and ultimately to be reduced to 50% of 1986 levels by 1998. Production and consumption of halons 1211 (CF_2BrCl), 1301(CF_3Br) and 2402 ($C_2F_4Br_2$) were to be restricted to 1986 levels.

Ozone depletion had, from the start, captured the US public imagination. The subject featured prominently in the media and in congressional hearings, and soon began to influence consumer behaviour. In 1978 the US, followed by Canada, Norway and Sweden, banned the use of CFCs as aerosol spray can propellants in non-essential applications.[18] Even before this ban, the US market for spray cans had fallen by nearly two thirds because of public commitment to protection of the environment. In the UK, such a ban was not introduced. Concerns about ozone depletion were defused by ministerial assurances that any terrestrial effects would be inconsequential.[19]

In the light of increasing scientific and public concern over ozone depletion, empirical evidence of global depletion, confirmation that CFCs and other man-made ozone depletion substances were the major factor in creating the Antarctic ozone hole, and generally improved prospects for replacing such substances, significant changes to the Protocol were approved in London (1990) and Copenhagen (1992). In London, the control measures were adjusted to provide for the phase-out of CFC and halon production and consumption by the year 2000. An intermediate cut of 50% of the 1986 level by 1995 was also agreed for CFCs and halons, together with an 85% cut, for CFCs but not halons, by 1997. Additionally, carbon tetrachloride and fully halogenated CFCs were to be phased-out by 2000 and methyl chloroform by 2005.

The 1990 Protocol Amendment introduced the concept of transitional substances, such as HCFCs. These are chemical substitutes for CFCs and other controlled substances, but unlike CFCs, are reactive in the lower atmosphere and so, tonne for tonne, transport less chlorine to the stratosphere. They are necessary in some applications, in the short to medium term, to enable a rapid phase-out of the controlled substances to take place. A non-binding resolution was approved with a view to replacing HCFCs by alternatives that do not deplete ozone by no later than 2040.

Further revisions to the Protocol were agreed in Copenhagen (1992). The phase-out dates were brought forward and controls on several new substances were imposed. For CFCs, carbon tetrachloride and methyl chloroform, the new phase-out date was 1 January 1996. Provision was made to allow production of these substances for 'essential use'. Within the European Union, Member states have adopted even tighter controls. The transitional substances were introduced into the Protocol and controls on HCFCs agreed. These controls set a cap on consumption, which will be reduced stepwise, leading to a phase-out by 2030. The numerical value of the cap is the sum of the quantity of all HCFCs that were produced and used during 1989 plus an amount equal to 3.1% of the calculated CFC consumption during 1989. Contributions to the 3.1% cap from individual HCFCs are adjusted by their ozone depletion potentials.

It is already clear that the Montreal Protocol and its amendments have been adhered to, and have worked.[20] Tropospheric abundance of long-lived ozone-depleting substances (CFCs *etc.*) has peaked and is now beginning to decline. In the stratosphere the time at which total chlorine peaks is a function of height, since the transport of air to the uppermost stratosphere, by the Brewer–Dobson circulation, is slow. The peak in chlorine therefore occurred later in the stratosphere than in the troposphere. At 22 km, for example, the peak was measured in 1998.

9.9 GEOENGINEERING

Geoengineering, which is sometimes referred to as climate engineering, is the deliberate modification of the Earth system to offset the negative effects of climate change.[21,22] There are two overarching categories of geoengineering:

(i) *Carbon dioxide removal* (CDR) which directly removes atmospheric carbon dioxide thereby reducing the amount of infrared radiation trapped within the atmosphere.
(ii) *Solar radiation management* (SRM) which reduces the amount of solar (UV/visible) radiation absorbed by the Earth.

Whilst SRM does not solve the root cause of climate change, namely rising greenhouse gas emissions, it is argued that its use could buy more time until the required greenhouse gas reductions are achieved. Many CDR and SRM schemes have been proposed, and each has their advantages and disadvantages. All schemes will affect the stratosphere in some way by perturbing the stratospheric radiation budget and hence energy budget. These perturbations in the solar radiation budget will affect photolysis rates and the ratio of direct to scattered (diffuse) radiation. Perturbations in the infrared budget will affect stratospheric heating and hence, dynamics.

One SRM scheme in particular would have a pronounced effect on the stratosphere – namely the deliberate stratospheric injection of aerosol particles. This scheme is modelled on its natural analogue of large explosive volcanic eruptions which inject sulfur dioxide (SO_2) into the stratosphere. Subsequently SO_2 is oxidised, *via* both gas and liquid phase reactions, to form sulfuric acid (H_2SO_4). The low vapour pressure of H_2SO_4 leads to the condensation of the acid into aerosol particles, and these particles scatter a proportion of the incoming solar radiation back to space. The last major eruption which put large amounts of sulfate aerosol into the stratosphere was the Mt Pinatubo eruption on the island of Luzon in the Philippines in 1991; the plume from this eruption reached altitudes well above the tropopause and in excess of 30 km (see section 9.2). This aerosol layer led to a reduction in average global temperature of $\sim$0.5 K for $\sim$2 years. Most studies to date which investigate the stratospheric injection of aerosols consider the aerosol to be composed of sulfate because of the volcanic analogue. However particle compositions of higher refractive index, and hence greater light scattering ability, could lead to lower volumes of aerosol being required for injection.[23]

The aerosol is injected into the stratosphere, rather than the troposphere, because of the much longer residence time within the stratosphere which is on the order of years compared to weeks within the troposphere. Most schemes involve injecting aerosol into the tropical stratosphere; from here the atmospheric circulation will spread the aerosol poleward (*cf.* Figure 9.2) creating a global distribution of aerosol. The aerosol particles are ultimately lost from the stratosphere by gravitational settling. However, the physical characteristics of the particles can be altered, within the stratosphere, *via* condensation, coagulation and chemical reaction. The stratosphere already contains significant aerosol as discussed earlier in the chapter.

The major pollution worry surrounding stratospheric aerosol injection is the impact it will have on the stratospheric ozone layer. After the Mt Pinatubo eruption, a reduction in the total ozone column concentration was observed in regions where there were large surface areas of sulfate aerosol.[24] Model calculations indicate that the Mt Pinatubo eruption resulted in a 2–3% reduction in the global mean ozone column, and this was largely attributable to increased heterogeneous surface area and hence chemistry (see section 9.2.2).[25] A geoengineered aerosol layer will have a significantly larger surface area than in the non-volcanic stratosphere, and will likely be in excess of the Mt Pinatubo generated surface area. Therefore ozone depletion resulting from aerosol injection will probably be similar or in excess of that seen after the Mt Pinatubo event. In particular, aerosol injection will affect ozone loss at the poles: increases will be seen in the extent of Arctic ozone depletion, and the recovery of the Antarctic ozone hole will be delayed.[26]

Another pollution threat posed by particle injection is the warming of the stratosphere and in particular the cold tropopause region. Absorption of infrared radiation by the particles would lead to localised warming. Under normal stratospheric conditions, the tropopause acts as a 'cold trap' which inhibits the transfer of water from the troposphere to the stratosphere. Therefore warming of this region could increase stratospheric water concentrations. The water can react with electronically excited oxygen atoms generated in reaction (2) *via* reaction (35) to generate OH which can go on to destroy ozone *via* the catalytic cycle shown in reactions (5) and (6).

$$O + H_2O \rightarrow 2OH \tag{35}$$

Geoengineering and stratospheric particle injection are relatively new concepts, and it is not yet clear whether they will ever be employed. However, as efforts to reduce global greenhouse emissions continue to flounder, the prospect of geoengineering is likely to become more persuasive. It is clear that the stratospheric injection of particles is likely to lead to stratospheric pollution and this should be considered carefully in any assessment of the technique.

9.10   CONCLUSIONS

Long-term depletion of ozone has been observed since the early 1980s in the Antarctic polar vortex and at mid-latitudes in both hemispheres, with most of the ozone loss occurring in the lower stratosphere. Intermittently since the spring of 1995, ozone depletion of similar relative magnitude to that observed in the Antarctic spring has been observed in the Arctic, with depletion equivalent to Antarctic ozone loss occurring in the winter 2010–11.

The present condition of the global ozone layer is much worse than we would wish, but it is easy to forget how close the world came to an environmental problem of much greater proportions. Had the Montreal Protocol and its amendments not been adopted, or if they are not adhered to, the stratosphere in 2050 would be set to look like this:[20]

   (i)   About five times more ozone-depleting gases would be in the stratosphere than are currently there. This corresponds to a total mixing ratio about 17 ppbv of ozone-depleting substances, having first weighted each substance for its ozone-depletion potential.
   (ii) Ozone depletion would be at least 50% over the middle latitudes of the Northern Hemisphere and 70% over the middle latitudes of the Southern Hemisphere. This is about ten times larger than present depletion.
   (iii) Surface ultraviolet (UV-B) fluxes would be at least double those in an unperturbed atmosphere in northern middle latitudes, and four times greater in southern middle latitudes. Currently UV-B levels are up by 5% and 8% in the northern and southern middle latitudes, respectively.

But work remains to be done. The coupling of the ozone layer and global climate is still very poorly quantified, and this makes forecasts of future ozone levels (and climate) uncertain.

ACKNOWLEDGEMENTS

The contributions of Ian Colbeck and Joe Farman to previous editions of this chapter are very gratefully acknowledged. The task of writing a short description of stratospheric ozone is made immeasurably easier by the existence of the WMO Scientific Assessments; we thank the many scientists who have laboured long and hard on them. We very grateful also to Terry Deshler, Roland Neuber and Michelle Santee for supplying figures and data for figures.

REFERENCES

1. J. C. Farman, B. G. Gardiner and J. D. Shanklin, Large losses of total ozone in Antarctica reveal seasonal ClOx–NOx interaction, *Nature*, 1985, **315**(6016), 207–210.
2. G. L. Manney, M. L. Santee, M. Rex, N. J. Livesey, M. C. Pitts, P. Veefkind, E. R. Nash, I. Wohltmann, R. Lehmann, L. Froidevaux, L. R. Poole, M. R. Schoeberl, D. P. Haffner, J. Davies, V. Dorokhov, H. Gernandt, B. Johnson, R. Kivi, E. Kyro, N. Larsen, P. F. Levelt, A. Makshtas, C. T. McElroy, H. Nakajima, M. C. Parrondo, D. W. Tarasick, P. von der Gathen, K. A. Walker and N. S. Zinoviev, Unprecedented Arctic ozone loss in 2011, *Nature*, 2011, **478**(7370), 469–475.
3. A. J. Weinheimer, *et al.*, Meridional distributions of NOx, NOy and other species in the lower stratosphere and upper troposphere during AASE II, *Geophys. Res. Lett.*, 1994, **21**(23), 2583–2586.
4. C. R. Webster, *et al.*, Hydrochloric acid and the chlorine budget of the lower stratosphere, *Geophys. Res. Lett.*, 1994, **21**(23), 2575–2578.
5. M. B. McElroy, *et al.*, Reductions of Antarctic ozone due to synergistic interactions of chlorine and bromine, *Nature*, 1986, **321**(6072), 759–762.
6. S. Solomon, R. R. Garcia and A. R. Ravishankara, On the role of iodine in ozone depletion, *J. Geophys. Res.: Atmos.*, 1994, **99**(D10), 20491–20499.
7. R. R. Garcia and S. Solomon, A new numerical model of the middle atmosphere: 2. Ozone and related species, *J. Geophys. Res.: Atmos.*, 1994, **99**(D6), 12937–12951.
8. A. R. MacKenzie, Stratospheric chemistry: aerosols and the ozone layer, in *Environmental Chemistry of Aerosols*, ed. I. Colbeck, Blackwell Publishing Ltd, Oxford, 2008, pp. 193–216.
9. L. W. Thomason and T. Peter, *Assessment of Stratospheric Aerosol Properties (ASAP)*, SPARC Report No. 4, SPARC (Stratospheric Processes And their Role in Climate), Zurich, Switzerland, 2006.
10. D. Lowe and A. R. MacKenzie, Polar stratospheric cloud microphysics and chemistry, *J. Atmos. Sol.-Terr. Phys.*, 2008, **70**(1), 13–40.
11. D. R. Hanson, A. R. Ravishankara and S. Solomon, Heterogeneous reactions in sulfuric acid aerosols: a framework for model calculations, *J. Geophys. Res.: Atmos*, 1994, **99**(D2), 3615–3629.
12. S. A. Montzka, *et al.*, Ozone Depleting Substances (ODS) and related chemicals, in *Scientific Assessment of Ozone Depletion: 2010*, World Meteorological Organization: Geneva, 2011.
13. A. R. Ravishankara, J. S. Daniel and R. W. Portmann, Nitrous oxide (N_2O): the dominant ozone-depleting substance emitted in the 21st century, *Science*, 2009, **326**(5949), 123–125.
14. P. Forster, *et al.*, Changes in atmospheric constituents and in radiative forcing, in *Climate Change 2007: The Physical Science Basis. Contribution of Working Group I to the Fourth Assessment Report of the Intergovernmental Panel on Climate Change*, ed. S. Solomon, D. Qin,

M. Manning, Z. Chen, M. Marquis, K. B. Averyt, M. Tignor and H. L. Miller, Cambridge University Press, Cambridge, UK and New York, NY, USA, 2007.

15. M. L. Santee, *et al.*, Interhemispheric differences in polar stratospheric HNO₃, H₂O, ClO, and O₃, *Science*, 1995, **267**(5199), 849–852.

16. P. Von der Gathen, *et al.*, Observational evidence for chemical ozone depletion over the Arctic in winter 1991–92, *Nature*, 1995, **375**(6527), 131–134.

17. R. E. Benedick, *Ozone Diplomacy*, Harvard University Press, Cambridge, MA, USA, 1991.

18. S. C. Zehr, Accounting for the ozone hole: scientific representations of an anomaly and prior incorrect claims in public settings, *Sociol. Quart.*, 1994, **35**, 603–619.

19. M. Purvis, M., *Yesterday in parliament: British politicians and debate over stratospheric ozone depletion, 1970–92, Environ. Plan. C: Gov. Policy*, 1994. **12**, p. 361–379.

20. O. Morgenstern, *et al.*, *The World svoided by the Montreal Protocol, Geophys. Res. Lett.*, 2008, **35**(16), L16811.

21. D. W. Keith, Geoengineering, *Nature*, 2001, **409**(6818), 420–420.

22. Royal Society, *Geoengineering the Climate: Science, Governance and Uncertainty*, London, UK 2009.

23. F. D. Pope, *et al.*, Stratospheric aerosol particles and solar-radiation management, *Nature Clim., Change*, 2012, **2**(10), 713–719.

24. S. Solomon, Stratospheric ozone depletion: a review of concepts and history, *Rev. Geophys.*, 1999, **37**(3), 275–316.

25. P. Telford, *et al.*, Reassessment of causes of ozone column variability following the eruption of Mount Pinatubo using a nudged CCM, *Atmos. Chem. Phys.*, 2009, **9**(13), 4251–4260.

26. S. Tilmes, R. Müller and R. Salawitch, The sensitivity of polar ozone depletion to proposed geoengineering schemes, *Science*, 2008, **320**(5880), 1201–1204.

CHAPTER 10

Atmospheric Dispersal of Pollutants and the Modelling of Air Pollution

MARTIN L. WILLIAMS

Science Policy, Environmental Research Group, King's College London, Franklin-Wilkins Building, 150 Stamford Street, London, SE1 9NH, UK
Email: martin.williams@kcl.ac.uk

10.1 INTRODUCTION

Considerable resources are often devoted to the measurement of air pollutant concentrations in the ambient atmosphere but measurements on their own provide little information on the origin of the pollutants in question, on the dispersal process in the atmosphere and on the impact of new sources or the benefits of controls. There is frequently the need, therefore, for detailed knowledge of the characteristics and quantities of pollutants emitted to the atmosphere and on the atmospheric processes which govern their subsequent dispersal and fate. This knowledge must then be built into an appropriate dispersion model, whether the problem to be addressed is the emission from a single stack or the emissions from a large multi-source urban/industrial area, or, on a larger scale, from a region or country.

Developments in air quality management systems to achieve air quality limit values, objectives or standards have created the need for an integrated approach to air pollution control involving monitoring networks, emission inventories and models. A wide variety of spatial scales are addressed currently, and a range of techniques are available, ranging from the most simple 'box' models through to numerical solutions of the basic equations of fluid flow, *etc.* Where chemical processes are important it is also necessary to incorporate the relevant atmospheric chemistry and current models are now capable of treating meteorology, dispersion and chemistry in single large sophisticated models. The spatial and temporal resolution and accuracy of the model output ideally must match the questions being posed. Where emissions vary greatly, both in space and time, or it is necessary to predict the temporally resolved time series of concentrations at specified locations, then the modelling task is difficult, even without the complications of a very chemically reactive pollutant species or substantial topographical effects on the dispersal pattern. On the other hand, if long term average concentrations (*e.g.* annual means) are required, then the modelling task is

Pollution: Causes, Effects and Control, 5th Edition
Edited by R M Harrison

generally much less demanding and uncertainties are lower. In general, long-period (*e.g.* annual) average concentrations can be modelled more accurately than can shorter averaging times where the turbulent fluctuations in the atmosphere can result in agreement only within factors of typically 2 to 3 with observed values. This chapter begins with a qualitative description of the main features of atmospheric boundary layers affecting pollutant dispersion so that these principles are made clear before embarking on a more mathematical (but still relatively simple) account of some methods used to perform dispersion modelling calculations. The advent of sophisticated freely-available open-source models means that there is a danger that they will be used as 'black boxes' without the user necessarily having a grounding in the basic principles of turbulence and modelling. This chapter attempts to provide an initial introduction to such principles.

10.2   DISPERSION AND TRANSPORT IN THE ATMOSPHERE

A pollutant plume emitted from a single source is transported in the direction of the mean wind. As it travels it is acted upon by the prevailing level of atmospheric turbulence which causes the plume to grow in size as it entrains the (usually) cleaner surrounding air. There are two main mechanisms for generating atmospheric turbulence. These are mechanical and convective turbulence, and will be discussed in sections 10.2.1 and 10.2.2.

10.2.1   Mechanical Turbulence

This is generated as the air flows over obstacles on the ground such as crops, hedges, trees, buildings and hills. The intensity of such turbulence increases with increasing wind speed and with increasing surface roughness and decreases with height above the ground. If there is only a small heat flux, either to or from the surface, in the atmosphere so that most of the turbulence is mechanically generated, the atmosphere is said to be 'neutral' or in a 'state of neutral stability'. In this case the wind speed will vary logarithmically with height z:

$$u(z) = (u_*/\mathrm{k})\ln(z/z_0) \tag{1}$$

Where k is von Karman's constant (~ 0.4), z_0 is the so called surface roughness length (~ 1 m for cities and ~ 0.3 m for 'typical' countryside in the UK), and u_* is the friction velocity and is a measure of the flux of momentum to the surfaces.

10.2.2   Turbulence and Atmospheric Stability

As solar radiation heats the earth's surface, the lower layers of the atmosphere increase in temperature and convection begins, driven by buoyancy forces. The motion of air parcels from the surface is unstable as a parcel in rising finds itself warmer than its surroundings and will continue to rise. Convective circulations are set up in the boundary layer and this form of turbulence is usually associated with large eddies, the effects of which are often visible in 'looping' plumes from stacks. At night when there is no incoming solar radiation and the surface of the earth cools, temperature increases with height, and turbulence tends to be suppressed. During calm clear nights, when surface cooling is rapid and little or no mechanical turbulence is being generated, turbulence may be almost entirely absent.

   When considering most dispersion problems it is convenient to classify the possible states of the atmosphere into what are usually referred to as 'stability' categories.

   The typing scheme developed by Smith from the original Pasquill formulation[1,2] is widely used because of its relative simplicity yet dependence on sound physical principles. Stability is classified according to the amount of incoming solar radiation, wind speed and cloud cover. A semi-quantitative guide is given in Table 10.1 and Figure 10.1 shows typical temperature profiles corresponding to the 'unstable', 'neutral' and 'stable' cases. The adiabatic lapse profile in Figure 10.1 is the vertical temperature gradient for the atmosphere in a state of adiabatic equilibrium when a parcel of air can rise and expand, or descend and contract, without gain or loss of heat; the temperature of the air parcel is always the same as that of the level surrounding air and the conditions correspond to neutral stability. The numerical value of the adiabatic lapse rate is $\sim 1\,^\circ$C per 100 m. For wind speeds in excess of about 6–8 m s^{-1}, mechanical turbulence dominates irrespective of the degree of insolation and neutral stability prevails. Similarly, in areas of high surface roughness like large cities, mechanical turbulence can dominate and the atmosphere can often be considered to be neutral. This can be a useful approximation in modelling studies in cities. Equally, in large cities with correspondingly large 'heat island' effects, stable boundary layers may well be very infrequent, or even absent. Table 10.2 gives typical annual frequencies of occurrence of the different stability categories in Great Britain. For other regions quite different frequencies might apply. In central Continental regions at lower latitudes, for example, the greater incidence of solar radiation would probably result in smaller incidence of neutral conditions and increased frequencies of unstable and stable categories.

More recently, a more quantitative measure of stability, the Monin–Obukhov length, has come into use. This essentially measures the balance between the mechanical and convective contributions to turbulence in a boundary layer, and the Monin-Obukhov length, L, is

Table 10.1 Pasquill's stability categories. (Based on Figure 6.10 in ref.).

Surface wind speed m s^{-1} ($\cong U_{10}$)	Insolation			Night	
	Strong	*Moderate*	*Slight*	*Thickly overcast or $\geq 4/8$ low cloud*	*$\leq 3/8$ cloud*
<2	A	A–B	B	–	G
2–3	A–B	B	C	E	F
3–5	B	B–C	C	D	E
5–6	C	C–D	D	D	D
>6	C	D	D	D	D

Strong insolation corresponds to sunny midday in midsummer in England. Slight insolation in similar conditions in midwinter. Night refers to the period from 1 hour before sunset to 1 hour after dawn. A is the most unstable category and G the most stable. D is referred to as the neutral category and should be used, regardless of wind speed, for overcast conditions during day or night.

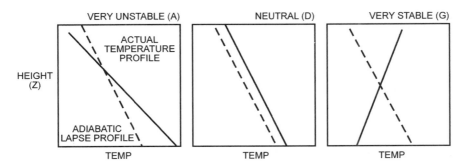

Figure 10.1 Typical atmospheric temperature profiles and corresponding stabilities.

Table 10.2 Typical annual frequency of occurrence of stability categories in Great Britain.

Stability category	Frequency of occurrence %
A	0.6
B	6.0
C	17.0
D	60.0
E	7.0
F	8.0
G	1.4

that height above the ground at which the two contributions are equal. In mathematical terms, L is given by:

$$L = -u_*^3(\rho c_p T_0) / \text{K}gF$$

where u_* is the friction velocity, K is the von Karman constant, g is gravitational acceleration, F is the surface heat flux, ρ is the air density, c_p is the specific heat of air at constant pressure and T_0 is the surface temperature. Recent developments in boundary layer models use the Monin-Obukhov length to classify stabilities and calculate dispersion. L takes on positive values in stable layers and is negative in convective boundary layers, approaching infinity as the heat flux (in either direction) approaches zero in a neutral boundary layer.

10.2.3 Boundary Layer and Mixing Heights

The stable atmosphere depicted in Figure 10.1 is an example of a ground based temperature inversion, *i.e.* the temperature increases with height unlike the normal decrease. An elevated inversion is often observed where a region of stable air caps an unstable layer below. Pollutants emitted below the inversion can be mixed up to, but generally not through, the inversion unless they are emitted with a high degree of buoyancy or momentum. The inversion in this case gives a clear example of a boundary layer height. However, in general the height of the boundary layer is less obvious and needs to be calculated from theories of boundary layer turbulence.[3–5] The boundary layer height is, in simple terms, the height over which the features of the surface-surface roughness, wind shear and heat flux are felt in the atmosphere. In the simple case of a neutral boundary layer, the height is given by:

$$h = cu_* / f$$

where c is a constant ~ 0.3–0.4 and f is the Coriolis parameter $\sim 10^{-4}$ s^{-1} at mid-latitudes. With a value of u_* typically of 0.3 m s^{-1}, we see that h is roughly 1000 m.

The term can be used more generally to describe the height of a boundary between two stability regimes. In the case of an emission above an elevated inversion, the pollutant will be prevented from reaching the ground so that for both surface and elevated sources, inversions can have a significant effect on ground level concentrations. The variation of boundary layer heights throughout the day and night due to solar heating and atmospheric cooling can have profound effects on ground level concentrations of pollutants. At night the atmosphere is typically stable with a shallow (~ 1–300 m) layer formed by surface cooling. As the sun rises the surface heating generates convective eddies and the turbulent boundary layer increases in depth, reaching a maximum in the afternoon at depths of ~ 1000 m or more in summer in strongly convective conditions. As the solar input decreases and stops, the surface cools and a shallow stable layer begins to form at the surface in the evening. In this idealized day, if emission rates remain constant, concentrations from surface sources will be at a maximum in the periods when the stable layers (with low wind speeds and mixing heights) are present and minimized during the afternoon. Sources emitting above the stable overnight layer will not contribute to ground-level concentrations until the height of the growing convective layer reaches the plume and brings the pollutants to ground level, a process known as fumigation. The patterns of ground-level concentrations from elevated sources can therefore be quite different from those of surface releases. In reality, the diurnal pattern of emissions can, of course, play a significant role.

It is important here to distinguish the terms boundary layer height and mixing height. As we have seen the boundary layer height measures the extent of the influence of the surface on the air flow. In general an emission of pollutant from a point at the surface will not necessarily be mixed up to the boundary layer height before impinging on a sensitive receptor. In general therefore, for a given source–receptor distance, the height to which a pollutant is mixed will be less than or equal to the boundary layer height.

In assessing air quality impacts, particularly of elevated sources such as power stations, estimates of boundary layer heights and their frequency of occurrence and variability throughout the day are therefore essential. An early study using acoustic sounding (SODAR) to determine boundary layer heights,[6] and an example of the diurnal and seasonal variation of mixing heights measured at a rural site in south-east England is given in Figure 10.2. The broad features described above are apparent in this diagram.

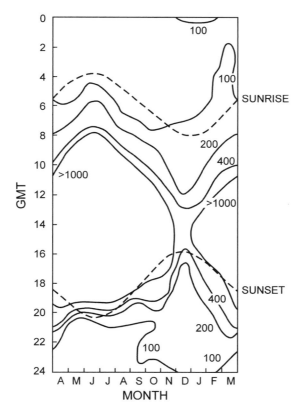

Figure 10.2   Annual and diurnal variation of boundary layer heights at Stevenage, UK from 1981–1983.

10.2.4 Building, Topographical and Street Canyon Effects

Hills or buildings can have significant adverse effects on plume dispersion if their dimensions are large in comparison with the dimensions of the plume or if they significantly deflect or disturb the flow of the wind. Figure 10.3 shows a simplified and idealized representation of the flow over a building. There is a zone on the immediate downwind (or leeward) side of the building which is to some extent isolated from the main flow and within which there is a reversal of the air flow. Further downstream the air flow is highly turbulent. Waste gases escaping through a relatively short chimney attached or adjacent to the building, will be entrained in this characteristic flow pattern and will not disperse according to the conventional Equations (2) and (3), shown in section 10.3.1. Wind tunnel studies demonstrate that up-wind buildings can have a significant effect on emissions from a chimney located within a few, say five, building heights; for example, to maintain the same maximum ground level concentration the chimney height required in the presence of one building type studied would be between 1.5 and 2 times the height of the chimney required if the buildings was not present. Further downwind, beyond roughly 10 building heights, the near-field effects of the buildings can be incorporated into dispersion models in a parameterized way as discussed in section 10.3.1. However, accurate numerical modelling in the near-field to buildings is difficult and most models treat this in an approximate way. Wind tunnel modelling is often the preferred solution, although improvements in computer power also mean that solutions of the detailed fluid flow equations (the Navier-Stokes equations) are now often feasible in Computational Fluid Dynamics packages.

Dispersion in street canyons is a particular case of wake effects from buildings. The recirculation zone in Figure 10.3 behind the windward side of the canyon will result in pollutants emitted in the street being advected towards the upwind side of the canyon at street level, leading to higher concentrations on that side of the canyon, as opposed to the downwind side which would be the case in the absence of the buildings forming the canyon.

Topographical features such as hills and sides of valleys can have similar effects on dispersion to those described above for buildings. Valleys are also more prone to problems arising from emissions close to the ground. The incidence of low level or ground based temperature inversions can be

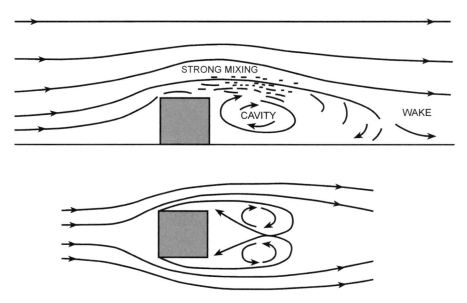

Figure 10.3   Simplified schematic flow patterns around a cubical building.

greater, either because solar heating of the ground is somewhat delayed in the morning or because during the night cold air drains down the valley sides (katabatic winds) thus creating a 'pool' of cold air on the valley floor. Any low-level emissions will therefore disperse very slowly and may even accumulate. The relatively undiluted emissions can also drift along and across the valley thus affecting areas other than the immediate surroundings of the source. Emissions from high chimneys located on the valley floor may not be detected at all on the valley floor while the ground based inversion persists. However, considerable horizontal spreading of the plume aloft can occur and during the morning fumigation period large parts of the valley may experience relatively high concentrations at much the same time.

The effects of hills on the flow of pollutant plumes and the resulting concentrations are complex and will not be discussed here.

10.2.5 Removal Processes – Dry and Wet Deposition

When considering pollution impacts from nearby sources, *e.g.* within say 10 km or so, the various losses of pollutants are generally not important (unless one happens to be interested specifically in such issues as the short range washout of a pollutant to a sensitive receptor or the deposition of a particularly toxic species in the near field). However, in considering impacts over longer ranges, and especially on the international scale, then consideration of the removal processes is essential.

Dry deposition takes place continuously in a turbulent boundary layer as a result of turbulent flux towards the surface. The efficiency of the process is determined by the overall deposition velocity which in general is a function of the prevailing level of turbulence (high levels of which result in increasing deposition other things being equal) and of the nature of the gas and the surface (for example a reactive gas such as HNO_3 will be deposited more readily than a less reactive species such as NO). The overall flux to the surface is given by the product of the surface concentration and the deposition velocity, v_g, so that the process is linear. The equivalent first-order rate constant for this process is given by (v_g/H) where H is the depth of the mixed layer through which deposition is taking place. Typical half-lives for this process are ~ 1–2 days for species such as SO_2 and NO_2, but ~ 5 days or more for sulfate and nitrate aerosols, which is one reason why 'acid rain' is a continental rather than purely a local phenomenon, and why transboundary transport of secondary inorganic aerosol components of particulate matter are important. The process of dry deposition is complex and a fuller account involving the resistance model of the atmospheric and surface components of the process is given in ref. 7.

Wet deposition is the term given to the removal of gases or particles from the atmosphere in clouds and/or in rain. For species such as SO_2 and aerosols this process is relatively efficient. Typical lengths of dry and wet periods in the UK are such that if transport times are of the order of the average dry period duration (~ 70 hours in the UK), wet removal processes should be included in the model.

10.3 MODELLING OF AIR POLLUTION DISPERSION

In section 10.2 we discussed the underlying physical principles of air pollution dispersion, transport and transformation in the atmosphere. In this section we summarize the methods used to apply these principles in a quantitative way to model the processes mathematically. The techniques used depend on the distance scales involved in the transport from the source to the receptor or 'target' area. If this distance is small, say of the order of tens or hundreds of metres, then very often buildings and local topography are important and the mathematical description of the ensuing complex flows and turbulence may not be tractable. In such cases (and others such as the dispersion of dense gases or longer distance problems in complex topography) a physical model in a wind tunnel or a complex Computational Fluid Dynamics calculation may be the only practicable solutions. These complications will be neglected in all that follows and which will deal with situations of uninterrupted flat terrain.

It is fairly clear that as a plume is transported in the direction of the mean wind, it grows through the effect of atmospheric turbulence and dilution producing, very roughly, a cone shaped plume with the apex towards the stack. Now clearly the plume will continue to expand until, in the vertical, it fills the atmospheric boundary layer ($\sim$1 km deep in neutral conditions). Beyond this point vertical dispersion has no further effect; concentrations are thence reduced only by horizontal dispersion, and by the deposition processes and, if appropriate, by chemical reactions. It can be shown that in neutral conditions this point is reached at downwind distances from a source of very roughly 50–100 km, so for source–receptor distances less than this value, vertical dispersion should be included in a model for an accurate representation of the dispersion. Beyond this region, plumes generally fill the mixing layer and uniformly mixed 'box- models' can be used with some confidence.

We will firstly discuss modelling on scales where vertical dispersion is important, before discussing problems involving longer range transport.

10.3.1   Modelling in the Near-field

In this section we will discuss modelling of pollutant dispersion from 0 to $\sim$100 km, using the Gaussian plume approach. This is not to condemn more sophisticated methods, but for illustrative purposes in an account such as this, the simple Gaussian approach has much merit, and is still often used with success in real-world applications. In many cases a sound knowledge of meteorology and aerodynamics can be used to parameterize the Gaussian model to simulate adequately, for example, the effects of buildings on dispersion.

In the Gaussian plume approach the expanding plume has a Gaussian, or Normal, distribution of concentration in the vertical (z) and lateral (y) directions as shown in Figure 10.4. The concentration C (in units of μg m^{-3} for example) at any point (x, y, z) is then given by:

$$C(x,y,z) = \frac{Q}{2\pi\sigma_y\sigma_z U}\exp\left[-\frac{y^2}{2\sigma_y^2}\right]\left\{\exp\left[-\frac{(z-H_e)^2}{2\sigma_z^2}\right] + \exp\left[-\frac{(z+H_e)^2}{2\sigma_z^2}\right]\right\} \qquad (2)$$

where Q is the pollutant mass emission rate in μg s^{-1}, U is the wind speed in m s^{-1}, x, y and z are the along wind, crosswind and vertical distances, H_e is the effective stack height given by the height of the stack plus the plume rise defined below. The parameters σ_y and σ_z measure the extent of plume growth and in the Gaussian formalism are the standard deviations of the horizontal and vertical concentrations, respectively, in the plume. When $y = z = 0$, this equation reduces to the familiar ground level concentration below the plume centre-line:

$$C(x) = \frac{Q}{\pi\sigma_y\sigma_z U}\exp\left(-\frac{H_e^2}{\sigma_z^2}\right) \qquad (3)$$

Equation (3) is less cumbersome to deal with and several important properties of Gaussian models are more clearly illustrated. Firstly, concentrations are directly proportional to the emission rate, Q, so it is essential that this is known accurately in any practical application. Secondly, unless $H_e = 0$ (*i.e.* unless the source is at ground level) the maximum concentration (C_{max}) will occur at some point downwind and this downwind distance will increase with increasing H_e and furthermore the value of C_{max} will decrease with increasing H_e. In fact, C_{max} is very roughly proportional to H_e^{-2}. This is the mathematical statement of the so-called 'tall stacks' policy which under-pinned air

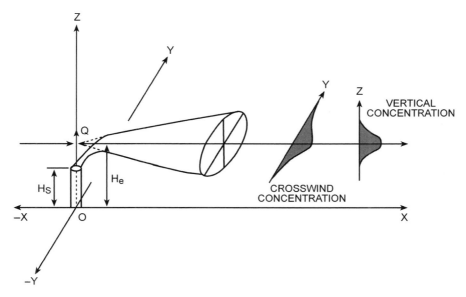

Figure 10.4   Gaussian plume distribution.

pollution control in the UK from the 1960s, and is still used, along with emission abatement techniques in granting approvals for industrial plant.

The specification of H_e involves calculating the plume rise, which is the height above the point of emission reached by the plume due to its buoyancy (if it is warmer than the surrounding air, as most combustion emissions are) or momentum (plumes may be driven up stacks at relatively high velocities). In the case of jet-engine exhaust, both buoyancy and momentum are important. For most plumes however, buoyancy is the dominating force and there have been a large number of studies of methods of determining plume rise. A widely used method however is that due to Briggs,[8] where the plume rise ΔH is given by:

$$\Delta H = 3.3(Q_H)^{1/3}(10H_s)^{2/3}U^{-1} \quad \text{for} \quad Q_H \geq 20\,\text{MW}$$

or   (4)

$$\Delta H = 20.5(Q_H)^{0.6}(H_s)^{0.4}U^{-1} \quad \text{for} \quad Q_H \leq 20\,\text{MW}$$

where Q_H is the sensible heat emission from the stack and U is the wind speed at the stack height (H_s). The dependence on Q_H is not strong and, if measured values of this quantity are not available, an approximation often used is to assume Q_H is equal to one-sixth of the total heat generated in the combustion of the fuel. An expression for ΔH due to Moore[9] has been developed for power stations in the UK and is given as:

$$\Delta H = aQ_H^{1/4}/U \tag{5}$$

where a = 515 in unstable and neutral conditions and a = 230 in stable atmospheres, with Q_H in MW and U in m s^{-1}. More recent and more complex formulations of Moore's and Briggs' equations have been given by the UK Dispersion Modelling Working Group.[10]

The standard deviations of the plume in the vertical and lateral directions, σ_z and σ_y, are extremely important quantities as they determine the dilution of the plume and hence resulting concentrations. They are determined by the prevailing atmospheric turbulence in the boundary layer.

Turbulent motions or eddies in the atmosphere vary in size and intensity; the greater their size and/or intensity the more rapid is the plume growth and hence the dilution of the pollutants. Small scale turbulent motions tend to dominate the plume growth close to the point of emission where the plume is still relatively small and the larger scale eddies dominate at greater distances. Furthermore, the small and larger eddies are associated respectively with short and longer time scales. Consequently, σ_y and σ_z increase in value with distance from the source (see Figure 10.4); their values (particularly for σ_y) also increase with the time or sampling period over which they have been measured. This latter point means that it is essential to state the sampling period to which σ_y and σ_z apply, especially if comparisons are being made between calculated and measured concentration; ideally the two periods should be identical. Most values of σ_y and σ_z to be found in the literature are for sampling periods in the range 3–60 minutes. It is also evident that σ_y and σ_z are dependent on atmospheric stability, being smallest when the atmosphere is most stable (category G) *i.e.* when atmospheric turbulence is least, increasing to their greatest values in highly turbulent very unstable conditions (category A). The underlying surface roughness elements also play a part, σ_y and σ_z increasing with increasing surface roughness so that for a given distance downwind of a chimney σ_y and σ_z will be larger in, for example, an urban area than in an area of open, relatively flat agricultural land.

In general, lateral (horizontal) motion is less constrained than vertical motion with the result that there are larger scale eddies in the horizontal than in the vertical. Fluctuations in wind direction also become important for longer sampling periods. Consequently, σ_y increases more rapidly with increasing sampling or averaging period than does σ_z. This dependence of σ_y on wind direction fluctuation also means that for longer sampling periods, say greater than one hour, σ_y values can increase with increasing atmospheric stability because during low wind speed stable conditions plume meandering can be significant.

Ignoring for the moment the plume meandering component of σ_y the parameters are often conveniently expressed in the form

$$\sigma_y = \sigma_{yo} + ax^b \tag{6}$$

$$\sigma_z = \sigma_{zo} + cx^d$$

where a, b, c and d are constants dependent on atmospheric stability, x is the downwind distance from the source and σ_{yo}, σ_{zo} are the initial plume spreads generated by, for example, building entrainment. To incorporate plume meander into σ_y, an extra term is added so that

$$\sigma_y^2 = \sigma_{yt}^2 + 0.0296Tx^2 / U \tag{7}$$

where σ_{yt} is given by Equation (6) and T is the averaging time in hours. A simple expression for σ_z based on Smith's[1] work is:

$$\sigma_z = \sigma_{zo} + 0.9(0.83 - \log_{10}P)x^{0.73} \tag{8}$$

which gives a good representation out to $\sim$30 km. Here P is Smith's stability parameter equal to 3.6 for neutral conditions and ranging from 0–1 (stability A) through to 6–7 in stability G. Values of coefficients specifying σ_{yt}, the so-called microscale σ_y, *i.e.* not including any plume meander effects, are given in Table 10.3. This table also includes typical values of boundary layer heights in the stability categories A–G. The effect of the boundary layer height on vertical plume dispersion can be taken into account in the following modification of Equation (2) for ground level concentrations:

$$C(x, y) = \frac{Q}{\pi \sigma_y \sigma_z U} \exp\left[-\frac{y^2}{2\sigma_y^2}\right] \left\{ \exp\left[-\frac{H_e^2}{2\sigma_z^2}\right] + \exp\left[-\frac{(2L - H_e)^2}{2\sigma_z^2}\right] \right\} \tag{9}$$

Table 10.3 Typical boundary layer heights and coefficients in $\sigma_y = cd^d$ (x in km) for different stabilities.

	Stability					
	A	B	C	D	E	F/G
Mixing height (m)	1300	900	850	800	400	100
c	213	156	104	68	50.5	34
d	0.894	0.894	0.894	0.894	0.894	0.894

where L is the mixing height. This equation is not valid for $H_e > L$ when the concentration is zero (*i.e.* the pollutant is emitted above the mixing height).

Where long period averages (*e.g.* annual) are of concern the detailed dependence on σ_y is of much less importance and the pollutant concentrations can be assumed to be uniformly distributed cross-wind within each wind sector. For a 30° sector the first two terms of Equation (9) become:

$$\frac{1.524Q}{U\sigma_z x} \tag{10}$$

The contribution of this wind sector to the overall annual average is then given by Equation (9) modified as in Equation (10) multiplied by the combined frequency of occurrence of that wind sector and stability category.

In section 10.2.1 we introduced the concept of the logarithmic wind speed profile with height in conditions of neutral stability. In different atmospheric stability conditions the variation will be different, but, in general, the wind speed will increase with height because of the surface drag. As would be expected intuitively this variation is smallest in unstable conditions (since in such boundary layers there is a considerable degree of vertical mixing) and greatest in stable conditions (for the opposite reason). In general one can write:

$$U(z) = U_{10}(z/10)^a \tag{11}$$

where U_{10} is the 10 metre wind speed and a is ~ 0.15 in unstable conditions, ~ 0.2 in neutral and ~ 0.25 in stable conditions.

There are several interesting derivations from the standard Gaussian equation and a particularly useful one is the formula giving the concentrations from a line source (of infinite length) obtained by integrating, for simplicity, Equation (2) over y to yield:

$$C(x) = \sqrt{\frac{2}{\pi}} \frac{Q}{\sigma_z U} \exp\left[-\frac{H_e^2}{2\sigma_z^2}\right] \tag{12}$$

which can be used to estimate the concentration downwind of roads, for example. Here, Q is the mass emission rate per unit length of road ($\mu g\ m^{-1}\ s^{-1}$).

10.3.2 Operational Models

Although much basic research has been, and continues to be, carried out in dispersion modelling, an essential feature of models is their use in practical operational situations. Such uses are increasingly made by non-specialists, and in the last decade or so there have been some significant developments in packaging models to facilitate their use in this way. These developments have been made possible by the rapid expansion in personal and desk-top computing over this period, so that quite major calculations are now possible on desk-based systems. It is worthwhile repeating, however, the crucial importance of high quality input data, particularly on emissions, as the numerical modelling calculations become easier through the use of standard packages.

Even without recourse to computers, some very useful estimates of air quality impacts of single sources can be made by the use of graphical workbooks such as the well-known *Workbook of Atmospheric Dispersion Estimates* by Turner,[22] and the *NRPB Report R91* referred to earlier.[2]

While such workbooks are probably most useful for screening calculations, more complex calculations require computer-based models, and the USEPA has produced a set of approved models which are freely available at http://www.epa.gov/ttn/scram/dispersion_prefrec.htm#aermod (accessed 24 January 2013) in its AERMOD and CALPUFF modelling systems. AERMOD is an extension of the Gaussian plume model, using boundary layer scaling where the boundary layer properties are determined by the Monin–Obukhov length and the boundary layer height. This represents an advance over the simple Gaussian approach in that it enables turbulence levels and hence plume dispersion (σ_y and σ_z) to vary with height in the boundary layer, whereas these are constant in the Gaussian approach. The models also have options to allow the influence of complicating factors such as building effects and the effects of complex terrain to be taken into account, and are supported by the USEPA, with full user guide literature freely available.

In the UK the most widely used urban scale dispersion model is ADMS (Atmospheric Dispersion Modelling System) originally developed by Cambridge Environmental Research Consultants (CERC) sponsored by a consortium of bodies including regulatory agencies, electricity generators, nuclear agencies and industrial companies. ADMS also uses the numerical description of the boundary layer based on the Monin–Obukhov length scale, discussed previously. ADMS is a commercially available model and has been extended to produce versions specifically directed to modelling in urban areas, near roads and airports. Full technical details are available on the ADMS website: http://www.cerc.co.uk/index.php (accessed 24 January 2013).

10.3.3 Emission Inventories

The importance of emission inventories in dispersion modelling cannot be overemphasized. We have already seen how important it is to specify the emission rate of a single source in order to model concentrations with confidence. In single stack applications this is often relatively straightforward. In general, the emission rate, Q, for a particular source is usually calculated as a product of some measure of activity, fuel consumption for example, or kilometres travelled in the case of a vehicle, and an emission factor. This latter could be a fuel sulfur content (allowing for some sulfur retained in the system, *e.g.* in the ash in coal combustion), an emission rate as measured on a vehicle as a mass emitted per kilometer travelled. Modelling of a single source can then be carried out. However, in multiple source applications, such as in the use of an urban air quality model, there can in principle be literally thousands of individual sources. It would be clearly impracticable to attempt to quantify the emission rate of every house, office, shop and car in, say, London so methods have to be devised of making the problem tractable yet retaining as accurate a description of reality as possible.

The usual way of achieving this is to apportion the area to be modelled into a grid and to combine all the numerous small emitters within each grid square (such as individual houses, cars of a given type-diesel or petrol for example,) into so-called 'area sources'. Major sources such as power stations are usually treated explicitly as individual point sources. The size of the grid square used will usually be determined by the size of the area, or domain, to be modelled and the computing resources available. Typical grid scales are 1 km (or smaller) for urban areas, 5–20 km for nationwide modelling in a country the size of the UK, and 25–50 km for European or other international scale long range transport models. However, the continuing increases in computer power mean that the spatial resolution of models is improving all the time.

The specification of emissions is therefore fundamental to modelling and, apart from some well-specified single sources, is a difficult task. The usual approach is to collect information on fuel consumption in particular sectors (such as power generation, domestic heating, *etc.*) and multiply this by appropriate emission factors which ideally will have been measured over a range of representative fuels, appliances and combustion conditions.

Very often such data on fuel consumption (or some other measure of industrial commercial activity) are available only on a large scale, *e.g.* at national level, when the area to be modelled is much smaller. The modeller then has to use some means of spatially disaggregating the total domain emissions over the individual grid squares of the model. This usually involves the introduction of another level of uncertainty as surrogate statistics have to be employed – for example domestic heating emissions may be assumed to have the same spatial pattern over the model grid as population for which data are often fairly readily available. Other surrogates which can be used are office floor space for emissions from the commercial sector and population for motor vehicle emissions. A detailed description of national emission inventories, disaggregated to small spatial scales (in this case 1 km×1 km) are to be found in the UK National Atmospheric Emission Inventory at http://naei.defra.gov.uk/ (accessed 22 January 2013).[11] Emission inventories for the countries of Europe are also produced as part of the EMEP (European Monitoring and Evaluation Programme) programme within the UNECE Convention on Long Range Transboundary Air Pollution. The data and related information are available from the Centre on Emission Inventories and Projections at http://www.ceip.at/ (accessed 23 January 2013).

As well as spatial resolution, temporal resolution is also an important issue for emission inventories and producing accurate hourly variations in emission rates for a wide variety of multiple sources can be a major challenge. Few measures of activity are available to assist in this task, although the emergence of traffic count measurements and automatic number plate recognition techniques has benefitted the calculation of vehicle emissions in urban areas considerably. Vehicle emissions are the most important source of population exposure to air pollutants in most cities in the developed world, and they are also one of the most difficult to quantify accurately. This is because they depend on a variety of factors such as the fuel and engine technology (petrol, diesel), exhaust after-treatment (three-way catalyst, particle filter), engine power, driving patterns and also road characteristics such as slopes and traffic congestion. Most of these are difficult to quantify accurately and improvements are still being made in vehicle emission inventories.[12]

Methodologies for producing emission inventories have been published by EMEP and the European Environment Agency.[13] The COPERT emissions model developed as part of co-operative European research on vehicle emissions is available at http://www.emisia.com/copert/General.html (accessed 23 January 2013) and provides reasonable estimates of vehicle emissions, although there may still be uncertainties surrounding modelled emissions and real-world values,[14] unless more up-to-date data are used.

10.3.4 Modelling beyond Urban Scales – Long Range Transport and Chemical Transport Models

Interest in pollutants other than primary combustion products has increased in the past few decades and considerations of atmospheric chemistry are now integral to many modelling studies. Examples include photochemical ozone, acid rain, and so-called secondary aerosols – particles containing sulfates, nitrates and organic aerosols which from parts of the PM (particulate matter) mix. Since the reactions forming these pollutants take place over times from minutes to days, pollutant transport distances of many hundreds, even thousands, of kilometres become important. Over these distances, simple single wind-direction Gaussian plume models are inappropriate, not least because wind directions can change significantly, and simple plume models are not able to handle complex chemistry well. There are several important features which must be considered in modelling long (in this context, greater than 100 km) as opposed to shorter range transport. Firstly the time scales of transport over these distances are such that the removal processes of wet and dry deposition must be incorporated. Secondly large scale meteorological features must be taken into account which involves specifying the movement of air masses on a synoptic scale. Thirdly chemical reactions will be important and must be included. To incorporate all these effects in detail imposes significant burdens on computer power, as well as raising questions over how accurately the input data on emissions and meteorology can be known and how well one can describe the detailed physics and chemistry of the processes. In addressing these issues, two model approaches are used. These are, Lagrangian, where a 'moving box' containing the pollutants moves over the domain in a series of air mass trajectories, or Eulerian, where the governing equations are solved for every grid point in the two-dimensional domain at each time step. In many ways Lagrangian models are the simpler in concept and application, as Eulerian models generally require greater complexity of input data and computing requirements, as well as suffering from numerical 'pseudodiffusion' if not appropriately constructed. Lagrangian models in general require fewer computer resources, allowing for example complex chemistry to be treated, and in the early days of addressing long-range transport, these were the model of choice. Lagrangian models are usually applied to a succession of air mass trajectories arriving at a receptor at relatively short intervals (*e.g.* every six hours). The specification of the trajectories is a fundamental step in using these models and they are usually obtained from the detailed models used in national meteorological services. In practice back-track trajectories up to 96 hours are used; longer timescales would introduce unacceptable errors. As it is, errors in trajectories increase with time back along the path, and particularly in slack pressure areas these can be very large even at relatively short times. These models work by moving the box or air parcel along the trajectory and at each time step the appropriate emissions are introduced from the underlying grid, pollutant is lost by dry deposition at the rate given by $v_g c$ where the deposition velocity, v_g, is appropriate to the under-lying surface type. Alternatively, more sophisticated dry deposition models may be used. Wet deposition removes pollutant according to the rainfall field at the particular location of the air parcel, and chemical transformations of reactive species are updated since the previous time step. A good summary of the use of Lagrangian and Eulerian long range transport models of acid rain has been given by Pasquill and Smith.[1]

Much use has been made of Lagrangian box models in modelling photo-chemical ozone formation in the UK. Because the physical and meteorological aspects of the problems are treated relatively simply in this application, complex chemical schemes can be used to describe the chemical processes involved. The model developed by Derwent,[15] for example, uses the Master Chemical Mechanism developed in the UK and which at that time involved 12 692 chemical reactions describing the fate of 4342 chemical species. Explicit chemistry is used rather than so-called 'lumped' schemes where for example one hydrocarbon is used as a surrogate for its class (*e.g.* propene could be used to describe all alkenes, *etc*).

Within the UK Meteorological Office, a regional Lagrangian model was developed following the nuclear accident at Chernobyl. This model, NAME, treats the pollutants as particles released into wind, temperature, and rainfall fields from the UK Meteorological Office's numerical weather prediction model. The model gives extremely accurate representations of the timing of pollution events at a given receptor, and has more recently been extended to incorporate chemical processes.[16] A further development in global modelling at the UK Meteorological Office has been *via* the STOCHEM model.[17] This is also a Lagrangian model but which operates in a global three-dimensional domain, incorporating chemistry as well as physical transport and dispersion. The model has been applied to problems in climate change, acidification and photochemical ozone at regional and global scales.

Another Lagrangian model which is widely used, particularly in the rest of Europe, is FLEX-PART,[18] see http://transport.nilu.no/flexpart for details (accessed 24 January 2013). This model only allows relatively limited treatment of chemical reactions but the trajectories can be calculated from detailed numerical weather forecasting models, most usually from the European Centre for Medium Range Weather Forecasting. Perhaps the most important thing to note about FLEX-PART however is that it is an open-source model, that is to say, the code is freely available in the internet. This is one example of a major development in numerical air pollution modelling in the past decade or so as models are increasingly being made available to users in this way.

Eulerian models were relatively sparsely used some years ago, largely because of their high demands on computer power. Advances in this regard over the past decade or so has meant that such models are now capable of handling complex meteorology and turbulence schemes as well as relatively complex chemistry, including both homogeneous gas phase and heterogeneous reactions. Moreover, as with FLEXPART above, several large, sophisticated Eulerian Chemistry-Transport Models (CTMs) are open source and freely available. They do nonetheless require a significant level of computing power.

The EMEP model, which provides scientific support to the UNECE Convention on Long Range Transboundary Air Pollution (CLRTAP), was originally developed as a Lagrangian box model in the early 1980s to treat acid rain problems in Europe. It grew to incorporate chemistry to deal with ozone, but more recently with the advent of more powerful parallel processing computers, the switch to an Eulerian framework was made.[19] This model is also an open source model available at http://www.emep.int/ (accessed 24 January 2013) and continues to underpin policy development in CLRTAP, not least through quantifying the impact of emissions in one country on the air quality in the others in Europe in the so-called 'blame matrices' published annually on the website.

Another European CTM which is becoming widely used is CHIMERE, developed in France and available at http://www.lmd.polytechnique.fr/chimere/ (accessed 24 January 2013). This has been widely applied by a variety of research groups and an extensive list of papers can be found on the CHIMERE website. One paper in particular, by Menut *et al.*, is of interest in the context of formulating the short time-scale variation in emission inventories discussed above.[20]

A further widely used model, developed initially in the US, is CMAQ (Community Multiscale Air Quality modelling system) another open source model which is now supported by the US Environmental Protection Agency available at http://www.cmaq-model.org/ (accessed 24 January 2013). This model has a wide user community who all contribute to scientific developments of the model, co-ordinated by a development group at the USEPA.

These large CTM models are now very sophisticated in that they are capable of handling relatively complex chemistry, sophisticated turbulence/deposition and solar radiation schemes, along with detailed wind fields. The models all need to be driven by detailed meteorological data from numerical weather models, and in fact the meteorological processing forms an integral part of the model and are inseparable from the CTM. Space does not permit a discussion of this aspect of the models and readers are referred to the model websites for further discussion.

The CTMs are generally applied over national to regional scales and with hourly time resolution over a year, spatial scales of $\sim 10–20\,km$ are feasible at the time of writing (2013), although computing improvements will continue.

The consequence of this spatial resolution is that modelling at urban scales is only approximate without further action. One approach has been to 'nest' a finer scale urban model into the larger CTMs .[21] Due care needs to be taken to minimize 'double counting' of emissions from the urban area which can contribute to the CTM and the urban scale models.

10.3.5 Uncertainty and Accuracy of Models

Outputs from all air quality dispersion models will be uncertain. Even with ideally accurate input data, the nature of atmospheric turbulence is inherently unpredictable. At best, a run of a model will give an ensemble mean of one set of atmospheric condition which is what an instantaneous measurement delivers. On top of this there are uncertainties in the emissions, in the meteorological inputs and in the description of the physical and chemical processes in the model.

Some models cope with this problem by scaling parameters in the model or the outputs to fit observations. While this may be defensible for obtaining full concentration fields in the areas where there are no monitors, it renders the model output questionable when used when projecting future concentrations. Far better in such cases to retain the deterministic nature of the physics and chemistry, but to obtain as good a measure of the uncertainty as possible. In this context therefore it is not helpful to think of 'validating' a model, but rather to evaluate quantitatively its performance and then to consider whether or not a model is acceptable for the use intended. In general, models perform better in reproducing measured annual average concentrations than they do for short-term (*e.g.* hourly) values. This is because long-term average emission rates and meteorology (particularly wind speeds) are generally reasonably well known. In very general terms, for annual averages in a model with well known emission rates, agreement with observations of the order of low 10s of percent could be expected, while factor of two accuracy would be typical for hourly values. In fact, the *EU Ambient Air Quality Directive (2008/50/EC)* requires modelling uncertainty to be no higher than 30% for annual averages of sulfur dioxide, NO_x, NO_2 and CO, but no higher than 50% for hourly averages of these pollutants. Recognising that emission rates are more uncertain for benzene, and both emission rates and formation mechanisms for PM_{10} and $PM_{2.5}$ are more uncertain, it allows an uncertainty of 50% for annual averages of these pollutants.

In carrying out the task of evaluating model performance, it is often helpful to compare not just one model against observations, but also an ensemble of models against each other. There have been several inter-comparison exercises in recent years. An international exercise known as AQMEII (Air Quality Model International Initiative) has recently taken place involving regional scale CTMs run by European and North American scientists and funded by the European Commission, the USEPA and Environment Canada. A good summary is given in a series of papers in a special edition of the journal Atmospheric Environment[23] which also covers the issue of the important meteorological pre-processing systems.

In the UK, Defra (Department for Environment, Food and Rural Affairs) has recently funded a model intercomparison exercise of models used in the UK to inform policy development. This exercise involved models in three groups – urban scale, regional scale and modelling of deposition. The intercomparison involved some novel techniques using the Openair software system[24] and has been published by Defra in three reports.[25] A range of statistical measures are used including the fraction of modelled values which are within a factor of two of the observations, mean bias (the average over the number of measurement sites of the difference between measured and modelled values), the normalized mean bias (mean bias divided by the mean observed value) and the root-mean-square error.

Finally, a very concise way of describing model performance is the Taylor diagram,[26] as shown in Figure 10.5. The correlation with observations is measured along the arc from the *y*-axis to the *x*-axis centred on the origin; the root-mean-square difference between modelled and observed values is measured in arcs centred on the 'observed' point on the *x*-axis, and the standard deviation of the simulated values is proportional to the radial distance from the origin. A 'good' model will therefore lie close to the 'observed' point on the *x*-axis, *i.e.* it would have a high correlation coefficient and a low RMS error.

In the context of the evaluation of air pollution control strategies, it should be noted that the Gaussian plume model and the behaviour of the primary pollutants to which it is usually applied are linear in emission rates, so that such models will predict concentration reductions proportional to emission reductions which might arise from any postulated control technology. This is very straightforward in single source problems; however, in urban areas a larger number of sources will be present and only one sector (*e.g.* motor vehicles under emission regulations) may be subject to controls. The accuracy of prediction of the effects of these controls will then depend on the accuracy

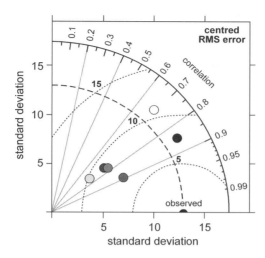

Figure 10.5   Taylor diagram showing the performance of six atmospheric models against measurements.

with which the 'source apportionment', *i.e.* the proportional contribution of the particular sources is predicted by the model, and this may be difficult to assess. It will depend to a large extent on how the model handles emissions at different release heights, which will in an Eulerian grid model depend on the vertical resolution of the grids, bearing in mind that in an emitting grid, the emission is uniformly spread over the whole grid, so if the grid volume is large, the calculated contribution can differ significantly from the true contribution of the sources in the grid.

Turning to the larger scale effects such as acid deposition, photochemical oxidant and secondary aerosol formation, the question of the linearity, or proportionality, is more important, that is, whether or not for a given reduction in emissions of a particular species (such as SO_2 for example) there is likely to be a proportional reduction in deposition of sulfur or the formation of sulfate particles. The formation of secondary sulfate aerosol, an important component of $PM_{2.5}$ and PM_{10}, has recently been shown to be significantly non-proportional.[27]

The problem of the evaluation of photochemical ozone formation is more complex in that ozone is formed from the atmospheric reactions of nitrogen oxides and many individual hydrocarbon species. The governing reaction schemes are overall non-linear and can be further complicated by the fact that NO_x (and some hydrocarbons) can act as both sources and sinks of ozone, on different scales.

With increasing interest in global problems, such as the effects greenhouse gases on climatic change, on the effects of man-made pollutants on stratospheric ozone, and on the recently recognized problem of inter-continental transport of air pollutants, modelling techniques are now being used to advise pollution control policies over a very wide range of atmospheric problems.

REFERENCES

1. F. Pasquill and F. B. Smith, *Atmospheric Diffusion*, Ellis Horwood, Chichester, 1983.
2. R. H. Clark, *A Model for Short and Medium Range Dispersion of Radionuclides Released to the Atmosphere*, First Report of a UK Working Group on Atmospheric Dispersion, NRPB Report R91, HMSO, London, 1979.
3. J. R. Garratt, *The Atmospheric Boundary Layer*, Cambridge University Press, 1992.
4. S. H. Derbyshire, Nieuwstadt's stable boundary layer revisited, *Q. J. R. Meteorol. Soc.*, 1990, **116**, 127–158.

5. A. P. van Ulden and A. A. M. Holtslag, Estimation of atmospheric boundary layer parameters for diffusion applications, *J. Climate Appl. Meteorol.*, 1985, **24**, 1196–1207.

6. A. M. Spanton and M. L. Williams, A comparison of the structure of the atmospheric boundary layers in Central London and a rural/suburban site using acoustic sounding', *Atmos. Environ.*, 1988, **22**, 211–223.

7. J. H. Seinfeld and S. N. Pandis, *Atmospheric Chemistry and Physics*, Wiley, New York, 1998.

8. G. A. Briggs, *Plume Rise*, US Atomic Energy Commission, Washington, DC, 1969.

9. D. J. Moore, A comparison of the trajectories of rising buoyant plumes with theoretical empirical models, *Atmos. Environ.*, 1974, **8**, 441–457.

10. J. A. Jones, *Models to Allow for the Effects of Coastal Sites, Plume Rise and Buildings on Dispersion of Radionuclides and Guidance on the Value of Deposition Velocity and Washout Coefficients*, Fifth Report of a UK Working Group on Atmospheric Dispersion, HMSO, London, 1983, NRPB Report 8157.

11. J. MacCarthy, G. Thistlethwaite, E. Salisbury, Y. Pang and T. Misselbrook, *Air Quality Pollutant Inventories for England, Scotland, Wales and Northern Ireland: 1990–2010*, AEA Technology, National Environmental Technology Centre, Culham, UK, 2012.

12. D. C. Carslaw, M. L. Williams, J. E. Tate and S. D. Beevers, The importance of high vehicle power for passenger car emissions, *Atmos. Environ.*, 2013, **68**, 8–16.

13. European Environment Agency, *EMEP/EEA Air Pollutant Emission Inventory Guidebook*, EEA Technical Report No. 9/2009, Copenhagen, ISSN 1725–2237, 2009.

14. D. C. Carslaw, S. D. Beevers, J. E. Tate, E. J. Westmoreland and M. L. Williams, Recent evidence concerning NO_x emissions from passenger cars and light duty vehicles, *Atmos. Environ.*, 2011, **45**, 7053–7063.

15. R. G. Derwent, C. S. Witham, S. R. Utembe, M. E. Jenkin and N. R. Passant, Ozone in Central England: the impact of 20 years of precursor emission controls in Europe, *Environ. Sci. Pol.*, 2010, **13**, 195–204.

16. D. B. Ryall and R. H. Maryon, *The NAME Dispersion Model: A Scientific Overview*, Met O APR Turbulence and Diffusion Note 217b, UK Meteorological Office, 1996.

17. W. J. Collins, D. S. Stevenson, C. E. Johnson and R. G. Derwent, Tropospheric ozone in a global-scale three-dimensional Lagrangian model and its response to NO_x emission controls, *J. Atmos. Chem.*, 1997, **26**, 223–274.

18. A. Stohl, Trajectory analysis – a new method to establish source–receptor relationships of air pollutants and its application to the transport of particulate sulfate in Europe, *Atmos. Environ.*, 1996, **30**, 579–587.

19. D. Simpson, A. Benedictow, H. Berge, R. Bergstrom, L. D. Emberson, H. Fagerli, C. R. Flechard, G. D. Hayman, M. Gauss, J. E. Jonson, M. E. Jenkin, A. Nyiri, C. Richter, V. S. Semeena, S. Tsyr, J.–P. Tuovinen, A. Valdebenito and P. Wind, The EMEP MSC-W chemical transport model-technical description, *Atmos. Chem. Phys.*, 2012, **12**, 7825–7865.

20. L. Menut, A. Goussebaile, B. Bessagnet, D. Khvorostiyanov and A. Ung, Impact of realistic hourly emissions profiles on air pollutants concentrations modeled with CHIMERE, *Atmos. Environ.*, 2012, **49**, 233–244.

21. S. D. Beevers, N. Kitwiroon, M. L. Williams and D. C. Carslaw, One way coupling of CMAQ and a road source dispersion model for fine scale air pollution predictions, *Atmos. Environ.*, 2012, **59**, 47–58.

22. D. B. Turner, *Workbook of Atmospheric Dispersion Estimates*, Lewis Publishers, Chelsea, MI, 2nd edn.,1994.

23. S. T. Rao, S. Galmarini and D. G. Steyn, Special issue: AQMEII: an international initiative for the evaluation of regional–scale air quality models – Phase 1, *Atmos. Environ.*, June 2012, 53.

24. D. C. Carslaw and K. Ropkins, Openair – an R package for air quality data analysis, *Environ. Modell. Softw.*, 2012, **27–28**, 52–61.

25. D. C. Carslaw, *Defra Urban Model Evaluation Analysis – Phase 1*, Defra, London, UK, 2011.

26. K. Taylor, Summarizing multiple aspects of model performance in a single diagram, *J. Geophys. Res.*, 2001, **106**(D7), 7183–7192.

27. R. M. Harrison, A. M. Jones, D. C. S. Beddows and R. G. Derwent, The effect of varying primary emissions on the concentrations of inorganic aerosols predicted by the enhanced UK Photochemical Trajectory Model, *Atmos. Environ.*, 2013, **69**, 211–218.

CHAPTER 11

Air Pollution and Health

ROBERT L. MAYNARD*[a,†] AND JON AYRES[b]

[a] School of Geography, Earth & Environmental Sciences, University of Birmingham, Edgbaston, Birmingham B15 2 TT, UK; [b] Institute of Occupational Health, School of Health and Population Sciences, University of Birmingham, Edgbaston, Birmingham B15 2 TT, UK
*Email: robertmaynard3@gmail.com

11.1 INTRODUCTION

Air pollution is a major cause of death and illness in both developed and developing countries around the world. It has been estimated that fine particulate air pollution contributes to about 3% of deaths from cardiovascular disease, to about 5% of mortality from cancer of the respiratory tract and to about 1% of deaths from acute respiratory infections in children under the age of five years.[1] Globally this amounts to about 0.8 million deaths and a loss of 6.4 million years of life every year: more than half of this occurring in Asia. In addition to the effect on mortality are the effects on morbidity or illness; these have not yet been estimated with the same accuracy as those on mortality.[1] That air pollution continues to play such an important part in causing death and illness is surprising to some. Air pollution levels in countries like the UK are probably as low as at any time since the industrial revolution and yet their impact on health continues. In 2010 the Department of Health Committee on the Medical Effects of Air Pollutants concluded that:

"Removing all anthropogenic ('human-made') particulate air pollution (measured as $PM_{2.5}$) could save the UK population approximately 36.5 million life years over the next 100 years and would be associated with an increase in UK life expectancy from birth, *i.e.* on average across new births, of six months. This shows the public health importance of taking measures to reduce air pollution.

A policy which aimed to reduce the annual average concentration of $PM_{2.5}$ by $1 \mu g \ m^{-3}$ would result in a saving of approximately 4 million life years or an increase in life expectancy of 20 days in people born in 2008.

The current (2008) burden of anthropogenic particulate matter air pollution is, with some simplifying assumptions, an effect on mortality in 2008 equivalent to nearly 29 000 deaths in the UK at

†This chapter is based upon an earlier contribution from Dr Sarah Walters which is gratefully acknowledged.

Pollution: Causes, Effects and Control, 5th Edition
Edited by R M Harrison
© Maynard and Ayres and The Royal Society of Chemistry 2014
Published by the Royal Society of Chemistry, www.rsc.org

typical ages and an associated loss of total population life of 340 000 life-years. The burden can also be represented as a loss of life expectancy from birth of approximately six months.

The uncertainties in these estimates need to be recognised: they could vary from about a sixth to double the figures shown."[2]

These are worrying findings and the more remarkable in that the days of the coal-smoke smogs are now almost historical. It was in 1952 that the London Smog focused attention on air pollution and triggered the expansion of research in this field. That expansion continues and in the short period since 2005 over a hundred time series epidemiological studies dealing with the effects on health of exposure to nitrogen dioxide have been published (personal communication: Inga Mills, 2012). So rapid is the current growth of the field that meta-analytical techniques are required to summarise what is now known.

In addition to these estimates of impact, new work has shown that the majority of the effects of air pollution are due to effects on the cardiovascular system and not, perhaps surprisingly, on the respiratory system. A major review by Brook *et al.*[3] has summarised the evidence. New work, for example by Tonne *et al.*[4] has linked levels of particulate matter measured as PM_{10} with changes in the thickness of the lining of the carotid artery. This work builds on earlier work by Künzli *et al.*[5] and on work on 'knock out' mice which are prone to atherosclerosis.[6,7] This body of evidence shows that particulate matter, by whatever mechanism(s), can affect the arterial system reducing blood flow to that part of the body which a particular artery may supply, thus increasing the risk of heart attack or stroke.

Remarkable as these findings are, there are still very large gaps in our knowledge. For example, we have little idea of which components of the ambient aerosol cause the effects noted above; we do not know whether nitrogen dioxide plays a direct role in causing effects on health or whether it acts as a surrogate for some fraction of the inhaled aerosol, and we are unable to identify those members of the population who are susceptible to the effects of air pollution on the cardiovascular system.

Perhaps more remarkable still is the finding that the effects of common air pollutants such as particulate matter and ozone do not seem to be characterised by a threshold of effect. Thus no ambient concentration can be regarded as without effects and no conventional standards offering a margin of safety can be set. An understanding of this has moved policy development away from attempts to meet standards and towards programmes of continuous reduction by cost–benefit tested measures. Standards continue to be set and efforts to meet them continue to be made but it might be better if such standards should be regarded as milestones along the road of progressive improvement. That some milestones are needed appears to be clear; that the reaching of milestones should not be misinterpreted as reaching the end of the journey is essential.

This chapter reviews the evidence for the effects of air pollutants on health. Inevitably the review is limited in scope and attention has been focused on the recent literature. The older literature has been extensively reviewed in the World Health Organisation's publication relating to *Air Quality Guidelines*[8–10] and in other monographs.[11,12]

11.2   EXPOSURE TO AIR POLLUTION

People are exposed to air pollutants every day of their lives and in all environments. Indeed, people are exposed to air pollutants, or at least to the effects of their mother's exposure to air pollutants, before they are born. Exposure varies with location: those working near busy roads are exposed to higher concentrations of traffic-generated air pollutants than those working in urban background or rural sites. Exposure occurs both indoors as well as outdoors. Air pollutants generated outdoors can filter into the indoor environment and affect indoor concentrations. For instance, in a non-smoking household, the main driver of indoor particle levels is the outdoor concentration of those particles. Air pollutants generated indoors may reach levels often substantially higher than those found outdoors; this is especially the case in households occupied by smokers or where cooking is undertaken in poorly ventilated rooms, notably in developing countries.

We know a great deal more of the relationships between outdoor concentrations of air pollutants and effects on health than we do of those between exposure to air pollutants and such effects. Personal exposure to air pollution can be monitored but this is expensive and the number of people who can be studied in this way, in any given programme of research, is perforce limited. Monitoring outdoor concentrations of air pollutants is, on the other hand, not too expensive and in countries such as the UK many outdoor monitoring stations have been established. This allows epidemiological studies to be undertaken and the expansion of outdoor monitoring has in large part been responsible for the surge of epidemiological studies in this field. The interpretation of the findings of such studies is predicated on the assumption that exposure, that is average exposure, is related to the outdoor, ambient concentrations of the pollutants in question. In general this is accepted though how exposure is distributed across the population is relatively unknown. This point is important; epidemiological studies that relate effects to a surrogate of exposure such as ambient concentration are susceptible to error.[13] The most common error occurs when a measurement of each person's exposure to a pollutant is incorrect. There is no reason to believe that the error is related to the true level: it is reasonable to assume that errors in measuring high levels of exposure will be no greater than those met in measuring low levels of exposure. Such an error blurs the exposure–response relationship and biases the results of regression analysis towards the null. Such an error is described as an example of a "classical measurement error".

The second type of error is less easy to grasp but it is potentially important in air pollution epidemiology. It occurs when a single measured value, for example of exposure, is assigned to all individuals in a study. It is likely that the true exposures for all individuals vary around the observed, monitored value. Thus the error is independent of the observed value. This is the difficult point: in the classical error model the error is independent of the *true* value; in the second type of error the error is independent of the *observed* value. The reason it is important is that if this type of error is present then ordinary linear regression is unbiased, *i.e.* it is not biased towards the null though the standard error of the regression coefficient is increased, but in the case of log-linear methods (including logistic regression, Cox regression and Poisson regression: methods widely used in the air pollution field) then the estimate can be biased away from, not towards, the null. This will not always occur; it will occur if the errors were greater for larger values of the observed variable than for smaller values of this variable.

In general, these problems are allowed for in the models that link concentrations of air pollutants with effects on health.

11.3   EPIDEMIOLOGICAL METHODS APPLIED IN THE AIR POLLUTION FIELD

A wide range of epidemiological methods have been applied in this field. Only three will be considered here: time series studies, cohort studies involving a number of locations with different long-term average concentrations of pollutants and intervention studies.

11.3.1   Time Series Methods

In essence these are easy to understand: the daily count of some variable related to ill-health (for example the daily death count) is plotted against the daily average concentration of some pollutant. If the death count varies with the concentration of the pollutant this suggests that there is some relationship between these two variables; specifically, that the daily variation in pollution might be causing the daily variation in deaths, the converse being unlikely. Such a straightforward approach was applied after the London Smog of 1952: daily death rates rose just after the initial increase in concentrations of sulfur dioxide and particulate matter (measured as Black Smoke) and followed the changes in pollution levels over the following week or so.[14] The evidence was so clear to the eye that everybody was convinced of the validity of the inference that air pollution was killing people. If the same approach were to be used today little linkage would be detected between the daily death count and, for example, daily average PM_{10} (for definition of PM_{10} see below) because the levels are so much lower. Nevertheless, using the more sophisticated statistical approaches available today, these variations have been consistently shown to be associated with daily health events.

It will be understood that factors other than concentrations of air pollutants may also vary from day to day (*e.g.* temperature) and some of these can also affect health. Some allowance for such factors is necessary if an effect of air pollution on health is to be detected. Factors such as temperature are known as "confounding" factors, adjustment for which is the essence of modern time series methods. A number of ways of adjusting are available: if we knew the exact relationship between temperature and the daily death count the adjustment would be straightforward.[15] Much more difficult are the adjustments needed when the effects of one pollutant are being dissected out from the effects of a mixture of pollutants. In the early days of air pollution time series studies the models used were single pollutant models. For example if nitrogen dioxide were being studied then only nitrogen dioxide concentrations, in addition to confounding factors such as temperature, would be entered into the statistical model that was being constructed. But imagine that the concentration of fine particles tracked with the concentration of nitrogen dioxide, it might appear that there was an association between nitrogen dioxide and deaths when in fact the real causal relationship was with the concentration of fine particles. Multi-pollutant models include more than one air pollutant in the statistical model of the relationship between concentration and response; this, of course, helps but is not always as helpful as one might wish. If two pollutants are very closely correlated then separating out their effects may well be impossible. In addition, as more and more variables are added to the models a loss of what is described as "stability" can occur and predictions become erratic and unreliable.

The confounding factors that are important in time series studies are those that vary from day to day. Thus temperature is important but cigarette smoking and socio-economic status are not. Standard statistical techniques have been developed and it is fair to say that time series analysis is now routine in the air pollution field. Perhaps because they are easy to do, the number of studies reported every year has risen to levels which make it very difficult to keep track of the literature. However, having such a large data base of studies of really rather similar design has allowed the development of meta-analytical techniques for combining their findings This has been a major step forward in this area and has been critically important is estimating the impact of air pollution on health in the UK and, more widely, in Europe (see, for example the detailed analysis of time series relating to effects on the cardiovascular system).[16]

As well as being confounders, air pollutants themselves and other factors such as temperature can also act as effect modifying factors. Suppose that the health response to fine particles is increased by co-exposure to nitrogen dioxide. It might be reasonable to infer that the effects of these pollutants were simply additive but it is also possible that nitrogen dioxide, though having no effects of its own, is in some way increasing the response to fine particles, *i.e.* acting as a permissive agent. These different types of effect modification need careful consideration when trying to dissect out the effects of air pollution on health.

Meta-analysis of literally hundreds of studies, and thousands of individual coefficients, has allowed summary coefficients to be derived for many associations between air pollutants and their effects. The power of meta-analysis lies in the combining of studies; summary coefficients are qualified by much narrower confidence intervals than individual studies alone thus increasing confidence in the accuracy of the size of the effect. The results of such analyses have been illustrated by "forest plots" which show the results of the studies included in the meta-analysis and the summary coefficient. An annotated example of a forest plot is shown in Figure 11.1.

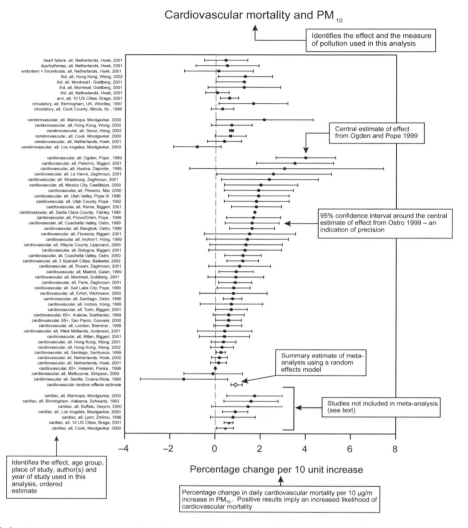

Figure 11.1 Forest plot showing the effects of PM₁₀ on cardiovascular mortality.

A range of such plots will be found in *Cardiovascular Disease and Air Pollution*, published by the Committee on the Medical Effects of Air Pollutants in 2006.[16] A glance at a selection of these plots will show that in the great majority of studies it was found that an increase in the concentration of particulate matter monitored as PM_{10} (see below) was associated with an increase in effects on health. The bottom line meta-analysis coefficient was expressed as (for instance) a cardiovascular mortality increase by 0.9% (CI: 0.7%–1.2%) for a $10\,\mu g\,m^{-3}$ increase in PM_{10}. The reader is referred to the report for a discussion of "random effects" *versus* "fixed effects" models, for an account of how publication bias was taken into account and for a discussion of sub-categories of the analysis. A selection of the bottom line results is provided in Table 11.1.

It will be seen that each pollutant is positively correlated with cardiovascular mortality. But even having allowed for confounding and effect modification of co-exposures the concentrations of some of these pollutants and the metrics of particulate matter may be closely correlated with each other which might contribute to some of results. The similarity between the coefficients for fine $PM_{2.5}$ and

Table 11.1 Cardiovascular mortality and air pollution: results of meta-analyses of time series studies.

Pollutant (all 24 hr average concentrations except ozone which is an 8 hour average)	Number of studies	Random effects estimate and confidence interval: percentage change per 10 μg m^{-3}
PM10*	40	0.9 (0.7, 1.2)
PM2.5*	9	1.4 (0.7, 2.2)
Black Smoke*	29	0.6 (0.4, 0.7)
TSP*	21	0.5 (0.3, 0.8)
SO_2	67	0.8 (0.6, 1.0)
O_3	26	0.4 (0.3, 0.5)
CO	12	1.1 (0.2, 2.1)
NO_2	44	1.0 (0.8, 1.3)

*Metrics of particulate matter are marked with an asterisk.

nitrogen dioxide might well be due to concentrations of these pollutants being closely correlated by virtue of their being produced by similar sources: in urban areas the major source is traffic. Nevertheless, the consistency of the findings is strong evidence of a real effect with, on average considering all countries or areas together, an accurate enough assessment of effect size to be helpful in informing policy development in air pollution control.

11.3.2 Cohort Studies

A second, very useful but expensive method of defining causation and effect size in this area is the use of cohorts of subjects. Imagine a cohort of a million people, all followed-up from birth, their personal habits all known, their occupational histories meticulously recorded, their health checked at intervals and their deaths fully investigated as they occur. If this cohort were then divided into ten large groups according to the levels of air pollution in the areas in which they lived their whole lives this would provide an excellent analytical framework to look at the effects of long-term exposure to air pollution on health. The confounding factors that would need to be considered in this sort of study are those which do not change from day to day but which are more likely to remain constant over long periods. Thus smoking habits, obesity, occupational history and socio-economic status are all key factors and have to be allowed for in the analyses. Setting up a cohort of a million people would be expensive but work using small cohorts set up deliberately to study the effects of air pollutants and work using existing cohorts has been done and has produced remarkable findings.

The first study of this kind was reported as part of the Six Cities Study set up by Harvard University. This prospective study involved 8111 people, aged 25–74, living in six rather different US cities.[17,18] Air pollutant monitors were installed in each city. The second, rather larger study capitalised on an existing cohort (the American Cancer Society cohort), involves 552 138 individuals from 151 US metropolitan areas from which routinely collected data on PM2.5 were available from 50 areas and data on sulfate concentrations from all areas.[19] The findings of this seminal study have been considered in depth, argued about and reanalysed and found to be reliable.[20,21] The bottom line findings are:

(i) Long term exposure to a pollutant increment of 10 μg m^{-3} PM2.5 is associated with a 6% increase in risk of death, at all ages, from cardiovascular disease.

(ii) The effect on the risk of death from lung disease, with the exception of lung cancer, is small (see ref. 16 for details, page 113). For lung cancer the increased risk is 14% per 10 μg m^{-3} PM2.5.

(iii) The effect on the risk of death from diseases other than cardiovascular and respiratory is zero.

Only a very brief summary of these important studies has been provided; the reader is urged to read not only the original papers published by Pope and colleagues but also the exceptionally detailed reports and reanalyses of the work that have been published both in the US and in the UK. More recent cohort studies have been undertaken at a smaller spatial scale than the original studies by Dockery and Pope (see above) and have reported larger effect sizes than those in the original work.[22] This may be due to a reduction in errors consequent upon equating ambient concentrations with exposure across a large area, or perhaps to the greater toxicity per µg of $PM_{2.5}$ in urban areas where the major source of $PM_{2.5}$ is traffic.

The results of the cohort studies discussed have been used to quantify the effects on health of long-term exposure to particulate air pollution.[2] The method used is, in principle, easy to understand; in practice it is requires a considerable amount of calculation. The essential starting point is the finding that long term exposure to unit $PM_{2.5}$ ($1\ \mu g\ m^{-3}$) increases the risk of death from cardiovascular diseases, at all adult ages, by a constant percentage. This is important to grasp: exposure to unit concentration increases the risk of death in a 25 year old by x% and in a 75 year old also by x%. Of course the underlying risk is less in the 25 year old than in the 75 year old and so the actual increases in risk are not identical. The second thing to understand is the Life Table. A Life Table is, for our purposes, simply a tabulation of the risk of death, either from all non-accidental causes or from specific causes such as cardiovascular disease, for all ages. Thus the table specifies the base-line risks (referred to by users of Life Tables as "base-line hazards") for each age group. It is a theoretical tool but one which can be populated by real data to extrapolate effects into the future.

Imagine a cohort of 1 million people aged 25 who could be followed through the life table, noting that a certain number die each year in accordance with the base-line risks specified by the table. If we followed the cohort for 100 years all would have died and we would be able to add up the total number of years lived by the cohort. Let us call this result "A" life years. Now repeat the process again but, this time, adjust all the baseline risks for a given level of pollution and let us call this new total of years lived "B" life years. B will be smaller than A and the difference will be due to the increased risk of death at each age caused by exposure to pollution. This is the principle; for the practice see the COMEAP report.[2] The adoption of this method has been a significant step forward in quantifying the effects of particulate matter on health. It has the great benefit that the results can be used for developing policy: the benefits of reducing levels of particulate matter can be calculated and weighed against the costs of the reductions.

11.3.3 Intervention Studies

Epidemiological studies of the effects of air pollutants on health began with "unhappy accidents" such as the London Smog of 1952; recent studies have capitalised on "happy accidents" involving sudden reductions in emissions of certain pollutants arising, usually, from policy interventions. Banning coal sales in Dublin, reducing the sulfur content of oil used by power plants and road vehicles in Hong Kong and changing transportation patterns before the Olympic Games in Atlanta (USA) all produced improvements in health.[23–25] Industrial action by workers cannot be regarded as a "happy accident", but Pope *et al.* studied the effects of a strike at a steel mill in the Utah Valley and showed a reduction in admissions of children to hospital for treatment of respiratory diseases during the period when the steel mill was inoperative.[26] It was also shown, uniquely, that the toxicological activity of the ambient aerosol decreased during the strike due to reduced emissions from the plant.[27]

11.4 INDIVIDUAL AIR POLLUTANTS

11.4.1 Particulate Matter

Particulate matter (PM) is the most intriguing of all the classical air pollutants. Epidemiological studies have shown clear associations between ambient concentrations of PM and a number of indices of effects on health. The number of studies, the range of methods used, the number of locations in which the studies have been undertaken, the consistency of the results and the evidence of coherence (if there are effects of deaths from heart disease there should be, and are, effects on admissions to hospital for treatment of heart disease) provide strong grounds for thinking that the associations are causal in nature. This conclusion is now widely accepted.[28]

PM comprises a mixture of materials and particles of different sizes. A complete analysis of the composition of, for example, particles comprising the ambient aerosol in urban areas would reveal dozens or more likely hundreds of chemical species. Attention has been focused on the major components: the soluble inorganic fraction comprising sulfates and nitrates; the insoluble organic fraction including, for example, polycyclic aromatic hydrocarbon compounds; the carbon particles produced by condensation and agglomeration of material produced by engines of motor vehicles; and the silicaceous crustal material contributed, in part, by the resuspension of road dust. A range of metals have been identified including iron, copper, nickel, vanadium, caesium, selenium, chromium, cobalt, lead, and mercury. It requires little imagination to realise that untangling the toxicology of such a mixture is challenging.

Attention has focused on those particles that are likely to reach the lung and to be deposited there. This has led to the characterisation of the ambient aerosol in terms of particle size. In general, particles with a diameter of more than $10\,\mu m$ are unlikely to reach the airways beyond the larynx, being deposited in the nose, naso-pharynx and larynx. Particles of less than $10\,\mu m$ diameter have a high probability of being carried beyond the larynx and into the lungs. These particles may be deposited within the lungs by a variety of mechanisms: the larger particles by impaction and sedimentation, the smaller ones by diffusion. Where along the pathways of the lung particles are most likely to be deposited is controlled by their size. Small particles are more likely to deposit in the gas exchange zone than the larger particles. Particles that can enter the upper airways are described as "inhalable" particles; whether a particle is in fact inhalable depends on its size and on whether the rate at which it will be carried into the nose exceeds the rate at which it will sediment in the ambient air around the nose. To help define the inhalable characteristics of particles, PM size is subdivided into two overlapping fractions: the "thoracic" fraction and the "respirable" fraction. The thoracic fraction is, in general, less than $10\,\mu m$ in aerodynamic diameter; the respirable fraction is, in general, of less than $4\,\mu m$ aerodynamic diameter.

The widely used indices of particle concentration, PM_{10} and $PM_{2.5}$, while similar to the thoracic and respirable fractions of inhalable particles are not identical in that PM_{10} and $PM_{2.5}$ are size specified fractions of the ambient aerosol, rather than of the inhalable fraction of the ambient aerosol. PM_{10} represents the mass concentration of particles that pass a size-selective filter that is characterised by passing 50% of particles of exactly $10\,\mu m$ aerodynamic diameter, rejecting all particles of more than about $15\,\mu m$ aerodynamic diameter and accepting essentially all particles of less than about $5\,\mu m$ aerodynamic diameter. $PM_{2.5}$ is defined in the same way except that the 50% cut point is at $2.5\,\mu m$ aerodynamic diameter. PM_{10} therefore includes $PM_{2.5}$. Setting a cut point at $2.5\,\mu m$ aerodynamic diameter has advantages: it allows separation of the accumulation mode particles from the coarse mode particles and includes most of those with the greatest probability of deposition in the gas exchange zone of the lung, namely in the respiratory bronchioles, alveolar ducts and alveoli. PM_{10} and $PM_{2.5}$ are metrics of concentration and are expressed in terms of $\mu g\ m^{-3}$. They differ in content by place and time and should not be referred to as "$PM_{10}s$" or "$PM_{2.5}s$".

Though some 70% of particles of about 2.5 µm aerodynamic diameter reach the proximal parts of the gas exchange zone (respiratory bronchioles: alveolar ducts), only about 20–30% of these particles are deposited in this zone: the remainder pass out with the expired air. Deposition by diffusion is related to the reciprocal of the square root of the diameter of the particle (real diameter, not aerodynamic diameter; shape is not important for diffusion); deposition by sedimentation is related to the square of the aerodynamic diameter of the particle. Thus the smaller the particle, the more effectively it is deposited by diffusion and the less effectively it is deposited by sedimentation. These conflicting mechanisms produce a minimum of deposition at about 0.5 µm diameter: at yet smaller diameters the percentage deposited increases and peaks at about 60% for particles of about 20 nm diameter. This means that very small particles deposit efficiently in the gas exchange zone. Inspired air flows rapidly in the large airways but very slowly in the small airways, and the final step of fresh inspired air reaching the alveolar walls is accomplished by diffusion rather than by bulk flow. Only very small particles diffuse sufficiently rapidly to be deposited efficiently in the alveoli.

11.4.2 Problems with Mass as a Metric of Dose

However, the mass of inhaled PM may not be the best measure of exposure when considering health effects. Let us assume that a typical $PM_{2.5}$ in urban air in the UK is 12 µg m^{-3}. Let us also assume that an individual breathes 20 m^3 of air per day and that all the particles measured as $PM_{2.5}$ are deposited in the lung. Every day some 240 µg of particles would be deposited. If all the particles that were deposited were evenly spread out on the alveolar surface (for the whole lung this is about 100 m^2) then the dose to that individual would be 2.4 µg m^{-2} or 2.4×10^{-12} µg µm^{-2}. Each human alveolus has an area of about 250 000 µm^2 and is covered by about 30 Type I alveolar cells.[29] The area of each Type I cell is thus about 8000 µm^2. The dose per cell would be $2.4 \times 10^{-12} \times 8000$ µg, *i.e.* about 20×10^{-9} µg per cell. Some toxicologists find it difficult to believe that such a dose of PM could do much harm. Suggesting that the mass would accumulate over time does not solve the problem; we know from time series studies that a small, transient change in ambient concentration produces an effect. Of course the distribution on the alveolar surface might well not be even; a denser deposit in the respiratory bronchioles and alveolar ducts is likely. But it remains difficult to explain the effects in terms of conventional toxicology. The answer is that we need to think beyond conventional toxicology.

11.4.3 Other Possible Metrics of "Dose"

11.4.3.1 *Particle Number or Surface Area*

Seaton *et al.*[30] and Oberdörster *et al.*[31] proposed in 1995 that ultrafine particles (particles of less than 100 nm diameter) might be playing an important role in causing the reported effects of exposure to ambient particulate matter especially as regards effects on the cardiovascular system. They stressed that the key might lie in the number of particles deposited in the lung rather than the mass of those particles. Table 11.2 makes the point that if mass is divided amongst very small particles the number concentration of particles will be high.

If we imagine, for a moment, that the mass concentration of 12 µg m^{-3} were contained *only* in 20 nm particles then the number concentration would be about 3 million per cm^3 or 3×10^{12} per m^3. Assuming 20 m^3 of air is taken in per day and that all the particles are deposited: the daily deposition would be 60×10^{12} particles. Or, 60×10^{12} spread over 100 m^2, 0.6×10^{12} per m^2, 0.6 per µm^2, $0.6 \times 8000 = 4800$ per Type I cell. This sounds more impressive than 20×10^{-9} µg per cell, as long as one can believe that cells respond to numbers of particles rather than to the mass of particles. Of course not all the mass is contained in 20 nm particles but the argument has some force in

Table 11.2   Particle number and surface area per $10\,\mu\mathrm{g}\ \mathrm{m}^{-3}$ airborne particles. (Adapted from Oberdörster).[32]

Particle diameter (nm)	Number of particles per cm³	Total surface area of particles (μm²) per cm³
5	153 000 000	12 000
20	2 400 000	3016
250	1200	240
5000 (5 μm)	0.15	12

that it suggests that the very large numbers of very small particles found in the ambient aerosol might be punching literally above their weight in producing effects on health.

A more subtle answer may lie in the importance of the surface area of the particles; this too will increase as a given mass is more and more finely subdivided, see Table 11.2. In short, the cell may respond to the amount of surface area of PM presented to its cell membrane rather than the number of particles taken into the cell. Oberdörster's important review of 2005 provides a detailed discussion of the evidence that supports this conjecture.[32] From a physiological standpoint, the suggestion makes excellent sense if it is assumed that some receptor mechanism is involved rather than a toxicological mechanism involving direct damage to cells. One might expect the "area-dose" rather than mass dose to be related to the number of receptors, using the term widely, that are activated. This idea of cells recognising particles and responding by releasing mediators has been followed up in detail by a number of workers.[32,33] Nano-toxicology is a fast-growing area of toxicology which grew out of air pollution studies.

11.4.3.2   Particle Specific Toxicity

It has also been suggested that particles are involved in the generation of free radicals in the lung.[34] Free radicals have long been known to play an important part in the toxicology of compounds including paraquat, ozone, oxygen, carbon tetrachloride and many other toxicants.[35] Metals could well be involved in this and might act as catalysts of free radical generation. If this turns out to be so, then an explanation of how such a small mass dose of metals could have significant effects may have been produced; catalysts, like enzymes, would not used up by reactions which they catalyse and thus even a small amount of metal could promote the generation of many free radicals. It has been suggested that hydrogen ions play a key role in the toxicity of the ambient aerosol.[12] This suggestion fits with the results of studies that link inorganic sulfate with effects on health. Inorganic sulfate (mainly ammonium sulfate) is not toxicologically active, *per se*, but may be acting as a marker for acidic species generated along the path, so to speak, from sulfur dioxide to sulfate.

Whatever the mechanisms of activity of particles in the lung there seems little doubt that these can have knock-on effects on the cardiovascular system. Changes in clotting factors, in the stability of atheromatous plaques and in the autonomic regulation of the heart beat have all been linked with exposure of both experimental animals and people to particles. A detailed discussion of the evidence is provided in the Department of Health *Cardiovascular Disease and Air Pollution* report.[16]

11.4.4   World Health Organisation Air Quality Guidelines for Particulate Matter

The recent and rapid developments in the area of particulate matter have led the World Health Organisation to review its *Guidelines on Air Quality*.[10] The current guidance acknowledges the absence of a threshold of effect and recommends Interim Targets in addition to more conventional

Table 11.3 Air Quality Guideline and Interim Targets for PM: Annual Mean. (Reproduced by kind permission of the World Health Organization).

Annual Mean Level	PM_{10} ($\mu g\ m^{-3}$)	$PM_{2.5}$ ($\mu g\ m^{-3}$)	Basis for the selected level
WHO Interim target 1 (IT-1)	70	35	These levels are estimated to be associated with about 15% higher long-term mortality than at AQG levels.
WHO Interim target 2 (IT-2)	50	25	In addition to other health benefits, these levels lower risk of premature mortality by approximately 6% (2–11%) compared to IT-1.
WHO Interim target 3 (IT-3)	30	15	In addition to other health benefits, these levels lower risk of premature mortality by approximately another 6% (2–11%) compared to IT-2.
WHO air quality guidelines (AQG)	20	10	These are the lowest levels at which total cardio-pulmonary and lung cancer mortality have been shown to increase with more than 95% confidence in response to $PM_{2.5}$ in the ACS study by Pope *et al.*[19,63] The use of the $PM_{2.5}$ guideline is preferred.

Table 11.4 Air Quality Guideline and Interim Targets for PM: 24-hour mean. (Reproduced by kind permission of the World Health Organization).

24-hour mean level[a]	PM_{10} ($\mu g\ m^{-3}$)	$PM_{2.5}$ ($\mu g\ m^{-3}$)	Basis for selected level
WHO Interim target 1 (IT-1)	150	75	Based on published risk coefficients from multicentre studies and meta-analyses (about 5% increase in short-term mortality over AQG).
WHO Interim target 2 (IT-2)	100	50	Based on published risk coefficients from multicentre studies and meta-analyses (about 2.5% increase in short-term mortality over AQG).
WHO Interim target 3 (IT-3)[b]	75	37.5	About 1.2% increase in short-term mortality over AQG.
WHO air quality guidelines (AQG)	50	25	Based on relationship between 24-hour and annual PM levels.

[a]99th percentile (3 days per year).
[b]For management purposes, based on annual average guideline values, the precise number to be determined on the basis of local frequency distribution of daily means.

guidelines. The recommendations are set out in Tables 11.3 and 11.4. Note that the 24 hour guideline is based on the annual guideline and that the latter does not posit a threshold of effect.

11.4.5 Nitrogen Dioxide

Nitrogen dioxide is a gas, the main sources in urban environments being vehicles and space heating. Nitrogen dioxide is not as soluble in water as sulfur dioxide and, like ozone, penetrates further into the respiratory tract when inhaled. Epidemiological studies have shown consistent associations between day-to-day variations in the concentration of nitrogen dioxide and both all-cause and cardiovascular mortality. A forest plot of time series studies is shown in Figure 11.2 (from ref 16).

The rather close similarity between Figure 11.1 and Figure 11.2 will be appreciated. Concentrations of NO_2 are usually closely correlated with concentrations of fine particles and this makes separation of the effects of these two pollutants difficult. Some experts regard NO_2 as *per se*

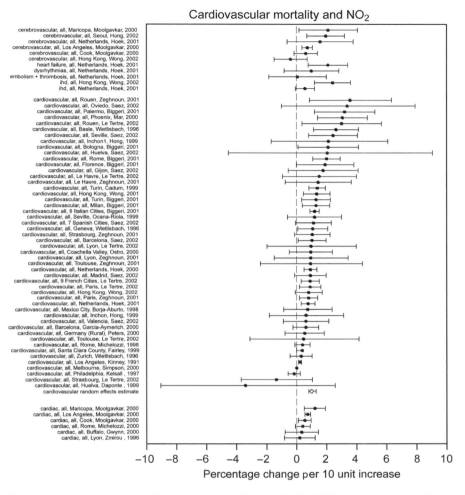

Figure 11.2 Forest plot showing cardiovascular mortality and daily NO2 concentrations. Figure shows the percentage increase in cardiovascular mortality per $10 \, \mu g \, m^{-3}$ increase in concentration of NO_2.

toxicologically active in low concentrations; others regard NO_2 merely as a marker for vehicle-generated pollutants: the effects being attributed to particles. This question has been examined by time-series studies which, in general, use single-pollutant models. A meta-analysis of 109 such studies was undertaken by Steib *et al*.:[36] 32 coefficients for NO_2 were included from single-pollutant models and 15 from multi-pollutant models. The "bottom line" result was expressed as:

(i) For single-pollutant models: a 24 ppb increment in NO_2 is associated with a 2.8% (CI: 2.1, 3.5) change in all-cause mortality.
(ii) For multi-pollutant models: a 24 ppb increment in NO_2 is associated with a 0.9% (CI: −0.1, 2.0) change in all-cause mortality.

It will be noted that the meta-analysis coefficient derived from multi-pollutant models is one third of that derived from single-pollutant models and, in formal statistical terms, lacks significance. This may be due to allowance being made in multi-pollutant models for the effects of other pollutants. If this finding is taken at face value, then it seems possible that nitrogen dioxide is acting as marker or surrogate for fine particles.

Effects on respiratory admissions to hospital have been reported.[37–39] In addition to these, studies of effects of short-term variations in exposure to NO_2 have reported that children growing up in areas with relatively high long-term average concentrations of NO_2 suffer from retarded lung development although these effects are reversible if exposure is reduced.[40–42] These results are supported by work on European adults.[43]

Explaining such effects in toxicological terms has not been successful. Exposure to high concentrations of NO_2 causes damage to the lung and damage to Type I alveolar cells, and ciliated airway epithelial cells has been shown, in rats, exposed to 340 ppb ($640\,\mu g\,m^{-3}$) NO_2.[44] Long-term exposure to NO_2 has been shown to produce emphysema-like changes in several species.[45] In man, changes in airway resistance (broncho-constriction) are seen on exposure to 300 ppb ($520\,\mu g\,m^{-3}$) NO_2 for 2–2.5 hours, but the changes tended to be small and of questionable clinical significance.[46–48] Airway inflammation has been recorded in man on exposure to 1 ppm ($1880\,\mu g\,m^{-3}$) NO_2 but not at 600 ppb ($1130\,\mu g\,m^{-3}$) – the latter being given as 4 separate 2 hour exposure over 6 days.[49–51] None of these studies provides an explanation for the effects of NO_2 on all-cause mortality reported in time-series studies. It is particularly difficult to account for the lack of a threshold of effect for nitrogen dioxide in that it is difficult to understand how exposure to very low concentrations could produce damage to the airways.

WHO has taken a cautious approach and has recommended the following guidelines:[10]

(i) Annual average concentration: $40\,\mu g\,m^{-3}$.
(ii) 1 hour average concentration: $200\,\mu g\,m^{-3}$.

Recent discussions of the effects on health of nitrogen dioxide have concluded that the evidence for an effect of nitrogen dioxide, *per se*, has strengthened. A summary of the discussions has been published (see *Review of Evidence on Health Aspects of Air Pollution* – REVIHAAP; www.euro.who.int).

11.4.6 Sulfur Dioxide

Sulfur dioxide is one of what might be regarded as the older air pollutants; it was a key component of the coal smoke smogs of the UK in the period before the *Clean Air Act of 1956*. In the UK today, power stations which burn fossil fuel are a major source of sulfur dioxide. Sulfur dioxide is converted into sulfur trioxide, a very hygroscopic gas, which condenses with water as sulfuric acid.[12] Sulfuric acid can react with ammonia to form the bisulfate (also a strong acid) and ammonium sulfate. Ammonium sulfate is an almost neutral salt, *almost* because the ammonium ion can act as a proton donor (drinking a solution of ammonium chloride leads to acidification of the blood).

Sulfur dioxide is very soluble in water and is removed from inspired air largely in the upper airways. Absorbed SO_2 reacts with water for form sulfite and bisulfite ions:

$$SO_2 + H_2O \rightarrow H_2SO_3$$

$$H_2SO_3 + H_2O \rightarrow HSO_3^- + H_3O^+ \, (pKa = 1.86)$$

$$HSO_3^- + H_2O \rightarrow SO_3^{2-} + H_3O^+ \, (pKa = 7.2)$$

Bisulfite ions are converted to sulfate ions under the influence, *in vivo*, of a molybdenum-containing oxidase enzyme. Bisulfite ions also react with oxidised glutathione; bisulfite ions are thus a reducing agent and contributed to the reducing activity of coal smoke smog. It is important to recall that the "smog" which occurs in many cities in developed countries today is an oxidant, or oxidative smog containing nitrogen dioxide and ozone. Sulfurous acid (H_2SO_3) can be converted to sulfuric acid (H_2SO_4) and this may occur at the surface of inhaled particles, especially if catalytically active metal species are present. This may explain the increased effects of SO_2 in the presence of an ultrafine zinc aerosol.[52] The idea that sulfur dioxide is transported into the lung in the form of acid formed at the surface of particles is appealing. Lippmann has made a strong case for the importance of hydrogen ions generated from acid which, in turn, has been produced from sulfur dioxide.[12]

Asthmatic subjects respond to inhaling SO_2 at concentrations of 100–400 ppb (286–1144 μg m^{-3}) with increased airway resistance (thus reducing airflow) though clinically significant effects at the lower end of this range are unlikely.[53] Non-asthmatic subjects are much less sensitive and > 1000 μg m^{-3} are needed to produce broncho-constriction. Inhaled SO_2 interacts with irritant receptors in the airway wall and this may lead to reflex changes in the control of heart rate.[54] The significance of these changes in subjects with normal myocardial function is unknown. Broncho-constrictor responses to SO_2 appear rapidly as is classically the case with irritants; the response is dependent on concentration to a much larger extent than on duration of exposure. In terms of the Haber relationship where the response to an inhaled toxicant is proportional to $C^n t$ where C is the concentration, t is the duration of exposure and n is a constant; for sulfur dioxide, n is approximately equal to 4. This means that the duration of exposure is rather unimportant in controlling the response to exposure. In many people the response passes off rapidly; in some it may persist for longer.[12] Sulfur dioxide impairs ciliary function, which may contribute to the induction of chronic bronchial inflammation in those exposed to high concentrations. It has been suggested that exposure to sulfur dioxide, and particulate matter, was a cause of chronic bronchitis in the UK in the coal-smog era.[55]

Time-series epidemiological studies show clear associations between all-cause mortality (excluding accidental deaths) and daily variations in SO_2 concentrations. A major European study conducted in 12 cities reported a 2.9% (CI: 2.3–4.5) increase in mortality per 50 μg m^{-3} increase in SO_2 concentration.[56] Similar findings have been reported by Samoli and Ballester.[57,58] The large US NNMAPS Study[59,60] reported a smaller coefficient: 1.1% (CI: 0.5–1.7) for the same pollutant increment. The mechanism by which short-term exposure to SO_2 increases all-cause mortality is unknown.

Hedley *et al.*[24] reported an important intervention study from Hong Kong. Reductions in sulfur content of fuel oil led to a fall in SO_2 concentrations from 44 to 21 μg m^{-3}; particle concentrations were unchanged. Remarkably, the increasing annual death rate (due to ageing of the population) for respiratory diseases fell by 3.9% and for cardiovascular diseases by 2.0%. This has been seen as evidence that long-term exposure to even low concentrations of SO_2 causes cardio-respiratory deaths. Interestingly, the vanadium content of particles also fell – the significance of this is unclear at present.

Time-series studies have also shown that SO_2 is associated with hospital admissions. Table 11.5 summarises the results of studies in this area and is taken, with permission, from reference 16.

The consistency of effects on the cardiovascular system is striking and unexpected; SO_2 has been regarded, classically, as a pollutant likely to affect the respiratory system.

Interestingly, a re-analysis of the major US cohort studies showed a clear association between long-term average concentrations of SO_2 and all-cause mortality.[20] All-cause mortality is, in these studies dominated by cardiovascular deaths. Remarkably, the effect of SO_2 remained statistically significant after adjustment for the co-variables $PM_{2.5}$ and sulfate (SO_4^{2-}) concentration. Once again, the question of how SO_2 causes such deaths is unanswered; a reflex effect on the control of the heart is at least possible.

Table 11.5 Summary table based on meta-analysis of time-series studies. (Reproduced by kind permission of the UK Department of Health).

Pollutant (24 hr average)	N	Outcome measure	Assessment	Random effects (95% CI) (% change per 10 μg m^{-3})
SO$_2$	67	CV mortality	+	0.8 (0.6, 1.0)
SO$_2$	7	CV admissions	+	0.6 (0.1, 1.2)
SO$_2$	18	Cardiac admissions	+	2.4 (1.6, 3.3)
SO$_2$	10	IHD admissions	+	1.2 (0.5, 1.9)
SO$_2$	5	Heart failure admissions		0.9 (−0.1, 1.8)
SO$_2$	7	Cerebrovascular admissions		0.3 (−0.5, 1.1)

11.4.7 Ozone

Concentrations of ozone in urban areas of the UK are rising slowly. In part, this is due to an increase in Northern Hemisphere background ozone concentrations and in part to a reduction in ozone scavenging by nitric oxide in urban areas as controls on emissions of oxides of nitrogen (NO$_x$) begin to bite. In the UK, ozone concentrations tend to be higher in rural than in urban areas because of the scavenging effect of nitric oxide; in California, on the other hand, the strong sunlight, high traffic density and production of reactive organic species (see below) leads to ozone concentrations being high in both rural and urban areas. This is also the case in parts of the Mediterranean area. Ozone reacts with nitric oxide to produce nitrogen dioxide:

$$NO + O_3 \rightarrow NO_2 + O_2$$

NO$_2$ breaks down under the action of sunlight and this leads to ozone formation:

$$NO_2 + h\nu \rightarrow NO + O^{\bullet}$$

$$O_2 + O^{\bullet} \rightarrow O_3$$

$$RO_2 + NO \rightarrow RO + NO_2$$

The final reaction in the series shows the reconstitution of nitrogen dioxide by reaction between organic oxidants, produced by motor vehicles, and nitric oxide. It will be noted that the reactions form a cycle; as long sunlight and the organic oxidants are present, ozone will continue to be formed. Concentrations of ozone close to the ground fall during the hours of darkness due to reaction with surface structures; down-mixing of air containing higher concentrations occurs in the morning.

Ozone is regarded as an outdoor air pollutant, though in summer, when windows and door are open, indoor levels rise towards those outdoors. Because the formation of ozone is dependent largely on the formation of traffic-generated precursors and on the action of sunlight, it is not surprising that ozone concentrations reach a peak in the afternoon. The toxicological response to ozone seems to be about equi-dependent on concentration and duration of exposure, and peak daily 8-hour average ozone concentrations have been used as a basis (metric) for air quality standards. Time series studies of the effects of ozone have often used the peak 8-hour average concentration of ozone as a metric of exposure and have related this to daily counts of deaths and hospital admissions. As in the case of particulate matter, effects on the cardiovascular system have

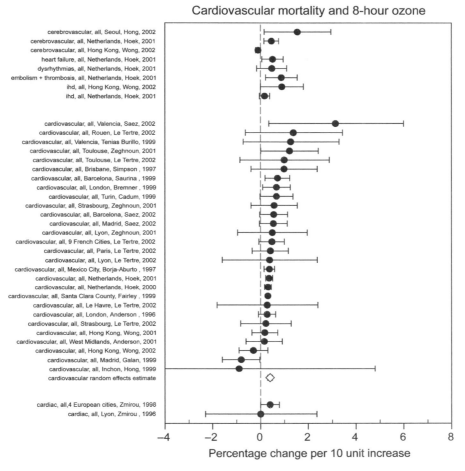

Figure 11.3   Effect of ozone (measured as maximum 8 hour average concentration of cardiovascular mortality. Figure shows the increase in risk of death from cardiovascular disease per $10 \, \mu g \, m^{-3}$ increase in ozone concentration.[16]

been reported. This makes one think that these effects may well not be specific to individual pollutants but, rather, dependent on reactions triggered in the respiratory system with "knock-on" effects on the cardiovascular system. It has been argued that particles, especially very small particles, could cross the air-blood barrier of the lung and reach the circulation and have effects on other organs such as the heart and the brain, although the evidence for that occurring in man in very limited. This argument will not do for ozone; such a reactive gas will not penetrate in an unreacted form beyond the lining of the lung.

The database on the effects of ozone on health is vast.[10] Time-series studies show consistent effects on cardiovascular mortality (see Figure 11.3). The WHO *Air Quality Guidelines*[9,10] should be consulted for references to the original studies.

Despite very clear associations with cardiovascular mortality, no consistent effect of ozone on admissions to hospital for treatment of cardiovascular diseases has been found.[16] This lack of coherence in findings is remarkable and unexplained. It is not easy to explain how a pollutant could cause deaths as a result of effects on the cardiovascular system without, perhaps at lower levels of exposure, causing damage that might lead to clinical symptoms and signs and hospital admissions.

One possible explanation is that exposure to ozone triggers fatal arrhythmias and these either kill before the individual reaches hospital or have no effect. This is possible but there is, as yet, no evidence to support this conjecture. For respiratory effects the picture is clearer: both daily deaths and admissions to hospital are associated with daily ozone concentrations.[10]

Long-term exposure to ozone has been shown to be related to a slowing of lung development, as indicated by poorer than expected performance in lung function tests in young people growing up in relatively high ozone areas.[61] Similar results have been reported from the Southern California Children's Health Study.[40,41]

Studies of volunteers exposed to ozone have reported reductions in indices of lung function and an increase in markers of inflammation (cytokines and inflammatory cells) in airway fluid sampled by broncho-alveolar lavage. A detailed account of these studies has been provided by Lippmann[12] and the World Health Organization.[10] Early work suggested a threshold of effect at about 80 ppb ($160 \mu g$ m^{-3}) ozone exposure for 6.6 hours with intermittent exercise. Responses at this level of exposure were small. As with SO_2 and NO_2, the lack of coherence between the results of volunteer studies and those of time-series epidemiological studies is a cause for concern. It may be that the range of sensitivity to these gases is much larger in the general population than in those subjects chosen for volunteer studies.

11.4.8   Carbon Monoxide

Carbon monoxide is perhaps the only air pollutant for which there can be no doubt about its capacity to kill.[12] Accidental exposure to carbon monoxide causes deaths and hospital admissions every year in the UK and, indeed, in all other countries where exposure occurs. Carbon monoxide is also the only common air pollutant for which there is a bio-marker that indicates both exposure and relates to effects. Carbon monoxide binds, rather remarkably, to haemoglobin (Hb) in precisely the same was as oxygen: the CO–Hb dissociation curve is identical in shape with the O_2–Hb dissociation curve. But the affinity of CO for haemoglobin is >200 greater than that of oxygen and thus low concentrations of CO compete effectively with ambient concentrations of oxygen for binding to haemoglobin. Oxygen transport is impaired and the O_2–Hb dissociation curve is shifted making release of O_2 at the tissues less effective.

The reason for displacement of the dissociation curve is easy to grasp. Normally, fully oxygenated Hb carries 4 molecules of oxygen and as each is released, the PO_2 (partial pressure of oxygen) falls, the "next to be released" is released more easily due to "cooperation" – a homely term to explain complex changes in conformation of the oxy-haemoglobin molecule. In CO poisoning, many molecules of Hb carry only one or two molecules of O_2 and thus the capacity for "cooperation" is reduced. In fact, the oxygen dissociation curve for Hb carrying a mixture of 2 CO molecules and 2 O_2 molecules looks like the rather flat upper part of the normal curve and the steep portion, showing the efficiency of release of oxygen as the P_{O2} falls, is "missing". The dissociation curve can be regarded as being shifted to the left. This so-called "left shift" of the dissociation curve is particularly important in the fetus: fetal Hb has an already left-shifted oxygen dissociation curve and further displacement to the left seriously impairs oxygen release.[62]

Exposure to CO that causes 2.3% of haemoglobin to occur as carboxyhaemoglobin (COHb) impairs oxygen delivery to the myocardium and patients suffering from poor coronary circulation experience chest pain earlier than expected during exercise. Studies in which volunteers suffering from angina due to impaired blood supply to the heart were exposed to low concentrations of carbon monoxide, and asked to exercise on a treadmill, have shown that chest pain occurred earlier during exercise than when ambient air was breathed. Furthermore, the characteristic signs of a reduction in oxygen supply to the heart seen on an electrocardiogram during exercise in such subjects occurred earlier than expected. For a review of these studies, see ref 11.

Despite this mechanism, whether exposure to ambient concentrations of CO damages health is unknown and regarded by many as toxicologically unlikely: concentrations of CO are generally too low to produce significant changes in COHb levels. Epidemiological studies have shown associations between daily variations in CO concentrations and effects on the cardiovascular system,[16] but the close correlation between ambient concentrations of CO and fine particles makes interpretation of these findings difficult. Long-term exposure to CO, *e.g.* in cigarette smokers, has been found to be associated with atheromatous changes in the coronary arteries, but there is little evidence that such an effect occurs at ambient concentrations.[9]

11.4.9 Carcinogenic Air Pollutants

Urban air contains a number of carcinogenic air pollutants. These include: benzene, 1,3-butadiene and the polycyclic aromatic hydrocarbon compounds (PAH). In addition, small amounts of inorganic carcinogens such as arsenic (produced by coal burning) are present. These compounds range in molecular weight and, at ambient temperature, from gases to solids. Some, including the higher molecular weight PAH compounds, are condensed onto the surface of particles and may account for the carcinogenicity of the urban aerosol monitored as $PM_{2.5}$.[63] It is generally accepted that no threshold of effect can be identified regarding genotoxic carcinogens; it is assumed that exposure to low concentrations causes a small increase in the risk of cancer. The fundamental mechanisms of action of genotoxic carcinogens, direct effects on the genetic material of cells where one molecule of carcinogen could, in theory, induce a critical mutation, support the view that no threshold of effect can be identified for these compounds. It should always be remembered that though exposure to a very low concentration *could* cause cancer, the *probability* of it causing cancer is very low. Only two of the major carcinogenic air pollutants will be discussed here: benzene and PAH compounds.

Benzene is classified by the International Agency for Research on Cancer (IARC) as a group 1 carcinogen, with proven causal association with acute non-lymphocytic leukaemia in humans. The main toxic effects occur on the bone marrow, with toxic exposures producing bone-marrow suppression, and reductions in red cell, white cell and blood platelet production (pancytopenia) which may lead to bone marrow failure (aplastic anaemia). It is important to adjust for smoking in epidemiological studies because benzene is present in high concentrations in cigarette smoke. Smokers may have up to 10 times the exposure of non-smokers, particularly in rural areas. In the longer term, studies in workers exposed to benzene have clearly demonstrated an excess risk of acute non-lymphocytic leukaemia, but in general this was not detectable in workers exposed to less than $1.5\,mg\,m^{-3}$ (500 ppb) over a working lifetime – an exposure considerably higher than any achieved by members of the general population.[64] There is evidence of chromosomal abnormalities in workers exposed to slightly lower levels (0.6 to $40\,mg\,m^{-3}$; 200 to 1300 ppb) over a long time period (over 11 years). This contrasts with ambient levels which are usually under $13\,\mu g\,m^{-3}$ (4 ppb) in the United Kingdom at the urban roadside and under $3\,\mu g\,m^{-3}$ (1 ppb) in rural areas.

Population epidemiological studies are extremely difficult to carry out because of the rarity of this type of leukaemia. Estimates of toxicity at low levels of exposure are therefore made from occupational studies. A combination of estimates of risk from a variety of studies suggest that for a lifetime exposure (70 years) to $1\,\mu g\,m^{-3}$ (0.3 ppb) benzene, the excess risk is between 3 to 30 cases per million population, with the World Health Organisation consensus estimate being an excess risk of around four.[8] However, acute non-lymphocytic leukaemia is extremely rare. There are only about 6 to 7 cases per million per year in the United Kingdom, or 420–490 per million over a lifetime of 70 years. An additional four cases resulting from ambient levels of $1\,\mu g\,m^{-3}$ (0.3 ppb) would be almost impossible to detect in epidemiological studies, and the potential risk at ambient levels of benzene remains difficult to prove at the population level, or indeed in workers with modest working lifetime exposures. In practice, the risk of leukaemia from ambient benzene exposure is so small as to be unmeasureable.

Polycyclic aromatic hydrocarbons (PAH) collectively describe a large number of chemicals, many of which, with their metabolites and nitro-derivatives are known to be animal or human carcinogens. The majority derive from the combustion or organic fuels, including wood, coal, oil, petrol and diesel as products of incomplete combustion. The best studied PAH is benzo[a]pyrene (BaP), which along with others is present in cigarette smoke. It is known that clearance of BaP is reduced if it is adsorbed onto particles, and that the dose required to produce tumours in animals is reduced if adsorbed onto particles.[65] This may therefore be relevant to the situation in ambient urban air. The fraction of urban particulates containing PAH is known to have carcinogenic effect in animals, and one study demonstrated dose–response relationship to concentration of BaP in extracts from urban particulates.[66]

11.4.9.1 *Setting Standards for Carcinogenic Air Pollutants*

Given that no threshold of effect can be identified, it is clearly impossible to recommend an air quality guideline or set an air quality standard that guarantees absolute safety. Standards have, however, been recommended in the UK; the thinking behind the approach used is explained by Maynard *et al.*[67] In essence, the approach depends on identifying a lowest observed adverse effect level (LOAEL) or a no observed adverse effect level (NOAEL) from the literature. If only a LOAEL could be identified this was divided by 10 to produce an estimated NOAEL. This value was then converted to an estimated lifetime exposure NOAEL:

If the NOAEL were to be expressed in terms of an occupational exposure (8-hour per day, 5 days per week) as

$$x \ \mu g \ m^{-3} \ \text{for} \ y \ \text{years}$$

then the estimated lifetime exposure NOAEL would be:

$$x \times \frac{y}{70} \times \frac{8}{24} \times \frac{240}{365} \tag{1}$$

i.e., 0.00313 xy

If a working lifetime (40 years) equivalent exposure is calculated then this can be divided by 10 to produce an approximate estimated whole life equivalent exposure (Department of the Environment, ref. 64). A further factor of 10 is then applied to allow for the possible range of sensitivity in the general population. For benzene this leads to a proposed standard of 5 ppb (15.95 $\mu g \ m^{-3}$) expressed as a running annual average concentration.[64]

The approach described above is similar to that used by the World Health Organisation in calculating Unit Risk Factors: *i.e.* the conditional risk associated with lifetime exposure to unit concentration of genotoxic carcinogens (in the case of benzene: 1 $\mu g \ m^{-3}$). This is expressed by the equation:

$$UR = \frac{PoR - Po}{X}$$

where:

P_o is the background lifetime risk of the effect (disease) in question in the specified population;

R is the Relative risk in the reference group (the group exposed occupationally to the carcinogen): *i.e.* R = Observed number of cases/Expected number of cases;

X is the correction factor need when considering lifetime exposure as compared with the exposure leading to Relative Risk, R.

$$X = x \times \frac{8}{24} \times \frac{240}{365} \times \frac{y}{70}$$

In the World Health Organization *Air Quality Guidelines*[9] the Unit Risk for benzene was given as, for a lifetime exposure to $1 \,\mu g \, m^{-3}$, 6×10^{-6}. Both approaches depend on good quality studies of the effects of occupational exposure to the compound concerned. Effects cannot be easily detected at environmental levels: very large and expensive studies would be needed. The two approaches described above are similar in that they ignore the possible non-linearity of the exposure response curve for carcinogens; it is assumed that long-term exposure to a low concentration is as dangerous as short-term exposure to a proportionately higher concentration. This is a not unreasonable assumption given the likely pattern of exposure to carcinogens in ambient air; it is less satisfactory when dealing with occupational exposures which may include occasional exposures to very high concentrations. Exposure to very high concentrations might well overwhelm defence mechanisms, or the usual metabolic pathways, and change the shape of the exposure–response curve. More sophisticated methods for predicting the response to a specified level of exposure are available, though proof of their greater reliability is, in general, lacking.

The approach adopted in the UK led to the recommendation of standards not Unit Risk Factors. It was argued that if the recommended standards were met then the risk of cancer would be very low, too low in fact to be detectable. Such an assertion is open to the challenge that if more sensitive methods for detecting the effects were to become available then effects might be detected despite the standard being met. This is conceded as a truism. However, the standards have provided useful targets for policy development and offer a very significant level of protection of health.

11.5 CONCLUSION

Levels of air pollution in developed countries are now low in comparison with levels recorded in the past. Levels in developing countries are, in some cases, rising as rapid industrialisation and the spread of urban areas continues. Air pollution continues to be responsible for much ill-health. Research into the effects of air pollutants on health has expanded rapidly in the past twenty years and a close understanding of the relationships between ambient concentrations of a number of pollutants and effects on health are now available. Our understanding of the mechanisms of effect of these pollutants has advanced less rapidly but, at least in the case of particulate air pollution, a number of promising conjectures and hypotheses are now being followed up. Quantification of the effects on health of a few air pollutants is now possible and this has been helpful in developing cost–benefit tested policies to reduce, further, emissions of air pollutants.

REFERENCES

1. A. J. Cohen, H. R. Anderson, B. Ostro, K. D. Pandey, M. Krzyzanowski, N. Künzli, Gutschmidt, A. Pope, I. Romieu, J. M. Samet and K. Smith, The global burden of disease due to outdoor air pollution, *J. Toxicol. Environ. Health*, 2005, Part A, 68, 1–7.
2. Department of Health, Committee on the Medical Effects of Air Pollutants, *The Mortality Effects of Long-term Exposure to Particulate Air ollution in the United Kingdom*, 2010.
3. R. D. Brook, S. Rajagopalan, A. C. Pope III, J. R. Brook, A. Bhatnagar, A. V. Diez-Roux, F. Holguin, Y. Hong, R. V. Luepker, M. A. Mittleman, A. Peters, D. Siscovick, S. C. Smith, L. Whitsel and J. D. Kaufman, Particulate matter air pollution and cardiovascular disease: an

update to the scientific statement from the American Heart Association, *Circulation*, 2010, **121**, 2331–2378.

4. C. Tonne, J. D. Yanosky, S. Beevers, P. Wilkinson and F. J. Kelly, PM mass concentration and PM oxidative potential in relation to carotid intima-media thickness, *Epidemiology*, 2012, **23**, 486–494.

5. N. Künzli, M. Jerrett, W. J. Mack, B. Beckerman, L. LaBree, F. Gilliland, D. Thomas, J. Peters and H. N. Hodis, Ambient air pollution and atherosclerosis in Los Angeles, *Environ. Health Perspect.*, 2005, **113**, 201–206.

6. M. Lippmann, T. Gordon and L. C. Chen, Effects of subchronic exposures to concentrated ambient particles (CAPs) in mice. I. Introduction, objectives and experimental plan, *Inhalation Toxicol*, 2005(a), **17**, 199–207.

7. M. Lippmann, T. Gordon and L. C. Chen, Effects of subchronic exposures to concentrated ambient particles in mice. IX. Integral assessment and human health implications of subchronic exposures of mice to CAPs, *Inhalation Toxicol*, 2005(b), **17**, 255–261.

8. World Health Organisation, *Air Quality Guidelines for Europe*, WHO Regional Publications, European Series Number 23, 1987.

9. World Health Organisation, *Air Quality Guidelines for Europe*, WHO Regional Publications, European Series Number 91, 2nd edn., 2000.

10. World Health Organisation, *Air Quality Guidelines*, Global Update 2005, World Health Organisation, 2006.

11. S. T. Holgate, J. M. Samet, H. S. Koren and R. L. Maynard, *Air Pollution and Health*, Academic Press, London UK, 1999.

12. M. Lippmann, *Environmental Toxicants*, Wiley-Interscience, 2nd edn., 2000.

13. K. Steenland and D. A. Savitz, *Topics in Environmental Epidemiology*, Oxford University Press, Oxford, UK, 1997.

14. UK Ministry of Health, *Mortality and Morbidity during the London Fog of December 1952*, Reports on Public Health and Medical Subjects Number 95, HMSO, 1954.

15. P. Brimblecombe and R. L. Maynard, *The Urban Atmosphere and its Effects*, Air Pollution Reviews, No. 1, Imperial College Press, London, 2001.

16. Department of Health, Committee on the Medical Effects of Air Pollutants (COMEAP), *Cardiovascular Disease and Air Pollution*, 2006.

17. D. W. Dockery, C. A. Pope III, X. Xu, D. Spengler, J. H. Ware, M. E. Fay, B. G. Ferris and F. E. Speizer, An association between air pollution and mortality in six US cities, *N. Engl. J. Med.*, 1993, **329**, 1753–1759.

18. J. Lepeule, F. Laden, D. Dockery and J. Schwartz, Chronic exposure to fine particles and mortality: an extended follow-up of the Harvard Six Cities Study from 1974 to 2009, *Environ. Health Perspect.*, 2012, **120**, 965–970.

19. C. A. Pope III, M. J. Thun, M. M. Namboodiri, D. W. Dockery, J. S. Evans, F. Speizer and C. W. Heath Jr., Particulate air pollution as a predictor of mortality in a prospective study of US adults, *Am. J. Respir., Crit. Care Med.*, 1995, **151**, 669–674.

20. D. Krewski, R. T. Burnett, M. S. Goldberg, K. Hoover, J. Siemiatycki, M. Jerrett, M. Abrahamowicz and W. H. White, *Reanalysis of the Harvard Six Cities Study and the American Cancer Society Study of Air Pollution and Mortality*, Health Effects Institute, Boston, A, 2000.

21. Department of Health, Committee on the Medical Effects of Air Pollutants, Long-term Exposure to Air Pollution: Effect on Mortality, 2009.

22. M. Jerrett, R. T. Burnett, R. Ma, C. A. Pope III, D. Krewski, K. Newbold, G. Thurston, Y. Shi, N. Finkelstein, E. E. Calle and M. J. Thun, Spatial analysis of the air pollution and mortality in Los Angeles, *Epidemiology*, 2005, **16**, 727–736.

23. L. Clancy, P. Goodman, H. Sinclair and D. W. Dockery, Effect of air-pollution control on death rates in Dublin, Ireland: an intervention study, *Lancet*, 2002, **360**, 1210–1214.

24. A. J. Hedley, C. M. Wong, T. Q. Thach, S. Ma, T. H. Lam and H. R. Anderson, Cardio-respiratory and all-cause mortality after restrictions on sulfur content of fuel in Hong Kong: an intervention study, *Lancet*, 2002, **360**, 1646–1652.

25. M. S. Friedman, K. E. Powell, L. Hutwagner, L. M. Graham and W. G. Teaque, *Impact of changes in transportation and commuting behaviors during the 1996 Summer Olympic Games in Atlanta on air quality and childhood asthma*, *JAMA*, 2001, **285**, 897–905.

26. C. A. Pope III, J. Schwartz and M. R. Ransom, Daily mortality and PM_{10} pollution in the Utah Valley, *Arch. Environ. Health*, 1992, **47**, 211–217.

27. A. J. Ghio, Biological effects of Utah Valley ambient air particles in humans: a review, *J. Aerosol Med*, 2004, **17**, 157–164.

28. C. A. Pope III and D. W. Dockery, Health effects of fine particulate air pollution: lines that connect, *J. Air Waste Manage. Assoc.*, 2006, **56**, 709–742.

29. K. A. Crapo, R. R. Mercer, P. Gehr, B. Stockstill and J. D. Crapo, Allometric relationships of cell numbers and size in the mammalian lung, *Am. Rev. Respir. Dis.*, 1989, A 289.

30. A. Seaton, W. MacNee, K. Donaldson and D. Godden, Particulate air pollution and acute health effects, *Lancet*, 1995, **345**, 176–178.

31. G. Oberdörster, R. M. Gelein, J. Ferin and B. Weiss, Association of particulate air pollution and acute mortality: involvement of ultrafine particles?, *Inhalation Toxicol*, 1995, **7**, 111–124.

32. G. Oberdörster, E. Oberdörster and J. Oberdörster, Nanotoxicology: an emerging discipline evolving from studies of ultrafine particles, *Environ. Health Perspect.*, 2005, **113**, 823–839.

33. N. A. Monteiro-Riviere and C. L. Tran, *Nanotoxicology*, Informa Health Care, 2007.

34. F. J. Kelly, Oxidative stress; its role in air pollution and adverse health effects, *Occup. Environ. Med.*, 2003, **60**, 612–616.

35. J. F. Nunn, *Nunn's Applied Respiratory Physiology*, Butterworth Heinemann, Oxford, 4th edn., 1993.

36. D. M. Stieb, S. Judek and R. T. Burnett, Meta-analysis of time-series studies of air pollution and mortality: effects of gases and particles and the influence of cause of death, age, and season, *J. Air Waste Manage. Assoc*, 2002, **52**, 470–484.

37. H. R. Anderson, C. Spix, S. Medina, J. P. Schouten, J. Castellsague, G. Rossi, D. Zmirou, G. Touloumi, B. Wojtyniak, A. Ponka, L. Bacharova, J. Schwartz and K. Katsouyanni, Air pollution and daily admissions for chronic obstructive pulmonary disease in 6 European cities: results from the APHEA project, *Thorax*, 1997, **52**, 760–765.

38. C. Spix, H. R. Anderson, J. Schwartz, M. A. Vigotti, A. LeTertre, J. M. Vonk, G. Touloumi, F. Balducci, T. Piekarski, L. Bacharova, A. Tobias, A. Ponka and K. Katsouyanni, Short-term effects of air pollution on hospital admissions of respiratory diseases in Europe: a quantitative summary of APHEA study results, *Arch. Environ. Health*, 1998, **53**, 54–56.

39. J. L. Peel, P. E. Tolbert, M. Klein, K. B. Metzger, W. D. Flanders, K. Todd, J. A. Mulholland, P. B. Ryan and H. Frumkin, Ambient air pollution and respiratory emergency department visits, *Epidemiology*, 2005, **16**, 164–174.

40. W. J. Gauderman, R. McConnell, F. Gilliland, S. London, D. Thomas, E. Avol, H. Vora, K. Berhane, E. B. Rappaport, F. Lurmann, H. G. Margolis and J. Peters, Association between air pollution and lung function growth in southern California children, *Am. J. Respir., Crit. Care Med.*, 2000, **162**, 1383–1390.

41. W. J. Gauderman, G. F. Gilliland, H. Vora, E. Avol, D. Stram, R. McConnell, D. Thomas, F. Lurmann, H. G. Margolis, E. B. Rappaport, K. Berhane and J. M. Peters, Association between air pollution and lung function growth in southern California children. Results from a second cohort, *Am. J. Respir. Crit. Care Med.*, 2002, **166**, 76–84.

42. R. McConnell, K. Berhane, F. Gilliland, J. Molitor, D. Thomas, F. Lurmann, E. Avol, W. J. Gauderman and J. M. Peters, Prospective study of air pollution and bronchitic symptoms in children with asthma, *Am. J. Respir. Crit. Care Med.*, 2003, **168**, 790–797.

43. C. Schindler, U. Ackerman-Liebrich, P. Leuenberger, C. Monn, R. Rapp, G. Bolognini, J. P. Bongard, O. Brändii, G. Domenighetti, W. Karrer, R. Keller, T. G. Medici, A. P. Perruchoud, M. H. Schoni, J. M. Tschopp, B. Villiger and J. P. Zellweger, Associations between lung function and estimated average exposure to NO$_2$ in eight areas of Switzerland. The SAL-PALDIA Team. Swiss Study of Air Pollution and Lung Diseases in Adults, *Epidemiology*, 1998, **9**, 405–411.

44. Department of Health, Advisory group on the Medical Aspects of Air Pollution Episodes, *Third Report: Oxides of Nitrogen*, HMSO, London, 1993.

45. G. L. Snider, J. Kleinerman and W. M. Thurlbeck, and Z. H. Bengali, The definition of emphysema: report of a National Heart, Lung, and Blood Institute, Division of Lung Diseases workshop, *Am. Rev. Respir. Dis.*, 1985, **32**, 182–185.

46. E. L. Avol, W. S. Linn, R. C. Peng, J. D. Whynot, D. A. Shamoo, D. E. Little and M. N. Smith, and J. D. Hackney, Experimental exposures of young asthmatic volunteers to 0.3 ppm nitrogen dioxide and to ambient air pollution, *Toxicol. Ind. Health*, 1989, **5**, 1025–1034.

47. M. A. Bauer, M. J. Utell, P. E. Morrow, D. M. Speers and F. R. Gibb, Inhalation of 0.30 ppm nitrogen dioxide potentiates exercise-induced bronchospasm in asthmatics, *Am. Rev. Respir. Dis.*, 1986, **134**, 1203–1208.

48. L. J. Roger, D. H. Horstman, W. McDonnell, H. Kerhl, P. J. Ives, E. Seal, R. Chapman and E. Massaro, Pulmonary function, airway responsiveness, and respiratory symptoms in asthmatics following exercise in NO$_2$, *Toxicol. Ind. Health*, 1990, **6**, 155–171.

49. R. Jörres, D. Nowak, F. Grimminger, W. Seeger, M. Oldigs and H. Magnussen, The effect of 1 ppm nitrogen dioxide on bronchoalveolar lavage cells and inflammatory mediators in normal and asthmatic subjects, *Eur. Respir. J.*, 1995, **7**, 1213–1220.

50. T. Sandström, R. Helleday, L. Brermer and N. Stjernberg, Effects of repeated exposure to 4 ppm nitrogen dioxide on bronchoalveolar lymphocyte subsets and macrophages in healthy men, *Eur. Respir. J*, 1992, **5**, 1092–1098.

51. H. A. Boushey, I. Rubinstein and B. G. Bigby, *Studies on Air Pollution: Effects of Nitrogen Dioxide on Airway Caliber and Reactivity in Asthmatic Subjects; Effects of Nitrogen Dioxide on Lung Lymphocytes and Macrophage Products in Healthy Subjects; Nasal and Bronchial Effects of Sulfur Dioxide in Asthmatic Subjects*, Report No ARB/R-89/384, California Air Resources Board, Sacramento, CA, 1988.

52. M. O. Amdur, Air Pollutants, in *Casarett and Doull's Toxicology: The Basic Science of Poisons*, ed. C. D. Klaassen,M. O. Amdur and J. Doull, Macmillan, London, New York, Toronto, 3rd edn., 1986, pp. 801–824.

53. W. S. Linn, T. G. Venet, D. A. Shamoo, L. M. Valencia, U. T. Anzar, C. E. Spier and J. D. Hackney, Respiratory effects of sulphur dioxide in heavily exercising asthmatics. A dose response study, *Am. Rev. Respir. Dis.*, 1983, **127**, 278–283.

54. W. S. Tunnicliffe, M. F. Hilton, R. M. Harrison and J. G. Ayres, The effect of sulfur dioxide on indices of heart rate variability in normal and asthmatic adults, *Eur. Respir. J.*, 2001, **17**, 604–608.

55. W. W. Holland and D. D. Reid, The urban factor in chronic bronchitis, *Lancet*, 1965, **285**, 445–448.

56. K. Katsouyanni, G. Touloumi, C. Spix, J. Schwartz, F. Balducci, S. Medina, G. Rossi, B. Wojtyniak, J. Sunyer, L. Bacharova, J. P. Schouten., A. Ponka and H. R. Anderson, Short-term effects of ambient sulfur dioxide and particulate matter on mortality in 12 European cities: results from time series data from the APHEA project, *BMJ*, 1997, **314**, 1658–1663.

57. E. Samoli, J. Schwartz, B. Wojtyniak, G. Touloumi, C. Spix, F. Balducci, S. Medina, G. Rossi, J. Sunyer, L. Bacharova, H. R. Anderson and K. Katsouyanni, Investigating regional differences in short-term effects of air pollution on daily mortality in the APHEA project: a

sensitivity analysis for controlling long-term trends and seasonality, *Environ. Health Perspect.*, 2001, **109**, 349–353.

58. F. Ballester, M. Saez, S. Perez-Hoyos, C. Iniguex, A. Gandarillas, A. Tobias, J. Bellido, M. Taracido, F. Arribas, A. Daponte, E. Alonso, A. Canada, F. Guillen-Grima, L. Cirera, M. J. Perez-Boillos, C. Saurina, F. Gomez and J. M. Tenias, The EMECAM project: a multicentre study on air pollution and mortality in Spain: combined results for particulates and for sulfur dioxide, *Occup. Environ. Med.*, 2002, **59**, 300–308.

59. J. M. Samet, S. L. Zeger, F. Dominici, F. Curriero, I. Coursac, D. W. Dockery, J. Schwartz and A. Zanobetti, The National Morbidity, Mortality and Air Pollution Study. Part II. Morbidity and mortality from air pollution in the United States, *Res. Rep. Health Eff. Inst.*, 2000, **94(Pt 2)**, 5–70(discussion), 71–79.

60. F. Dominici, A. McDermott, M. Daniels, S. L. Zeger and J. M. Samet, Mortality among residents of 90 cities, in *Revised Analyses of Time-Series Studies of Air Pollution and Health*, Health Effects Institute Special Report, Capital City Press, Montpelier, VT, 2003, 9–24.

61. A. Galizia and P. Kinney, Long-term residence in high areas of ozone: associations with respiratory health in a nationwide sample of nonsmoking young adults, *Environ. Health Perspect*, 1999, **107**, 675–679.

62. L. D. Longo, Carbon monoxide in the pregnant mother and fetus and its exchange across the placenta, in Biological Effects of Carbon Monoxide, R. F. Coburn. *Ann. N. Y. Acad. Sci.*, 1970, **174**, 1–430.

63. C. A. Pope, R. T. Burnett, M. J. Thun, E. E. Calle, D. Krewski, K. Ito and G. D. Thurston, Lung cancer, cardiopulmonary mortality, and long-term exposure to fine particulate air pollution, *JAMA*, 2002, **287**, 1132–1141.

64. Department of the Environment, Expert Panel on Air Quality Standards, *First Report: Benzene*, HMSO, London, 1994.

65. R. Sellakumar, R. Montesano, U. Saffioti and D. G. Kaufman, Hamster respiratory carcinogenesis induced by benzo(a)pyrene and different levels of ferric oxide, *J. Natl. Cancer Inst.*, 1973, **50**, 507–510.

66. F. Pott, R. Tomingas, A. Brockhaus and F. Huth, Studies on the tumorigenic effects of extracts and their fractions of atmospheric suspended particulates in the subcutaneous test of the mouse, *Zbl. Bakt. I Abt. Orig. B*, 1980, **170**, 17–34.

67. R. L. Maynard, K. M. Cameron, R. Fielder, A. McDonald and A. Wadge, Setting air quality standards for carcinogens: an alternative to mathematical quantitative risk assessment – discussion paper, *Regulat. Toxicol. Pharmacol*, 1997, **26**, S60–S70.

CHAPTER 12

Impacts of Air Pollutants on Crops, Trees and Ecosystems

MIKE ASHMORE[†]

Stockholm Environment Institute, University of York, York YO10 5DD, UK
Email: mike.ashmore@york.ac.uk

12.1 INTRODUCTION

When we look at old photographs of smoke-ridden industrial cities in Europe and North America from around the beginning of the 20th century, we are not surprised to discover that plants did not grow well. In 1928, W. W. Pettigrew[1] delivered a lecture which included a graphic description of the problem of growing ornamental plants in Manchester: 'While a smoke-laden atmosphere is inimical to both animal and vegetable life, its effects are undoubtedly more apparent if not more deadly in the case of vegetation...'. The best of the early experiments on air pollution and plant life were those conducted in Leeds by Cohen and Ruston.[2] They found that plants grew three to four times larger on the outskirts of the conurbation than in the polluted air of the city centre, and they found a remarkably good negative correlation between the estimated annual sulfur deposition and the stunting of plant growth (see Figure 12.1).

It is also clear that, when present in high enough concentrations, air pollutants can not only reduce plant growth, but may cause the complete elimination of all major plant species. For example, near the large Sudbury smelter in Ontario in the early 1970s, areas devoid of vegetation occurred up to 8 km from the source, and species numbers and productivity were reduced up to 20–30 km from the smelter.[3] Figure 12.2 illustrates the relationships between tree stem basal area, tree species number, and an index of the diversity of the forest canopy, and the distance from the smelter; similar relationships were found for ground vegetation, although the trend with distance was not as strong, and for insect communities. These effects were primarily attributed to the combined effects of sulfur dioxide and metal emissions from the smelter.

Effects of the kind illustrated in Figures 12.1 and 12.2 were identified throughout Europe and North America in regions where industrial activity or large urban centres produced localized

---

[†]This chapter is based upon an earlier contribution by Terry Mansfield and Peter Lucas which is gratefully acknowledged.

Pollution: Causes, Effects and Control, 5th Edition
Edited by R M Harrison

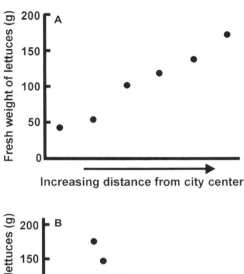

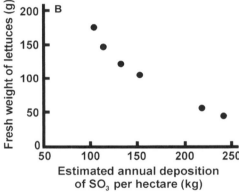

Figure 12.1 (A) Average fresh weights of lettuce plants grown at six different sites in Leeds and its suburbs between July and September, 1911. (B) Annual deposition of SO_3 at the six sites showing an apparently clear inverse relationship with the growth of the crop. (Plotted from data in the book by Cohen and Ruston).[1]

sources of pollutants from fossil fuel combustion. However, even in Victorian England, the effects of air pollution were not confined to urban and industrial zones. For example, a paper published in the *Halifax Naturalist* in 1901 commented on the effects of coal smoke on bark-loving moss species, stating that they 'have mostly disappeared around the large towns, and even in country districts seem to be much scarcer than formerly'.[4] In the Pennine hills of Northern England, examination of peat profiles and historical records has shown vegetation decline over the 19th and 20th centuries, with loss of the dominant peat-forming *Sphagnum* mosses following the appearance of soot particles in the peat profiles.[5,6] Over the last fifty years, we have gained a much improved understanding of impacts on vegetation in rural areas of pollutants such as ozone, nitrogen deposition and sulfur deposition, which are widely distributed at a regional scale.

Over the same period, the smoke and sulfur dioxide smogs that enveloped European and North American cities a hundred years ago have disappeared. However, the effects of air pollution from local industry and urban areas have become increasingly apparent in many other parts of the world.[7] For example, many studies in India have demonstrated the effects of local industrial sources and urban air pollution on plant performance and crop yield.[8] In the holy city of Varanasi, yields of mung bean were reduced by 75–80% at sites with the highest levels of sulfur dioxide and nitrogen dioxide,[9] an effect of similar size to that reported a century earlier in Leeds.[2]

This chapter aims to provide an overview of current evidence of effects of major air pollutants on crop yields, forest health and ecosystem services, drawing primarily on examples from studies in

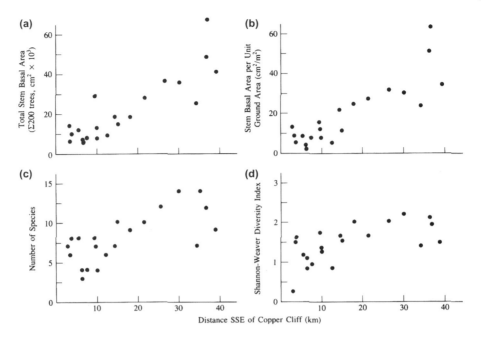

Figure 12.2   Forest overstorey characteristics, expressed as (a) total stem basal area, (b) total stem basal area per unit ground area, (c) number of species, and (d) Shannon-Weaver diversity index, with distance south of the Sudbury smelter complex. (Published with permission from Freedman, 1995).

Europe, but also in North America and Asia. It highlights some of the key mechanisms underlying these effects of air pollution, and the interactions between effects of air pollutants and those of other environmental stresses, The focus of the chapter is on the gaseous pollutants sulfur dioxide (SO_2), nitrogen oxides (NO_x, composed of NO and NO_2), ammonia (NH_3) and ozone (O_3), and the deposition of sulfur and nitrogen, as these are now the most important air pollutants in terms of effects on vegetation at a regional scale. Particulates are of much less concern than is the case for effects on human health, and deposition of particles onto vegetation can reduce urban particulate concentrations.[10] However, they can have effects such as reducing the light energy available for photosynthesis and blocking stomata, hence disrupting regulation of water loss from leaves. A range of other air pollutants may be significant around specific industrial sources but these are not considered in any detail here; for example, hydrogen fluoride emissions from aluminium smelters or brickworks may have local impacts on both plant growth and the health of grazing animals.[11,12]

12.2   METHODS OF INVESTIGATION

The evidence of effects of air pollution on vegetation has been obtained using a range of methods, each with their own strengths and weaknesses. These range from controlled laboratory chambers, which clearly link cause and effect, but which often do so under highly unrealistic conditions, through to field observations, which provide no certainty about the link between cause and effect. The most important information over recent decades has come from field experiments, which fall into two major categories.

   (i) *Field chambers*, in which gaseous pollutant exposures can be controlled, either by filtration of ambient pollutants or by the addition of pollutants in temporal patterns that simulate ambient concentration profiles. The most commonly used type of chamber, the open-top chamber (OTC), provides microclimatic conditions that are comparable, but not identical, to those outside.[13] However, OTCs do modify the microclimate, and significantly change the interactions between plants, insects and pathogens, and size prevents their use for mature trees.[14] Experiments involving such chambers have also been used to define exposure–response relationships, for example between pollutant concentrations and crop yield,[13] by assigning chambers to a range of different pollutant levels.

(ii) *Open-air field manipulation experiments*, in which gaseous pollutants are released over outdoor plots, or in which pollutant deposition is increased in artificial rainwater. This approach allows multi-year exposure of ecosystems under unmodified climatic conditions, with exposure to stress events and realistic interactions with insects and pathogens.[15,16] Such experiments have provided important information on how long it takes for different components of the ecosystem to respond to increases in pollutant exposure, and can be designed to assess interactions, for example with management intensity, elevated CO_2 concentrations, and temperature, or between pollutants.[17] Experiments with a range of concentrations or deposition rates have also been used to identify thresholds for effects, for example on vegetation composition or the decline of a key plant species.[18] However, such systems cannot be used to assess the effects of reducing pollutant exposures, and it is difficult to identify thresholds from experiments in areas with a relatively long history of elevated pollutant deposition, where there may already have been significant impacts, for example on biodiversity.

All of these experiments have important constraints in terms of spatial scale and timescale. There are enormous technical challenges, for example, in exposing more mature trees, especially to gaseous air pollutants, over long periods of time under field conditions. However, two experiments with ozone have successfully done so in recent years, and have provided important new insights into the ecosystem-level effects of exposure to this pollutant over periods of 5–10 years.[19,20] Less obviously, experiments in semi-natural ecosystems are typically based on replicated sampling units of limited size (*e.g.* areas of 1 m^2), within which, for example, the species richness (total number of species found) or the cover of individual plant species is determined. However, a reduction in species richness within the experimental area in these studies does not necessarily indicate the loss of species over a wider area, although it may mean that fewer individuals of some species occur in the wider landscape.[21] Rare species, which may be of particular concern, can be absent, or present at very low cover or frequency in the small experimental unit, so that no useful conclusions can be drawn about their responses.

A further major limitation of experimental studies is that they typically assess relatively short-term responses (even the longest experiments seldom exceed 10–20 years), and it is impossible to design experiments which include all the factors which may be involved, for example in the long-term decline in forest health. This means that field observations may be the only feasible way of assessing the long-term impacts of air pollution. For example, long-term field studies in the San Bernadino mountains, close to Los Angeles, have elucidated complex interactions between ozone exposure, attacks by bark beetle, fire management policies, and droughts that have contributed to the decline of the dominant ponderosa pine and Jeffrey pine trees.[22]

There are two main approaches to field surveys. The first is through *targeted field surveys* of sites covering a gradient of pollutant exposure. Where there is a short but intense gradient of pollution (as in Figures 12.1 and 12.2), there is little doubt that the observed effects are primarily caused by variation in pollutant exposure. However, such studies may also have a wider regional or national scale, in which case gradients in the pollutant of interest may be correlated with those of other potential drivers (*e.g.* other pollutants, climate, soil characteristics or site management).[23] These other drivers need to be measured or estimated, and considered in statistical analyses and interpretation. For this reason, targeted surveys cannot prove causality, but can only determine the statistical significance of pollutant exposure as a possible driver of changes in ecosystem response.

Furthermore, although such field surveys can provide information on longer-term responses, it is not clear when in the past the observed effects actually occurred unless a long-term series of observations is available. This is important for once-off surveys when, as is often the case, present pollutant exposures show a spatial correlation with historical, and often higher, pollutant concentrations.

An alternative is the use of broader *ecological surveillance networks*, which cover a wider range of plant communities representative of the region of concern. For example, many countries have networks of forest monitoring plots which can be used to assess the impacts of air pollution and other stress factors,[24] or which record the presence or absence of species in larger areas (*e.g.* 10 × 10 km squares). Such networks are not specifically targeted to identify pollution impacts, and reflect the influence of land use and a range of climatic, edaphic and management factors, making attribution of any effect to pollutant exposure even more problematic than for targeted surveys. However, when major changes in forest health, or in the distribution of a plant species, have occurred at this scale, they can provide convincing evidence of the importance of air pollution across the landscape, *e.g.* in showing the loss of lichen species in response to elevated concentrations of SO_2 and the subsequent recovery as concentrations fall again (see section 12.3).

12.3   SULFUR DIOXIDE AND SULFUR DEPOSITION

Sulfur is one of the important mineral nutrients of plants. It is normally taken into the roots from the soil in the form of sulfate, and then transported to the leaves, where it is reduced and used in the synthesis of organic S compounds, including amino acids.[25] However, gaseous forms of sulfur, such as SO_2, can also be taken up directly by the leaves. Exchange of carbon dioxide and water vapour between leaves and the atmosphere is fundamental to many physiological processes such as photosynthesis, respiration and transpiration. These gaseous exchanges are regulated by the opening and closing of stomata, and the uptake of air pollutants such as SO_2, O_3, and NO_x occur primarily *via* these stomatal pores at the surface of the leaf.[26] However, some gaseous pollutants may be sorbed by the leaf cuticle which could have important implications in reducing cuticular integrity, leading, for example, to enhanced water loss.[27]

When SO_2 enters the leaves *via* the stomata, it dissolves in the apoplast, the liquid phase of the cell wall, forming sulfite and bisulfite ions.[28] The apoplast is part of the pathway by which water and minerals from the soil are distributed to individual cells of the leaf, and it normally contains sulfate ions, but not sulfite or bisulfite, in significant concentrations. The presence of the enzyme sulfite oxidase in the apoplast can act to detoxify sulfite by converting it into sulfate,[29] but, if not, elevated concentrations of bisulfite and sulfite can cause a range of physiological effects in the leaf, including damage to the cell membrane and chloroplast structures, reductions in photosynthetic rate and, at a high enough concentration, characteristic visible damage to the affected leaf.

As for all pollutants, there is a large degree of variation between species in their sensitivity to SO_2, as well as genetic variation within species.[30] This variation broadly relates to the ability of the plant either to restrict pollutant uptake or, once it has been taken up, to detoxify, metabolize, or sequester the pollutant. Sensitivity to SO_2, and other air pollutants, may also depend on environmental conditions, as discussed in more detail in section 12.7. It is well documented that evolution of resistance to SO_2 can occur. For example, a classic early study showed that the yield of two ryegrass clones that were indigenous to a heavily polluted area in the UK were much more tolerant to experimental SO_2 exposure than commercial cultivars.[31] Studies of emission sources in both the United States and the United Kingdom have shown that tolerance to SO_2 in annual and perennial species increases with increasing concentrations closer to the source.[32] Decreasing urban SO_2 concentrations have been associated with a loss of this tolerance in grass species, suggesting that the SO_2-tolerant genotypes are at a selective disadvantage in the absence of the pollutant.[33]

The most sensitive group of species to SO_2 are recognised to be lichens. The disappearance of many lichen species from European and North American cities during the 19th and 20th centuries has been well documented; in the United Kingdom, between 30% and 90% of species were lost from areas in which air pollution was a dominant factor.[34] Many of these species were highly sensitive to direct effects of relatively low concentrations of SO_2. In contrast, other lichen species are relatively tolerant, and several different scales have been developed to map SO_2 concentrations based on the occurrence of particular groups of lichen species.[35]

In many European and North American cities, SO_2 concentrations have decreased dramatically in recent decades, but the re-invasion of lichens in response to this decline has been patchy and variable. In London, for example, it has been found that some relatively pollution-sensitive lichens have re-invaded more quickly than species which are more tolerant.[36] There are several possible explanations for these different rates of recovery, some of which have been investigated using transplant experiments.[37] One factor is the speed of dispersal of vegetative propagules into areas of decreasing SO_2; certain species (so-called "zone-skippers") are able to disperse much more rapidly than others. Another important factor is changes in bark chemistry. Recovery of epiphytic lichen species seems to be more rapid on tree species such as ash and willow, which have a relatively high pH bark, than on species with acidic bark, such as oaks. A third factor is the importance of other air pollutants, especially ammonia and nitrogen oxides, in influencing lichen biodiversity.[38,39]

There has long been extensive evidence of the adverse effects of SO_2 on crop yield and forest growth from experimental studies, transects away from point sources, and field observations,[40,41] and threshold concentrations for effects on sensitive species have been established (see section 12.8). On the other hand, for crops with a high sulfur demand (*e.g.* oilseed rape), and where the soil is deficient in sulfur, it is possible that atmospheric deposition of sulfur can make a significant contribution to the crop's sulfur requirement.[42] Long-term datasets from the Rothampsted Experimental Station in the UK show that the sulfur content of grasses increased from 2.5 mg g^{-1} in 1860 to a peak of approximately 4 mg g^{-1} in the 1970s, tracking the long-term changes in sulfur emissions and deposition in the UK. However, there has been a rapid decline in foliar sulfur from 1980 to the present day, when values are similar to those of the mid-1860s.[43] Similar trends in foliar sulfur have been observed in cereal crops, which now require sulfur fertilizer applications in some areas of the UK to achieve optimum yield.[44]

12.4   NITROGEN OXIDES, AMMONIA AND N DEPOSITION

Nitrogen, like sulfur, is an essential element for plant growth, as it is a major constituent of proteins and nucleic acids; indeed, plant productivity in many terrestrial ecosystems is limited by nitrogen supply from the soil.[45] Plants have developed a diverse range of mechanisms to enhance their acquisition of this important nutrient, including fixation of atmospheric nitrogen by symbiotic bacteria in legumes, associations with mycorrhizal fungi, and insectivory. The majority of plants obtain nitrogen through root absorption of ammonium and nitrate ions from the soil solution. In the leaves, nitrate is reduced to ammonium, which is then incorporated into amino acids in the chloroplasts. However, as for sulfur, gaseous forms of nitrogen oxides can also enter the leaf *via* the stomatal pores, and, if present in high enough concentrations, can cause direct toxicity, primarily through accumulation of nitrite ions. The susceptibility of different plant species to NO_x may depend on where they normally carry out nitrate and nitrite reduction.[46] In some woody plants, for example, reduction of soil-derived nitrates primarily occurs in the roots, and these species might be particularly sensitive to NO_x because the reduction pathway is absent or poorly developed in the leaves. The exposure of plants to NO_x leads to marked increases in nitrate and nitrite reductase activities in leaves,[47] a clear indication that the plant has some ability for metabolic adjustment to detoxify the nitrite ions as they enter the cells. In the case of NH_3, direct toxicity is more related to changes in acid/base regulation,[48] especially as the bi-directional exchange of NH_3 is controlled by the ammonium concentration in the apoplast, which is pH-dependent.[49]

Lichens and bryophytes are among the most sensitive groups of organisms to direct effects of NO_x and NH_3,[50–52] as is the case for SO_2. There is evidence of adverse effects on these species, as well as on vascular plants, close to major sources, such as roads for NO_x,[53,54] and intensive livestock units for NH_3.[55,56] These adverse effects of NH_3 have been substantiated also by a field release experiment, which has demonstrated that dry-deposited NH_3 has much greater effects (more rapid accumulation of foliar N and loss of membrane integrity leading to a decline in lichen cover) than wet deposition of ammonium or nitrate on adjacent plots.[57]

However, it is the effects of total N deposition, in the form of gaseous NO_x and NH_3, but also wet-deposited nitrate and ammonium, that is now of greatest concern, because of its wider regional impacts.[58] In Europe, N deposition is now considered the most important air pollution driver of loss of biodiversity, as atmospheric concentrations of SO_2 have dramatically decreased, and effects of N deposition and the resulting eutrophication appear to be much more widespread in terrestrial ecosystems than effects of S deposition and resulting acidification.[59] Effects of N deposition are now recognised in nearly all terrestrial ecosystems with low nutrient levels in Europe, including aquatic habitats, forests, grasslands, tundra, mires and bogs, heathlands, and coastal and marine habitats.[58,60,61] In these systems, nitrogen is generally the most important growth-limiting element, and species are adapted to a nitrogen-deficient environment. However, effects of N deposition on sensitive ecosystems are not confined to Europe, but are well documented in North America,[62] with additional evidence of effects in Africa, Latin America, and Asia, where N deposition rates are predicted to increase further over the next few decades.[63]

The mechanisms by which long-term exposure to elevated N deposition can affect vegetation are complex, with many ecological processes interacting at different temporal and spatial scales, but the key processes are summarised in Figure 12.3. The importance of these different mechanisms varies

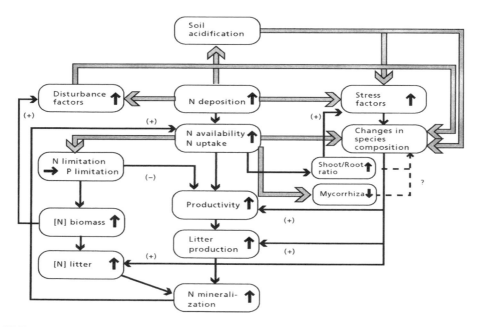

Figure 12.3 Scheme of the main impacts of increased N deposition on terrestrial ecosystems. ↑indicates increase; ↓indicates decrease; solid arrow: effect will occur in the short term (<5 yrs); smaller arrow indicates long-term impact; (+): positive feedback; and (−): negative feedback. (Published with permission from Bobbink and Lamers, 2002).

with the species or community of concern and the physical and chemical form of the deposition. Where N is the limiting nutrient for plant growth, increased N deposition, at least initially, increases plant growth and productivity and thus leads, over the longer term (years to decades), to higher annual litter production.[64] Because of this, N mineralisation in the soil will gradually increase, which may further increase plant productivity.[65] This is a positive feedback, because higher N mineralisation leads to higher N uptake by plants.

However, this increase in N status can also change species composition, and may reduce species richness, as it favours species adapted to quick exploitation of available nutrients. Above a certain level of primary productivity, local species diversity declines, as the growth of those species able to exploit the available N increases, due to competitive exclusion of characteristically slow-growing species of nutrient-poor habitats by the relatively fast-growing nitrophilic species. Such changes in species richness and composition are generally long-term effects, although they may be induced by relatively large doses of nitrogen applied over a few years.[66–69]

When N is no longer the limiting nutrient in the ecosystem, plant growth becomes limited by other resources, such as phosphorus (P) or water, and plant productivity will not increase with further increases in N, indicating N saturation.[70] Under long-term N inputs in P-limited ecosystems, or after a shift from N to P limitation, anomalous N : P ratios can gradually lead to changes in plant species composition and plant nutrient imbalance.[71,72] N concentrations in plant tissues will also tend to increase because N availability still increases. This accumulation tends to increase the sensitivity of the plant to stress factors, including increasing the palatability of the vegetation for herbivores and its sensitivity to fungal pathogens (see section 12.7).

Finally, deposition of both oxidised and reduced N can acidify soils; oxidised N through acting as a mobile anion accompanying basic cations leached from soil, and reduced N through the acidifying effects of both nitrification and root exchange of NH_4^+ for H^+.[73] Soil acidification triggers many long-term changes, such as increasing levels of potentially toxic metals, especially reactive aluminium, in the soil solution.[74] Furthermore, as pH declines, the ecosystem capacity to remove reduced N is also compromised, resulting in the accumulation of ammonium in soil solution, which is toxic to the roots of sensitive plant species.[75,76] As a result of this cascade of changes, species composition can change, with acid-tolerant plant species becoming dominant, and species typical of intermediate and high pH soil disappearing.

The evidence of widespread changes in plant biodiversity in Europe caused by N deposition, as recently collated for the *European Nitrogen Assessment*,[61] includes: reduced numbers of forb species in grasslands; reduced cover of lichens and increased cover of vascular plants in tundra; decreased cover of ericaceous shrubs in boreal forests; loss of lichen species and reduced overall species diversity in coastal dunes; loss of characteristic lichen and bryophyte species and invasion by grasses in heathlands; and decline of carnivorous species and characteristic bryophyte species in bogs and mires. Much of the evidence for these effects in Europe and North America comes from field experiments.[77,78] However, such experimental findings may be site-specific and important evidence of widespread long-term ecological effects of N deposition across Europe has also come from field surveys. For example, a targeted survey of one specific acid grassland community across the UK showed a large reduction in species richness as N deposition increased (see Figure 12.4).[21] In particular, forb species richness was reduced by about 75% over this gradient of N deposition.[79] Further analysis of the study plots clearly identified N deposition, alongside management, climate and ozone, as a major driver of species composition in this particular community across the UK.[80] Detailed analysis of the survey results suggests that acidification is a stronger driver of the observed effects than eutrophication in these poorly buffered soils.[81]

Supporting evidence for this effect has now come from two important sources. Firstly, comparison of the results of this survey with those from a long-term surveillance survey showed a similar relationship between N deposition and species richness in all UK acid grasslands,[82] even though, as expected given the greater range of community types, there was more scatter between

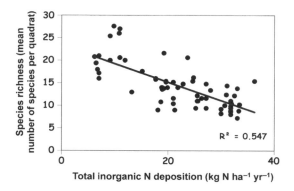

Figure 12.4   Species richness (number of vascular plant species) in 2×2 m quadrats in acid grassland communities across the UK expressed as a function of nitrogen deposition. (Published with permission from Stevens *et al.*, 2004).

individual sites. Secondly, the UK targeted survey has been extended to include acid grassland sites across western Europe, and, despite the greater climatic variation, a similar decrease in species richness, especially of forbs, was found with increasing N deposition.[83] Further evidence is provided by analysis of data from historic and more recent large-scale ecological surveys. These studies have consistently shown either: (i) a decline in species richness and, more specifically, a decline in species associated with nutrient-poor habitats; or (ii) little change or even an increase in species richness, but a decline in species adapted to low nutrient availability, including species of high conservation importance.[84,85]

Therefore, there is now a large body of experimental and field evidence to suggest that N deposition has had a significant adverse effect on biodiversity in many habitats of conservation importance in western and central Europe, and there is increasing evidence of such effects elsewhere in the world.[58]

12.5   OZONE

Considerations of the impacts of O_3 are not complicated by its providing a source of an essential nutrient, as is the case with SO_2 and NO_x. Ozone entering the leaf through the stomata reacts primarily in the apoplast to form reactive oxygen species, such as hydrogen peroxides, superoxide and hydroxyl radicals.[86,87] The parts of cells that are most vulnerable to these reactive species are the cell membranes, where the double bonds of unsaturated fatty acids are very vulnerable to attack, leading to loss of osmotic control and ultimately cell death. However, it is also now well established that the apoplast provides an important line of defence against incoming O_3, due to the presence of antioxidants such as ascorbate.[87,88] and there is evidence of a correlation between levels of apoplastic antioxidants and ozone tolerance.[89,90] At high concentrations (above 100 ppb for most plant species), and when the apoplastic defence mechanisms are inadequate, ozone causes characteristic injury to the foliage, typically in the form of brown or white flecking; these injury symptoms can be widespread, especially on irrigated crops in regions with high ozone exposures.[91]

The most important long-term effect of ozone exposure within the leaf is a reduction in net CO_2 fixation as a result of reduced photosynthetic efficiency, increased rates of respiration which are associated with maintenance and repair of damaged tissue, and reduced stomatal conductance.[92] Ozone also accelerates the rate of leaf senescence, which in turn reduces the period of leaf growth with a positive carbon balance.[93] Ozone often reduces the translocation of assimilates to other parts of the plant, such as the roots and grains, either as a result of diversion of carbon to support new leaf growth or the synthesis of protective chemicals, or through a direct effect on translocation efficiency in the phloem.[94] Ozone has also been shown to have direct effects on reproductive organs, for example reducing flower production or increasing grain abortion.[95] All these mechanisms may contribute to reduced plant growth and reductions in crop yields.

These effects of ozone within the plant have a wider range of ecological implications. These include effects on species composition and diversity, biogeochemical cycles, soil chemistry and soil biology. Some of these effects are a direct consequence of the reduced photosynthesis and carbon assimilation in the leaf, but other mechanisms, such as interference with hormonal signalling by ozone, are also important.[96,97] To provide one example of the range of effects that might occur within a simple experimental ecosystem, a study of a constructed meadow in Finland reported that ozone caused visible leaf injury, delayed the onset of spring growth, reduced leaf and root growth, delayed flowering, reduced the number of berries, reduced rust infection of the leaves, reduced nitrogen levels in the soil, reduced emissions of carbon dioxide (CO_2) and nitrous oxide from the soils, and reduced the microbial biomass in the soil.[98,99] The implications of some of these responses to ozone are discussed further in sections 12.6, 12.7 and 12.9.

However, rather than cover all these effects of ozone, this section will focus on the impacts of ozone on crop production and food security. For sensitive crop species that are harvested for their leaves, such as spinach, visible injury can directly reduce economic returns for farmers, but for most crops it is the effect on the harvested yield that is important. But how large is this loss of crop yield due to ozone exposure? The approach used to answer this question has been empirical rather than mechanistic. Ozone exposure–yield response relationships for major crop species have been derived from experimental studies, and then applied to measured or modelled ozone concentration fields and data on crop distributions and yields to estimate the overall impacts on crop production at a regional or national scale. The first major experimental programme to establish exposure–response relationships used open-top chambers to provide a range of different ozone exposures over a crop growing season.[13,14] This National Crop Loss Assessment Network (NCLAN) used 10 crops, representing 85% of the cropped area of the USA, and identified the yield of soybean and cotton as being particularly sensitive to ozone, while that of rice, barley and sorghum was relatively insensitive.[100]

A key question in deriving these exposure–response relationships when ozone concentrations vary from hour to hour is how best crop exposure to ozone over a growing season is characterised as a single exposure index. In the NCLAN programme, this was done using the seasonal mean ozone concentration during daylight hours, and comparing the crop yield with a baseline seasonal mean concentration of 20 ppb or 25 ppb.[100] A different exposure index was adopted in Europe, in an attempt to recognise the capacity of plants to detoxify ozone and the cumulative effect of long-term ozone exposure. This new cumulative index, termed AOT40, sums all the hourly mean concentrations above a threshold of 40 ppb during daylight hours over the growing season, and has consistently been found to result in linear relationships with crop yield.[101]

However, both these indices use the ozone concentration immediately above the crop canopy. Since the primary sites of ozone damage are within the plant, it has been argued that an exposure index based on the cumulative flux into the leaves through the stomata would provide a stronger mechanistic approach for risk assessment.[102,103] This is because it takes into account the effect of factors such as high vapour pressure deficits or soil water deficits causing reductions in stomatal conductance that lead to reduced ozone flux into the leaf.[104] This theoretical argument is supported by experimental evidence; for example, recent data on effects on grass growth under two watering regimes showed that differences in response can be explained by the differences in modelled cumulative ozone flux with and without watering over the experimental period.[105] Because the flux approach accounts, at least to some extent, for the influence of climatic, edaphic and developmental factors, it is better suited to modelling ozone effects over large areas with a range of different climatic conditions than an approach which assumes that exposure–response relationships from chamber experiments can be applied without any modification. Use of a flux-based risk assessment leads to a very different spatial distribution of risks of ozone impacts across Europe compared with use of AOT40,[106] and there is evidence that this spatial distribution of flux is more representative of the actual risk of ozone impacts on crops.[107] Using this flux-based approach, one study has estimated that there is an average 14% yield loss for wheat across Europe, corresponding to a total economic loss of over three billion euro.[108]

Mean ozone concentrations in the background troposphere of the northern hemispheric increased over the last century, and concentrations are expected to increase further through the first half of this century.[109,110] Models and meta-analyses using concentration-based exposure indices suggest that there are substantial global yield losses due to current levels of ozone.[111,112] For example, one study estimated global yield losses of 2–5% for maize, 4–15% for wheat and 8–14% for soybean.[113] These estimates are consistent with a meta-analysis of filtration experiments from three continents, which suggested that the mean yield improvement of wheat from removing ambient ozone was 9%.[114] Furthermore, these crop yield losses are likely to increase further by 2030, assuming precursor emissions continue to increase as predicted.[112,115] Measures to reduce these precursor emissions or introduce more ozone-resistant cultivars could significantly reduce yield losses in 2030.[116]

These global models predict that some of the largest reductions in the yield of staple crops occur in south and east Asia. However, are these model predictions consistent with observations and experiments in this region? In the case of south Asia, where regional models also show that critical levels for ozone are exceeded over large areas of the densely populated Indo-Gangetic plain,[117] experimental studies involving filtration of ambient air provide clear evidence of large yield losses, often in the range 30–50%, to staple cereals and legumes.[118,119] In the Pearl River region of China, mean yield losses of 12% and 20% for local cultivars of rice and wheat, respectively, have been reported from open-air release experiments increasing ambient ozone levels by 25–30%.[120,121] A comparison of exposure–yield responses derived in North America and Europe with experimental results with Asia (see Figure 12.5) suggested that they may underestimate the actual size of the

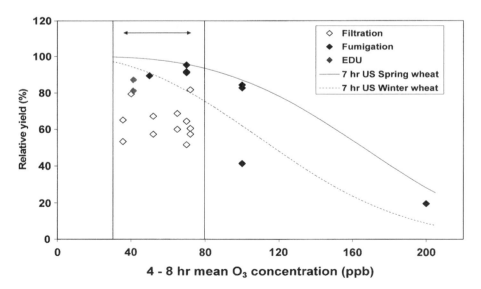

Figure 12.5   Wheat yield loss data plotted against 4–8 hr growing season mean O_3 exposure, from experiments in Asia involving filtration ($\diamond$), fumigation ($\blacklozenge$) and EDU applications ($\blacklozenge$). Dose–response relationships from North America for spring wheat and winter wheat based on 7 hour growing season mean O_3 exposures are also shown. The range of the local ambient pollutant concentrations in the experimental investigations is also indicated as the arrowed gap between the two vertical lines. (Published with permission from Emberson *et al.*, 2009).

effects in the field in the region derived from experiments involving fumigation with ozone, filtration of ambient air, or use of the ozone-protectant chemical EDU.[122] This may be because environmental conditions enhance the sensitivity of the crops or because local cultivars are relatively sensitive to ozone. A flux-based comparison of Chinese and European wheat cultivars suggested that differences remain when the effects of local environmental conditions on stomatal flux of ozone have been accounted for, and hence cultivar differences may be more important.[123] Hence it is possible that the global crop yield losses predicted using European and North American exposure–response relationships significantly underestimate yield losses in parts of the world such as south and east Asia where food security is a critical issue.

Quantifying the large-scale long-term impact of ozone on managed forests or semi-natural ecosystems is more challenging. Relatively short-term experiments on experimental saplings, primarily in OTCs, have consistently shown reductions in height and trunk diameter,[124] and one meta-analysis of these studies suggests that current global ozone concentrations would reduce tree biomass by 7%, a comparable effect to that estimated for crop yields.[125] However, the longer-term effects are less certain, and free-air release experiments have provided variable answers. For instance, in the AspenFACE study in the northern USA, although there was a significant reduction in aspen trunk growth in the first few years, most of this effect was lost after 12 years, primarily because more tolerant clones began to dominate the ozone treatment.[126] In contrast in the Kranzberg experiment in Bavaria, elevated O_3 caused a 50% reduction in trunk growth in beech (although not in spruce), a much larger effect than the early effects on stem diameter suggested.[127]

Many herbaceous natural plant communities contain species with high ozone sensitivity;[128] with a mix of species differing in sensitivity to ozone, species composition tends to be affected before total production. However, long-term field experiments have also produced conflicting results on the effects of ozone. For example, a study in Switzerland showed very few significant effects of ozone on species composition of a species-rich hay meadow,[129] while a study in the UK showed large shifts in response to ozone in species composition of an upland meadow undergoing a change in management to improve species diversity.[130]

12.6   INTERACTIONS BETWEEN POLLUTANTS

Air pollutants rarely occur singly, yet much of the literature covers their individual effects rather than their joint action in realistic mixtures. The evidence that has emerged from studies of pollutant mixtures is complex. The effects of combinations of pollutants are sometimes greater and sometimes less than we would predict from their separate effects, and only rarely are responses simply additive; furthermore the nature of the interactions may depend on the concentrations used and experimental conditions. Furthermore, responses when plants are exposed to two pollutants simultaneously may differ from those when they are applied sequentially,[131] which, for example for ozone and NO_x, is often more representative of ambient pollutant patterns. Detailed reviews are available,[132] but here only two specific pollutant combinations are considered, for which there are clear mechanistic explanations for the observed interactions.

12.6.1   Sulfur Dioxide and Nitrogen Oxides

Sulfur dioxide and NO_x occur together in many situations, because they are often produced simultaneously by the same source, for example during the combustion of coal. Short-duration fumigations with both SO_2 and NO_2 have often led to far greater visible leaf injury than found with the pollutants separately.[133] Studies of long-term effects of SO_2 and NO_2 on grasses and trees have shown similar synergistic effects on plant growth.[134,135] There is a fairly simple biochemical explanation of these synergistic effects of SO_2 and NO_2, one of the few clear explanations we have of the mechanistic basis of pollutant interactions at the metabolic level. This relates to the enzyme nitrite reductase, which is considered to play a critical part in the metabolism of the solution products of NO_2 after its entry into cells. Experiments have shown that, while fumigation with NO_2 on its own leads to a statistically significant increase in the activity of nitrite reductase, no such increase is found when the NO_2 is accompanied by SO_2. This clearly suggests that SO_2 prevents the normal induction of greater nitrite reductase activity in the presence of NO_2. so that the plants may be unable to detoxify the solution products of NO_2.[136] However, these synergistic effects of SO_2 and NO_2 have been reported at relatively high concentrations, and responses to this pollutant mixture at lower concentrations are less clear-cut.[131]

12.6.2   Interactions between Ozone and Elevated CO_2 Concentrations

The impacts of O_3 together with elevated CO_2 concentrations are receiving much attention in current research programmes, since it is predicted that the two pollutants will increase alongside one another in the future. In general, elevated CO_2 has been found to reduce the adverse effects of O_3 on the growth of plants, but O_3 exposure in turn can reduce the beneficial effects of additional CO_2. Several mechanisms have been identified to explain these interactive effects. Increasing concentrations of CO_2 generally cause reductions in stomatal conductance, and hence reduce ozone flux into the leaf,[137] providing one simple mechanism for any decreased toxicity of O_3, although some studies report more complex interactions on stomata.[138] Other studies have identified compensatory effects (*i.e.* elevated CO_2 reduces the adverse effects of O_3, while O_3 reduces the beneficial effects of elevated CO_2) directly on photosynthetic processes within the leaf,[139] while a recent meta-analysis suggests that elevated CO_2 alleviates the effect of O_3 in reducing root biomass in woody plants, but not in herbaceous species.[140]

Two major studies in North America have examined the interactive effects of elevated ozone and elevated CO_2 under field conditions. The AspenFACE field fumigation study in the northern US applied elevated ozone and elevated CO_2 concentrations for about a decade to communities of three important North American trees – aspen, birch and maple – that were grown from seedlings. Elevated CO_2 prevented any adverse effects of ozone on tree growth and many other variables, but effects of O_3 on leaf senescence, timing of bud burst, insect and fungal damage, and regulation of water loss, were not prevented by elevated CO_2, suggesting different mechanisms of ozone impacts that are not directly related to changes in the leaf such as reduced stomatal conductance or increased leaf antioxidant potential.[141] The SoyFACE experiment revealed similar interactions between elevated CO_2 and ozone in terms of crop yield,[142] with elevated CO_2 preventing the adverse effects of O_3. More complex effects were found below-ground; for example, elevated CO_2 decelerated nitrogen cycling below-ground, whereas O_3 enhanced it – an effect that was lost in the presence of elevated CO_2,[143] although another study has suggested no significant interaction between elevated CO_2 and O_3 effects on nitrogen fixation by soybean.[144]

12.7   INTERACTIONS WITH BIOTIC AND ABIOTIC FACTORS

This chapter to date has considered the effects of air pollutants in isolation from other abiotic and biotic environmental factors. Yet the evidence from experiments, supported by many observations in the field, suggests that such interactions can at least play a significant role in determining plant responses to pollutants, and that in some cases they may be of overriding importance. Some of the most important of these interactions are discussed below.

12.7.1   Climate

In terms of abiotic (primarily climatic) factors, the two most important effects are: (i) growth conditions modifying plant response to a given air pollutant exposure,[145] and (ii) pollutant exposure modifying the sensitivity of the plant to an extreme climatic event.[146] This can be illustrated by the interactions between irradiance and temperature and exposure to SO_2. In terms of (i), a study of timothy grass showed that when growth was rapid, in high irradiance and long days, and at favourable temperatures, 120 ppb SO_2 had no detectable effect on the plants, but when the same concentration was applied to plants in low irradiance and short days, and at low temperatures, large reductions in growth were found.[147,148] It is likely that these results can, at least partly, be explained by a reduced capacity of the plants to metabolise and detoxify the incoming SO_2. However, experiments in which grass species were grown over winter and then through to a harvested yield in the summer, showed a recovery in spring from the effect of SO_2 over winter, leading to no final significant effect.[149] In terms of (ii), there is considerable evidence that plants exposed to SO_2 become more sensitive to frost injury. There is, for example, a substantial reduction in survival of SO_2-polluted ryegrass upon subsequent exposure to sub-zero temperatures.[150] Experiments with coniferous tree species have similarly found that exposure to SO_2 increases the sensitivity to freezing in the spring,[151] and large-scale forest dieback in SO_2-polluted mountain areas in central Europe has been associated with increased severity of frost damage. Studies have also suggest that conifers exposed to elevated concentrations of ozone are more sensitive to freezing damage, but these effects may reflect an increased sensitivity to winter dessication, through a loss of stomatal control, rather than a direct effect on cold tolerance.[152] Greater winter dessication of heather plants has also been found to be a response to elevated N deposition.[153]

Interactions with water availability are also very important, especially in the case of ozone, as the highest concentrations typically occur in hot dry summers. At low vapour pressure deficits and low soil moisture deficits, partial closure of the stomata may reduce the ozone flux into the leaf, and hence the impacts of ozone,[154] although species vary in the extent of stomatal closure, for example in response to drought, and hence the extent to which ozone effects are reduced.[155] However, ozone exposure may also reduce the capacity of the plant to withstand periods of drought stress; for example, if ozone reduces carbon allocation to the roots, there may be a reduced capacity for water uptake under drought conditions. Recent studies have suggested that ozone exposure can also lead to a reduced sensitivity of stomata to environmental stimuli such as changes in light intensity,[156] or reduced soil water availability.[157] Of particular significance is the effect of ozone in reducing or preventing the closure of stomata in response to drought stress, an effect that will both reduce the protective effect of water stress in reducing the ozone dose received by vegetation, and increase plant vulnerability to drought episodes.[158–160] There is now persuasive evidence that this loss of stomatal control is caused by ozone preventing the normal stomatal response to the hormone abscisic acid,[97] probably through increased production of ethylene.

12.7.2 Interactions with Pests and Diseases

In terms of interactions with biotic factors, it is the effects of air pollution on important insect pests, and fungal and viral diseases, that has received the most attention.[161,162] In understanding these interactions, it is important to distinguish between direct effects of air pollutants on the pest or disease organism, and those which modify plant responses, while tertiary interactions (*e.g.* through effects on predators or parasites of insects, or insect vectors of viral diseases) may also be significant.[163] A nice illustration of these first two effects is provided in Figure 12.6. There is a linear increase in the growth rate of aphids feeding on pea plants in response to increasing SO_2 concentration, but above about 100 ppb there is a steep decline, which is thought to due to exceeding the threshold concentration at which SO_2 begins to be toxic to the aphids themselves.[164]

At ambient concentrations, experimental studies with SO_2 and NO_2, and studies in which ambient air in city centres has been filtered, consistently show that air pollution predisposes plants to attack by insects.[165] Recent studies have demonstrated significant effects of ambient urban air pollution in an Indian city increasing infestation of local crops by both an aphid and a moth pest.[166]

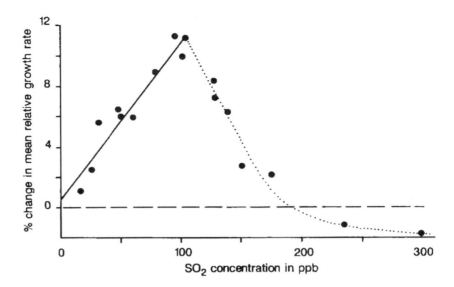

Figure 12.6   Effects of atmospheric SO_2 concentration on the mean relative growth rate of pea aphids (*Acyrthosiphon pisum*). The aphids fed for four days on garden peas growing in the various SO_2 concentrations to which the plants had also been exposed for about 20 days prior to the placement of the aphids. Percentage changes are shown compared with aphids on plants in clean air (From Warrington[102] and used with permission of Applied Science Publishers Ltd).

This effect is primarily mediated through an alteration in the quality of the host plant by the pollutant. Aphids, for example, feed specifically on the phloem sap of plants, which is the main transport pathway for sugars and amino acids, and it has been demonstrated that changes in amino acid composition of the phloem are the main mechanism which is responsible for increased aphid growth rates in response to SO_2.[167]

Similar effects of elevated N deposition on insect performance have been identified as part of a sequence of events leading to major ecological change; for example, heather beetle outbreaks are much more frequent in heathlands exposed to higher levels of N deposition, reflecting higher growth rates on foliage with higher N contents.[167,168] The gaps created in the heather canopy by beetle attacks can be invaded by other species, accelerating a change from heathland to acid grassland. In contrast, the evidence for ozone is much less consistent, with both positive and negative effects on major insect pests being reported, as well as studies showing no significant effect.[169] This may reflect the variety of mechanisms involved, which besides effects on the insects and through primary nutrients, may include a change in secondary metabolites that provide defence against insect attack and the aerial release of compounds that act as attractants or repellents of insects.[170]

In the case of fungal pathogens, a similar range of direct and indirect mechanisms is at work. Air pollutants can have direct effects on spore germination or fungal development, while chemical and physical changes in the leaf surface (*e.g.* through changes in foliar exudates or leaf wettability) may alter the likelihood of successful infection.[161,162] Fungal pathogens induce a range of defence mechanisms within the leaf, which may be similar to those induced by pollutant exposure; for example exposure to O_3 can induce a range of plant defence reactions that are maintained for significant periods after ozone exposure, and which would also provide enhanced defence against fungal pathogens.[171] However, when O_3 increases the wettability of leaf surfaces, increased fungal infection can result.[172]

There are well-known cases of fungal diseases, such as blackspot on roses and tarspot on sycamore, that are less frequent in urban areas, an effect that has been ascribed to air pollution. For example, *Diplocarpon rosae*, which is responsible for black spot on roses, has been shown experimentally to be sensitive to SO_2.[173] However, caution is required before hasty conclusions are drawn. For example, although the frequency in urban areas of *Rhytisma acerinum*, the fungus responsible for tar spot on the leaves of sycamore (*Acer pseudoplatanus*), has been reported to decrease with increasing concentrations of SO_2,[174] other factors have been identified as leading to lower frequencies in urban areas, including elevated concentrations of NO_x, and the low abundance of overwintered sycamore leaves infested with tar spot, most of which are removed in cities, hence eliminating the source of infection.[167,175]

As for insect pests, elevated N deposition can enhance N contents of tree foliage and increase the prevalence and frequency of fungal disease outbreaks.[176,177] Such fungal outbreaks can also contribute to ecological change in response to N deposition. For example, the decline of the ericaceous shrub *Vaccinium myrtillus*, and its replacement by grass species, in Swedish forests has been associated with a greatly increased frequency of fungal pathogens in areas with higher N deposition.[178] Air pollution is also known to have significant effects on biotic interactions below-ground. The most important of these relate to the mycorrhizal fungal associations in the roots, which have vital roles in nutrient acquisition, and which have been shown to respond to both elevated nitrogen and ozone exposure as a result of processes such as soil eutrophication and acidification, and reduced carbon allocation below ground.[179,180]

12.8 CRITICAL LOADS AND LEVELS

One of the major objectives of research of the kind discussed in this chapter is to provide an evidence base for policy to protect crops, forests, and both natural and managed ecosystems, from the adverse effects of air pollution. Quantification of the impacts of air pollutants, and the benefits of policy measures to reduce pollutant concentrations and deposition rates, is only possible when robust exposure–response relationships are available. However, these are only available for a limited range of plant species or ecosystems, and for a limited range of response variables, and even when they do exist there are major uncertainties in their application at a regional scale (as discussed for ozone in section 12.6). Therefore, policy assessment has more often been based on identifying critical thresholds above which adverse ecological effects may occur, and then aiming to reduce emissions to minimise exceedance of these thresholds.[181]

Within Europe, this policy approach has been used effectively to develop multilateral agreements on control of air pollution to protect ecosystems within the Convention on Long-Range Transboundary Air Pollution (CLRTAP).[182] This relies on two concepts – the 'critical load' and the 'critical level'. The critical load for ecosystems refers to long-term cumulative effects of total N and S deposition, and is defined as 'a quantitative estimate of exposure to one or more pollutants below which significant harmful effects on sensitive elements of the environment do not occur according to present knowledge'. The critical level refers to shorter-term effects of pollutant gases, such as SO_2, NO_2 and O_3, and is defined as 'the concentration in the atmosphere above which direct adverse effects on receptors such as plants, ecosystems or materials, may occur according to present knowledge'. Both critical loads and critical levels are based on the precautionary principle, with the aim of protecting sensitive elements of any ecosystem from long-term damage. Given the range in sensitivity between plant species, and between ecosystems, a key element of the approach has been to define different thresholds corresponding to ecosystems or plant species of different sensitivity, and hence to optimise emissions control measures to maximise ecological and societal benefits.[183,184]

But how have appropriate values of critical loads and critical levels been determined? In the case of critical loads for acidification and eutrophication, steady-state values are typically calculated to prevent long-term critical changes in soil or water chemistry, rather than directly from observed effects on vegetation,[185] while, more recently, dynamic models have been used to complement this steady state approach.[186] Both these approaches require a large number of important simplifying assumptions when calculations are applied across the continent of Europe,[187] but these issues are outside the scope of this section, which rather is focussed on values that have been derived directly from experimental data or from field observations of plant responses, as described in previous sections.

Since the recognition of N deposition as a major driver of loss of biodiversity in Europe (see section 12.4), a number of expert workshops have taken place to reach agreement on so-called 'empirical critical loads' of nitrogen, for those ecosystems that are recognised as being relatively sensitive to N deposition.[60,88,189] In this approach, long-term experimental studies of the effects of nitrogen addition to vegetation, ideally under field conditions, play a central role, with the critical load being the highest addition of nitrogen that does not lead to adverse changes in ecosystem function or loss of species. In some cases, experimental results are supplemented by field studies to provide confirmation of the threshold for effects. However, appropriate dose–response relationships are only available for a limited number of broadly defined ecosystems, many studies have been conducted at sites with deposition rates above the threshold for effects, and there is a need to extrapolate from studies of limited duration to the long-term effects on which the critical load concept is based.[188]

Given these uncertainties in interpreting this limited evidence base to provide appropriate critical load values, a degree of expert judgement and interpretation is needed. Table 12.1 identifies

Table 12.1 Empirical critical loads of nitrogen deposition as currently applied in Europe, adapted from Bobbink and Hettlingh, 2011.[60] For each ecosystem, the critical load range (expressed as an annual mean deposition rate), the level of certainty placed on the value and the major ecological effects that are observed when the critical load is exceeded are shown. Only a limited number of ecosystems of those for which critical loads have been proposed are tabulated.

Ecosystem	Critical load ($kg\ ha^{-1}\ yr^{-1}$)	Level of certainty	Ecological indicators of critical load exceedance
Raised and blanket bogs	5–10	reliable	Altered growth and species composition of bryophytes Increase in vascular plants Increased N in peat and peat water
Arctic and alpine heaths	5–15	quite reliable	Decline in lichens and bryophytes Decline in evergreen dwarf shrubs
Broadleaved deciduous woodland	10–20	reliable	Altered species composition of ground flora Altered composition of mycorrhizae Nutrient imbalance Changes in soil processes
Dry heath	10–20	reliable	Transition from heather to grass dominance Decline in lichens Increased sensitivity to abiotic stress
Stable coastal dune grasslands	8–15	quite reliable	Increase in tall grasses Decrease in prostrate plants Loss of typical lichen species Increased N leaching and soil acidification
Non-Mediterranean dry acidic and neutral closed grasslands	10–15	reliable	Increase in grass cover Decline in typical species Decreased total species richness
Sub-Atlantic semi-dry calcareous grasslands	15–25	reliable	Increased in tall grasses Decline in species diversity Increased N leaching and surface acidification
Mid-upper salt marshes	20–30	expert judgement	Increase in domination of grasses

empirical critical loads for N deposition in Europe that have recently been agreed, for a selected range of ecosystems only,[189] to illustrate the approach used to communicate this uncertainty. For each ecosystem, the value is given as a range, representing the likely variation in sensitivity within a given ecosystem, based on the available evidence and a degree of expert judgement; this variation reflects differences in factors such as climate, soil type, and management across Europe, and guidance is also provided on how an appropriate value within the range for a particular location might be selected. The strength of the evidence is also indicated on a simple three-point scale, from expert judgement to reliable (*i.e.* with strong supporting data). Defining what constitutes an adverse effect is also a very important element in setting appropriate values of critical loads and levels. Table 12.1 shows for each ecosystem, the criteria used to set the critical load range. There may be multiple criteria used, and these vary between ecosystems, based on knowledge of the key mechanisms of impact and the nature of the evidence that is available.[189] In general, these are confined to loss of characteristic species or major changes in plant species composition that would clearly be recognised as adverse by conservationists, although some (*e.g.* soil acidification, increased N leaching) are chemical indicators of wider ecosystem perturbation. This empirical approach involves a considerable degree of expert judgement in interpretation of the available evidence. Empirical critical loads that have been proposed for the United States[62] are consistently lower than those identified for similar ecosystems in Europe. A number of reasons for this difference have been identified, including the much lower background N deposition in pristine environments, and the use of biogeochemical indicators of ecosystem response as response criteria, rather than significant change in species composition.

In contrast, formal statistical analysis of exposure–response relationships, identifying the concentration at which a specific percentage change in the chosen response occurs, has been more commonly used to define critical levels, which, unlike critical loads, are based on a specific duration of exposure. A selection of the currently recommended values is provided in Table 12.2,[190] indicating the evidence base on which each value has been quantified. While the critical levels for SO_2 rely primarily on field observations,[191] and should be regarded as only approximate values, those for O_3 and NH_3 are based on more rigorous statistical analysis. In the case of ozone effects on crops and trees, exposure–response relationships based on collated experimental datasets from across Europe have been used to quantify the exposure that corresponds to a reduction in annual yield or growth of 5%, a criteria which is based on the minimum effect that is typically significantly different from zero in experimental studies.[192] In contrast, for NH_3, robust statistical analysis of data from field release experiments and field observations has been used to identify 'no observed effect thresholds' for effects on plant species composition.[193]

In the case of ozone, the AOT40 index was initially used to define critical levels,[194,195] although there was some evidence that use of an AOT30 index rather than AOT40 (*i.e.* using a threshold of 30 ppb rather than 40 ppb) would have provided a better fit to experimental data.[196] As described in Section 12.5, there is now increasing evidence that the effects of ozone are better related to the cumulative flux through the stomata into the leaf. A new exposure index, the AFstx (Accumulated Flux through the stomata above a critical flux threshold of x mmol m^{-2} s^{-1}), has been quantified based on standardised flux models, and used to derive flux–response relationships over critical periods of plant development. Critical levels have then been deduced following a similar approach to that adopted for AOT40.[197] Critical levels for some of the limited range of species for which appropriate flux–response relationships are available are included in Table 12.2.

12.9 EFFECTS ON ECOSYSTEM SERVICES

The previous sections have described some of the effects of major air pollutants, discussed the underlying mechanisms, and identified critical exposure thresholds above which

Table 12.2 Summary of critical levels for major gaseous pollutants, indicating the rationale for the recommended values, adapted from Mills *et al.* 2010.[190] Note that this is not a comprehensive list of all critical levels and plant groups or species but provides illustrative values for sulfur dioxide, ammonia and ozone.

Pollutant	Plant species/ group	Recommended critical level	Averaging time	Major evidence base
SO_2	Lichens	$10 \ \mu g \ m^{-3}$	annual mean	Field observations of locations where the most sensitive lichen groups are eliminated
SO_2	Forests	$20 \ \mu g \ m^{-3}$	annual mean and winter mean	Field observation of threshold for declines of a sensitive tree species (Norway spruce)
SO_2	Crops	$30 \ \mu g \ m^{-3}$	annual mean	Experimental dose–response data for the most sensitive species (ryegrass)
NH_3	Lichens and bryophytes	$1 \ \mu g \ m^{-3}$	annual mean	Field observations of threshold for changes in species composition
NH_3	Vascular plants	$3 \ \mu g \ m^{-3}$	annual mean	Field observations of threshold for changes in species composition
O_3	Agricultural crops	3 ppm. h AOT40	3 months	Exposure causing 5% yield reduction, derived from OTC experiments
O_3	Wheat	1 mmol m^{-2} AFst6	approx. 55 days around anthesis	Exposure causing 5% yield reduction, derived by applying a wheat stomatal flux model to OTC experimental data
O_3	Potato	5 mmol m^{-2} AFst6	approx. 70 days around anthesis	Exposure causing 5% yield reduction, derived by applying a potato stomatal flux model to OTC experimental data
O_3	Forest trees	5 ppm.h AOT40	growing season mean	Exposure causing a 5% growth reduction each growing season, derived from OTC experiments
O_3	Beech and birch	4 mmol m^{-2} AFst1.6	one growing season	Exposure causing 5% growth reduction, derived by applying a stomatal flux model for birch and beech to OTC experimental data
O_3	Semi-natural communities dominated by perennial crops	5 ppm.h AOT40	6 month growing season	Estimated from effects on species composition, individual species biomass and senescence in OTC experiments

adverse effects of these pollutants may be found, primarily on plant vitality, growth and yield. However, it is now recognised that both managed and unmanaged ecosystems provide a much wider range of services for human society. Over the past decade, ecologists have developed methods to categorise these so-called ecosystem services (defined as 'the services that the environment provides that people need'), and have begun to quantify and value the impacts of anthropogenic stresses on them.[198,199] These ecosystem services can be divided into:

(i) Provisioning services, *e.g.* the food, fuel and fibre provided by ecosystems;
(ii) Regulating services, *e.g.* the ways in which ecosystems contribute to regulation of climate, water flow and quality, and pollination; and
(iii) Cultural services, *e.g.* aesthetic and educational qualities of ecosystems, and their contribution to tourism and recreation.

Air pollution can both enhance and reduce the delivery of these ecosystem services. Some of the major effects of air pollution on ecosystem services are summarised in Table 12.3. This summary does not include every type of ecosystem service, but focuses on those for which there is clear

Table 12.3 Potential impact of air pollutants on major ecosystem services. The nature of the ecosystem service is indicated as: P = provisioning; R = regulating; C = cultural; The mechanisms of impact column provides reference either to the chapter sections in which this is covered or supporting references.

Pollutant	Ecosystem service	Mechanism of impact
S and N deposition	Enhanced food and fibre production (P)	Increase nutrient availability for arable crops and grasslands (sections 12.3 and 12.4).
Ozone	Decreased food and fibre production (P)	Decreased growth rate of sensitive crops and trees and reduced forage quality in grasslands (section 12.5).
N deposition	Both increased and decreased forest fibre production (P)	Moderate N deposition increases growth but long-term high deposition can reduce growth through acidification and eutrophication (section 12.4)
All pollutants	Increased or decreased air quality regulation (R)	Changes in growth rate, species composition and surface characteristics can alter rates of pollutant deposition.[200]
N deposition	Increased climate regulation (R)	Increased primary production (section 12.4), above ground C sequestration, and soil C storage in N-limited ecosystems.[204,206]
N deposition	Decreased climate regulation (R)	Nitrogen-induced increase in N_2O emissions.[208,209]
Ozone	Decreased climate regulation (R)	Decreased primary production (section 12.5), decreased above ground C sequestration, and decreased soil C storage.[205,207]
N deposition, S deposition and ozone	Changes in climate regulation (R)	Soil methane fluxes from wetlands altered by pollutant exposure.[210–212]
Ozone	Altered water flow and quality (R)	Changes in ecosystem transpiration losses (section 12.5), altering water retention in summer.[213]
N deposition, ozone	Altered pollination (R)	Changes in species composition (sections 12.4 and 12.5) and both decreased and increased flowering.[77,95,214,215]
All pollutants	Loss of genetic resources (C)	Loss of individual species or of sensitive genotypes, decreasing overall fitness (sections 12.3, 12.4 and 12.5).
N deposition	Aesthetic (C)	Loss or decline of sensitive and attractive species (section 12.4).
S and N deposition	Aesthetic (C)	Soil acidification results in the clarifying of peat stained waters through reductions in dissolved organic carbon.[216,217]

evidence of the adverse effects of air pollution. Effects on provisioning services, and in particular food and fibre production, have been described in previous sections, but those on regulating and cultural services deserve more discussion here.

The ecosystem service of air quality regulation is linked to the exchange of trace gases and particles between ecosystems and the atmosphere; ecosystems can be both a source and a sink of air pollutants. Therefore, ecosystems can have positive effects on air quality, primarily through interception, deposition and removal of pollutants,[200] or have negative effects (*e.g.* VOC emissions from vegetation contribute significantly to ozone formation). This air quality regulation can itself also be modified by pollutant exposure,[201] Air quality regulation has important implications for human exposure to air pollution and effects on human health. For example, increases in local particle deposition due to urban tree planting[202] and decreases in national-scale O_3 deposition in drought summers[203] have been shown, respectively, to decrease and increase the impacts of air pollution on human health.

Air pollution exposure can also alter the fluxes of greenhouse gases between ecosystems and the atmosphere, and the long-term storage of carbon, and hence the ecosystem service of climate regulation. The changes in primary production and growth that are caused by air pollution have been modelled to cause significant changes in CO_2 uptake and above-ground sequestration of carbon at a European scale, with the effects being positive for N deposition and negative for ozone.[204,205] In addition, exposure to these pollutants is associated with significant changes in respiration rates and soil carbon storage.[206,207] Effects on fluxes of other greenhouse gases also need to be considered. Increased N deposition, and associated increase in soil N pools, is associated with increased N_2O emissions,[208,209] while increased N deposition, S deposition and ozone have all been shown to alter fluxes of methane to the atmosphere.[210–212]

As discussed in section 12.5, ozone can reduce above-ground biomass and leaf area; reduce root/shoot ratio; and interfere with the response of stomata to soil water deficits, all of which can modify the ecosystem water balance at a landscape scale.[159] How these different effects actually alter the ecosystem service of water regulation in catchments is unclear, although one study in the eastern USA suggested that periods of elevated ozone exposure led to a more rapid depletion of catchment soil water and reduced late-summer stream flow.[218]

Exposure to air pollution can also alter cultural services, for example through loss of sensitive species and genotypes, which will have implications at a larger scale for genetic resources. Flowering is also affected, although there is evidence of both positive and negative effects; for example, N deposition increases flowering in experiments with heathlands but decreases it in experiments on grasslands.[77,214] The loss of attractive species, and reduced flowering, can also reduce the aesthetic value of ecosystems, a cultural service. The implications for the important ecosystem service of pollination are less clear; although ozone has a number of relevant effects, *e.g.* changes in flowering, interference with detection of scents by insects, and reduced pollen growth,[95,215] there have been no studies to measure or model the overall effects on pollination of key commercial species. Finally, effects on air pollutant deposition below-ground may have many implications for ecosystem services, besides those on carbon sequestration and the microbial communities responsible for N_2O and CH_4 production. For example, acidification caused by S and N deposition may suppress production of dissolved organic carbon,[216,217] which is important for ecosystem carbon budgets, and causes brown drainage and drinking water in peatland regions, reducing its aesthetic value.

So, for every air pollutant considered in this brief analysis, there are processes that degrade ecosystem services and processes that enhance them; furthermore these effects may be concentration- or ecosystem-dependent. Although a comprehensive assessment of the impacts of air pollution on national ecosystem services has been advocated, there are enormous challenges in such an approach, particularly if it is extended to an economic valuation of the effects on ecosystem services that could be compared with the cost of pollution control. Nevertheless, for some specific policy interventions, the effects of air quality management policies in terms of ecosystem services could be comparable in size to those for human health; for example, the economic value of scenarios to reduce UK ammonia (NH_3) emissions are comparable in magnitude in economic terms for effects on human health, ecosystem carbon sequestration and ecosystem nitrous oxide (N_2O) emissions.[218] However, no study has yet quantified the effect of air quality on a full range of ecosystem services, and very significant barriers to such an endeavour exist in terms of both data and models.

This is a key challenge for the future. As this chapter has shown, although there is much detail still to be learnt, we have a good understanding of the mechanisms and effects of air pollutants on individual plants. However, the cascade of effects, and the interactions with other biotic and abiotic factors, that lead to effects at an ecosystem and landscape scale are far from fully understood. Furthermore, the greatest effects of air pollution over this century are likely to be in regions of the world (Asia, Africa, Latin America), where we know almost nothing about the response to air pollution of the unique local plant communities and ecosystems, and the interactions with other elements of global change.[7,219]

REFERENCES

1. W. W. Pettigrew, *The Influence of Air Pollution on Vegetation*, lecture to the Smoke Abatement League of Great Britain, 1928.
2. J. B. Cohen and A. G. Rushton, *Smoke: a Study of Town Air*, Edward Arnold, London, 1912.
3. B. Freedman and T. C. Hutchinson, *Can. J. Bot.*, 1980, **58**, 2123.
4. A. Wilson, *Halifax Naturalist*, 1901, 57.
5. J. H. Tallis, *J. Ecol.*, 1964, **52**, 207.
6. M. C. Press, P. Ferguson and J. A. Lee, *Naturalist*, 1983, **108**, 125.
7. L. D. Emberson, M. R. Ashmore and F. Murray, eds., *Air Pollution Impacts on Crops and Forests: A Global Assessment*, Imperial College Press, London, 2003.
8. M. Agrawal, in *Air Pollution Impacts on Crops and Forests: A Global Assessment*, ed. L. D. Emberson, M. R. Ashmore and F. Murray, Imperial College Press, London, 2003, pp. 165–187.
9. M. Agrawal, B. Singh, S. B. Agrawal, J. N. B. Bell and F. Marshall, *Water Air Soil Pollut.*, 2006, **169**, 239.
10. M. Tallis, G. Taylor, D. Sinnett and P. Freer-Smith, *Landscape Urban Plan.*, 2011, **103**, 129.
11. M. N. Ahmad, L. J. L. van den Berg, H. U. Shah, T. Masood, P. Buker, L. Emberson and M. Ashmore, *Environ. Pollut.*, 2012, **162**, 319.
12. L. H. Weinstein and A. W. Davison, *Fluorides in the Environment*, CABI Publishing, Wallingford, 2004.
13. W. W. Heck, O. C. Taylor and D. T. Tingey, *Assessment of Crop Loss from Air Pollution*, Elsevier Applied Science, London, 1988.
14. A. S. Heagle, L. W. Kress, P.J. Temple, R. J. Kohut, J. E. Millar and H. E. Heggestad, in *Assessment of Crop Loss from Air Pollution*, ed. W. W. Heck, O. C. Taylor and D. T. Tingey, Elsevier Applied Science, London, 1988, pp. 141–179.
15. D. M. Eastburn, M. M. Degennaro, E. H. Delucia, O. Dermody and A. J. McElrone, *Global Change Biol.*, 2010, **16**, 320.
16. G. K. Phoenix, B. A. Emmett, A. J. Britton, S. J. M. Caporn, N. B. Dise, R Helliwell, L. Jones, J. R. Leake, I. D. Leith, L. J. Sheppard, A. Sowerby, M. G. Pilkington, E. C. Rowe, M. R. Ashmore and S. A. Power, *Global Change Biol.*, 2012, **18**, 1197.
17. S. Bassin, M. Volk, N. Buchmann and J. Fuhrer, *New Phytol.*, 2007, **175**, 523.
18. P. Redbo-Tortensson, *Acta Bot. Neerl.*, 1994, **43**, 175.
19. D. F. Karnosky, D. R. Zak, K. S. Pregitzer, C. S. Awmack, J. G. Bockheim, R. E. Dickson, G. R Hendry, G. E. Host, J. S. King, *et al.*, *Function. Ecol.*, 2003, **17**, 289.

20. H. Werner and P. Fabian, *Environ. Sci. Pollut. Res.*, 2002, **9**, 117.
21. C. J. Stevens, N. B. Dise, J. O. Mountford and D. J. Gowing, *Science*, 2004, **303**, 1876.
22. P. Miller and J. McBride, *Oxidant Air Pollution Impacts in the Montane Forests of Southern California: The San Bernadino Mountain Case Study*, Springer-Verlag, New York, 1999.
23. S. Braun, B. Rihm, C Schindler and W. Fluckiger, *Water Air Soil Pollut.*, 1999, **116**, 357.
24. N. Clarke, R. Fischer, W. De Vries, L. Lundin, T. Vesela, O. Merila, G. Matteucci, M. Mirti, D. Simpson and E. Paoletta, *iForest – Biogeosci. For.*, 2011, **4**, 162.
25. H. Marschner, *Mineral Nutrition of Higher Plants*, Academic Press, San Diego, 1995.
26. S. Cieslik, K. Omasa and E. Paoletti, *Plant Biol.*, 2009, **11**, 24.
27. K. J. Lendzian, *Aspects Appl. Biol.*, 1988, **17**, 97.
28. U. Takama, S. Veljovic-Iovanovic and U. Heber, *Plant Physiol.*, 1992, **100**, 261.
29. H. Rennenberg and A. Polle, in *Plant Responses to the Gaseous Environment*, ed. R. G. Alscher and A. R. Wellburn, Chapman and Hall, London, 1994, p. 165.
30. G. E. Taylor Jr., L. F. Pitelka and M. T. Clegg, eds., *Ecological Genetics and Air Pollution*, Springer-Verlag, New York, 1991.
31. J. N. B. Bell and W. S. Clough, *Nature*, 1973, **241**, 47.
32. G. E Taylor and W. H. Murdy, *Bot. Gaz.*, 1975, **136**, 212.
33. J. N. B. Bell, M. R. Ashmore and G. B. Wilson, in *Ecological Genetics and Air Pollution*, ed. G. E. Taylor Jr., L. F. Pitelka and M. T. Clegg, eds., Springer-Verlag, New York, 1991, pp. 33–59.
34. B. W. Ferry, M. S. Baddeley and D. L. Hawksworth, eds., *Air Pollution and Lichens*. Athlone Press, London, 1973.
35. D. L. Hawksworth and F. Rose, *Lichens as Pollution Monitors*, Arnold, London, 1976.
36. J. W. Bates, J. N. B. Bell and A. M. Farmer, *Environ. Pollut.*, 1990, **68**, 81.
37. K. Batty, J. W Bates and J. N. B Bell, *Can. J. Bot.*, 2003, **81**, 439.
38. L. Davies, J. W. Bates, J. N. B. Bell, P. W. James and O. W. Purvis, *Environ. Pollut.*, 2007, **146**, 299.
39. C. M. van Herk, *Lichenologist*, 1999, **31**, 9.
40. J. N. B. Bell, in *Effects of Gaseous Air Pollution in Agriculture and Horticulture*, M. H. Unsworth and D. P. Ormrod, eds., Butterworth Scientific, London, 1982, pp. 225–246.
41. R. Guderian, *Air Pollution Phytoxicity of Acidic Gases and its Significance in Air Pollution Control*, Ecological Studies 22, Springer-Verlag, Heidelberg, 1977.
42. S. P. McGrath and F. J. Zhao, *Soil Use Management*, 1995, **11**, 110.
43. F. J. Zhao, J. S. Knights, Z. Y. Hu and S. P. McGrath, *J. Environ. Qual.*, 1998, **32**, 33.
44. F. J. Zhao, B. Spiro, P. R. Poulton and S. P. McGrath, *Environ. Sci. Technol.*, 2003, **32**, 2288.
45. J. J. Elser, M. E. S. Bracken, E. E. Cleland, D. S. Gruner, W. S. Harpole, H. Hillebrand, J. T. Ngai, E. W. Seabloom, J. B. Shurin and J. E. Smith, *Ecol. Lett.*, 2007, **10**, 1135.
46. A. R. Wellburn, *New Phytol.*, 1990, **115**, 395.
47. A. R. Wellburn, J. Wilson and P. H. Aldridge, *Environ. Pollut.*, 1980, **22**, 219.
48. J. Pearson and G. R. Stewart, *New Phytol.*, 1993, **125**, 283.
49. M. A. Sutton, D. Fowler, J. K. Burkhardt and C. Milford, *Water Air Soil Pollut.*, 1995, **31**, 2615.
50. P. A. Wolseley, P. W. James, M. R. Theobold and M. A. Sutton, *Lichenologist*, 2006, **38**, 161.
51. C. M. van Herk, E. A. M. Mathijssen-Spiekman and D. de Zwart, *Lichenologist*, 2003, **35**, 347.
52. M. A. Sutton, S. Reis and S. M. H. Baker, eds., *Atmospheric Ammonia*, Springer Science, 2009.
53. L. Davies, J. W. Bates, J. N. B. Bell, P. W. James and O. W. Purvis, *Environ. Pollut.*, 2007, **146**, 299.

54. K. L. Bignal, M. R. Ashmore, A. D. Headley, K. Stewart and K Weigert, *Appl. Geochem.*, 2007, **22**, 1265.
55. M. A. Sutton, P. A. Wolseley, I. D. Leith, N. van Dijk, Y. Sim Tang, P. W. James, M. R. Theobold and C. Whitfield, in *Atmospheric Ammonia*, ed. M. A. Sutton, S. Reis and S. M. H. Baker, Springer Science, 2009, pp. 71–86.
56. C. E. R. Pitcairn, I. D. Leith, A. Crossley, N. van Dijk, J. N. Cape, D. Fowler and M. A. Sutton, in *Atmospheric Ammonia*, ed. M. A. Sutton, S. Reis and S. M. H. Baker, 2009, Springer Science, pp. 59–69.
57. L. J. Sheppard, I. D. Leith, T. Mizunuma, J. N. Cape, A. Crossley, S. Leeson, M. A. Sutton, N van Dijk and D. Fowler, *Global Change Biol.*, 2011, **17**, 3589.
58. R. Bobbink, K. Hicks, J. Galloway, T. Spranger, R. Alkemade, M. Ashmore, M. Bustamante, S. Cinderby, E. Davidson, F. Dentener, B. Emmett, J.-W. Erisman, M. Fenn, F. Gilliam, A. Nordin, L. Pardo and W. de Vries, *Ecol. Appl.*, 2010, **20**, 30.
59. M. A. Sutton, C. M. Howard, J. W. Erisman, G. Billen, A. Bleeker, P. Grennfelt, H. van Grinsven and B. Grizzetti, eds., *The European Nitrogen Assessment*, Cambridge University Press, Cambridge, 2011.
60. B. Achermann and R. Bobbink, *Empirical Critical Loads for Natural and Semi-natural Ecosystems: 2002 Update*, Swiss Agency for Environment, Forestry and Landscapes, Berne, 2003.
61. N. B. Dise, M. Ashmore, S. Belyazid, A. Bleeker, R. Bobbink, W. de Vries, J. W. Erisman, T. Spranger, C. J. Stevens and L. van den Berg, in *The European Nitrogen Assessment*, ed. M. A. Sutton, C. M. Howard, J. W. Erisman, G. Billen, A. Bleeker, P. Grennfelt, H. van Grinsven and B. Grizzetti, Cambridge University Press, Cambridge, 2011, pp. 463–494.
62. L. H. Pardo, M. E. Fenn, C. L. Goodale, L. H. Geiser, C. T. Driscoll, E. B. Allen, J. S. Baron, R. Bobbink, W. D. Bowman, C. M. Clark, *et al.*, *Ecol. Appl.*, 2011, **21**, 3049.
63. G. K. Phoenix, W. K. Hicks, S. Cinderby, J. C. I. Kuylenstierna, W. D. Stock, F. J. Dentener, K. E. Giller, A. T. Austin, R. D. B. Lefroy, B. S. Gimeno, M. R. Ashmore and P. Ineson, *Global Change Biol.*, 2006, **12**, 470.
64. R. Bobbink and L. Lamers, in *Air Pollution and Plant Life*, ed. J. N. B. Bell and M. Treshow, John Wiley and Sons, Chichester, 2nd edn., 2002, pp. 201–236.
65. S. A. Power, M. R. Ashmore and D. A. Cousins, *Environ. Pollut.*, 1998, **102**, 27.
66. J. A. Carroll, S. J. M. Caporn, D. Johnson, M. D. Morecroft and J. A. Lee, *Environ. Pollut.*, 2003, **121**, 363.
67. C. M. Clark and D. Tilman, *Nature*, 2008, **451**, 712.
68. R. Bobbink and J. N. Willems, *Biol. Conserv.*, 1987, **40**, 301.
69. K. N. Suding, S. L. Collins, L. Gough, C. Clark, E. E. Cleland, K. L. Gross, D. G. Milchunas and S. Pennings, *Proc. Natl. Acad. Sci. U. S. A.*, 2005, **102**, 4387.
70. B. Emmett, *Water Air Soil Pollut., Focus*, 2007, **7**, 99.
71. G. K. Phoenix, R. E. Booth, J. R. Leake, D. J. Read, J. P. Grime and J. A. Lee, *New Phytol.*, 2004, **161**, 279.
72. S. J. T. Arens, P. F. Sullivan and J. M. Welker, *J. Geophys. Res.*, 2008, **113**, G03209.
73. W. D. Bowman, C. C. Cleveland, L. Halada, J. Hresko and J. S. Baron, *Nature Geosci.*, 2008, **1**, 767.
74. N. B. Dise, E. Matzner, M. Armbruster and J. Mcdonald, *J. Environ. Qual.*, 2001, **30**, 1747.
75. J. G. M. Roelofs, R. Bobbink, E. Brouwer and M. C. C. de Graaf, *Acta Bot. Neerl.*, 1996, **45**, 517.
76. D. Kleijn, R. M. Bekker, R. Bobbink, M. C. C. de Graaf and J. G. M. Roelofs, *J. Appl. Ecol.*, 2008, **45**, 680.

77. G. K. Phoenix, B. A. Emmett, A. J. Britton, S. J. M. Caporn, N. B. Dise, R. Helliwell, L. Jones, J. R. Leake, I. D. Leith, L. J. Sheppard, A. Sowerby, M. G. Pilkington, E. C. Rowe, M. R. Ashmore and S. A. Power, *Global Change Biol.*, 2012, **18**, 1197.

78. C. M. Clark, E. E. Cleland, S. L. Collins, J. E. Farglone, L. Gough, K. L. Gross, S. C. Pennings, K. N. Suding and J. B. Grace, *Ecol. Lett.*, 2007, **10**, 596.

79. C. J. Stevens, N. B. Dise, D. J. G. Gowing and J. O. Mountford, *Global Change Biol.*, 2006, **12**, 1823.

80. R. J. Payne, C. J. Stevens, N. B. Dise, D. J. Gowing, M. G. Pilkington, G. K. Phoenix, B. A. Emmett and M. R. Ashmore, *Environ. Pollut.*, 2011, **159**, 2602.

81. C. J. Stevens, K. Thompson, J. P. Grime, C. J. Long and D. J. G. Gowing, *Function. Ecol.*, 2010, **24**, 478.

82. L. C. Maskell, S. M. Smart, J. M. Bullock, K. Thompson and C. J. Stevens, *Global Change Biol.*, 2010, **16**, 671.

83. C. J. Stevens, C. Dupre, E. Dorland, C. Gaudnik, D. J. G. Gowing, A. Bleeker, M. Diekmann, D. Alard, R. Bobbink, D. Fowler, E. Corcket, J. O. Mountford, V. Vandvik, P. A. Aarrestad, S. Muller and N. B. Dise, *Environ. Pollut.*, 2011, **159**, 2243.

84. C. Dupre, C. J. Stevens, T. Ranke, A. Bleeker, C. Peppler-Lisbach, D. J. G. Gowing, N. B. Dise, E. Dorland, R. Bobbibk and M. Diekmann, *Global Change Biol.*, 2010, **16**, 344.

85. W. L. M. Tamis, M. van't Zelfte, R. van der Meijen, C. L. G. Groen and U. de Haes, *Biol. Conserv.*, 2005, **125**, 211.

86. E. L. Fiscus, F. L. Booker and K. O. Burkey, *Plant Cell Environ.*, 2005, **28**, 997.

87. M. Plochl, T. Lyons, J. Ollernshaw and J. Barnes, *Planta*, 2000, **210**, 454.

88. P. L. Conklin and C. Barth, *Plant Cell Environ.*, 2004, **27**, 959.

89. M. Frei, J. P. Tanaka, C. P. Chan and M. Wissuwa, *J. Exp. Bot.*, 2010, **61**, 1405.

90. T. Lyons, J. H. Ollerenshaw and J. D. Barnes, *New Phytol.*, 1999, **141**, 253.

91. I. Fumagalli, B. S. Gimeno, D. Velissariou, L. De Temmerman and G. Mills, *Atmos. Environ.*, 2001, **35**, 2583.

92. P. B. Morgan, E. A. Ainsworth and S. P. Long, *Plant Cell Environ.*, 2003, **26**, 1317.

93. I. F. Mckee and S. P. Long, *New Phytol.*, 2001, **152**, 41.

94. D. A. Grantz, S. Gunn and H. B. Vu, *Plant Cell Environ.*, 2006, **29**, 1193.

95. C. P. Leisner and E. A. Ainsworth, *Global Change Biol.*, 2012, **18**, 606.

96. M. R. Ashmore, *Plant Cell Environ.*, 2005, **28**, 949.

97. S. Wilkinson and W. J. Davies, *Plant Cell Environ.*, 2010, **33**, 510.

98. K. Ramo, T. Kanerva, K. Ojanpera and S. Manninen, *Environ. Pollut.*, 2007, **145**, 850.

99. T. Kanerva, S. Palojarvi, K. Ramo and S. Manninen, *Soil Biology Biochem.*, 2008, **40**, 2502.

100. R. M. Adams, J. D. Glyer, S. L. Johnson and B. A. McCarl, *J. Air Pollut. Control Assoc.*, 1989, **39**, 960.

101. J. Fuhrer, L. Skarby and M. R. Ashmore, *Environ. Pollut.*, 1997, **97**, 91.

102. L. D. Emberson, M. R. Ashmore, H. M. Cambridge, D. Simpson and J.-P. Tuovinen, *Environ. Pollut.*, 2000, **109**, 403.

103. J. Fuhrer, *Naturwissenschaften*, 2009, **96**, 173.

104. H. Pleijel, H. Danielsson, L. D. Emberson, M. R. Ashmore and G. Mills, *Atmos. Environ.*, 2007, **41**, 3022.

105. F. Hayes, S. Wagg, G. Mills, S. Wilkinson and W. Davies, *Global Change Biol.*, 2012, **18**, 948.

106. D. Simpson, L. D. Emberson, M. R. Ashmore and J.-P. Tuovinen, *Environ. Pollut.*, 2007, **146**, 715.

107. G. Mills, F. Hayes, D. Simpson, L. Emberson, D. Norris, H. Harmens and P. Buker, *Global Change Biol.*, 2011, **17**, 592.

108. G. Mills and H. Harmens, *Ozone Pollution: A Hidden Threat to Food Security*, ICP Vegetation, Centre for Ecology and Hydrology, Bangor, 2011.

109. Royal Society, *Ground-level Ozone in the 21st Century: Future Trends, Impacts and Policy Considerations,* The Royal Society, London, 2008.

110. F. Dentener, D. Simpson, K. Ellingson, T. van Noije, M. Schultz, M. Amann, C. Atherton, N. Bell, D. Bergmann, I. Bey, *et al., Environ. Sci. Technol.,* 2006, **40**, 3586.

111. Z. Z. Feng and K. Kobayashi, *Atmos. Environ.,* 2009, **43**, 1510.

112. R. van Dingenen, F. J. Dentener, F. Raes, M. C. Krol, L. Emberson and J. Cofala, *Atmos. Environ.,* 2009, **43**, 604.

113. S. Averny, D. L. Mauzerall, J. Liu and L. W. Horowitz, *Atmos. Environ.,* 2011, **45**, 2284.

114. H. Pleijel, *Environ. Pollut.,* 2011, **159**, 897.

115. S. Averny, D. L. Mauzerall, J. Liu and L. W. Horowitz, *Atmos. Environ.,* 2011, **45**, 2297.

116. S. Averny, D. L. Mauzerall and A. M. Fiore, *Global Change Biol.,* 2013, **19**, 1285.

117. M. Engardt, *Atmos. Phys. Chem.,* 2008, **59**, 61.

118. S. B. Agrawal, A. Singh and D. Rathore, *Chemosphere,* 2005, **61**, 218.

119. A. Wahid, *Sci. Total Environ.,* 2006, **371**, 304.

120. G. Y. Shi, L. X. Yang, Y. X. Yang, K. Kobayashi, J. Zhu, H. Tang, S. Pan, T. Chen, G. Liu and Y. Wang, *Agri. Ecosyst. Environ.,* 2009, **131**, 176.

121. X. K. Zhu, Z. Z. Feng, T. F. Sun, X. Liu, H. Tang, J. Zhu, W. Guo and K. Kobayashi, *Global Change Biol.,* 2011, **17**, 2697.

122. L. D. Emberson, P. Buker and M. R. Ashmore, *Atmos. Environ.,* 2009, **43**, 1945.

123. Z. Feng, H. Tang, J. Uddling, H. Pleijel, K. Kobayashi, Z. Jianguo, H. Oue and W. Guo, *Environ. Pollut.,* 2012, **164**, 16.

124. L. Skarby, *Atmos. Environ.,* 2004, **38**, 2225.

125. V. E. Wittig, E. A. Ainsworth, S. L. Naidu, D. F. Karnosky and S. P. Long, *Global Change Biol.,* 2009, **15**, 396.

126. D. R. Zak, K. S. Pregitzer, M. E. Kubiske and A. J. Burton, *Ecol. Lett.,* 2011, **14**, 1220–1226.

127. H. Prestzch, J. Dieler, R. Matyssek and P. Wipfler, *Environ. Pollut.,* 2010, **158**, 1061.

128. A. W. Davison and J. D. Barnes, *New Phytol.,* 1998, **139**, 135.

129. S. Bassin, M. Volk, M. Suter, N. Buchmann and J. Fuhrer, *New Phytol.,* 2007, **175**, 523.

130. K. V. Wedlich, N. Rintoul, S. Peacock, J. N. Cape, M. Coyle, S. Toet, J. Barnes and M. Ashmore, *Oecologia,* 2012, **168**, 1137.

131. J. Bender, H.-J. Weigel and H.-J. Jager, *New Phytol.,* 1991, **119**, 261.

132. A. Fangmeier, J. Bender, H.-J. Weigel and H.-J. Jager, in *Air Pollution and Plant Life,* ed. J. N. B. Bell and M. Treshow, John Wiley and Sons, Chichester, 2nd edn., 2002, pp. 251–272.

133. D. T. Tingey, R. A. Reinert, J. A. Dunning and W. W. Heck, *Phytopathology,* 1971, **61**, 1506.

134. T. W. Ashenden and T. A. Mansfield, *Nature,* 1978, **273**, 142.

135. P. H. Freer-Smith, *New Phytol.,* 1984, **97**, 49.

136. A. R. Wellburn, C. Hugginson, D. Robinson and C. Walmesley, *New Phytol.,* 1981, **88**, 223.

137. I. F. Mckee, M. Eiblmeier and A. Polle, *New Phytol.,* 1997, **137**, 275.

138. J. Uddling, A. J. Hogg, R. M. Teclaw, M. A. Carroll and D. S. Ellsworth, *Environ. Pollut.,* 2010, **158**, 2023.

139. H. Kobayakawa and K. Imai, *Photosynthetica,* 2011, **49**, 227.

140. X. Z. Wang and D. R. Taub, *Oecologia,* 2010, **163**, 1.

141. D. Karnosky, K. S. Pregitzer, D. R. Zak, M. E. Kubiske, G. R. Hendrey, D. Weinstein, M. Nosal and K. Percy, *Plant Cell Environ.,* 2005, **28**, 965.

142. O. Dermody, S. P. Long, K. McConnaughhay and E. H. DeLucia, *Global Change Biol.,* 2008, **14**, 556.

143. C. Decock, H. Chung, R. Venterea, S. B. Gray, A. D. B. Leakey and J. Six, *Soil Biol. Biochem.,* 2012, **51**, 104.

144. L. Cheng, F. L. Booker, K. O. Burkey, C. Tu, H. D. Shew, T. W. Rufty, E. L. Fiscus, J. L. Deforest and S. Hu, *PLoS ONE,* 2011, **6**, e21377.

145. G. Mills, in *Air Pollution and Plant Life*, ed. J. N. B. Bell and M. Treshow, John Wiley and Sons, Chichester, 2nd edn., 2002, pp. 343–358.
146. A. W. Davison, J. D. Barnes, in *Air Pollution and Plant Life*, ed. J. N. B. Bell and M. Treshow, John Wiley and Sons, Chichester, 2nd edn., 2002, pp. 359–377.
147. T. Davies, *Nature*, 1980, **284**, 483.
148. T. Jones and T. A. Mansfield, *Environ. Pollut.*, 1982, **27**, 57–71.
149. M. E. Whitmore and T. A. Mansfield, *Environ. Pollut.*, 1983, **31**, 217.
150. A. W. Davison and I. F. Bailey, *Nature*, 1982, **297**, 400.
151. T. Keller, *Environ. Pollut.*, 1978, **16**, 243.
152. J. D. Barnes, M. R. Hull and A. W. Davison, in *Plant Response to Air Pollution*, ed. M. Yunus and M. Iqbal, John Wiley and Sons, London, 1996, pp. 135–166.
153. L. J. Sheppard, I. D. Leith, A. Crossley, N. van Dijk, D. Fowler, M. A. Sutton and C. Woods, *Environ. Pollut.*, 2008, **154**, 404–413.
154. M. Fagnano, A. Maggio and I. Fumagalli, *Environ. Pollut.*, 2009, **157**, 1438.
155. P. Bungener, G. R. Balls, S. Nussbaum, M. Geissman, A. Grub and J. Fuhrer, *New Phytol.*, 1998, **142**, 271.
156. E. Paoletta and N. E. Grulke, *Environ. Pollut.*, 2010, **158**, 2664.
157. G. Mills, F. Hayes, S. Wilkinson and W. J. Davies, *Global Change Biol.*, 2009, **15**, 1522.
158. S. B. Mclaughlin, S. D. Wullschleger, G. Sun and M. Nosal, *New Phytol.*, 2007, **174**, 125.
159. S. Wilkinson, G. Mills, R. Illidge and W. J. Davies, *J. Exp. Bot.*, 2011, **63**, 527.
160. S. Wagg, G. Mills, F. Hayes, S. Wilkinson, D. Cooper and W. J. Davies, *Environ. Pollut.*, 2012, **165**, 91.
161. W. Fluckiger, S. Braun and E. Hiltbrunner, in *Air Pollution and Plant Life*, ed. J. N. B. Bell and M. Treshow, John Wiley and Sons, Chichester, 2nd edn., 2002, pp. 379–406.
162. J. N. B. Bell, S. McNeill, G. Houlden, V. S. Brown and P. J. Mansfield, *Parasitology*, 1993, **106**, S811.
163. I. M. Gate, S. McNeill and M. R. Ashmore, *Water Air Soil Pollut.*, 1995, **85**, 1425.
164. S. Warrington, *Environ. Pollut.*, 1987, **43**, 155.
165. G. Houlden, S. McNeill, M Aminu-Kano and J. N. B. Bell, *Environ. Pollut.*, 1990, **67**, 305.
166. J. N. B. Bell, S. A. Power, N. Jarraud, M. Agrawal and C. Davies, *Int. J. Sustain. Dev. World Ecol.*, 2011, **18**, 226.
167. G. Houlden, S. Mcneill, S. Craske and J. N. B. Bell, *Environ. Pollut.*, 1991, **72**, 45.
168. S. A. Power, M. R. Ashmore, D. A. Cousins and L. J. Sheppard, *New Phytol.*, 1998, **138**, 663.
169. V. C. Brown, in *Insects in a Changing Environment*, ed. R. Harrington and N. E. Stork, Academic Press, London, 1995, p. 219.
170. Q. S. McFrederick, J. D. Fuentes, T. Roulston, J. C. Kathilankal and M. Lerdau, *Oecologia*, 2009, **160**, 411.
171. H. Sandermann, *Environ. Pollut.*, 2000, **108**, 327.
172. D. F. Karnosky, K. E. Percy, B.X. Xiang, B. Callan, A. Noormets, B. Manovska, A. Hopkin, J. Sober, W. Jones, R. E. Dickson and J. G. Isebrands, *Global Change Biol.*, 2002, **8**, 329.
173. P. J. W. Saunders, *Ann. App. Biol.*, 1966, **58**, 103.
174. R. J. Bevan and G. N. Greenhalgh, *Environ. Pollut.*, 1976, **10**, 271.
175. I. D. Leith and D. Fowler, *New Phytol.*, 1988, **108**, 175.
176. J. G. Roelofs, A. J. Kempers, A. L. Houdijk and J. Jansen, *Plant Soil*, 1985, **84**, 45.
177. W. Fluckiger and S. Braun, *Water Air Soil Pollut.*, 1999, **116**, 99.
178. J. Strengbom, A. Nordin, T. Nasholm and L. Ericson, *J. Ecol.*, 2002, **90**, 61.
179. T. Wallenda and I. Kottke, *New Phytol.*, 1998, **139**, 187.
180. I. P. Edwards and D. R. Zak, *Global Change Biol.*, 2011, **17**, 2184.
181. M. R. Ashmore, in *Air Pollution and Plant Life*, ed. J. N. B. Bell and M. Treshow, John Wiley and Sons, Chichester, 2nd edn., 2002 pp. 417–430.

182. UBA, *Manual on Methodologies and Criteria for Mapping Critical Levels/Loads and the Geographical Areas where they are Exceeded*, Umweltbundesamt, Berlin, 2004.
183. J.-P. Hettelingh, M. Posch and P. A. M. de Smet, *Water Air Soil Pollut.*, 2001, **130**, 1133.
184. M. Posch, J.-P. Hettelingh and P. A. M. de Smet, *Water Air Soil Pollut.*, 2001, **130**, 1139.
185. M. Posch, J. Aherne and J.-P. Hettelingh, *Environ. Pollut.*, 2011, **159**, 2223.
186. M. Posch and G. J. Reinds, *Environ. Model. Software*, 2009, **24**, 329.
187. M. S. Cresser, *Sci. Total Environ.*, 2000, **249**, 51.
188. R. Bobbink, D. Boxman, E. Fremsted, G. Heil, A. Houdijk and J. Roelofs, in *Critical Loads for Nitrogen*, ed. P. Grennfelt and E. Thornelof, Nordic Council of Ministers, Copenhagen, 1992, p. 111.
189. R. Bobbink and J.-P. Hettelingh, *Review and Revision of Empirical Critical Loads and Dose–Response Relationships*, Coordination Centre for Effects, National Institute for Public Health and the Environment, Bilthoven, 2011.
190. G. Mills, H. Pleijel, P. Buker, L. Emberson, H. Harmens, D. Simpson, L. Grunhage, P.-E. Karlsson, H. Danielsson, V. Bermejo and I. G. Fernandez, *Chapter 3, Mapping Critical Levels for Vegetation, 2010, Revision of UBA, Manual on Methodologies and Criteria for Mapping Critical Levels/Loads and the Geographical Areas where they are Exceeded*, Umweltbundesamt, Berlin, 2004.
191. WHO, *Air Quality Guidelines for Europe*, World Health Organisation, Copenhagen, 2000.
192. P. E. Karlsson, J. Uddling, S. Braun, M. Broadmeadow, S. Elvira, B. S. Gimeno, D. Le Thiec, E. Oksanen, K. Vandermeiren, M. Wilkinson and L. Emberson, *Atmos. Environ.*, 2004, **38**, 2283.
193. J. N. Cape, L. J. van der Eerden, L. J. Sheppard, I. D. Leith and M. A. Sutton, *Environ. Pollut.*, 2009, **157**, 1033.
194. G. Mills, A. Buse, B. Gimeno, V. Bernejo, M. Holland, L. Emberson and H. Pleijel, *Atmos. Environ.*, 2007, **41**, 2630.
195. P.-E. Karlsson, S. Braun, M. Broadmeadow, S. Elvira, L. Emberson, B. S. Gimeno, D. le Thiec, K. Novak, E. Oksanen, M. Schaub, J. Uddling and M. Wilkinson, *Environ. Pollut.*, 2007, **146**, 608.
196. H. Pleijel, K. Ojanpera and L. Mortensen, *Acta Agri. Scand., B Soil Plant Sci.*, 1997, **47**, 20.
197. G. Mills, H. Pleijel, S. Braun, P. Buker, V. Bermejo, E. Calvo, H. Danielsson, L. Emberson, I. G. Fernandez, L. Grunehage, H. Harmens, F. Hayes, P.-E. Karlsson and D. Simpson, *Atmos. Environ.*, 2011, **45**, 5064.
198. Millenium Ecosystem Assessment. *Ecosystems and Human Welfare: Synthesis*. Island Press, Washington, DC, 2005.
199. UK National Ecosystem Assessment, *The UK National Ecosystem Assessment Technical Report*. UNEP-WCMC, Cambridge, 2011.
200. D. Fowler, K. Pilegaard, M. A. Sutton, P. Ambus, M. Raivonen, J. Duyzer, D. Simpson, H. Fagerli, S. Fuzzi, J. K. Schjoerring, *et al.*, *Atmos. Environ.*, 2009, **43**, 5193.
201. P. Smith, M. R. Ashmore, H. I .J. Black, P. J. Burgess, C. D. Evans, T. A. Quine, A. M. Thomson, K. Hicks and H. G. Orr, *J. App. Ecol.*, 2013, **50**, 812.
202. A. Tiwary, D. Sinnett, C. Peachey, Z. Chalabi, S. Vardoulakis, T. Fletcher, G. Leonaride, C. Grundy, A. Azapagic and T. R. Hutchings, *Environ. Pollut.*, 2009, **157**, 2645.
203. L. D. Emberson, N. Kitwiroon, S. Beevers, P. Buker, H. Cambridge and S. Cinderby, *Atmos. Chem. Phys.*, 2013, **13**, 6741.
204. W. de Vries and M. Posch, *Environ. Pollut.*, 2011, **159**, 2289.
205. H. Harmens and G. Mills, *Ozone Pollution: Impacts on Carbon Sequestration in Europe*, ICP Vegetation, Centre for Ecology and Hydrology, Bangor, 2012.
206. P. Olsson, S. Linder, R. Giesler and P. Hogberg, *Global Change Biol.*, 2005, **11**, 1745.

207. K. S. Hofmockel, D. R. Zak, K. K. Moran and J. D. Jastrow, *Soil Biol. Biochem.*, 2011, **43**, 1518.
208. K. Butterbach-Bahl, E. Nemitz and S. Zaehle, in *The European Nitrogen Assessment*, ed. M. A. Sutton, C. M. Howard, J. W. Erisman, G. Billen, A. Bleeker, P. Grennfelt, H. van Grinsven and B .Grizzetti, Cambridge University Press, Cambridge, 2011, pp. 434–462.
209. K. Pilegaard, U. Skiba, P. Ambus, C. Beier, N. Breuggemann, K. Butterbach-Bahl, J. Dick, J. Dorsey, J. Duyzer, M. Gallagher, *et al.*, *Biogeosciences*, 2006, **3**, 615.
210. V. Gauci, E. Matthews, E, N. Dise, B. Walter, D. Koch and G. Granberg, *Proc. Natl. Acad. Sci. U. S. A.*, 2004, **101**, 12583.
211. Z. P. Wang and P. Ineson, *Soil Biol. Biochem.*, 2003, **35**, 427.
212. S. Toet, P. Ineson, S. Peacock and M. R. Ashmore, *Global Change Biol.*, 2011, **17**, 288.
213. G. Sun, S. B. Mclaughlin, J. H. Porter, J. Uddling, P. J. Mulholland, M. B. Adams and N. Pederson, *Global Change Biol.*, 2012, **18**, 3395.
214. A. J. Britton and J. M. Fisher, *Environ. Pollut.*, 2008, **153**, 564.
215. V. J. Black, C. R. Black, J. A. Roberts and C. A. Stewart, *New Phytol.*, 2000, **147**, 421.
216. C. D. Evans, P. J. Chapman, J. M. Clark, D. T. Monteith and M. S. Cresser, *Global Change Biol.*, 2006, **12**, 2044.
217. C. D. Evans, C. L. Goodale, S. J. M. Caporn, N. B. Dise, B. A. Emmett, I .J. Hernandez, C. D. Field, S. E. G. Findlay, G. M. Lovett, H. Meesenburg, F. Moldan and L. J. Sheppard, *Biogeochemistry*, 2008, **91**, 13.
218. J. R. Smart, K. Hicks, T. Morrissey, A. Heinemeyer, M. Sutton and M. Ashmore, *Environmetrics*, 2011, **22**, 649.
219. R. S. Sokhi, *World Atlas of Atmospheric Pollution*, Anthem Press, London, 2008.

CHAPTER 13

Control of Pollutant Emissions from Road Transport

CLAIRE HOLMAN

Brook Cottage Consultants, Elberton, South Gloucestershire, BS35 4AQ, UK
Email: claireholman@brookcottage.info

13.1 INTRODUCTION

Road traffic is a major source of air pollution in urban areas around the world, despite efforts to control emissions since the 1960s when the first legislation to control vehicle emissions came into force in California. High concentrations of nitrogen dioxide (NO_2) and particulate matter (PM) occur not only in the major cities of the world, but also in a large number of smaller cities and towns, particularly those with narrow streets that restrict the dispersion of emissions from vehicles.

There are many ways of reducing emissions from traffic such as improving standards of vehicle maintenance, reducing the volume of traffic and smoothing traffic flow. These are all important elements in an emissions control strategy. However, this chapter focuses on the role of technology to improve air quality. Specifically, it discusses cleaner engines and fuels and after-treatment systems such as catalysts and diesel particle filters.

Over the last decade or so, with the growing emphasis on reducing carbon dioxide (CO_2) emissions in many countries around the world there have been a number of initiatives to increase the use of renewable energy and electricity for road transport. For example, the United Kingdom (UK) government requires petrol and diesel to contain 5% biofuel under its Renewable Transport Fuels Obligation, typically achieved using ethanol and biodiesel and offers fiscal incentives for the purchase of electric vehicles (EVs). However, petrol and diesel are forecast to remain the major transport fuels for the foreseeable future,[1,2] and much of this chapter discusses technology to improve emissions from vehicles using these fuels.

Emissions from motor vehicles have reduced significantly since 1972 when the first exhaust emissions standards for passenger cars were introduced into Europe. Emissions of the regulated

Pollution: Causes, Effects and Control, 5th Edition
Edited by R M Harrison
© The Royal Society of Chemistry 2014
Published by the Royal Society of Chemistry, www.rsc.org

Table 13.1 Summary of EU car emission limits.

Standard	Date of introduction		Emission Limit (mg km^{-1})						Number of particles (PN) km^{-1}
	New models	All models	CO	NOx + HC	NOx	THC	NMHC	PM	
Petrol									
Euro 1	01.07.92	31.12.92	2720	970					
Euro 2	01.01.96	01.01.97	2200	500					
Euro 3	01.01.00	01.01.01	2300		150	200			
Euro 4	01.01.05	01.01.06	1000		80	100			
Euro 5	01.09.09	01.01.11	1000		60	100	68	5.0^a	
Euro 6	01.09.14	01.09.15	1000		60	100	68	4.5^a	6.0×10^{11}
Diesel									
Euro 1	01.07.92	31.12.92	2720	970				140	
Euro 2 IDI	01.01.96	01.01.97	1000	700				80	
Euro 2 DI			1000	900				100	
Euro 3	01.01.00	01.01.01	640	560	500			50	
Euro 4	01.01.05	01.01.06	500	300	250			25	
Euro 5a	01.09.09	01.01.11	500	230	180			5	
Euro 6	01.09.14	01.09.15	500	170	80			4.5^b	6.0×10^{11}

^{a}Applies to GDI vehicles only, 4.5 mg km^{-1} using the PMP measurement procedure.
^{b}Using PMP measurement procedure (equivalent to 5 mg km^{-1} under old test protocol).
Euro 1 to Euro 4 Cars >2.5t GVW were type approved as small vans.
EU emission legislation is very complex and this table only provides a brief summary of the car emission limits; there are many additional requirements. For full details see the relevant EU Directives/Regulations on the European Commission web site (www.ec.europea.eu/environemnt/air/transpirt/road.htm).

gaseous pollutants – nitrogen oxides (NO_x), carbon monoxide (CO) and hydrocarbons (HC)[†] – per vehicle kilometre are less than one per cent of their unregulated levels when measured over the legislative test cycles (which have changed over time to better reflect real world driving conditions). These improvements have been in response to progressively more stringent emission limits for new vehicles in the European Union (EU) and United States (US) which have been widely adopted in other jurisdictions. There have also been significant reductions in the permitted emissions of PM per vehicle kilometre.

Table 13.1 provides a summary of the exhaust emission limits for passenger cars since Euro 1[‡] standards were introduction in the EU in the early 1990s. Although four emission limits had already been adopted by the United Nations Economic Commission for Europe (UNECE) prior to Euro 1 they were not mandatory for all new cars, and not all EU Member States adopted them.

The Euro 1 standard introduced the first mandatory emission limit values for new cars, which applied to virtually all new cars first registered in EU Member States from 1993. The most recently adopted standard, Euro 6, will come into effect from 2014. Over time the legislation has become increasingly complex. Emissions are tested on a chassis dynamometer using the New European Driving Cycle (NEDC). This test consists of an urban driving cycle repeated four times followed by a higher speed extra-urban driving cycle. There is concern that this test is not representative of real world driving conditions where fast acceleration occurs and that manufacturers produce vehicles that meet the emissions only under the legislative test conditions but not in real world driving conditions. As a result a series of new test procedures have been and continue to be developed

[†]The road vehicle emissions legislation regulates hydrocarbons (HC) emissions. In this context hydrocarbons are assumed to be equivalent to volatile organic compounds (VOCs). The most recent legislation includes emission limits for total HC (THC) and non-methane HC (NMHC).
[‡]The convention is to use Arabic numbers (Euro 1, Euro 2, Euro 3, *etc.*) for the EU light duty and Roman numbers (Euro I, Euro II, Euro III, *etc.*) for the heavy duty emission standards.

under the auspices of the UNECE, in co-operation with the European Commission, for both light duty vehicles (LDVs) and heavy duty vehicles (HDV). These include cold start and evaporative emissions, off-cycle, and in-service tests.

For the most recent standards an additional test procedure has been adopted, known at the Particle Measurement Programme (PMP) procedure, to measure the number of particles (PN). This new test was introduced to ensure that manufacturers fit diesel particle filters, particularly to heavy duty engines.

Vehicles are used for many years before they are scrapped (typically around 10 years for a passenger car, longer for other vehicles in the UK, although this varies with the economy). It therefore takes some years for the benefits of cleaner and more fuel-efficient new vehicles to have maximum impact on the environment. To encourage a more rapid turnover of the fleet, and the purchase of cleaner vehicles ahead of their mandatory introduction, fiscal incentives have been used in many EU countries.

Controls have also been introduced on fuel quality in the EU to complement engine and exhaust after-treatment technologies, but in other jurisdictions, where fuel does not meet the same exacting standards, certain after-treatment technologies cannot be used effectively.

Whilst emissions per vehicle kilometre have declined in the four decades since emission controls were first introduced in Europe, the rapid increase in vehicle mileage offset the progress made in introducing cleaner technology, and total emissions increased in the 1980s. However, in the early 1990s road transport emissions peaked and have since declined. This is shown for the EU 27 countries in Figures 13.1, 13.2 and 13.3 for NO_x, PM_{10} (PM with an aerodynamic diameter less than 10 μm) and CO_2, respectively. These figures show the total estimated EU emissions as well as that for road transport.[3,4]

Between 1990 and 2010 road transport NO_x emissions fell by 46% and PM_{10} by 33%. The NO_x reduction was primarily due to the fitting of three-way catalysts to petrol cars, whilst the reduction in PM_{10} was primarily due to improvements to heavy duty vehicle engines.

However, ambient urban concentrations of NO_2 have not fallen by as much as the reported emissions. According to the European Environment Agency from 2001 to 2010, annual mean NO_2

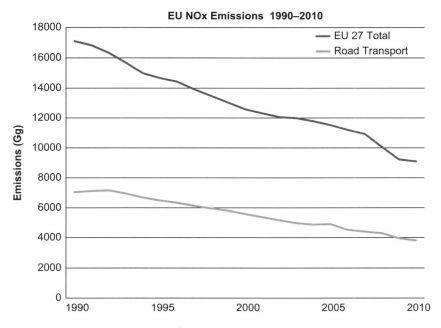

Figure 13.1 EU NO_x emissions 1990–2010.[3]

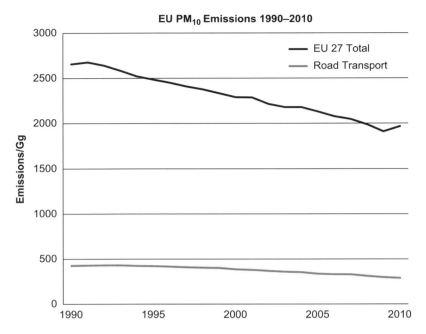

Figure 13.2   EU PM$_{10}$ emissions 1990–2010.[3]

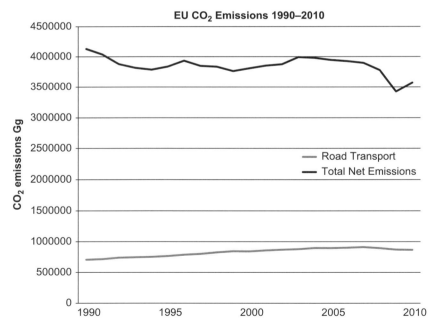

Figure 13.3   EU CO$_2$ emissions 1990–2010.[4]

concentrations at urban background sites fell by just 10.6% on average during which time the reported NO$_x$ emissions for the EU-27 decreased by 24.9%.[5] The disparity between these trends is thought, at least in part, to be due to optimistic emissions factors used in emission inventories as

Table 13.2 Contribution of road transport to total EU-27 emissions, 2010.

Pollutant	Road Transport Contribution (%)
Nitrogen oxides (NO_x)	42
Carbon monoxide (CO)	29
Non methane volatile organic compounds (NMVOC)	16
Particulate Matter less than 10 μm (PM_{10})	15
Particulate Matter less than 2.5 μm ($PM_{2.5}$)	16
Carbon dioxide (CO_2)	20

(Source: EU submission to UNECE).[3]

'real-world' emission performance of modern diesel vehicles has not shown the improvements that were anticipated (see section 13.8). As a result a number of Member States' NO_x emissions could be higher than the estimates shown in the figures. Another reason is that the proportion of NO_x emitted from road transport as NO_2 has increased.[6]

For CO_2 the picture is quite different. Over the period 1990 to 2010 EU road transport emissions increased by 22% while the total net emissions (*i.e.* taking account of land use, land-use change and forestry) decreased by 13%. The EU has a policy target of reducing its 1990 greenhouse gas emissions by 20% by 2020. Reducing emissions from road transport is more difficult than for some other sectors of the economy, due to the increasing demand for mobility and the difficulty, as discussed later, in replacing fossil fuels with low carbon alternatives.[4]

Table 13.2 shows the contribution of road transport to EU emissions in 2010, which depending on the pollutant, ranges from 15% (PM_{10}) to 42% (NO_x). At the urban scale, away from major industrial areas, road transport makes a much greater contribution to total emissions than the national/EU figures suggest, particularly for NO_x and PM. For example in 2008 road transport contributed 39% to the UK national NO_x emissions but 49% in inner and central London.[7,8] In the centre of large conurbations there are likely to be pollution 'hot-spots' where the road traffic contribution is even higher.

At the end of 2011 there were approximately 34 million vehicles licensed for use on the roads in Great Britain, of which 29 million were cars. Although HDVs (*i.e.* vehicles greater than 3.5 tonne gross vehicle weight) were only 2% of the vehicles on the roads they have significantly higher emissions per vehicle kilometre than LDVs and therefore can be a major source of local emissions, for example in urban centres close to roads used by a large number of buses.[9]

It should be noted that all emission estimates are uncertain. For the UK National Atmospheric Emissions Inventory it has been estimated that the uncertainties are within ±10 per cent for NO_x and −20 to +30 per cent for PM_{10}.[10] Future forecasts are even more uncertain as they have to rely on forecasts of activity in different economic sectors, which are also subject to much uncertainty. In estimating how much further emissions need to be reduced to meet air quality objectives these uncertainties need to be taken into account.

Emission inventories only provide information on the source contribution within its boundaries. In reality sources outside an urban area are often also very important. In London it is thought that emissions from outside the capital may account for around 40 per cent of nitrogen dioxide (NO_2) concentrations.[11]

13.2 ENGINES

13.2.1 Introduction

Petrol and diesel are the two main automotive fuels currently used worldwide. In 2011, 70% of cars in Great Britain were fuelled using petrol and 30% diesel, and these fuels are likely to continue to dominate in the foreseeable future.[9] The vast majority of light duty commercial vehicles and virtually all heavy duty vehicles (HDVs) use diesel engines because of their greater fuel efficiency, durability and torque compared with petrol engines.

Due to the different physical characteristics of petrol and diesel these fuels are burnt in different types of internal combustion engines (ICE). Petrol requires a spark to ignite the fuel, and thus is used in spark ignition (SI) engines, also known as positive ignition (PI) engines. Compressing the fuel ignites diesel, and therefore the engines using this fuel are known as compression ignition (CI) engines.

In ICEs the fuel is either mixed with air in the combustion chamber or before the combustion chamber, known as direct injection (DI) or indirect injection (IDI) respectively. In an IDI engine the fuel is mixed with air in a pre-chamber (diesel) or in the intake ports (petrol).

The sophisticated electronic control of direct fuel injection, at high pressures, facilitates precise timing of when the fuel enters the combustion chamber and has resulted in the manufacture of engines with increased power and lower emissions. Formerly only diesel used DI, but in the past 15 years petrol DI engines, known as GDI (gasoline direct injection) engines, have come onto the market.

A range of other fuels such as natural gas (compressed or liquefied), ethanol and biodiesel can also be used in ICEs and they are discussed in the section on alternative fuels.

13.2.2   Spark Ignition/Petrol Engines

Until relatively recently four stroke SI engines were used in most cars and small vans, because of their superior power to weight ratio and their wider operational range compared with CI engines, but also for reasons such as lower cost, less noise and more refined operation.

Most SI engines run under stoichiometric conditions, that is where the amount of oxygen available is that required to just fully combust the fuel. When more oxygen is available the combustion becomes lean. SI engines typically use a homogenous air–fuel mixture with early fuel introduction for good fuel vaporization. Fuel delivery systems have evolved from carburettors to multipoint fuel injection (MPFI). The latest evolutionary step, stoichiometric direct injection, represents a significant improvement for SI engines and when combined with turbocharging and engine downsizing makes them more competitive with CI engines in terms of fuel economy and performance.

Control of the air to fuel ratio has a major impact on the formation of hydrocarbons (HC), or unburned fuel, and carbon monoxide (CO), which is partially oxidized fuel. In contrast, NO_x is a by-product of combustion, created when nitrogen and oxygen in the air, and to a lesser extent, the fuel, combine during the combustion process. The higher the cylinder temperature, the more NO_x is formed. Thus, the primary strategy to reduce the formation of NO_x in the engine is to reduce combustion temperature.

With the growing pressure to reduce passenger car CO_2 emissions GDI engines were introduced into the European market from the late 1990s and by 2010 were responsible for about 20% of the new petrol cars sold in the EU. For some manufacturers, such as BMW and Audi, about 80% of their petrol cars sold in 2010 had GDI engines.[12] These tend to be fitted to medium or large cars.

The advantages of a GDI engine are increased fuel efficiency and high power output, achieved through the precise control over the amount of fuel, and injection timing that is varied according to engine load, and no throttling losses. Engine speed is controlled by the engine control unit/management system which regulates fuel injection and ignition timing, instead of having a throttle plate that restricts the incoming air supply. But the major fuel efficiency improvements with GDI engines occur when they are running lean. The stoichiometric air to fuel ratio for petrol is 14.7 : 1 by weight, but in ultra-lean mode the air to fuel ratio can be as high as 65 : 1 or for very limited periods, even higher. The engine management system continually chooses between ultra-lean burn, stoichiometric, and full power output, depending on the driving conditions and engine load. This is discussed later in the section on fuel injection. The real world emissions benefits depend on the driving conditions and driver behaviour.

13.2.3 Compression Ignition/Diesel Engines

Modern diesel engines use turbocharging with inter-cooling to increase power. This has effectively closed the gap in performance between petrol and diesel passenger cars. Turbochargers use the exhaust gas to turn a turbine to compress the engine's intake air. This effectively increases the amount of oxygen available in the combustion chamber increasing volumetric efficiency. It is possible to reduce emissions by matching the level of turbocharging to the vehicle operating conditions and by cooling the air input into the engine using electronic control of the intercooler.

The rated maximum power output of diesel cars has increased from about 70 kW for Pre-Euro to Euro 2 cars, to about 113 kW for Euro 5 cars. On the other hand it has remained about 80 kW for petrol vehicles, irrespective of the Euro class.[13]

Diesel engines are heavier than the equivalent petrol engine due to the need to withstand much higher combustion pressures and therefore need more power for similar performance.

This has resulted in the popularity of diesel cars and vans increasing in recent years in many European countries. Diesel cars accounted for over 50% of new car sells in the EU in 2010.[12] The use of small diesel vehicles has largely been a European and Asian phenomenon, with few models on the North American market. However, in 2012 there were signs of that changing as more manufacturers launched diesel cars onto the US market.

Light duty diesel vehicles used to have either DI or IDI engines, whereas only DI engines were used in heavy duty vehicles due to their superior fuel efficiency. However, most diesel LDVs, until the late 1990s, used IDI engines because DI engines could not meet the emission limits for NO_x, PM and noise (EU emission legislation has separate requirements for vehicles below and above 3.5 tonnes gross vehicle weight). However, improvements to the technology resulted in the range of DI cars on the market widening and DI has replaced IDI in the light duty diesel market.

13.3 CONTROLLING REGULATED EMISSIONS

13.3.1 Introduction

If complete combustion of the fuel in an internal combustion engine (ICE) were possible vehicle exhaust would contain only CO_2 and water vapour. However, as a result of a number of factors including the short time available for combustion in the engine, the incomplete mixing of the fuel and air, and the high temperature and pressure of combustion, vehicle exhaust also contains CO, HC, PM, NO_x and a range of other compounds depending on the fuel.

Emissions from road transport depend on a large number of factors. The main ones are the vehicle technology, including the fuel used, and the mileage driven. Other factors such as fuel quality, vehicle maintenance, driver behaviour, congestion, ambient temperature and the scrappage rate of older vehicles can also be important.

For each type of vehicle there is a compromise during engine design between optimising for fuel consumption, emissions or performance. For luxury vehicles, fuel consumption has tended to be a lower priority than performance, whilst for small compact cars the emphasis has been more on fuel consumption. However, during the 2000s, but particularly since 2007 when the first mandatory targets for new car CO_2 emissions were agreed, there has been renewed interest in fuel consumption amongst car manufacturers selling products in Europe.

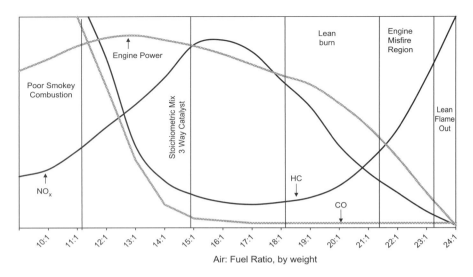

Figure 13.4 The effect of air : fuel ratio on emissions and engine power.

In the early phases of EU emission control, the emphasis was on improving the precision and timing of the injection of fuel into the combustion chamber and the design of the combustion chamber itself to improve combustion and minimise the deposit of fuel on the walls of the chamber, which leads to exhaust HC emissions.

Increasingly, as emission limits became more stringent, new techniques for reducing emissions have emerged. Both petrol and diesel vehicles now depend on the use of 'after-treatment', that is the removal of the pollutants in the exhaust system as well as controlling its formation in the engine. This has been an evolutionary process of gradual improvement to the combustion process, engine management system and catalytic after-treatment.

For petrol engines one of the most important factors influencing emissions is the ratio of air to fuel in the engine. Figure 13.4 shows that there is no ideal ratio at which all the main emissions are low and the engine power is at an acceptable level. Indeed where CO and HC emissions are at their lowest, the NO_x emission is at its maximum. The best compromise is found in the lean burn region. Lower emissions, however, can be achieved using a three-way catalyst, which as explained below, needs a stoichiometric air to fuel ratio. The main advantage of lean burn engines is that they are more fuel-efficient than those running richer.

The greatest advance in controlling emissions from road vehicles was the introduction of the closed loop (or controlled) three-way catalyst for petrol vehicles, used in essentially all new petrol vehicles sold in the world's major markets. This technology can remove more than 99% of emissions of CO, HCs and NO_x and is described below in the section on after-treatment devices.

From Euro 3 (2000) new petrol cars have been fitted with on-board diagnostic systems (OBD), which monitor the performance of a number of components of the exhaust, and evaporative emission control systems. When the system detects a fault a warning comes on and the engine performance may be reduced. As engine management systems have developed, an increasing number of components can be monitored, and on-board systems directly measuring exhaust emissions are used on some heavy duty vehicles (HDVs).

Engine-out CO and HC emissions are lower from diesel cars compared to similar modern stoichiometric petrol cars. On the other hand emissions of NO_x and PM are greater. Typical NO_x type approval emissions from new EU petrol cars meeting the Euro 3 emission limits were an order of magnitude lower than those from diesel cars. This gap is also closing; for Euro 5 cars the difference is about a factor of five, and for early Euro 6 cars about a factor of two.[12]

Petrol direct injection (GDI) cars tend to have emission characteristics more similar to diesel cars, with higher NO_x and PM emissions, but lower CO_2 emissions than a stoichiometric petrol car.

The types of HC emitted from petrol and diesel engines are very different, reflecting the differences in the chemical composition of the fuels. Typically, petrol cars have greater emissions of low molecular weight HC (*e.g.* benzene) whilst diesel have greater emissions of high molecular weight compounds (*e.g.* polycyclic aromatic compounds).

European emissions legislation has essentially treated light duty vehicles (LDVs) differently from heavy duty vehicles (such as buses and lorries), with LDVs being tested on a chassis dynamometer while just the engines used in HDVs being tested on an engine dynamometer. As a result the test cycles are very different. For LDVs different emission limits have applied to diesel and petrol vehicles reflecting their emission characteristics.

13.3.1.1   *Fuel Injection*

One of the major changes in engine design, driven by increasingly stringent exhaust emission requirements, has been the way fuel enters the combustion chamber. Mechanical devices, that is carburettors for petrol and pumps for diesel engines were widely used in Europe until the early 1990s. However the need to reduce engine-out emissions and to control the air to fuel ratio in petrol cars fitted with three way catalysts, resulted in the widespread deployment of fuel injection linked to electronic engine management systems.

For IDI petrol engines the main difference between fuel injection and a carburettor is that fuel injection atomizes the fuel by pumping it through a small nozzle under high pressure, while a carburettor relies on suction to draw the fuel into the airstream. Today multi point fuel injection (MPFI) systems inject fuel into intake ports just upstream of the intake values, rather than at a central point in an intake manifold. Typically sequential systems are used which inject fuel during the engine cylinder intake stroke.

In direct injection (DI) engines common rail fuel injection is used. On earlier diesel engines a fuel rail fed individual solenoid values at pressures greater than 10 kPa, but now piezoelectric injectors are used at pressures double that and with greater precision.

These sophisticated electronically controlled fuel injection systems allow much more precise timing of when the fuel enters the combustion chamber and have resulted in the manufacture of engines with increased power, lower fuel consumption and lower emissions. They have also facilitated the use of GDI engines which require particularly complex engine management systems to manage the fuel injection under all operating modes. Under light load conditions such as when driving at constant or reduced road speeds, the fuel is injected later than during stoichiometric conditions – during the compression stroke. This allows a small amount of air–fuel mixture to be injected and placed near the spark plug. This is surrounded mostly by air and residual gases, which keeps the fuel and the flame away from the cylinder walls, and is known as 'stratified charge'. The lower combustion temperature produces lower NO_x emissions and heat losses. When in the stoichiometric mode, during moderate load conditions, the fuel is injected during the intake stroke, creating a homogeneous fuel–air mixture in the cylinder. The full power mode is used for rapid acceleration and heavy loads, and again the air–fuel mixture is homogeneous but the air and fuel mixture is slightly richer than stoichiometric.

It is also possible to inject more than once during a single combustion cycle. After the first fuel charge has been ignited, it is possible to add fuel as the piston descends. The benefits are more power and economy, but there may be adverse effects on the durability of exhaust valves.

13.3.1.2   *Exhaust Gas Recirculation*

Exhaust gas recalculation (EGR) is used to control NO_x emissions. It returns a proportion of the exhaust gas to the combustion chamber. In diesel engines this reduces the amount of oxygen in the combustion chamber, while in a petrol engine it displaces combustible matter resulting in lower combustion chamber temperatures and reduced NO_x emissions. In some EGR systems the exhaust gas is cooled prior to further reduce NO_x emissions. This technology was commonly used to reduce NO_x emissions from diesel engines, but is no longer sufficient to meet the current emission limits on its own.

13.3.2 Exhaust After-treatment

13.3.2.1 Introduction

Modern spark ignition (SI) and compression ignition (CI) engines require the use of exhaust after-treatment devices to meet emission limits. These include auto catalysts, absorbers and diesel particle filters (DPF). These devices when used with good quality fuels and lubricants, and sophisticated engine management systems can reduce new vehicle emissions to low levels under the legislative test cycle. They can also be retrofitted to existing vehicles, although their pollutant removal efficiency may be lowered.

The first after-treatment devices to be fitted to cars were simple oxidation catalysts that reduced CO and HC emissions and were introduced into the US in 1974. These were first used in Europe, about a decade later but are no longer used with petrol engines, having been replaced by three-way catalysts.

13.3.2.2 Three-way Catalysts

Virtually all petrol cars in the EU have been fitted with three way catalyst (TWC) since the end of 1992. In other parts of the world this technology has a much longer history. TWCs were widely used from 1981 and 1978 in the United States and Japan, respectively, when new emission legislation for cars was introduced. Today the EU has caught-up with these other countries and is now among the world leaders in emission control.

A TWC uses a ceramic or metal substrate with an active coating incorporating aluminium, cerium or other metal oxides, and combinations of the precious metals platinum, palladium and rhodium. The metal oxides effectively store oxygen for the oxidation reactions and the precious metals facilitate the reduction of the NO_x.

A TWC simultaneously removes CO, HC and NO_x from a vehicle's exhaust through chemical processes on the catalyst surface. The reactions are:

Oxidation reactions

$$2CO + O_2 \rightarrow 2CO_2 \tag{1}$$

$$HC + O_2 \rightarrow CO_2 + H_2O \tag{2}$$

Reduction reactions

$$2CO + 2NO \rightarrow 2CO_2 + N_2 \tag{3}$$

$$HC + NO \rightarrow CO_2 + H_2O + N_2 \tag{4}$$

For the effective removal of all three pollutants the air to fuel ratio needs to be close to the stoichiometric ratio (*i.e.* 14.7 : 1). Cars fitted with these catalysts require one or more oxygen sensors to monitor the exhaust gas composition and electronically manage the fuel system to control the air to fuel ratio. The effect on catalyst efficiency of moving away from the stoichiometric

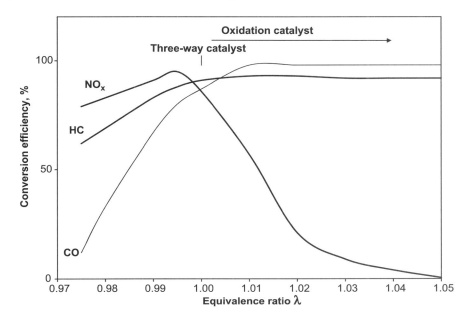

Figure 13.5 The effect on catalyst efficiency of moving away from the stoichiometric air : fuel ratio.

air to fuel ratio (*i.e.* the equivalence ratio $\lambda = 1$) is shown in Figure 13.5. Conversion efficiency falls very rapidly when the engine is operated outside of the optimum air fuel ratio.

For other fuels, such as liquefied petroleum gas (LPG), natural gas and ethanol, the optimum air fuel ratio is slightly different, requiring modification to the engine settings when using those fuels.

Under lean engine operation there is excess oxygen making the reduction of NO_x difficult. Under rich conditions, the excess fuel consumes the available oxygen prior to the catalyst, reducing the amount of oxygen available for oxidation. Closed loop control systems are necessary because of the conflicting requirements for effective NO_x reduction and HC oxidation. The control system must prevent the NO_x reduction catalyst from becoming fully oxidized, yet replenish the oxygen storage material to maintain its function as an oxidation catalyst.

In modern vehicles two oxygen sensors are fitted, upstream and downstream of the three way catalyst to monitor the O_2 levels. The on-board computer compares the readings to identify whether it is malfunctioning or has been removed, and if so will illuminate an indicator light and reduce engine performance ('limp home mode'). However, there are after-market devices available to circumvent this on-board diagnostic system that may allow the engine to run in a more fuel-economical lean burn mode.

Most of the emissions from petrol cars fitted with three way catalysts occur while the catalyst is warming up. The development of fast light-off catalysts has reduced the exhaust temperature required for the catalyst to operate, reducing emissions over the legislative test cycle. Moving the TWC closer to the engine, made possible by the development of more thermally durable catalyst substrates and changes to the type and composition of the precious metals used, has brought light-off times down from a few minutes to a few seconds. These catalysts also have a longer life particularly during demanding driving when there are high exhaust temperatures.

The catalytic material is coated onto a honeycomb structure, to minimise the weight of the substrate. Early catalysts used ceramic substrates with a density of about 30 cells cm^{-2} and a wall thickness of 0.3 mm. Today's catalysts can have a wall thickness of 0.03 mm and cell densities of over 150 cells cm^{-2}. These developments have allowed a larger catalyst surface area to be incorporated into a given converter volume, giving better conversion efficiency and durability.

Alternatively the same performance can be incorporated into a smaller converter volume, making the catalyst easier to fit close to the engine, the smaller size being important for compact cars.

Three way catalysts cannot be used in the oxygen rich exhaust of a lean burn petrol or diesel engine.

13.3.2.3 Diesel Oxidation Catalysts

Carbon monoxide (CO) and HC emissions from diesel vehicles can be reduced using a simple oxidation catalyst, developed specially to operate efficiently in the relatively low exhaust temperatures of these vehicles. This type of catalyst also oxidises the soluble organic fraction of the PM in the exhaust of diesel engines, reducing the total PM emission by up to 50%. It can reduce CO and HC emissions by over 90%, and has the benefit of also reducing the characteristic diesel odour. However the number of particles is unchanged.

Similar oxidation catalysts are also used on some buses running on compressed natural gas and motorcycles.

13.3.2.4 Diesel Particle Filters

The first production car with a particle filter (DPF) was introduced to the EU market in 2000. Today DPFs are standard equipment for new diesel cars in the EU and are also fitted to some HDVs. It is thought that DPFs will be required to meet the Euro VI particle number (PN) limits for heavy duty vehicles (HDVs) and the Euro 6 PN limits for gasoline direct injection (GDI) light duty vehicles (LDVs). One of the motivators for the introduction of the PN standard was that earlier legislation, which is technology neutral, had failed to result in the widespread use of DPFs on HDVs as manufacturers found alternative methods to pass the emission tests.

However, a large number of trucks and buses have been retrofitted with DPF. Transport for London, for example, has fitted over 8000 DPFs to London buses and plan to fit a further 1000. Due to the relatively high cost of retrofitting this has generally occurred only where fiscal incentives have been provided or where there are low emissions zones that restrict entry to certain specified vehicles.

The most common type of DPF is a wall-flow filter in which PM is filtered using a ceramic honeycomb structure with the channels blocked at alternative ends. The exhaust is forced to flow thought the walls between the channels and the PM is deposited on the walls. Ceramic wall-flow filters can remove more than 95% by weight of the particulate matter (PM) and particle number over a wide range of operating conditions.

To avoid the DPF walls becoming saturated with PM, the filter needs to be regenerated by burning the collected PM on a regular basis. In the presence of oxygen regeneration occurs when the exhaust temperature exceeds about 550 °C. However, diesel exhaust temperatures are generally considerably lower. PM can burn off at normal exhaust temperatures using the oxidation properties of NO_2 or in oxygen when the exhaust temperature is periodically increased.

The methods to achieve regeneration include:

(i) Incorporating an oxidation catalyst upstream of the DPF to increases the proportion of NO_2 in the exhaust.

(ii) Coating the surface of the DFP with a catalyst to lower the temperature at which the PM burns. This can contain precious metal catalysts (palladium or platinum) to replicate the action of an oxidation catalyst in forming NO_2, or be combined with zirconium or cerium oxides which can act directly to make oxygen react with PM at the filter wall at lower temperatures.

(iii) Fuel-borne catalysts such as cerium or iron oxides added to the fuel using an on-board dosing system. These compounds also reduce the temperature the PM burns at. Very small quantities are added so that a tank of less than 2 litres can last the life of a diesel light duty vehicle (240 000 km).

(iv) Electrical heating of the trap either on or off the vehicle. Some DPFs are made from metal fibres woven into a monolith. An electrical current can be passed through the monolith to heat and regenerate the filter, allowing it to occur at low exhaust temperatures. Metal fibre cores tend to be more expensive than ceramic filters.

(v) Engine management used to increase exhaust temperature through late fuel injection or injection during the exhaust stroke.

(vi) A fuel burner after the turbocharger but before the DPF to increase the exhaust temperature.

DPFs require low sulfur fuel (see section 13.5).

All on-board active systems use extra fuel, whether through burning to heat the DPF, or providing extra power to the DPF's electrical system, although the use of a fuel borne catalyst reduces the energy required very significantly. The latter has been used in cars for over a decade.

Typically the engine management system monitors one or more sensors that measure back pressure and/or temperature, and based on pre-programmed set points the computer makes decisions on when to activate the regeneration cycle. Running the regeneration cycle too often while keeping the back pressure in the exhaust system low will result in high fuel consumption. Not running the regeneration cycle soon enough increases the risk of engine damage and/or uncontrolled regeneration (thermal runaway) and possible DPF failure. Some DPF require periodic cleaning to remove ash from the combustion of the PM that can build up on the filter walls.

Some applications use off-board regeneration. Off-board regeneration requires operator intervention (*i.e.* the machine is either plugged into a wall/floor mounted regeneration station, or the filter is removed from the machine and placed in the regeneration station). Off-board regeneration is not suitable for on-road vehicles, except in situations where the vehicles are parked in a central depot when not in use such as buses. Off-board regeneration is mainly used in industrial and mining applications.

13.3.2.5 Selective Catalytic Reduction

With the development of petrol direct injection (GDI) engines and the increased use of diesel engines, controlling NO_x emissions in lean combustion has been one of the greatest challenges for the automotive catalyst industry. The conventional three-way technology used with petrol engines needs a richer environment to reduce NO_x, so a different approach was required.

Selective catalytic reduction (SCR) was originally developed to reduce emissions from stationary emission sources such as power stations. In automotive applications it continuously removes NO_x by creating a rich microclimate where NO_x is converted to nitrogen using ammonia, while the overall exhaust remains lean. It is called selective catalytic reduction because the ammonia reacts with NO_x in preference to oxygen. It is now fitted to most new HDVs in the EU, and is being introduced for LDVs (diesel and GDI).

The conversion efficiencies for SCR fitted to HDVs are typically 60–85% during transient operations and up to 99% under steady state conditions. Exhaust temperatures tend of be lower in diesel light duty vehicle exhaust and the SCR conversion efficiencies are also lower.

Its use also provides a small fuel efficiency benefit, as engine manufacturers can take advantage of the trade-off between NO_x, PM and fuel consumption, and calibrate the engine to have higher engine-out NO_x emissions but lower PM emission and improved fuel consumption, with the overall NO_x emissions being substantially reduced by the SCR.

The ammonia is stored on-board as urea and sprayed into the exhaust ahead of the SCR catalyst where it hydrolyses into ammonia. The primary reactions are:

$$CO(NH_2)_2 + H_2O \rightarrow 2NH_3 + CO_2 \tag{5}$$

$$4NO + 4NH_3 + O_2 \rightarrow 4N_2 + 6H_2O \qquad (6)$$

$$2NO_2 + 4NH_3 + O_2 \rightarrow 3N_2 + 6H_2O \qquad (7)$$

$$NO + NO_2 + 2NH_3 \rightarrow 2N_2 + 3H_2O \qquad (8)$$

Since the introduction of Euro IV engines in 2008 a distribution network for urea solution, known as AdBlue®, has developed. The consumption of AdBlue® is typically 3–4% of the fuel consumption for Euro IV engines and slightly higher for a Euro V engines. On-board systems alert the driver when to re-fill the AdBlue® tank.

The system needs to be tuned to ensure that little ammonia passes through the catalyst and is emitted from the vehicle, known as ammonia slip. On-board diagnostics with NO_x or ammonia sensors are used to control the system. At low exhaust temperatures SCR works best when about 50% of the NO_x is NO_2. On the other hand more NO is needed to oxidise any ammonium nitrate condensed onto the catalyst.

Early SCR catalysts were based on vanadium pentoxide (vanadia), however as Euro VI HDV engines are likely to require a DPF, new SCR systems using zeolites (micro-porous aluminium silicate minerals) have been developed as they are better able to withstand the high exhaust temperatures associated with DPF regeneration. Typically copper-zeolites and iron-zeolites, which operate best in high and low exhaust temperatures, respectively, are used together to produce the best overall results.[14]

New combined DPFs and SCR systems, with the NO_x catalyst incorporated in the DPF have been developed; however, these tend to be slightly less efficient at reducing NO_x. In addition, alternatives to urea for storing ammonia on-board are being developed.

13.3.2.6   Lean NO_x Traps or Adsorbers

Lean NO_x traps (LNTs) store NO_x under lean engine operations and release it under richer conditions. A brief return to stoichiometric or rich operation, for one or two seconds, is enough to remove the stored NO_x for reduction to nitrogen using a conventional three-way catalyst mounted downstream. This approach is suitable for use with petrol direct injection (GDI) engines which operate under both rich and lean conditions.

Over 90% NO_x removal has been demonstrated with new LNTs. The main problem with NO_x absorbers is their sensitivity to the effects of fuel and lubricant sulfur which rapidly reduces the removal efficiency.[14]

13.3.2.7   Lean DeNO$_x$ Catalysts

Lean De-NO_x catalysts (also known as hydrocarbon selective catalytic reduction or HC-SRC) remove NO_x in the same way as SCR but uses hydrocarbons as the reductant. Passive systems use hydrocarbons in the exhaust as the reducing agent, whilst active systems use additional fuel added upstream of the catalyst.

The NO_x reduction achievable with a DeNO$_x$ catalyst system is significantly lower than using an ammonia SCR system and the active system can give rise to an undesirable increase in HC, CO and PM emissions and fuel consumption. In some lean DeNO$_x$ systems there may also be an increase in the proportion of NO_x emitted as N_2O, a potent greenhouse gas. A large number of factors influence the performance of these systems. These include the concentration of HC and NO_x in the exhaust, the exhaust temperature, the type and loading of the catalyst material, fuel sulfur content and the concentration of certain HCs in the exhaust.

13.4 REDUCING CARBON DIOXIDE EMISSIONS

Improving vehicle efficiency is the most cost effective means of reducing CO_2 emissions in the transport sector in the short term, and is likely to play a larger part than biofuels and electricity together.[13] Electricity can be used for cars and small vans, either in hybrid or full plug-in electric vehicles (EVs), and hybrid EVs have the potential to reduce CO_2 emissions from buses. Biofuels are able to replace liquid fossil fuels used in heavy goods vehicles and other transport modes, such as aeroplanes, that cannot be easily electrified. Therefore these fuels have the potential to play a major part in de-carbonising transportation fuels, although there remain issues with the use of full EVs as discussed in the section 13.6.

In 1998 the European motor manufacturers association (ACEA) agreed with the European Commission voluntary CO_2 emissions targets. For new cars these were an average of 140 g km^{-1} by 2008, a 25% reduction from 1996 levels. This was followed by similar agreements with the Asian motor manufacturers. However this voluntary approach did not prove to be very effective and it was not until the 2007 when mandatory limits were agreed that there were significant reductions in new car CO_2 emissions.

For cars, manufacturers are now obliged to ensure that the new car fleet does not emit more than an average of 130 g km^{-1} by 2015 and 95 g km^{-1} by 2020. This compares with an average of almost 160 g km^{-1} in 2007 and 136 g km^{-1} in 2011. For vans the equivalent limits are 175 g km^{-1} in 2017 and 147 g km^{-1} in 2020.[15] Further limits are to be introduced for new vehicles to apply after 2020, in order to continue to exert pressure on the motor manufacturers to innovate.

As a consequence of these legally binding average emission limits, the CO_2 emissions from new cars first registered in 2011 in Great Britain was 138 g km^{-1}, down 21% since 2001, and down 13% since 2008. About 3.4 m cars had CO_2 emissions below 130 g km^{-1}.[9]

Vehicle CO_2 emissions can be reduced by using different fuels which is discussed in section 13.6 on alternative fuels. This section discusses methods to improve the fuel economy of conventionally fuelled vehicles to reduce CO_2 emissions.

There is no add-on technology, such as a catalyst, that can be fitted to reduce CO_2 emissions from motor vehicles. A number of factors affect the fuel consumption and hence CO_2 emissions. These include the weight of the vehicle, its aerodynamic resistance and the efficiency of the engine and transmission.

Much of the emphasis today is on improving engine efficiency. However, as the manufacturers' legal requirement relates to average new vehicle CO_2 emissions, measures also need to ensure that trends towards larger and more powerful passenger cars are halted. There are indications that the previous long-term rising trend may have stabilised as the average engine size of cars in Great Britain remained around 1750 cm^3 between 2009 and 2011, but this could also be due to economic conditions rather than the start of a long term trend.[9]

Conventional petrol engines requiring stoichiometric air to fuel mixtures are likely to become increasingly incompatible with the need to *significantly* improve fuel consumption of new vehicles to meet the legislative CO_2 limits. There is a thermodynamic limit to the efficiency of IDI SI engines, and with a combination of higher compression ratios, and fewer energy losses GDI engines are inherently more energy efficient. Information from the International Council on Clean Transportation illustrates the benefits of petrol direct injection (GDI) in terms of CO_2 emissions for certain car models. A new Ford Focus, one of the top five most popular cars in Europe, had about 30% lower CO_2 emissions in 2012 than it had in 2010, without any reduction in vehicle weight and while maintaining all vehicle performance characteristics. It has a smaller, turbocharged GDI engine and uses a start-stop system, and has NEDC emissions of 99 g km^{-1}. The Audi A3 has also improved by a similar amount using similar technologies plus a dual-clutch automated manual transmission.[16]

There are trade-offs between the technology that might be applied to reduce fuel consumption and that required to reduce the regulated emissions. For example, if improved fuel consumption were the only concern lean burn engines would be in widespread use now. However, because of greater concern during the last two decades over the regulated emissions this technology has only recently been more widely used. Meeting increasingly stringent emission standards for the regulated pollutants has, in general, led to small fuel consumption penalties over the last ten years. These have typically been 2 to 3% over the legislated test cycle for each new standard. On the other hand, the use of selective catalytic reduction may result in a small fuel consumption benefit.

Vehicles with GDI engines require both a three-way catalyst and NO_x absorbers to control all regulated emissions when running stoichiometrically and NO_x emissions when running lean. The greater the time spent running stoichiometrically, the lower the fuel economy benefits of the technology. GDI engines need to periodically run stoichiometrically to remove NO_x from the low NO_x traps (see section 13.3.2.6).

Spark and compression ignition engines have different emission characteristics, as reflected in the emissions legislation shown in Table 13.1. Diesel cars are more fuel-efficient, and have lower CO_2 emissions despite the higher carbon content of the fuel. However over the last two decades this gap has closed significantly. There are both petrol and diesel cars with CO_2 emissions less than $100 \, \mathrm{g \, km}^{-1}$. The petrol cars typically have an engine capacity of less than 1 litre, whereas the diesel cars have an engine capacity larger, typically around 1.4 litres. There are also a number of hybrid cars on the market with these low CO_2 emissions.[9]

The difference in average CO_2 emissions from new petrol and diesel cars in the EU in 2010 was about 3%, whereas ten years earlier it was about 10%. However this data is sales weighted and reflects the fact that there were more smaller petrol cars sold and thus, on a like for like basis, the gap is larger than indicated by these figures.[12]

13.5   FUEL QUALITY

The chemical and physical characteristics of automobile fuels have a significant impact on vehicle emissions, and as a result, limits have been put on many fuel parameters. Improving fuels can reduce emissions directly but also indirectly by affecting the performance of after-treatment devices.

Many studies have showed that several fuel parameters, for example, volatility, distillation characteristics and the sulfur and aromatics content, have significant effects on emissions. As a result fuel quality standards in Europe, and elsewhere, were extended to include a wide range of parameters in the 2000s.

In general, the effect of fuel quality changes on emissions is relatively small, compared to the improvements achievable by engine and after-treatment technologies. However for certain pollutants (*e.g.* lead, sulfur and benzene) their emission is related to their content in the fuel. The aromatic content of petrol also affects the benzene emission as it is formed in the combustion process from other aromatic compounds.

Reducing the content of these substances in the fuel can have a large impact on emissions. For example, lead emissions in the EU reduced by over 90% in 1990s as legislation was introduced in the 1980s that required all new cars to be able to use unleaded petrol. The availability of unleaded petrol allowed the use of three-way catalysts for petrol cars, as well as reducing emissions of tetraethyl lead. This compound was added to petrol from the 1920s when it was discovered that it was a cheap way of increasing the octane rating. Now, it is widely banned as the lead binds to active sites within the catalyst and oxygen sensor, greatly reducing their performance. It has also been shown to be a neurotoxin.

Octane number is a measure of petrol's ability to resist auto-ignition which can cause engine knock and damage. Vehicles are designed and calibrated for a certain octane rating. When petrol with an octane rating lower than required is used knocking may result. Several compounds have been used to replace tetraethyl lead including methylcyclopentadienyl manganese tricarbonyl

(MMT), ferrocene (Fe(C$_5$H$_5$)$_2$), toluene and isooctane (2,2,4-trimethylpentane). There are concerns that organometallic compounds and metallic contaminants, such as calcium, copper, phosphorus, sodium and zinc, can adversely affect the operation of the emission control systems of petrol cars, and the motor industry's World Fuel Charter recommends that their use is avoided in markets such as the EU where there are stringent emission requirements.[17]

Studies have shown that most of the manganese from the MMT remains in the engine, catalyst and exhaust system, while a small portion is emitted from the exhaust. The oxidized manganese coats exposed surfaces throughout the system, including spark plugs, oxygen sensors and the catalytic converter. These effects result in higher emissions and lower fuel economy. MMT has not been widely used within the EU, and since 2011 its use in petrol has been limited. Similar issues have been found with the use of ferrocene.[17]

Oxygenated organic compounds, such as methyl tertiary butyl ether (MBTE) and ethanol, are often added to petrol to increase octane. This can affect vehicle emissions (exhaust and/or eva-porative), performance and durability. Adding ethanol affects the distillation of the petrol. The vapour pressure and distillation of ethanol–petrol blends must be carefully regulated to ensure proper vehicle operation and emissions control. Ethanol also will make vapour lock more likely. Ethanol by itself has a very low vapour pressure, but when added to petrol has a non-linear and synergistic effect. Importantly, the final vapour pressure of the blend could be either higher or lower than the base petrol, depending on temperature and ethanol concentration. At lower ethanol concentrations (below about 10% by volume) and typical temperatures, ethanol will cause the blend's vapour pressure to exceed that of the base petrol.[17]

Methyl tertiary butyl ether (MTBE) has been a controversial component of petrol since it was first used in the US in the 1990s. It reduces CO and HC emissions from catalyst equipped cars. For NO$_x$ there is a greater variability of response with an increase in emissions in some studies and a decrease in others. MTBE was thought to have a low toxicity, especially when compared to other components of petrol. The main concern regarding its use is that it is extremely soluble in water and has been found in drinking water. Concerns over its health impacts have led to it being largely replaced by ethanol.

The motor industry believes that increasing the minimum octane rating of petrol would help vehicle CO$_2$ emissions. Octane rating is becoming a limiting factor in future efficiency improve-ments as more efficient engine designs, such as smaller capacity turbo-charged engines, are approaching their theoretical knock limits when using lower octane rated petrol.[17]

Sulfur is naturally present in crude oil and the level in petrol and diesel depends on the sulfur content of the crude oil feedstock, the refinery processes and the extent to which sulfur is removed. If the sulfur is not removed it will have a significant impact on vehicle emissions by reducing the efficiency of catalysts and exhaust gas oxygen sensors.

Stringent exhaust emission limits and requirements for long-life compliance, demand efficient and durable after-treatment systems. For example, catalyst efficiency at removing HCs at 160 000 km must be about 97% for a vehicle meeting Euro 4 petrol standards.

Three way catalysts and low NO$_x$ traps (LNT) are both highly sensitive to sulfur and once exposed to sulfur in the fuel their efficiency may never fully recover. In a LNT sulfur oxides are more strongly trapped than NO$_x$, and, therefore the NO$_x$ capacity of the device is reduced. Sulfur removal requires prolonged rich operating conditions, which is possible for petrol engines but negates the fuel efficiency benefits of the GDI engine. Sulfur-free petrol and diesel is therefore necessary to ensure that these technologies can be used effectively.

Diesel vehicles also benefit from lower PM emissions with low fuel sulfur. The sulfur is oxidised first to sulfur dioxide and then to sulfates. For vehicles fitted with oxidation catalysts, a large proportion of the engine-out sulfur dioxide is oxidized to sulfate, and therefore reducing sulfur in the fuel reduces PM emissions. The US Diesel Emission Control – Sulfur Effects Program has shown that with 3 parts per million by weight (ppm) sulfur diesel the traps tested removed over

95% of the PM over a transient test cycle, but with 30 ppm sulfur fuel the removal efficiency declined to just under 75%. With 150 ppm sulfur fuel there is virtually no PM removal.[18]

In the EU the maximum permitted sulfur content fell from 350 and 150 ppm (by mass) for diesel and petrol, respectively, from 2000, to 10 ppm for both fuels from 2009.

Many different types of additives are used in automotive fuels and lubricants. For example, detergents are used to reduce the build up of combustion generated deposits over time in engines which can impair driveability and emission control systems. Other additives may include an ignition improver, cold flow improver, anti-corrosion additive, anti-foam compounds, and a lubricant.

13.6 ALTERNATIVE FUELS

13.6.1 Introduction

In the 1990s and early 2000s alternative transport fuels were promoted to improve air quality. For example, a 1995 Department of Transport publication stated "Alternative fuels have traditionally been considered for social, economic or strategic reasons. However, in recent years there has been growing interest in their environmental benefits and the contribution they might make to improving air quality in urban areas".[19] In particular, liquefied natural gas (LNG) and compressed natural gas (CNG) were promoted to replace diesel for heavy duty applications, such as buses, because the combustion of these fuels produces significantly less PM, but can also emit less NO_x depending on whether the vehicle is manufactured for gas or is a conversion, and whether or not a catalyst is fitted to control emissions. As noted earlier the political drive for alternative fuels is now the need to reduce CO_2 emissions from road transport.

Gas is also used in cars and in 2011 there were about 50 000 gas cars on British roads, most of which use liquid petroleum gas (LPG), a fuel mainly comprising propane. In general motorists opted for LPG for economic reasons as the fuel was cheaper. In Europe, Italy has the highest proportion of natural gas vehicles with a market share of 15–20% for new cars and light commercial vehicles.[12] In Sweden the use of ethanol has been actively promoted by the government and in 2008 accounted for more than 20% of the new car market, although it has fallen since.[12]

As methane, a potent greenhouse gas (GHG), is the main component of natural gas, CNG and LNG are not regarded as a low carbon solution for transport fuels. Today the key fuels are seen to be biofuels (including biogas), electricity and in the longer term hydrogen fuel cells.[20]

In Britain, less that 1% of cars were fuelled by pure alternative fuels in 2010, more than 99% used petrol or diesel.[9] However petrol and diesel is required to contain 5% biofuels which is not reflected in this figure. The biofuel content was scheduled to increase as a result of the EU mandatory requirement for renewable energy to contribute at least 10% to transport energy consumption by 2020, which is expected to be met mainly through using biofuels. However, the European Commission has proposed that the contribution of biofuels produced from crops be capped at 5%.

Electrifying the car fleet would reduce exhaust emissions from individual vehicles to zero, a substantial benefit in terms of local air quality, although would not reduce the non-exhaust emissions, for example of PM from brake and wheel wear. Its impact on wider air quality and GHG emissions would depend on the fuels used to generate the electricity. However, hybrid cars, using either petrol or diesel ICEs, have proved more popular accounting for about 1% of all new cars sold in EU in 2010, but almost 10% of the Toyota's car sales.[12]

When comparing emissions from vehicles using different fuels it is important to compare similar vehicles. Comparisons should be between vehicles of similar function and performance. The emissions from a petrol car cannot realistically be compared to those of a 32 tonne lorry. Nor should comparisons be made between a high performance car operating on one fuel, and an economy model on another. Similarly comparisons must be made with similar generations of vehicle as technological advances have changed the emissions. In addition, comparisons must be made for vehicles tested using the same test conditions, as some vehicles have lower emissions under cruising conditions while others have advantages under the stop start driving conditions of urban areas.

There is evidence that the measured benefits of different fuels may depend on the test method.[21] Emission measurement methods typically use engine dynamometers, vehicle dynamometers, remote sensing or on-board (or portable) emissions monitoring systems. Tunnel studies have also been used. Each method has advantages and disadvantages. Historically heavy duty emission testing has largely been undertaken on engine dynamometers, whereas cars have been tested on chassis dynamometers. This was due to the high cost of heavy duty chassis dynamometers, and for regulatory purposes engine dynamometers are still used. Chassis dynamometers can be used to replicate a wide range of driving conditions. On-board measurements have the advantage of measuring under real-world driving conditions, but this means that the conditions often vary from test to test, making comparisons difficult. Remote sensing has the advantage that it allows a very large number of in-use vehicles to be tested under real-world driving conditions but only provides a snapshot of the emissions, and there are limitations as to where it can be used. As it is a spectroscopic method it can only measure where there is a single line of traffic and the vehicle needs to be under load. It is therefore not able to capture data for all operating conditions.

13.6.2 Natural Gas

Although petroleum derived gas is no longer being actively promoted as a clean fuel in the UK, as it is not low carbon, it is currently the most commonly used alternative fuel. The main emission advantage of gaseous fuels is their significantly lower PM emissions compared to diesel vehicles. They can also have lower NO_x emissions, but not always.

Both SI and CI engines can be used to run on natural gas (NG). Vehicles with SI engines can also be used as bi-fuel vehicles, operating on gas and petrol. Where the engine is not optimised for gaseous fuels, for example in conversions or less sophisticated dual fuel vehicles, the vehicle is likely to give higher emissions than a dedicated vehicle. Where a catalyst is fitted, it might not be optimised for NG operation.

Although gaseous fuel technology has improved, it has not kept up with progress in emission control for conventionally fuelled vehicles, and some heavy-duty engine manufacturers believe that meeting stringent emission standards may be easier with diesel engines than gaseous engines. Therefore gas vehicles may fill a niche market for a relatively short period of time. Certainly the LDV natural gas market in the EU declined since its peak in 2009.[12]

Bi-fuelled vehicles have lower emissions of CO_2, CO, and PM emissions when operating on compressed natural gas (CNG) compared with petrol. However, HC emissions can be greater or lesser than when using petrol, depending on the driving conditions. Natural gas mainly consists of methane, and this is a significant component of the HC emissions from CNG vehicles. Under CNG operation the NO_x emission can be the same or less than with petrol operation.

Dedicated CNG vehicles offer greater benefits than bi-fuelled vehicles. Compared to petrol vehicles emissions of all the regulated pollutants are either lower or similar. Compared to diesel vehicles emissions of CO, NO_x and PM are generally lower with CNG vehicles. In the case of particulate matter, engine-out emissions may be up to 90% lower for smaller vehicles such as vans, and 50 to 80% for large vehicles such as buses. Total HC emissions tend to be higher with CNG under all driving conditions.

CNG buses have higher CO_2 emissions compared to similar diesel buses. This increase is greater than the expected loss of efficiency resulting from switching a CI to a SI engine and may be due to poor matching of the engine to the bus operating cycle.

Liquid petroleum gas (LPG) vehicles can have lower CO and HC emissions than petrol vehicles, but the NO_x emissions may be slightly higher under some driving conditions. Compared to diesel vehicles LPG vehicles have about 80% lower engine-out PM emissions.

The CO_2 emissions from LPG vans are lower than the equivalent diesel vehicles except under heavily congested driving conditions. For buses CO_2 emissions are greater when using LPG than with diesel, probably for the same reason as for the compressed natural gas buses.

13.6.3   Electric Vehicles

Electric vehicles have started coming onto the British market albeit in very small numbers. Only 250 were registered in 2010 and 1200 in 2011, and the UK Low Carbon Vehicle Partnership believe that they are unlike to dominate the market in the UK until 2035 at the earliest.[22]

These vehicles have zero emissions at the point of use, but their life-cycle emissions depend on the fuel used at the power station. They are best suited to urban operation because of their relatively limited range (typically around 150 km). Widespread use of these vehicles will depend upon the development of more advanced cheaper batteries, with greater range, and the development of the re-charging infrastructure. The latter is evolving, but does not approach the density of conventional filling stations, and together with purchase cost is perceived to be a barrier to the purchase of EVs. On the other hand running an EV is significantly cheaper than an internal combustion engine (ICE), to a large extent because there is no tax on the electricity used to charge car batteries. Other barriers to their use include motorists' perception regarding their safety and speed of recharging.

Unlike ICEs, electric motors are efficient with exceptionally high power to weight ratio providing adequate torque over a wide speed range. There is evidence from BMW's trial with its Mini-E that consumers enjoy the driving experience. Although a number of motor manufacturers are planning to introduce EVs into their product ranges they are likely to remain a niche product until the purchase price falls.

13.6.4   Hybrid Electric Vehicles

More promising are hybrid electric vehicles, the first of which came onto the UK market in 2000. In the UK by the end of 2011 there were around 100 000 hybrid cars registered in Great Britain, mainly using a petrol engine.[9] although more recently diesel hybrids have come onto the market.

There are three types of hybrid vehicle:

  (i) Mild hybrids typically are a parallel system with start-stop only or in combination with modest levels of engine assist or regenerative braking features.

 (ii) Full hybrids (HEV) are vehicles that can run using just the internal combustion engine (ICE), just the batteries, or a combination of both. A large, high-capacity battery pack is needed for battery-only operation.

(iii) Plug-in hybrids (PHEV) are full hybrid vehicles able to run in electric-only mode, with larger batteries and the ability to recharge from the electricity grid. Their main benefit is that they can be used as EVs for daily commuting, but also can have an extended range using the ICE for longer trips. The first PHEV cars came into the UK market in 2011.

HEV and PHEV can either be series or parallel hybrids. A series hybrid uses electricity to drive the wheels and the ICE is used to generate electricity on-board. The Honda Insight is an example of a series hybrid. A parallel hybrid can use either electricity or the ICE, separately or together, to drive the wheels. Toyota hybrid vehicles are examples of this type of HEV. According to the International Council for Clean Transportation, the 2012 hybrid model of the Toyota Yaris emits about 50% less CO_2 than the 2010 non-hybrid version, illustrating the potential of this technology, however it is likely that the 95 g km^{-1} CO_2 limit for 2020 can be achieved without the use of hybrids.[16]

The efficiency and emissions of PHEVs compared to petrol hybrids depends on the fuels used to generate the grid electricity.

13.6.5 Biofuels

Biofuels are liquid and gaseous fuels produced from organic matter derived from plants or animals. The first-generation biofuels currently in commercial production are ethanol, biodiesel (fatty acid methyl ester known as FAME), and biogas derived through anaerobic digestion. Typical feedstocks include sugarcane and sugar beet, starch-bearing grains like corn and wheat for ethanol, and oil crops like rape, soybean and oil palm, and in some cases animal fats and used cooking oils for biodiesel. Anaerobic digestion is a natural process in which microorganisms break down organic matter, in the absence of oxygen, into biogas (a mixture of CO_2 and methane) and digestate (a nitrogen-rich fertiliser). The biogas can replace natural gas as a vehicle fuel. Anaerobic digestion is considered to be a low carbon fuel because the emissions would occur anyway when the waste was disposed into landfill, and using biogas replaces the greenhouse gas (GHG) emissions from the combustion of fossil fuels.

The GHG emissions depend on the biofuel feedstock and processes used. Advanced biofuels, which are still at their development phase include: hydro-treated vegetable oil (HVO), which is based on animal fat and plant oil, as well as biofuels based on lignocellulosic biomass, such as cellulosic ethanol, biomass-to-liquids diesel (BtL diesel) and bio-synthetic gas (bio-SG), algae biofuels and the conversion of sugar into diesel-type biofuels using biological or chemical catalysts.

Biofuels have a long history, having been used from the late 19th century and until the 1940s until falling fossil fuel prices stopped their further development. Interest in commercial production of biofuels for transport rose again in the mid-1970s, when ethanol began to be produced from sugarcane in Brazil and then from corn in the United States, largely to support the agricultural sector and rural economies. However, the fastest growth in biofuel production has taken place since the beginning of the 21st century, supported by government policies to de-carbonise transport fuels.

One of the most common support measures is what is known as a 'blending mandate' – which defines the proportion of biofuel that must be used in road transport fuel – often combined with other measures such as tax incentives. More than 50 countries, have adopted blending targets. As a result, global biofuel production grew from 16 billion litres in 2000 to more than 100 billion litres in 2010. Today, biofuels provide around 3% of road transport fuel globally (on an energy basis).[23]

Over the last few years, there has been much debate about the extent to which biofuels lead to GHG reductions. Research has suggested that when the emissions associated with direct and indirect land-use changes (LUC) caused by biofuel production are taken into account that the GHG advantage over conventional fuels diminishes. Direct LUC occurs when biofuel feedstocks are grown on land that was previously forested; indirect LUC occurs when they are grown on previously cultivated land displacing the production of other crops, which are then produced on land converted elsewhere. In particular, growing energy crops on carbon rich soils or where there was carbon rich vegetation increases the direct carbon emissions as some of the stored carbon will generally be released into the atmosphere, leading to the formation of CO_2. The resulting negative GHG impact can offset the positive impact of the biofuels, in some cases by a wide margin.

Life-cycle analysis (LCA) studies that compare the GHG emissions associated with different biofuels against the replaced fossil fuel suggest that there is a large range for each biofuel, depending on the process and way the feedstock is produced, including the amount of fertilisers used. In general, ethanol produced from sugar cane (*e.g.* in Brazil or Thailand) shows the greatest benefits if no indirect land-use change occurs. For other commercial biofuels the benefits appear to be generally less. However there are gaps in knowledge of the indirect LUC impacts and there is currently no consistent approach to modelling these impacts. Further work is needed to develop data for use in LCA (*e.g.* on fugitive emissions from biogas vehicles).[24]

There is some controversy over the potential economic and social impacts of biofuel production and use. For example whether biofuels and food production competing for land results in decreased food production and associated increased prices.

The European Union has introduced regulations under the *Renewable Energy Directive* that lay down sustainability criteria that biofuels must meet before being eligible to contribute to the binding national targets that each Member State must attain by 2020. In order to count towards the Directive's target, biofuels must provide 35% greenhouse gas emissions saving compared to fossil fuels. This threshold will rise to 50% in 2017 and to 60% in 2018 for new biofuel plants. Other sustainability criteria include not growing the energy crop on land with high biodiversity value or land with a high carbon stock.

The *Renewable Energy Directive* includes a 10% renewable energy mandate for transportation fuels. To achieve the target advanced biofuels production technologies will be required as well as petrol and diesel blends with higher percentages of biofuels than currently mandated. There remain gaps in knowledge on the emissions from higher biofuel percentage transport fuels. Whilst new vehicles can be built to run on these fuels there may be issues of compatibility for existing vehicles similar to those that occurred when lead was removed from petrol and sulfur drastically reduced in diesel, which were addressed using fuel additives. A revision to the Directive to limit the amount of crop derived biofuels to 5% is currently under discussion.

13.6.5.1   Biodiesel

Biodiesel either blended with petroleum diesel or as pure biodiesel can be used in diesel engines without modification, although, particularly higher biodiesel blends benefit from engine optimisation to take account of the different characteristics of the fuel.

Pure biodiesel contains no aromatic hydrocarbons, has a higher cetane value, better lubricity, and a higher flash point than diesel. These characteristics lead to lower emissions of some pollutants, but higher emissions of others.

In general, it appears that 20% blends of biodiesel with petroleum diesel (known as B20) and pure biodiesel (B100) reduces CO, HC and PM emissions from heavy duty vehicles. However, the NO_x emissions may increase. A 2012 review of the available data from heavy duty chassis dynamometer tests from around the world shows NO_x emission increased by 3.5% and 9.0% for B20 and B100, respectively, but on-road tests showed NO_x emissions using B20 reduced by 3.3% (no data was provided for B100).[21]

For light duty vehicles no significant change was found for HC. For CO there was no significant effect for all biodiesel blends, except B20 for which there was a 5.5% reduction, while PM emission were lower (5.8–16.0%) and NO_x emissions were higher (1.1 to 7.3%) except for B5 when there was no significant change.[21]

Biodiesel manufactured from different feedstocks have different characteristics that can affect emissions. For example in one study of the impact of B10 blends on emissions showed that NO_x increased by up to 20% for two out of five blends, decreased by up to 15% for two other blends, and remained unchanged for one blend. PM emissions were reduced for all blends by up to 25%.[21]

Late technology engines, which are optimized to achieve low NO_x emissions, seem to lead to higher NO_x emissions with biodiesel.

It is thought that the increase in NO_x emissions is mainly due to the higher cetane number, which leads higher combustion temperature and pressure. The unsaturated organic compounds that make up biodiesel and its higher oxygen content also means higher flame temperatures and oxygen in the flame front which also may lead to higher NO_x formation. In addition, the high oxygen content means more fuel is injected, earlier injection to counterbalance the energy loss and, again, higher temperature combustion. In Euro 2 and later engines, equipped with electronic control, more fuel injected is translated as higher load which means lower exhaust gas recirculation (EGR) rate, again giving rise to higher NO_x emissions.[24]

Although most studies have shown a decrease in PM emission with the use of biodiesel blends, there is evidence that the particle number may increase.[25]

There is also limited data on the effect of the use of biodiesel on unregulated pollutants. Emissions of polyaromatic hydrocarbons (PAH) and nitro-PAH appear to be significantly lower than those observed with petroleum derived diesel, which probably reflects their absence in the biodiesel. This trend appears to be consistent on both light- and heavy-duty engines and on different test cycles.[24]

13.6.5.2 Bioethanol

Bioethanol can be used directly either pure in specially modified petrol engines, as a blend in petrol (from 5 to 85%), or transformed into ETBE (ethyl tertiary-butyl ether). It can also be used as a blend with diesel (E-diesel) or biodiesel (BE-diesel) in diesel engines. The advantage of ETBE is that it reduces some of the historical impediments to the greater use of ethanol, such as increased volatility of petrol and incompatibility with petrol pipelines.

In Europe, a few manufacturers have produced flexible fuel vehicles that run on blends up to E85 or on pure petrol, particularly for the Swedish market where government policy favours the use of bioethanol. However none are currently on the UK new car market and it is likely that in the UK in the foreseeable future, bioethanol use will be limited to low percentage blends, although this may be higher than the current 5% (E5).

Petrol cars need to be modified to use bioethanol blends higher than E5. Ethanol degrades certain types of rubber and several metals and some engine components that come in contact with ethanol may need to be replaced. Bioethanol has a higher octane number than petrol that enables its use at high compression ratios, thus increasing engine efficiency. However, it has a lower energy density and this requires an adjustment of the ignition timing and the fitting of a larger tank to achieve the same distance between refuelling. There can be difficulties in starting vehicles using E100 in cold weather and therefore the fuel is usually blended with a small amount of petrol to improve ignition (E85 is a common high percentage blend).

Studies investigating the impact of the use of E10 on NO_x emissions have found no consistent pattern with some studies showing an increase, some no impact and others a decrease. The average increase of NO_x emissions was about 1%. For E20 the average increase in NO_x emissions in tests was about 25%, and there is some evidence that as mileage increases, NO_x emissions also increase compared to pure petrol. Limited evidence also suggests that modern (Euro 4) flexible fuel cars have large reductions in NO_x emissions when using E85.[24]

Petrol engine emissions are very dependent on the performance of the three way catalyst. Small variations of the combustion stoichiometry may have significant effects on the catalyst efficiency (see section 13.3.2.2). If the ethanol oxygen content in the fuel is not properly compensated for by the engine, this will lead to a lean exhaust, which inhibits the NO_x reducing efficiency of the catalyst and will lead to higher emissions.

The use of ethanol–petrol blends appears to also reduce PM emissions, albeit pure-petrol emissions are significantly lower than those from diesel engines. The use of E10 ethanol–diesel blends reduced PM emissions from passenger cars by 5% and from heavy duty vehicles by about 23%.[24]

13.6.6 Hydrogen

In the longer term, hydrogen either burnt in a spark ignition engine but more likely used in fuel cells may become important alternatives to petrol and diesel. The widespread use of hydrogen is likely to depend on the development of a clean and economical production method for the fuel. Hydrogen vehicles have very low emissions during use, but currently more than 90% of the hydrogen produced comes from fossil fuels. For hydrogen to have low life-cycle emissions clean production using the electrolysis of water by electricity generated by renewable energy sources is likely to be needed.

Hydrogen can be used in spark ignition engines, re-tuned to take account of its different ignition and flammability characteristics. The advantage is that the only emissions are NO_x and water vapour. However its use requires the development of a hydrogen infrastructure and therefore its use is, at least in the short to medium term, likely to be restricted to fixed fleets, such as buses, which return to the same depot for refuelling.

It is generally considered that hydrogen would be better used to power fuel cells rather than being burnt in a SI engine, as these are essentially zero emission vehicles with the only emission being water vapour.

Fuel cells generate electricity on-board the vehicle, typically from hydrogen and atmospheric oxygen. The Ballard fuel cell, which was developed for automotive applications with Daimler Benz and Ford Motor Company, uses two electrodes separated by a polymer membrane electrolyte (proton exchange membrane). The electrodes are coated on one side with a thin platinum catalyst layer. Hydrogen dissociates into free electrons and protons in the presence of the platinum catalyst at the anode. The electrons are conducted in the form of useable electric current through the external circuit. The protons migrate through the membrane electrolyte to the cathode. At the cathode the combination of oxygen from the air, electrons from the external circuit and protons produces water and heat. Individual fuel cells produce about 0.6 volt and are combined into a fuel cell stack to provide the amount of electrical power required.

Transport for London first introduced hydrogen fuel cell buses in 2003 as part of a wider European trial, and the success of this trial led to the introduction of the new hydrogen powered buses and a refuelling station in east London. Currently there are five hydrogen fuel cell buses being operated in a route in central London with a further three expected to join the fleet in 2013. The 'well to wheel' CO_2 emissions for the fuel cell buses compared to diesel buses are expected to be about 50% less for the fuel cell buses.

13.7 PARTICLE EMISSIONS

Until very recently only the mass of particulate matter (PM) emitted from diesel motor vehicles was limited by legislation. There were no controls on specific sizes of particles or the number of particles emitted, or legislation to control the emissions of PM from petrol vehicles. Given the importance of the health impacts of PM, particularly of smaller particles, this is perhaps surprising.

Different vehicle technologies give rise to different mass of PM emission. Diesel light duty engine-out emissions are perhaps one to two orders of magnitude greater than from an equivalent petrol vehicle. In general, diesel engines produce much greater number of particles than petrol engines. However, there may be little difference under high speed/load driving conditions. This is due to an increase in the petrol emissions rather than a decrease in the diesel emissions. The new petrol direct injection (GDI) technology is thought to have PM emissions more similar to a diesel than a conventional petrol engine.

In terms of mass more than 85% of light duty diesel PM emissions are less than 1 μm. This corresponds to about 99% by number. There is some evidence that petrol vehicles emit a higher proportion of smaller particles than diesel vehicles. That is, more than 99% of the number of particles are less than 1 μm.

There has been concern that, as the mass of PM emitted from diesel vehicles has declined in response to increasingly more stringent limit values, the number of very small particles increases. Since particle mass is approximately proportional to diameter cubed, bigger particles are more significant to total mass than smaller ones. However, the results of a test programme run by the European Automobile Manufacturers Association suggests that measures taken to reduce PM mass emissions from light duty diesel vehicles also reduce PM number emissions, and that there is little difference in the size distribution of PM emitted from 'conventional' and 'advanced' diesel vehicles .

The number of particles measured in vehicle exhaust is very sensitive to the dilution technique used. Significant gas to particle conversion takes place as the exhaust dilutes and cools. It is thought that about 90% of the number and 30% of the mass is formed during dilution. New particles are formed by nucleation while pre-existing particles grow by adsorption or condensation. Nucleation and adsorption are competing processes, and are dependent on the dilution rate. There is evidence that high dilution ratio and high relative humidity both favour the formation of nanoparticles.

The majority of work on the measurement of vehicle particle size and number has been carried out using a standard dilution tunnel (*i.e.* as in the legislated emission tests). On the road dilution of a vehicle exhaust can reach 1000 to 1 in 1 to 2 seconds; that is considerably more rapid that in a dilution tunnel. This has raised questions as to how adequately the size/number of particles measured in the dilution tunnel relate to the real world.

An UNECE programme, known as the Particle Measurement Programme (PMP), led by the UK Government, has developed light duty and heavy duty PN test procedures for type approval purposes. This test procedure will be used from Euro 6/VI and will apply to GDI and diesel vehicles.

13.8 NON-EXHAUST PARTICLES

Non exhaust emissions of particulate matter (PM) are becoming increasingly important as exhaust emissions are controlled. In 2010 emissions from the wear of brakes, tyres and road surfaces was estimated to contribute 45% to the total EU road transport PM_{10} emissions.[3] A number of studies using ambient measurements have confirmed that non-exhaust and exhaust emissions can contribute approximately equal amounts to elevated PM_{10} concentrations close to roads.[27] The use of diesel particle filters (DPFs) on all new diesel vehicles and petrol direct injection (GDI) cars by 2015 will increase the relative contribution from these non-exhaust sources.

Emissions of PM from brake and tyre wear are caused by abrasion between surfaces, either the brake lining and disc/drum or the tyre and road surface. The rate of wear is determined largely by the composition of the materials used and how vehicles are driven. These are complex processes influenced by many factors including the vehicle speed, vehicle weight, severity of braking, vehicle condition and maintenance history. Non exhaust PM tends to be larger than exhaust particles but a significant proportion are in the $PM_{2.5}$ size.[28]

Another important source of non-exhaust PM is the resuspension of dust on road surfaces. The main source is soil and other crustal material but it also includes brake, tyre and road surface particles, some of which are sufficiently large to settle under gravity and deposit on the road, exhaust particles, de-icing salt and grit, and biogenic material deposited on the road. This source is very difficult to quantify due to the lack of a source specific tracer, varies significantly and is site specific.[27]

Calcium magnesium acetate (CMA) dust suppressant has been used successfully to reduce PM_{10} concentrations in locations where there is a particularly high resuspension component to the ambient PM_{10} concentrations such as in Scandinavia due to the use of studded tyres in winter. The CMA solution is sprayed onto a surface, and binds with particles on the surface. This prevents the

PM from becoming airborne when agitated by wind, tyre action or vehicle turbulence. Transport for London have undertaken trials of its effectiveness and have shown that CMA application is most effective in locations with unusually high local levels of PM_{10} due to resuspension, such as near industrial sites, and that application had no identifiable effect in more typical roadside locations, even where total PM_{10} levels were elevated. However, it could be useful in reducing the deposition of PM onto roads from adjacent industrial and construction sites, and major road works.[29]

13.9   IN-SERVICE EMISSIONS

Although the motor industry has invested heavily in the development of increasingly sophisticated emission control technology over the last few decades, urban air quality, particularly ambient concentrations of NO_2 and PM, remains a widespread issue. There is evidence that emission control systems are not working under real world driving conditions as well as intended by the regulators.

From the 1990s several European governments, particularly the Swedish and Dutch governments, have undertaken testing of in-service vehicles to ensure that emissions reflected the standard to which the vehicles were constructed. The evidence obtained suggested that in-service performance was not as good as expected. This was due to a number of factors, one of which was that the use of electronic engine controls led to an aftermarket industry selling 'performance chips' that increase the performance and torque of cars, but with increased emissions and fuel consumption.

The poorer than expected emissions in-service led to a series of mandatory requirements for the use of OBD to indicate the malfunctioning of key components of the emission control system. This initially focused on measuring the performance of the oxygen sensor, but has been extended over time to include the performance of diesel particle filters (DPFs) and selective catalytic reduction (SCR). Other measures that have been introduced into emission control legislation have included increased durability requirements, regulation of the performance of replacement after-treatment devices, the prohibition of the use of emission "defeat devices", and in-service testing.

However, one of the reasons may be that emission data obtained using the legislative engine and chassis dynamometer tests can be very different to those obtained from on-board emissions testing or in-service testing. This is because manufacturers have no choice but to design their engines and after treatment strategies to meet the mandatory regulations, otherwise their products would be banned from use on Europe's roads. To avoid excessive emissions under conditions not explicitly tested, the Euro VI emission requirements includes off-cycle emission measurements with a maximum emission requirement and in-service testing using portable emissions monitoring. These approaches are under consideration for light duty vehicles.

Work undertaken for Defra has shown that ambient concentrations of NOx and NO_2 in the UK declined up until around 2003. Since then concentrations have been much more stable, with only a small reduction of 1 to 2% between 2004 and 2009. This is much less than suggested by emission inventories. This work has also identified that the fraction of primary NO_2 in vehicle exhaust has increased from around 5–7% in 1996 to 15–16% in 2009 in the UK, and 21–2% in London. Most of the increase occurred before 2004. This is thought to be largely due to the increase in the diesel light duty fleet and the retrofitting of pollution control devices, such as diesel particle filters (DPFs) fitted to London buses. An analysis of the measurements at the London Marylebone Road monitoring site, and comparison with Transport for London's retrofit scheme for London buses, shows that the fitting of DPFs provides a plausible explanation of the observations, although emissions from other diesel vehicles probably also contribute to the increase in primary NO_2.[26] As a consequence the European Commission intends to introduce legislation controlling NO_2 emissions.

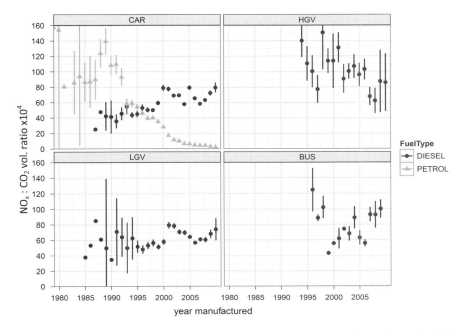

Figure 13.6 $NO_x : CO_2$ ratio for major classes of vehicle from Leeds University and Enviro Technology plc remote sensing data.[6] (The equipment used measured NO and not NO_2, Emissions of total NO_x were calculated using literature values of the proportion of NO_x as NO_2).[6]

Analysis of remote sensing data from 72 000 vehicles in the UK has provided very useful data on the in-service performance of vehicles. This suggests that a *new* Euro 5 petrol car emits about 96% less NO_x than a pre-catalyst vehicle, and that NO_x, CO and HC emissions from Euro 3–5 vehicles are very well controlled. For older cars there is some evidence of failed or inefficient catalysts.[6]

Figure 13.6 shows the $NO_x : CO_2$ ratios for different types of vehicle from the remote sensing. There is uncertainty in the estimated mean emissions for very old or new vehicles due to smaller sample sizes. However, there is a clear benefit of the EU legislation for petrol cars, as there is a steep drop in the $NO_x : CO_2$ ratio from Euro 0 to Euro 1 (1992–3) and again from Euro 2 (1996–7) to Euro 3 (2000–1). For diesel vehicles the situation is very different. Emissions of NO_x appear to increase from Euro 1 to Euro 3 and then remain more or less the same. Euro 4 vehicles (2005–6) emit similar or more NO_x than vehicles prior to 1995, suggesting that real-world NO_x emissions from diesel cars have changed little over the last 20 years despite the more stringent emission limits, summarised in Table 13.1. In addition, there is evidence that Euro 3 to 5 diesel cars emit up to twice the amount of NO_x under higher engine load conditions compared with older generation vehicles, probably due to the increased use of turbo-charging in modern diesel cars. For HC and CO there is some evidence that emissions have decreased by modest amounts although 'smoke' emissions from diesel cars show a larger decrease since Euro 2, thought to be due to the increased use of oxidation catalysts and particle filters.[6]

The heavy duty vehicle (HDV) trend in NO_x emissions is also relatively flat but there is evidence of a decrease in emissions from 2006–2007 when Euro IV vehicles were beginning to enter into service. For buses there has been a steady increase in emissions over time, but are likely to be affected by specific fleet characteristics. However, it should be noted that the sample size was small.[6]

A potentially important issue to emerge is that early selective catalyst reduction (SCR) used on most Euro V HDVs appears to be ineffective under urban driving conditions *i.e.* slow speeds resulting in low engine temperature). SCR can be calibrated to work under these conditions, for example for use in urban buses, and there is evidence that HDVs fitted with later SCR systems perform better.[30]

13.10 CONCLUSIONS

Controlling emissions from road vehicles is very complex. Whilst the manufacturers have to ensure their products meet the limit values under the legislative test procedures this has not always resulted in real world vehicle emissions reducing in parallel.

The nature of the three way catalyst means that the engine has to be closely controlled to ensure that both the oxidation and reduction reactions can take place. The evidence suggests that this has been effective with emissions being generally tightly controlled, particularly from Euro 3 onwards, provided the catalyst is working. However, for diesel vehicles, despite decades of emissions legislation, the real world NO_x emissions do not appear to have improved. In addition, the heavy duty engine manufacturers have succeeded in meeting legislation without using diesel particle filters which are proven to be affective at reducing both PM mass and numbers, although this is likely to change with the Euro VI requirements.

A number of new tests have been introduced to reduce emissions, with increasingly global harmonisation. In the future the focus will be on in-service testing using portable emissions monitoring to ensure that the promises of legislation are fully achieved in reality.

ACKNOWLEDGEMENTS

I wish to thank my former employer, ENVIRON UK Ltd., for supporting me in writing this chapter and Nafeezah Padamsey for assistance with figures.

REFERENCES

1. UKPIA, *Fuelling the UK's Future – The Role of our Refining and Downstream Oil Industry*, London, 2011.
2. L. Ntziachristos and P. Dilara, *Sustainability Assessment of Road Transport Technologies*, Joint Research Centre, European Commission, Ispra, 2012.
3. European Environment Agency, *Air Pollutant Emissions Data Viewer (LRTAP Convention)*; http://www.eea.europa.eu/data-and-maps/data/data-viewers/air-emissions-viewer-lrtap (accessed 31/01/2013).
4. European Environment Agency, *National Emissions Reported to the UNFCCC and to the EU Greenhouse Gas Monitoring Mechanism*; http://www.eea.europa.eu/data-and-maps/data/ national-emissions-reported-to-the-unfccc-and-to-the-eu-greenhouse-gas-monitoring-mechanism-6 (assessed 31/01/2013).
5. European Environment Agency, *Nitrogen Oxides (NO_x) Emissions (APE 002) Assessment* 2012; www.eea.europa.eu/data-and-maps/indicators/eea-32-nitrogen-oxides-nox-emissions-1/ assessment.2010-08-19.0140149032-2 (assessed 09/01/2013).
6. D. C. Carslaw, S. D. Beevers, E. Westmoreland, M. L. Williams, J. E. Tate, J. Stedman, Y. Li, S. Grice, A. Kent and I. Tsagatakis, *Trends in NOx Emissions and Ambient Measurement in the UK: Version July 2011*, Prepared for Defra, 2011.
7. EIONET Central Data Repository, *2012 LRTAP Submission UK*; http://cdr.eionet.europa.eu/ gb/un/cols3f2jg/envtzp7xq/2008.xls/manage_document (assessed 25/01/2013).
8. Greater London Authority, *London Atmospheric Emission Inventory*, London, 2008.
9. Department for Transport, *Vehicle Licensing Statistics 2011*, HM Government, London, 2012.

10. N. R. Passant, T. A. Murrells, A. Misra, Y. Pang, H. L. Walker, R. Whiting, C. Walker, N. C. J. Webb and J. MacCarth, *UK Informative Inventory Report (1980 to 2010)*, AEA Group, Harwell, 2012.

11. *Mayor's Air Quality Strategy*, Greater London Authority, London, 2010.

12. M. Mock and P. Campestrini, *European Vehicle Market Statistics*, The International Council on Clean Transportation, Washington DC, 2011.

13. International Energy Agency, *Improving the Fuel Economy of Road Vehicles – A Policy Package*, OCED/IEA, Paris, 2012.

14. Association for Emissions Control By Catalyst, *Emission Control Technoogies to Meet Current and Future European Vehcile Emissions Legislation*, undated, Brussels.

15. European Commission, *Road Transport: Reducing CO_2 Emissions from Vehicles*; http://ec.europa.eu/clima/policies/transport/vehicles/index_en.htm (accessed 21/01/2013).

16. The International Council for Cleaner Transportation, *Reducing CO_2 and Fuel Consumption: A Summary of the Technology Potential for New Cars in the EU*, Brussels, 2013.

17. ACEA, Alliance of Automobile Manufacturers, Truck and Engine Manufacturers Association, Japan Automobile Manufacturers Association, *Worldwide Fuel Charter: Draft. 2012.*

18. Diesel Emission Control – Sulfur Effects (DECSE) Program, *Program Phase 1 Interim Data Report No 4: Diesel Particulate Filters, Final Report*, 2000.

19. The Department of Transport, *Trials of Alternative Fuels in the UK*, London, 1995.

20. Department of Energy and Climate Change, *The Carbon Plan: Delivering our Low Carbon Future*, December 2011, HM Government, London, 2011.

21. G. L. Anderson, Effects of Biodiesel Fuels Use on Vehicle Emissions, *J. Sustain. Energy Environ.*, 2012, **2**, 35.

22. G. Archer, *EV, Fuel Cells and Biofuels – Competitors or Partners?* Presentation to the Institute of Engineering and Technology, 19th November 2011, Low Carbon Vehicle Partnership, 2011.

23. International Energy Agency, *Technology Roadmap – Biofuels for Transport*, OCED/IEA, Paris, 2011.

24. M. Kousoulidou, G. Fontaras, G. Mellios and L. Ntziachristos, *Effect of Biodiesel and Bioethanol on Exhaust Emissions*, ETC/ACC Technical Paper 2008/5, The European Topic Centre on Air and Climate Change, 2008.

25. P. Kumar, A. Robins and H. ApSimon, Nanoparticle Emissions from Biofuelled Vehicles - Their Characteristics and Impacts on the Number Based Regulation of Atmospheric Particles, *Atmos. Sci. Lett.*, 2010, **11**, 327.

26. Air Quality Expert Group, *Trends in Primary Nitrogen Dioxide in the UK*, London, 2007, Defra, London.

27. R. M. Harrison, A. M. Jones, J. Gietl, J. Yon and D. C. Green, Estimation of the contributions of brake dust, tire wear and resuspension to nonexhaust traffic particles derived from atmospheric measurements, *Environ Sci. Technol.*, 2012, **46**, 6523–6529.

28. A. Thorpe and R. M. Harrison, Sources and properties of non-exhaust particulate matter from traffic: a review, *Sci. Total Environ.*, 2008, **400**, 270–282.

29. B. Barratt, D. Carslaw, G. Fuller, D. Green and A. Tremper, *Evaluation of the Impact of Dust Suppressant Application on Ambient PM_{10} Concentrations in London*, King's College London, Environmental Research Group, Prepared for Transport for London, 2012, London.

30. TNO, *Real World NO_x Emissions – Euro V*, 2012, Delft.

CHAPTER 14

Climate Change

KEITH P. SHINE

Department of Meteorology, University of Reading, Earley Gate, Reading RG6 6BB, UK
Email: k.p.shine@reading.ac.uk

14.1 HISTORICAL AND POLITICAL BACKGROUND

It is now abundantly clear that human activity has been a major driver of global climate change over the past 100 years. In addition, it is apparent that climate change impacts on, and is impacted by, many of the other forms of environmental pollution discussed in the various chapters of this volume. Climate change has wide-ranging consequences, perhaps more so than any other form of pollution.

The roots of quantitative understanding of climate and climate change date back to the 19th century. What was recognizable as "modern" climate science began around the mid-1960s, when the direct forerunners of today's methodologies began to be used.[1] It was at about the same time that the potential seriousness of human-induced climate change for society began to be widely recognised. Two influential and multi-author studies, *Man's Impact on the Climate*[2] and *Inadvertent Climate Modification*,[3] appeared in 1971. The then available temperature records indicated that the northern hemisphere had cooled between 1940 and 1970. The authors of these studies noted that this cooling could have either natural or human causes. It was also recognised that a doubling of the concentration of carbon dioxide (CO_2) (which would eventually result from a continued burning of the fossil fuels such as coal, oil and natural gas) might increase surface temperatures by 2 °C. The many uncertainties were well appreciated. These reports also recognised that the human influence on climate went well beyond emissions of CO_2 and included, for example, the roles of aerosol particles and aviation.

During the late 1970s and 1980s, international assessments of the climate change issue became more common,[4,5] focusing more closely on the role of CO_2 and other greenhouse gases emitted by human activity. Observations indicated a clear increase in their concentrations and global temperatures were showing indications of a sustained increase.

A major landmark was in 1988, when the World Meteorological Organization and the United Nations Environmental Programme established the Intergovernmental Panel on Climate Change (IPCC). In 1990, and every five or six years since then, the IPCC has produced major assessments of understanding, covering three working groups (WG): WG I covers the physical science of climate

Pollution: Causes, Effects and Control, 5th Edition
Edited by R M Harrison
Published by the Royal Society of Chemistry, www.rsc.org

change; WG II covers the impacts of climate change on both natural and human systems, and possible adaption strategies; and WG III considers possible options for mitigation of climate change by limiting or reducing emissions by human activity. This chapter will draw heavily on the *Fourth Assessment Report* (which will henceforth be referred to as "IPCC AR4") published in 2007,[6] particularly emphasizing physical science aspects. At the time of writing, the IPCC's *Fifth Assessment Report* is being produced, with the main conclusions of WG I due for release in September 2013.

This chapter aims to provide a brief and broad overview of the science of climate change, and the nature and extent of the evidence linking climate change to human activity. The focus is on climate change over the past 150 years and over the next 100 years. Readers can find more in-depth discussion elsewhere.[6–8]

Since 1990, the IPCC has helped inform political developments in the area. In 1992, the United Nations Framework Convention on Climate Change (UNFCCC) (http://unfccc.int) was signed and is now ratified by 194 countries. The UNFCCC's objective is the "stabilization of greenhouse gas concentrations in the atmosphere at a level that would prevent dangerous anthropogenic interference with the climate system". It goes on to say that this "should be achieved within a time-frame sufficient to allow ecosystems to adapt naturally to climate change, to ensure that food production is not threatened and to enable economic development to proceed in a sustainable manner". The exact meaning of this objective has been much scrutinized (in particular, how the word "dangerous" should be interpreted), but the intent was clear.

The UNFCCC itself does not set any binding targets on greenhouse gas emissions – it was the Kyoto Protocol to the UNFCCC that first did this. The Kyoto Protocol was signed in 1997, and entered into law in 2005, once it had been ratified by sufficient signatories. The Protocol covered emissions of CO_2 and other relatively long-lived greenhouse gases: methane, nitrous oxide, sulfur hexafluoride and a number of hydrofluorocarbons and perfluorocarbons. These emissions were placed on a common "equivalent CO_2" scale using a metric called the "Global Warming Potential" which will be briefly described in section 14.4. The Protocol committed industrialized countries as a whole to reducing their equivalent CO_2 emissions, averaged over the period 2008–2012, by 5% compared to 1990. Individual industrialised countries had quite diverse targets – some were allowed to grow their emissions, while others were committed to reduce them. The Kyoto Protocol target has been met (http://unfccc.int/ghg_data/ghg_data_unfccc/items/ 4146.php); it is widely accepted that much of this was to do with the consequences of the reorganisation of the economies of countries making up the former Soviet Union and its neighbours (the so-called "economies in transition").

Extending the Kyoto Protocol beyond its so-called first commitment period, which ended in 2012, has proven difficult. This is not least because of the status of countries such as China and India; their emissions were not limited under the Kyoto Protocol but their economies (and hence emissions) have grown greatly since the original Kyoto Protocol was signed. As an interim measure, the UNFCCC conference which met in Doha in late 2012 agreed a second commitment period for the Kyoto Protocol up to 2020, when it is intended that a new global climate agreement will come into force. In this second commitment period, individual countries or groups of countries have made their own pledges for emission cuts. In the case of the European Union, the aim is to cut emissions by 20%, relative to 1990 levels, by 2020 (it was committed to an 8% reduction by 2012, relative to 1990 levels, in the first commitment period). Many countries have, so far, declined to take part in this second commitment period: those that have done so constitute about 14% of total world equivalent-CO_2 emissions.

In addition to these multilateral agreements, some individual countries, and even states within countries, have their own targets. In the United Kingdom, the Climate Change Act of 2008 commits the Government to reducing the UK's equivalent-CO_2 emissions by 80%, relative to 1990 levels, by 2050, with an interim target of at least 34% by 2020.

14.2 SCIENTIFIC BACKGROUND

A number of overarching concepts are developed here to focus the later discussion on the role of human activity in climate change. "Climate" is conveniently defined as the long-term average of a given parameter (*e.g.* near-surface temperature, precipitation, cloudiness) on scales from local, through regional, to global; the average could be for a given month, season or for the annual mean. A conventional averaging period for such averages is 30 years. "Climate change" can then be defined as some statistically significant deviation from the long-term average. Climate science distinguishes between the *detection* of climate change (*i.e.* a demonstration that a statistically significant change in climate has occurred) and the *attribution* of that change to a particular cause or causes. This distinction is sometimes lost in popular discussions.

To understand why climate change could occur, we start by considering the planetary energy balance.[9] The sun is the primary source of energy to drive the climate system. The global and annual average incoming solar radiation at the top of the atmosphere is about 340 W m^{-2}, and is concentrated in the ultraviolet, visible and near-infrared portions of the spectrum (wavelengths less than 4 μm). About 30% of this incoming radiation is reflected back to space (this reflected fraction is referred to as the planetary albedo), especially by clouds, but also by the surface, atmospheric gases and particulates; the remaining 70% (about 240 W m^{-2}) is absorbed by the earth and its atmosphere. The atmosphere is relatively transparent to solar radiation and about 165 W m^{-2} is absorbed by the surface itself.

For the planet to remain in energy balance it must emit 240 W m^{-2}, on a global and annual average, back to space – it does so mostly in the mid and far infrared (wavelengths from 4 to 200 μm). The surface emits *much* more than this (about 400 W m^{-2}). Clouds and atmospheric gases (notably water vapour and CO_2) absorb much of this radiation and emit infrared radiation both upwards and downwards. As temperature decreases with height in the troposphere (the bottom 10 km or so of the atmosphere), the amount of upward infrared radiation decreases gradually, reaching 240 W m^{-2} at the top of the atmosphere, where it balances the incoming solar radiation. Clouds and atmospheric gases emit about 340 W m^{-2} down to the surface; this energy, which is about double the amount of radiation that reaches the surface directly from the sun, is one measure of the greenhouse effect. This natural greenhouse effect is crucial for keeping the earth's surface as warm as it is, making life as we know it possible.

From the above numbers, it can be seen that the surface receives about 505 W m^{-2} either directly from the sun, or by emission by the atmosphere, and yet emits "only" 400 W m^{-2}. The surface energy balance is maintained mostly by evaporation (which cools the surface by 80–90 W m^{-2}), the remainder coming from direct loss of so-called "sensible heat" in the form of small-scale thermals.[9]

Global-scale climate change can be driven by processes that perturb the energy balance at the top of the atmosphere – such perturbations are called radiative (or sometimes climate) forcings. These could be natural – for example, changes in the energy emitted by the sun, or by particulates resulting from large volcanic eruptions which alter the planetary albedo. Most of the focus here will be on forcings due to human activity. For example, human activity has led to an enhancement of the greenhouse effect (by increasing the concentrations of greenhouse gases such as CO_2, methane and nitrous oxide) which tend to warm the planet (by convention, these are referred to as positive forcings); it has also led to increases in the concentrations of sulfate and other particulates, which increase the planetary albedo and hence tend to cool the planet (*i.e.* they cause a negative forcing).

Once a forcing has been imposed on the climate system, from either natural or human causes, the planet as a whole is no longer in energy balance. The climate system responds and seeks to restore the balance by, for example, surface warming in response to a positive forcing. It is often useful in discussing climate change to distinguish between the forcing, which initiates the climate change, and the eventual response.

To give one idealised example, imagine that the amount of CO_2 is suddenly doubled. Detailed calculations indicate that the radiative forcing would be about 3.6 W m^{-2}.[6] If the energy balance were to be restored simply by the surface and atmosphere warming up equally, the resulting surface temperature response would (eventually) be about 1.2 K. However, the very act of warming triggers a number of feedbacks which alter this response. A warmer atmosphere contains more water vapour; water vapour is a powerful greenhouse gas, so this leads to an enhancement of the response. Similarly, the extent of snow and ice would be expected to decrease in a warmer climate; as snow and ice contribute to the planetary albedo, a loss of snow and ice decreases the planetary albedo, leading to more absorption of solar radiation and a further enhancement of the response. The types and amounts of clouds can also respond to climate change, although there is only limited confidence in our ability to predict these responses. Clouds play a complicated role in both the solar and infrared energy budget, and so, depending on how the cloud properties change, the surface temperature response could either be reduced or amplified – this is one major uncertainty in climate science.

The current consensus based on results from both climate models and analyses of observations, is that a doubling of CO_2 amount would likely lead to a global average surface warming of between 2 and 4.5 K;[6] these values are sometimes used as measures of the *climate sensitivity*, although it can also conveniently be reported instead as the temperature change divided by the radiative forcing, which gives 0.6–1.25 K (W m^{-2})$^{-1}$.

Note that this 2 to 4.5 K warming assumes that the forcing acts for sufficient time for the earth and atmosphere to reach a new equilibrium. The timescale for the climate system to respond is determined largely by the time it takes to warm the upper layers (and to some extent the deep layers) of the ocean – a useful order of magnitude for an exponential timescale is about 10 years. Hence if a radiative forcing exists for a period shorter than a few decades, the climate system will only partially respond to it. One important example concerns volcanic eruptions. An individual powerful volcanic eruption can increase the amount of sulfate particulates in the lower stratosphere (*i.e.* at altitudes of around 10–20 km), causing a negative radiative forcing of several W m^{-2}. However, the atmospheric lifetime of such particulates is about a year, and so the climate system can only partially respond to their presence, and the climate response is muted.

In the above discussion, it was assumed that climate change has to be "forced" by some mechanism. It is important to appreciate that climate can change on many timescales in the absence of a forcing, due to interactions between different components of the climate system, such as the atmosphere, oceans and sea ice. This is sometimes referred to as "internal climate variability". An important part of the process of attributing climate change to a particular forcing or forcings is an assessment of whether the observed climate change is larger than could reasonably be expected to have occurred due to such internal variability. This variability is particular important on small scales (such as the British Isles) where relatively small changes in the strength, position and frequency of occurrence of weather systems can lead to marked and persistent changes in climate. It is also an important consideration at the global average scale.

14.3 OBSERVED CHANGES IN CLIMATE

Figure 14.1 shows a time series of global average (near) surface temperature over the past 160 years,[10] plotted as anomalies from the mean temperature between 1961 and 1990. This series combines measurements made at conventional meteorological stations on land, and sea–surface temperature measurements. There are many challenges in producing such time series.[6] The number and geographical distribution of observations has changed markedly over time, the techniques used to measure temperature have changed, particular for the oceans, and local conditions around individual stations can change, for example, due to urbanisation. Hence a number of adjustments

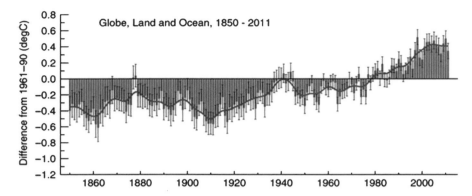

Figure 14.1   Global and annual average combined land–surface air and sea–surface temperature for the period 1850–2011, shown as anomalies from the 1961–1990 mean (°C).[10] The error bars show the 95% confidence interval of the annual averages. The dark solid line shows the annual values after smoothing with a 21-point binomial filter. (Crown Copyright).

have to be made to the raw data to produce a time series that is homogeneous. Nevertheless, confidence in the overall features shown in Figure 14.1 is high, as several research groups have performed their own independent analyses and produced closely similar results. The mid-19th century is normally taken as the starting point of such series, as it is judged that the quality of the measurements and their distribution was, by then, sufficiently good to perform a global analysis.

Figure 14.1 shows the temperature anomalies for individual years, the uncertainty in these anomalies and the running mean. Several features are of note. Over the 160 years there has been an overall warming of the planet by about 0.8 °C and such is the confidence in this result, that the IPCC AR4 labelled this warming as "unequivocal".[6] The warming has not been gradual, but appears to have been concentrated in two periods, from about 1910 to 1940, and 1980 to 2000. The past 15 years has been exceptionally warm, and even relatively cold years (such as 2011) would have seemed unprecedentedly warm prior to the 1990s. There is significant inter-annual variability, and many of the years that are warm relative to their near neighbours (such as 1998) are associated with marked changes in ocean currents in the tropical Pacific Ocean that occur every few years, which are part of the El Niño phenomenon.[6]

The period since 1998 is the subject of much discussion.[11] The period is sometimes referred to as a "hiatus" in the overall warming trend. However, it is not clear that this period is especially unusual, as the period of warming since 1980 is characterised by several periods of flattening. Further, given the uncertainties in the measurements (which is indicated on Figure 14.1), it is not easy to assert how much flattening has actually occurred; this uncertainty means that in some analyses 2005 and 2010 are warmer than 1998, which is the warmest year in the particular record shown in Figure 14.1. Although attribution of temperature changes to specific causes is the subject of a later section, there has been much discussion as to whether this apparent flattening is the result of a particular human forcing (such as increased sulfate particles resulting from emissions in East Asia, with their associated negative forcing); natural forcing (such as changes in the sun's output); or simply internal climate variability related to, for example, variations in the rate at which the deep ocean takes up the additional energy associated with the positive radiative forcing due to human activity (see section 14.4).

The global mean temperature record (see Figure 14.1) is unusual in both the overall quality of the available temperature data, and the representativeness of the available measurements. This makes the construction of such a long-period global data series feasible. For other climate parameters, either coherent observations have not been available for such a long period (especially those from satellites, which are, at best, available from the late 1970s) or the quality and/or coverage of data is not adequate. As an example, direct rainfall measurements are essentially absent over the ocean. Over land, each individual rainfall station is generally representative of a smaller area than a temperature measurement, and there has been less consistency in design and exposure of rain gauges than temperature measurements. In addition, rainfall is generally more variable from year to year, and hence it is less easy to discern long-term trends.

Nevertheless, the overall picture presented by these additional data is consistent with expectations of a warming world. Measurements from satellites and balloon-borne instruments show a warming of the troposphere and an increase in water vapour amount.[6] Variability is an important characteristic of the climate and, in a warming world, it would be expected that the incidence of extreme high temperature events would exceed those of extreme cold temperatures. Analyses for the United States indicate that in recent years, warm extremes are occurring at double the rate of cold extremes.[12] A global analysis of temperature anomalies for the June–July–August period has shown that warm anomalies that would have been extremely rare between 1950 and 1980 (*i.e.* they are more than 3 standard deviations from the mean, and hence would be expected to cover less than 0.2% of the globe) covered between 4 and 13% of the globe during each year between 2006 and 2011.[13] The occurrence of cold anomalies was found to be correspondingly less common.

One of the most striking trends is the decrease in Arctic sea ice extent.[14] Arctic sea ice extent goes through a marked annual cycle, and is at its greatest extent at the end of winter in March, and at its smallest extent in September. Trends in sea ice extent are most marked in September (see Figure 14.2) where there is only about 50% of what was observed (from satellites) in the late 1970's. The sea ice that remains is also believed to be thinner, and thus more prone to summertime melting. Interestingly, the Antarctic sea ice extent has overall increased slightly, although with marked regional variations in this trend. Sea ice is moved by the winds, and generally drifts north away from the Antarctic continent. The changes in extent are believed to be due to changes in the strengths of winds around Antarctica.[15]

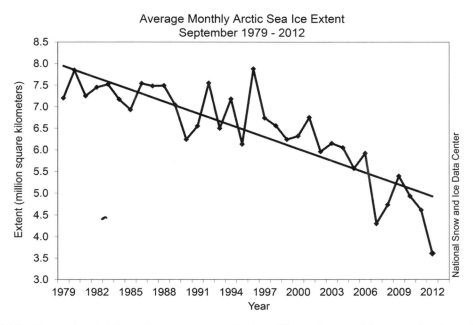

Figure 14.2 September Arctic sea ice extent anomalies in millions of square kilometres for the period 1979 to 2012. Individual years are shown by diamonds, with the linear trend line also shown. (From the US National Snow and Ice Data Center; http://nsidc.org/).

Rainfall trends are much more difficult to discern.[6] While long-term increases have been found in some regions, they have not been found in other regions. Since intense rainfall events feed off water vapour already in the atmosphere, the increase in the amount of that vapour leads to an expectation of more heavy precipitation events. There is some evidence of such an increase in some regions, but adequate data are not available in many regions to assess changes.[16] Similarly, there is some evidence for an increase in the intensity of tropical cyclones in some regions, but not in the overall number, but again, data reliability is a serious issue.[16]

Sea-level rise is one further consequence of climate change, with a potentially large impact on coastal communities. As with other indicators, there are many challenges in producing a homogeneous time series from the available data, and there has been a transition from the use of *in situ* tide gauges to the use of satellite altimeter data, which provide an essentially global view of the world's oceans. After accounting for the continued adjustments in sea level as a result of "recovery" from the last ice age, it is believed that global sea-level increased by about 210 mm between 1880 and 2009,[17] and that there has been a significant acceleration of the trend – between 1900 and 2009 the average trend was about 1.7 mm year^{-1} but from 1993 it has been about 2.8 mm year^{-1}. Over the period since the 1970s, when better quality data are available, it is believed that the rise has been due to a roughly equal mixture of thermal expansion of the oceans, and the melting of glaciers and ice sheets (including Greenland and Antarctica).[18]

14.4 CHANGES IN ATMOSPHERIC COMPOSITION AND RADIATIVE FORCING

The concentrations of several greenhouse gases have increased, and there is compelling evidence that these changes are largely due to emissions from human activity, including the burning of fossil fuels and changes in land use and agriculture. Many of the greenhouse gases are relatively long-lived once in the atmosphere (10s to 100s of years). This means that once emitted, they can be spread relatively evenly over the globe by the winds – therefore, a relatively sparse network of measurements can quite accurately characterise the global concentration changes. For many gases, the atmospheric lifetime can be well characterised by a simple exponential timescale,[6] but CO_2 is different. Because of the nature of interchanges in carbon between the atmosphere, oceans and land, if a sudden pulse of CO_2 was emitted into the atmosphere, part of the CO_2 increase would get taken up by vegetation very quickly (timescales of a few years); but because of changes in the chemistry of the upper oceans, and other mechanisms, even after a 1000 years, it is expected that 20% of this pulse would effectively remain in the atmosphere, and maybe 10% would persist for many thousands of years afterwards.[7]

CO_2 has been directly measured at Mauna Loa (Hawaii) since the late-1950s (see Figure 14.3); a more global network has been in place since the 1970s. Longer period changes can be deduced from air trapped in bubbles of ice in the large continental ice caps on Antarctica and Greenland[6] – cores of ice drilled from these ice caps yield a history of greenhouse gas concentration stretching back more than half a million years in the case of CO_2. Between 1750 and 2010, CO_2 mixing ratios have increased from about 285 ppm (parts per million) to 389 ppm; methane increased from 715 ppb (parts per billion) to 1799 ppb; nitrous oxide increased from 270 ppb to 324 ppb (all 2010 values from http://data.giss.nasa.gov/modelforce/ghgases/). Changes in the concentrations of some CFCs (which are powerful greenhouse gases, as well as gases that deplete stratospheric ozone) are discussed in Chapter 9. The radiative forcing due to the changes in these long-lived greenhouse gases (LLGHGs) are relatively straightforward to calculate, and the forcing due to changes between 1750 and 2010 is about 2.8 W m^{-2} (see Figure 14.4).[6,19]

Several other important radiative forcings arise from human activity, due to emissions of short-lived species, with atmospheric lifetimes of weeks to months.[6] These include pollutants such as

Mauna Loa Observatory, Hawaii
Monthly Average Carbon Dioxide Concentration
Data from Scripps CO_2 Program   Last updated March 2013

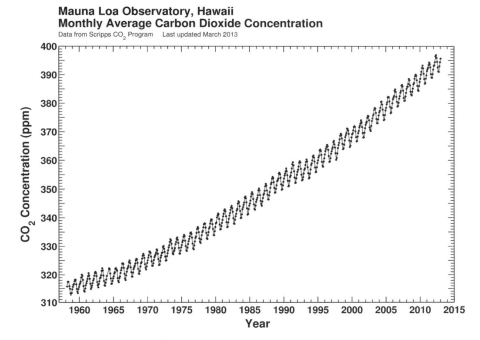

Figure 14.3 Monthly average atmospheric CO_2 concentration *versus* time at Mauna Loa Observatory, Hawaii (20°N, 156°W) where CO_2 concentration is in parts per million. (Data from the Scripps CO_2 Program; http://scrippsco2.ucsd.edu).

oxides of nitrogen, carbon monoxide and non-methane hydrocarbons. These gases increase tropospheric ozone concentrations (as does methane), ozone being an important greenhouse gas. The abundance of tropospheric particulates has also increased due to emissions of gases that lead to particulate formation (such as sulfur dioxide and organic carbon) and by direct emission of particles (for example, so-called black carbon or soot). Since the emissions of such species are very regional (*e.g.* from industrialised regions or regions of biomass burning), and because their atmospheric lifetime is short compared to the time required by the winds to spread them globally, their distribution is much more inhomogeneous than the long-lived greenhouse gases, and hence local measurements are not globally representative. Furthermore, the observational techniques employed, especially for past concentrations, are generally less reliable, and there is no equivalent to the ice-core observations of the long-lived greenhouse gases. Consequently, there is much less robust knowledge of how their concentrations have changed and more heavy reliance on results from numerical models.

For particulates, the situation is more complex still.[6] Most particulates in the atmosphere are mixtures of various components which have contrasting impacts on radiative forcing. Sulfate particles, on their own, cause a negative radiative forcing by reflecting solar radiation to space, while black carbon particles on their own, cause a positive radiative forcing, as they absorb solar radiation that would otherwise be reflected back to space (the "direct aerosol effect" shown in Figure 14.4). In addition, these same particulates play a fundamental role in the formation of clouds, and changes in their concentrations are believed to increase the reflectivity of clouds, hence causing an additional negative radiative forcing – these forcings are labelled as the "cloud albedo effect" and "cloud lifetime effect" indicated in Figure 14.4, where they can be seen to be particularly uncertain.

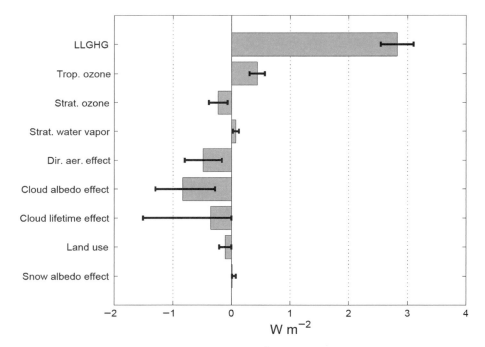

Figure 14.4 Estimates of the radiative forcing (in W m^{-2}) in 2010 relative to 1850 due to human activity.[19] Where LLGHG: long-lived greenhouse gases (including CO_2, methane, nitrous oxide and a number of fluorocarbons); trop. ozone: tropospheric ozone; strat. ozone: stratospheric ozone; strat. water vapour: stratospheric water vapour; dir. aer. effect: the direct impact of particulates (aerosols); cloud albedo effect: the effect of particulates on the albedo of clouds; cloud lifetime effect: the effect of particulates on the thickness of clouds; land use: the effect of land use changes on the surface albedo; and snow albedo effect: the effect of the deposition of black carbon particulates on snow surfaces. The error bars represent the 90% confidence interval.

From Figure 14.4, the net forcing due to human activity since pre-industrial times is estimated to be around 1.4 W m^{-2} with an uncertainty of about 0.5 Wm^{-2}, indicating the expectation that human activity will have caused a warming of the climate system.[18]

Radiative forcing is a metric that looks at the effect of changes over a given period of time (for example between 1750 and 2010) and includes the effect of all emissions that have occurred during that period. For policymaking, and other purposes, it is useful to have some measure of the impact of different present-day emissions on future climate. Such a measure needs to take into account both the radiative forcing due to these emissions and the lifetimes of the emitted substance in the atmosphere.

A particular example concerns the Kyoto Protocol, which covers emissions of a range of different greenhouse gases (see section 14.1) and requires some metric to place the emissions of different gases on to a "CO_2-equivalent" scale. There is no unique way of defining such a metric, and a number of possibilities are available, which may be driven by user-choice and the needs of the climate policy that is being pursued.[6] The Kyoto Protocol has chosen to use one particular metric, the Global Warming Potential (GWP). The GWP considers the emission of a pulse of a greenhouse gas, which then decays with time as it is removed from the atmosphere. The time-integrated radiative forcing is then calculated for a given period (the Kyoto Protocol chose 100 years); this value is then divided by the time-integrated radiative forcing due to an equal mass emission of CO_2. If emissions of a given gas (*e.g.* in kg year^{-1}) are multiplied by the GWP for that gas, this yields an estimate of the CO_2-equivalent emission of that gas. Hence, for example, during the first phase of the Kyoto Protocol, a 1 kg emission of methane was deemed equivalent to an emission of 21 kg of CO_2 (http://unfccc.int/ghg_data/items/3825.php). For some heavily-fluorinated greenhouse gases, a 1 kg emission was deemed equivalent to many thousand kg of CO_2. GWPs, and other metrics, are subject to frequent updates, as a result of changes in understanding.[6]

14.5 MODELLING CLIMATE CHANGE

While radiative forcing indicates the tendency of climate change, it does not directly indicate how much temperature (or any other climate parameter) will vary. To obtain such an estimate it is necessary to use some kind of "climate model" although this is a rather vague term. There is a vast hierarchy of climate models, all trying to incorporate the physics of the climate system.[8] These can range from very simplified "back-of-the-envelope" considerations, which can be used to estimate the change in global and annual mean surface temperature, through slightly more complicated models that can be incorporated in basic computer spreadsheets, all the way to state-of-the-art models that require some of the largest computer resources on the planet to produce their results.

One example of a back-of-the-envelope calculation is to take the estimated radiative forcing of 1.4 W m^{-2} due to human activity (see section 14.4) and multiply this by a mid-range value of the climate sensitivity (say, 0.9 K (W m^{-2})$^{-1}$, from section 14.2), which yields an eventual warming (given sufficient time for the oceans to warm, and assuming the radiative forcing remains fixed) of about 1.2 K – a value that is not at odds with the observed warming shown in Figure 14.1. (It should be stressed that such back-of-the-envelope calculations are still reliant on knowing values for the radiative forcing and climate sensitivity, which are themselves derived from detailed calculations.)

The state-of-the-art models represent climate variables (*e.g.* wind, temperature, humidity, cloudiness, *etc.*) on a three-dimensional grid covering the whole globe and often stretching from the surface to above the stratopause at 50 km, with typical horizontal resolutions of 100 km (a number that has decreased with time, as computer power has increased). The winds are simulated using numerical representations of Newton's Laws; the laws of thermodynamics are used to simulate the temperature; and the principles of conservation of water are used in modelling humidity and cloudiness. As well as simulating the atmospheric circulation, the models will include sophisticated representations of the oceans, sea ice, the land–surface, and chemical, and increasingly biological, processes that interact with climate. The top-of-the-range climate models are so encompassing in their ambition that they are now sometimes referred to as "earth system models".

Sophisticated as they are, there are many processes in the earth system (such as cloud droplet growth) that occur on such small scales that they cannot be explicitly represented by the models; and there are others (such as, for example, the temperature and humidity dependence of forest fires, which is a process beginning to be incorporated in some models) for which the governing equations are not fully understood. Such processes have to "parameterised" in terms of the model variables that are explicitly represented. The uncertainties in these parameterizations are one reason why climate model predictions are uncertain. The scale of this uncertainty is captured by comparing results from climate models developed in different institutes around the world – these climate models may differ in for example, the techniques used to represent the physical laws on a computer, their horizontal and vertical resolution, and the details of the processes they attempt to capture in the parameterisations. It is also possible to capture some of the uncertainty by using slightly different versions of the same climate model.

Climate models can be used in various ways – two are described here in more detail. One is the attempt to simulate past climate, with the results compared with observations. The second is to use them to simulate future climate using some assumptions as to how emissions due to human activity will change in the future. A third is to use the models to do more "what-if" experiments (*e.g.* "what if we suddenly double the amount of CO_2 in the atmosphere" or "what if we take all the ozone from the stratosphere?") which can be useful in building up a more generic understanding of how the climate system works.

It is important to appreciate that our understanding of the effect of, for example, increases in CO_2 on climate change does not come from a simple correlation of the observed changes in CO_2 with the observed climate change, as appears to be sometimes thought. Rather, it arises from simulations, using models of greater or lesser simplicity, of the impact of CO_2 changes on the fluxes of energy (*i.e.* radiative forcing) and the subsequent impact of those energy changes on surface temperature and, in the more complex models, on other climate parameters. Hence the models represent our best understanding of the physics of the climate system.

14.6 ATTRIBUTION OF CLIMATE CHANGE OVER PAST 150 YEARS

One of the most important pieces of evidence that links human activity to climate change arises from experiments with models that attempt to simulate the changes in past climate, first using just natural forcing mechanisms, and then with natural and human forcings. Figure 14.5 shows the results presented in IPCC AR4,[6] for the global mean changes in surface temperature. The black solid line shows a smoothed version of the observed temperatures. The dark grey band shows results from a range of climate models (the spread is one indication of the uncertainty in the simulations) using only the natural forcings, while the lighter grey band shows the effects when natural and human forcings are included. Up until the 1950s, the curves are pretty much indistinguishable but increasingly the "human plus natural" curve departs from the "natural only" curve, and tracks well the actual changes in temperature, unlike the natural curve which shows little systematic change. The model results can also be examined for each continent individually – for those continents for which we have good long-term climate records, the results are broadly the same, indicating that the observed temperature changes are unlikely to be due to natural forcings alone.[6]

The output from the sophisticated models has also been used to provide additional evidence for a human impact by looking at more subtle changes in the vertical and horizontal patterns of temperature change. Further, when these models are used to simulate temperature variations in the

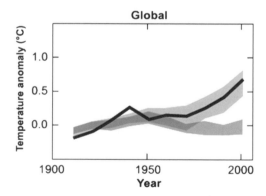

Figure 14.5 Comparison of observed global-mean surface temperature anomaly with results simulated by climate models using natural and anthropogenic forcings.[6] Decadal averages of observations are shown for the period 1906 to 2005 (black line) plotted against the centre of the decade and relative to the corresponding average for 1901–1950. The darker shaded band shows the 5–95% range for simulations from climate models using only the natural forcings due to solar activity and volcanoes. The lighter shaded band shows the 5–95% range for simulations from climate models using both natural and anthropogenic forcings.

absence of *any* forcing, they provide evidence of the size of the internal climate variability. These simulations indicate that it is very unlikely that the observed temperature increases, both globally and on continental-scales, could have originated without some forcing mechanism.[6]

14.7 MODELLING FUTURE CLIMATE CHANGE

Projection of future climate change is difficult, not only because of the remaining uncertainties in the underlying science of the climate system but also because we can have no clear idea how human emissions will change over the next, say, hundred years. Imagine trying to make an accurate prediction of emissions due to human activity in 2013 even 50 years ago. Instead, climate scientists use a range of possible scenarios of future emissions, taking into account factors such as population growth, economic development (including the degree to which the developing world converges on the developed world's use of fossil fuels) and technological and political factors. In the *Fifth Assessment Report* of the IPCC, a range of so-called "Representative Concentration Pathways" (RCPs) have been constructed that lead to a given radiative forcing in the future,[20] each of which could result from different combinations of future factors. In IPCC AR4,[6] so-called SRES (Special Report on Emission Scenarios) emissions were used to give a range of possible future emissions.

These RCPs are now being widely used in climate models of varying complexities to investigate possible future climate changes. The projected global mean warmings by 2100, relative to pre-industrial, range from 1.5 K to 4.5 K (hence, even in the lowest case, this constitutes almost double the warming than has been observed to date).[20,21] The RCPs have been used in an extensive inter-comparison of more complex climate models.[21] Such inter-comparisons are valuable for not only understanding the range of possible future climate change, but also for understanding which aspects of climate change are robust amongst the models and which are not, and which changes are statistically significant and which are not.

A summary of these inter-comparison calculations shows that predictions of regional surface temperature changes are relatively robust across different climate models;[21] there is a consistent pattern of greater warming over land compared to the oceans, with greatest warming at high latitudes (exceeding 11 K in winter time in the Arctic by 2100, compared to the global mean of 4.5 K, for the highest emission scenarios). In regions of the ocean (notably the north Atlantic and in the southern oceans) the warming is predicted to be much slower, as these are areas where heat is mixed by the oceans to greater depths than elsewhere. By contrast, predictions of precipitation change are significantly less robust, in terms of inter-model agreement (particularly in the tropics). It also takes much larger amounts of warming for the predicted changes to become significantly different from the present-day rainfall, because rainfall is inherently more variable than tempera-ture. For example, even a 2 K global average warming relative to present-day conditions would lead to statistically significant changes in rainfall over only about 50% of land areas. The general flavour of the predicted precipitation changes is that dry regions will get drier and wet regions will get wetter,[6,21] but this is by no means a general rule at all locations and in all models.

14.8 CONCLUSIONS

The development of our understanding of the role of human activity in climate change can be illustrated by the changes in the "attribution" statements that have been made in the main sum-maries of the successive IPCC assessments. In its *First (1990) Assessment Report*,[21] IPCC was equivocal, stating that "the size of [the] warming is broadly consistent with predictions of climate models [which simulate the impact of human activity], but it is also of the same magnitude as natural climate variability". During the 1990s and 2000s, the increased length of the temperature record (and in particular the marked warming that was occurring, as shown in Figure 14.1), and improvements in both modelling and the statistical techniques used to interpret model output, led to an increasingly confident set of statements. The *Second Assessment Report* (1995) stated that "The balance of evidence suggests that there is a discernible human influence on climate change".[23] The *Third Assessment Report* (2001) concluded that "most of the observed warming over the last 50 years is likely to have been due to the increase in greenhouse gas concentrations".[24] The *Fourth Assessment Report* (2007) used a modified version of this statement, saying "most of the observed increase in globally averaged temperatures since the mid-20th century is very likely due to the observed increase in anthropogenic greenhouse gas concentrations".[6] The IPCC reports use a "lexicon" of terminology that it can use to signal the level of confidence – the change in wording from "likely" to "very likely" between the *Third* and *Fourth Assessment Reports* was deliberate and significant.

At the time of writing, we await the conclusions from the *Fifth Assessment Report*, but I think it unlikely that it will be any less confident than the *Fourth Assessment Report* about the role of human activity on the observed climate change over recent decades. This is not to say, though, that there are not many outstanding and challenging issues in the science of climate change. There are. Some are chronic ones, such as understanding the size of cloud feedbacks; others are more recent, such as understanding how the natural cycle of carbon between the atmosphere, land and oceans is disturbed by climate change; and others are perplexing, such as why different state-of-the-art climate models produce such a range of predictions of some aspects of regional climate change (such as the change in the strength of the Indian monsoon) over the coming decades.

There are, of course, divergent views of the nature of the climate change, and this chapter has sought to present the mainstream view of climate science, as encapsulated by the IPCC assessments, which are written by teams of active climate scientists. Equally, there are divergent views on how society should respond to the prospect of climate change, which range from those who believe it is easier/cheaper to let climate change happen and adapt to it, to those who believe we should make dramatic changes in the way we live and the way we generate and use energy. If there is a mainstream view on how humanity should respond to climate change, it is probably encapsulated in the Copenhagen Accord (http://unfccc.int/meetings/copenhagen_dec_2009/items/5262.php) which was agreed by a majority of parties at the 2009 Conference of the Parties to the UNFCCC. This stated that "We agree that deep cuts in global emissions are required according to science ... with a view to reduce global emissions so as to hold the increase in global temperature below 2 degrees Celsius [above pre-industrial levels], and take action to meet this objective consistent with science and on the basis of equity".

REFERENCES

1. S. Manabe and R. T. Wetherald, *J. Atmos. Sci.*, 1967, **24**, 241.
2. W. H. Matthews, W. W. Kellogg and G. D. Robinson, *Man's Impact on Climate*, Massachusetts Institute of Technology Press, Cambridge, MA, 1971.
3. *Inadvertent Climate Modification*, Massachusetts Institute of Technology Press, Cambridge, MA, 1971.
4. J. D. Charney, *et al.*, *Carbon Dioxide and Climate: A Scientific Assessment*, National Academy of Sciences, Washington DC, 1979.
5. Atmospheric Ozone 1985: Global Ozone Research and Monitoring Project – Report No. 16, World Meteorological Organization, Geneva, 1986.
6. *Climate Change 2007: The Physical Science Basis*, Working Group I Contribution to the Fourth Assessment Report of the Intergovernmental Panel on Climate Change, Cambridge University Press, Cambridge, 2007.
7. D. Archer, *Global Warming: Understanding the Forecast*, John Wiley and Sons, 2nd edn., 2011.

8. J. D. Neelin, *Climate Change and Climate Modeling*, Cambridge University Press, Cambridge, 2010.
9. G. L. Stephens, J. Li, M. Wild, C. A. Clayson, N. Loeb, S. Kato, T. L'Ecuyer, P. W. Stackhouse Jr, M. Lebsock and T. Andrews, *Nature Geosci.*, 2012, **5**, 691.
10. J. Kennedy, C. Morice and D. Parker, *Weather*, 2012, **67**, 212.
11. G. A. Meehl, J. M. Arblaster, J. T. Fasullo, X. Hu and K. E. Trenberth, *Nat. Clim. Change*, 2011, **1**, 360.
12. K. E. Trenberth, *Clim. Change*, 2012, **115**, 283.
13. J. E. Hansen, M. Sato and R. Ruedy, *Proc. Natl. Acad. Sci. U. S. A.*, 2012, **109**, 14726.
14. J. C. Stroeve, M. C. Serreze, M. M. Holland, J. E. Kay, J. Malanik and A. P. Barrett, *Clim. Change*, 2012, **110**, 1005.
15. P. R. Holland and R. Kwok, *Nature Geosci.*, 2012, **5**, 872.
16. *Managing the Risks of Extreme Events and Disasters to Advance Climate Change Adaptation*: *Special Report of the Intergovernmental Panel on Climate Change*, A Special Report of Working Groups, Cambridge University Press, Cambridge, 2012.
17. J. A. Church and N. J. White, *Surv. Geophys.*, 2011, **32**, 585.
18. J. M. Gregory, N. J. White, J. A. Church, M. F. P. Bierkens, J. E. Box, M. R. Van den Broeke, J. G. Cogley, X. Fettweis, E. Hanna, P. Huybrechts, L. F. Konikow, P. W. Leclercq, B. Marzeion, J. Oerlemans, M. E. Tamisiea, Y. Wada, L. M. Wake and R. S. W. Van de Wal, *J. Clim.*, 2013, **16**, 4476.
19. R. B. Skeie, T. K. Berntsen, G. Myhre, K. Tanaka, M. M. Kvalevåg and C. R. Hoyle, *Atmos. Chem. Phys.*, 2011, **11**, 11827.
20. M. Meinshausen, S. J. Smith, K. Calvin, J. S. Daniel, M. L. T. Kainuma, J.–F. Lamarque, K. Matsumoto, S. A. Montzka, S. C. B. Raper, K. Riahi, A. Thompson, G. J. M. Velders and D. P. P. van Vuuren, *Clim. Change*, 2011, **109**, 213.
21. R. Knutti and J. Sedláček, *Nature Clim. Change*, 2013, **3**, 369.
22. *Climate Change: The IPCC Scientific Assessment*, Cambridge University Press, Cambridge, 1990.
23. *Climate Change 1995: The Science of Climate Change*, Working Group I Contribution to the Second Assessment Report of the Intergovernmental Panel on Climate Change, Cambridge University Press, Cambridge, 1995.
24. *Climate Change 2001: The Scientific Basis*, Working Group I Contribution to the Third Assessment Report of the Intergovernmental Panel on Climate Change, Cambridge University Press, Cambridge, 2001.

CHAPTER 15

Soil Pollution and Risk Assessment

CHRIS D. COLLINS[†]

Soil Research Centre, Dept. Geography and Environmental Science, University of Reading, Reading, RG6 6DW
Email: c.d.collins@reading.ac.uk

15.1 INTRODUCTION

Soil is an essential component of terrestrial ecosystems because the growth of plants and biogeochemical cycling of nutrients depend upon it. Soil can be viewed as natural capital which supplies a number of ecosystem services as shown in Tables 15.1(a) and (b) (adapted from ref. 1–3).

Of the total area of the world's land mass (13.07×10^9 ha): 11.3% is cultivated for crops; permanent grazing occupies 24.6%; forest and woodland 34.1%; and 'other land' including urban/industry and roads accounts for 31%.[4] From a resource perspective, soil is vitally important for the production of food and fibre crops and timber and it is therefore essential that the total productive capacity of the world's soils is not impaired. Additionally soil is a major store of carbon containing *over 2700 Gt*, larger than the amount of carbon retained in biomass and the atmosphere, and as such has a critical role to play in future climate change.[5] Pollution, along with other types of degradation, such as erosion, and the continuing spread of urbanization poses a threat to the sustainability of soil resources and hence its ability to operate in biogeochemical cycles and support the human functions described above. Soil pollution can have a direct and negative impact on human health when potentially toxic substances move through the food chain or reach groundwater subsequently used for potable supplies. There is also the potential for the direct ingestion of soils either inadvertently through work and play activities or deliberately as a consequence of the condition known as 'pica'.[‡]

[†]This chapter is based upon an earlier contribution from Brian Alloway which is gratefully acknowledged.
[‡]Pica – pattern of eating non-food materials such as soil or paper.

Pollution: Causes, Effects and Control, 5th Edition
Edited by R M Harrison

Table 15.1a Soil natural capital.

MASS	
Solid	Inorganic material: (i) mineral stock and (ii) nutrient stock
	Organic material: (i) OM/carbon stock and (ii) organisms
Liquid	Soil water content
Gas	Soil air
ENERGY	
Thermal Energy	Soil temperature
Biomass Energy	Soil biomass
ORGANISATION	
Physico-chemical structure	Soil physico-chemical organisation, soil structure
Biotic structure	Biological population organisation, food webs and biodiversity
Spatio-temporal structure	Connectivity, patches and gradients

Table 15.1b Soil ecosystem services.

SUPPORTING
Physical stability and support for plants
Renewal, retention and delivery of nutrients for plants
Habitat and gene pool

REGULATING
Regulation of major elemental cycles
Buffering, filtering and moderation of the hydrological cycle
Disposal of wastes and dead organic matter

PROVISIONING
Building material

CULTURAL
Heritage sites, archaeological preserver of artifacts
Spiritual value, religious sites and burial grounds

In comparison with air and water, soil is more variable and complex and as such it is difficult to define. The Soil Science Society of America uses the following definition for soil:

'The unconsolidated mineral or organic matter on the surface of the Earth that has been subjected to and shows effects of genetic and environmental factors of: climate (including water and temperature effects), and macro- and microorganisms, conditioned by relief, acting on parent material over a period of time. A product-soil differs from the material from which it is derived in many physical, chemical, biological, and morphological properties and characteristics.'

Despite this variable composition, soils function as a sink for pollutants, a filter which retards the passage of chemicals to the groundwater and a bioreactor in which many organic pollutants can be decomposed. It is generally accepted that most of the soil in industrialised regions of the world is polluted (or contaminated), at least to a slight extent and determining the impacts of this background pollution is a challenge. However, in many cases the relatively small amounts of pollutants involved may not have a significant effect on either soil fertility or animal and human health. It should also be noted that in certain regions of the world *e.g.* Cornwall, UK and Bangladesh, high concentrations of pollutants occur in their soils as a consequence of their geology.

More severe 'chemical pollution' which poses a greater hazard has been estimated by a Global Assessment of Soil Degradation (GLASOD) to affect a total of 21.8×10^6 ha of land in Europe, Asia, Africa and Central America.[6] Realistic estimates of areas affected by soil pollution are difficult owing to unreliable official figures and inadequate data for many regions of the world. Industrially contaminated land tends to contain higher concentrations and a wider diversity of pollutants than other sources of pollution. It has been estimated that there are between 50 000 and 100 000 contaminated sites in the United Kingdom which are estimated to occupy around 300 000 ha[7] and would cost approximately £20–40 billion to remediate.[8] In the rapidly developing economies of China and India, soil pollution has worsened according to incomplete statistics. About 10 million hectares of arable land in China has been polluted with potentially toxic elements (PTEs), and the mining regions of India have been shown to have elevated levels.[9] Industrial sites, which act as point sources, are generally very heterogeneous; both with regard to the distribution of pollutants and also to the properties of the soil materials. Industrial soils are usually characterized by low organic carbon and high pH which will influence the behaviour of the pollutants within them (see section 15.3).[10] Depending on the pollutant and the type of spill they can penetrate the whole soil profile and contaminate the underlying groundwater or remain in the surface layers.

In contrast, atmospherically deposited pollutants tend to have a more even distribution at the field scale with gradual changes in concentrations, which tend to decrease with distance from the source, for example there are significant differences between rural and urban concentrations of PAH in UK soils.[11] The upper horizons of the soil are contaminated to the greatest extent by atmospheric deposition and a number of persistent organic pollutants (or POPs) contaminate soil *via* this route.[12] For further discussion of this topic see POPs Chapter 18.

All contamination/pollution situations comprise the following components: (i) a source of pollutant; (ii) the pollutant itself; (iii) a transport mechanism by which the pollutant is dispersed; and (iv) the receptor where the transport phase terminates. Within regulatory frameworks it is only if these four components are connected that there is a risk to the receptor, otherwise the pollutant is just a hazard. Within the UK this is referred to as a source–pathway–receptor (S–P–R) linkage and this has to be present for land to be classed as contaminated.[13] When exploring those factors where we can alleviate or remediate soil pollution, the contractor can decide whether to remove the pollutant or block the pathway in order to break the S–P–R linkage.

Providing the source–pathway–receptor linkage is maintained, a suite of negative impacts can result from polluted soils. In many cases these are specific to the individual contaminant. For example, BaP (a PAH) is a known Class 1 human carcinogen and has similar impacts on other mammals. Other PAHs have known negative impacts on earthworms which are known as sentinel organisms within the soil. In Bangladesh, a region with high arsenic in the geology (source), the installation of wells in shallow aquifers to provide clean drinking and irrigation water (pathway) resulted in significant impacts on human health (receptor). It has been referred to as the 'largest mass poisoning of a population in history' according to the World Health Organization and up to 30 000 people have been diagnosed with arsenicosis.[45]

Although the S–P–R is a very simple conceptual model it does provide a useful basis to focus our energies in the context of soil pollution and provides the structure for this chapter.

15.2 SOURCES OF SOIL POLLUTANTS

15.2.1 Potentially Toxic Elements

Many workers have previously used the term 'heavy metals', but this is not useful because many of the contaminants of concern are not 'heavy' *e.g.* atomic weight of chromium = 52, unlike lead = 207. These are also considered to be 'trace elements' because they collectively comprise < 1% of the rocks in the earth's crust. All trace elements are toxic to living organisms at excessive concentrations, but some are essential for the normal healthy growth and reproduction by either plants or animals at low but critical concentrations. This was first noted by Paracelsus 500 years ago:

'All things are poison, and nothing is without poison. The dose alone makes a thing not poison'.

Deficiencies of 'essential trace elements' or 'micronutrients' can lead to disease and even death of plants and animals. The essential trace elements include: Co (for bacteria and animals), Cr (animals), Cu (plants and animals), Mn (plants and animals), Mo (plants), Ni (plants), Se (animals) and Zn (plants and animals). In addition, B (plants), Cl (plants), Fe (plants and animals), I (animals) and Si (plants and animals – probably) are also essential trace elements.

Other elements, including: Ag, As, Ba, Cd, Hg, Tl, Pb, Sb, have no known essential function and, like the essential trace elements, cause toxicity above a certain tolerance level. The most important PTEs with regard to potential hazards and risks in contaminated soils are: As, Cd, Cu, Cr, Hg, Pb and Zn.

15.2.1.1 Sources of PTEs

In 2002 the Department of Environment, Food and Rural Affairs (Defra) in the UK produced a guide to those industries where specific toxins might be expected at legacy sites;[14] the principal present sources for PTEs are summarized in Table 15.2. Atmospheric deposition would result from a range of industries and may be high at the national level, but it would be low at the field scale when compared with point sources *e.g.* sewage sludge. A number of the other sources relate to agricultural operations and this is because of their direct application to land. Consequently guidelines have been developed for the application rates of sewage sludges and composts.

Source of PTEs include

(i) *Metalliferous Mining.* This is an important source of contamination by a wide range of metals, especially As, Cd, Cu, Ni, Pb and Zn, because ore bodies generally include a range of minerals containing both economically exploitable metals (in ore minerals) and uneconomic elements (in gangue minerals). Most mine sites are contaminated with several metals and accompanying elements (*e.g.* sulfur). Wind-blown tailings (finely ground particles of ore) and ions in solution from the weathering of ore minerals in heaps of tailings tend to be the major sources of pollution from abandoned metalliferous mine sites.

(ii) *Metal Smelting.* This is the process of producing metals from mined ores and so can be a source of many different metals. These pollutants are mainly transported in air and can be in the form of fine particles of ore, aerosol sized particles of oxides (especially important in the case of the more volatile elements, such as As, Cd, Pb and Tl) and gases (SO_2). In some cases, pollution is directly traceable in soil up to 40 km downwind of smelters. The slags from these industries can also cause a significant pollution source.

(iii) *Metal-using Industries.* These can be a source of metals in gaseous/particulate emissions to the atmosphere, effluents to drains and solid wastes. These include: the electronics industry where metals are used in semiconductors, contacts, circuits, solders and batteries; plating (Cd, Ni, Pb, Hg, Se and Sb); pigments and paints (Pb, Cr, As, Sb, Se, Mo, Cd, Co,

Table 15.2 Sources of PTEs into soil (g ha^{-1} yr^{-1}). (Adapted from ref. 15).

Source	Zn	Cu	Pb	Cd	Cr	As	Hg
Atmospheric deposition	2485	638	611	22	84	35	11
Livestock manures	1666	541	44	4	32	15	<1
Sewage sludge	385	271	106	2	78	3	1
Industrial 'wastes'	65	25	7	1	6	nd	<1
Inorganic fertilisers	199	67	13	9	94	6	<1
Composts	52	13	28	<1	6	nd	<1
Lead shot	–	–	18000	–	–	–	–

nd = no data

Ba and Zn); the plastics industry (polymer stabilizers such as Cd, Zn, Sn and Pb); and the chemical industry which uses metals as catalysts and electrodes including Hg, Pt, Ru, Mo, Ni, Sm, Sb, Pd and Os.

(iv) *Waste Disposal.* Municipal solid waste, special wastes and hazardous wastes from many sources can contain a wide suite of different metals (see Table 15.2).

(v) *Corrosion of Metals in Use.* Corrosion and chemical transformation of metals used in structures are the source of PTEs, *e.g.* Cu and Pb on roofs and in pipes; Cr, Ni and Co in stainless steel; Cd and Zn in rust preventative coatings on steel; Cu and Zn in brass fittings; and Cr and Pb from the deterioration of painted surfaces.

(vi) *Agriculture.* This mainly includes As, Cu, and Zn which are (or have been) added to pig and poultry feeds; Cd and U contaminants in some phosphatic fertilizers; and metal-based pesticides (historic and current) such as As, Cu, Mn, Pb and Zn.

(vii) *Forestry and Timber Industries.* Wood preservatives containing As, Cr and Cu have been widely used for many years and have caused contamination of soils and waters in the vicinity of timber yards. Several organic chemicals, including tar derivatives (creosote) and pentachlorophenol, are also used as wood preservatives.

(viii) *Fossil Fuel Combustion.* Trace elements present in coals and oils include Cd, Zn, As, Sb, Se, Ba, Cu, Mn and V and these can be present in the ash or gaseous/particulate emissions from combustion. In addition, various metals are added to fuels and lubricants to improve their properties (Se, Te. Pb, Mo and Li).

(ix) *Sports and Leisure Activities.* Game and clay pigeon shooting involves the use of pellets containing Pb, Sb and As but alternatives to these metals such as steel, Mo and Bi are also being introduced.

(x) *Crematoria.* There are significant emissions of Hg during cremation and emission control measures have being introduced to reduce these.

(xi) *Transport.* Platinum group elements from catalytic converters and other traffic derived elements *e.g.* Zn and Pb decrease rapidly away from roadside >12m.

15.2.2 Organic Pollutants

Fuel spills are possibly the most common point sources of organic pollution worldwide. A United Nation Development Programme report states that there have been a total of 6817 oil spills between 1976 and 2001, which account for a loss of three million barrels of oil, of which more than 70% was not recovered.[4] Acts of vandalism are also significant causes of uncontrolled discharges in many producing countries. One advantage of the hydrocarbons is that many of them can be biodegraded and are therefore subject to natural attenuation, while the more volatile components such as BTEX can be treated by vapour extraction (see section 15.5). The persistent organic chemicals (POPs) contain the 'dirty dozen' which are subject to the UNEP Stockholm Convention as a consequence of their high toxicity. POPs accumulate in soils as they bind tightly to the organic fraction as a consequence of their high K_{OW} values and are generally resistant to microbial degradation (see Chapter 18). Sources of organic pollutants include:

(i) *Diffuse sources.* These include many of the Stockholm Convention 'dirty dozen' which are now banned *e.g.* PCBs, hexachlorobenzene, dioxins and furans, DDT.

(ii) *Agriculture.* Pesticides are widely used in agriculture and include many POPs *e.g.* mirex and DDT. Some pesticides breakdown in the environment and poison unintended targets *e.g.* atrazine.

(iii) *Combustion sources.* Polycyclic aromatic hydrocarbons (PAHs) are produced whenever combustion occurs. They are principally associated with gasworks sites and other combustion sources *e.g.* coal ash and are a small component of fuel. PAHs are routinely observed at elevated levels in urban soils. Dioxins arise when chlorine is present in the feedstock of incinerators and there is inadequate control technology.

(iv) *Transport*. In the UK, oil and fuel spills were the second most frequent incidents with *ca.* 3000 spills in 2010. The majority of these were from fuel storage and distribution *e.g.* leaks from underground storage tanks. These may then penetrate the soil and pollute groundwaters.

(v) *Cleaning agents*. Solvents are usually used for degreasing activities in the electronics industry and a number of these compounds are dense non-aqueous phase liquids (DNAPLs) which are not readily soluble, but can result in considerable impacts on water supplies because of their toxicity *e.g.* chlorinated solvents, such as trichloroethylene, tetrachloroethene and carbon tetrachloride.

(vi) *Sewage sludge*. The profile of contaminated sludges in many industrialised countries has changed as control technologies have been introduced and domestic sources now dominate *e.g.* antibiotics, pharmaceuticals, flame retardants.

15.2.3 Nanoparticles

Nanoparticles are a relatively new potential pollutant of soils. They are being introduced to many materials because of the beneficial properties and alteration in chemistry offered by their size. For example, nano silver is now routinely used in many clothing products for its disinfection properties which prevent the build-up of odours. Sources of nanoparticle pollution include:

(i) *Sewage sludge*. Nanoparticles may enter the soil *via* sewage sludge application due to their presence in paints, fabrics, personal health care products and cosmetics.

(ii) *Remediation*. Nanoscale zero-valent iron is directly introduced for the clean-up of polluted soils and ground waters.

15.3 PATHWAYS OF POLLUTANTS IN SOILS

Following contact with the soil *e.g.* accidental spillage, pollutants can partition between several different solid, liquid and gas phases. For example, metals in soil can occur in three solid phases: (i) a component of the mineral lattice; (ii) a precipitate; or (iii) an exchangeable ion on mineral surfaces.[16] Chemicals may also dissolve in the soil water or volatilise into the air contained within the spaces between soil aggregates. The relationship between these different phases is often complex, depending on a range of different soil conditions and chemical properties (see Figure 15.1).

The partitioning between these different phases is usually calculated by a series of partition coefficients *e.g.* K_{AW} describes the partition between air (A) and water (W) for organic compounds. This coefficient describes the concentration in the air phase relative to the water phase when the two phases are in chemical equilibrium. The soil to water partition coefficient (K_d) is the one

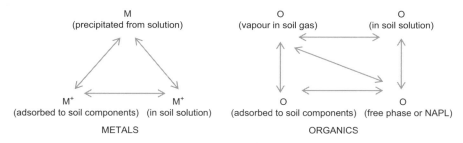

Figure 15.1 Chemical partitioning of contaminants in soils. (Adapted from ref. 7).

most important when considering the release of pollutants from the soil. The K_d describes the distribution between the solid and liquid phase as shown in Equation (1). Such that the pore water concentration can be calculated and hence the subsequent potential leaching into the groundwater.

$$K_d = M_{solid}/M_{solution} \qquad (1)$$

Where:

K_d = partition coefficient of specific chemical (cm^3 g^{-1})
M_{solid} = concentration of chemical in soil (µg g^{-1})
$M_{solution}$ = concentration of chemical in water (µg cm^{-3})

15.3.1 PTEs

Sorption is the dominant process for a number of the metals in polluted soils *e.g.* Cd, Co, Ni, Pb, Zn. The sorption is strongly influenced by pH because of its impact on the binding sites of different soil components. For the majority of the PTEs of interest the sorption to soil decreases with decreasing pH. This is because H$^+$ ions saturate the binding sites on organic matter, clays and ferric hydroxides so that the metal ions (M$^+$) are kept in soil solution.

Models of sorption describe the partition of the metal between the solid phase and solution. One frequently used model divides the solid phase in two: with a non-labile phase and labile phase. The labile phase freely exchanges with the soil solution and can be replenished from the non-labile phase. There can also be transport into the non-labile or inert phase from the labile phase (see Figure 15.2).

The degree of association depends on the contaminant of interest. It is important that models developed to predict the K_d use routinely, hence easily measured, soil characteristics so they frequently take the form:

$$\log M_{solid} = \log k + n \log M_{solution} + a\,pH + b\,\log(OM) \qquad (2)$$

The calculated value for $M_{solution}$ derived from incubation studies in water, or a weak electrolyte such as CaCl$_2$ to represent soil solution, is then added to Equation (1) along with M_{solid}, derived from a strong acid digestion, to calculate the K_d. This K_d will clearly not be linear as M_{solid} and $M_{solution}$ are not linearly related in Equation (2), but it will vary over a range of environments. Estimates of the labile pool can be determined using different mild extractants *e.g.* EDTA (ethylenediaminetetraacetic acid). Alternatively, the isotopic exchange technique can be used where the release of the 'native' isotope *e.g.* Cd-112 is contrasted with the sorption of an 'introduced' isotope *e.g.* Cd-109.[17]

Diffusive gradient thin films (DGT) have been used to determine speciation and/or *in situ* concentrations of toxic elements.[18] DGTs are best described as an *in situ* sampler. It consists of a resin gel overlain by a hydrogel layer and filter membrane that is in contact with the solution. The resin has a high affinity for metal ions and acts as an infinite sink. After a fixed time, the metal accumulation on the resin is measured, and the flux can be calculated.

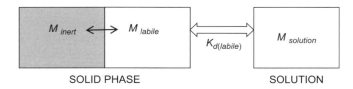

Figure 15.2 Schematic diagram of metal distribution in soil.

15.3.2 Organic Pollutants

The behaviour of organic pollutants in the environment can be illustrated by the use of evaluative triangle phase diagrams (see Figure 15.3). The accumulation of compounds in different phases is controlled by a number of their physico-chemical properties, specifically the vapour pressure, water solubility and octanol[§]–water partition coefficient (K_{OW}), and values for these can be found in the literature or calculated using quantitative structure–activity relationships (QSARs) using programmes such as EPI Suite.[19]

The accumulation of mass in a simple 3 phase (air, water, soil) environment as described in Figure 15.3 is described by:

$$\text{Mass} = C_W V_W + (K_{AW} C_W) V_A + (K_{SW} C_W) V_S$$

Where

C_W = concentration in water
V_W = volume of water
V_A = volume of air
$K_{SW} = K_d$ = soil to water partition coefficient
V_S = volume of soil or sediment

The are many calculations which can be undertaken for the derivation of K_d, but the simplest and most widely used is that derived by Karickhoff:[20]

$$K_d = K_{OW} \times 0.41 \times f.o.c.$$

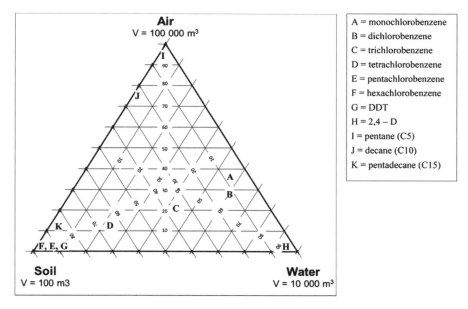

A	= monochlorobenzene
B	= dichlorobenzene
C	= trichlorobenzene
D	= tetrachlorobenzene
E	= pentachlorobenzene
F	= hexachlorobenzene
G	= DDT
H	= 2,4 – D
I	= pentane (C5)
J	= decane (C10)
K	= pentadecane (C15)

Figure 15.3   Phase diagram to illustrate distribution of a range of organic pollutants in the environment.

[§]Octanol is used as a surrogate of body lipid.

The f.o.c. (fraction of organic carbon) is of importance because the soil carbon acts as a sink for many organic pollutants particularly the POPs. In contrast to the PTEs, the clay content of the soil has little impact on the K_d for organics.

Fuel spills would include both aromatics (*i.e.* containing benzene ring structures) and aliphatics (non-aromatic structures). In order to derive fate and transport characteristics for these mixtures the Equivalent Carbon Number (ECN) concept has been developed.[21] The ECN refers to the boiling point (BP) of a chemical normalized to the boiling point of the *n*-alkanes. The relationship is described by:

$$ECN = 4.12 + 0.02(BP) + 6.5 \times 10^{-5}(BP)^2$$

This alteration in the boiling point can be seen in the following petroleum hydrocarbon fractions: benzene, toluene, ethyl benzene and xylene (BTEX) (ECN 6-8), petrol (ECN 4-12), diesel (ECN 7-24) and mineral oils (ECN 12-50). It allows for the range of petroleum constituents to be compared under a unified physical property. The carbon number has a considerable bearing on boiling point but also their fate in the environment (see Figure 15.3).

While it is easy to position the majority of the organic compounds of interest on a phase diagram, the pesticides are considerably more difficult because they differ considerably in their composition and structure which results in a wide range of physico-chemical properties *e.g.* the high water solubility of 2,4 D in contrast with that of DDT (see Table 15.3 and Figure 15.3).

Table 15.3 Physico-chemical properties of frequently detected pesticides in soils.

Pesticide	Structure	Log K_{OW}	Vapour pressure (Pa)	Water solubility (mg l^{-1})
DDT		6.79	0.0010	0.0055
Mirex		7.01	0.00039	0.00048
2,4-D		2.62	0.0037	680
Chlorpyrifos		5.11	0.0027	1.12

In parallel with the PTEs, the organics have labile and non-labile pools. There has been considerable interest in the carbohydrate, cyclodextrin, to determine the labile pool because it has a hydrophobic cavity to determine the labile pool,[22,23] but mild solvents *e.g.* butanol have also been used.[22,24]

15.3.3 Nanoparticles

One consequence of their relative novelty is that less is known about the long term fate of nanoparticles in soils. It is difficult to generalise because they are engineered for each specific application.[25] For example metallic nanoparticles maybe coated to maintain their stability as colloidal solutions and these coatings will affect their behaviour in the environment.[26] Once in the environment they usually aggregate particularly when they come into contact with large molecular fractions of natural organic matter and their reactivity is reduced.[27] Although even in this condition they can travel long distances depending on the hydrological conditions.[15]

15.4 CONSQUENCES OF SOIL POLLUTION – RISK ASSESSMENT

Risk assessment estimates the amount of exposure to chemicals of receptors such as humans and/or environment, and the potential for harm as a consequence of the toxicity of the pollutant.[28–30] Chemical pollutants are emitted from various sources and distribute through a range of media based on the physico-chemical characteristics as described above. The receptor, human and/or environment, becomes exposed through contact with the pollutant *via* the contaminated medium (see Figure 15.4).

The US Environmental Protection Agency explained that exposure assessment of a chemical can be divided into several phases:[30] (i) identifying pollutant sources and transfer media; (ii) determining the concentration of the chemical pollutant; (iii) determining the exposure scenarios, exposure routes and pathways; (iv) determining the exposure factors; and (v) identifying the exposed population. Phases (i)–(iii) are completed by establishing the source–pathway–receptor linkages described previously using the information outlined in sections 15.2 and 15.3.

There are various regulatory modelling tools for estimating chemical exposure as a part of risk assessment. For example, the European Chemicals Agency recommends EUSES (European Union System for the Evaluation of Substances) as the tool of environmental exposure estimation in the Chemical Safety Assessment mandated by REACH regulation.[29] RAIDAR is used for the rapid

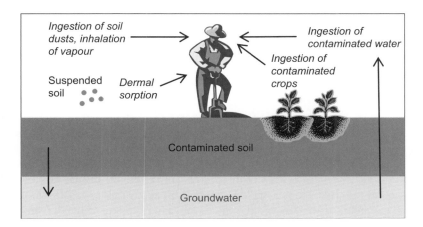

Figure 15.4 Exposure pathways to humans from contaminated soils.

Table 15.4 Soil Guideline Values derived the UK Environment Agency.

	Soil guideline value (mg kg^{-1} dry weight)		
	Residential	Allotment	Commercial
Cadmium	10	1.8	230
Arsenic	32	43	640
Benzene	0.33	0.03	95
Dioxins (sum)	0.008	0.008	0.024

screening of chemicals identifying their potential to cause harm under the *Canadian Environmental Protection Act, 1999*.[31] Additionally, there are the CLEA and CSOIL exposure assessment tools developed by the regulatory agencies in the UK and Netherlands. These are used to produce standard guidance values for the acceptable levels of soil pollutants to enable the protection of human health (see Table 15.4).

The difference in the guideline values between scenarios in Table 15.4 relates to the different exposure pathways which are active. For example, ingestion of vegetation is more important for cadmium than arsenic, resulting in a lower guideline value on allotments where people potentially eat more home grown produce. This derives from the differences in the plant uptake of PTEs from polluted soils. For the PTEs of interest this is in the order: Cd (0.55) = Hg (0.33) > As (0.025) > Pb (0.015), so the former is more likely to contaminate crops. The figures in brackets refer to the bio-concentration factor (mg kg^{-1} fresh weight (f.wt.) plant/mg kg^{-1} dry weight (d.wt.) soil).[32]

15.4.1 Fine Tuning the Risk Assessment

In the case of soil pollution it is usually the direct ingestion of soil or the ingestion of contaminated crops that dominate the exposure (see Figure 15.4). Consequently there is considerable interest in determining the quantity of pollutants transported from soils to these matrices. In many models this is considered to be 100% of the total in either the soil (ingestion) or soil water (plant uptake). The total concentrations are measured by harsh extractions *i.e.* strong acids for the PTEs and aggressive solvents for organics. However, the labile fraction is often better correlated with measures in biota. In order to quantify these processes better, a range of methods that mimic uptake processes in target organisms have been developed to fine tune the risk assessment

15.4.1.1 Bioaccessibility and Bioavailability Testing

When soil is ingested it passes into the gastro-intestinal tract where the pollutants maybe desorbed from the soil matrix. Bioaccessibility measures the quantity of pollutant desorbed into the gut medium; it is only this desorbed material that can pass across the gut lining and into the bloodstream. Once material has crossed a biological membrane it is termed 'bioavailable' (this could be the gut lining in humans or the cell wall of a bacterium). To measure actual blood levels following ingestion is costly and has associated ethical issues hence the need to develop bioaccessibility tests. There are a number of physiologically based extraction tests (PBETs) that mimic the stomach and small intestine. When a range of soils were measured for As, bioaccessibility values were lower in geologically contaminated soils (5–9%) than in soils impacted by diffuse urban pollution (19–28%).[33] In soils contaminated from a nickel smelter, bioaccessibility values ranged from 3.6 to 19%,[34] while bioaccessibility of benzo[a]pyrene at a gasworks site ranged from 1–23%.[35] The controls on the desorption are complex; acidity increases metal desorption in the gut for PTEs, while bile salts and transit time increase desorption of organics. Should the bioaccessibility be below 20%, this will have a profound impact on the risk assessment, effectively reducing the exposure by a factor of five.

Bioavailability of pollutants to crops, another ingestion pathway, has also been investigated. When reviewing the potential for DGT to predict plant uptake, Degryse *et al.*[36] reported that uptake in DGT and crops would increase proportionally if uptake is limited by diffusion of the metal ion from the soil (high affinity uptake by the plant, low concentration in soil solution). This is probably the case for Cd and Zn in most soils. However, when the soil solution concentration was high and the plant requirement was low, as may occur in contaminated soils, the prediction by DGT was no better than the total soil concentration. Uptake of organics by crops has also been correlated with passive samplers *e.g.* TECAM,[37] solid phase microextraction fibres and butanol.[38]

There is also the potential to predict impact to ecosystems. Similar techniques have been used to predict uptake into earthworms, which are a sentinel species involved in the soil processing of soil carbon. These have involved simulated gut fluids as in the bioaccessibility tests previously described,[39,40] and the prediction of bioavailability for both PTEs,[41,42] and organics.[38,43]

15.5 REMEDIATION OF CONTAMINATED SOILS

Where the concentrations of pollutants exceed statutory or environmental quality standards following the risk assessment, it will be necessary to carry out some form of clean-up or remediation. All the methods will attempt to break the S–P–R linkage. This may be by removing the source (bioremediation), blocking the pathway (physical barriers) or removing the receptor (reducing exposure times, *e.g.* commercial not residential buildings). In former times the principal remediation method involved excavating the most polluted soil and either disposing of it safely in a licensed landfill (often referred to as 'dig and dump'), or cleaning it up off-site (*ex situ* remediation). However, with the high cost of haulage and increasing taxes on landfilling of wastes to encourage recycling, greater use is now made of *in situ* remediation methods. The two largest remediation projects in the UK in the last 15 yrs were the Millennium Dome site and the Olympic Park. The greatest difference in the remediation approach between these sites was the adoption of a more sustainable approach at the Olympic Park. Treatment centres were constructed on site, *e.g.* bio-piling for organic pollutants and soil sieving to separate the fines,¶ where the majority of the contamination resides, for PTEs. This significantly reduced the environmental 'footprint' of the remediation, *e.g.* fuel usage, and had other benefits such as reducing conflicts with the local community as transport movements were substantially decreased.

The main remediation methods in use are briefly outlined in Table 15.5. These methods include:

(i) *Containment*. The contamination is retained by physical barriers (such as trenches filled with impermeable clay, *e.g.* bentonite) or hydraulic barriers based on maintaining fluid pressure differentials by the extraction or injection of water to confine the contaminant plume in aquifers (permeable underground strata).

(ii) *'Pump and Treat'*. This method is applied to contaminated groundwaters which are pumped to the surface where they are treated to remove the contaminants, and the cleaned water is then usually returned to the aquifer. In the case of volatile organic compounds, such as chlorinated solvents, this can include bubbling air through the extracted water to enhance the volatilization of the organic pollutants ('air sparging'). When water is injected into contaminated soil to dissolve soluble contaminants and then extracted, this technique is referred to as *'in situ* soil washing' which can be successfully applied to the removal of a wide range of contaminants (not all necessarily water soluble), including salts such as sulfates. Surfactants or other chemicals may be added to the pumped liquid to enhance the clean-up process.

¶Fines include the clay fraction – see section 15.3.

Table 15.5   Techniques available for the remediation of contaminated land.

Method	Contamination problem
Civil engineering based methods	
(1) Excavation and safe disposal (usually to landfill)	'Hot spot' contamination at concentrations too high for safe *in situ* treatment, especially inorganic chemicals such as heavy metals
(2) Soil solidification and stabilisation	As above
(3) Physical containment- using in-ground barriers and covers	Organic NAPLs, landfill leachates
(4) Hydraulic controls	Supporting containment and/or for treatment of contaminated surface or ground waters
Process-based remediation	
(5) Thermal treatment- to remove, stabilize or destroy contaminants	Incineration of soils contaminated with POPs such as PAHs, PCBs and PCDDs
(6) Physical treatment to separate contaminants from soils or different fractions of contaminated media (usually *ex situ*)	Soil washing to remove sulfates, metals and other pollutants adsorbed on silt and clay particles
(7) Chemical treatment- reactions to remove, or modify contaminants (*in situ*, or *ex situ*)	Dechlorination of PCB contaminated soils. Acid leaching of metal-contaminated soils
	Organic contaminants, including hydrocarbons, pesticides, solvents, tars, PAHs

(Adapted from ref. 44).

(iii) *In situ enhanced recovery*. This technique includes soil vapour extraction by soil venting, where air is pumped into soil which has sufficient macropores to provide adequate air permeability and is collected from extraction wells from which the volatile compounds can be either incinerated on site, or adsorbed onto activated carbon.

(iv) *Soil extraction*. Soil pollutants adsorbed to the soil solids are generally more difficult to treat. In the case of organic compounds, bioremediation is used which involves optimizing the conditions for the microbial decomposition of the chemicals (*e.g.* pH, temperature). For PTEs, chemical treatment, involving the use of extractants such as acids or chelating agents, may be used.

(v) *Phytoremediation*. This method uses plants with a high capacity to accumulate metals to deplete the plant-available fraction of metals in the soil. However, although several hyperaccumulator plant species have been identified and show considerable potential, most do not produce sufficient biomass to effectively remove significant amounts of metals from contaminated soil. Plants may also be used to stabilise polluted soils preventing dispersion of the contaminants.

(vi) *Natural attenuation*. Some organic pollutants undergo natural processes of adsorption and degradation. There is an increasing body of evidence to show that these natural processes account for a significant decrease in the amount of pollutant available and therefore a reduction of risk from soil pollution at many types of sites. However, the pollutant levels need to be carefully monitored to ensure that improvements are taking place over an appropriate period of time.

15.6 CONCLUSIONS

Soils are important for a whole range of environmental processes and pollution may impact upon these. Therefore we need to control and remediate pollution sources. When considering the risk to humans and biota from soil pollution we can use the source–pathway–receptor paradigm. This model allows us to bring together our knowledge of the fate of contaminants in soils and their toxicity to develop regulatory guidelines to protect human and environmental health. Chemical tools are allowing us to fine tune the inputs into these models so we can focus the remediation of soil where it is necessary and identify appropriate techniques, thereby providing a more sustainable solution to contaminated land restoration.

REFERENCES

1. G. C. Daily, P. A. Matson and P. M. Vitousek, *Ecosystem Services Supplied by the Soil.*, in *Nature's Services: Societal Dependence on Natural Ecosystems*, ed. G. C. Daily, Island Press, Washington, DC, 1997, pp. 113–132.
2. D. A. Robinson, I. Lebron and H. Vereecken, On the definition of the natural capital of soils: a framework for description, evaluation, and monitoring, *Soil Sci. Soc. Am. J.,* 2009, **73**(6), 1904–1911.
3. Centre for Ecology and Hydrology, *Soil Natural Capital and Ecosystem Services,* 2013; http://www.ceh.ac.uk/sci_programmes/soil-capital-ecosystem-services.html (accessed 10/07/2013).
4. B. J. Glass, World Resources 1994–1995: A Guide to the Global Environment, World-Resources Institute, *J. Gov. Informat.,* 1996, **23**(4), 534–535.
5. Y. P. Wang, R. M. Law and B. Pak, *A global model of carbon, nitrogen and phosphorus cycles for the terrestrial biosphere,* Biogeosciences, 2010, **7**(7), 2261–2282.
6. L. R. Oldeman, R. T. A. Hakkeling and W. G. Sombroek, *World Map of the Status of Human Induced Soil Degradation*, CIP-Gegevens Koningklijke Bibliothek, Den Haag, 1991.
7. J. Jeffries and I. Martin, Updated technical background to the contaminated land exposure assessment model, *Sci. Report,* 2008, SC050021/SR3.
8. English Partnerships, *Contamination and Dereliction Remediation Costs,* Best Practice Note 27, 2008.
9. K. S. Patel, *et al.*, A survey of lead pollution in Chhattisgarh State, central India, *Environ. Geochem. Health,* 2006, **28**(1–2), 11–17.
10. B. J. Alloway and D. C. Ayres, *Chemical Principles of Environmental Pollution*, Blackie Academic & Professional, London, 2nd edn., 1997, 395.
11. Environment Agency, *UK Soil and Herbage Pollutant Survey UKSHS Report No. 9. Environmental Concentrations of Polycyclic Aromatic Hydrocarbons in UK Soil and Herbage,* Environment Agency, Bristol, 2007.
12. J. L. Barber, A. J. Sweetman, D. van Wijk and K. C. Jones, Hexachlorobenzene in the global environment: emissions, levels, distribution, trends and processes, *Sci. Total Environ.,* 2005, **349**(1–3), 1–44.
13. Defra, *Environmental Protection Act 1990: Part 2A. Contaminated Land Statutory Guidance,* 69, 2012.
14. DEFRA and Environment Agency, *Potential Contaminants for the Assessment of Land,* Report CLR8, available from The R&D Dissemination Centre, WRc plc, Swindon, Wilts, UK, 2002, p. 34.
15. F. A. Nicholson and B. J. Chambers, *Sources and Impacts of Past, Current and Future Contamination of Soil,* 2007, Publication SP0547 DEFRA, London.
16. B. J. Alloway, *Heavy Metals in Soils*, Blackie Academic & Professional, London, 2nd edn., 1995, 368.
17. I. A. M. Ahmed, S. D. Young and N. M. J. Crout, Ageing and structural effects on the sorption characteristics of Cd^{2+} by clinoptilolite and Y-type zeolite studied using isotope exchange technique, *J. Hazard. Mater.,* 2010, **184**(1–3), 574–584.

18. S. Mongin, *et al.*, Key role of the resin layer thickness in the lability of complexes measured by DGT, *Environ. Sci. Technol.*, 2011, **45**(11), 4869–4875.

19. US Environmental Protection Agency, *Estimation Program Interface Suite 4.1*, 2011.

20. S. W. Karickhoff, Semiempirical estimation of sorption of hydrophobic pollutants on natural sediments and soils, *Chemosphere*, 1981, **10**(8), 833–846.

21. J. B. Gustafson, J. Griffith-Tell and D. Orem, *Selection of representative TPH fractions based on fate and transport considerations*, Total Petroleum Hydrocarbon Criteria Working Group Series, Vol. 3, Amherst Scientific Publishers, MA, 1997, 102.

22. J. L. Gomez-Eyles, C. D. Collins and M. E. Hodson, Relative proportions of polycyclic aromatic hydrocarbons differ between accumulation bioassays and chemical methods to predict bioavailability, *Environ. Pollut.*, 2010, **158**(1), 278–284.

23. J. D. Stokes, *et al.*, Prediction of polycyclic aromatic hydrocarbon biodegradation in contaminated soils using an aqueous hydroxypropyl-beta-cyclodextrin extraction technique, *Environ. Toxicol. Chem.*, 2005, **24**(6), 1325–1330.

24. H. H. Liste and M. Alexander, Butanol extraction to predict bioavailability of PAHs in soil, *Chemosphere*, 2002, **46**(7), 1011–1017.

25. I. Bhatt and B. N. Tripathi, Interaction of engineered nanoparticles with various components of the environment and possible strategies for their risk assessment, *Chemosphere*, 2011, **82**(3), 308–317.

26. F. Mafune, *et al.*, Structure and stability of silver nanoparticles in aqueous solution produced by laser ablation, *J. Phys. Chem. B*, 2000, **104**(35), 8333–8337.

27. E. Navarro, *et al.*, Environmental behavior and ecotoxicity of engineered nanoparticles to algae, plants, and fungi, *Ecotoxicology*, 2008, **17**(5), 372–386.

28. US Environmental Protection Agency, *Guidelines for Exposure Assessment*, 1992.

29. US Environmental Protection Agency, *Guidance on Information Requirements and Chemical Safety Assessment, Part A: Introduction to the Guidance Document*, 2011.

30. US Environmental Protection Agency, *Exposure Factors Handbook*, 2011.

31. Environment Canada, Health Canada, *Rapid Screening of Substances of Lower Concern*, 2013; http://www.ec.gc.ca/ese-ees/2A7095CD-A88C-4E7E-B089-486086C4CBC4/RSI%20Final%20-%20EN.pdf (accessed 01/07/2013).

32. M. R. Thorne, R. C. Walke and P. Maul, *The PRISM Foodchain Modelling Software: Parameter Values for the Soil Plant Model*, 2005.

33. J. D. Appleton, M. R. Cave and J. Wragg, Anthropogenic and geogenic impacts on arsenic bioaccessibility in UK topsoils, *Sci. Total Environ.*, 2012, **435**, 21–29.

34. L. Vasiluk, M. D. Dutton and B. Hale, *In vitro* estimates of bioaccessible nickel in field-contaminated soils, and comparison with *in vivo* measurement of bioavailability and identification of mineralogy, *Sci. Total Environ.*, 2011, **409**(14), 2700–2706.

35. E. L. Tilston, G. R. Gibson and C. D. Collins, Colon Extended Physiologically Based Extraction Test (CE-PBET) increases bioaccessibility of soil-bound PAH, *Environ. Sci. Technol.*, 2011, **45**(12), 5301–5308.

36. F. Degryse, *et al.*, Predicting availability of mineral elements to plants with the DGT technique: a review of experimental data and interpretation by modelling, *Environ. Chem.*, 2009, **6**(3), 198–218.

37. Y. Wang, *et al.*, Reducing the bioavailability of PCBs in soil to plant by biochars assessed with triolein-embedded cellulose acetate membrane technique, *Environ. Pollut.*, 2013, **174**, 250–256.

38. J. L. Gomez-Eyles, *et al.*, Passive samplers provide a better prediction of PAH bioaccumulation in earthworms and plant roots than exhaustive, mild solvent, and cyclodextrin extractions, *Environ. Sci. Technol.*, 2012, **46**(2), 962–9.

39. W. K. Ma, *et al.*, Development of a simulated earthworm gut for determining bioaccessible arsenic, copper and zinc from soil, *Environ. Toxicol. Chem.*, 2009, **28**(7), 1439–1446.

40. B. A. Smith, B. Greenberg and G. L. Stephenson, Comparison of biological and chemical measures of metal bioavailability in field soils: test of a novel simulated earthworm gut extraction, *Chemosphere*, 2010, **81**(6), 755–766.
41. R. Bade, S. Oh and W. S. Shin, Diffusive gradients in thin films (DGT) for the prediction of bioavailability of heavy metals in contaminated soils to earthworm (*Eisenia foetida*) and oral bioavailable concentrations, *Sci. Total Environ.*, 2012, **416**, 127–136.
42. P. S. Fedotov, *et al.*, Extraction and fractionation methods for exposure assessment of trace metals, metalloids, and hazardous organic compounds in terrestrial environments, *Crit. Rev. Environ. Sci. Technol.*, 2012, **42**(11), 1117–1171.
43. M. T. O. Jonker, *et al.*, Predicting PAH bioaccumulation and toxicity in earthworms exposed to manufactured gas plant soils with solid-phase microextraction, *Environ. Sci. Technol.*, 2007, **41**(21), 7472–7478.
44. M. Harris and S. Herbert, *Contaminated Land*, Institution of Civil Engineers, Thomas Telford Publishers, London, 1994.
45. A. Heikins. *Arsenic Contamination of Irrigation Water, Soil and Crops in Bangladesh: Risk Implications for Sustainable Agriculture and Food Safety in Asia*. RAP PUBLICATION 2006/20. Food and Agriculture Organisation 2006.

CHAPTER 16

Solid Waste Management

GEV EDULJEE

SITA UK, SITA House, Grenfell Road, Maidenhead SL6 1ES, UK
Email: gev.eduljee@sita.co.uk

16.1 INTRODUCTION

Waste management and recycling industries in the EU generate a turnover of €95 billion and provide between 1.2 and 1.5 million jobs.[1] The recycling industry alone represents a turnover of €24 billion, provides 500 000 jobs and involves 60 000 companies, 95% of which are small and medium sized companies.

Waste is not a unique material in terms of its constituents: the main distinguishing feature relative to the products from which it derives is its perceived lack of value. Dictionary definitions of waste include the descriptions "useless" or "valueless". In the UK, the legal definition of waste in the *Waste (England and Wales) Regulations 2011* follows the definition given in Article 3(1) of the *European Union Directive on Waste (2008/98/EC)*[2] which states that "any substance or object the holder discards, or intends or is required to discard" is waste. The Directive also defines two further categories of material that need not be regarded as waste provided certain stipulated conditions are met: by-products (Article 5) and materials conferred so-called "end-of-waste" status (Article 6). This latter designation applies to specified waste materials that have undergone a recovery operation, including recycling, such that they cease to be waste and attain a product-like status.

Up to the 1970s, the perception of waste as unwanted, "useless" material with no intrinsic value shaped society's approach to waste management. The ultimate disposal of waste was the overriding priority. Waste generators (domestic, commercial and industrial) sought disposal at the lowest cost (overwhelmingly in landfills) and had little or no incentive to "manage" waste as opposed to merely dispose of it.

However, the past three decades have witnessed a sea change in attitudes to waste. Public sensitivity towards waste disposal outlets has put enormous pressure on operators and regulators alike; the siting of new facilities has at best proceeded after lengthy delays and in the teeth of intense opposition. Waste generators and policy makers have perforce turned their attention to upstream waste-related activities in an effort to minimise waste production and hence make more efficient use of ultimate disposal capacity.

Pollution: Causes, Effects and Control, 5th Edition
Edited by R M Harrison
© The Royal Society of Chemistry 2014
Published by the Royal Society of Chemistry, www.rsc.org

Another key driver has been the concept of sustainable development,[3] defined as "development that meets the needs of the present without compromising the ability of future generations to meet their own needs". The slogan "more with less" encapsulates the current thinking: extracting the maximum value and benefit from products and services, using the minimum of energy and rejecting the minimum of waste materials or emissions to the environment. Arising out of this is the concept of resource efficiency and sustainable waste management.[4,5] In essence, waste is given value. Viewed against the principle of the conservation and nurturing of natural resources, the production of waste is in itself seen as a manifestation of the inefficient management of the earth's raw materials. Commencing with waste reduction during the production of goods and the provision of services, sustainable waste management calls for the recovery and reuse of materials so as to conserve raw materials, the use of waste as a source of energy in order to conserve non-renewable natural resources, and finally for the safe disposal of unavoidable waste. A new policy approach has been developed to instil a sense of common responsibility towards waste, and to create overall strategies for waste and resource management that do more than pay mere lip service to the concept of sustainability.

The aim of this chapter is to illustrate how this approach has been applied to the management of municipal solid waste (MSW). For the purpose of this chapter, we define MSW as solid waste collected from households, and waste of similar composition collected from commercial and industrial premises. In many ways MSW presents the greatest challenge to waste management. Waste arising from domestic, commercial and industrial premises is extremely heterogeneous in nature, and as a result components of potential value (such as glass, metals or biodegradable material) cannot be beneficially used without significant effort being put into segregation and sorting schemes. Even with such systems, care needs to be taken to achieve products of value due to cross-contamination and fragmentation of the constituents. In many countries, the poor standard of MSW disposal also leads to a significant threat to public health.

In keeping with the importance governments now place on resource efficiency and sustainable development, this chapter emphasises the preventative aspects and the resource potential of waste management as opposed to the impact of MSW disposal *per se*; other publications have examined this latter aspect of waste management in some detail.[6-8] The chapter commences with an introduction to the so-called "waste management hierarchy" and the need to integrate various strands of policy into a coherent waste management strategy. Next, technical options for waste prevention and recycling are presented, followed by a discussion of the policy options that can be applied to encourage or to mandate waste prevention and reuse. Bulk waste reduction technologies, in particular incineration and composting, are then presented. Finally, some of the issues concerning the development of an integrated waste management strategy are discussed.

16.2   AN INTEGRATED APPROACH TO WASTE MANAGEMENT

16.2.1   The Waste Management Hierarchy

Options for waste management are often arranged in a hierarchical manner to reflect their desirability.[2,4] The first priority is waste prevention, that is not producing the waste in the first place. If the waste must be produced, then the quantities should be minimised. Once that has been achieved, the next priority is to maximise reuse, followed by recycling and recovery of suitable waste materials. Taken together, these three options are often called waste prevention, although strictly speaking only the first two are prevention whereas the third is already an end-of-pipe solution. Once the possibilities for waste prevention have been exhausted, the next priority is to reduce the volume of residual wastes being passed on for final disposal, extracting resources in the form of products and/or energy in the process. The waste management hierarchy is illustrated in Figure 16.1.

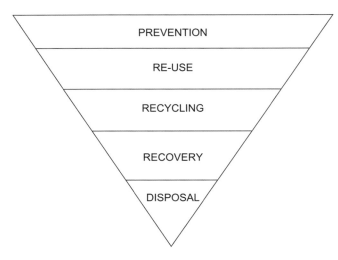

Figure 16.1   Components of the waste hierarchy.

To move from the current situation in which the majority of wastes are dealt with by final disposal or other end-of-pipe solutions to one where waste prevention becomes dominant, it is necessary for governments to provide some support through policy measures. Policies which encourage waste avoidance and waste minimisation are to be preferred over those which focus purely on further encouraging present recycling, recovery and reuse.

The waste management hierarchy is discussed further in section 16.6.

16.2.2   An Integrated Approach

If waste management is to change significantly, the behaviour of individuals and groups in society has to change. Three groups in society are key to this process: government, industry and commerce, and individuals. A fourth group, lobby groups and NGOs, is effective as a conduit of ideas and energy. Policies need to be designed to change the behaviour of all these groups in order to reduce or even reverse the growth in waste generation which accompanies increasing wealth – decoupling waste generation from economic growth.

Effective policies generally operate on a two pronged approach. Where possible, governments prefer that waste reduction policies should be implemented on a voluntary basis, in order to reduce the regulatory burden on the competent authorities as well as on the regulated sectors. In order to kick-start some initiatives, government can provide an incentive or "carrot" in the form of financial or other support. Combined, these form the first prong. For the second prong, governments implement policies to tackle the underlying causes relating to waste generation and positively discourage it. Some such measures need to be implemented alongside those from the first prong. Others may only be required if measures from the first prong, for whatever reason, prove to be ineffective. This second prong of policies can be viewed as the disincentives, sanctions or "sticks" which complement the "carrots" above.

In addition, it is essential that policies build on, rather than undermine, existing strengths. Where there are already efforts to reduce waste generation or recover materials for recycling and reuse on a voluntary basis or using current market mechanisms, policies must aim to reinforce these efforts, and not be so invasive as to undermine such existing initiatives.

Finally, even with all these policies and measures in place and operational, some wastes will still be generated and require disposal. The provision of bulk waste reduction technologies will help to reduce the volume and weight of these remaining residues prior to final disposal.

An integrated strategy for waste reduction requires a combination of all the types of measure detailed above. The following sections address these measures, along with technical options for waste prevention and segregation.

16.3 TECHNICAL OPTIONS FOR WASTE PREVENTION AND RECYCLING

16.3.1 Opportunities for Waste Avoidance and Minimization

Waste avoidance and minimization are the most desirable options in the waste management hierarchy as discussed in section 16.2.1 above. This section explores the ways in which waste avoidance and minimization have been or may be practised to reduce the amounts of domestic, commercial and industrial wastes that arise. Table 16.1 lists some options which may be appropriate and these are discussed in greater detail under each sector. Some of these waste avoidance and minimization options may be applicable to all waste generators while some would be applicable only to industry (or even to specific industry sectors).

16.3.1.1 Domestic Sector

There are many ways in which individuals can avoid or minimise the amount of waste they put out for disposal. As consumers, they may select product types, packaging types and material types that would lead to the generation of less waste. For example, waste can be reduced by buying in bulk, utilising reusable shopping bags, buying reusable and more durable products and by buying equipment that generates less waste (for example electrical equipment which runs off mains power or uses rechargeable batteries to reduce the quantity of primary cells used). Thoughtless household practices can also lead to waste generation. In 2006/7 UK households discarded 8.3 million tonnes of food and drink waste, 5.3 million tonnes of which were judged to be avoidable (*i.e.* food and drink that could have been eaten). [9] A repeat study found that by 2010 domestic food and drink waste generation had reduced by 13%,[10] with avoidable waste down to 4.4 million tonnes. The environmental impact of avoidable household food and drink waste amounts to 17 million tonnes of CO_2 equivalents and 4% of the UK's total water footprint.

A major factor is to change the public's perception of waste and how to deal with waste materials. Education and communication programmes therefore need to be built into any

Table 16.1 Ways to achieve waste avoidance and minimization.

Waste Management Options	*Opportunities for Avoidance and Minimization*
Product design change	• Product design with less waste • Increase product life/durability of product
Package change	• Product in bulk or concentrate form • Reusable or recyclable pack
Materials change	• Substitution of less toxic materials • Use of reusable or recyclable materials
Technological change	• Improved/more efficient equipment • Cleaner technology
Management practices	• Good housekeeping • Proper operating procedures and regular maintenance • Inventory control • Training and clear instructions • Waste segregation

approach aimed at the domestic sector before significant take-up of the options offered could be achieved.[11–13] Inducements and incentives can be deployed to support information dissemination and communication, encouraging communities to change their perception of waste.[14] However, lack of attention to practical obstacles such as the provision of recycling bags or adequate collection facilities can negate the drive towards waste avoidance, even where there is a general motivation to avoid, minimize and recycle waste.[15,16]

Some waste avoidance and minimization measures taken within the domestic sector would not be possible without the supporting actions of both industry and commerce. The manufacturing, packaging and labelling of products and the provision of some services that would lead to the generation of less waste are necessary to enable householders to have some choice. This type of responsibility relating to producers is examined in later sections.

16.3.1.2   *Commercial and Industrial Sectors*

As manufacturers, companies may produce longer life products, requiring less maintenance. Manufacturing processes could use less toxic substitutes and recycling materials and may be improved in terms of reduction in material wastage. Proper operating procedures and regular maintenance would also reduce wastage as well as reducing other emissions. Such measures are likely to be industry specific, *i.e.* their application and potential to reduce wastes will vary industry by industry. For example, in the plastics industry, internal recycling of segregated clean plastics may contribute 10% to 15% of total consumption of plastic materials. In metal products fabrication, improved maintenance of cutting machines has been identified as a way of preventing contamination of scrap metal waste which may then be recycled. In the wearing apparel industry, use of wider roll of cloth and re-design of garments may significantly reduce cutting wastes. Improving management and control of processes can lead to significant waste reduction, which in turn can lead to considerable savings in costs due to less wastage and more efficient procedures.[17]

It should be noted that waste avoidance and minimization measures are already adopted by many individuals and companies, partly for environmental concern and partly for financial concern. Experience from running cooperative waste minimisation programmes around the world shows that less waste leads to higher profit. However, there are barriers and constraints identified in these programmes and government policies and measures to encourage the adoption of these methods would be required, as discussed in section 16.4.

16.3.2   Collection and Sorting

A prerequisite for the cost-effective reuse and/or recycling of potentially valuable components in MSW is that these materials be separated out from the bulk waste. A number of options are available, as discussed in the following sections.

16.3.2.1   *Source Separation*

Collections of waste materials at source are termed "kerbside collection systems". This method involves the householder putting out recyclable materials for collection separate from the normal refuse. These recyclable materials may either be mixed, all materials being placed into one container for future sorting either by the collector at a reclamation facility, or separated into individual materials.

In the "mixed at source" scheme the householder places all recyclable materials into one container which is either separated at the kerbside by the waste collection operative using a specially adapted vehicle, or emptied into the collection vehicle and taken to a central facility for sorting. This central facility is usually referred to as a Materials Reclamation Facility (MRF). In MRFs, co-mingled materials are re-separated, stored and perhaps some initial processing is carried out, prior to selling on to a manufacturer or into the secondary materials markets.[18] Separate processing lines are usually set up; for the separation of co-mingled containers (aluminium cans, steel cans, plastic bottles and glass bottles) and for co-mingled paper (newspaper, domestic paperboard, white paper). The process separates these and prepares them for onward sale to recycling companies. The initial processing of waste at a MRF (for example, washing, baling, *etc.*) can add value to the separated materials.

Source separated materials which reach the MRF should already be fairly clean and uncontaminated. Hence, a MRF should not give rise to significant air or water pollution so long as the civil works of the facilities are properly designed. Proper enclosure and management of MRFs would reduce the impact of noise and odour.

To a large extent, the economics of source separation and MRFs are determined by the price for recovered materials, which can fluctuate widely in international markets. The relative quantities of recyclable collected and sorted are also key, because of different prices and different sorting regimes for different materials. A study found that co-mingled and two-stream collection systems could be expected to achieve higher yields of collected dry recyclables, relative to kerbside sorting.[19] However, when material rejected at the MRF, by secondary processors and by reprocessors was taken into account, differences in the tonnage actually recycled between the systems appeared to be marginal.

16.3.2.2 Bring Systems

Bring systems are widely used in many parts of the world, and are typically employed for the recovery of glass or paper. In the UK, the recycling centres at Civic Amenity sites (also called Household Waste Recycling Centres) and recycling facilities such as bottle banks and paper banks outside supermarkets and in town centres, are examples of "bring systems". HWRCs permit the deposit of recyclable materials in separate containers. Typically separate containers are used for green/garden waste, glass, metals, aluminium and steel cans, paper, cardboard, oils and textiles. Containers for hardcore, wood, car and domestic batteries, *etc.* are also provided. Refrigerators containing CFCs can also be placed in HWRCs for recycling, as can smaller electrical and electronic items.

The overall diversion rate of materials from the waste stream destined for direct landfill varies widely, but is generally higher for kerbside collection, presumably because this involves less effort on the part of the householder.

16.4 POLICY OPTIONS TO MAKE WASTE PREVENTION AND RECYCLING WORK IN PRACTICE

16.4.1 Introduction

For recycling to become more effective, it is necessary to introduce measures to "level the playing field" so that secondary materials can compete more fairly with virgin stock, and to provide direct support to increase the size of local markets. These measures can be *via* voluntary participation, *via* positive encouragement, generally in the form of some sort of government support (the carrots referred to in section 16.2.2), or *via* mandatory measures, enacted by government to more forcefully encourage the adoption of waste reduction measures (the sticks discussed in section 16.2.2). The central objective of the application of these policy measures is to encourage waste avoidance, minimization, reuse and recycling. In general, they each seek to change behaviour by making waste avoidance, minimization, reuse and recycling more attractive options than disposal. Most successful schemes for waste prevention and recycling work by combining the "carrot" of financial

Table 16.2 Policy measures for managing municipal solid waste in the EU.[21]

	Estonia	Finland	Flemish Region	Germany	Hungary	Italy
User charge for waste collection and management	√	√	√	√	√	√
Environmental product charges					√	
Landfill tax	√	√	√			√ (regional)
Incineration tax			√			
Landfill ban	√	√	√	√	√	√
Separate collection of biowaste	√	√	√	√	√	√ (regional)
Producer responsibility deal for waste paper		√	√	√		
Producer responsibility deal for packaging waste	√	√	√	√	√	√

incentives and other measures to encourage positive behaviour with the "stick" of either financial penalties or legal requirements to discourage negative behaviour. Policies and other measures may target either those who provide the goods (*e.g.* manufacturers and importers) or those who use and dispose of the goods (*e.g.* householders), or both.

Five types of policy options that have emerged as key policy drivers are discussed in this section:

   (i) Producer responsibility.
  (ii) Eco-labelling.
 (iii) Charges and economic incentives.
 (iv) Persuasion measures.
  (v) Integrated product policy.

To illustrate the mix of measures that can be adopted, Table 16.2 provides examples of how six waste management authorities in the EU have approached the problem of diverting waste from landfill, and raising recycling and recovery rates.[20]

16.4.2 Producer Responsibility

16.4.2.1 Principles

Waste avoidance, waste minimization and the separation of waste materials at source to facilitate recycling, all require the active participation of the waste producers and waste generators, including householders and commercial/industrial companies. The concept of "producer responsibility" is one which has become the norm rather than the exception around the world for the management of such wastes. The concept is that the manufacturer or importer of the products giving rise to the waste should take responsibility for those wastes. These groups are thereby encouraged to consider the implications of disposal of their product and are given an incentive to investigate methods of reducing, reusing or recycling their wastes. The producer either levies a charge on the product to finance the cost of recovery and collection of materials, or uniquely in the UK waste packaging legislation, system costs are financed by reprocessing of waste material (as discussed later). The concept is illustrated in Figure 16.2.

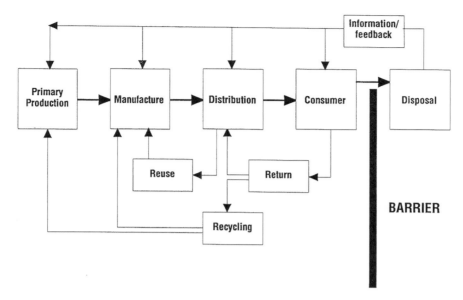

Figure 16.2 The concept of producer responsibility.

This concept has been implemented in a number of countries, often on a voluntary basis, whereby industry negotiates agreed targets for waste prevention and recycling with government, and is then left to implement these in the most cost-effective manner. All voluntary schemes worldwide have been negotiated between industry and government on the understanding that, if a satisfactory agreement is not reached or if agreed targets are not met, a mandatory scheme will be introduced. However, within the European Union the introduction of legislation such as the *Packaging and Packaging Waste Directive* has generally resulted in voluntary quotas being replaced by national targets (see the following section). The concept has been applied to a range of products, including batteries, waste paper, end-of-life vehicles, *etc.* Three examples relating to waste packaging, end-of-life electrical and electronic equipment and batteries are described in the following sections.

16.4.2.2 Waste Packaging

One of the first and most studied producer responsibility schemes is that which applies to waste packaging, following the EU-wide *Packaging and Packaging Waste Directive 94/62/EC* (as amended in *Directive 2004/12/EC*), which came into force on 31 December 1994. The Directive set recovery targets (60%) and recycling targets (55%) for packaging waste, as well as material-specific recycling targets, to be met by 31 December 2008. After 2008 member states must continue to meet these minimum targets, but they have the freedom to set higher targets if they so choose. The UK responded with the *Producer Responsibility Obligations (Packaging Waste) Regulations* which came into force in 1997 (subsequently amended in 1999 and in 2007) and is currently consulting on higher targets to run from 2013 to 2017.[21] The Regulations establish a shared responsibility for each participant in the packaging chain, from supply to waste creation, as well as an incentive to reduce the cost of compliance by reducing the amount of packaging handled. Obligated businesses can implement their obligations independently, or join one of a number of registered compliance schemes which certify compliance on behalf of its members and meet their obligations in the aggregate. These compliance certificates are called Packaging Waste Recovery Notes (PRNs) or Packaging waste Export Recovery Notes (PERNs). Surplus PRNs and PERNs can be traded, the intention being that the resale value of the certificates would provide reprocessors with an incentive to increase capacity.

Rather than relying on a market-led mechanism to deliver the *Packaging and Packaging Waste Directive* targets, other European Union member states have taken a different approach to that of the UK by adopting a levy-based system. Typical of these schemes are the Duales System Deutschland (DSD) packaging waste system in Germany, the FOST-Plus system in Belgium, and the Eco-Emballages scheme in France. While it is claimed that the costs of these latter systems are greater than in the UK, robust country comparisons are in fact very difficult to make due to the different structures and obligations placed by national bodies on the various stakeholders along the supply and waste management chain.

16.4.2.3   *The WEEE Directive*

The *EU Directive on Waste Electrical and Electronic Equipment 2002/96/EC* (the *WEEE Directive*) came into force on 13 February 2003. The Directive has three aims: to prevent waste from end-of-life electrical and electronic equipment; to set out measures for re-use, recycling and other forms of recovery; and to minimize the risks and impacts associated with the production and the treatment and disposal of end-of-life electrical and electronic equipment. The Directive sets targets for re-use, recycling and recovery. In seeking to ensure that products are designed to make recycling easier by removing hazardous substances, such as heavy metals, from the waste stream, it is supported by two further Directives: *2002/95/EC on the Restriction of the use of Certain Hazardous Substances in Electrical and Electronic Equipment* – the "*RoHS Directive*" – and the *2005/32/EC Directive on Establishing a Framework for the Setting of Eco-design Requirements for Energy-Using Products* – the "*EuP Directive*". The *WEEE Directive* places the onus for the collection, recycling and reuse of equipment on the electrical and electronic equipment producers. The Directive affects all players within the electronics industry including producers, importers and distributors. Producers are responsible for the development of collection procedures associated with end-of-life goods. The recycling targets range from 70–90% of end-of-life goods (by weight) depending on the product category.

16.4.2.4   *The Batteries Directive*

The *EU Directive on Batteries and Accumulators (2006/66/EC)* came into force on 26 September 2006. Its aim is to reduce the negative environmental impacts of waste batteries and accumulators by encouraging their collection and recycling, harmonising labeling requirements and by imposing limits on certain hazardous substances. The operating principles, mandatory in all member states, include the banning of all batteries and accumulators containing greater than 0.002% cadmium (subject to some exceptions) and the establishment of collection schemes. Take-back of portable batteries must be free of charge for consumers. Collection targets have been set for portable batteries only – 25% by weight by 2012, and 45% by weight by 2016. Minimum recycling targets (by weight) have also been set: 65% for lead–acid, 75% for nickel cadmium and 50% for all other battery types. In the UK, battery compliance schemes register their members with the competent authority and collect, treat and recycle batteries on behalf of large producers (those placing more than one tonne of portable batteries on the UK market each year). Distributors of portable batteries/accumulators who sell over 32 kilogrammes of portable batteries per year in an individual store must provide a free in-store battery take-back facility for any member of the public who has batteries to dispose of.

16.4.3   Eco-labelling

Eco-labelling schemes for products are designed to encourage consumers to purchase environmentally friendly (including low waste) products. Labels may help to facilitate recycling for materials which would otherwise be difficult to segregate (*e.g.* different types of plastics). Eco-labels seek to give a product a rating for environmental impact and assist consumers in making environmentally based purchasing decisions and to educate consumers about environmental issues. Eco-labelling helps to support and reinforce other policy measures. For example, if a product charge were introduced, an eco-label could briefly explain the purpose of the product charge. The earliest eco-labelling schemes date from the late 1970s – for example, the Blue Angel scheme in Germany, introduced in 1977. Eco-labelling schemes can be found in most regions, ranging from the Green Seal scheme in the USA, to the Nordic Swan scheme in the Nordic countries, the ECOMARK scheme in India, the Green Label scheme in China, the Green Mark Program in Taiwan, and the Eco-Mark scheme in Japan. The Forest Stewardship Council (FSC) scheme, launched in 1993, applies internationally. The FSC evaluates, accredits and monitors forest products and their sources. Currently, forests covering 165 million hectares in 80 countries have been certified, including 1.6 million hectares in the UK.

Within the EU, *Council Regulation EEC 880/92* set up an Eco-Label Award Scheme in 1992, with the objective of establishing an easily recognizable label for all member states. A major revision was proposed in *Regulation EC 1980/2000*, with the aim of establishing a voluntary eco-label award scheme intended to promote products with a reduced environmental impact during their entire life cycle and to provide consumers with accurate, non-deceptive, science-based information on the environmental impact of products. A further revision was introduced through *Regulation EC 66/2010*, which established a European Union Ecolabelling Board (EUEB) consisting of the representatives of the competent bodies of all the member states, and of other interested parties such as consumer groups. The EUEB advises the Commission and issues recommendations on minimum environmental performance requirements. The European Eco-label scheme is currently applied across 23 product and service sectors, with licences awarded to more than 1300 companies, covering 17 000 different product types.

In order to control dubious and confusing environmental information supplied to consumers on products, the Standards community have specified different types of environmental claim. Under the umbrella of the International Standards Organization (ISO), national experts have developed standards related to both self declared environmental claims and the formal eco-labels. A terminology has developed related to the various types of environmental claim:

(i) Type I claims are based on criteria set by a third party. The awarding body may be either a governmental organization or a private non-commercial entity.
(ii) Type II claims are based on self-declaration by manufacturers.
(iii) Type III claims are based on product information on environmental aspects/impacts, but without comparing or weighting against other products.

In addition to labeling, the standards framework covers environmental claims made on product or packaging labels, through product literature, advertising, or other applications. International Standard *ISO 14021:1999* covers Type II claims, *ISO 14024:1999* covers Type 1 claims, while *ISO 14025:2006* covers Type III claims.

16.4.4 Charges and Economic Incentives

Four types of economic incentive are examined in this chapter:

(i) Grants and subsidies.
(ii) Recycling credits.
(iii) Preferential purchase policies.
(iv) Taxes and tradeable permits.

16.4.4.1   Grants and Subsidies

Grants in the form of low interest loans or one-off cash payments and/or tax allowances for investments may be used to encourage innovative projects on waste avoidance, waste minimisation, collection and processing of recyclable. This type of support is compatible with funding schemes for demonstration projects introduced at the inception stage before the innovation becomes a commercially viable option in the free market system. Grants and subsidies tend to be aimed at supporting the initial capital outlay for the introduction of new technologies or for "kick starting" other waste reduction initiatives. Normally, grants would not exceed 50% of the total capital required, though in some cases up to 80% of the project costs can be funded.

Several countries use grants or subsidy systems to encourage the development of recycling schemes or initiatives. Subsidy systems can be used to encourage both supply (encouragement of collection schemes) and demand (support for processing facilities and innovative project development) for materials for recycling. Examples include the subsidised introduction of home composting units in Canada, grants towards the establishment of separate collection programmes for household recyclables in Luxembourg, and grants to cover the capital cost of thermal recycling facilities for waste plastic in Japan. Ireland has a system of grants which cover up to 50% of the cost of approved recycling developments. Several countries also provide assistance to industry to adopt cleaner, low-waste, technologies.

16.4.4.2   Recycling Credits

A recycling credit can be defined as a payment to those who divert materials from final disposal for recycling. The recycling credit would reflect the saving in reduced collection, transfer and disposal costs. The aim of recycling credits is to encourage the recovery of materials in situations where the economics of doing so would be marginal. Typically, these situations include operations where the amounts of materials recovered are small compared with the effort required and where the value of the recovered material is sufficiently low to make the recovery process marginal in financial terms. Recycling credits are not intended to support recovery activities that are already economically viable or those that are completely untenable, in that, for the example, there is no market for the recovered material.

Recycling credits are also aimed at the initial recovery end of the recycling process. Their role is to increase the amount of material removed from the waste stream in marginal situations, so that later in the process, they are in sufficient quantity or are of sufficient quality to make the remainder of the process viable. They are not intended to support bulking up operations or, indeed, recycling industries.

Recycling credit schemes have been introduced in the UK, Canada and Australia. In the UK, a scheme was introduced under Section 52 of the *Environmental Protection Act 1990*, and implemented by Waste Disposal Authorities (WDAs) paying a credit to Waste Collection Authorities (WCAs) operating recycling schemes, and also to third parties collecting waste for recycling. WCAs in turn can make a discretionary payment to recyclers of a collection credit commensurate with the saving in household waste collection costs.

16.4.4.3   Preferential Purchase Policies

A concern with many recycling initiatives relates to identification of reliable markets for the recovery materials. Many recycling initiatives in the past have failed because either markets for the recovered materials could not be found or because they collapsed shortly after the scheme was initiated. Importance must therefore be given to stimulating the demand side of the recycling process to ensure that sufficient markets exist to absorb the recovered materials. A preferential purchasing scheme is one of the options for achieving this. So-called Green Public Procurement (GPP), which takes advantage of the huge collective purchasing power of the public sector to stimulate the market for goods and services with a lower environmental impact, is one such initiative being examined by the European Commission.

The need for a preferential purchasing policy implies that the products targeted cost more than other products. If the products receiving the preference could operate under normal market conditions, the preference would not be needed. For example, if recycled paper is more expensive than paper produced from virgin pulp because there is greater demand for the latter, then until the demand for recycled paper increases, the small market for recycled paper means that the prices are higher (due to diseconomies of scale). The objective is to encourage greater purchases which will stimulate price reductions and, in the long run, encourage further purchases of recycled paper.

Other examples of demand side initiatives include guarantes of a certain minimum price for waste paper, as practised by local authorities in the Netherlands, and the supporting of renewable energy from the combustion of waste under the Renewables Obligation, Renewable Heat Incentive and Feed-in Tariff schemes in the UK.

16.4.4.4 *Taxes and Tradeable Permits*

These economic measures are used both to discourage specific waste management practices, and to influence environmentally beneficial alternatives. Landfill taxes are intended to raise landfilling costs to better reflect the environmental impact of this form of waste disposal, while also discouraging producers from producing waste. The UK introduced a two-tier landfill tax in 1996: £2 per tonne for inactive or inert waste and £7 per tonne for all other "active" waste. The charge is borne by waste producers, and to offset their costs, revenues from the Landfill Tax are used to reduce the level of employers' National Insurance Contribution. The tax is collected by Her Majesty's Revenue and Customs (HMRC) through the landfill operators. Since 1998 an annual escalator (currently standing at £8 per tonne per year) has been applied to the higher tax rate, but the lower tax rate for inert/inactive waste has remained unchanged at its 1998 level of £2.50 per tonne. At current and projected escalations at the higher rate, the UK will arguably have one of the highest differential landfill tax rates of the twenty EU member states that apply a tax – £2.50 per tonne against £80 per tonne at the higher rate in 2014/15, by which time the total amount raised by the tax is estimated to exceed £1 billion. The Netherlands has a comparable differential (€16.79 per tonne for inert waste against €107.49 per tonne for active waste) but with only 5% of the country's waste being landfilled and with a landfill ban in place, the tax is in the process of being withdrawn.

In parallel with this tax the UK operates a Landfill Communities Fund, through which landfill operators can claim credit against their landfill tax payments if they make a voluntary contribution of up to 5.6% of their annual landfill tax bill to an approved environmental body. Up to 90% of the contribution can be reclaimed. The contribution to the environmental body is intended to finance environmental improvement schemes approved by the Environmental Trust Scheme Regulatory Body Limited (ENTRUST). Landfill operators may bear the remaining 10% themselves, or else an independent third party can make up this 10% difference.

The concept of a tradeable permit was introduced in section 16.4.2.2 in relation to the PRN and PERN system associated with the packaging waste regulations in the UK. This market-based approach is intended to incentivise reprocessors to provide more facilities, and raise the collection, recovery and recycling rate of waste packaging. Revenue from the scheme varies according to the demand for compliance certificates to meet company obligations relative to the quantity of recycled packaging placed on the market. However, insufficient funds from PRN sales appear to have been directed into investment in recycling projects. Other trading schemes within the UK have included the Landfill Allowance Trading Scheme, which commenced in 2003 and placed landfill diversion targets on local authorities, which were to be met by the sale or purchase of landfill allowance certificates. The scheme will terminate at the end of the 2012/13 scheme year, with trading concluding on 31 September 2013. The UK's Carbon Reduction Commitment Energy Efficiency Scheme introduced in 2008 is another trading-based mandatory scheme aimed at improving energy efficiency and cutting greenhouse gas emissions in large public and private sector organisations. Waste management facilities qualify as participants if, during the 2008 calendar year, they had at least one half hourly electricity meter, settled on the half hourly market and if they consumed at least 6000 megawatt hours through all half hourly meters. A half hourly meter registers how much electricity is used by a site every half hour of every day, helping the company to better understand and regulate its energy use. The readings are typically transmitted electronically in real time to the energy supplier, allowing for more accurate billing. The European Commission is currently examining the feasibility of designing recycling certificates to incentivise companies to use recycled materials in their processes as an alternative to virgin raw materials.

16.4.5   Persuasion Measures

Several types of persuasion measures are available, either based wholly on economic incentives, or linking economic measures with mandatory "command" policies.

16.4.5.1   Take Back Requirements

In the "take back" system, householders are required to return the waste to the retailers to enter a parallel, private waste collection, recycling or disposal system rather than the public system (*e.g.* the Austrian scheme); or it could be a compulsory deposit-refund system (*e.g.* as in Taiwan); or it could involve a charge levied on either raw materials used or on products sold.

16.4.5.2   Product Charges and Product Taxes

Product charges are levied at the point of consumption on products that are harmful to the environment. The charges can be set at such a level as to achieve the desired reduction in usage, or to incorporate some or all of the costs of recycling or disposing of the product. The former is preferable.

  Italy, Austria and Ireland are among a number of countries to have introduced product charges on disposable carrier bags in an effort to reduce waste and encourage consumers to use durable bags. The Republic of Ireland introduced a charge of €0.15 per bag in 2002, which led to a 95% reduction in plastic bag litter. Within a year, 90% of shoppers were reported to be using long-life bags. The levy was raised to €0.22 in 2007 after evidence showed that the number of plastic bags used annually had risen from 21 per person immediately after the ban to 30 (compared with 328 previously). The levy raised €75 million in revenue, which was put into an environment fund and used to reduce waste and research new ways of recycling. Wales introduced a levy of 5p per bag in 2011, and Northern Ireland is considering introducing a plastic bag levy.

  Product taxes or levies are used in many European countries as a means of raising the price of disposal or non-recyclable goods relative to less environmentally demanding alternatives. Taxes are commonly used as an incentive to set up deposit/refund schemes, as is the case in Norway and Finland, where tax exemptions are allowed if a suitable return rate is achieved.

16.4.5.3   Refunds

Deposit refund schemes are another policy instrument for waste reduction. For example, there are several such schemes in Scandinavia, while in Germany and the Netherlands deposit schemes for plastic beverage containers were introduced. There are also examples of deposit refund schemes used for a number of other products including car bodies (Sweden, Norway), batteries (Denmark, Netherlands, the US) and disposable cameras (Japan). In Korea an industrial deposit refund scheme has been set up whereby manufacturers are required to pay a deposit to the government which is refunded if, after customer use, the company collects and treats the product itself.

16.4.5.4   Compulsory Collection and Recycling

Legislation to force local authorities to collect and recycle materials is becoming a widely used means of achieving a reduction in the quantity of MSW which is sent for disposal. Measures of this type usually focus on particular components of the MSW stream. For example, legislation enacted in the Netherlands and Austria obliges municipal authorities to set up source-separated organics collection programmes to collect and compost household organic waste. Similar legislation applies in the UK through the *Household Waste Recycling Act 2003*, which requires all English waste collection authorities to collect at least two types of recyclable waste from all households in their area. The *Directive on Waste 2008/98/EC* requires member states to separately collect paper, glass, metal and plastic wastes from households, and to meet specified recycling targets for these materials. These obligations have in turn been passed down to local authorities responsible for household waste collection.

16.4.6   Integrated Product Policy (IPP)

Integrated Product Policy (IPP) represents an important paradigm shift in policy development within the EU and its member states. The concept of IPP differs from traditional policy approaches (which are fundamentally process focused, being source, substance and media specific) in that it is based on a life cycle perspective, and seeks to avoid shifting environmental problems from one stage of the life cycle to another. It further avoids the risk of conflicting life cycle legislation. IPP has been defined as "public policy which explicitly aims to modify and improve the environmental performance of product systems", and in the context of waste management, integrates different, but related aspects of product consumption:[22]

(i)   Preventing and managing wastes generated by the consumption of products.
(ii) Introducing more environmentally sound products.
(iii) Creating markets for environmentally sound products.
(iv) Transmitting information up and down the product chain.
(v)   Allocating responsibility for managing the environmental burden of product systems.

Thus, many of the initiatives discussed individually above, such as waste minimization, waste prevention, chemicals policy, eco-labeling, and standardization, and producer responsibility can be integrated under the umbrella of IPP.

In 2003 the European Commission adopted *Communication COM (2003) 302 final* on IPP, which set out an implementation strategy based on five key principles:

(i)   Life cycle thinking – considering a product's life cycle "from cradle to grave".
(ii) Working with the market – setting incentives to encourage the supply and demand of greener products.
(iii) Stakeholder involvement – encouraging all those who come into contact with the product to act on their sphere of influence.
(iv) Continuous improvement – in design, manufacture, use and disposal.
(v)   A variety of policy instruments – from voluntary agreements to regulation.

Since 2003 the European Commission has been reviewing its environmental and product-related legislation to set the right economic and legal framework for IPP. Current initiatives include a review of the *Ecodesign Directive 2009/125/EC* to extend its provisions beyond energy using products to include non-energy intensive products, public procurement legislation to stimulate the supply and demand of greener products (GPP – see section 16.4.4.3) and (in the *Directive on Waste 2008/98/EC*) deregulating waste-derived materials that can be demonstrated to have attained product-like properties.

IPP initiatives are not restricted to the EU. In the UK a voluntary responsibility deal called the Courtauld Commitment was brokered in 2005 with the grocery retail sector, with the aim of improving resource efficiency and reducing the carbon and wider environmental impact of the sector. Phase 2, launched in 2010, moved away from solely weight-based targets and aims to achieve more sustainable use of resources over the life cycle of products, throughout the whole supply chain. The responsibility deal currently has over 50 signatories. Similar voluntary responsibility deals and codes of practice have been set up with the waste management and construction sectors.

16.5 BULK WASTE REDUCTION TECHNOLOGIES AND FINAL DISPOSAL

Whatever success is achieved in reducing waste arisings and in separating materials for recycling, some waste will always remain. To achieve high waste reduction rates in terms of landfill demand, a technology component is required. Some options are listed in Table 16.3.

Other than the physical size and weight reduction technologies such as baling and separation, of the options listed in Table 16.3, waste fired power generation, anaerobic digestion and composting are perhaps the most widely used. These three options are discussed in this section, along with landfilling, which remains the most common waste disposal option.

Table 16.3 Selected bulk waste reduction technologies.

Options	*Technologies*
Size reduction technologies	• Baling • Pulverization/shredding • Homogenization/wet pulping
Weight reduction technologies	• Separation • Materials recycling facilities • Waste derived fuels
Waste to energy generation	• Mass burn incineration • Fluidised bed incineration • Combustion of prepared waste derived fuels
Other combustion technologies	• Aggregate/block production • Cement kiln firing • Wood burning power/CHP stations • Tyre burning power/CHP stations • Gasification • Pyrolysis
Biological systems	• Composting • Vermiculture • Hydrolysis • Anaerobic digestion
Combined systems	• Mechanical biological systems combining shredding and other separation processes with biological processes such as composting

16.5.1 Combustion/Incineration

Waste incineration has been practised as a waste management and volume reduction technique since the 1890s. It is an extremely effective bulk waste reduction technology, typically reducing waste volume by 90% and mass by around 70%. In terms of waste processing, incineration is a relatively simple option, with residual waste (after removal of recyclables) being fed into a furnace and, by burning in an excess of air, reduced to one-tenth of its original volume. Typically the only materials removed from the waste stream prior to burning are large bulky objects or potentially hazardous materials such as gas bottles.

The combustion gases then typically pass through a boiler system to recover energy. The most flexible means of recovering energy from the hot gases is to produce steam for direct use (at lowered temperature) or for electricity generation. To generate electricity, superheated steam is passed from the boiler system through a turbine generator. Depending on the nature of the turbine, the resulting temperature of the steam can still be sufficient for its subsequent use in district or process heating. Around 550 kilowatt hours of electricity can typically be produced from a tonne of waste input, compared to 2300 kilowatt hours of electricity from industrial coal. Where both steam and electricity are produced by waste burning, the thermal efficiency of such schemes is between two and three times that of electricity-only production, although such co-generation schemes requires specialised turbine technology.[8]

The gases are then cleaned prior to discharge to atmosphere through a tall stack, in order that the discharge conforms to the strict emission limits laid down by national governments. Gas cleaning strategies aim to remove the following components:

(i) Acid gases such as nitrogen and sulfur oxides, and hydrogen chloride. A variety of wet and dry scrubbing systems can be applied, but in essence the gas cleaning method involves neutralisation of the acidity by dosing with an alkaline reagent such as lime. Chemical methods of controlling nitrogen oxide emissions (so called De-NOx systems) have also been developed.

(ii) Particulate matter, often associated with trace metals and semivolatile organic micro-pollutants. Fabric filters are generally used for this purpose. The particulate matter is retained in the electrostatic precipitator or the filter, and is periodically removed from the system as ash which requires land disposal.

(iii) Organic micropollutants such as dioxins. Control of these emissions is achieved by control of combustion conditions, as well as by dosing in the gas cleaning train with activated carbon or other adsorbents. A fabric filter located further downstream removes the adsorbent material along with the reaction products of acid gas with the neutralizing agent. Simultaneous reduction of emissions of metals such as mercury and cadmium is also achieved.

The impact of prior materials separation and recycling on the calorific value of waste sent for incineration has been studied. UK research has indicated that the calorific value will fall if large quantities of only paper and plastics are recovered. However, if putrescibles are also separated from the waste stream, there is little effect on the overall calorific value.[8]

The EU-27 incinerated 51 million tonnes of municipal solid waste in 2009 (20% of the total amount treated) in approximately 400 combustion facilities.[23]

16.5.2 Other Thermal Processes

Gasification and pyrolysis are thermal treatment techniques that are being developed as alternatives to mass burn incineration. Unlike conventional combustion, pyrolysis involves heating the waste in the complete absence of air. The waste decomposes to produce a char-like solid, liquid hydrocarbons and a hydrocarbon-rich gas.[8] In contrast, combustion of the waste in excess of air in a conventional incinerator will produce a highly oxidised and degraded gas with very little residual hydrocarbons, and an inert solid residue in the form of ash.

Pyrolysis offers some advantages over conventional incineration. Firstly, because no air is fed into the combustor, far less waste gases are produced and therefore the gas cleaning system can be smaller and hence less costly. Secondly, the waste itself must be pre-prepared to make it homogeneous and to remove bulky materials. This results in a smaller volume of waste being treated, and hence overall, a smaller plant. Thirdly, in theory, the solid, liquid and gaseous streams can be further processed into useful products (oil, soil conditioner, *etc.*) and hence there should be less material to discard to landfill. While these merits are attractive, the technology is still relatively unproven insofar as commercial size and applicability to MSW are concerned.

Gasification operates along similar lines to pyrolysis, except that oxygen is fed in sub-stoichiometric quantities into the combustor. The waste is partially oxidised at high temperature to again produce a hydrocarbon-rich gas, and an effluent.[8] The potential benefits and disbenefits of gasification units are similar to those for pyrolysis.

16.5.3 Composting

Composting is essentially the controlled aerobic decomposition of putrescible material. Of the various methods of aeration, windrowing (mechanical or manual turning of the material) and forced aeration of the static pile are the most common methods used. However, closed systems in which the composting process is regulated in rotating drums or packed towers are increasingly being applied in order to reduce environmental nuisance from dust and odours. During composting, putrescible material is progressively broken down by microorganisms in a series of distinct stages. In the mesophilic stage, microorganisms begin to actively break down the organic material, the temperature of the composting material rising to around 50 °C in about two days. During the second, or thermophilic stage, temperatures begin to rise so that only the most temperature resistant microorganisms survive. As the microorganism population reduces, the composting material cools and anaerobic conditions may develop unless sufficient air is introduced. In the third stage, the material continues to cool and microorganisms begin to compete for the remaining organic material, in turn leading to breakdown of cellulose and lignin in the waste. During the final, maturation stage, levels of microbial activity continue to fall as the remaining organic material is broken down, and the microorganisms die off as their food sources deplete.[24]

Overall, maintaining the correct balance of oxygen, and therefore temperature, is crucial for the successful degradation of wastes. If the process becomes anaerobic, odour problems can be severe and microbial activity can cease. Temperature and moisture content can be monitored by electronic sensors. Depending on the type of organics and moisture levels in the material to be composted, volume and weight reduction levels for that waste fraction are typically in the range of 40–60% and 40–50%, respectively.

Maintaining a consistently high quality, and as such a marketable compost product, and producing it in an efficient and environmentally acceptable manner are critical to the success of composting schemes. A key factor has been a switch to the composting of uncontaminated source separated organic wastes rather than of mixed MSW. Operational experience in the US and elsewhere with mixed MSW feedstock has indicated that the end product is often of poor quality and has limited end use. Present composting schemes have therefore focused on processing source segregated kitchen and garden wastes, "green" wastes from parks and gardens, and food wastes from the food processing industry and large commercial generators. A segregated waste input is a pre-requisite for compost to achieve end-of-waste status in the EU (see section 16.1).

16.5.4 Anaerobic Digestion

Anaerobic digestion (AD) is the microbiological conversion of organic matter in the absence of oxygen. AD generates two main products:

(i) Biogas – a mixture of carbon dioxide and methane, which can be used to generate heat and/or electricity.
(ii) Digestate – a mixture of fibre (which can be used as a nutrient-rich soil conditioner, and liquor (which can be used as liquid fertilizer).

The feedstock for AD can range from wastewater sludge to food waste, agricultural manures and residues, and industrial organic wastes from the pharmaceutical, textile or rendering sectors. The process takes place in a sealed airless digester into which the prepared feedstock is pumped either as a low solids (3% total solids by weight) or medium solids (3–12% total solids by weight) "wet" mixture, or as a high solids (up to 25% by weight) so-called "dry" mixture depending on the characteristics of the waste. There are typically two types of plant, operating under either mesophilic or thermophilic conditions. In the former, most commonly used process for anaerobic digestion, decomposition of the volatile suspended solids is around 40% over a retention time of 15 to 40 days at a temperature of 30–40 °C. In thermophilic digestion the digester is heated to 55 °C and held for a period of 12–14 days. Mesophilic digestion is generally more robust than the thermophilic process, but the biogas production tends to be less, and additional sanitisation is usually required. Thermophilic digestion systems provides higher biogas production, faster throughput and an improved pathogen and virus "kill", but the technology is more expensive, more energy is needed and it is necessary to have more sophisticated control and instrumentation.[25,26]

During the digestion process 30–60% of the organic material is converted into biogas, which can be burned in a conventional gas boiler for heat or in a combined heat and power (CHP) system to generate heat and electricity. The digestate is stored and can be applied straight to land or it can be separated to produce fibre and liquor.

The devolved governments of the UK are actively supporting the wider uptake of AD. For example, the administration in England appointed an AD Task Group in 2009 to develop an implementation plan for AD, *Accelerating the Uptake of Anaerobic Digestion in England: An Implementation Plan*, which was published in 2010,[27] followed by an *Anaerobic Digestion Strategy and Action Plan* in 2011.[28] The Plan sets out actions in creating a long-term economic framework, a robust regulatory framework and incentives to build capacity – such as the £10 million AD Demonstration Programme. An infrastructure study estimated that by 2011 the UK had 214 AD facilities with an overall capacity of 5 million tonnes of material per year and with a total installed generating capacity of over 170 MW of electricity.[29] Two of these plants injected biogas directly into the gas grid.

In 2009 the governments of England and Wales launched an *Anaerobic Digestate Quality Protocol (BSI PAS110)* for the production and use of products from the anaerobic digestion of source-segregated waste.[30] Digestate which meets certain set criteria can be classified as a product rather than a waste, so that the material would no longer be subject to waste management controls. The aim of the Quality Protocol is to stimulate the market for the use of anaerobic digestate products as a fertiliser, soil conditioner and a material in land restoration and reduce the amount of organic waste being sent to landfill. In 2012 the governments of England and Wales consulted on a quality protocol setting out proposed criteria for the use of biomethane as a fuel. This applies where it is supplied through the gas grid and used as fuel for vehicles. The quality protocol will allow biomethane produced from biogas generated at landfills and AD treatment plants to make a contribution to renewable power, and help to ease the regulatory burden on businesses by defining when the material ceases to be waste and is no longer subject to waste management controls.

16.5.5   Mechanical Biological Treatment (MBT)

Mechanical Biological Treatment (MBT) is a term used to describe a particular waste treatment concept for the management of municipal and non-hazardous industrial and commercial waste. MBT plants combine individual mechanical and biological processes such as shredding, crushing and milling, screening and other mechanical waste classification processes, magnetic separation of metals, and ballistic sorting systems, biological drying, composting, AD and pelletising in different ways depending on the output required.[31] An MBT plant combines mechanical processes to separate out the dry recyclables such as glass and metals, with biological processes to drive out moisture and to handle the organic-rich fraction of the incoming waste. In addition to the separation of dry recyclables from the incoming waste stream, the plant can be designed to produce:

(i)   An energy-rich refuse derived fuel (RDF) comprising paper, plastics and other combustible fractions, that can be combusted in an energy from waste plant or in an industrial furnace.
(ii) An organic-rich fraction that is suitable for composting or anaerobic digestion.
(iii) A biologically stable material that can be landfilled.

As an example, in a typical RDF application incoming waste is pre-sorted to remove glass and bulky objects, crushed, and biologically dried (aerobic fermentation). The dried material is then subjected to mechanical separation to produce three fractions: RDF, ferrous metals, and an inert residual fraction. Depending on the nature of the incoming waste, 20–30% of the original weight is lost as water during the drying stage, 45–55% is converted into RDF, 10–15% is recovered as metals and glass, and the remainder is residue that is landfilled. In cases where an MBT plant produces a recyclable or marketable product (an RDF, biological material that is subsequently composted, *etc.*) the issue of product quality and consistency is paramount if a long-term market is to be secured. Products developed from mixed waste inputs are often too contaminated to permit their use in anything other than basic, low value applications, be it as compost or RDF. This restricts the commercial viability of an MBT plant designed to handle unsorted municipal waste, or mixed residual waste from households.

16.5.6   Landfilling

Landfills are the final destination for the residues from incineration and other treatment and processing options, as well as for the primary waste stream – in the UK about 55% of domestic waste and 50% of commercial waste was dispatched directly to landfill in 2011.[4] It is therefore a critical element in a waste management strategy since, despite the best attempts at waste minimisation, recycling and recovery, its use is unavoidable. The challenge is to design and manage the process of landfilling in a sustainable manner so as not to leave a long term potential for environmental damage.

The process of landfilling consists of the following steps:[7]

(i)   Preparation – waste may be processed prior to landfilling, for example by shredding, baling or compaction, which often takes place as part of the transfer system or in a MRF rather than at the landfill site.
(ii) Waste placement – waste is deposited in phases or cells by tipping from the delivery vehicles, followed by spreading over the area by on-site plant. The deposited waste is often compacted by specialist plant or by the movements of the bulldozers to increase the density of the waste, to reduce subsidence and to preserve valuable void space. At regular intervals, typically daily, the waste is covered with an inert material to protect it from scavenging birds and animals, and to reduce odours and airborne dispersal.

(iii) Landfill completion – after the void space within the site is filled, the landfill is capped with a low permeability layer to minimise the ingress of rainwater, and covered with soil to return the area to the surrounding landscape.

In practice, landfill restoration is a continuing process throughout the lifetime of the site; void space is typically utilised in a phased manner, with progressive contouring, capping and restoration.

Within the landfill, the constituents of the waste undergo biodegradation and stabilisation. The infiltration of rainfall and surface waters into the waste mass, coupled with the biochemical and physical breakdown, produces a leachate which contains soluble components of the waste.

The breakdown products also include so-called "landfill gas", which can be harnessed for its energy content. The maximum gas volume which is generated from the decomposition of organic matter is in the region of 350–400 m^3 per tonne of MSW, amounting to an average of about 6 m^3 per tonne per year. In reality gas generation peaks in the early years and then gradually decreases as the waste ages, diminishing to non-commercial recovery levels after 30–40 years. The gas typically consists of 50–70% methane and 30–50% carbon dioxide with traces of nitrogen, hydrogen, oxygen, hydrogen sulfide and a range of trace organic compounds. Landfill gas utilisation schemes can capture 50–60% of the gas released by the site; its combustion can potentially generate about 350–370 kWh of energy per tonne of MSW.

16.5.7 Environmental Considerations

16.5.7.1 Waste Incineration

All bulk waste reduction technologies have the potential to produce gaseous, liquid and solid contaminants. For waste incineration, the key environmental issues are:

(i) Dust and odour from waste handling and storage.
(ii) Ash management (grate ash and fly ash from the gas cleaning system).
(iii) Atmospheric emissions from the stack.

The potential effects of dust and odour from waste handling can be controlled by the following measures:

(i) Location of the waste unloading area within an enclosed building.
(ii) Extraction of air from above the storage bunker for use as combustion air in the incinerator, where any odorous compounds and dust entrained in the air will be destroyed. This also results in a slight negative pressure within the building which draws air inwards, thus minimising the escape of dust or odours.
(iii) The use of dust suppression waste sprays at the waste tipping bays.

The first two of these measures are standard in all modern incinerator plants, and are not identifiable as separate environmental mitigation within a scheme.

Grate ash (also called Incinerator Bottom Ash or IBA) and fly ash are produced at a rate of 20–30% and 3–4% of the waste input, respectively.[8] IBA is essentially inert; fly ash has relatively high levels of heavy metals and is classed as a hazardous waste in some countries. Due to its inert characteristics, the disposal of IBA is not generally considered a significant pollution risk and it is generally used for beneficial purposes such as road building, in construction materials, for intermediate cover at landfill sites, *etc.* Prior to disposal, the ash is normally scavenged for ferrous metals by magnetic separation; up to 10% of the ash may be recoverable ferrous material.[8]

The cleaning of combustion gases prior to their release to atmosphere has been discussed in section 16.5.1. Additional information of the environmental impact of atmospheric releases can be obtained from other references.[32,33]

16.5.7.2 Composting

The key environmental issues of composting are as follows:[7,34]

(i) Fugitive emissions of litter, dust and bioaerosols. These emissions may arise from wind dispersal of the waste feedstock during preparation and from compost piles.
(ii) Odour, arising from trace organics produced if anaerobic conditions are permitted to occur in the composting process.
(iii) Leachate generation, since the requirement for moisture in the composting process may lead to excess water and run-off from the composting pile, which may be contaminated by materials present in the waste.

Dust, litter, odour and bioaerosol emissions from waste material preparation can be mitigated by enclosure of the operation, both with respect to individual items of equipment (shredders, *etc.*) and by locating the process within a building with appropriate ventilation systems. Dust, litter and odour from the composting process can also be mitigated by enclosure of the activity, or by in-vessel composting. However, since these measures are relatively expensive due to the large volume of material which needs to be processed, windrow composting is most commonly applied. This is normally conducted out of doors, although Dutch barn buildings are sometimes used (*i.e.* roofed buildings with open sides). Regular wetting of the windrows and proper management of the composting process assist in minimisation of dust, odour and litter.

The requirement to keep the compost pile wet increases the potential for leachate generation from excess water and run-off. The control of this leachate is important for the protection of water resources. Typical design features include organisation of the windrows or compost piles to retain water inflow and avoid run-off, or the use of a concrete plinth equipped with a controlled drainage system to collect run-off.[35]

16.5.7.3 Anaerobic Digestion

Being an enclosed process, the environmental impact of an AD plant is generally related to pre- and post-digestion activities and discharges. The main impacts are as follows:

(i) Odour from waste delivery, waste handling and feedstock preparation activities, and from storage of digestate.
(ii) Release of bioaerosols from waste handling and feedstock preparation activities.
(iii) Release of leachate from run-off in the waste handling area, and during processing/thickening of the digestate.
(iv) Releases to atmosphere of combustion gases during electricity generation.

Odours from the waste handling, feedstock preparation and digestate storage areas are controlled by maintaining process buildings at negative pressure, and installing scrubbers or biofilters at the exits of the pump exhausts. Design measures include equipping the incoming waste building with quick-opening doors. Bioaerosols are also controlled by these means, with the additional precaution of discharging air discharged from the biofilters at a height sufficient to ensure adequate dispersion in the ambient environment. Leachate from the digestion process is collected and treated to the appropriate standard prior to discharge to the receiving water body. Being a relatively clean fuel, the combustion of biogas does not generate appreciable levels of airborne pollutants such as dust, heavy metals or organic micropollutants. Sulfur, nitrogen and halogens present in the waste will be combusted to their respective acid gases, though scrubbing and cleansing of the exhaust gases is rarely necessary. Biogas destined for direct injection into the gas grid has to be thoroughly cleaned of sulfur and materials such as siloxanes, an issue addressed in the biomethane quality protocol discussed in section 16.5.4.

16.5.7.4 Landfilling

Of the potential releases and environmental effects of landfilling, the following have raised most concern: [36–38]

(i) Leachate – leachate arises from the moisture contained in the deposited waste, from the infiltration of water into the site, and from the biodegradation process itself. Escape of leachate, for example due to engineering failures of landfill caps, covers and liners, has been linked to contamination of water resources.

(ii) Landfill gas – consisting of a mixture of methane and carbon dioxide, landfill gas can cause damage to vegetation and is also an explosion hazard. Methane and carbon dioxide are also greenhouse gases.

(iii) Trace organics – a variety of trace organic compounds can be entrained with landfill gas, for example vinyl chloride, benzene, toluene, alkanes, organosulfur compounds, *etc*. While many of these compounds are potentially toxic, their concentration in offsite air is generally too low to pose a threat to public health. Odour nuisance is potentially a more common problem.

(iv) Litter, vermin, noise, *etc.* – the nuisance aspects of landfills and their operation are potentially the most intrusive in terms of disturbance and disruption to the amenities enjoyed by the surrounding population. Careful consideration is given to operational work plans, traffic movements on and off site, the fitting of screens to reduce the visual intrusion and dispersal of litter, daily cover of the waste, *etc.* to mitigate against the possibility of nuisance.

Landfilling provides a method of reclaiming existing excavations as well as the development of new landforms. Hence, it can be used to return unproductive land to beneficial use. However, after landfilling operations are complete and the site has been capped, the *in situ* processes of bio-degradation continue for a significant length of time, measured in decades. Hence, the generation of leachate and landfill gas also continues, as does the potential for offsite migration of these releases. Post closure management of the landfill is therefore a key consideration; typically this takes the form of the installation of a landfill gas utilisation/control scheme, provision for leachate collection and treatment, and regular monitoring of releases from the site.

16.6 INTEGRATED WASTE MANAGEMENT STRATEGIES

16.6.1 Revisiting the Waste Management Hierarchy

In order to achieve an integrated approach to waste management, the following need to be included in any strategy:

(i) Policy and other measures to encourage the avoidance and minimisation of waste.

(ii) Policy and other measures to encourage the recovery, recycling and reuse of materials that would otherwise enter the waste stream.

(iii) Adoption of bulk waste reduction technologies which will effectively reduce the volume of materials remaining in the waste stream after the above measures have been put into place, thus minimising the amount of materials requiring final landfill.

While there are options within each of these categories, a strategy which omits the inclusion of measures from any one complete category is likely to achieve much lower levels of overall waste reduction than one which uses a more integrated approach. In conceptual terms, what an integrated strategy seeks to achieve is to convert a linear transference of materials through the producer–user–disposer chain, to a series of circular and as far as possible, closed systems where beneficial aspects of the waste (materials, energy, *etc.*) are drawn out at each stage.

However, the phrase "waste management hierarchy" may suggest that after waste prevention and minimisation, the remaining options invariably represent more environmentally acceptable solutions and conversely, options such as landfilling invariably represent less environmentally acceptable and less sustainable solutions. This view could lead to anomalous and unbalanced waste management strategies which take insufficient account of local situations and economic sustainability. For example, in a predominantly rural area in which MSW is generated in sparsely populated and widely separated conurbations, long transport hauls to a central composting or waste to energy incineration facility may not represent an environmentally or economically preferable option relative to local landfilling, if the cost, energy and emissions relating to transportation are taken into account. Reusable containers need to withstand repeated cleaning and handling, and therefore require more material and energy in their manufacture than do single-trip containers.[39] The overall energy consumption of a bring system which involves the householder in a 2 km car ride to a Civic Amenity site to deliver recyclables is over double that of a basic system involving non-segregated kerbside collection followed by landfilling.[40]

In order to optimise the overall waste management strategy, both with respect to the environment and to economics, it is necessary to address the entire strategy in a holistic sense. The final section of this chapter discusses the concept of life cycle assessment (LCA) and its role in the optimization of waste management strategies.

16.6.2 Principles of Life Cycle Assessment (LCA)

Life cycle assessment (LCA) is an environmental management technique in which the inputs and outputs of an activity are systematically identified and quantified "from cradle to grave"; that is, from the extraction of raw materials from the environment to their eventual assimilation back into the environment. The inputs and outputs include raw material and energy consumption, emissions to air and water and the production of solid waste. These environmental flows are then assessed in terms of their potential to contribute to specific environmental impacts which might include global warming, acidification, ozone depletion, eutrophication and toxic effects. By taking the comprehensive life cycle approach, one can identify the environmental advantages and disadvantages of alternatives with the assurance that all their upstream and downstream consequences have been taken into account. Comparisons made between competing systems can highlight counter-productive initiatives, which might at first sight appear to be worthwhile, but in reality merely shift an environmental burden to a different part of the waste management chain.

The process of life cycle assessment is generally regarded as consisting of four distinct activities:[41]

(i) *Goal Definition and Scoping* which defines the purpose and scope of the study and sets out the framework in which it will be carried out.

(ii) *Life Cycle Inventory (LCI)* which is the quantification of environmental burdens throughout the life cycle of the product, process or activity.

INPUTS
OUTPUTS

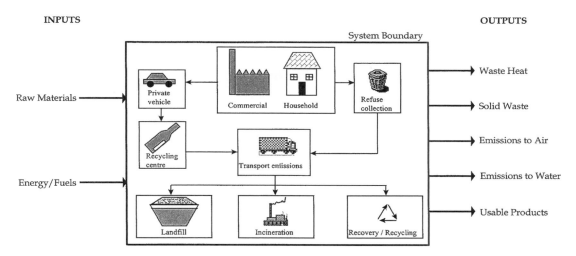

Figure 16.3   Environmental effects of waste management systems.

(iii) *Life Cycle Impact Analysis (LCIA)* which involves assessing the effects of the burdens identified in the inventory, and is often divided into:
   a. *Classification* – grouping burdens into impact categories.
   b. *Characterization* – quantifying the significance of burdens in each category.
   c. *Valuation* – assessing the relative importance of the categories.
(iv) *Life Cycle Interpretation* in which the results are assessed and applied.

The life cycle process is iterative, with preliminary results prompting a reassessment of goals and fresh data collection.

Typically a waste management system would be described in terms of the management of a quantity of waste of a given composition, which allows the comparison of alternative systems which might perform this service in very different ways (*e.g.* systems based on recycling, incineration or landfill). A schematic diagram of the elements of a life cycle inventory of a generic system is shown in Figure 16.3.

The system is defined as part of the Goal Definition and Scoping Phase of LCA, but in waste management studies might include, for example, the collection, transport, treatment and disposal of wastes, together with the recovery of materials and energy and there replacement for virgin materials and fuels.

The procedures of LCA have been codified at an international level in the *ISO 14000* series of Environmental Management Standards: in *ISO 14040:2006, 14044:2006, ISO 14048:2002* and *ISO 14049:2000*.

16.6.3   Selecting the Best Environmental Option for Individual Materials

LCA has been used to assess which option (typically, between landfilling, energy recovery and recycling) is preferred from an environmental standpoint. A comprehensive meta-analysis conducted in 2006[42] and updated in 2010, undertook an international review of LCA studies on key materials that are often collected for recycling – paper/cardboard, plastics, aluminium, steel, glass, wood and aggregates, extended to food waste and textiles in 2010. Several hundred studies were screened to arrive at 55 methodologically robust LCAs comprising over 200 different scenarios. In general, recycling consistently offered more environmental benefits and lower environmental impacts than other waste management options. For materials such as paper/cardboard and plastics, the results were sensitive to assumptions such as the energy mix used in virgin paper and plastics manufacture. LCA-based guidance on applying the five-step waste hierarchy (discussed in section 16.2.1) has been published in England for 11 individual materials or product streams.[43,44]

16.6.4 Waste Management and Climate Change

Waste management activities release greenhouse gases that contribute to global warming and hence to climate change. In England the government's *Waste Review 2011* estimated that direct emissions from the waste management sector in the UK (which includes wastewater treatment) accounted for 3.2% of the UK's total estimated emissions of greenhouse gases in 2009, or 17.9 million tonnes of CO_2 equivalents, of which landfill emissions contributed 89%. However, the net benefits of diverting waste from landfill into recycling and energy recovery were indicated by a comparison with sector emissions in 1990 of 59 million tonnes of CO_2 equivalents. *England's Waste Strategy 2007* envisaged a further reduction of 9 million tonnes of CO_2 equivalents per annum by raising recycling targets,[3] agreeing responsibility deals with key sectors, and potentially imposing landfill restrictions.

A number of studies have examined the greenhouse gas balance of waste management scenarios, and the reduction potential of recycling and energy recovery.[45–47] At an EU level, PROGNOS[47] compared the greenhouse gas emissions from three scenarios – business as usual with landfill, recycling and energy recovery at the EU average as practiced in 2004, against three scenarios with enhanced recycling and energy recovery. The study considered 18 individual waste streams comprising a total of 2.4 billion tonnes, of which 1.3 billion tonnes were landfilled in 2004. The study concluded that with ambitious recycling and energy recovery targets, the EU could save between 146 and 234 million tonnes of CO_2 equivalents annually, contributing 19–30% of the EU's greenhouse gas reduction target by 2020. Similar conclusions were reached by the European Environment Agency,[46] with recycling and energy recovery contributing 75% and 25% to avoided emissions, respectively. In the UK, the *Low Carbon Transition Plan*[48] requires the waste management sector to cut England's yearly emissions by the equivalent of 1 million tonnes of CO_2 by 2020, on top of the reductions already predicted, thereby reducing UK waste emissions to 13% below 2008 levels.

16.6.5 Waste Management Strategy for Northern Ireland

An example of the consideration of environmental criteria (through the application of LCA) alongside cost, social aspects such as employment, and feasibility criteria in aiding the decision-making process, is the study of alternative waste management options conducted by Environmental Resources Management (ERM) on behalf of the Environment and Heritage Services of Northern Ireland.[49] The base case and alternative waste management options are described in Table 16.4.

A stakeholder workshop was held to discuss the outcome of the LCA (covering environmental impacts) alongside the three other decision criteria listed above. The workshop elicited an average relative ranking of the decision criteria as follows (rounded to 100%):

  (i) Feasibility: 34%.
 (ii) Cost: 26%.
(iii) Environment: 24%.
 (iv) Social aspects: 16%.

On the basis of the analysis, the Best Practicable Environmental Option for municipal waste management in Northern Ireland was defined as follows:

  (i) Recycling and composting growing to stretching but achievable levels – 35%, 40% and 45% by 2010, 2013 and 2020, respectively.

Table 16.4 Waste management scenarios for Northern Ireland.[46]

Option	Description
Base case	• 65% of total waste to directly landfill in 2020, starting from 85% in 2003; • 1% annual growth in recycling and composting, starting from 12% in 2003; • 50 000 tonnes per annum AD plant planned by 2020
Option 1	• Recycling and composting increases to 45% by 2020 • Thermal treatment and AD handles the balance of the waste • Residues to landfill
Option 2	• Recycling and composting increases to 55% by 2020 • Thermal treatment and AD handles the balance of the waste • Residues to landfill
Option 3	• Recycling and composting increases to 45% by 2020 • AD reduced in favour of thermal treatment for the balance • Residues to landfill
Option 4	• Recycling and composting increases to 55% by 2020 • AD and MBT for the balance of waste • Residues to landfill
Option 5	• Recycling and composting increases to 45% by 2020 • AD and MBT for the balance of waste • Residues to landfill
Option 6	• Recycling and composting increases to 45% by 2020 • Balance of waste treated by AD, thermal treatment and MBT • Residues to landfill
Option 7	• Recycling and composting increases to 45% by 2020 • Balance of waste going to MBT first, with output streams sent to AD and thermal treatment • Residues to landfill
Option 8	• Recycling and composting increases to 45% by 2020 • MBT and thermal treatment for the balance of the waste • Organic fraction would be put to beneficial use • Residues to landfill

Table 16.5 Required infrastructure for Northern Ireland waste arisings in 2020.

Technology	Capacity (tonnes year^{-1})	% of Waste Treated
Recycling	325 000	21
Composting	366 000	24
AD	110 000	7
MBT	200 000	13
Thermal	150 000	10
Landfill	385 000	25

(ii) New AD and mechanical biological treatment facilities early on, later followed by thermal treatment facilities to deal with the balance of the waste.

(iii) Diminishing levels (50%, 40% and 25% in the same years as above) going directly to landfill.

On the basis of the assessment, infrastructure capacity requirements were computed as shown in Table 16.5,[50] for consideration in the sub-regional waste strategies.

16.6.6 Recycling and Recovery of Plastics in Germany

LCA can also be applied to specific components of the waste stream. For example, a German study[51] assessed the following plastics recycling and recovery methods in the context of the Duales System Deutschland (DSD) discussed in section 16.4.2.2:

(i) Mechanical recycling (production of bottles, cable conduits and waste sacks).
(ii) Feedstock recycling (gasification, thermolysis, utilisation in blast furnaces, hydrogenation).
(iii) Energy recovery *via* combustion.

The recovery methods included collection close to the consumers, transport, sorting, preparation, and recovery through to production of useful end products. The sorting stage yielded a range of plastics: film, containers, mixed plastics and residues. Each recovery and recycling process was offset against an equivalent process for plastics manufacture from primary resources, which allowed comparisons of resource consumption and environmental pollution from different recovery methods producing different products. The life cycle inventory took into consideration primary resources, energy requirements, emissions and waste disposal, while the impact categories assessed were global warming potential, acidification, eutrophication, resource depletion, energy use and waste generation.

The study was able to deliver a detailed analysis of the relative merits of different operating systems for the DSD scheme. For example, the additional collection, sorting and transport required for mechanical sorting of plastics, while incurring a high cost, did not pose a sufficiently significant environmental impact to rule out this process, since compensatory gains were made in terms of reduced consumption of resources, and reduced emissions in comparison to feedstock recycling and energy recovery. The critical factor was the degree of substitution of mechanically recycled regrind product for highly refined primary material. A substitution rate of 1 kg of regrind for 1 kg of primary granules provided an improvement in the recovery operation for consumption of energy and primary resources, global warming potential, acidification and eutrophication. However, if the substitution rate was significantly less, then mechanical recycling resulted in a net deterioration in environmental performance relative to the best feedstock recycling or energy recovery processes.

This raised the issue of the expenditure required within the waste management system, to achieve the critical levels of substitution associated with break-even environmental performance. On the basis that plastics from DSD collections could not be completely recovered by mechanical recycling, a balanced recovery strategy recommended combining mechanical and feedstock recycling with energy recovery of the waste plastics that could not be recovered.

REFERENCES

1. General Secretariat of the Council of the European Union, *Note 6918/09*, Brussels, 2009.
2. European Commission, *Directive on Waste 2008/97/EC*, Brussels, 2008.
3. World Commission on Environment and Development, *Our Common Future*, Oxford University Press, Oxford, 1987.
4. Department for the Environment, Food and Rural Affairs, *Waste Strategy for England*, Defra, London, 2007.
5. European Commission, *Roadmap to a Resource Efficient Europe*, COM(2011)571 final, Brussels, 20 September 2011.
6. R. E. Hester and R. M. Harrison, *Issues in Environmental Science and Technology. No. 2 Waste Incineration and the Environment*, The Royal Society of Chemistry, Cambridge, 1994.

7. R. E. Hester and R. M. Harrison, *Issues in Environmental Science and Technology. No. 3 Waste Treatment and Disposal*, The Royal Society of Chemistry, Cambridge, 1995.
8. P. T. Williams, *Waste Treatment and Disposal*, John Wiley & Sons, Chichester, 1998.
9. Waste and Resources Action Programme, *Household Food and Drink Waste in the UK*, WRAP, Banbury, 2009.
10. Waste and Resources Action Programme, *New Estimates for Household Food and Drink Waste in the UK*, WRAP, Banbury, 2011.
11. F. L. Margai, *Environ. Behaviour*, 1997, **29**, 769–792.
12. W. J. Bryce, R. Day and T. J. Olney, *J. Consumer Affairs*, 1997, **31**, 27–52.
13. Waste and Resources Action Programme, *Love Food Hate Waste Campaign*, WRAP, Banbury, 2007.
14. London Assembly, *Carrots and Sticks: A Review of Waste Financial Reward and Compulsory Recycling Schemes*, London, 2011.
15. D. Birley and J. Hummel, *Householder Participation in Recycling Schemes: Experiences from ERRA Programmes*, European Recovery and Recycling Association, Brussels, 1996.
16. P. Tucker, *Environ. Waste Manage.*, 1999, **2**, 55–63.
17. The Industry Council for Packaging and the Environment, *Packaging Reduction,* Incpen, London, 2012.
18. Waste and Resources Action Programme, *Materials Recovery Facilities*, WRAP, Banbury, 2006.
19. Waste and Resources Action Programme, *Kerbside Collections Options: Wales*, WRAP, Banbury, 2011.
20. Department for the Environment, Food and Rural Affairs, *Consultation on Recovery and Recycling Targets for Packaging Waste for 2013–2017*, Defra, London, 2011.
21. European Environment Agency, *Diverting Waste from Landfill*, EEA Report No. 7/2009, Copenhagen, 2009.
22. European Commission, *Workshop on Integrated Product Policy*, Directorate-General XI, Brussels, 8 December, 1998.
23. European Commission, *IPPC Reference Document on the Best Available Techniques for Waste Incineration*, Brussels, 2006, *Generation and Treatment of Municipal Waste*, Eurostat Issue No. 31/2011, Brussels, 2011.
24. L. F. Diaz, G. M. Savage, L. L. Eggerth and C. G. Golueke, *Composting and Recycling of Municipal Solid Wastes*, Lewis Publishers, Boca Raton, FL, 1993.
25. Juniper Consultancy Services Limited, *Anaerobic Digestion Technology Evaluation,* Uley, 2007.
26. C. A. de Lemos Chemicharo, *Anaerobic Reactors*, IWA Publishing, London, 2007.
27. Department for the Environment, Food and Rural Affairs, *Accelerating the Uptake of Anaerobic Digestion in England: An Implementation Plan*, Defra, London, 2010.
28. Department for the Environment, Food and Rural Affairs, *Anaerobic Digestion Strategy and Action Plan,* Defra, London, 2011.
29. Waste and Resources Action Plan, *Anaerobic Digestion Infrastructure in the UK: September 2011*, WRAP, Banbury, 2012.
30. Waste and Resources Action Plan, *Anaerobic Digestate Quality Protocol (Draft)*, WRAP, Banbury, 2009.
31. Juniper Consultancy Services Limited, *Mechanical Biological Treatment: A Guide for Decision Makers – Processes, Policies and Markets*, Uley, 2005.
32. G. H. Eduljee, in *Issues in Environmental Science and Technology. No. 2 Waste Incineration and the Environment*, R. E. Hester and R. M. Harrison, The Royal Society of Chemistry, Cambridge, 1994.
33. P. T. Williams, in *Issues in Environmental Science and Technology. No. 2 Waste Incineration and the Environment*, R. E. Hester and R. M. Harrison, The Royal Society of Chemistry, Cambridge, 1994.

34. P. R. Bardos, *Survey of Composting in the UK and Europe*, Paper No. W93036, Warren Spring Laboratory, Stevenage, 1993.

35. P. Stenbro-Olsen and P. Collier, *J. Waste Manage. Resour. Recovery*, 1994, **1**, 113–118.

36. K. Westlake, in *Issues in Environmental Science and Technology. No. 3 Waste Treatment and Disposal*, R. E. Hester and R. M. Harrison, The Royal Society of Chemistry, Cambridge, 1995.

37. D. J. Lisk, *Sci. Total Environ.*, 1991, **100**, 415–468.

38. G. H. Eduljee, in *Issues in Environmental Science and Technology. No. 9 Risk Assessment and Risk Management*, R. E. Hester and R. M. Harrison, The Royal Society of Chemistry, Cambridge, 1998.

39. J. Bickerstaffe, *J. Waste Manage. Resour. Recovery*, 1994, **1**, 91–96.

40. F. McDougall, P. R. White, M. Franke and P. Hindle, *Integrated Solid Waste Management – A Lifecycle Inventory*, Blackwell Science, Oxford, 2nd edn., 2000.

41. Society for Environmental Toxicology and Chemistry (SETAC), *A Technical Framework for Life-Cycle Assessment*, and *Guidelines for Life-Cycle Assessment: A Code of Practice*, SETAC-Europe, Brussels, 1993.

42. Waste and Resources Action Programme, *Environmental Benefits of Recycling*, WRAP, Banbury, 2006 (Updated 2010).

43. Department for the Environment, Food and Rural Affairs, *Applying the Waste Hierarchy: Evidence Summary*, Defra, London, 2011.

44. Department for the Environment, Food and Rural Affairs, *Guidance on Applying the Waste Hierarchy,* Defra, London, 2007.

45. Greater London Authority, *Greenhouse Gas Balances of Waste Management Scenarios*, London, 2008.

46. European Environment Agency, *Better Management of Municipal Waste will Reduce Greenhouse Gas Emissions,* EEA Briefing 2008/01, Copenhagen, 2008.

47. PROGNOS AG, *Resource Savings and CO2 Reduction Potential in Waste Management in Europe and the Possible Contribution to the CO2 Reduction Target in 2020,* Report 2008/10, Berlin, 2008.

48. H. M. Government, *The UK Low Carbon Transition Plan*, The Stationery Office, London, 2009.

49. ERM (Environmental Resources Management), *Assessment of the Best Practicable Environmental Option for Waste Management in Northern Ireland*, Oxford, 2005.

50. Department of the Environment, Northern Ireland, *Best Practicable Environmental Option for Waste Management in Northern Ireland*, Belfast, June 2005.

51. Association of Plastics Manufacturers in Europe (APME), *Life Cycle Analysis of Recycling and Recovery of Household Plastics Waste Packaging*, APME, Brussels, 1996.

CHAPTER 17

System Approaches: Life Cycle Assessment and Industrial Ecology

ROLAND CLIFT

Centre for Environmental Strategy, University of Surrey, Guildford, Surrey GU2 7XH, UK
Email: r.clift@surrey.ac.uk

17.1 INTRODUCTION: CHANGING PARADIGMS

The emphasis in managing the environmental impacts of human activities is increasingly moving from limiting emissions of pollutants or mitigating their impacts to avoiding forming the pollutants in the first place. Closely associated with this shift in focus is the realisation that some resources are becoming increasingly scarce. This shift leads to increasing attention to the whole system within which materials and energy are used rather than focussing on processing or manufacturing. This chapter gives an introduction to the system approaches used in the developing paradigm of *Industrial Ecology* and the analytical tools used, of which the most widespread is *Life Cycle Assessment*. Industrial Ecology is also concerned with the social impacts and benefits of human activities[1,2] and the concepts of *sustainability* and *sustainable development* necessarily embrace the social equity dimension, but this aspect is not explored here.

Figure 17.1 gives a highly aggregated view of the industrial ecology perspective. Human society is maintained by food and other products obtained by agriculture and from natural ecosystems, and by goods and services provided by industrial production. The energy flows needed to bring about the material transformations derive in part from renewable sources (ultimately from incident solar energy), and in part from non-renewable fossil sources. From the industrial revolution, the process of industrial development has involved a shift from renewable to non-renewable energy sources. This applies even to agricultural production, where the use of fossil energy has become large, mainly for production of synthetic fertilisers. A shift back to carbon-neutral energy sources, primarily to mitigate global environmental change due to accumulation of carbon dioxide in the atmosphere, is an imperative (see Chapter 14). Material products are derived from agricultural production (including silviculture), which may be more or less sustainable, and from non-renewable abiotic resources (primarily mineral bodies). Material resource efficiency depends on the systematic

Pollution: Causes, Effects and Control, 5th Edition
Edited by R M Harrison

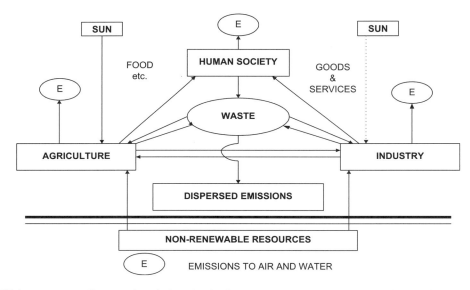

Figure 17.1 Resource flows and emissions in the human economy. (Reproduced from refs. 1, 3, 4).

use and re-use of materials, so that the industrial ecology paradigm sees waste as remaining part of the economy unless it is emitted and dispersed (see Figure 17.1).

Figure 17.1 also illustrates the problems inherent in attempting to use economic market approaches to regulate pollutant emissions (see Chapters 21 and 22). For resource inputs, scarcity leads to price rises which promote the development of alternatives through conventional market mechanisms. Whereas material supply represents a real physical resource constraint, the market economy is not constrained in the same direct way by the limited *carrying capacity* of the biosphere for emissions from human activities. Emission trading schemes amount to attempts to construct an artificial market to limit pollutant emissions. Whether this approach will be effective and adequate remains to be seen; the failure of the European Trading System (ETS) for greenhouse gases to maintain a sufficient price to drive structural change is not encouraging.

17.2   ENVIRONMENTAL SYSTEM ANALYSIS

17.2.1   Economy and Environment

Systematic study of the interactions between the human economy and the environment (used here in the sense of "that which surrounds the economic system")* is the subject of the emerging field of Environmental System Analysis. The general approach is summarized in Figure 17.2. Environmental interventions – *i.e.* resource inputs and emissions – pass between the environment and the part of the economic system which delivers a human benefit. Following the definition of waste implicit in Figure 17.1, material products remain within the economic system unless emitted and dispersed. Environmental System Analysis includes analysing the relationship between the economic activities providing a service and the associated environmental interventions, along with

---

*The word "environment" has acquired a range and depth of meanings, including the one used here: physical surroundings and conditions. This is the original meaning. The word entered the English language *via* the thermodynamic literature as a translation or transliteration of the French "environnement" in the work of Sadi Carnot. "Environment" is about 20 years younger than another word which entered the language *via* thermodynamics: "Energy", which appears to have been first used by Young in a lecture to the Royal Society in 1805.

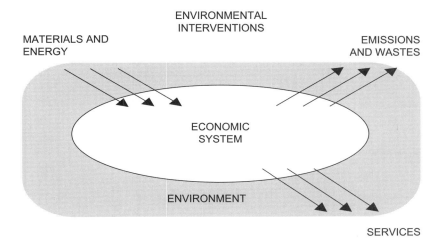

Figure 17.2 Environmental system analysis.

the significance of the interventions. Different forms of Environmental System Analysis differ in the way in which the system boundary between the economic activity and its environment is drawn.

17.2.2 Life Cycle Assessment

17.2.2.1 Product Systems

In Life Cycle Assessment[†] (LCA), the system boundary between the system and the environment is drawn around the supply chain providing a product or service: the system comprises all the operations along the material and energy supply chains, including transport, to obtain a complete "cradle to grave" assessment of the environmental interventions in providing the service or the function delivered by the product. According to the International Organisation for Standardization (ISO),[5] LCA is "a systematic tool for identifying and evaluating the environmental aspects of products and services from extraction of resource inputs to the eventual disposal of the product or its waste". The significance of LCA is that it provides a structured approach to defining and evaluating the total environmental load associated with providing a service.[6]

The scope of a life cycle assessment for a simple product system is shown in Figure 17.3. Rather than considering solely the material and energy use and emissions associated with producing a material product or a service (system boundary 1 in Figure 17.3), LCA considers the complete supply chain (*i.e.* the life cycle) of energy and materials needed to provide a service (system boundary 2), including transportation steps (which, for simplicity, are not shown in Figure 17.3). Extending the system analysed from the process to the full life cycle gives a complete picture of the environmental and resource impacts of the product system, and avoids the risk of *burden shifting*, *i.e.* apparently improving one part of the life cycle but actually transferring impacts to some other point in the material or energy supply chain. A full LCA, like that shown by System 2, is sometimes termed a *cradle-to-grave assessment*. When a material product is not controlled following production, for example because it is sold to a different organisation for use in producing a final product, the system boundary may be cut off following the manufacturing process; the assessment, which may provide the basis for an Environmental Product Declaration (EPD), is then termed

---

[†]The terms "Life Cycle Assessment" and "Life Cycle Analysis" are in practice used interchangeably although some authors have argued that there is a distinction between them.

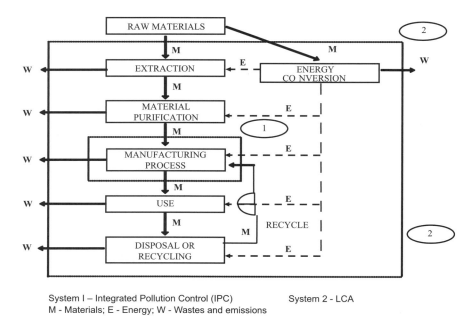

Figure 17.3 Life Cycle Assessment for a simple product system. (Reproduced from ref. 7). System boundaries – 1: process or plant; 2: cradle-to-grave; M: material flow; E: energy flow; and W: wastes and emissions.

cradle-to-gate. Increasingly, products and materials are being recovered for re-use or recycling (see section 17.3). The analysis then needs to consider a "closed loop" supply chain, as shown schematically in Figure 17.4.

LCA was originally developed for manufacturing and processing systems; in fact the first recognisable LCA studies were carried out for beverage packaging systems.[6] However, LCA is increasingly being applied to production, processing and delivery of food and other biotic products. The basic concept is the same as in Figures 17.3 and 17.4 but, in this kind of application, agricultural or silvicultural production and subsequent processing replace manufacturing. The inputs and emissions of concern now include agrochemicals and any water used for irrigation. A significant concern in LCA of biotic production is variability between different producers, which is much wider for agriculture, horticulture and silviculture than for manufacturing and processing.

Because LCA aims to describe the full supply chain, the locations of some of the operations will be unknown (see section 17.2.2.3). Therefore LCA has developed primarily as a way of comparing technological routes; *i.e.* the classic application of LCA is to decisions like A in Figure 17.5, where a product system is to be selected rather than a manufacturing or processing site. LCA should therefore be seen as complementary to other tools such as Environmental Impact Assessment (EIA) which have been developed to aid in selecting sites for a preselected technology; *i.e.* decisions like B in Figure 17.5. However, decisions which involve selecting a technology for a pre-identified site (C in Figure 17.5) or, in the most difficult case (D in Figure 17.5), selecting both technology and site from amongst several options, require both LCA and a site-specific tool such as EIA.

In recent years, a distinction has arisen between two types of LCA: *attributional* and *consequential*.[10–12] Attributional (or "accounting") LCA describes an existing supply chain; it is used, for example, in estimating the "carbon footprint" of a product (see section 17.3.4). Consequential (or "prospective") LCA attempts to explore the system effects of changes in economic activities; a conspicuous use of consequential LCA is in exploring the effects of changes in land use due to switching from production of food to fuel crops in one location, compensated by increased food

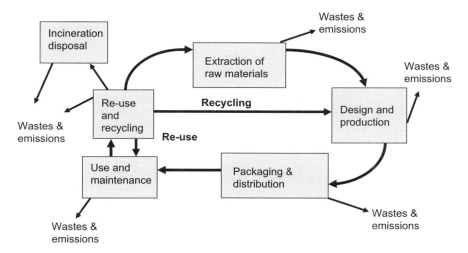

Figure 17.4 Product system with closed-loop. (Adapted from UNEP).[8]

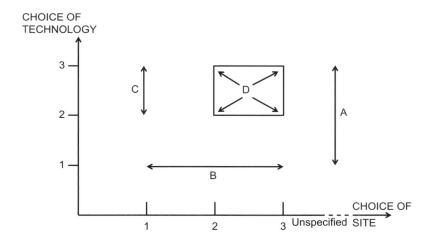

Figure 17.5 Application of LCA and site-specific approaches such as Environmental Impact Assessment (EIA). (Reproduced from ref. 9). A: established application of LCA. B: established application of EIA. C: decisions requiring both LCA and EIA.

production elsewhere. The methodological differences between the two forms of LCA are a matter of current debate but, for food products and bioenergy in particular, consequential LCA usually gives significantly higher impacts, particularly for climate change, because it includes changes in carbon stock resulting from indirect land use change.

The methodology of LCA has developed so that a study conventionally passes through four phases, shown in Figure 17.6.[5,6]

17.2.2.2 Goal and Scope Definition

It is essential at the outset of the LCA to define precisely why the study is to be carried out, and what decision is to be informed by the results. This involves deciding whether to use an attributional or consequential approach, and then defining the system boundary, whether cradle-to-grave or cradle-to-gate, to ensure that no relevant processes are omitted.

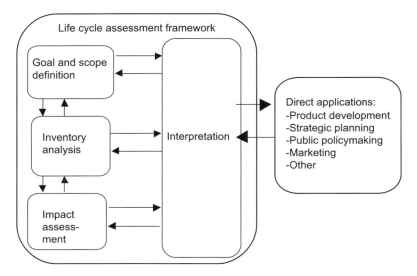

Figure 17.6 Phases in a Life Cycle Assessment. (Reproduced from ref. 5, 6).

When LCA is used to compare alternative ways of providing a service, the basis for comparison is the Functional Unit, defined as a number of units of service delivered by the product system. To take an obvious example, in comparing alternative packaging materials or systems, the basis for comparison would be a common number of units of specified volume of a beverage delivered to the consumer rather than a common mass of packaging material.

17.2.2.3 *Inventory Analysis*

The next phase identifies and quantifies the materials and emissions crossing system boundary 2 in Figure 17.3, *i.e.* the environmental interventions in this embodiment of Figure 17.2. The complete set of interventions, in the form of quantified exchanges with their chemical compositions, constitutes the *Inventory Table*. In process engineering terms, Inventory Analysis amounts to compiling material and energy balances over the processes and operations making up the Life Cycle. However, Life Cycle Inventory Analysis differs from conventional process analysis in certain respects. It is necessary to include trace emissions of substances with large actual or potential impacts even when the material flows are trivial (see section 17.2.2.4).

A further difference arises from the frequent need to include multi-functional processes. Some processes within the system boundary may be shared with other supply chains, most commonly when a process has two or more outputs which are used in different product systems or, in the case of waste management, where waste streams from more than one source are processed. It is then necessary to *allocate* environmental interventions and the associated impacts between the different outputs or inputs. A related problem arises when materials or components are re-used or recycled (see Figure 17.4); interventions, for example associated with production of the primary material, must then be allocated between successive uses. Alternative approaches to allocation have been discussed since the earliest days of LCA.[13,14] The relevant international standard, ISO 14044,[5] specifies a hierarchical approach to allocation:

"Allocation should firstly be avoided through process division or system expansion where possible. Where not possible, the physical relationships between products or functions should be used to partition inputs and outputs. When physical relationships cannot be established, other relationships shall be used instead (e.g. economic)."

However, this approach is at the level of general guidance. Detailed application of the ISO guidelines remains an area of contention between different practitioners: it is unfortunately the case that there are no approaches to allocation which are universally accepted as correct.[11] Allocation by economic value is probably the approach most commonly used,[15,16] even though the ISO standard recommends that this should be the last resort if other approaches cannot be applied.

Compiling the inventory table is usually the most labour-intensive phase in a LCA. To confine the effort needed, it is common practice to distinguish between:[9,17]

(i) *Foreground:* the set of processes whose selection or mode of operation is affected directly by decisions based on the study.
(ii) *Background:* all other processes which interact with the Foreground, usually by receiving or supplying material or energy. A sufficient (but not necessary) condition for a process or group of processes to be in the Background is that the exchange with the Foreground takes place through a homogeneous market.

Foreground processes are described by primary plant- and process-specific data, whereas Background processes are represented by generic industry-average data, usually taken from one of the several commercially available LCA databases. However, the geographical location of the Background processes is undefined, leading to the limitations on the use of LCA noted in section 17.2.2. Modelling of the exchanges between Foreground and Background (see Figure 17.7) are a further methodological contention, with some authors maintaining that different approaches should be used for attributional and consequential analysis.

The distinction between Foreground and Background underlies another recent development in LCA methodology. Conventionally, the secondary data used to describe the environmental interventions and activities associated with processes in the Background system correspond to the averages over all technologies and operators; this is termed the *process-based approach* to LCA. An alternative is to represent the Background economic activities by *Environmentally Extended Input–Output Analysis* (EEIO). This approach, derived from economic analysis,[18] treats the Background as a linear homogeneous system and represents it by a matrix which relates changes in output back to change in activities and the associated environmental burdens. The Background may be modelled by EEIO and the Foreground by process-based LCA; in this case, the approach is known as *hybrid LCA*.[19,20] Data for EEIO are usually only available at a highly aggregated level: industrial sectors rather than specific materials or components. Therefore the EEIO approach is valuable for

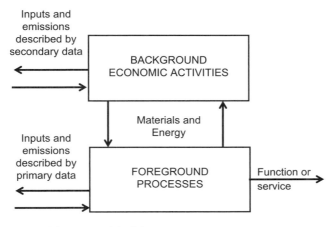

Figure 17.7 Background and foreground (sub-) systems.

analyses such as the environmental impacts embodied in trade[21,22] and the environmental consequences of different lifestyles and consumption patterns.[23] However, for more conventional applications of LCA, particularly involving selection or comparison of supply chains and processes, the process-based approach is still preferred.

17.2.2.4 Impact Assessment

The Inventory Table comprises a great body of numerical information which is usually too detailed to display clearly the most important environmental impacts. Life Cycle Impact Assessment (LCIA) aims to evaluate and clarify the magnitude and significance of the potential environmental impacts of the system within boundary 2 of Figure 17.3. LCIA and Interpretation include several elements, some of which are held by ISO to be mandatory.[5]

The first mandatory element is selection of a manageable number of categories defining resource usage and environmental Impact Categories, together with indicators for the categories and models to quantify the contribution of the different interventions to these categories. Two approaches have developed,[11,12,24] shown schematically in Figure 17.8. The longer established approach uses *mid-point impacts*, each of which refers to a specific physico-chemical effect. Table 17.1 lists the set of mid-point impact categories most commonly used in LCA. These impacts are selected to lie on the source-impact path at a point where "further modelling is not feasible or involves too large uncertainties, or where a relative comparison can be made without the need for further modelling".[11] The alternative approach uses *end-points*, corresponding to impacts on areas of social and economic concern. Proponents of the end-point approach argue that expressing results in these terms – quantified impacts on human health, for example[25] – makes them more accessible to policy makers and the lay public. Set against this, there is the unavoidable problem that proceeding from mid-points to end-points introduces further uncertainties into an assessment which is already subject to wide uncertainty even at the mid-point level. The two approaches are not inherently incompatible,[24] and there are current efforts to develop a common harmonised modelling framework.[11]

Following the mid-point approach, the next mandatory element, Classification identifies which of the interventions in the Inventory Table contribute to which of the classified impacts (see Figure 17.9). For example, carbon dioxide emissions only contribute to global climate change, while chlorofluorocarbons (CFCs) and hydrochlorofluoro-carbons (HCFCs) contribute both to global warming and to depletion of stratospheric ozone. Characterization is the ensuing mandatory step,

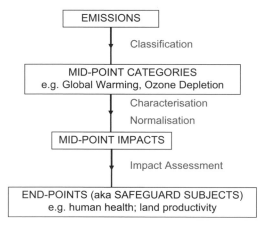

Figure 17.8 Life Cycle Impact Assessment: mid-point and end-point impacts.

Table 17.1 Principal mid-point impact categories used in LCA.[14–17,26,27]

Impact Category	Description
Global	
Abiotic depletion potential	A measure of the potential of the system to contribute to the depletion of non-living natural resources that are not renewable within human lifetimes. The measure is based on remaining reserves and rates of extraction and expressed as equivalents to antimony (Sb).
Global warming potential (GWP 100)	The potential of a system to contribute to global warming, based on the radiative forcing capacities of atmospheric emissions. Expressed as equivalents to carbon dioxide.
Ozone layer depletion potential	The potential of a system to contribute to the loss of stratospheric ozone, based on the change in the stratospheric ozone column due to gases emitted to atmosphere. Expressed as equivalents to CFC-11.
Regional and Local	
Acidification potential	The potential of the system to contribute to the acidification of freshwaters through emissions of acidifying substances to air. Expressed as equivalents to sulfur dioxide.
Eutrophication potential	The potential of the system to contribute to the eutrophication of freshwater through release of primary nutrients to water. Expressed as equivalents to phosphate.
Human toxicity potential	The potential for a system to impact on human health as a result of toxic substances released to the environment. It is the product of a number of factors covering the transfer, intake, fate and effect of different substances. Expressed as equivalents to 1,4-dichlorobenzene.
Freshwater aquatic ecotoxicity potential	The potential for a system to impact on the freshwater environment as a result of toxic substances released. It is the product of a number of factors covering the transfer, intake, fate and effect of different substances. Expressed as equivalents to 1,4-dichlorobenzene.
Marine aquatic ecotoxicity potential	The potential of a system to impact on the marine environment as a result of toxic substances released, expressed as equivalents to 1,4-dichlorobenzene.
Terrestrial toxicity potential	The potential for a system to impact on the soil environment as a result of toxic substances released. It is the product of a number of factors covering the transfer, intake and effect of different substances. Expressed as equivalents to 1,4-dichlorobenzene.
Photochemical oxidation potential (POCP)	The potential of a system to create tropospheric ozone through emissions of VOCs, expressed in equivalents to ethene.

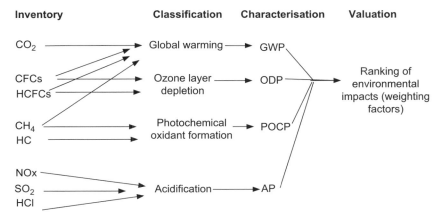

Figure 17.9 Life Cycle Impact Assessment using mid-point impacts.

in which the contribution to each impact category is quantified by expressing the effect of each intervention in terms of an equivalent quantity of a reference species (see Table 17.1). Thus Greenhouse Warming Potential (GWP), for example, is assessed by expressing the effect of each emission contributing to this impact category in terms of an equivalent mass of carbon dioxide, and summing these to obtain the total contribution to GWP per Functional Unit. For this specific case, the contribution of any species to global climate change is assessed as the integral of the contribution to radiative forcing of the species and any products of atmospheric reactions (*e.g.* carbon dioxide from methane) over some specified time period. Because different species have different decay profiles, the relative GWP values depend on the accounting period used (see Chapter 14). Solely as a matter of convention, LCIA normally uses the 100-year values for GWP.

The mid-point approach is in principle straightforward for those impacts which are truly global – *i.e.* resource depletion, greenhouse warming and stratospheric ozone depletion – so that the nature and extent of the impact are independent of the location of the activities in the product system. It is more problematic for the other impact categories in Table 17.1, which are site-dependent. It was noted above that it is inevitable in a full LCA that the locations of some parts of the life cycle cannot be defined. Therefore established practice in Life Cycle Impact Assessment is to estimate potential global rather than actual localized impacts. Thus, for example, Acidification Potential (AP) is assessed as the stoichiometric production of hydrogen ions by complete reaction of the emission, referred to sulfur dioxide as the reference species to give a value for AP in SO_2 equivalents. For some impacts, notably acidification, approaches have been proposed for site-dependent impact assessment, allowing for air movement, sensitivity of the receiving population and ecosystem, and even on the height of the stack from which a pollutant is emitted.[28,29] Although the impacts can vary widely according to local conditions,[30] an agreed methodology for site-dependent LCIA has yet to emerge.[11] On the positive side, the approach of comparing emissions on a common globalized basis lends itself logically to measuring and reporting the environmental performance of multinational companies with global operations.[31]

Other impacts are increasingly being considered, in addition to those listed in Table 17.1. For farming and forestry, land use is a concern,[32–34] both because land is a constrained resource and because changes in land cover and use can lead to emissions, for example resulting from changes in the stock of carbon in the soil and vegetation.[35] This is one of the principal differences between attributional and consequential LCA: an attributional study normally considers a specific established land use, whereas consequential analysis includes both direct foreground land use change and indirect background changes due to displacement or expansion of agriculture or silviculture elsewhere.[36,37] Biodiversity is a related topic;[38] whether impacts on biodiversity can be expressed in terms of ecosystem services is a current research question.[39] Dissipative water use is also an increasing focus of Life Cycle Impact Assessment, again particularly for agriculture.[40,41]

17.2.2.5 *Interpretation*

Even within the mid-point approach, what is done with the estimated impacts differs between different studies and different practitioners. A common approach to interpreting mid-point impact estimates is to express them as fractions of reference values, such as the total contribution to each impact category by all the economic activity in the relevant country or economic area. This approach, known as *normalisation*, can help to show the significance of the different impacts and reveal which are of greatest significance. However, appropriate reference values for normalisation are a matter of debate.[42] Other forms of normalization can be used, for example to compare different products or different parts of the supply chain.[2,43]

Other elements, summarized in Figure 17.9, are more contentious. *Grouping* entails sorting and possibly ranking the (normalized) impacts in order of importance. *Weighting* aims to attach to each

Impact a factor indicating its importance, opening the possibility of aggregating the different impacts to a single score. However, the weights must be obtained by a carefully constructed process, such as Multi-Criterion Decision Analysis, if the results are to have any value.[44,45] Weighting is sometimes expressed in terms of monetary damage costs, in which case the process may be called *Valuation*; this amounts to a form of Cost–Benefit Analysis. Proponents of the end-point approach argue that it makes weighting simpler and more reliable by reducing the number of impacts to be compared and making them more intelligible to a lay public. However, this is to be set against the uncertainty introduced by going from mid-point to end-point, and the possibility that the model used may overlook some end-point impacts.

The usefulness of Weighting and Valuation are much debated; critics regard such a process of aggregation as conceptually and theoretically ill-founded,[46–48] misleading[44,45,48] and unnecessary.[49,50] The view taken here is that LCA is to be seen as a tool for structuring and presenting information in a way that is intelligible and accessible to those making decisions; *i.e.* it is one decision support tool amongst others, not an algorithm which generates the "right answer". The kind of gross aggregation represented by Weighting and Valuation obscures information. Furthermore, LCA results are unavoidably subject to substantial uncertainty (see section 17.2.2.7). Moreover, some of the bases for aggregation, notably the *ecological footprint* which attempts to express impacts and resource impacts in terms of the land area needed to absorb the emissions and supply the inputs, are at best of questionable scientific validity.[51] Therefore aggregation beyond normalized impacts is to be avoided. Instead, it is preferable to recognize that environmental improvement usually requires balancing performance in non-commensurable impact categories; multi-objective optimization can then be used to inform decisions.[49,50]

17.2.2.6 An Application of LCA: Product Labelling

One of the most conspicuous public uses of LCA to date has been in providing information to consumers and retailers on the environmental impacts of purchases in the form of environmental labels on products. Table 17.2 summarises the principal types of environmental product labels or declarations. Environmental Product Declarations (see section 17.2.2.1) are Type III declarations. Because environmental labels are usually provided as the basis for implicit or explicit comparative assertions, the methodology needs to be much more closely prescribed and standardised than the more general standards on LCA.

Ecolabelling is an approach intended to identify products with improved environmental performance compared with equivalent products or services, typically the "best" 10 to 15% in their product group. Amongst the various national and multinational schemes are the European "Eurodaisy",[53] the German "Blue Angel" label[54] and the Nordic "Swan".[55] Taking the European label as typical, ecolabels have two objectives:

(i) "Benchmarking": to guide and encourage development of environmentally sound products.
(ii) Consumer Information: objective identification of products whose environmental performance goes beyond compliance with regulations.

While there is some evidence that the former objective is realistic, evidence that labelling has significant effect on consumer purchasing is scant.[56]

In the taxonomy of ISO,[52] ecolabels are a Type I label (see Table 17.2). Establishing the criteria to identify products and services eligible for an ecolabel entails carrying out life cycle assessment of a representative range of products on the market; estimating the range of the principal impacts and, usually, where they arise in the supply chain; and establishing the environmental performance that must be met to be recognised as eligible for award of a label of excellence. Given that, in most cases,

Table 17.2   Types of Life Cycle-based Product Declaration.[6,52]

Type I environmental labelling – a voluntary, multiple-criteria-based third party programme that awards a license which authorises the use of environmental labels on products indicating overall environmental preferability of a product within a particular product category based on life cycle considerations.

Type II environmental labelling (self-declared environmental claims) – environmental claims that are made, without independent third-party certification, by manufacturers, importers, distributors, retailers or anyone likely to benefit from such a claim.

Type III environmental declaration – quantified environmental data for a product with pre-set categories of parameters based on the ISO 14040-series of standards, but not excluding additional environmental information within a type III environmental declaration programme.

Type III environmental declaration programme – voluntary process by which an industrial sector or independent body develops a type III environmental declaration, including setting minimum requirements, selecting categories of parameters, defining the involvement of third parties and the format for external communications.

more than one impact emerges as significant, award of a single label for good performance across all environmental categories inevitably requires some form of aggregation (see section 17.2.2.5). Agreement on the impacts to be covered and the values constituting "excellent performance" has been one of the problems in the development of ecolabelling schemes.

More recently, attention in product labelling has turned to single-impact labels, usually self-declared by the producer or retailer (Type II) but sometimes with independent third-party certification (Type III). These labels report the total life cycle contribution of a product or service to a single impact category expressed in terms of the quantity of the reference species (see Table 17.1). To date, the most popular, known as *carbon labels* or *carbon footprints,* report the life cycle GHG emissions.[57–60] Other labels are in development, notably covering water use in the supply chain. The functional unit for the LCA is the labelled consumer product, usually at the point of sale to the consumer. The drive for introduction of product labels has come from retailers and some producers;[56,61] as with ecolabels, there is little evidence that specific consumer purchases are influenced by labelling.

Treatment of recycling in product labelling illustrates how the LCA methodology must be specified closely for use in the kind of claim represented by product labels. Figures 17.10 and 17.11 illustrate the basis for the equations which are specified (but not explained) in the British standard for carbon labelling, PAS 2050.[60] If "the recycled material does not maintain the same inherent properties as the virgin material input", so that recycled material is drawn from a recyclate pool and recovered material is returned to the pool (see Figure 17.10), the total GHG emissions associated with putting a product into use are calculated as:

$$E = (1 - R_1)E_V + R_1 E_R + (1 - R_2)E_D \tag{1}$$

where

R_1 = proportion of recycled material input;
R_2 = proportion of material in the product that is recycled at end of life;
E_R = GHG emissions arising from recycled material per functional unit;
E_V = GHG emissions arising from producing virgin material per functional unit;
E_D = GHG emissions arising from disposal of waste per functional unit.

Equation (1) implicitly assumes that GHG emissions associated with material recovery following use are negligible. However, it recognises that the proportion of the product recovered after use

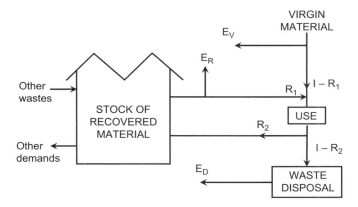

Figure 17.10 "Recycled Content" approach to recycling. (Reproduced from ref. 62).

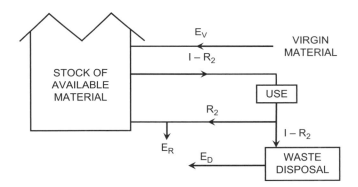

Figure 17.11 "Closed-loop Approximation" approach to recycling.

(R_2) may be different from the proportion of secondary material in the product (R_1) because R_1 is determined by the product manufacturer whereas R_2 is normally determined by the body responsible for management of post-consumer waste.

As an alternative, "if the recycled material maintains the same inherent properties as the virgin material input",[‡] the approach shown in Figure 17.11 is used, leading to:

$$E = (1 - R_2)E_V + R_2 E_R + (1 - R_2)E_D \qquad (2)$$

Both of the approaches illustrated by Figures 17.10 and 17.11 contain a rather stronger assumption than the difference in properties between virgin and secondary material. In both approaches, material recovered to be recycled or re-used is taken to be free of "inherited" impacts from previous uses; *i.e.* environmental interventions are allocated to the current use cycle. This is an example of the long-running methodological arguments for which practicality demands a simple solution.

[‡]On a thermodynamic basis this assumption can never be strictly true (see section 17.3.3.1) but in some cases it may be a justifiable approximation.

17.2.2.7   A Cautionary Note on Precision and Uncertainty

Even when the methodology is tightly prescribed, Life Cycle Assessment is not a precise science or art; the specific problem of variability in agricultural production was noted previously, but even industrial processes are subject to some unavoidable uncertainty, while the precision of data on inputs and emissions over the whole life cycle is uncertain and often completely unknown. It is therefore considered good practice to include uncertainty analysis in an LCA (although "good practice" is by no means universally followed). The commonest approach to analysing uncertainty is stochastic, using Monte Carlo simulation (although the resulting estimates are of questionable value, given that the range of variability in the data is usually only a guess and the probability density functions are unknown).[63,64] Analytic uncertainty analysis has been less used[65] although it might offer potential advantages.[66,67]

Recognising the problem of inherent variability and uncertainty, some labelling standards impose limits on the precision which can be claimed in the results, with "rounding rules" that imply 5 to 10% residual uncertainty.[60] Two-figure accuracy is indeed all that can be expected, dropping to one significant figure for many agricultural products, while for some products (notably biofuels subject to consequential analysis) the sign of the final result may be contested. Estimates for smaller differences between life cycles may be meaningful provided that the differences are confined to a few well-characterised stages in the supply chain.

17.2.3   Material Flow Accounting

Referring to Figure 17.2, whereas in Life Cycle Assessment the boundary defining the "system" studied is drawn around the supply chain delivering a product or service, in Material Flow Accounting (MFA) it is drawn around a geographical area or an economic sector.[§] Graedel and Allenby[68] define MFA as "an analytical approach to how materials are used by an (industrial) ecosystem". When applied to an economic sector, MFA provides what amounts to a flowsheet for the flows of a particular element or material between operations in that sector. When applied to a national economy, it leads to National Material Accounts. Material flows embodied in international trade can conveniently be represented in the form of input-output tables (see section 17.2.2.3).

Figure 17.12 shows the basic approach of MFA. A material flows into the stock of material in use *via* manufacturing processes or in imported manufactured goods. Following use, the material may be re-used or recycled (see section 17.3.3) or exported, or it may enter the stock of waste from which it is lost when it is dissipated (see Figures 17.1 and 17.13). One of the uses of MFA is to estimate the stock of material in an economy. For sector i, the stock of material (S_i) is related to the input and output flows by

$$\frac{dS_i}{dt} = p_i - q_i \tag{3}$$

where p_i is the flow of the material into the sector and q_i is the flow out of the sector when the stock reaches the end of its service life. From Equation (3), to estimate S_i by integration of p_i and q_i over time, it is necessary to have historical time-series data for p_i and qi.[69] Data for p_i are usually compiled from national statistical data. Data for q_i may be harder to obtain, but are sometimes inferred from historical p_i data allowing for the service life (*i.e.* the residence time) of the stock in sector i. Different sectors are characterised by different product lives; this is one of the principal reasons for distinguishing between different sectors in compiling the MFA. It is also possible to allow for the distribution of service lives within each sector,[71,72] although this only affects the results when dS_i/dt is significant.

[§]Some authors prefer the term Material Flow Analysis.[68] Others, *e.g.* Brunner,[69,70] make a distinction between Substance Flow Analysis (SFA) which focuses on individual substances and Material Flow Analysis (MFA) which includes "economic entities such as consumer or investment goods".[70] This semantic distinction is ignored here.

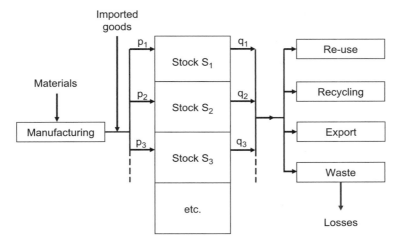

Figure 17.12 Material flow accounting.

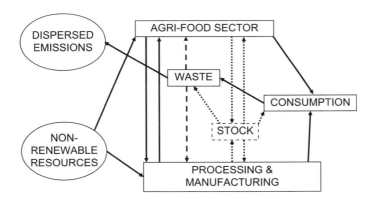

Figure 17.13 Linear, circular and performance economies. ——— linear economy, — — — circular economy, ••••••• performance economy

   Whereas Life Cycle Assessment has its origins in industrial practice,[6] Material Flow Accounting developed originally as an academic analytical tool.[73] However, MFA is increasingly being used for public policy support and also planning in the private sector. Uses include planning the material requirements to maintain economic activities and projecting future waste arisings, recovery and management.[74,75] With increasing tendency for post-consumer waste to be returned to the producer, MFA is starting to be used within companies to predict material recovery in end-of-life products and hence plan material input requirements allowing for re-use and recycling (see section 17.3.3.1).

17.3 APPLICATIONS AND ASPIRATIONS

17.3.1 Industrial Ecology and the Circular Economy

Figure 17.13 expands on the idea behind industrial ecology, introduced in Figure 17.1, which views "an industrial system... not in isolation from its surrounding systems, but in concert with them. It is a systems view in which one seeks to optimise the total materials cycle from virgin material, to component, to product, to obsolete product, and to ultimate disposal. Factors to be optimised include resources, energy, and capital".[76]

The flows which have characterised the "linear" industrialised economies since the industrial revolution are shown by the solid black lines in Figure 17.13. However, there is a widespread consensus that this approach to consuming material and energy resources cannot be maintained.[77-80] Constraints arise from increasing scarcity of some critical materials and from the effect of anthropogenic emissions on the environment, most notably through climate change. A radical rethink of resource use is arguably needed.

Waste generation is unavoidable – the notion of a "waste free" economy is a thermodynamic fantasy – but what is done with waste is an important part of resource efficiency. Part of the waste stock can be re-used or recycled, as shown by the chain-dotted arrows in Figure 17.13. Clearly this not only reduces demand for primary material but also saves the energy (or, more meaningfully, the exergy)[79,81] used and the waste and emissions generated in producing primary material. As a specific example, recycling is particularly beneficial for aluminium: recycling requires around 5% of the energy for primary production and avoids problematic wastes such as "red mud".[77,82] One, recently popularised, form of the basic industrial ecology idea of closed-loop use of materials is that of the circular economy: "The circular economy refers to an industrial economy that is restorative by intention... and eradicates waste through careful design".[83]

A further step in thinking is to focus on the stock of materials in use in infrastructure, buildings, plant, vehicles, appliances *etc.* The argument is that the quality of the stock of goods is at least as important as material flows in maintaining quality of life,[84] so that the focus should be on the most materially efficient ways to maintain and improve the capital stock. Stahel has termed this approach "the performance economy" and has explored its implications for resource use and the structure of the economy.[85]

Ensuing sections illustrate applications of the system approach of industrial ecology.

17.3.2 Clean Technology and Pollution Prevention

17.3.2.1 Clean vs. Clean-up Technology

One of the longest established embodiments of the system approach to pollution and resource use is *clean technology*, defined as "a means of providing a human benefit which, overall, uses less resources and causes less environmental damage than alternative means with which it is economically competitive".[7,86] As in Figures 17.1 and 17.13, the focus is on the service rather than the material product. "Overall" means considering the whole life cycle, as in Figures 17.3 and 17.4. Clean Technology is to be distinguished from clean-up technology, which controls the emissions of pollutants whereas the clean technology approach attempts to avoid forming the pollutants in the first place. In the USA, the equivalent approach is called *pollution prevention*, defined as "source reduction and other practices that reduce or eliminate the creation of pollutants".[87]

The distinction between Clean Technology and Clean-up Technology is illustrated by Figure 17.14. Each of the curves in Figure 17.14 represents the trade-off between the economic cost of providing a specified benefit or service and the associated environmental impact. The economic cost represents the conventional cost of providing the service – the concept of Clean Technology can be developed without invoking the neo-classical economic concept** of "external costs". The environmental impact is estimated over the whole set of economic activities needed to provide the

** Conventional economics has difficulties with the whole concept of Clean Technology, on the argument that if it were possible to introduce a technology which complies with existing regulations and meets demand for a service more cheaply, then it would necessarily be implemented and would displace older technology. In reality, as distinct from economic theory, there are barriers to the adoption of Clean Technology which have been the subject of several careful studies (*e.g.* refs. 89 and 90).

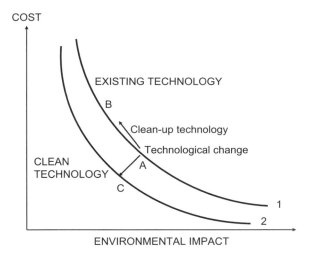

Figure 17.14   Clean technology and clean-up technology. (Reproduced from ref. 7). 1: Established technology for providing a service. 2: Alternative 'cleaner' technology.

service, established by Life Cycle Assessment. Representation of environmental impact by a single parameter is purely for explanatory purposes; the arguments against reducing non-commensurable impacts to a single index were adumbrated in section 17.2.2.5.

In fact, the curves are multi-dimensional surfaces.[88] They form *Pareto surfaces* (also known as *decision frontiers* or *non-inferior surfaces*) which represent the set of design or operating conditions such that it is impossible to improve one objective describing economic or environmental performance without worsening at least one other objective.

Current standard practice might correspond to technology 1 (see Figure 17.14) operated at the conditions corresponding to point A. It is possible in principle to reduce environmental impact by adding some end-of-pipe abatement device (bearing in mind that the end-of-pipe approach may introduce a new environmental problem to replace the emission being contained). This Clean-up approach by definition involves adding to the process, or treating or reprocessing the product which provides the benefit or service. Thus it involves increasing the economic cost to reduce the environmental impact, thus moving around the curve to point B shown in Figure 17.14. Following the definition given above, introducing Clean Technology involves shifting to a new way of providing the service which has less environmental impact and is also less expensive. The Clean Technology is represented by curve 2 in Figure 17.14, so that introduction of a Clean Technology corresponds to shifting from point A to point C.

17.3.2.2   Process Selection and Design

Decisions over process selection, design and operation necessarily involve trade-offs between cost and the different environmental impacts. The problem in terms of process design is shown schematically in Figure 17.15, in the two-dimensional simplification of trading off cost against one measure of environmental performance. The performances of technologies 1–3 are each represented by a space which represents the possible range of performance. For each technology, there is a decision frontier which represents the set of designs for which it is impossible to improve one performance parameter without making the other parameter worse. Technology 4 is clearly less effective than technologies 1–3, and is therefore considered no further.

The overall decision envelope is tangential to the decision frontiers representing the individual technologies. The optimal design point lies on the decision envelope but at a point determined by

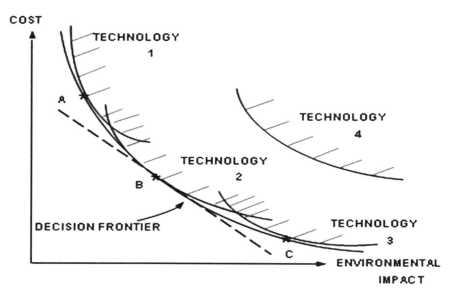

Figure 17.15 Process selection and design. (Reproduced from ref. 88).

the trade-off between economic and environmental performance (or, in the general case, between different measures of environmental performance). For example, if technology 2 is selected, the design point would be at point B. The negative gradient of the tangent at B gives the marginal cost of abating the environmental impact, and thus helps in selection of the preferred design. The Pollution Prevention and Control regime, brought in by the European Directive on Integrated Pollution Prevention and Control (IPPC) and intended to act as a driver to promote cleaner technologies, requires explicit trade-offs between different categories of environmental impact assessed on a lifecycle basis.[91] In some European member states, IPPC is interpreted in a way which is equivalent to the approach shown schematically in Figure 17.15.[92,93]

17.3.2.3 An Example: Phosphorus

Phosphorus provides an informative example of the distinction between the end-of-pipe and industrial ecology approaches. Phosphorus is an essential element for almost all life-forms, and cannot be substituted by any other element. It is therefore used very widely as a fertiliser for producing food and animal fodder. It is obtained from phosphate rock. In terms of Figure 17.1, it is one of the key inputs to agriculture produced from a non-renewable resource. Current agricultural practices mainly use phosphorus once before it is emitted from the economic system and dispersed, ultimately into the oceans. In addition to the questions over depletion of the resource, phosphorus emissions cause environmental problems in water bodies (see Chapters 1 to 4); for example, it is one of the principal causes of eutrophication (see section 17.2.2.4).

A common response to depletion of reserves and phosphorus pollution[††] to adopt the end-of-pipe approach: to recover phosphorus from waste water and recycle it as a fertiliser, for example in the form of struvite. However, mapping the overall system in which phosphorus is used shows that

---

[††]Whether resource depletion or pollution is the more urgent problem is a matter for debate, although opinion is converging on the conclusion that phosphorus reserves are sufficient to meet long-term demand albeit concentrated in a few countries that will continue to dominate the market.[94–99]

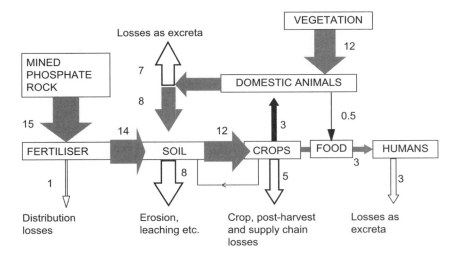

Figure 17.16 Global phosphorus flows in the agricultural sector. (Reproduced from ref. 96).

the situation is more complex. Figure 17.16 shows in broad outline the global industrial ecology of phosphorus; it represents a global MFA for phosphorus in the agricultural sector, in the form of a Sankey diagram which is sometimes recommended for presenting MFA results. The total phosphorus input to agriculture is approximately 15 Mtpa (*i.e.* million tonnes per annum) from mined phosphate rock and 12 Mtpa from vegetation consumed by grazing animals but only 3 Mtpa (*i.e.* about 10% of the total input) enters waste water in human excreta. The majority of the losses are from agriculture, particularly animal husbandry, plus significant accumulation (*i.e.* stock change, in the terms of Figure 17.12) in soil. Thus "closing the loop" on phosphorus depends on improving the efficiency of use in agriculture, much more than on recovering phosphorus in waste water treatment.[94–96] This will become more critical as global diets shift from vegetarian to dairy and meat diets which are much more phosphorus-intensive.[94,96,100]

17.3.2.4 Cleaner Production

It is useful to distinguish between three levels of change in producing any particular product by a cleaner route:[101,102]

(i) *Level 1 change* – minimize waste arisings, effluent and energy consumption within the process or operating unit. This approach – source waste reduction – essentially remains within system boundary 1 in Figure 17.3.

(ii) *Level 2 change* – modify the technology to use the same materials but more efficiently and with less energy. In effect, the system boundary in Figure 17.3 is extended to include the process along with upstream material preparation and energy supply.

(iii) *Level 3 change* – redesign the process completely and change input materials. This implies changing the entire cradle-to-gate supply chain.

This classification is not sharp, but it provides a useful framework for discussing Cleaner Production. Substitution of the product by a different way of delivering its function is a more radical step, and is discussed in section 17.3.3.

(i) *Waste Reduction.* Because waste reduction at source can both improve environmental performance and save money, it usually constitutes a Clean Technology change in the sense of Figure 17.14. The financial risks are usually small, so that the likely return can in principle be estimated.[101] Therefore the barriers to embarking on a waste reduction programme should be low. However, there is little evidence that many companies do estimate the financial benefits and then embark on such a programme.[89,90] Nevertheless for many companies, systematic waste reduction is the first step in implementing Clean Technology. It can lead on to the other levels of change, particularly when an appropriate systematic approach is used.[102–105] For example, a waste audit can prioritize waste streams for further attention according to their environmental impacts, treatment costs and safety aspects rather than merely their quantity.[101,105] Losses during loading and unloading and, particularly for batch operations, from cleaning operations can be particularly significant, and sometimes relatively straightforward to contain.[105]

(ii) *Technology and Raw Material Changes.* If experience with the modified technology or something resembling it is available, then it may be possible to estimate the financial implications of a level 2 change.[101] However, the uncertainties are likely to be greater than for waste reduction, so that the barriers may be greater. Particularly if life cycle considerations are introduced at the stage of process selection and design, the overall environmental improvements can be substantial and the risk small. However, different changes to avoid, reduce or enable partial recycling of process waste need to be considered in a life cycle perspective, even if a full LCA is not carried out, to identify off-site agents, such as waste contractors, who need to be involved in the change.[102] Conventional approaches, such as integration of heat and mass exchange networks and use of catalysts with improved selectivity and re-usability, also represent process improvements at this level.

(iii) *Redesigning the Life Cycle.* Level 3 changes usually involve substantial financial uncertainty and require significant capital investment. Both of these aspects represent significant barriers to this level of Cleaner Production,[89] particularly where there has been relatively recent investment in updating existing plant.[90] Change at this level may involve rethinking the product completely, including whether the service it delivers could be provided in some other way. Service provision and dematerialisation are considered further in section 17.3.3.

17.3.2.5   *Industrial Symbiosis*

Symbiosis in natural ecosystems refers to the interdependence of species for mutual or unilateral benefit. Extending the concept to industrial ecology, *industrial symbiosis* is an approach to turning waste from one industrial activity into a beneficial input to another.[106,107] Some authors extend the definition of industrial symbiosis to include shared infrastructure or utilities, and to non-material services such as training and management practices.[108] Such symbiotic relationships are sometimes informal and unplanned but nevertheless real. At the opposite extreme, exchanges may occur between activities or companies located close together; in this case the area within which the interrelated activities are located is commonly termed an *Eco-Industrial Park* (EIP). EIPs are usually a result of planning and design; they are therefore a feature of industrialising economies with a tradition of economic planning.[109] In other cases, the interrelated activities may not be contiguous but may be located in the same broad area or region. Symbiotic relationships may also be established between industrial activities which are remote from each other with normally no direct relationship; this manifestation of industrial symbiosis most commonly takes the form of a brokered waste exchange.

The archetypal example of industrial symbiosis is provided by Kalundborg in Denmark, outlined in Figure 17.17. At the core of the Kalundborg system are a refinery and a CHP station. Other activities have developed around them, including production of chemicals, pharmaceuticals and building products and also food production – fruit, vegetables and fish farming – using the waste

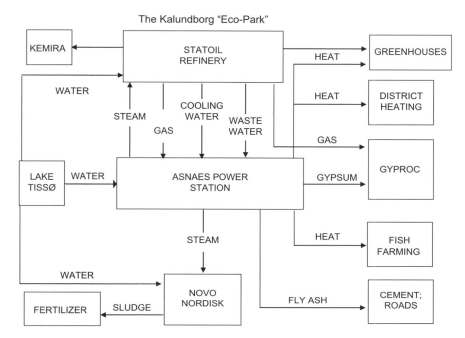

Figure 17.17 Industrial symbiosis at Kalundborg.

heat. At a purely technological level, there is nothing particularly remarkable about Kalundborg; it shows the kind of process integration which is normal in, for example, a petrochemicals complex. What is more interesting is the way in which mutually beneficial relationships have developed between companies who have no common management and might even be competitors under other circumstances.[110] Dependence on relationships, sometimes personal relationships, is recognised as one of the drivers for or barriers to developing industrial symbiotic systems.

In developed industrial or post-industrial economies where construction of new plants is infrequent, waste exchanges represent the most significant opportunity for industrial symbiosis. Because the organisations which can benefit from this kind of relationship are usually unaware of each other, some form of mediation or brokering is usually considered essential;[111] an example is the National Industrial Symbiosis Programme (NISP) in the UK. European Policy is advocating industrial symbiosis as a key aspect of the "green economy" (see section 17.4).[112]

17.3.3 Life Cycle Management

17.3.3.1 Re-use and Recycling

Whereas clean technology, cleaner production and industrial symbiosis consider the life cycles of inputs to and wastes from production and manufacturing, life cycle management extends to material products which are not inherently dissipated when used so that subsequent use of the material or product is possible. The industrial ecology approach, including closed-loop material use, is represented from the perspective of the consumer in Figure 17.18 and in terms of material flows in Figure 17.19.

On its passage through the economy, a material may pass through a succession of applications, usually with progressively lower performance requirements to allow for degradation or contamination of the material. Within each application, the product may be *re-used*; *i.e.* it may be put through the same application with no mechanical, physical or chemical reprocessing. Refilling a

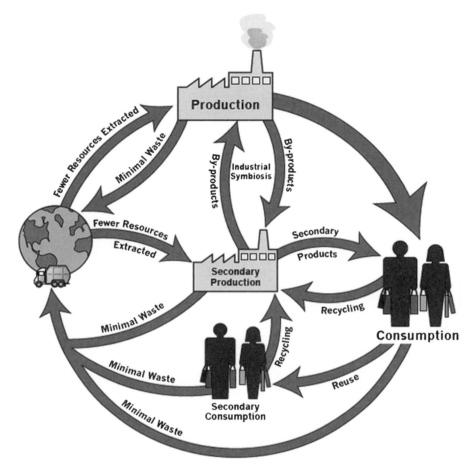

Figure 17.18 Industrial ecology and the consumer. (Reproduced with permission from Department of Natural Resources and Parks, King Country, WA, USA).[113]

container is an obvious example of re-use: it may be cleaned but is not re-formed. Re-use is emphasised because, even within a closed-loop economy, the energy requirements for reforming materials may be unsustainable.[77]

If the artefact cannot be re-used, the material may be recycled, *i.e.* reformed, commonly mixed with virgin material (*cf.* Figures 17.10 and 17.11). Because materials inevitably become contaminated or degraded through successive use cycles, they cannot be recycled in the same application indefinitely. They must therefore eventually leave the loops around Use 1 but may then be "cascaded" to the next use (see Figure 17.19). An example is glass, which may be first used as clear glass and then pass on to progressively darker coloured bottles.

These principles are illustrated by the use of plastics. Figure 17.20 shows, in very broad terms, the industrial ecology of hydrocarbon-based plastics.[‡‡] The material inputs are hydrocarbons, which must be extracted and processed to provide the monomer feedstock for polymerisation. The raw polymer must then be blended, usually with a range of additives including plasticisers, fillers and

---

[‡‡]Plastics from biomaterials are somewhat different. Life cycle assessment shows that biopolymers in particular are not necessarily better than conventional polymers on environmental and resource grounds.

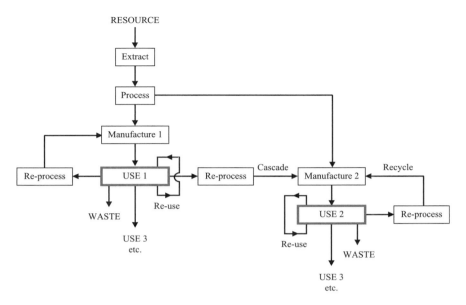

Figure 17.19   Re-use, recycling and cascaded use.

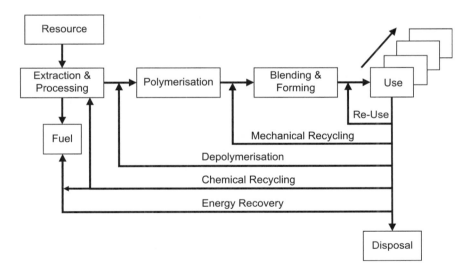

Figure 17.20   Industrial ecology of hydrocarbon-based polymers. (Reproduced from ref. 86).

pigments, before being formed into a plastic artefact. Re-use of scrap within the supply chain is not shown in Figure 17.20.

Following use, a number of options are available, shown by the different loops in Figure 17.20. The most environmentally benign loops are in the top right corner of the diagram: moving out from that corner, successive loops are less desirable in terms of resource efficiency and environmental impact. Consistent with the waste hierarchy (see Chapter 16), re-use of the plastic artefact is preferred. For packaging (which includes uses such as printer toner cartridges as well as packaging for food and beverages), this means returning the container for refilling. Deposit/return and reverse vending systems, such as the RETURPACK system used in Sweden,[114] can favour re-use.

If the artefact is recycled, then contamination with other materials must be avoided or at least minimised. This requires separation of different materials, including caps and labels, and may be aided by separation of different packages at the point of collection, as in reverse vending systems. The material is then "chipped" or "granulated". The resultant flakes may be cleaned, particularly if they are to be used for food and beverage containers, and then reformed into a recycled product. This may be identical to the previous artefact (closed-loop recycling) or may have lower performance specifications (down-cycling). If the recovered stream includes a number of commingled products, even bottles with caps made from a different material, then some separation of flakes of different materials can be achieved (see Chapter 16). Some of the additives, notably pigments and carbon black, can interfere with material separation; thus even post-use recycling is influenced by product design. There is a current debate over whether overall recovery of high-specification recyclate is favoured by sorting at the point of recovery or by separation of commingled material. The balance differs between different countries: the UK seems firmly set on a course of mixed collection and subsequent separation, whereas other countries in Europe have attached more emphasis to source separation.

As an alternative to mechanical recycling, it may be possible to depolymerise some materials (see Figure 17.20). However, depolymerisation still requires a single-material feedstock. The next option, applicable to a mixed feedstock, is chemical recycling by some form of thermal processing – pyrolysis or gasification – usually to produce a synthesis gas which can be used either as a chemical feedstock or as a fuel. Alternatively, mixed waste can be used directly as a fuel in electricity generation, heat production or combined heat and power (CHP). This at least has the merit of offsetting demand for primary fuels; for this reason, energy recovery from waste appears in the waste hierarchy, albeit below re-use and recycling, and the energy is sometimes classified as "renewable". Given the low density and value of mixed plastic waste, transport over long distances is not justified on either economic or environmental grounds. Therefore, as for all facilities for treating solid waste, viable waste-fuelled plants are limited in scale, determined by the quantities arising from the local catchment area.

While the inner loops in Figure 17.20 are associated with improved resource efficiency and environmental impact, they are also generally associated with increasing economic cost, particularly investment in reverse-vending and material separation. Development of a resource-efficient ecology for plastics therefore depends on demand for the recyclate, which in turn depends on ensuring that the recyclate meets accepted performance and quality standards.

17.3.3.2 Product Life

Figure 17.13 pointed out the importance of the capital stock of buildings, plant, vehicles, appliances *etc.* in providing the services which underpin quality of life. Figure 17.21 applies the approach of Material Flow Accounting summarised in Figure 17.12 to the maintenance of a particular stock by re-use and recycling.[115] For this case, Equation (4) takes the simple form:

$$dS/dt = p - q \tag{4}$$

For the common cases where stock is roughly constant over time (*e.g.* infrastructure, buildings and clothing in a mature economy) or where change in stock is small compared to the throughput of material, p and q are roughly equal. If the mean service life (or residence time) of the product group in question is T, then

$$p = S/T \tag{5}$$

i.e. the throughput of materials and good is strongly dependent on service life.

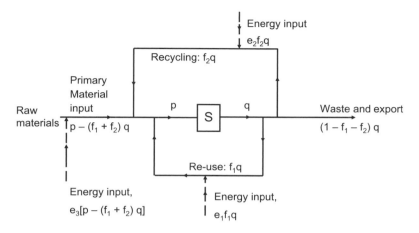

Figure 17.21 Closed-loop recycling and re-use. (Reproduced from ref. 115).

The flow p to maintain the quality of the stock is, in general, provided by a mix of re-use (fraction f_1), recycling (fraction f_2) and virgin material (fraction $1-f_1-f_2$). Usually the energy required per tonne or unit for re-use (e_1) is much lower than for recycling (e_2), while the energy for recycling is smaller than that for providing virgin material (e_3). The annual energy used to support the stock S is, from Figure 17.21 and Equation (5):

$$E = [e_1f_1 + e_2f_2 + e_3(1 - f_1 - f_2)]p$$
$$= [e_1f_1 + e_2f_2 + e_3(1 - f_1 - f_2)]S/T \qquad (6)$$

Similar relationships apply for emissions and resource use (see Figures 17.10 and 17.11).

The energy use terms (e) are normally driven down by commercial pressures. Therefore the greatest scope for reducing emissions and use of energy and resources lies in the other terms in Equation (6). The options for improving efficiency are therefore, in priority order:

(i) Extend service life, T, to reduce p;[§§]
(ii) Increase the proportion of post-use product re-used, f_1;
(iii) Increase the proportion of post-use product recycled, f_2;
(iv) Reduce the energy required for recycling, e_2;
(v) Reduce the energy required for re-use, e_1;
(vi) Reduce the energy required for primary material production, e_3;

This simple application of MFA supports the case for the closed-loop and performance economies (see section 17.3.1) in which product life is extended; as Stahel puts it "Do not repair what is not broken; do not remanufacture something that can be repaired; do not recycle a product that can be remanufactured".[85] This vision of the performance economy, with energy use (usually non-renewable) reduced by increasing the use of skilled labour (which is a renewable resource), runs counter to the current economic model but could be promoted by changes in fiscal structures.[117] The remaining sections of this chapter outline some of the implications of a shift from disposable products to service delivery.

[§§]Where the efficiency of the stock improves over time, there may be an optimal service life at which the impacts of stock replacement are offset by improvements in performance; see, for example, Smith and Keoleian.[116]

17.3.3.3 Product Design and Use

Life Cycle Product Design is an application of Life Cycle Assessment to guide selection of materials and design of components and sub-assemblies for reduced resource use and emissions.[118–120] The approach has developed from concentrating on design and manufacturing to considering the use and possible subsequent lives of a product. As shown in the preceding section (17.3.3.2), product and component life are important. Design for dismantling and re-use should also therefore be important, but trends towards miniaturisation in IT and use of a wider range of materials pull in the opposite direction.[121] Measures such as the European Directives on End-of-Life Vehicles (ELV) and Waste Electrical and Electronic Equipment (WEEE) were originally intended to return used products to the manufacturer to provide an incentive for design for re-use and recycling, but they have arguably been implemented in ways which do not have this effect.[122,123] Leasing and shared use of manufactured products can also provide incentives to extend product life and promote re-use at the end of life.[85] This approach, termed *servicisation*, is currently attracting interest.

Photocopiers provide a particularly instructive example.[7] The great majority of photocopiers are leased, and are therefore returned to the supplier after use. The return loop is therefore already closed, and the problems associated with reverse logistics have already been addressed. This also enables multiple-trip packaging to be used. A used machine is not treated just as scrap, but is recovered for re-use and recycling as shown in Figure 17.22. Lightly used machines may be partially dismantled, to remove worn or damaged parts, and returned for re-engineering. Machines which have received heavier use are disassembled completely. Components can then be fed back into the assembly line for re-use, after quality checking to ensure that they meet the same specifications as new components. Components which cannot be re-used, for example because they are more heavily worn or because they contain materials which are no longer used or permitted, may be recycled, typically passing back to the manufacturer who can blend the material with fresh material or reprocess it for other "cascaded" applications. However, major changes in technology disrupt this kind of product system; in the case of photocopiers, the shift from analog to digital devices was disruptive.

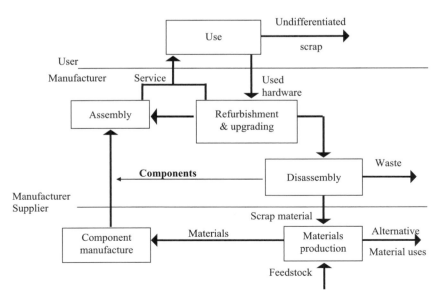

Figure 17.22   Recovery and recycling of a manufactured product. (Reproduced from ref. 7).

17.3.3.4 Waste Management

Even in a circular or performance economy, waste is inevitable. Management of solid waste is discussed in detail in Chapter 16. The system approach to environmental management can be applied to waste management, as a specific application of Life Cycle Assessment. The basic approach introduced in Figure 17.7 is applied to waste management as shown in Figure 17.23. For waste management, the Foreground comprises the waste management activities themselves, from the location where the waste arises, through the beneficial recovery of materials and/or energy (to the point where the material or energy is returned to the Background economy) to final emissions to the environment. The environmental interventions arising from the Foreground activities are termed *Direct Burdens*. They include emissions from vehicles, from thermal treatment/ combustion (gasification or pyrolysis) or from composting, uncontained landfill gas, and long-term leachate emissions from landfill. The resource usages and emissions arising from the Background activities are termed *Indirect Burdens*.

Particular care is needed in defining the benefits arising from recovery of materials or energy from the waste. The explicit assumption in the attributional approach is that recovery of materials and/or energy in the Foreground does not affect the demand for goods and services in the Background (except for materials and energy supplied to the Foreground). Thus paper produced from recycled paper, for example, is treated as a direct substitution for virgin fibre. The environmental interventions displaced by material or energy recovery are termed *Avoided Burdens*. The total Life Cycle Inventory for the waste management scheme is thus:

> *Direct Burdens* – arising in the Foreground waste management system
> *plus Indirect Burdens* – arising in the supply chains of materials and energy provided to the Foreground
> *minus Avoided Burdens* – associated with activities displaced by material and/or energy recovered from the waste.

In this way, LCA can provide a systematic framework for comparing different strategies for Integrated Waste Management. How the results are used for waste management planning is more contentious because, particularly in English-speaking countries, waste policy attracts much public

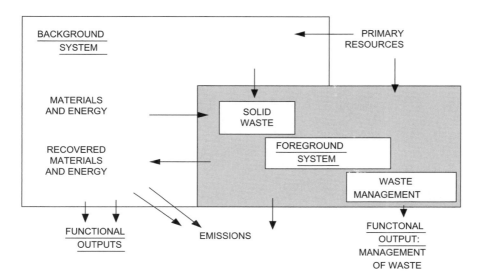

Figure 17.23 Application of Life Cycle Assessment to waste management. (Reproduced from ref. 9).

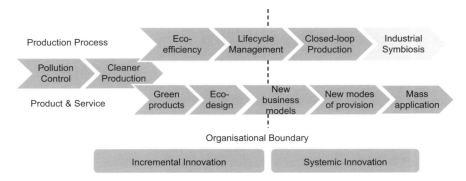

Figure 17.24 Economic transformations for a green economy. (Reproduced from ref. 125).

attention and debate. Application of LCA results in this context is a classic case where technological analysis has to be seen as an input to a public deliberation process.[45,88,124]

17.4 THE GREEN ECONOMY

The theme of this chapter has been avoiding or minimising pollution and dissipation of resources, rather than containing pollution or mediating its consequences. The industrial ecology approach goes beyond clean technology and cleaner production to consider system- or economy-wide changes to improve resource efficiency and reduce pollution, *i.e.* the "Green Economy" agenda which is now part of national and international policy discourse. Figure 17.24 summarises the changes envisaged by the Organisation for Economic and Co-operation and Development (OECD) in moving to a "green economy", involving the distinction introduced in section 17.3.2.4 between changes at Level 2: incremental innovation and transformations at Level 3: systemic innovation. How fast and effectively these industrial transformations will be achieved is an open and important question.

REFERENCES

1. C. A. Mitchell, A. L. Carew and R. Clift, *Sustainable Development in Practice: Case Studies for Engineers and Scientists*, ed. A. Azapagic, S. Perdan and R. Clift, John Wiley & Sons, Chichester, 2004, ch. 2, pp. 29–55.
2. R. Clift, S. Sim and P. Sinclair, *Treatise in Sustainability Science and Engineering*, ed. I. S. Jawahir, S. K. Sikhdar and Y. Huang, Springer Publishers, Dordrecht, 2013, pp. 291–309.
3. R. Clift, *Trans. Inst. Chem. Eng. B*, 1998, **6**, 151.
4. R. Clift, *J. Chem. Technol. Biotechnol.*, 1997, **68**, 347.
5. ISO (International Organisation for Standardisation), *Environmental Management – Life Cycle Assessment – Principles and Framework* (ISO 14040) *and – Requirements and Guidelines* (ISO 14044), ISO, Geneva, 2006.
6. H. Baumann and A.-M. Tillman, *The Hitch Hiker's Guide to LCA*, Studentlitteratur, Lund, 2004.
7. R. Clift and A. J. Longley, *Clean Technology and the Environment*, ed. R. C. Kirkwood and A. J. Longley, Blackie Academic and Professional, London, 1995, 174.
8. United Nations Environment Programme (Industry and Environment), *Life Cycle Assessment: What it is and how to use it*, Technical Report, Paris, 1996.
9. R. Clift, A. Doig and G. Finnveden, *Trans. Inst. Chem. Eng. B*, 2000, **8**, 279.

10. A.-M. Tillman, *Environ. Impact Assess. Rev.*, 2000, **20**, 113.
11. G. Finnveden, M. Z. Hauschild, T. Ekvall, J. Guinée, R. Heijungs, S. Hellweg, A. Koehler, D. Pennington and S. Suh, *J. Env. Manage.*, 2009, **91**, 1.
12. A. Zamagni, J. Guinée, R. Heijungs, P. Masoni and A. Raggi, *Int. J. LCA*, 2012, **17**, 904.
13. G. Huppes and F. Schneider, *Proceedings of the European Workshop on Allocation in LCA*, SETAC-Europe, Brussels, 1994.
14. B. Weidema, *J. Ind. Ecol.*, 2001, **4**, 11.
15. S. Lundie, A. Ciroth and G. Huppes, *Inventory Methods in LCA: towards Consistency and Improvement – Final Report*, UNEP-SETAC Life Cycle Initiative, Paris, 2007.
16. F. Ardente and M. Cellura, *J. Ind. Ecol.*, 2012, **16**, 387.
17. R. Clift, R. Frischknecht, G. Huppes, A.-M. Tillman and B. Weidema, *SETAC-Eur.News*, 1999, **10**(3), 14.
18. R. E. Miller and P. D. Blair, *Input-output Analysis: Foundations and Extensions*, Cambridge University Press, Cambridge, 2nd edn., 2009.
19. Y. Moriguchi, Y. Kondo and H. Shimuzu, *Ind. Environ.*, 1993, **16**, 42.
20. S. Suh, M. Lenzen, G. Treloar, H. Hondo, A. Horvath, G. Huppes, O. Jolliet, U. Lann, W. Krewitt and Y. Moriguchi, *Environ. Sci. Technol.*, 2004, **38**, 657.
21. G. Peters and E. Hertwich, *Environ. Sci. Technol.*, 2008, **42**, 1401.
22. House of Commons Energy and Climate Change Committee, *Consumption-based Emissions Reporting – Twelfth Report of Session 2010–2012*, **1**, The Stationery Office, London, 2012.
23. A. Druckman and T. Jackson, *Ecol. Econom.*, 2009, **68**, 2066.
24. J. C. Bare, D. Hofstetter, D. Pennington and H. A. Udo de Haes, *Int. J. LCA*, 2000, **5**, 319.
25. M. Goedkoop and R. Spriensma, *The Eco-Indicator 99 – A Damage-oriented Method for Life Cycle Impact Assessment: Methodology Report*, Pré Consultants, Amersfoort, 2nd edn., 2000.
26. J. B. Guinee, R. Heijungs, H. A. Udo de Haes and G. Huppes, *J. Cleaner Prod.*, 1993, **1**, 81.
27. L.-G. Lindfors, K. Christiansen, L. Hoffman, Y. Virtanen, V. Juntilla, O.-J. Hanssen, A. Ronning, T. Ekvall and G. Finnveden, *Nordic Guidelines on Life-Cycle Assessment*, Nordic Council of Ministers Report Nord 1995:20, Copenhagen, 1995.
28. G. Finnveden and M. Nilsson, *Int. J. LCA*, 2005, **10**, 235.
29. R. Van Zelm, M. A. J. Huijbregts, H. A. den Hollander, H. A. van Jaarsveld, F. J. Sauer, J. Struijs, H. J. van Wiljnen and D. van de Meent, *Atmos. Environ.*, 2008, **42**, 441.
30. J. Potting, W. Schöpp, K. Blok and M. Z. Hauschild, *J. Ind. Ecol.*, 1998, **2**, 63.
31. M. Wright, D. Allen, R. Clift and H. Sas, *J. Ind. Ecol.*, 1997, **1**, 117.
32. L. Milà i Canals, C. Bauer, J. Depestele, A. Dubneuil, K. R. Freiermuth, G. Guillard, O. Michelsen, R. Müller-Wenk and B. Rydgren, *Int. J. LCA*, 2007, **12**, 5.
33. T. Koellner and R. Scholz, *Int. J. LCA*, 2008, **13**, 32.
34. M. Brandao, R. Clift and L. Milà i Canals, *Food, Feed, Fuel, Timber on Carbon Sink? – Towards Sustainable Land-use Systems*, Springer, Dordrecht, 2013.
35. P. Smith, D. Martino, Z. Cai, D. Gwary, H. Janzen, P. Kumar, B. McCarl, S. Ogle, F. O'Mara, C. Rice, B. Scholes, O. Sirotenko, M. Howden, T. McAllister, G. Pan, V. Romanenkov, U. Schneider, S. Towprayoon, M. Wattenbach and J. Smith, *Philos. Trans. R. Soc. London, Ser. B*, 2008, **363**, 789.
36. T. Searchinger, R. Heinlich, R. A. Houghton, F. Dong, A. Elobeid, J. Fabiosa, S. Tokgoz, D. Hayes and T.-H. Yu, *Science*, 2008, **319**, 1238.
37. J. Kløverpris, H. Wenzel and P. Nielsen, *Int. J. LCA*, 2008, **13**, 13.
38. R. Schenck, *Int. J. LCA*, 2001, **6**, 114.
39. Millenium Ecosystem Assessment, *Ecosystems and Human Well-Being: Synthesis*, Island Press, Washington DC, 2005.

40. L. Milà i Canals, J. Chenoweth, A. Chapagain, S. Orr, A. Antón and R. Clift, *Int. J. LCA*, 2009, **14**, 28.
41. S. Pfister, A. Koehler and S. Hellweg, *Environ. Sci. Technol.*, 2009, **43**, 4098.
42. J. Kim, Y. Yang, J. Bae and S. Suh, *J. Ind. Ecol.*, 2013, **17**, 385.
43. R. Clift and L. Wright, *Technol. Forecasting Social Change*, 2000, **65**, 281.
44. J. Seppälä, L. Basson and G. Norris, *J. Ind. Ecol.*, 2002, **5**, 45.
45. L. Elghali, R. Clift, K. G. Begg and S. McLaren, *Eng. Sustainability*, 2008, **161**, 7.
46. J. Foster, *Valuing Nature? – Economics, Ethics and Environment*, Routledge, London, 1997.
47. J. O Neill, *Ambio*, 1997, **26**, 546.
48. M. Banner, *Christian Ethics and Contemporary Moral Problems*, Cambridge University Press, Cambridge, 1999, 5.
49. A. Azapagic. and R. Clift, *J. Cleaner Prod.*, 1999, **7**, 135.
50. A. Azapagic and R. Clift, *Comput. Chem. Eng*, 1999, **23**, 1509.
51. J. Van den Bergh and F. Grazi, *Environ. Sci. Technol.*, 2010, **44**, 4843.
52. ISO (International Organisation for Standardisation), *Environmental Labels and Declarations – General Principles* (ISO 14020), *– Self-declared Environmental Claims (Type II Environmental Labelling)* (ISO 14021), *– Type I Environmental Labelling – Principles and Procedures* (ISO 14024), *– Type III Environmental Declarations* (ISO 14025), 1999 and 2000.
53. *Regulation (EC) No 66/2010 of the European Parliament and of the Council of 25 November 2009 on the EU Ecolabel.*
54. *The "Blue Angel" label*; http://www.blauer-engel.de/en/blauer_engel/index.php (accessed 03/07/2013).
55. *Nordic "Swan" label*, Regulations for Nordic Ecolabelling, Nordic Ecolabelling Board, 12 December 2001.
56. R. Clift, R. Malcolm, H. Baumann, L. Connell and G. Rice, *J. Ind. Ecology*, 2005, **9**, 4.
57. M. Finkbeiner, *Int. J. LCA*, 2009, **14**, 91.
58. G. Sinden, *Int. J. LCA*, 2009, **14**, 195.
59. L. Draucker, S. M. Kaufman, R. ter Kuile and C. J. Meinrenken, *J. Ind. Ecol.*, 2011, **15**, 169.
60. BSI, *Specification for the Assessment of the Life Cycle Greenhouse Gas Emissions of Goods and Services (PAS 2050)*, British Standards Institution, London, 2011.
61. WRI, Greenhouse Gas Protocol: Product Life Cycle Accounting and Reporting Standard, World Resources Institute, Washington DC, 2011.
62. R. Clift, L. Basson and D. Cobbledick, *Chem. Eng.*, 2009 Sept, 35.
63. B. Steen, *J. Cleaner Prod.*, 1997, **5**, 255.
64. S. Ross, D. Evans and M. Webber, *Int. J. LCA*, 2002, **7**, 47.
65. R. Heijungs, *J. Cleaner Prod.*, 1996, **4**, 159.
66. C. J. Meinrenken, S. M. Kaufman, S. Ramesh and K. S. Lackner, *J. Ind. Ecol.*, 2012, **16**, 669.
67. H. Imbeault-Tétreault, O. Jolliet, L. Deschênes and R. K. Rosenbaum, *J. Ind. Ecol.*, 2013, **17**, 485.
68. T. E. Graedel and B. R. Allenby, *Industrial Ecology and Sustainable Engineering*, Pearson, Boston, MA, 2010.
69. P. H. Brunner and H. Rechberger, *Practical Handbook of Material Flow Analysis*, Lewis Publishers, Boca Raton, FL, 2004.
70. P. H. Brunner, *J. Ind. Ecol.*, 2012, **16**, 293.
71. A. Van Schaik and M. A. Reuter, *Mineral Eng.*, 2002, **15**, 1001.
72. J. Davis, R. Geyer, J. Ley, J. He, R. Clift, A. Kwan, M. Sansom and T. Jackson, *Resour., Conserv. Recycl.*, 2007, **51**, 118.
73. C. R. Binder, E. van der Voet and K. S. Rosselot, *J. Ind. Ecol.*, 2009, **13**, 643.
74. P. T. Jones, *J. Ind. Ecol.*, 2009, **13**, 843.
75. U. Arena and F. di Gregorio, *Resour., Conserv. Recycl.*, in press.

76. T. E. Graedel and B. R. Allenby, *Industrial Ecology*, Prentice Hall, Upper Saddle River, NJ, 2003.
77. J. M. Allwood, M. F. Ashby, T. G. Gutowski and E. Worrell, *Resour., Conserv. Recycl.*, 2011, **55**, 362.
78. J. M. Allwood and J. M. Cullen, *Sustainable Materials with Both Eyes Open*, UIT, Cambridge, 2012.
79. B. R. Bakshi, T. G. Gutowski and D. P. Sekulić, *Thermodynamics and the Destruction of Resources*, Cambridge University Press, Cambridge, 2011.
80. World Economic Forum, *More with Less: Scaling Sustainable Consumption and Resource Efficiency*, World Economic Forum, Geneva, 2012.
81. R. U. Ayres, L. T. Peiró and G. V. Méndez, *Environ. Sci. Technol.*, 2011, **45**, 2011.
82. J. M. Allwood, J. M. Cullen and R. L. Milford, *Environ. Sci. Technol.*, 2010, **44**, 1888.
83. McKinsey & Company, *Towards the Circular Economy: Economic and Business Rationale for an Accelerated Transition*, Ellen MacArthur Foundation, London, 2012.
84. T. Jackson, *Prosperity without Growth*, Earthscan, London, 2009.
85. W. R. Stahel, *The Performance Economy*, Palgrave MacMillan, Basingstoke, 2nd edn. 2010.
86. R. Clift, *J. Chem. Technol. Biotechnol.*, 1995, **62**, 321.
87. S. K. Sikdar, J. Drahos and E. Drioli, *Tools and Methods for Pollution Prevention*, ed. S. K. Sikdar and U. Diwekar, Kluwer Academic Publishers, Dordrecht, 1999, 1.
88. R. Clift, *Chem. Eng. Sci.*, 2006, **61**, 4179.
89. I. Christie, H. Rolfe and R. Legard, *Cleaner Production for Industry*, Policy Studies Institute, London, 1995.
90. A. M. H. Clayton, G. Spinardi and R. Williams, *Policies for Cleaner Technology – A New Agenda for Government and Industry*, Earthscan, London, 1999.
91. M. J. Nicholas, R. Clift, A. Azapagic, F. C. Walker and D. E. Porter, *Trans. Inst. Chem. Eng.*, 2000, **78B**, 193.
92. N. Emmott and N. Haigh, *J. Environ. Law*, 1996, **8**, 301.
93. J. Geldermann, C. Jahn, T. Spengler and O. Rentz, *Int. J. LCA.*, 1999, **4**, 94.
94. D. Cordell, J.-O. Drangert and S. White, *Global Environ. Change*, 2009, **19**, 292.
95. D. Cordell, A. Rosemarin, J.-J. Schröder and A. L. Smit, *Chemosphere*, 2011, **84**, 747.
96. R. Clift and H. Shaw, *Procedia Eng.*, 2012, **46**, 39.
97. D. Cordell and S. White, *Sustainability*, 2011, **3**, 2027.
98. US Geological Survey, Mineral Commodity Summaries for Phosphate Rock, January 2013.
99. P.-N. Giraud, *Procedia Eng.*, 2012, **46**, 22.
100. G. S. Metson, E. M. Bennett and J. J. Elser, *Environ. Res. Lett.*, 2012, **7**, 044043.
101. G. Allen, *Philos. Trans. R. Soc. London*, 1997, **355**, 1467.
102. N. Weston, R. Clift, L. Basson, A. Pouton and N. White, *Can. J. Chem. Eng.*, 2008, **86**, 302.
103. M. Overcash, *Philos. Trans. R. Soc. London*, 1997, **355**, 1299.
104. R. Bahu, B. Crittenden and J. O'Hara, *Management of Process Industry Waste*, IChemE, Rugby, 1997.
105. D. T. Allen and K. S. Rosselot, *Pollution Prevention for Chemical Processes*, John Wiley & Sons, New York, 1997.
106. M. R. Chertow, *Ann. Rev. Energy Environ.*, 2000, **23**, 313.
107. M. R. Chertow, *J. Ind. Ecol.*, 2007, **11**, 11.
108. D. R. Lombardi, D. Lyons, H. Shi and A. Agarwal, *J. Ind. Ecol*, 2012, **16**, 2.
109. H. Shi, J. Tian and L. Chen, *J. Ind. Ecol.*, 2012, **16**, 8.
110. J. Ehrenfeld and N. Gertler, *J. Ind. Ecol.*, 1997, **1**, 67.
111. R. L. Paquin and J. Howard-Grenville, *J. Ind. Ecol.*, 2012, **16**, 83.
112. P. Laybourn and D. R. Lombardi, *J. Ind. Ecol*, 2012, **16**, 11.

113. W. Collins, *pers. comm.*, Department of Natural Resources and Parks, King Country, WA, 2009.
114. RETURPAK website; http://www.pantamera.nu/en/ (accessed 03/07/2013).
115. R. Clift and J. M. Allwood, *The Chemical Engineer*, 2011 (March), 30.
116. V. M. Smith and G. Keoleian, *J. Ind. Ecol.*, 2004, **8**, 193.
117. W. R. Stahel, *Philos. Trans. R. Soc.*, London, in press.
118. G. A. Keoleian and D. Menerey, *Life Cycle Design Guidance Manual – Environmental Requirements and the Product System*, US EPA Report EPA/600/R-92/226, Office of Research and Development, Washington, DC, 1993.
119. L. Alting, M. Hauschild and H. Wenzel, *Phil. Trans. Roy. Soc.*, 1997, **355**, 1373.
120. H. Wenzel, M. Hauschild and L. Alting, Environmental Assessment of Products – Methodology, Tools and Techniques and Case Studies in Product Development, Chapman and Hall, London, 1997.
121. R. Lifset and M. Eckelmann, *Philos. Trans. R. Soc.*, *London*, 2013, **371**, 1471.
122. A. Castell, R. Clift and C. France, *J. Ind. Ecol.*, 2004, **8**, 4.
123. K. Mayers, R. Peagam, C. France, L. Basson and R. Clift, *J. Ind. Ecol.*, **15**, 4.
124. R. Clift, *Ambiente Rischio Communicazione*, 2012, **2**, 74.
125. OECD, Sustainable Materials Management: Making Better Use of Resources, 2012, OECD, Paris.

CHAPTER 18

The Environmental Behaviour of Persistent Organic Pollutants

STUART HARRAD*[a] AND MOHAMED ABOU-ELWAFA ABDALLAH[b]

[a] Division of Environmental Health & Risk Management, University of Birmingham, Edgbaston, Birmingham B15 2TT, UK; [b] Department of Analytical Chemistry, Faculty of Pharmacy, Assiut University, 71526 Assiut, Egypt
*Email: S.J.Harrad@bham.ac.uk

18.1   INTRODUCTION

The last half century has seen tremendous growth in public, scientific and governmental interest in the environmental effects of persistent organic pollutants (POPs). This is attributable to a number of factors, principally: the increasing weight of scientific evidence related to their adverse effects; the dramatic advancement of the analytical technology required to measure such compounds; and enhanced public awareness. The latter was promoted initially by media coverage of incidents such as the 1976 explosion at a chemical plant in Seveso, Italy, alongside books such as Rachel Carson's *The Silent Spring*,[1] that publicized the potentially detrimental effects of pesticide use. Over the ensuing decades, international concern grew substantially; culminating on 22nd May 2001 with the adoption of the United Nations Environment Programme (UNEP) Stockholm Convention on Persistent Organic Pollutants (POPs). By 2004, the Convention entered into force when the 50th party ratified, and at the time of writing there are a total of 178 participants (www.pops.int). The Convention's objective is "to protect human health and the environment from persistent organic pollutants." It initially listed 12 contaminants (or groups of contaminants) nicknamed "the dirty dozen" to be targeted for action. These are listed below:

- Aldrin
- Chlordane
- DDT (+DDD and DDE)
- Dieldrin
- Polychlorinated dibenzo-*p*-dioxins (PCDDs)
- Polychlorinated dibenzofurans (PCDFs)

Pollution: Causes, Effects and Control, 5th Edition
Edited by R M Harrison
© Harrad and Abou-Elwafa Abdallah and The Royal Society of Chemistry 2014
Published by the Royal Society of Chemistry, www.rsc.org

- Endrin
- Heptachlor
- Hexachlorobenzene
- Mirex
- Polychlorinated biphenyls (PCBs)
- Toxaphene

The Convention seeks the elimination or restriction of production and use of all intentionally produced POPs (*i.e.* industrial chemicals and pesticides). It also seeks the continuing minimization and, where feasible, ultimate elimination of unintentionally produced POPs (*e.g.* dioxins and furans). Moreover, stockpiles of POPs (*e.g.* pesticides) must be managed and disposed of in a safe, efficient and environmentally sound manner.

The Convention also allows for the inclusion of additional chemicals under its scope. Parties to the Convention may propose such additions, and the case for their inclusion is considered by the POPs Review Committee (POPRC). For a chemical to be listed under the Convention, it must display the following:

- Persistence
- Bioaccumulation
- Potential for long-range environmental transport, and
- Adverse effects

Full details of the evidence required for these criteria to be fulfilled are outlined in the next section, and at the time of writing, a further 10 chemicals are deemed to have met these and have been listed, see below:

- α-hexachlorocyclohexane
- β-hexachlorocyclohexane
- γ-hexachlorocyclohexane (lindane)
- Chlordecone
- Hexabromobiphenyl
- Commercial pentabromodiphenyl ether (Penta-BDE)
- Commercial octabromodiphenyl ether (Octa-BDE)
- Pentachlorobenzene
- Perfluorooctane sulfonic acid (PFOS)
- Technical endosulfan

Such high-profile international efforts serve to highlight the importance of POPs. This chapter addresses their environmental impact and the factors influencing that impact.

18.1.1   Definition of Persistent Organic Pollutants (POPs)

To be listed under the Stockholm Convention, the proposed substance should demonstrate the following:

   (i) *Potential for long range environmental transport.* Evidence for this must consist of one of the following: (a) measured levels of the chemical in locations distant from the sources of its release that are of potential concern; (b) monitoring data showing that long-range environmental transport of the chemical, with the potential for transfer to a receiving environment, may have occurred *via* air, water or migratory species; or (c) environmental fate properties and/or model results that demonstrate that the chemical has a potential for long-range environmental transport through air, water or migratory species, with the potential for transfer to a receiving environment in locations distant from the sources of its release. For a chemical that migrates significantly through the air, its half-life in air should exceed two days.

(ii) *Adverse effects.* Such evidence should comprise either: (a) evidence of adverse effects to human health or to the environment that justifies consideration of the chemical within the scope of the Convention; or (b) toxicity or ecotoxicity data that indicate the potential for damage to human health or to the environment.

(iii) *Persistence.* This should take the form of evidence that the substance's half-life in water is greater than two months, or that its half-life in soils is greater than six months, or that its half-life in sediments is greater than six months. Alternatively, evidence should be provided that the chemical is otherwise sufficiently persistent to be of concern within the scope of the Convention.

(iv) *Bioaccumulation.* To meet this criterion, the following must be provided: (a) evidence that the bioconcentration factor (BCF) or bioaccumulation factor (BAF)[†] in aquatic species exceeds 5000 or in the absence of such data, that the log K_{OW} is greater than 5 (see sections 18.5.1 and 18.6.2); or (b) evidence that a chemical presents other reasons for concern, such as high bioaccumulation in other species, high toxicity or ecotoxicity; or (c) monitoring data in biota indicating that the bio-accumulation potential of the chemical is sufficient to justify its consideration within the scope of the Convention.

Other factors such as production, uses, socio-economic factors, alternatives, cost and benefits are also to be considered. Summarised, POPs are chemicals that are resistant to environmental degradation, that can accumulate through food chains, and induce toxic effects in humans and wildlife.

18.1.2   Scope

This chapter will consider the following compounds already listed as POPs under the Stockholm Convention; *viz*: PCDDs and PCDFs, (collectively known as PCDD/Fs), PCBs, and poly-brominated diphenyl ethers (PBDEs). In addition, we will discuss hexabromocyclododecane (HBCD), which has been recommended by POPRC for listing under the Stockholm Convention at the 6th Conference of the Parties in Spring 2013.

For each class of compounds, the following key areas pertinent to their environmental impact are addressed: their toxicology – with particular emphasis on their human effects; the methods used to monitor their presence in the environment; their major sources; their physicochemical properties and the influence of these on their environmental fate and behaviour; and finally, a brief examination of the use of theoretical models to predict such behaviour.

18.1.3   Chemical Structure and Nomenclature

The environmental fate and behaviour of groups of POPs is strongly structurally-dependent, and a brief discussion of their structures and nomenclature is deemed worthwhile.

[†]Bioconcentration factor (BCF) is the ratio of the concentration of a chemical in a biological tissue to the concentration of the same chemical in the water surrounding that tissue. The BCF is usually measured under laboratory conditions. Bioaccumulation Factor (BAF) is the ratio of the contaminant level in a certain organism to its concentration in the ambient environment at a steady state. BAF is closely related to the BCF but applies usually to field measurements and takes into account multiple exposure routes (*i.e.* dietary uptake as well as passive uptake from the surrounding environment).

18.1.3.1 *Polychlorinated Dibenzo-p-dioxins and Polychlorinated Dibenzofurans (PCDD/Fs)*

The basic chemical structures of PCDD/Fs are illustrated in Figure 18.1. The numbers indicate sites of chlorination – for example, 2,3,7,8-tetrachlorodibenzo-*p*-dioxin (2,3,7,8-TCDD) contains four chlorine atoms, one each at the 2, 3, 7 and 8 positions; in total, there are 75 possible PCDDs and 135 possible PCDFs. Each individual PCDD/F is referred to as a congener, whilst those congeners possessing identical empirical formulae are isomers of each other. Each group of isomers (there exists one for each degree of chlorination) constitutes a homologue group.

18.1.3.2 *Polychlorinated Biphenyls (PCBs)*

Figure 18.2 illustrates the basic chemical structure of PCBs. As with PCDD/Fs, the numbers denote sites of chlorination – *e.g.* 3,3′,4,4′,5-pentachlorobiphenyl possesses five chlorines, three attached to one biphenyl ring at the 3, 4 and 5 positions, the remainder on the other ring in the 3′ and 4′ sites. 209 PCB congeners are possible, and the definitions and distinctions between congeners, isomers and homologues are identical to those given above for PCDD/Fs. Unlike PCDD/Fs, a IUPAC system[2] assigning a single unique number to each possible PCB congener is widely – indeed almost universally – utilized and this greatly simplifies reference to individual PCBs. For example, 3,3′,4,4′-tetrachlorobiphenyl is referred to as PCB No. 77.

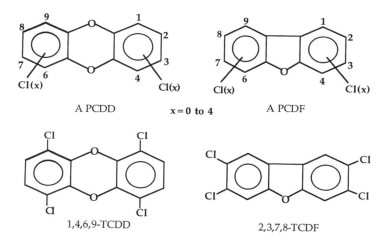

Figure 18.1   PCDD/F nomenclature.

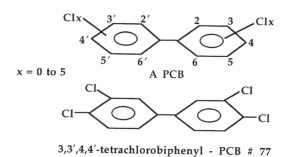

Figure 18.2   PCB nomenclature.

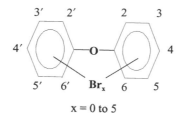

x = 0 to 5

Figure 18.3 Structures of PBDEs.

18.1.3.3 Polybrominated Diphenyl Ethers (PBDEs)

The structures of PBDEs are illustrated in Figure 18.3. Note that the numbering system employed for the 209 possible PBDEs is identical to that of PCBs, *i.e.* 3, 3′, 4, 4′-tetrabromodiphenylether is known as PBDE No. 77.

18.1.3.4 Hexabromocyclododecanes (HBCDs)

Unlike PBDEs, HBCD is a hexabrominated aliphatic compound giving rise to 16 possible isomers. However, the HBCD technical mixture is composed mainly of three diastereomers denoted as alpha (α), beta (β) and gamma (γ). Each of these three diastereomers are chiral (see also section 18.3.2.4) and therefore exist as two closely-related chemical forms known as enantiomers denoted respectively as (−) and (+). The structures of these enantiomers of α-, β-, and γ-HBCD are illustrated in Figure 18.4. Note that the collective term that covers all the isomeric forms of HBCD is stereoisomers.

18.2   ADVERSE EFFECTS

An in-depth treatment of the toxicology and ecotoxicology of POPs is beyond the scope of this chapter and only a brief summary of the adverse effects of each compound class is presented. A general consideration in relation to the toxic effects of compounds present at extremely low concentrations in the environment, is the difficulty in ascribing health or reproductive abnormalities to contamination by a specific chemical alone, in the presence of other similarly potentially toxic pollutants, often at much higher concentrations. In particular, one cannot discount the possibility of synergistic and/or antagonistic[‡] toxicological effects arising from contamination by a "cocktail" of pollutants like PCBs and PCDD/Fs, all of which are known to evoke similar bio-chemical responses. This area, which has important health implications, is now receiving much closer scientific scrutiny. In addition to these difficulties, there are considerable problems with assessing the effects in humans, in the understandable absence of clinical data. Instead, evidence for adverse effects in humans is reliant on epidemiological surveys and extrapolations from animal studies, with all the problems inherent in such indirect measurements of human toxicity. A final general point, is the tendency of all of the compounds considered here to partition into fatty tissues. As a result, their adverse effects are compounded by their bioaccumulation in species at the head of food chains, such as marine mammals, birds of prey and humans.

18.2.1   PCDD/Fs

Probably no other group of chemicals has been subjected to greater scrutiny with respect to their environmental effects than PCDD/Fs. It is to the Seveso incident in 1976, that present interest may be

---

[‡]Synergism is defined as a situation in which the combined effect of two or more substances exceeds the sum of their separate effects. Antagonism is the opposite phenomenon, whereby the combined presence of two or more substances lessens the effects of those substances acting independently.

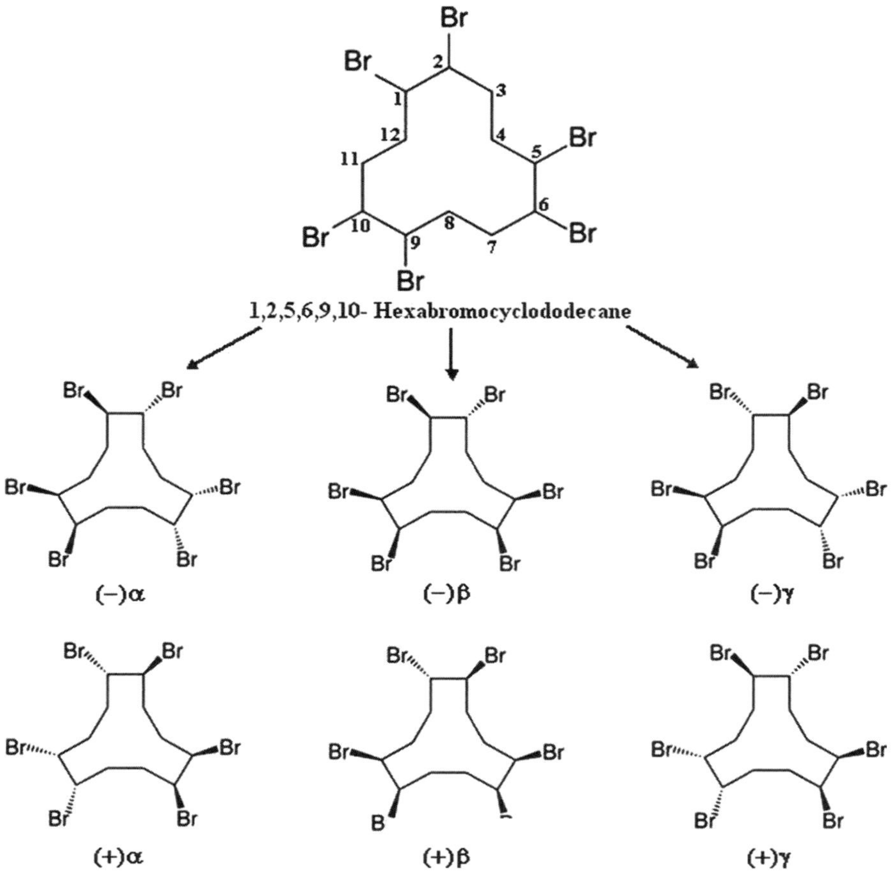

Figure 18.4 Structure and nomenclature of the three most abundant HBCD diastereomers.

traced. The episode occurred on July 10th, 1976 at the ICMESA chemical plant in Seveso, near Milan, Italy. The alkaline hydrolysis of 1,2,4,5-tetrachlorobenzene to produce 2,4,5-trichlorophenol (2,4,5-TCP) – an intermediate in the production of the bactericide, hexachlorophene – went out of control, and the resultant explosion distributed large quantities of 2,3,7,8-TCDD (formed *via* the self-condensation of 2,4,5-TCP) over the surrounding area. Although the exact environmental impact of the incident has yet to be fully evaluated – whilst widespread animal mortality occurred, the only undisputed adverse effect in humans remains the induction of chloracne (an extremely disfiguring skin complaint) – the psychological effect on the general public was immense. Widespread media coverage both in Italy and beyond – ranging from the responsibly factual to the hysterical (an assertion was made that dioxin could facilitate "the end of Western civilization") – propelled the "dioxin issue" onto the political agenda of the industrialized world. Since then, a number of incidents have maintained a high public profile for these compounds. Examples include: the apparent severe chloracne suffered by the then Ukrainian presidential candidate Viktor Yushchenko following his alleged poisoning with dioxin-contaminated soup; the detection of high concentrations of dioxin-like PCBs in Belgian eggs and chickens as a result of contamination of animal feed; and an incident in Ireland whereby pig feed contaminated with dioxin-like compounds led to contamination of pork and the withdrawal from sale of all Irish pork products.

It is important to note that amongst the range of 210 compounds known as PCDD/Fs, there exist wide differences in toxicological potency. In summary, only those PCDD/Fs chlorinated at the

Table 18.1 PCDD/F and PCB Congeners for which Toxic Equivalent Factors (TEFs) are defined.

Congener	TEF
2,3,7,8-tetrachlorodibenzo-*p*-dioxin (2,3,7,8-TCDD)	1
1,2,3,7,8-pentachlorodibenzo-*p*-dioxin (1,2,3,7,8-PeCDD)	1
1,2,3,4,7,8-hexachlorodibenzo-*p*-dioxin (1,2,3,4,7,8-HxCDD)	0.1
1,2,3,6,7,8-hexachlorodibenzo-*p*-dioxin (1,2,3,6,7,8-HxCDD)	0.1
1,2,3,7,8,9-hexachlorodibenzo-*p*-dioxin (1,2,3,7,8,9-HxCDD)	0.1
1,2,3,4,6,7,8-heptachlorodibenzo-*p*-dioxin (1,2,3,4,6,7,8-HpCDD)	0.01
octachlorodibenzo-*p*-dioxin (OCDD)	0.0003
2,3,7,8-tetrachlorodibenzofuran (2,3,7,8-TCDF)	0.1
1,2,3,7,8-pentachlorodibenzofuran (1,2,3,7,8-PeCDF)	0.03
2,3,4,7,8-pentachlorodibenzofuran (2,3,4,7,8-PeCDF)	0.3
1,2,3,4,7,8-hexachlorodibenzofuran (1,2,3,4,7,8-HxCDF)	0.1
1,2,3,6,7,8-hexachlorodibenzofuran (1,2,3,6,7,8-HxCDF)	0.1
1,2,3,7,8,9-hexachlorodibenzofuran (1,2,3,7,8,9-HxCDF)	0.1
2,3,4,6,7,8-hexachlorodibenzofuran (2,3,4,6,7,8-HxCDF)	0.1
1,2,3,4,6,7,8-heptachlorodibenzofuran (1,2,3,4,6,7,8-HpCDF)	0.01
1,2,3,4,7,8,9-heptachlorodibenofuran (1,2,3,4,7,8,9-HpCDF)	0.01
octachlorodibenzofuran (OCDF)	0.0003
3,3′,4,4′-tetrachlorobiphenyl (PCB # 77)	0.0001
3,4,4′,5- tetrachlorobiphenyl (PCB # 81)	0.0003
3,3′,4,4′,5-pentachlorobiphenyl (PCB # 126)	0.1
3,3′,4,4′,5,5′-hexachlorobiphenyl (PCB # 169)	0.03
2,3,3′,4,4′-pentachlorobiphenyl (PCB # 105)	0.00003
2,3,4,4′,5-pentachlorobiphenyl (PCB # 114)	0.00003
2,3′,4,4′,5-pentachlorobiphenyl (PCB # 118)	0.00003
2′,3,4,4′,5-pentachlorobiphenyl (PCB # 123)	0.00003
2,3,3′,4,4′,5-hexachlorobiphenyl (PCB # 156)	0.00003
2,3,3′,4,4′5′-hexachlorobiphenyl (PCB # 157)	0.00003
2,3′,4,4′,5,5′-hexachlorobiphenyl (PCB # 167)	0.00003
2,3,3′,4,4′,5,5′-heptachlorobiphenyl (PCB # 189)	0.00003

2, 3, 7 and 8 positions are considered of toxicological significance. The seventeen congeners fitting this criterion are listed in Table 18.1, along with their toxic equivalent factors (TEFs).[§] This structural-dependence of toxicity is due to the fact that the toxic effects of PCDD/Fs are mediated by initial interaction with a cellular protein – the Ah-receptor. Such binding is subject to the "lock and key" principle commonly associated with compound–receptor interactions and hence, only those compounds able to assume a coplanar molecular configuration similar to that of 2,3,7,8-TCDD are capable of binding to the Ah-receptor. This receptor-mediated theory of toxicity also lends a rationale to the considerable species-specific variations in the toxicity of 2,3,7,8-TCDD (guinea-pigs have an LD_{50} of 0.6 µg kg^{-1} body weight (bw) *c.f.* hamsters which have an LD_{50} of 3500 µg kg^{-1} bw),[4] as the precise nature of the Ah-receptor varies widely between species.

Scientific opinion concerning the exact toxicological potency of PCDD/Fs continues to differ widely. Despite this, there is a significant body of opinion that PCDD/Fs may well present a human cancer hazard, and that they are potent toxins with the potential to elicit a range of non-cancer effects, starting with binding to the Ah-receptor. These effects may be occurring in humans at very

---

[§]TE (or TEQ) is an acronym for 2,3,7,8-TCDD equivalents. It is a means of expressing the toxicity of a complex mixture of different PCDD/Fs and related compounds (such as coplanar PCBs – see section 18.2.2) in terms of an equivalent quantity of 2,3,7,8-TCDD. Each compound is assigned a Toxic Equivalency Factor (TEF) based on its toxicity relative to that of 2,3,7,8-TCDD, which is universally assigned a TEF of 1. Multiplication of the concentration of a compound by its assigned TEF gives its concentration in terms of TE and the toxicity of a mixture is the sum of the TEs calculated for all congeners. The weighting scheme referred to in this chapter is that devised by the WHO.[3]

low levels comparable with the upper limit of background (*i.e.* non-occupational) exposure, and include reproductive, immunological, and developmental effects. To put this into context, the average background exposure to PCDD/Fs of UK adults is 0.4 pg ΣTE kg^{-1} bw per day, compared with the UK government's tolerable daily intake (TDI) for dioxin-like compounds of 2 pg ΣTE kg^{-1} bw per day.[5]

18.2.2 PCBs

PCBs have been linked to a number of toxic responses, including the impairment of immune responses in biota. Other cited effects include hepatotoxicity, carcinogenicity, teratogenicity, and reproductive toxicity. In evaluating the significance of these effects, it is important to recognise that – as for PCDD/Fs – considerable species- and congener-specific variations exist. In particular, there are indications that their potency in humans is markedly less than in other animals, and it is very important that caution is exercised when extrapolating toxicological data obtained from laboratory animals to humans, and that PCB behaviour is studied on an individual congener basis. Especially relevant in this context, is the fact that some PCBs (especially those lacking chlorine substituents at the *ortho* 2, 2', 6 and 6' positions), are capable of eliciting similar toxicological effects to the 2,3,7,8-chlorinated PCDD/Fs. This is due to their ability to adopt the coplanar molecular configuration necessary to interact with the Ah-receptor responsible for mediating "dioxin-like" effects, and has led to the assignation of TEFs to such "coplanar" PCBs.[3] When the toxic equivalent contribution of PCBs is included with that from PCDD/Fs for the purposes of comparing exposure with the TDI for dioxin-like toxicity of 2 pg ΣTE kg^{-1} bw per day, this has the effect of further narrowing the gap between background human exposure and the TDI. Those PCBs assigned TEFs are included in Table 18.1. In terms of UK adult human exposure, PCBs and PCDD/Fs make an approximately equal contribution on a ΣTE basis – average UK background exposure to PCDD/Fs *and* PCBs together is estimated at 0.9 pg ΣTE kg^{-1} bw per day of which PCBs contribute 0.5 pg ΣTE kg^{-1} bw per day.

   More recently, concerns have also been raised about the adverse neurodevelopmental impacts of non-dioxin-like PCBs.[6] Although the levels at which such effects are thought to occur are much higher than those required to elicit dioxin-like effects, this is offset by the fact that exposure to such PCBs is also much greater.

18.2.3 PBDEs

While evidence of adverse effects arising from PBDE exposure is to date limited, the widely reported presence in human tissues, together with their structural similarity to PCDD/Fs and PCBs has raised concerns. Several synthetic chemicals can mimic, block or synergize the response to natural hormones in our bodies. These chemicals are generally labeled endocrine disruptors. In general, the acute toxicity of PBDEs is low and lower brominated PBDEs can cause adverse effects at comparatively lower doses than higher brominated congeners. All PBDE technical products have been shown to have thyroid disrupting properties *via* competitive binding of PBDEs to thyroid hormone receptors.[7] PBDEs can also alter the liver functions, leading to changes in thyroid hormones homeostasis, often resulting in increased elimination of the thyroid hormone, T4.[8] Animal studies have shown reduced levels of thyroid hormones in serum following exposure to PBDEs. At high doses, BDE-209 was reported to induce tumors in rats and mice;[9] while penta-BDEs have been shown to interfere with sexual development in rats causing delayed onset of puberty and decreased follicle formation.[10] However, the most critical toxicological endpoint for PBDEs appears to be developmental neurotoxicity which has been reported for a range of congeners (including BDE-209) in mice.[11] Of particular interest are the effects of neonatal exposure, which has been demonstrated to adversely affect learning and memory functions in adult animals.[12] If replicated in humans, such effects are of potential concern.

18.2.4 HBCDs

The toxicological database for HBCD is limited. The direct acute and chronic toxicity of HBCD appears to be low.[13] However, HBCD has an antagonistic effect on detoxification enzymes that may increase the toxicity of other compounds.[14] It has also been suggested that HBCD may induce cancer by a non-mutagenic mechanism,[14,15] and can disrupt the thyroid hormone system resulting in lower circulating concentrations of thyroxine.[16]

Neonatal exposure to HBCD can induce developmental neurotoxic effects, such as aberrations in spontaneous behaviour, learning, and memory function.[13,17] HBCD can also alter the normal uptake of the neurotransmitters in rat brains[18] *via* inhibition of the glutamine and dopamine uptake mechanism. This effect is additive to the effects of other brominated flame retardants (BFRs) and methyl mercury.[19] Recent studies have shown the importance of oxidative stress and initiation of apoptotic cell death as mediators of HBCD induced adverse effects at the cellular level.[20]

18.3 MEASUREMENT TECHNIQUES

Table 18.2 indicates "typical" levels of selected POPs in soil, air, human milk and freshwater. From this, it is evident that sensitivity is an important prerequisite of any measurement technique for such compounds, particularly for PCDD/Fs, where the quantities involved may justifiably be described as "ultra-trace". Given the increasing level of legislation relating to the presence of such compounds in the environment, there is also a clear need for the techniques to be both as accurate and precise as is possible. In achieving these aims, the crucial role of sampling methodology must be recognized. Finally, the measurement techniques employed must be able to distinguish the target compounds from the complex "soup" of other chemicals present in the sample – in short, they must be selective.

18.3.1 Sampling Methodology

Even using the most sophisticated analytical instrumentation, a measurement will be severely compromised if the sample taken for analysis is unrepresentative. Hence, a good sampling method provides a sample that accurately reflects the levels of the chosen analyte(s) in the matrix studied. It must also provide sufficient sample for the purposes of the study (in this respect, the sensitivity of

Table 18.2 Typical environmental levels of selected POPs.

"Typical" Concentration of...				
Compartment	*PCDD/Fs[a]*	*PCBs[b]*	*PBDEs[c]*	*HBCDs[d]*
Soil	$0.3\ \mu g\ kg^{-1}$	$5\ \mu g\ kg^{-1}$	$100\ \mu g\ kg^{-1}$	$1\ \mu g\ kg^{-1}$
Outdoor Air	$0.003\ ng\ m^{-3}$	$0.2\ ng\ m^{-3}$	$0.005\ ng\ m^{-3}$	$0.06\ ng\ m^{-3}$
Human milk	$0.015\ \mu g\ kg^{-1\,e}$	$10\ \mu g\ kg^{-1\,e}$	$5\ \mu g\ kg^{-1\,f}$	$5\ \mu g\ kg^{-1\,f}$
Freshwater	$0.003\ ng\ dm^{-3}$	$1\ ng\ dm^{-3}$	$0.065\ ng\ dm^{-3}$	$0.15\ ng\ dm^{-3}$

[a]Reported as the sum of all tetra- through octachlorinated congeners.
[b]Expressed as the sum of the most commonly occurring PCBs.
[c]Expressed as the sum of the most commonly occurring PBDEs.
[d]Expressed as the sum of α-, β- and γ-diastereomers.
[e]Data on a whole milk basis.
[f]Data on a lipid weight basis.

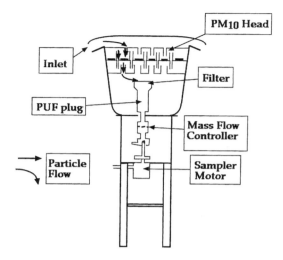

Figure 18.5   Schematic of hi-vol apparatus for sampling POPs in ambient air.

the analytical instrumentation employed must also be taken into account). Another important consideration for both soil and water sampling is that the suitable containers for sample storage must be cleaned thoroughly prior to use, to minimise potential cross-contamination.

In terms of sampling methodology, the major differences occur according to the nature of the matrix under scrutiny – *e.g.* soil, air or water – rather than on a compound-specific basis. As a result, the following sections will discuss the types of sampling strategies used for the determination of trace levels of organic chemicals in air, soil and water, with allusion to compound-specific refinements where necessary.

18.3.1.1   Air Sampling

There are two principal categories of "air" sampling. One involves sampling ambient air, *i.e.* that to which humans and other animals are exposed in their normal working and recreational lives, whilst the other involves the study of the gaseous and/or dust emissions from a pollutant source such as a waste incinerator. For the purposes of this chapter, we shall limit our focus to ambient air.

A general consideration for all air sampling is the monitoring of sampling efficiency – *i.e.* the proportion of analyte retained by the sampler. This may be achieved by the addition of "sampling evaluation standards" to the sampler prior to commencement of the sampling campaign. The fraction remaining at the end of sampling affords a measure of the sampling efficiency.

Figure 18.5 shows a schematic of the sampling apparatus used to sample POPs in ambient air. The equipment basically comprises: a PM_{10}-selective impactor head (designed to remove particles above 10 μm diameter); followed by a smaller sampling head housing a filter paper (commonly glass fibre or PTFE; the latter is more expensive but minimizes analyte reaction on the filter during sampling); followed by one or two cylindrical "plugs" of polyurethane foam (PUF). POPs present in ambient air will exist in both gaseous and aerosol form (vapour and particulate phase, respectively). The aerosol fraction is collected by the filter, with the vapour phase adsorbed onto the PUF "plugs". Owing to the low levels of POPs (particularly PCDD/Fs) in ambient air, very high volumes (100–3000 m^3) must be sampled in order to procure measurable quantities of the analyte(s) of interest.

As will be evident from Table 18.2, atmospheric concentrations of the other POPs considered in this chapter are considerably higher than those of PCDD/Fs, especially indoors where there are

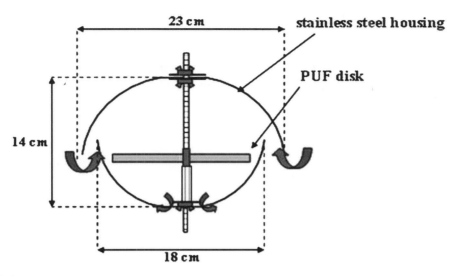

Figure 18.6 Typical configuration of a passive air sampler.

often substantial sources. Because of this, and the need for more portable and less noisy air samplers; passive air samplers (PAS) are now used widely. PAS provide several other advantages over active samplers including: ease of use, inobtrusiveness during deployment, low cost, no electricity requirements, as well as the provision of time weighted average concentrations which make PAS ideal for indoor air sampling. Typical sampling rates (which vary between different compounds) are 1–4 m^3 day^{-1} and thus longer deployment periods ($\sim$28 days) are standard for PAS. Classic configurations of PAS were applied mainly to sample the vapour phase of air.[21–23] However, recent modifications have resulted in passive samplers capable of sampling POPs associated with atmospheric particles, albeit at a different sampling rate, as well as those associated with the vapour phase.[24] Figure 18.6 illustrates a typical PAS configuration designed to sample the vapour phase.

18.3.1.2 Soil Sampling

Procuring soil samples is relatively straightforward. A cylindrical metal corer of *ca.* 10 cm diameter is sunk into the soil, with the material removed being placed into a clean glass jar, which is sealed to minimise adsorption and volatilization of analytes from and to the air.

An important aspect of any soil sampling campaign is the sampling depth. This will depend on the nature of the analyte (most POPs are not prone to movement through the soil column, thus largely restricting contamination to the surface), and on whether the soil has been ploughed or otherwise recently disturbed, as disturbance will enhance analyte distribution throughout the soil column. Typical sampling depths for POPs are 5–10 cm on undisturbed land, as concentrations fall dramatically below such depths. Obtaining a representative sample is also very important, and is commonly achieved by combining sampled soil cores from various points in the vicinity of the sampling site. These points are situated along either an 'X' or a 'W' pattern covering an area typically 10 m by 10 m. The combined samples are homogenized and appropriately sized sub-samples taken for analysis.

18.3.1.3 Water Sampling

Hydrophobicity is a common property of all POPs dealt with in this chapter. As a result, the levels of most POPs present in aqueous media are low (see Table 18.2), and much of the burden of such chemicals associated with "water" samples, is in fact bound to particulate matter. Consequently, large sample volumes are necessary (*e.g.* 40 L for the determination of HBCDs in lake water),[25] and the particulate fraction must be separated from the aqueous fraction if the true "dissolved phase" concentration is to be measured. Such separation is carried out by pulling the sample through a fine mesh filter (usually, but not always, of 0.45 µm porosity), to trap particle-bound pollutants. The filtrate is subsequently either solvent extracted, or passed through an adsorptive resin-filled tube or PUF plug to concentrate the dissolved phase, which is subsequently solvent extracted.

18.3.2 Analytical Methodology

Once sampling is complete, a vast array of analytical techniques is available to the analyst. That chosen depends on both the nature of the matrix (the determination of POPs in sewage sludge is extremely difficult owing to the high levels of potential interferences that are present), the expected concentrations of the target analyte, and the nature of the analyte itself. In this section therefore, the range of available techniques will be examined on a compound-specific basis.

18.3.2.1 PCDD/Fs

The universal method of choice for PCDD/F determination is GC-MS. The selectivity afforded by mass spectrometry is essential, given that the levels of potential chemical interferences present in samples will usually be far in excess of those of PCDD/Fs, whilst the requisite sensitivity is partly achieved *via* use of selected ion monitoring (SIM). In general, and particularly for the determination of PCDD/Fs in ambient air where individual congener concentrations may be as low as a few fg m^{-3} (1 fg = 10^{-15} g), the use of high resolution mass spectrometers capable of providing "on-column" detection limits of 25 fg is favoured.

Quantification of PCDD/Fs is made *via* reference to $^{13}C_{12}$-labelled PCDD/F internal standards (IS) added to the sample prior to extraction. These standards – which are identical to the corresponding "native" PCDD/F except for their mass – are used to correct analyte concentrations for losses throughout the extensive extraction and purification procedures employed prior to mass spectrometric analysis, as well as fluctuations in mass spectrometric response during an analysis. As noted in later sections, the use of such IS is commonly employed for the determination of many POPs. The quantification method employed is referred to as "isotope dilution".

18.3.2.2 PCBs

There are two principal selectivity requirements for the analysis of PCBs. The first, the separation of individual PCB congeners from each other, is crucial, given the congener-specific variations in toxicity, and is achieved by the use of GC. The second selectivity requirement – *viz* the ability to differentiate PCBs from co-eluting interferences – is less pivotal than for PCDD/Fs, given the fact that environmental concentrations of individual PCBs are much greater. As a result, while mass spectrometric detection – especially when employed alongside the use of $^{13}C_{12}$-labelled PCB IS for the determination of dioxin-like congeners – is now the method of choice for the determination of PCBs, bench-top unit mass resolution mass spectrometers are usually adequate. The selectivity of mass spectrometry facilitates differentiation of PCBs from co-eluting interferences such as β-HCH and DDE, as well as the resolution of PCB congeners with closely similar GC retention times but different molecular masses (*e.g.* PCBs 77 and 110).

18.3.2.3 PBDEs

To date, the most widely used technique for analysis of PBDEs is GC-MS. In general, low resolution MS with an EI (electron ionization) source can be used for detection of mono- to hexa- brominated congeners with good sensitivity (~ 50 pg on-column) in SIM mode. The major ions formed from PBDEs in EI-MS are the M^+ and $[M-2Br]^+$, which can be used for their identification, and quantification.[26] The EI source provides the added advantages of separate quantification of co-eluting mass homologues with a different number of Br substituents and possible application of $^{13}C_{12}$-labelled isomers as IS.[27]

On the other hand, an ECNI (electron capture negative ionization) source is required to attain a good sensitivity for hepta- to deca-BDEs. The chief benefit of ECNI is higher sensitivity than EI. However – with one exception – only the Br isotopes (m/z 79 and 81) can be monitored using this ionization technique which precludes the use of ^{13}C-labelled isomers as IS. This is especially problematic for the determination of nona- and deca-BDEs where considerable thermal degradation on the GC occurs, that cannot be compensated for without isotopically labeled IS. To overcome such problems for BDE-209, the use of $^{13}C_{12}$-BDE 209 as IS in ECNI-MS mode (which can be monitored *via* the ion $[C6Br5O]^-$) has found broad application.[28]

Most recently, an isotope-dilution LC-MS-MS method has been developed for the analysis of tetra- to deca BDE congeners. The method is advantageous over GC-MS methods as it not only permits use of isotopically-labelled IS for all congeners but avoids on-column thermal degradation of higher brominated congeners, as well as offering the high sensitivity (~ 40 pg on-column) and selectivity inherent to tandem mass detection.[29]

18.3.2.4 HBCDs

HBCDs can be analyzed by GC-MS, typically using an ECNI source for which the monitoring of $[Br]^-$ provides excellent sensitivity.[30] However, HBCD diastereomers cannot be resolved using GC columns and, until recently, environmental data have been reported as total HBCD (*i.e.* the sum of α-, β- and γ-HBCDs). This is mainly because at temperatures > 160 °C (commonly used in GC analyses), thermal rearrangement of the HBCD diastereomers occurs leading to isomeric interconversions, with HBCD decomposition occuring at temperatures > 240 °C. Hence, GC methods are prone to errors regarding the accurate quantification of the total HBCD concentration. Moreover, co-elution of HBCD and some PBDEs during GC analyses is also possible, and this will result in biased data during the MS quantification of these chemicals when monitoring the bromine isotope ions with m/z 79 and 81.[31]

Reliable HBCD stereoisomer-specific analysis is essential for further understanding of the environmental fate and behaviour of this group of pollutants. Therefore, for isomer-specific analysis of HBCDs, reversed phase, high-performance liquid chromatography (RP-HPLC) operating at ambient temperatures is a better alternative to GC. When LC is interfaced with an atmospheric pressure ionization source, such as electrospray ionization (ESI), the mass spectrometer can then selectively detect the chromatographically separated stereoisomers. Currently LC-MS-MS is the method of choice for isomer-specific analysis of HBCDs in environmental samples. Isotope labelled IS for the three main HBCD diastereomers are essential for accurate diastereomer and /or enantiomer- specific quantification.[32,33]

18.4   SOURCES

Source apportionment is an important research area, the purpose of which is to identify and rank the major sources of pollutants to the environment. Such ranking tables permit the identification of major release pathways, and hence the prioritization of emission control strategies. Indeed, one of the principal aims of the Stockholm Convention is to identify the key sources of the targeted POPs.

The basic strategy of a source inventory is to derive an emission factor for a specific source activity (*e.g.* 10 µg per t of waste burnt), and subsequently to multiply this by an activity factor – *i.e.* the extent to which the activity is practiced (*e.g.* 3 million t waste burnt per year). In this way, an estimate of annual pollutant emissions for a specific source can be derived – in the case illustrated, annual emissions would amount to 30 g. The degree of difficulty involved in the construction of a source inventory – and hence its accuracy – varies widely according to the nature of the emission source (*e.g.* quantifying pesticide releases using well-documented industry production figures is more reliable than estimating emission factors for unintentional releases of dioxins from municipal waste incineration) and, as a result, the accuracy of such source inventories varies widely, with the primary benefit of many proving to be the identification of areas requiring further investigation. The section that follows summarizes present knowledge of the magnitude and sources of current and past releases of POPs.

18.4.1 PCDD/Fs

PCDD/Fs have never been intentionally produced – other than on a laboratory scale. Figure 18.7 illustrates the temporal variation of PCDD/F concentrations in archived soil taken from the Rothamsted experimental crop station in South England. This study – which covers the period 1844–1986 – shows there to have existed a low, essentially constant "background" of PCDD/Fs throughout the latter part of the 19th century, after which, there has been a dramatic rise.[34] The inference of this, is that – whilst some natural sources exist, notably forest fires (which may arguably be of significance in some countries) – the origins of the environmental burden of these compounds are principally as by-products of anthropogenic activities, in particular, the manufacture and use of organochlorine chemicals (such as chlorophenols) and combustion processes (such as waste incineration and steel manufacture).

The mechanisms *via* which PCDD/Fs form during combustion activities are complex, and are still not wholly understood. It appears that the mechanism of formation is *via de novo* synthesis – *i.e.* formation from the basic chemical "building blocks" of carbon, chlorine, oxygen and hydrogen. In summary, this means that, provided the conditions of temperature *etc.* outlined below are met, PCDD/F formation can occur from the combustion of any "fuel", providing sources of these four elements are present, although formation is enhanced if levels of so-called "precursor" compounds like chlorophenols and chlorobenzenes are present in the "fuel". The evidence of laboratory experiments and studies on working waste incinerators, is that PCDD/F formation during combustion occurs in post-combustion zones – *i.e.* oxygen-rich regions where temperatures are in the region 250–350 °C – such as electrostatic precipitators. In these regions, a series of reactions, catalyzed by the presence of metal chlorides, occur on the surface of fly ash particles and as a result PCDD/Fs are formed.

The presence of PCDDs in chlorophenols and products produced *via* chlorophenol intermediates (such as chlorophenoxy acetic acid derivatives; one of the principal constituents of

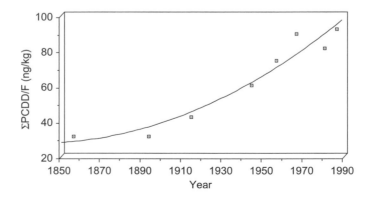

Figure 18.7 Temporal trends in ΣPCDD/F soil concentrations.[34]

"Agent Orange" – a defoliant used in the Vietnam War, and heavily contaminated with 2,3,7,8-TCDD – was the *n*-butyl ester of 2,4,5-trichlorophenoxyacetic acid) is more easily explained. PCDD formation occurs *via* the inadvertent reaction of two chlorophenol molecules (or chlorophenate ions), which yields PCDD(s), the exact identity(ies) of which are dependent on the chlorination pattern of the reactants, and the possibility of Smiles rearrangement products.

The existence of other sources of dioxins is highlighted by source inventories of atmospheric emissions of PCDD/Fs that fail to account for substantial proportions of deposition to the surface. The exact identity and contribution of the various sources is subject to uncertainty, but essentially centre on combustion activities and inadvertent contamination of other chlorinated chemicals (*e.g.* the fungicide and dyestuff intermediate, chloranil).

18.4.2 PCBs

In stark contrast to PCDD/Fs – while combustion sources are a known source of dioxin-like PCBs – the origin of the environmental burden of PCBs is almost exclusively their deliberate manufacture, primarily for use as dielectric fluids in electrical transformers and capacitors, but also for use in *inter alia,* carbonless copy papers and inks. Industrial manufacture commenced in the USA in 1929 (UK production in 1954) and between then and the late 1970s, when production (although not use of existing stocks) ceased in most western nations, an estimated total of 1.2 million t were produced (67 000 t in the UK). Temporal variations in PCB levels in archived soils from the so-called "Woburn Market Garden" experiment over the period 1942 to 1992 revealed UK PCB levels to have essentially reflected temporal trends in use, with the restrictions on such use producing a precipitous decline in soil concentrations.[35] The principal loss mechanism from soils (which has been estimated to hold >90% of the UK's burden of PCBs) is considered to be volatilization – degradation in soils is known to be slow – and until relatively recently, such volatilization was thought to represent the principal source of PCBs to the contemporary atmosphere. However, an increasing body of research shows that emissions of PCBs from remaining applications (such as fluorescent light ballasts and building sealants) has led to substantial contamination of indoor air and dust, and *via* ventilation of contaminated indoor air is driving outdoor contamination. The evidence that indoor air ventilation is maintaining present-day outdoor air concentrations rather than volatilization from soil consists of three strands. Firstly, studies in both the late 1990s and the mid-2000s showed concentrations of ΣPCB in indoor air in the UK West Midlands were on average 30 times higher than those in outdoor air.[36,37] Secondly, examination of spatial variation in PCB concentrations in outdoor air along an urban–rural transect across the West Midlands reveals a clear "urban pulse", whereby concentrations in the city centre (where there is a greater density of contaminated indoor environments) exceed substantially those in rural and suburban locations (see Figure 18.8).[38]

The final and most conclusive strand of evidence involves exploitation of the chiral properties of some PCBs. Nineteen PCBs containing three or four chlorine atoms exist as stable atropisomers – that is, they exist in two forms (known as enantiomers) which are essentially physically and chemically indistinguishable, except that they rotate the plane of polarised light in opposite directions, and may biodegrade at different rates. In commercial PCB formulations, these 19 PCBs are present as racemates – *i.e.* concentrations of the two enantiomers are equal, such that the enantiomer fraction (EF) is racemic *i.e.* = 0.5 (EF = concentration of the (+) enantiomer divided by the sum of the concentrations of both enantiomers). By comparison, due to enantiomeric differences in resistance to biodegradative processes, the EFs of chiral PCBs found in soils – and which are preserved upon volatilisation – deviate from 0.5, in theory varying between close to zero to *ca* 1. Hence knowledge of EFs of chiral PCBs in samples of outdoor air, indoor air, and soil provides an indicator of the extent to which the contemporary ambient atmospheric burden is due to volatilisation from soil, and how much arises from the ventilation of indoor air. Measurements in the

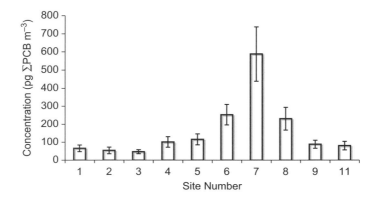

Figure 18.8 Spatial variation of average concentrations (pg m^{-3}) of ΣPCB in air samples along a transect in the UK's West Midlands going from rural South-West (Whitbourne, near Worcester, Site No. 1) to urban (Birmingham City Centre, Site No. 7) to rural North-East (Newton Regis near Tamworth, site No. 11).[38] (Reprinted with kind permission from A. Jamshidi *et al.*, *Environ. Sci. Technol.*, 2007, **41**, 2153; Copyright 2007 American Chemical Society).

West Midlands of the EF of the most abundant chiral PCB (PCB 95) have demonstrated that the racemic signature observed in almost all outdoor air samples is significantly different from the non-racemic value observed in soil, but matches the racemic values observed in indoor air.[38]

18.4.3 PBDEs

PBDEs do not occur naturally in the environment; but are produced synthetically as mixtures applied as additives (*i.e.* not covalently bonded to the polymer matrix) to flame-retard a wide range of consumer products as follows:[39]

  (i) PentaBDE mixture: used in polyurethane foam products such as furniture and upholstery in domestic furnishing, and in the automotive and aviation industries.
 (ii) OctaBDE mixture: used in plastic products, such as housings for computers, automobile trim, telephone handsets and kitchen appliance casings.
(iii) DecaBDE mixture: used in the high impact polystyrene (HIPS) housing for electrical and electronic equipment, and in various applications in the textile, automotive, and aviation industries.

Information is limited on how PBDEs incorporated in plastics and foams are released from products to the environment. However, several studies have shown that PBDEs can be released during its production, application to consumer products, emissions from waste disposal/recycling and in small amounts from landfills (by leaching). PBDEs can also be released to the air during the life-span of flame retarded products.[40] In this regard, the greater the vapour pressure, and the lower the octanol–air partition coefficient (see section 18.5.1), the more likely the BDE congener will volatilize from the plastic product Another release mechanism reported recently is the flaking off, crumbling, and/or general abrasion of PBDE-containing particles/fibres from treated polymers. Although such a loss mechanism may be significant, there is currently no method to quantify this avenue of loss, or compare it to other emission mechanisms.[41,42]

Sewage treatment operations can also result in environmental release of PBDEs. Both sewage sludge (from the treatment process) as well as the effluent from sewage treatment plants (STP) near PBDE point sources contain appreciable amounts of PBDEs.[43]

A limited life cycle analysis can provide some information on the magnitude of PBDEs releases to the environment. Following the releases from PBDE production processes, volatilization from products, and waste disposal, the Toxics Release Inventory (TRI) reports a total release of 32.2 t of decaBDE to the air, land, and water of the United States in 2007. According to the TRI, total environmental releases peaked in 1999, with a release of 53.9 t, dropped in 2003 to 36.3 t, followed by an increase in 2004 to 44.8 t, and then declined in 2005, 2006, and 2007.[40]

For comparison, a mathematical model was used to estimate the emissions of octaBDE to the European environment in 2003. Total EU emissions estimated from the use of plastics were 7–15 t year^{-1} to air, 0.2–0.9 t year^{-1} to waste water treatment plants, 7–14 t year^{-1} direct to surface water and 20–42 t year^{-1} to urban/industrial soil. These total emissions were dominated by estimates over the service life of polymers and from waste disposal.[44]

The application of sewage sludge for soil amendment resulted in 0.26, 2.69, 2.72, and 8.29 t year^{-1} of BDE 153, BDE 47, BDE 99, and BDE 209, respectively, being applied to agricultural land in the United States in 2009. Estimates of life cycle analysis reported that 37, 46, 82, 498, and 531 kg year^{-1} of BDE 154, BDE 153, BDE 209, BDE 47, and BDE 99, respectively, are discharged annually into surface waters from sewage treatment plant effluent.[39]

A final important observation is that, despite the current EU complete ban on the pentaBDE and octaBDE formulations and the restrictions on the decaBDE formulation, environmental release and subsequent human exposure to these chemicals is likely to continue for some time, given the persistence and ubiquity of flame retarded products accentuated by the environmental persistence of PBDEs.[45]

18.4.4 HBCD

HBCD is an additive flame retardant applied mainly in the building industry where it is incorporated (typically at <3% by weight) into extruded (XPS) or expanded polystyrene (EPS) foam materials. Secondary applications include upholstered furniture, automotive interior textiles, electric and electronic equipment. With global production of 16 700 t in 2001 and an EU consumption of 11 000 t in 2007, HBCD is the most widely used cycloaliphatic additive brominated flame retardant.[20]

Like PBDEs, HBCD can be released to the environment *via* a number of different pathways including: emission during production, the manufacture of flame retarded products, by leaching from consumer products during their life-span, or following product disposal/recycle.[46]

Very little is known as to what extent end-products containing HBCDD is landfilled, incinerated, left in the environment or recycled. An environmental assessment study of a HBCD production site (5000 t year^{-1}) was carried out in 2001. Results showed releases to air of 3400 kg year^{-1} and to wastewater from the onsite STP of 2000 kg year^{-1}. The annual release of HBCD in solid waste including sludge from the onsite STP was estimated to 1136 t year^{-1}.[47]

A life cycle study of HBCD emissions from its various sources to the European environment in 2007 reported total emissions of ~2 t year^{-1}. These total emissions were dominated by estimates over industrial use (textiles and insulation boards) and formulation of EPS and HIPS.[47]

18.5 IMPORTANT PHYSICOCHEMICAL PROPERTIES AND THEIR INFLUENCE ON ENVIRONMENTAL BEHAVIOUR

The following sections deal with the influence of a number of important physicochemical properties of POPs on their environmental behaviour; in particular, their availability for uptake by plants and animals, routes of human exposure, their relative partitioning between different environmental compartments, and their transport throughout the environment. Note that the practical difficulties** of accurately measuring such properties, has resulted in considerable uncertainty surrounding values reported for the compounds studied here. As a result, a table of definitive values for physicochemical properties is not included, and selected values are instead quoted to illustrate general trends, and to indicate the order of magnitude of these properties for individual pollutants.

18.5.1 Equilibrium Partitioning Coefficients

18.5.1.1 K_{OW}

Strictly, K_{OW} is a measurement of the equilibrium partition coefficient of a compound between water and octan-1-ol. The significance of this – at first sight, seemingly rather obscure parameter – is greater than might be expected, as it approximates to the lipid–water partition coefficient (K_{LW}). As a result, K_{OW} (the log form is most commonly quoted for convenience) is considered an expression of a compound's hydrophobicity, with implications for its bioaccumulative tendencies and movement through the food chain. It is important to note that the "solubility" of organic chemicals in octan-1-ol only varies within a very narrow range (*i.e.* 200 to 2000 mol m^{-3}). As a consequence, K_{OW} for an organic chemical is predominantly dependent on water solubility, and the common assumption that K_{OW} is an expression of lipophilicity is misleading, as most organic chemicals possess an equal affinity for lipids, whilst having very different affinities for water. As a general rule, K_{OW} increases with increasing molecular weight for a given class of compounds, and hence, OCDD (8.2) and BDE-183 (8.27) have higher log K_{OW} values than 2,3,7,8-TCDD (6.8) and BDE-28 (5.9), respectively.

18.5.1.2 K_{OC}

The exact definition of K_{OC} is the equilibrium partition coefficient of a chemical between water and natural organic carbon. The wider implications of this parameter (which has units of dm^3 kg^{-1}), is as an indication of the tendency of a chemical to sorb to organic matter. This can have important consequences for – amongst other things – a chemical's ability to leach through soil columns, its availability for uptake by plants and animals from soil and sediments, and its tendency to volatilize from soil. As a general rule for a particular group of chemicals, K_{OC} increases with increasing molecular weight and hence, OCDD and BDE-183 have higher K_{OC} values than 2,3,7,8-TCDD and BDE-28, respectively. As a very general approximation, $K_{OC} = 0.41$ K$_{OW}$, although the dependence of K_{OC} on the chemical form of organic carbon must not be overlooked, and the margin of error associated with this approximation is about a factor of two.

18.5.1.3 K_{OA}

This represents the equilibrium partition coefficient of a chemical between octan-1-ol and air. It has found particular utility in describing the equilibrium partitioning of POPs between plant lipids and the air in which they are immersed (directly analogous to the use of K_{OW} to represent the partitioning of POPs between fish lipid and the water in which they are immersed), and has been incorporated into mathematical modelling frameworks designed to describe POPs transfer through

---

**To illustrate, accurate measurement of K_{OW} for highly hydrophobic POPs is extremely difficult. Measurements of pollutant concentration in the octan-1-ol phase are relatively simple, but accurate determination of the extremely low levels in the water layer is severely hampered by the difficulty in achieving complete separation of the two layers. Given the much higher levels present in the octan-1-ol phase, the presence of even a very small fraction of this phase in the aqueous layer, will lead to an exaggeration of the pollutant concentration in the latter. To illustrate, if 50 μg of a substance with a log K_{OW} of 8, is partitioned between 1 dm^3 of both octan-1-ol and water, the equilibrium concentrations in the two layers will be 50 μg dm^{-3} and 0.5 pg dm^{-3}, respectively. Inadvertent contamination of the aqueous layer with just 1 μl of the organic phase will raise the apparent aqueous phase concentration to 50.5 pg dm^{-3}, giving rise to an apparent log K_{OW} value of 6.

terrestrial food chains (see section 18.6.2). K_{OA} is the ratio of the octanol–water and air–water partition coefficients *i.e.*

$$K_{OA} = K_{OW}/K_{AW} \tag{1}$$

where $K_{AW} = H/RT$, H is the Henry's Law constant (see section 18.5.1.4), R is the ideal gas constant, and T the absolute temperature. Although K_{OA} values are calculable from a knowledge of K_{OW} and H, direct measurements are now possible, with values of Log K_{OA} (measured at 25 °C) ranging from *ca* 6 for the trichlorinated PCBs, through to 12.7 for OCDD, with values for PBDEs falling in the range 7–13. K_{OA} is strongly temperature-dependent, with values inversely related to temperature – *i.e.* low temperatures result in higher K_{OA} values.

18.5.1.4 Henry's Law Constant

Henry's Law constant (H) is a measure of a compound's tendency to partition between a solution and the air above it and can be expressed in different ways. Its dimensionless form H' is directly equivalent to the air–water equilibrium partition coefficient (K_{AW}), and relates the concentration of a chemical in the gas phase Csg to its concentration in the liquid phase Csl,

$$H' = Csg/Csl \tag{2}$$

Henry's Law constant can also be written thus,

$$H = P_{vp}/S \tag{3}$$

where Pvp is the vapour pressure of the chemical (in Pa) and S the aqueous solubility or saturation concentration in mol m^{-3}. H therefore has units of Pa m^3 mol^{-1} and can be related to H' by,

$$H' = H/RT \tag{4}$$

where R is the gas constant (8.31 Pa m^3 mol^{-1} K^{-1}) and T is the temperature (K). The utility of Henry's Constant, is that it describes the tendency of a chemical to volatilize from water/plant/soil surfaces into the atmosphere, and compounds with a high H value are more prone to volatilization from aqueous solutions. In practical environmental terms, knowledge of H affords an insight into the movement of a chemical into and out of aquatic ecosystems such as ponds, lakes or oceans. Examples of H values (expressed as Pa m^3 mol^{-1}) at 25 °C are: 2,2′,3,3′-tetrachlorobiphenyl (22), 2,2′,4,4′,5,5′-hexachlorobiphenyl (43), 2,3,7,8-TCDD (3.3), OCDD (0.68), BDE-99 (0.36), HBCD (0.17) and BDE-183 (0.06).

18.5.2 Aqueous Solubility

As would be expected, a nonpolar organic compound's aqueous solubility is inversely related to its K_{OW} and K_{OC} and a good approximation is that solubility at neutral pH and a given temperature, will decrease with increasing molecular weight for a given class of compounds. To illustrate, the water solubilities at 25 °C of BDE-28, BDE-47, and BDE-183 are, respectively: 70, 11, and 2 µg L^{-1}. The environmental significance of this property is obvious; not least because more hydrophobic POPs are more prone to food chain transfer.

18.5.3   Environmental Persistence

An exact measure of environmental persistence is impossible to define, as persistence will vary tremendously between different environmental media (*e.g.* atmospheric persistence is appreciably lower than in other environmental compartments – reaction with the hydroxyl radical and dry and wet deposition constituting the principal removal mechanisms) and there are so many confounding variables. Even if environmental persistence is arbitrarily defined as the half-life in soil, measurements of this one parameter vary wildly, depending on the binding of the compound to the soil (which is, in turn, dependent on factors like soil organic matter content, available surface area, K_{OC}, soil moisture content and temperature) and the depth of incorporation below the surface (this influences the intensity of UV radiation received by the chemical and hence its photolysis rate). Even where a definitive half-life can be obtained for a compound, there are grave difficulties in reliably assessing the relative importance of each contributory loss mechanism – *e.g.* bacterial degradation, photolysis, leaching, chemical reaction, uptake by biota and volatilization – although experimental evidence strongly suggests that for tri- through pentachlorinated PCBs, volatilisation is the major pathway of loss from soils.[48] Despite these caveats, knowledge of the range of a compound's persistence is extremely important in assessing its environmental impact, not least because a longer soil half-life will allow more time for the chemical to transfer into the food chain, and elicit any adverse effects in biota. Illustrative values for soil half-lives range from a few months for lower molecular weight PCBs, PBDEs and OCPs, and up to 10 years or more for PCDD/Fs, and higher molecular weight PBDEs (estimated half-life of 28 years for BDE-209 in agricultural soils).[49]

18.5.4   Vapour Pressure

As a general rule, the vapour pressure of a chemical decreases with increasing molecular weight, and hence decachlorobiphenyl possesses a lower vapour pressure (5×10^{-8} Pa) than 2,2′,3,3′-tetrachlorobiphenyl (2.3×10^{-3} Pa). Other illustrative measurements of vapour pressure at 25 °C are: 2,3,7,8-TCDD (2×10^{-7} Pa) and OCDD (1.1×10^{-10} Pa). As with all of the other properties dealt with in this section, the influence of vapour pressure on environmental behaviour cannot be considered in isolation, and it is strongly dependent on environmental factors such as temperature. The environmental significance of vapour pressure is tremendous; *inter alia* it influences a compound's partitioning between air and other environmental compartments, thus affecting its rate of atmospheric transport, availability for uptake by biota, and susceptibility to photolytic degradation (compounds spending more time in the vapour phase are generally more susceptible to this). Vapour pressure also exerts a powerful influence on the relevance of inhalation as a human exposure route. While the relatively low vapour pressures of most POPs means inhalation of outdoor air is generally a negligible pathway of human exposure, the presence of substantial numbers of indoor sources of some POPs (*e.g.* PCBs, PBDEs, and HBCDs) leads to elevated indoor air concentrations rendering inhalation in some instances a significant exposure pathway. For higher molecular weight POPs with significant indoor sources (*e.g.* PBDEs), the lower vapour pressures of such higher molecular weight compounds means that they partition preferentially to indoor dust on floors and other surfaces. This results in contact with indoor dust being identified as the most important pathway of exposure of the American population to PBDEs.[50] Moreover, owing to their greater propensity for hand-to-mouth behaviour; toddlers have been identified as substantially more exposed than adults and older children.[51]

18.5.5   General Comments

The combined influence of the above physicochemical properties of individual chemicals, together with environmental factors such as meteorology (temperature, rainfall) and other factors like soil properties *etc.*, governs the environmental fate and behaviour of such chemicals in a predictable fashion. Thus, detailed knowledge of a compound's physicochemical properties, and the range of prevailing environmental conditions which it may experience, affords important insights into how an individual compound may behave once released into the environment.

For example, most POPs display common characteristics in their environmental behaviour. They reside primarily in soil, accumulate through the food chain, and food represents an important pathway of human exposure. In addition, they are capable of volatilizing from soils and water and subsequently undergoing long-range atmospheric transport either in the vapour phase or bound to aerosol. It is *via* such atmospheric transport, which occurs in a repeated cycle sometimes referred to as "the grasshopper effect", that these chemicals distribute in a ubiquitous fashion. This universal distribution is illustrated by the presence of POPs in polar regions, which is a matter of concern, for although inputs to such locations are small, the low temperatures minimise volatilization losses, leading to a steady accumulation of the overall pollutant burden. This accumulation is compounded by the reduced biomass to surface area ratio of polar regions, with the result that the pollutant burden is distributed amongst a much smaller biomass than in industrialized areas. These factors account for the observations of PCB concentrations in the breast milk of Canadian Inuit mothers, that exceed those found in women from urban Canada.

18.6 MODELLING ENVIRONMENTAL BEHAVIOUR

The environmental behaviour of a chemical under uniform meteorological and other environmental conditions is largely predictable in terms of a relatively small number of physicochemical properties. Without doubt, the most challenging aspect of predicting the environmental behaviour of a chemical lies in accounting for the uncertainties introduced by fluctuations in meteorological and other environmental parameters. For example, a hot, still and sunny day will result in enhanced volatilization from soil into the atmosphere, where a greater proportion will partition into the vapour as opposed to the aerosol phase, with concomitant susceptibility to reaction with the hydroxyl radical. Similarly, the influence of both the magnitude and chemical composition of the organic content of a soil will exert a profound influence on the environmental fate of a given pollutant. To illustrate, PCBs present in peat-like soils will bind much more strongly to the soil than if they were present in a sandy soil. As a result, movement of pollutants from peat soils – either by volatilization or leaching – is minimal; there is also evidence to suggest that uptake of pollutants from such soils by biota is similarly reduced – in other words, the "bioavailability" of the chemical is diminished.

18.6.1 The Fugacity Concept

These simple examples, serve as illustrations of the complex requirements that a successful environmental modelling technique must meet. Whilst a completely satisfactory model has yet to be developed – there are too many real-world variables for that – work pioneered by Prof. Donald Mackay and co-workers has provided an excellent framework within which the environmental fate of organic contaminants may be predicted with some accuracy. Their approach is essentially to construct a "model world"; for example a small lake of defined dimensions and with clearly defined values for properties like suspended particulate matter (SPM), the organic content of such matter, and the "volume" of biota like fish and plants that are present. Assumptions must also be made concerning the depth and physical properties – such as organic content – of lake sediment, and the height of the air layer above the lake surface deemed relevant. Combined with data on the "model world" dimensions and composition, are mass loadings and calculated values of fugacity for the compound(s) of interest; fugacity essentially describing the potential for a substance to move from

one environmental compartment to another. Incorporation of all these data into a single computational package, allows the construction of a reasonably realistic model, capable of predicting the
distribution of a chemical, throughout the various environmental compartments of the "model
world". The model can be further enhanced by inclusion of transport factors relating to processes
such as leachability and volatilization, along with reaction properties to provide information on
loss processes such as biodegradation and hydrolysis, and this basic approach can be modified to
consider smaller or larger "model worlds" as appropriate, in whatever degree of detail necessary.
To illustrate, the concept has been used to assess the global fate and distribution of toxaphenes –
dividing the globe into nine separate "model worlds" such as the arctic, the antarctic, northern
hemisphere temperate, tropical *etc.* and modelling chemical distribution within and between each of
the "model worlds".[52] One use for this global model has been to assess the relative propensity for
chemicals to transfer from temperate zones and "condense out" at the polar regions. On a smaller
scale, the approach can be utilized to predict chemical movement within a soil column following
amelioration with sewage sludge, and assess the potential for leaching to groundwater or uptake by
crops. Other applications include the prediction of the fate and behaviour of PBDEs within an
office.[53]

18.6.2 Equilibrium Partitioning Modelling Approaches

In addition to the more sophisticated modelling approaches represented by *inter alia* the fugacity
approach, it is possible to predict POPs behaviour to an acceptable degree of accuracy by simple
equilibrium partitioning approaches. The most widely employed example is in predicting POPs
transfer from water to fish. Ignoring intake *via* the diet, and assuming that POPs are at equilibrium
between fish lipid and the surrounding water, the relationship between POPs concentration in fish
lipid (C_f) and that in water (C_w), may be described in terms of K_{OW} thus:

$$\mathrm{Log}(C_f/C_w) = \mathrm{Log}\, K_{OW} \qquad (5)$$

C_f/C_w is referred to as the bioaccumulation factor (*BAF*). In reality, while experimental observations have demonstrated a linear relationship between the BAF and K_{OW}, the relationship is of the
type shown below, with values of m and c varying on both a location and species-specific basis:

$$\mathrm{Log}\, BAF = \mathrm{m}\, \mathrm{Log}\, K_{OW} + \mathrm{c} \qquad (6)$$

As an illustration, the relationships between *BAF* and K_{OW} for eels and pike from the River Severn
are:[54]

$$\mathrm{Log}\, BAF = 0.97 \times \mathrm{Log}\, K_{OW} - 0.52 \ (\text{eels}) \qquad (7)$$

$$\mathrm{Log}\, BAF = 0.94 \times \mathrm{Log}\, K_{OW} + 0.04 \ (\text{pike}) \qquad (8)$$

A similar approach has been taken to describing POPs partitioning between air and plants, leading
to relationships of the form:

$$\mathrm{Log}(C_P/C_A) = \mathrm{m}\, \mathrm{Log}\, K_{OA} + \mathrm{c} \qquad (9)$$

where C_P and C_A are the POPs concentrations in plant lipid and air, respectively. As with the
analogous relationship between fish lipid and water, such relationships assume POPs to exist in
an equilibrium between plant lipid and air. This has been shown to be true for the tri- through
pentachlorinated PCBs, but it appears that other POPs such as the higher chlorinated PCBs and

PCDD/Fs do not reach such an equilibrium. As with the relationship between *BAF* and K_{OW}, there is evidence of appreciable species- and location specific variation in both m and c.

18.6.3 Pharmacokinetic Models of Human Exposure

Recently, an interesting modelling approach was adopted to further understand the relationship between external exposure to PBDEs[50] and HBCDs[55] and their human body burdens. The model uses a simple, one-compartment (where the target compounds are hypothesized to accumulate only in lipids) pharmacokinetic approach to convert exposure data from various pathways to human body burdens of the studied brominated flame retardants (BFR). The change in BFR lipid concentration over time can be expressed by Equation (10).

$$\frac{\delta C_{BFR}}{\delta t} = \frac{I_{BFR}(t) x A F_{BFR}}{BL(t) - K_{BFR} x C_{BFR}(t)} \tag{10}$$

Where C_{BFR} is the compound specific concentration in lipids (ng g^{-1} lipid weight); I_{BFR} is the daily intake of the target BFR (ng day^{-1}); AF_{BFR} is the absorption fraction; *BL* is body lipid mass (g) and K_{BFR} is the compound specific first order dissipation rate (day^{-1}).

If K_{BFR} is assumed to be constant over time then Equation (11) can be resolved into:

$$C_{BFR}(t) = C_{BFR}(0) x e^{(-K_{BFR} \cdot t)} + \left[\frac{(I_{BFR}(t) x A F_{BFR})}{BL(t)} \right] x \left[\frac{(1 - e^{(-K_{BFR} \cdot t)})}{K_{BFR}} \right] \tag{11}$$

Where $C_{BFR}(0)$ is the studied BFR body lipid concentration at time 0 (initial concentration before intake).

Assuming a constant dose over time at constant body lipid mass, the steady state BFR lipid concentration can be calculated from Equation (12).

$$C_{BFR} = \frac{(I_{BFR} x A F_{BFR})}{BL x K_{BFR}} \tag{12}$$

While Equation (12) can be used to predict the body burdens of the target BFRs, it is stressed that the assumption of steady state conditions is an inherent uncertainty with this approach.

In conclusion, it is important to note that although mathematical models are extremely useful in comparing the behaviour of different chemicals under identical environmental conditions, they are not yet sufficiently reliable to provide accurate predictions of actual concentrations.

REFERENCES

1. R. Carson, *The Silent Spring*, Hamish Hamilton, London, 1963.
2. D. E. Schulz, G. Petrick and J. C. Duinker, *Environ. Sci. Technol.*, 1989, **23**, 852.
3. M. Van den Berg, L. S. Birnbaum, M. Denison, M. De Vito, W. Farland, M. Feeley, H. Fiedler, H. Hakansson, A. Hanberg, L. Haws, M. Rose, S. Safe, D. Schrenk, C. Tohyama, A. Tritscher, J. Tuomisto, M. Tysklind, N. Walker and R. E. Peterson, *Toxicol. Sci.*, 2006, **93**, 223.
4. F. W. Karasek and O. Hutzinger, *Anal. Chem.*, 1986, **54**, 309.
5. Committee on Toxicity of Chemicals in Food, Consumer Products and the Environment, *Statement on the Tolerable Daily Intake for Dioxins and Dioxin-like Polychlorinated*

Biphenyls, October 2001, Reference COT/2001/07; http://cot.food.gov.uk/cotstatements/cotstatementsyrs/cotstatements2001/dioxinsstate (accessed 21/12/2012).

6. D. Yang, K. H. Kim, A. Phimister, A. D. Bachstetter, T. R. Ward, R. W. Stackman, R. F. Mervis, A. B. Wisniewski, A. L. Klein, P. R. S. Kodavanti, K. A. Anderson, G. Wayman, I. N. Pessah and P. J. Lein, *Environ. Health Perpect.*, 2009, **117**, 426.
7. L. S. Birnbaum and D. F. Staskal, *Environ. Health Perspect.*, 2004, **112**, 17.
8. R. G. Ellis-Hutchings, G. N. Cherr, L. A. Hanna and C. L. Keen, *Toxicol. Appl. Pharmacol.*, 2006, **215**, 135.
9. P. O. Darnerud, *Environ. Int.*, 2003, **29**, 841.
10. H. Lilienthal, A. Hack, A. Roth-Harer, S. W. Grande and C. E. Talsness, *Environ. Health Perspect.*, 2006, **114**, 194.
11. A. P. Vonderheide, K. E. Mueller, J. Meija and G. L Welsh, *Sci. Total Environ.*, 2008, **400**, 425.
12. P. Eriksson, H. Viberg, E. Jakobsson, U. Örn and A. Fredriksson, *Organohalogen Compds.*, 1999, **40**, 333.
13. R. J. Law, M. Kohler, N. V. Heeb, A. C. Gerecke, P. Schmid, S. Voorspoels, A. Covaci, G. Becher, K. Janak and C. Thomsen, *Environ. Sci. Technol.*, 2005, **39**, 281A.
14. D. Ronisz, E. F. Finne, H. Karlsson and L. Forlin, *Aquat. Toxicol.*, 2004, **69**, 229.
15. T. Helleday, K. L. Tuominen, A. Bergman and D. Jenssen, *Mutat. Res.*, 1999, **439**, 137.
16. P. O. Darnerud, *Int. J. Androl.*, 2008, **31**, 152.
17. P. Eriksson, N. Johansson, H. Viberg, C. Fischer and A. Fredriksson, *Organohalogen Compds.*, 2004, **66**, 3119.
18. M. M. L. Dingemans, H. J. Heusinkveld, A. de Groot, A. Bergman, M. van den Berg and R. H. S. Westerink, *Toxicol. Sci.*, 2009, **107**, 490.
19. H. Lilienthal, L. T. van der Ven, A. H. Piersma and J. G. Vos, *Toxicol. Lett.*, 2009, **185**, 63.
20. C. H. Marvin, G. T. Tomy, J. M. Armitage, J. A. Arnot, L. McCarty, A. Covaci and V. Palace, *Environ. Sci. Technol.*, 2011, **45**, 8613.
21. T. Harner, M. Shoeib, M. Diamond, G. Stern and B. Rosenberg, *Environ. Sci. Technol.*, 2004, **38**, 4474.
22. S. Harrad and M. A. Abdallah, *J. Environ. Monit.*, 2008, **10**, 527.
23. S. Hazrati and S. Harrad, *Chemosphere*, 2007, **67**, 448.
24. M. A. Abdallah and S. Harrad, *Environ. Sci. Technol.*, 2010, **44**, 3059.
25. S. Harrad, M. A. Abdallah, N. L. Rose, S. D. Turner and T. A. Davidson, *Environ. Sci. Technol.*, 2009, **43**, 9077.
26. H. M. Stapleton, *Anal. Bioanal. Chem.*, 2006, **386**, 807.
27. A. Covaci, S. Voorspoels and J. de Boer, *Environ Int.*, 2003, **29**, 735.
28. J. Bjorklund, P. Tollback and C. Ostman, *J. Mass Spectrom.*, 2003, **38**, 394.
29. M. A. Abdallah, S. Harrad and A. Covaci, *Anal. Chem.*, 2009, **81**, 7460.
30. E. Eljarrat and D. Barcelo, *TraC,Trends Anal. Chem.*, 2004, **23**, 727.
31. A. Covaci, S. Voorspoels, L. Ramos, H. Neels and R. Blust, *J. Chromatogr., A.*, 2007, **1153**, 145.
32. C. H. Marvin, G. MacInnis, M. Alaee, G. Arsenault and G. T. Tomy, *Rapid Commun. Mass. Spectrom.*, 2007, **21**, 1925.
33. G. T. Tomy, T. Halldorson, R. Danell, K. Law, G. Arsenault, M. Alaee, G. MacInnis and C. H. Marvin, *Rapid Commun. Mass Spectrom.*, 2005, **19**, 2819.
34. L.-O. Kjeller, K. C. Jones, A. E. Johnston and C. Rappe, *Environ. Sci. Technol.*, 1991, **25**, 1619.
35. R. E. Alcock, A. E. Johnston, S. P. McGrath, M. L. Berrow and K. C. Jones, *Environ. Sci. Technol.*, 1993, **27**, 1918.
36. G. M. Currado and S. Harrad, *Environ. Sci. Technol.*, 1998, **32**, 3043.
37. S. Harrad, S. Hazrati and C. Ibarra, *Environ. Sci. Technol.*, 2006, **40**, 4633.
38. A. Jamshidi, S. Hunter, S. Hazrati and S. Harrad, *Environ. Sci. Technol.*, 2007, **41**, 2153.

39. US EPA 2010, An Exposure Assessment of Polybrominated Diphenyl Ethers, EPA/600/R-08/086F.
40. EU Risk Assessment Report 2004, Update Risk Assessment of Bis (Pentabromophenyl) Ether. CAS number: 1163-19-5, EINECS Number: 214-604-9.
41. G. Suzuki, A. Kida, S. Sakai and H. Takigami, *Environ. Sci. Technol.*, 2009, **43**, 1437.
42. T. F. Webster, S. Harrad, J. R. Millette, R. D. Holbrook, J. M. Davis, H. M. Stapleton, J. G. Allen, M. D. McClean, C. Ibarra, M. A. Abdallah and A. Covaci, *Environ. Sci. Technol.*, 2009, **43**, 3067.
43. EU European Union Risk Assessment Report (2000), *Diphenyl Ether, Pentabromo Derivative. CAS Number: 32534-81-9*, EINECS Number: 251-084-2.
44. EU European Union Risk Assessment Report (2002), *Diphenyl Ether, Octabromo Derivative. CAS No: 32536-52-*0, EINECS No: 251-087-9.
45. S. Harrad and M. Diamond, *Atmos.Environ.*, 2006, **40**, 1187.
46. A. Covaci, A. C. Gerecke, R. J. Law, S. Voorspoels, M. Kohler, N. V. Heeb, H. Leslie, C. R. Allchin and J. De Boer, *Environ. Sci. Technol.*, 2006, **40**, 3679.
47. KEMI (National Chemicals Inspectorate), *Final draft of the EU Risk Assessment Report on Hexabromocyclododecane*; R044_0710_env_hh.doc. R044_0710_env_hhdoc; Sundbyberg, Sweden 2007.
48. S. Ayris and S. Harrad, *J. Environ. Monit.*, 1999, **1**, 395.
49. N. A. Andrade, L. L. McConnell, A. Torrents and M. Ramirez, *J. Agric. Food Chem.*, 2010, **58**, 3077.
50. M. Lorber, *J. Exp. Sci. Environ. Epidemiol.*, 2008, **18**, 2.
51. H. A. Jones-Otazo, J. P. Clarke, M. L. Diamond, J. A. Archbold, G. Ferguson, T. Harner, G. M. Richardson, J. J. Ryan and B. Wilford, *Environ. Sci. Technol.*, 2005, **39**, 5121.
52. F. Wania and D. Mackay, *Chemosphere*, 1993, **27**, 2079.
53. X. Zhang, M. L. Diamond, C. Ibarra and S. Harrad, *Environ. Sci. Technol.*, 2009, **43**, 2845.
54. S. Harrad and D. Smith, *Environ. Sci. Pollut. Res.*, 1997, **4**, 189.
55. M. Abdallah and S. Harrad, *Environ. Int.*, 2011, **37**, 443.

CHAPTER 19

Radioactivity in the Environment

C. NICHOLAS HEWITT

Lancaster Environment Centre, Lancaster University, Lancaster LA1 4YQ, UK
Email: n.hewitt@lancaster.ac.uk

19.1 INTRODUCTION

Pollution of the natural environment by radioactive substances is of concern because of the considerable potential that ionising radiation has for damaging biological material and because of the very long half-lives of some radionuclides. Although both the benefits of controlled exposure for medical purposes and the catastrophic effects of large doses of radiation (for example, those received by the inhabitants of Hiroshima and Nagasaki in 1945, and in the vicinity of Chernobyl in 1988) are well understood, what is less clear are the effects of small doses on the general population. For this reason it is necessary to evaluate in detail the exposure received by humans, collectively and as individuals, from each of the multiplicity of natural and artificial sources of radioactivity in the environment. In order to do this, an understanding is required of the actual and potential source strengths, the pathways and cycling of radioactivity through the environment and their flux rates, and the possible routes of exposure for humans.

19.2 RADIATION AND RADIOACTIVITY

19.2.1 Types of Radiation

Radiation arises from a spontaneous rearrangement of the nucleus of an atom. Whilst some nuclei are stable, many are not and these can undergo a change, losing mass or energy in the form of radiation. Some unstable nuclei are naturally occurring while others are produced synthetically. The most common forms of these radiations are alpha particles, beta particles, and gamma rays, the physical properties of which are shown in Table 19.1 and described below.

   (i) Alpha (α) particles consist of two protons and two neutrons bound together and so are identical to helium nuclei. They have a mass number of 4 and a charge of $+2$. Because they are so large and heavy, alpha particles travel slowly compared with other types of radiation

Pollution: Causes, Effects and Control, 5th Edition
Edited by R M Harrison
© The Royal Society of Chemistry 2014
Published by the Royal Society of Chemistry, www.rsc.org

Table 19.1 Types of radioactive particles and rays.

Radiation	Symbol	Composition	Charge	Mass Number	Approximate Tissue Penetration (cm)
Alpha	α	particle containing two protons and two neutrons	$2+$	4	0.01
Beta	β	particle of one electron	$1-$	0	1
Gamma	γ	very short wavelength electromagnetic radiation	0	0	100

(at maximum, about 10% the speed of light) and can be stopped relatively easily. Their ability to penetrate into living tissue is therefore limited and damage occurs to animals only when alpha-emitting isotopes are ingested or inhaled. However, their considerable kinetic energy and the double positive charge which attracts and pulls away electrons from atoms belonging to tissue, means that alpha particles can cause the formation of ions and free radicals and hence cause severe chemical change along their path. The emission of alpha particles is common only for nuclides of mass number greater than 209 and atomic number 82, nuclides of this size having too many protons for stability. An example of α-decay is:

$$^{210}_{84}\text{Po} \rightarrow {}^{206}_{82}\text{Pb} + {}^{4}_{2}\text{He}$$

Note that the sum of the superscripts (mass numbers or sum of neutrons and protons) and the sum of the subscripts (atomic numbers or sum of protons) remain unchanged during the decay.

(ii) Beta (β) particles are simply electrons emitted by the nucleus during the change of a neutron into a proton. They have minimal mass (1.36×10^{-4} times that of an alpha particle) but high velocity (typically 40% the speed of light) and a charge of -1. They may penetrate through skin or surface cells into tissue and may then pass close to the orbital electrons of tissue atoms, where the repulsion of the two negative particles may force the orbital electron out of the atom, ionizing the tissue, and forming radicals. Because beta decay involves the change of a neutron into a proton the atomic number of the nuclide increases by one, but the mass number does not change. Beta decay is a common mode of radioactive disintegration and is observed for both natural and synthetic nuclides. An example is:

$$^{90}_{38}\text{Sr} \rightarrow {}^{90}_{39}\text{Y} + {}^{0}_{-1}\text{e}$$

(iii) Gamma (γ) radiation is very short wavelength electromagnetic radiation. It travels at the speed of light, is uncharged, but is highly energetic and so has considerable penetration power. As it passes through biological tissue the electric field surrounding a gamma ray may eject orbital electrons from atoms and so can cause ionisation of the tissue and formation of radicals along its path. The emission of gamma radiation does not lead to changes in mass or atomic numbers, and may occur either on its own from an electronically excited nucleus or may accompany other types of radioactive decay. Examples of these two processes are:

$$^{125}_{52}\text{Te}^* \rightarrow {}^{125}_{52}\text{Te} + \gamma$$

the asterisk signifying the excited state of the nucleus

and:

$$^{137}_{55}\text{Cs} \rightarrow\,^{137}_{56}\text{Ba} +\,^{-1}_{0}\text{e} \rightarrow\,^{137}_{56}\text{Ba} + \gamma$$

19.2.2 The Energy Changes of Nuclear Reactions

The energy changes associated with nuclear reactions are considerably greater than those associated with ordinary chemical reactions. The sum of the mass of the products of the nuclear reaction is invariably less than the sum of the mass of the reactants, and the amount of energy released (ΔE) is equivalent to this difference in mass (Δm). Most of this energy is released as kinetic energy, although some may be used to promote the nucleus to an excited state from where it will lose energy in the form of irradiation and return to the ground state.

The energy equivalent of a given mass m can be calculated from the velocity of light c by means of Einstein's equation:

$$\Delta E = (\Delta m)c^2$$

The energy equivalent of one atomic mass unit ($1\ \text{u} = 1.660566 \times 10^{-27}$ kg) is:

$$\Delta E = (1.660566 \times 10^{-27}\ \text{kg})\,(2.99792 \times 10^{8}\ \text{m s}^{-1})^2$$
$$= 1.49244 \times 10^{-10}\text{J}$$

This is usually expressed in units of electron volts (eV) where:

$$1\,\text{eV} = 1.60219 \times 10^{-19}\text{J}$$
and
$$1\,\text{MeV} = 1.60219 \times 10^{-13}\text{J}$$

The energy equivalent of 1 u is therefore 931.5 MeV. The amount of energy released by a decay process can now be calculated. For example, in the alpha decay of polonium-210:

$$\Delta\text{m} = (\text{mass}\,^{210}_{84}\text{Po}) - (\text{mass}\,^{206}_{82}\text{Pb} + \text{mass}\,^{4}_{2}\text{He})$$
$$= 209.9829\,\text{u} - (205.9745\,\text{u} + 4.0026\,\text{u})$$
$$= 0.0058\,\text{u}$$

The energy released by the decay is therefore:

$$\Delta E = 0.0058\,\text{u} \times 931.5\,\text{MeV/u} = 5.4\,\text{MeV}$$

19.2.3 Rates of Radioactive Decay

The rates of decay of radioactive nuclides are first-order and independent of temperature. This implies that the activation energy of radioactive decay is zero and that the rate of decay depends only on the amount of radioactive substance present. If N is the number of atoms present at time t the rate of change of N is given by

$$\text{d}N/\text{d}t = -\lambda N$$

Table 19.2 Half-lives of some environmentally important radionuclides.

Radionuclide	Half-life
^{131}I	8.1 d
^{85}Kr	10.8 y
^{3}H	12.3 y
^{90}Sr	28 y
^{137}Cs	30 y
^{239}Pu	2.4×10^4 y
^{238}U	4.5×10^9 y

where λ is the characteristic (or disintegration or decay) constant for that radionuclide. Integrating between times t_1 and t_2 gives:

$$N_2 = N_1 \exp[-\lambda(t_2 - t_1)]$$

where N_1 and N_2 are the number of atoms of the radionuclide present at times t_1 and t_2 respectively. If t_1 is set to zero then:

$$N = N_0 \exp(-\lambda t) \tag{1}$$

where t is the elapsed time and N_0 is the number of atoms of the radionuclide present when $t_1 = 0$.

When N/N_0 is equal to 0.5 (*i.e.* half the atoms have decayed away) t is defined as being the half-life, $t_{1/2}$. Then:

$$N/N_0 = 0.5 = \exp(-\lambda t_{1/2})$$
$$t_{1/2} = 0.693/\lambda$$

Equation (1) can now be written in terms of the more readily available t_1, rather than λ, to give

$$N = N_0 \exp(-0.693 t_1/t_{1/2}).$$

The half-lives of some selected radionuclides are given in Table 19.2.

19.2.4 Activity

The amount of radiation emitted by a source per unit time is known as the activity of the source, expressed in terms of the number of disintegrations per second. The unit of activity, the becquerel (Bq), is defined as being one disintegration per second. The activity of a source is proportional to the number of radioactive atoms present and so diminishes with time according to first order kinetics.

19.2.5 Radioactive Decay Series

Some radioactive decay processes lead in one step to a stable product but frequently a disintegration leads to the formation of another unstable nucleus. This can be repeated several times, producing a radioactive decay series that only terminates on the formation of a stable nuclide. There are three naturally occurring decay series, headed by ^{232}Th, ^{238}U and ^{235}U. Each of these

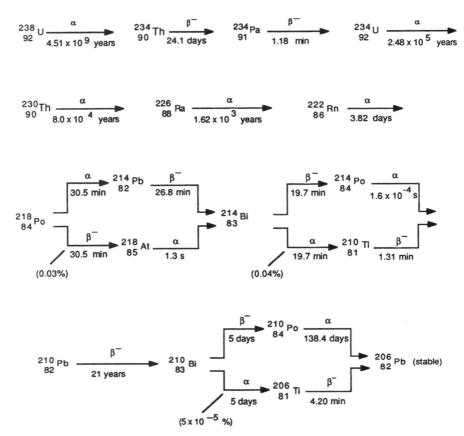

Figure 19.1 Disintegration series of $^{238}_{92}$U (half-lives of isotopes are indicated).

nuclides has a half-life which is long in relation to the age of the Earth and each finally produces stable isotopes of lead, ^{208}Pb, ^{206}Pb and ^{207}Pb, respectively. The 14 steps of the ^{238}U series are shown in Figure 19.1 and it will be seen that at several points branching occurs as the series proceeds by two different routes which rejoin at a later point.

19.2.6 Production of Artificial Radionuclides

The first artificial transmutation of one element into another was achieved in 1919 when Ernest Rutherford passed α-particles (produced by the radioactive decay of ^{214}Po) through nitrogen:

$$^{14}_{7}\text{N} + ^{4}_{7}\text{He} \rightarrow ^{17}_{8}\text{O} + ^{1}_{1}\text{H}$$

Subsequently the first artificial radioactive nuclide, ^{30}P, was produced:

$$^{27}_{13}\text{N} + ^{4}_{2}\text{He} \rightarrow ^{30}_{15}\text{P} + ^{1}_{0}\text{n (neutron)}$$

$$^{30}_{15}\text{P} \rightarrow ^{30}_{14}\text{Si} + ^{0}_{1}\text{e (positron)}$$

Bombardment reactions of this type have been used to produce isotopes of elements which do not exist in nature and which are particularly important as pollutants of the environment, for example:

$$^{238}_{92}U + ^{1}_{0}n \rightarrow ^{239}_{92}U + \gamma \quad \text{uranium}$$

$$^{239}_{92}U \rightarrow ^{239}_{93}Np + ^{0}_{-1}e \quad \text{neptunium}$$

$$^{239}_{93}Np \rightarrow ^{239}_{94}Pu + ^{0}_{-1}e \quad \text{plutonium}$$

$$^{239}_{94}Pu + ^{2}_{1}H \rightarrow ^{240}_{95}Am + ^{1}_{0}n \quad \text{americium}$$

19.2.7 Nuclear Fission

Many heavy nuclei with mass numbers greater than 230 are susceptible to spontaneous fission, or splitting into lighter fragments, as a result of the forces of repulsion between their large number of protons. Fission can also be induced by bombarding heavy nuclei with projectiles such as neutrons, alpha particles, or protons. When the fission of a particular nuclide takes place the nuclei may split in a variety of ways, producing a number of products. For example, some of the possible fission reactions of ^{235}U are:

$$^{235}_{92}U + ^{1}_{0}n \rightarrow ^{95}_{39}Y + ^{138}_{53}I + 3^{1}_{0}n$$

$$^{235}_{92}U + ^{1}_{0}n \rightarrow ^{97}_{39}Y + ^{137}_{53}I + 2^{1}_{0}n$$

$$^{235}_{92}U + ^{1}_{0}n \rightarrow ^{90}_{36}Kr + ^{144}_{56}Ba + 2^{1}_{0}n$$

$$^{235}_{92}U + ^{1}_{0}n \rightarrow ^{90}_{35}Br + ^{143}_{57}La + 3^{1}_{0}n$$

In each of these reactions neutrons are formed as primary products and these neutrons can, in turn, cause the fission of other ^{235}U nuclei, so causing a chain reaction of the fissile uranium.

The number of nuclei of a particular daughter product formed by the fission of 100 parent nuclei is defined as the yield of the process. Fission yields vary, fission induced by slow neutrons (as in a nuclear power plant) having a different set of yields to that induced by fast neutrons (as in many nuclear weapons). Table 19.3 shows examples of the percentage yields of the fast-neutron induced fission of ^{239}Pu. It also shows some of the yields of ^{235}U fission, which are similar for both slow and fast neutron induced processes.

Table 19.3 Yields of some long-lived radionuclides following uranium and plutonium fission.

Radioisotope	Half-life	Yield (%) ^{235}U	^{239}Pu
Strontium-90	28 y	5.8	2.2
Iodine-131	8 d	3.1	3.8
Caesium-137	30 y	6.1	5.2
Cerium-141	33 d	6.0	5.2

19.2.8   Beta Decay of Fission Products

The primary daughter products of nuclear fission are almost always β-radioactive and often these also quickly produce β-radioactive products. Only after several such decays are products with long half-lives formed. For example the fission of ^{235}U produces ^{90}Br:

$$^{235}_{92}U + ^1_0n \rightarrow ^{90}_{35}Br + ^{143}_{57}La + 3^1_0n$$

which quickly decays to ^{90}Kr with a half-life of 1.4 seconds. This in turn has a half-life of 33 seconds and decays to ^{90}Rb ($t_{1/2} = 2.7$ minutes), and this to ^{90}Sr:

$$^{90}_{35}Br \rightarrow ^{90}_{36}Kr + ^0_{-1}e \quad (t_{1/2} = 1.4\,s)$$

$$^{90}_{36}Kr \rightarrow ^{90}_{37}Rb + ^0_{-1}e \quad (t_{1/2} = 33\,s)$$

$$^{90}_{37}Rb \rightarrow ^{90}_{38}Sr + ^0_{-1}e \quad (t_{1/2} = 2.7\,min)$$

Strontium-90 decays to yttrium-90:

$$^{90}_{38}Sr \rightarrow ^{90}_{39}Y + ^0_{-1}e \quad (t_{1/2} = 28\,years)$$

and has a half-life of 28 years, which is sufficiently long for it to be widely circulated through the environment, in contrast to its short-lived precursors. After one further beta decay, a stable product, ^{90}Zr, is formed. It is therefore radioactive strontium rather than its precursors that is the important environmental pollutant in this series.

Examples of other environmentally-important nuclides formed by beta decay chains are iodine-131 and caesium-137. ^{131}I is formed from ^{131}Sn, itself produced by the fission of ^{235}U, and forms the stable nuclide ^{131}Xe:

$$^{131}_{50}Sn \rightarrow ^{131}_{51}Sb + ^0_{-1}e \quad (t_{1/2} = 3.4\,min)$$

$$^{131}_{51}Sb \rightarrow ^{131}_{52}Te + ^0_{-1}e \quad (t_{1/2} = 23\,min)$$

$$^{131}_{52}Te \rightarrow ^{131}_{53}I + ^0_{-1}e \quad (t_{1/2} = 24\,min)$$

$$^{131}_{53}I \rightarrow ^{131}_{54}Xe + ^0_{-1}e \quad (t_{1/2} = 8\,days)$$

^{137}Cs begins as the uranium fission daughter ^{137}I and ends as the stable barium isotope, ^{137}Ba:

$$^{137}_{53}I \rightarrow ^{137}_{54}Xe + ^0_{-1}e \quad (t_{1/2} = 24\,s)$$

$$^{137}_{54}Xe \rightarrow ^{137}_{55}Cs + ^0_{-1}e \quad (t_{1/2} = 3.9\,min)$$

$$^{137}_{55}Cs \rightarrow ^{137}_{56}Ba + ^0_{-1}e \quad (t_{1/2} = 30\,years)$$

19.2.9   Units of Radiation Dose

The amount of biological damage caused by radiation and the probability of the occurrence of damage are directly related to dosage and so before discussing the effects of radiation it is necessary to consider the units of dose.

19.2.9.1 Absorbed Dose

The amount of energy actually absorbed by tissue or other material from radiation is known as the *absorbed dose* and is expressed in a unit called the *gray* (Gy). One gray is equal to the transfer of one joule of energy to one kilogram of material. Table 19.4 shows the relationship of this and other SI units with the older units they replace.

19.2.9.2 Dose Equivalent

Because the physical and chemical properties of α, β, and γ radiations vary, equal absorbed doses of radiation do not necessarily have the same biological effects. In order to equate the biological effects of one type of radiation with another it is therefore necessary to multiply the absorbed dose by a radiation weighting factor (w_R) that accounts for the differences in biological damage caused by radioactive particles having the same energy. This is known as the *dose equivalent* and is expressed in units of *sieverts* (Sv). For γ-rays, X-rays, and β-particles w_R equals 1, for protons 2, and for α-particles 20. The dose equivalent resulting from an absorbed dose of 1 Gy of alpha radiation therefore equals 20 Sv.

19.2.9.3 Effective Dose Equivalents

Having established the dose equivalent for each tissue type (H_T) it is now necessary to weight this to take account of the differing susceptibilities of different organs and types of tissue to damage. For example the testes and ovaries are more easily damaged than are the lungs or bones. The risk weighting factors (WT) currently recommended by the International Commission for Radiological Protection (ICRP) are shown in Table 19.5. The effective dose equivalent (H_E) for the body is then expressed as the sum of the weighted dose equivalents:

$$H_E = \Sigma_T H_T W_T$$

Table 19.4 SI and old radiation units.

Quantity	SI Unit	Old (non-SI) Unit	Relationship
Activity	becquerel	curie	1 Ci $= 3.7 \times 10^{10}$ Bq
Absorbed dose	gray	rad	1 rad $= 0.01$ Gy
Dose equivalent	sievert	rem	1 rem $= 0.01$ Sv

Table 19.5 ICRP risk weighting factors (2007).

Tissue or organ	Tissue weighting factor
Bone-marrow, colon, lung, stomach, breast, remaining tissue	0.12
Gonads	0.08
Bladder, oesophasus, liver, thyroid	0.04
Bone surface, brain, salivary glands, skin	0.01
Whole body total	1.00

19.2.9.4   Collective Effective Dose Equivalents

As well as quantifying the effective dose equivalent (or *dose*) received by individuals it is also important to have a measure of the total radiation dose by a group of people or population. This is the *collective effective dose equivalent* (or *collective dose*) and is obtained by multiplying the mean effective dose equivalent to the group from a particular source by the number of people in that group, to give units of person Sieverts (person Sv).

19.3   BIOLOGICAL EFFECTS OF RADIATION

19.3.1   General Effects

The formation of ions and free radicals in tissue by radiation and the subsequent chemical reactions between these reactive species and the tissue molecules causes a range of short-term and long-term biological effects. As a frame of reference, acute exposure can cause death, an instantaneously absorbed dose of 5 Gy probably being lethal. The range of effects include cataracts, gastrointestinal disorders, blood disorders including leukaemia, damage to the central nervous system, impaired fertility, cancer, genetic damage and changes to chromosomes producing mutations in later generations. Of the long-term effects of radiation to the exposed generation, cancer and especially leukaemia is probably the most important. The number of excess (or extra) cancers observed in an exposed group compared with a non-exposed control group divided by the product of the exposed group size and the mean individual dose gives the *risk factor*. This is the risk of the effect occurring per unit dose equivalent.

Risk factors have been estimated by the United Nations Scientific Committee on the Effects of Atomic Radiation (UNSCEAR) by studying various groups of exposed people, including those affected by the Chernobyl disaster (see section 19.7.2.4). The risk factor found for fatal leukaemia, for example, is $\sim 2 \times 10^{-2}$ Sv^{-1}, so there is approximately a 1 in 500 chance of dying of leukaemia following an exposure of 1 Sv. When fatal cancers and serious hereditary defects in the first two generations are considered together the latest ICRP (2007) estimate of the risk factor is $\sim 5 \times 10^{-2}$ Sv^{-1}. This is based on the assumption that there is no lower threshold of dose below which the probability of effect is zero. In other words it is assumed that any exposure to radiation carries some risk and that the probability of the effect occurring is proportional to the dose. It is worth noting that the historical trend has been for estimates of risk to increase with time, and hence the estimates of acceptable dose have decreased with time.

19.3.2   Biological Availability and Residence Times

The rate of uptake of a radionuclide into the body depends upon its concentrations in the various environmental media (air, water, food, dust, *etc.*), the rates of intake of these media and the efficiency with which the body absorbs the nuclide from the media. This latter parameter largely depends upon the physical and chemical properties of the radioactive substance in question. For example, the chemical properties of strontium are similar to those of another Group II element, calcium, which is of vital biological significance. Calcium enters the body to form bones and to carry on other physiological functions and ^{90}Sr can readily follow calcium into the bones where it will remain. In the same way, the chemical similarity of caesium to potassium, which is present in all cells in the body, allows caesium to be readily transported throughout the entire body. On the other hand, the chemically inert noble gases krypton, argon, xenon and radon, which are all radioactive air pollutants, are not readily absorbed by the body (although they can still contribute to the exposure of an individual by external irradiation and by forming radioactive daughters with different physico-chemical properties).

In any given organism there exists a balance between the intake of an element and its excretion and this controls the concentration of the element in the organism. For radioactive isotopes this balance will include on the debit side the radioactive decay of the isotope. The effective half-life, t_{eff}, of a nuclide in an organism, *i.e.* the time required to reduce the activity in the body by half, is a function of both the biological half-life, t_{biol}, and the radioactive half-life, t_{rad}:

$$t_{eff} = t_{rad} t_{biol}/(t_{rad} + t_{biol})$$

Whilst t_{rad} is fixed for any given nuclide, t_{biol} will vary from species to species and between individuals with age, sex, physical condition, and metabolic rate. For ^{90}Sr, which has $t_{rad} = 28$ y and $t_{biol} =$ about 35 y, the effective half-life for humans is about 15.5 y. Obviously those nuclides of most concern as pollutants are those present in the environment in the highest concentrations, with the most energetic emissions and with the longest effective biological half-lives.

19.3.3 Radiation Protection of Terrestrial Ecosystems

Although most thought has been given to the dose assessment and protection of humans from radioactivity, the need to protect wildlife and natural ecosystems has recently become apparent. The multitude of potentially exposed plant and animal species and the difficulties of estimating exposure to individuals within ecosystems has lead to the ICRP adopting a 'reference organism' approach. The concept of hypothetical 'Reference Plants and Animals' (RAPs) which represent broad species groups (*e.g.* large terrestrial mammals, small terrestrial mammals, freshwater fish *etc.*) is designed to complement the approach used for human radiological protection. ICRP use the concept of a 'reference human' to help manage and assess the many different situations in which human beings would or could be exposed to ionising radiation and the same approach is now used for wildlife with RAPs. More information on this topic is available in the chapter by Howard and Beresford in *Nuclear Power and the Environment* (ed. Hester and Harrison, 2011).

19.4 NATURAL RADIOACTIVITY

19.4.1 Cosmic Rays

The Earth's atmosphere is continuously bombarded by highly energetic protons and alpha particles emitted by the sun and of galactic origin. Their energies range from about 1 MeV to about 10^4 MeV, with a flux rate at the outer edge of the atmosphere of about 2×10^7 MeV m^{-2} s^{-1}. These primary particles have two effects. They cause some radiation exposure directly to humans, the magnitude of which varies with altitude and latitude, and they interact with stable components of the atmosphere causing the formation of radionuclides. Of these, ^{3}H and especially ^{14}C are important to the biosphere, while several of the shorter-lived nuclides (*e.g.* ^{7}Be, $t_{1/2} = 53$ days; ^{39}Cl, $t_{1/2} = 55$ min) have applications as tracers in the study of atmospheric dispersion and deposition processes.

Carbon-14 is formed from atmospheric nitrogen by:

$$^{14}_{7}N + ^{0}_{1}n \rightarrow ^{15}_{7}N$$

$$^{15}_{7}N + ^{14}_{6}C \rightarrow ^{1}_{1}p$$

The ^{14}C is then oxidized to CO_2 and enters the global biogeochemical cycle of carbon, being incorporated into plants by photosynthesis and into the oceans by absorption. Most natural production of tritium is also by the reaction of cosmic ray neutrons (with energy >4.4 MeV) with nitrogen:

$$^{14}_{7}N + ^{1}_{0}n \rightarrow ^{12}_{6}C + ^{3}_{1}H$$

Tritium is then oxidized or exchanges with ordinary hydrogen to form tritiated water and enters the global hydrological cycle. The Radiation Protection Division of the UK government agency Public Health England (PHE, formerly known as the Health Protection Agency, HPA) estimates that the annual effective dose equivalent from cosmic rays in the UK is about 300 µSv y^{-1} on average. However, people that live at high altitudes and high latitudes or who spend significant time in aircraft will receive rather greater doses.

19.4.2 Terrestrial Gamma Radiation

The Earth's crust contains three elements with radioactive isotopes which contribute significantly to human exposure to radiation. These are: ^{40}K, with an average concentration in the upper crust of ~3 µg g^{-1}; ^{232}Th, which is present in granitic rocks at 10–15 µg g^{-1}; and the three isotopes of uranium which total on average 3–4 µg g^{-1} in granite. These latter nuclides have relative abundances of 99.274% ^{238}U, 0.7205% ^{235}U, and 0.0056% ^{234}U in the crust.

PHE estimate that the gamma rays emitted by these radionuclides and their daughters in soil, sediments, and rocks and in building materials give an average annual effective dose equivalent in the UK of about 400 µSv y^{-1}, with the whole body being irradiated more or less equally. The local geology and types of building material used lead to considerable variation about this figure.

19.4.3 Radon and its Decay Products

The decay series of ^{238}U and ^{232}Th both contain radioactive isotopes of the element radon, ^{222}Rn in the former case and ^{220}Rn (sometimes known as thoron) in the latter. Radon, being a noble gas, readily escapes from soil and porous rock and diffuses into the lower atmosphere. There the nuclides decay with half-lives of 3.8 days (^{222}Rn) and 55 s (^{220}Rn) producing a series of shortlived daughter products. The full ^{238}U decay series is shown in Figure 19.1. These daughters attach themselves to aerosol particles in the atmosphere. These particles are efficiently deposited in the lungs if inhaled. Their subsequent α and β emissions can then irradiate and damage the lung tissue. The mean flux rate of radon from soil to the atmosphere is about 1200 Bq m^{-2} d^{-1} but there are large variations about this value at very small spatial scales.

Ambient outdoor concentrations of radon are low, typically 1–100 Bq m^{-3}, but less (0.1 Bq m^{-3}) above the oceans. The average outdoor concentration in the UK, weighted for population, is about 4 Bq m^{-3}. However when this gas enter a building, either through the floor following soil emissions or from the building's construction materials or by desorption from the water supply, the low air exchange rates in modern buildings causes a substantial increase in concentrations. Considerable variations in indoor concentrations have been observed, with maximum concentrations exceeding 2000 Bq m^{-3}. In the UK, a positively-skewed distribution with a mean of ~20 Bq m^{-3} is found, with about 100 000 homes having concentrations above 200 Bq m^{-3}. The concentrations in the outside air, in the soil, and in soil pore water, the rate of emanation from building materials and the building ventilation rate will all affect the indoor concentration, and in a large scale measurement programme a few individual dwellings with concentrations two or three orders of magnitude above the mean might be observed.

Because of the large variations in radon concentration and differences in individual's personal activity patterns, the calculation of an individual's exposure to radiation from radon is rather uncertain. The average annual effective dose equivalent in the UK is estimated by PHE to be $\sim 1300~\mu Sv~y^{-1}$, or $\sim 50\%$ of the total annual average individual dose of 2700 μSv, although some individuals may receive very much higher doses from radon than this. For example, the average dose to the population of Cornwall is about 7800 μSv. In any case, radon and its daughters generally give a greater dose of radioactivity to populations than any other sources of radiation. In the UK an 'Action Level' has been proposed by PHE (200 Bq m^{-3}) above which remedial action to lower radon concentrations in buildings is recommended. Areas with $> 1\%$ of dwellings with radon levels above the Action Level are designated as 'Affected Areas'.

In the USA in particular, but increasingly in other countries as well, attempts are made to reduce indoor radon exposure in the worst affected areas and buildings. This can be done by preventing radon entering the building, by increasing the building ventilation rate or by removing the radon decay products from the air by, for example, using an electrostatic precipitator. Since most radon in buildings comes from the ground below the best method is to reduce the ingress of radon into the building by sealing the floor or by creating or increasing under floor ventilation.

19.4.4 Radioactivity in Food and Water

The most important naturally occurring radionuclides in food and water are ^{226}Ra, formed from ^{238}U, and its α-emitting daughter products, ^{222}Rn, ^{218}Po and ^{214}Bi. There is a wide variation in the ^{226}Ra content of public water supplies, it usually being very low and representing a minor source of intake, but some well and spring water may contain 0.04–0.4 Bq L^{-1}. As previously mentioned, the domestic water supply, particularly water used in showers, may act as a source of ^{222}Rn into the indoor environment. For the population in general the main source of radium intake is from food. Wide variations in concentration are found but a typical daily intake from the average diet is about 0.05 Bq per day. Some foods, for example Brazil nuts and Pacific salmon, accumulate radium in preference to calcium and can have very much higher concentrations than the average for food. Unbalanced diets based on these foods could lead to enhanced intakes of radium. Both ^{210}Pb and ^{210}Po can enter food, both from the soil but also by wet and dry deposition from the atmosphere where they are present as daughters of ^{222}Rn. For the average individual in the UK, the PHE estimate of the total effective dose equivalent of alpha activity from the diet is about 200 $\mu Sv~y^{-1}$.

An additional source of exposure to α-particles for some individuals is cigarette smoke. The decay products of radon emitted from the ground under tobacco plants can adsorb onto the growing plant leaves and hence be incorporated into cigarettes. Dose rates have been estimated to be as high as 6–7 mSv per year from this source for some individuals.

Most of the naturally occurring β activity in food is due to ^{40}K. The availability of potassium to plants is subject to wide variations and hence the ^{40}K activity of foods also varies. However potassium is an essential element for plants and animals, constituting about 0.2% of the soft tissue of the body. This leads to a total ^{40}K content of ~ 4 kBq for the average person, although the individual amount depends upon age, weight, sex, and proportion of fat. PHE estimate that the combined average effective dose equivalent from food and drink to individuals in the UK is about 260 $\mu Sv~y^{-1}$.

19.5 MEDICAL APPLICATIONS OF RADIOACTIVITY

Although not generally regarded as being pollutants, radiations used for medical purposes make a significant contribution to an individual's exposure. Indeed they provide the major artificial route of exposure to most individuals and so warrant mention here. X-rays are used for a wide range of diagnostic purposes, a typical chest X-ray giving an effective dose equivalent of about 20 μSv. Short-lived radionuclides are also used diagnostically. For example, ^{99}Tc is used for bone and brain scans. The use of radiation for diagnostic purposes gives an average effective dose equivalent in the UK of about 250 μSv per year, of this about 120 μSv y^{-1} is considered to be genetically significant, compared with about 1000 μSv y^{-1} genetically significant dose from natural sources. Restrictions on the use of X-rays during pregnancy, limiting the X-ray beam area to the minimum needed, reducing radiation leakage from the source and improving the sensitivity and reliability of the measuring methods used are all aimed at reducing the genetically significant dose to reduce the probability of future genetic effects.

The diagnostic use of CT (computerised axial tomography) scans has increased dramatically in the last two decades, with a typical CT scan of the chest giving a dose of about 6 mSv and a whole body scan about 10 mSv. These are very much greater than the doses resulting from conventional two-dimensional diagnostic X-rays.

Externally administered beams of X-rays, gamma rays (from ^{60}Co sources) and neutrons and radiations from internally administered radionuclides (*e.g.* ^{131}I in the thyroid) are all used for therapeutic purposes. Some of the doses involved are very high but the potential adverse effects have to be weighed against the benefits accrued to the individual patient.

As much as 15% of the effective dose of radiation received by the UK population may now be due to medical applications (410 μSv y^{-1}), although this may be higher in some countries where radiation-based diagnostics and treatment are more common (*e.g.* the USA). Regular and stringent calibration and testing of medical radiation sources is vital if patients are to receive the correct dose in the correct place.

19.6 POLLUTION FROM NUCLEAR WEAPONS EXPLOSIONS

Since 1945 the products of nuclear weapons explosions, both of fission and fusion devices, have caused considerable pollution of the globe. The major bomb testing programmes were held in 1955–8 and 1961–1990, with the number of documented test explosions now around 2000 (45 by the United Kingdom, including 21 atmospheric tests) and totalling at least 1000 megatons of explosive force. The most recent tests have been held by India, Pakistan and North Korea (in 2006 and 2009). Both underground and atmospheric tests have been used, with most of the explosive yield being from explosions in the atmosphere. Following an atmospheric explosion there will be some local deposition (or fallout) of activity but the majority of the products are injected into the upper troposphere and stratosphere, allowing their dispersion and subsequent deposition to the Earth's surface on a global scale. About 10^{29} Bq of fission products have been injected into the atmosphere by these tests.

Fission bombs, such as those used against Japan, depend upon the rapid formation of a critical mass of ^{235}U or ^{239}Pu from several components of subcritical size. The critical mass for ^{235}U is about 10 kg which on fission produces a very large amount of radioactivity: the total activity released by the 14 kiloton device at Hiroshima was about 8×10^{24} Bq. Most of the released activity is in the form of short-lived nuclides, but considerable amounts of environmentally important fission products, including ^{85}Kr, ^{89}Sr, ^{90}Sr, ^{99}Tc, ^{106}Ru, ^{137}Cs, ^{140}Ba, and ^{144}Ce, are also produced. Of particular biological significance are ^{89,90}Sr, ^{131}I, and ^{137}Cs.

As mentioned above, strontium is of importance because it has similar chemical properties to those of calcium. It enters the body by inhalation and ingestion, particularly through milk and vegetables. Iodine differs from strontium in that it is present in the atmosphere in both the gas and aerosol phases. Exposure may be by inhalation and ingestion, but of particular significance is the concentration of iodine in milk caused by cows grazing on contaminated grass and its subsequent consumption by humans. Estimates of the amount of ^{137}Cs released into the atmosphere by bomb tests vary but it is probably of the order of 10^{18} Bq, giving rise to human exposure mainly through grain, meat, and milk.

Fusion bombs (also known as thermonuclear or hydrogen bombs) use the reaction of lithium hydride with slow neutrons to generate tritium:

$$^{6}_{3}\text{Li} + ^{1}_{0}\text{n} \rightarrow ^{3}_{1}\text{H} + ^{4}_{2}\text{He}$$

which then reacts with deuterium releasing energy:

$$^{2}_{1}\text{H} + ^{3}_{1}\text{H} \rightarrow ^{4}_{2}\text{He} + ^{1}_{0}\text{n}$$

The slow neutrons are initially supplied by the fission of ^{235}U or ^{239}Pu and the deuterium by the use of lithium-6 deuteride. Fusion bombs therefore release fission products as well as large amounts of tritium.

Tritium is a pure beta emitter with a half-life of 12.3 y. It is readily oxidized in the environment forming tritiated water and hence enters the hydrological cycle. The amount of tritium injected into the atmosphere by weapons tests probably exceeds 10^{20} Bq giving rise to estimated dose commitments over an average lifetime of 2×10^{-5} Gy in the northern hemisphere and 2×10^{-6} Gy in the southern hemisphere.

Various other radionuclides are produced by weapons explosions by the neutron activation of other elements in the soil or surface rocks, the air and the bomb casings. Of these ^{14}C is the most significant pollutant, being formed from stable nitrogen in the air. It is a pure beta emitter of mean energy 49.5 keV and a half-life of 5730 years. The rate of natural production of ^{14}C by cosmic ray interactions is about 1×10^{15} Bq y^{-1} compared with a mean of $\sim 5 \times 10^{15}$ Bq y^{-1} produced by weapons testing since 1945.

The other important group of pollutant nuclides produced by weapons testing are the transuranics (elements with atomic numbers >92), including plutonium. Of these ^{239}Pu is the most significant as it has been produced in large quantities ($\sim 1.5 \times 10^{16}$ Bq) and has a long half-life (24 360 y). It is formed by capture of a neutron by ^{238}U:

$$^{238}_{92}\text{U} + ^{1}_{0}\text{n} \rightarrow ^{239}_{92}\text{U}^{\beta-} \rightarrow ^{239}_{93}\text{Np}^{\beta-} \rightarrow ^{239}_{94}\text{Pu}$$

The most important route to humans for plutonium from weapons tests is by inhalation. The UNSCEAR estimates for the population-weighted dose, up to the year 2000, from tests carried out to 1977, are 1×10^{-5} Gy in the northern hemisphere and 3×10^{-6} Gy in the southern hemisphere.

The current PHE estimate for the average effective dose equivalent in the UK from weapon test activity is about 6 pSv per year at present, compared with about 80 pSv per year in the early 1960s. About half the current effective dose equivalent is received *via* external radiation, mainly from ^{137}Cs, and about half from ingestion, mainly of ^{90}Sr and ^{14}C.

19.7 POLLUTION FROM ELECTRIC POWER GENERATION PLANT AND OTHER NUCLEAR REACTORS

19.7.1 Emissions Resulting from Normal Reactor Operation

There are a large number of nuclear reactors in operation world-wide of many different sizes, designs, and applications. As well as 434 large reactors operational in 31 countries around the world, with an installed generating capacity of 370 GW, and 62 reactors in 14 countries under construction (totals for 2012), there are many other smaller reactors used for research, isotope production, education, materials testing, and military purposes. Included in this latter category are reactors used to power submarines and other naval vessels. However the greatest use remains the generation of electricity. The share of electricity generated by nuclear power varies from country to country. Only in France, Belgium and Slovakia is more than half of all electricity produced by nuclear power. In the UK the fraction is 16% and in the USA 19% (2012).

The production of electric power by using thermal energy derived from nuclear fission involves a chain of steps from the mining and preparation of fissionable fuel through to the disposal of radioactive wastes. There are actual and potential emissions of radioactivity into the environment at each of these steps. An up-to-date review of the effects of nuclear power generation on the environment is available in *Nuclear Power and the Environment* (ed. Hester and Harrison, 2011).

19.7.1.1   *Uranium Mining and Concentration*

Uranium ores contain, typically, about 0.15% U_2O_3 and require milling, extraction and concentration before shipping to the consumer. At several stages in this process dusts bearing radioactivity are produced which may lead to contamination of the environment in the vicinity of the plant. Large volumes of mill tailings are also produced and as these contain virtually all the radium and thorium isotopes present in the ore they give rise to emissions of radon. Also of environmental significance are liquid releases from the mills which, if uncontrolled, may lead to contamination of surface and ground waters. It should also be mentioned that the mining and milling of uranium ore inevitably leads to the occupational exposure of workers to uranium and its daughters, especially *via* inhaled ^{222}Rn.

19.7.1.2   *Purification, Enrichment, and Fuel Fabrication*

The concentrated and purified ore extract, known as 'yellowcake', contains about 75% uranium and is converted to usable forms of uranium metal and uranium oxide by several processes. First uranium tetrafluoride, UF_4, is produced. The yellowcake is digested in nitric acid, insoluble impurities removed by filtration and soluble impurities extracted with an organic solvent. The uranyl nitrate solution is concentrated, oxidized to UO_2 and reacted with hydrogen fluoride to form uranium tetrafluoride, UF_4. The UF_4 so produced contains the relative natural abundances of the various uranium isotopes, including about 0.7% of ^{235}U. This is adequate for some reactors, such as the first generation British Magnox reactors, and unenriched uranium metal is produced for them by reaction of the UF_4 with magnesium. The metal is cast and machined into fuel rods, heat treated, and then inserted into magnesium–aluminium alloy ('Magnox') cans.

Other designs of reactors, including the Advanced Gas-cooled and Pressurized Water Reactors, require fuel in the form of enriched uranium oxide containing 2–3% of ^{235}U. This is produced by first reacting UF_4 with fluorine gas to form uranium hexafluoride, UF_6. The UF_4 gas is then repeatedly centrifuged or diffused through porous membranes, the small mass differences between the isotopes resulting in their eventual fractionation. Following enrichment the UF_6 is hydrolysed and reacted with hydrogen to form uranium oxide, UO_2, in powder form. This is pressed into pellets and packed in helium filled stainless steel or zirconium–tin alloy cans.

Radioactive releases into the environment during all these steps should be minimal, although as with all chemical and mechanical processes some emissions are inevitable. The current PHE estimates of the maximum annual effective dose equivalents arising from fuel preparation for members of the public living near to the relevant plants in the UK are less than 5 μSv from discharges to the air and about 50 μSv from discharges to water. These each give estimated *collective* effective dose equivalents for the UK of about 0.1 person Sv per year.

19.7.1.3   *Reactor Operation*

An operational nuclear reactor contains, and potentially may emit, radioactivity from three sources: from the fuel, from the products of fission reactions, and from the products of activation reactions. The amount of activity present as fuel itself is of course variable but a Pressurized Water Reactor with 100 tonnes of 3.5% enriched uranium would contain $\sim 0.25 \times 10^{12}$ Bq of ^{235}U and 1.1×10^{12} Bq of ^{238}U. Unless a catastrophic accident occurs, releases of fuel into the environment should not occur.

Fission of the fuel produces a large number of primary products which in turn decay, producing a large number of secondary decay products. Because of their very wide range of half-lives the relative composition and amounts of fission products present in a reactor varies with time in a complex manner. As each nuclide has a different rate of decay the rate of accumulation will vary. However there reaches a point when the rate of production of a nuclide equals its rate of removal by decay and an equilibrium arises between parent and daughter. If the rate of formation is constant then for all practical purposes equilibrium may be assumed to have been reached after about seven half-lives. For a short-lived nuclide such as ^{85}Kr, equilibrium will be reached about 10 weeks after the start of the reactor, whereas for ^{137}Cs with a half-life of 30.1 y it will take more than 200 years of continuous operation to reach equilibrium. As the reactor continues to operate with a given set of fuel rods the long-lived fission products become relatively more important and eventually the reactor might contain 10^{20} Bq of activity in total.

Most of the fission products are contained within the fuel cans themselves but total or partial failure of the fuel cladding will allow contamination of the coolant. In addition, contamination of the outer surfaces of the fuel rods by uranium fuel will allow fission products to form in the coolant. Most of the fission product activity is in the form of gaseous elements and although various scrubbing systems are used to remove them from the coolant, leakage into the environment can and does occur. The amount of gaseous fission products released will depend upon the number of fuel cladding failures, the design of the ventilation, cooling, and coolant purification systems and the length of operation of the reactor. Of the fission products, ^{85}Kr, which has a half-life of 10.8 y, makes the greatest contribution to the global dose commitment from reactor operation.

The third source of activity in the reactor results from the neutron activation of elements present in the fuel and its casing, the moderator, the coolant, and other components of the reactor itself. Those areas most subjected to neutron activation are those where the incident neutron flux is highest, and include the fuel casings and the coolant system. From an environmental point of view the most important activation product is probably tritium. This is formed by various routes, especially by the activation of deuterium present in the water of Heavy Water Reactors but also by the activation of deuterium present in the water of light water-cooled reactors, the activation of boron added as a regulator to the primary coolant of Pressurized Water Reactors and by activation of boron present in the control rods of Boiling Water Reactors. The small size of the tritium nucleus allows it to diffuse through the cladding materials and so be released into the environment where it will be inhaled and absorbed through the skin. A smaller dose will also be received from drinking and cooking water, following the incorporation of released tritium into the water cycle. Estimates have been made of the annual dose equivalent for individuals living 1, 3, and 10 km from a Canadian CANDU Heavy Water Reactor and are of the order of 8, 3, and 2 µSv, respectively.

Other important activation products are ^{58,60}Co, ^{65}Zn, ^{59}Fe, ^{14}C and the actinides. This latter group, which includes the three long lived isotopes of plutonium: ^{238}Pu, ^{239}Pu, and ^{240}Pu (with half-lives of 86.4, 24360, and 6580 y, respectively), are produced by the neutron activation of fuel uranium. Plutonium has a very low volatility and is not significantly released during normal reactor operation. However, it can potentially enter the environment following a reactor accident and as a result of fuel reprocessing.

Various estimates are available for the total dose arising from these various discharges of radioactivity to the environment during the normal operation of power generation reactors. The current PHE estimate of the maximum effective dose equivalents resulting from discharges to the atmosphere for the most exposed individuals living near to power stations in the UK is about 100 µSv per year, giving a collective effective dose equivalent in the UK of about

4 person Sv per year. In addition to discharges to the atmosphere there are also discharges of radioactivity from power stations to natural waters. These result from the temporary storage of used fuel under water in specially constructed ponds. PHE estimate that these discharges to natural waters give a maximum effective dose equivalent to local individuals of less than 350 μSv per year. The collective effective dose equivalent in the UK, mainly through the eating of contaminated seafood, is about 0.1 person Sv per year.

In addition to the above, members of the public who live close to nuclear power stations, especially older Magnox stations, may be exposed to direct radiation from parts of the plant themselves. In the worst case, small numbers of the population might receive up to 500 μSv per year from this, although annual doses at the site fence are typically less than 10 μSv per year.

19.7.2 Pollution Following Reactor Accidents

With the large number of reactors operational world-wide it is inevitable that accidents with environmental consequences should occur. To date six major incidents have taken place, in the UK, USA, the former USSR and Japan.

19.7.2.1 *Windscale, UK, 1957*

In October 1957, a graphite-moderated reactor at Windscale, UK, overheated, rupturing at least one fuel can and causing the release of fission products into the environment. An estimated 740×10^{12} Bq of ^{131}I, 22×10^{12} Bq of ^{137}Cs, 3×10^{12} Bq of ^{89}Sr, and 3×10^{12} Bq of ^{90}Sr were released. The cause of the accident was the sudden release of a huge amount of Wigner energy from the graphite in the core. When graphite is bombarded by fast neutrons, as in a reactor, an increase in volume and a decrease in thermal and electrical conductivity results. The graphite can then efficiently store energy and, if heated above 300 °C, this is released in the form of further heat. Normally this Wigner energy is released by slowly heating the graphite, allowing the stored energy to be released under control and reversing the radiation damage. During 7–8 October 1957, however, an unsuccessful attempt was made to release the Wigner energy at Windscale, following which the reactor rapidly overheated releasing radioactive material into the atmosphere.

Following the accident the inhalation of aerosol and gaseous nuclides probably represented the largest source of exposure to the general population, although the drinking of milk contaminated by ^{131}I was also significant, despite a restriction on the sale of milk from farms in the worst affected area. The total committed effective dose equivalent in the UK has been estimated at about 2000 person Sv, resulting in about 30 excess deaths from cancer in the UK during the first 40 y after the accident.

19.7.2.2 *Kyshtym, Russia, 1957*

In 1957 a chemical explosion in a high-level radioactive waste tank caused the release of about 7×10^{17} Bq into the environment.

19.7.2.3 *Idaho Falls, USA, 1961*

In January, 1961, during maintenance work on an enriched uranium-fuelled boiling water reactor at the National Reactor Testing Station, Idaho Springs, USA, the central rod was accidentally removed causing a large explosion, killing three men. Most of the coolant water was ejected together with an estimated 5–10% of the total fission products in the core. About 4×10^{12} Bq of activity escaped from the reactor building.

19.7.2.4 Three Mile Island, USA, 1979

Whilst operating at full power on 28 March, 1979, the cooling water system failed on the pressurized water reactor at Three Mile Island reactor 2 at Harrisburg, Pennsylvania. The reactor automatically shut down and three auxiliary pumps started to provide cooling water. However valves in this circuit had inadvertently been left closed and the resultant increase in temperature and pressure in the primary circuit caused an automatic relief valve to open, allowing about 30% of the primary coolant to escape. Fresh cooling water was injected into the primary circuit by the emergency cooling system but, following an error in their analysis of the accident, this and the main coolant pumps were deactivated by the operators. It was only when cooling water was later added that the overheating of the core ceased.

Despite the reactor core sustaining serious damage, the amount of activity released into the environment was limited to about 10^{17} Bq, almost entirely as short-lived noble gases, especially ^{133}Xe (which has a half-life of 5.2 days). The resultant total committed effective dose equivalent to the population was estimated to be about 20 person Sv with the maximum individual dose equivalent being about 1 mSv.

19.7.2.5 Chernobyl, Ukraine, 1986

The most serious reactor accident to date began on 26 April 1986, when an explosion and fire occurred at the Chernobyl number 4 reactor near Kiev in the Ukraine. The reactor came into service in 1984 and was one of 14 RBMK boiling water pressure tube reactors operational in the USSR. These are graphite-moderated reactors of 950–1450 MW electrical power generation capacity fuelled with 2% enriched uranium dioxide encased in zirconium alloy tubing. The normal maximum temperature of the graphite is about 700 °C and in order to prevent this being oxidized the core is surrounded by a thin-walled steel jacket containing an inert helium–nitrogen mixture.

Prior to the accident the reactor had completed a period of full power operation and was being progressively shut down for maintenance when an experiment was begun to see whether the mechanical inertia of one of the turbo-generators could be used to generate electricity for a short period in the event of a power failure. The reactor core contained water at just below boiling point but when the experiment began some of the main coolant pumps slowed down causing the core water to boil vigorously. The bubbles of steam so formed displaced the water in the core and, because steam absorbs neutrons much less efficiently than water, the number of neutrons in the core began to rise. This increased the power output of the reactor, so increasing the heat output and the amount of steam in the core, which in turn led to a further rise in the neutron density. This positive feedback mechanism led to a rapid surge in power causing the fuel to melt and disintegrate. As the fuel came into contact with the surrounding water, steam explosions occurred destroying the structure of the core and the pile cap, causing radioactive material to be ejected into the atmosphere. The core fires allowed a continuing release of activity that was slowly reduced by the dumping of clay and other materials onto the core debris. However the core temperature again began to rise and a second peak in activity release occurred on 5 May. After this the core was progressively buried and finally sealed in a concrete sarcophagus.

Estimates of the amount of activity released from the reactor vary but Ukranian measurements suggest that all of the noble gases, 10–20% of the volatile fission products (mainly iodine and caesium), and 3–4% of the fuel activity, giving a total of 1.85×10^{18} Bq, escaped into the environment. On the basis of air concentration and deposition measurements the UK AEA's Harwell Laboratory initially estimated that about 7×10^{16} Bq of ^{137}Cs was released.

The immediate casualties of the accident were 31 killed and about two hundred diagnosed as suffering from acute radiation effects. About 135 000 people and a large number of animals were evacuated from a 30 km radius area surrounding the plant. However, in 1995 the Ukranian Health

Ministry announced that a total of 120 000 fatalities had occurred as a result of the accident but the basis of this claim is not clear.

The meteorological conditions prevailing over Europe at the time of the accident were rather complex, leading to the dispersion of activity over a very wide area. In the UK peak air concentrations occurred on 2 May when ~ 0.5 Bq m^{-3} of ^{137}Cs were recorded at Harwell. No Chernobyl radioactivity was immediately detected in the southern hemisphere. In the year following the accident the annual mean ^{137}Cs concentration in the northern hemisphere was about the same as that of 1963 when weapons testing activity was at its highest. The amount of ^{137}Cs deposited on the ground surface obviously varied with air concentration, rainfall, and other parameters but close to Chernobyl (within 30 km) as much as 104 kBq m^{-2} of ^{137}Cs was deposited. The average for Austria was 23 kBq m^{-2}, for the UK 1.4 kBq m^{-2}, and for the USA 0.04 kBq m^{-2}. The very heavy but localized rainfall which occurred in parts of Europe and the UK during the time when the plume was overhead led to a very patchy distribution of deposited activity on the ground. In the UK for example it varied from > 10 kBq m^{-2} of ^{137}Cs in parts of Cumbria to < 0.3 kBq m^{-2} in parts of Suffolk.

Following the wet and dry deposition of activity from the atmosphere some contamination of foodstuffs was inevitable and outside of the immediate area around Chernobyl this was the major consequence of the accident. In general the most vulnerable foods in the UK were lamb and milk products. In the UK a ban on the movement and slaughter of lambs was imposed within specified areas until the meat consistently contained less than 1000 Bq kg^{-1} of radiocaesium. No restrictions were placed on the sale of milk.

The main exposure pathways to humans following the accident were direct gamma irradiation from the cloud and from activity deposited on the ground (groundshine), inhalation of gaseous and particulate activity from the air and, most importantly, the ingestion of contaminated foods. The Organization for Economic Co-operation and Development's Nuclear Energy Agency estimate that the average individual effective dose equivalents received in the first year after the accident range from a few microsieverts or less for Spain, Portugal, and most countries outside of Europe, to about 0.7 mSv for Austria. However, hidden within these averages are the higher peak doses received by the most exposed individuals, or critical group, in each country. These vary from a few microsieverts outside Europe, to an upper extreme of 2–3 mSv for the Nordic countries and Italy. In the UK, a person consuming 8 kg of sheep meat per year, contaminated with 10^3 Bq kg^{-1} of ^{137}Cs, would receive an additional dose of about 100 µSv y^{-1}.

The total collective dose in Eastern and Western Europe has been estimated to be about 1.8×10^5 person Sv. The PHE estimate that the average effective dose equivalent received in the UK during the first year was about 40 µSv with a total of about 20 µSv being received over subsequent years. The total committed effective dose equivalent for the UK population is estimated to be about 2100 person Sv. When compared with the other doses of radioactivity normally received by individuals, for example the 2 mSv per year individual effective dose equivalent received on average in Europe from natural background radiation, these doses are small and probably insignificant. Using the ICRP estimate of risk it is possible to calculate the additional mortality likely to result from Chernobyl. In the 40 years following the accident about 50 excess deaths are likely in the UK (compared with about 145 000 non-radiogenic cancers per year in the UK), suggesting that the impact of the accident on future mortality statistics will be small.

The Chernobyl accident presented a unique opportunity for experiments and studies in a wide range of environmental studies. It allowed the validation and refinement of atmospheric dispersion models, the calculation of washout ratios and deposition velocities, and the study of the behaviour of caesium and iodine in food chains, natural waters, sediments, and in the urban environment. It demonstrated the necessity of better international harmonization of scientific databases and public health protection policies and it provided a vivid example of the long-range trans-boundary transport of pollutants. In 1994 and 1995, a very large European Union-funded tracer release experiment (ETEX) was carried out to simulate a nuclear reactor accident on the scale of Chernobyl with the aim of testing the predictive capabilities of long range transport models and forecasts. Inert tracers were released from a site in Brittany, western France, and the resultant plume detected across Europe by a network of monitoring sites and by aircraft.

19.7.2.6 *Fukushima Daiichi, Japan, 2011*

On 11 March 2011 a magnitude 9 undersea earthquake with an epicentre east of the Oshika Peninsula of Honshu Island (the main island of Japan) triggered a massive tsunami, with waves reaching heights of more than 30 m. In total the tsunami inundated over 500 sq km of land and directly caused around 20 000 deaths in Japan, with over a million buildings destroyed or damaged. The affected region of Japan contained four nuclear power plants, with eleven reactors in total. None of these were significantly damaged by the earthquake itself, and all shut down automatically as planned. However, at the Fukushima Daiichi nuclear power plant the subsequent tsunami submerged and damaged the main and auxiliary condenser circuits used to cool the General Electric-designed boiling water reactors (first commissioned in 1971) at the site as well as backup generators, batteries and switchgear. Of the six reactors, one was empty of fuel and two were already shut down for planned maintenance, but the flooding caused the three operational reactors to overheat. Deliberate flooding of the reactors with seawater did not prevent the overheating from continuing, causing meltdown of the three cores. Several explosions caused by production of hydrogen then caused very significant further damage to the reactors, releasing radioactivity to the atmosphere. Fuel rods stored in a pond at reactor 4 also overheated as the cooling water level fell due to pump failure and this caused further releases of radioactivity to the environment.

After the first hydrogen explosion on 12 March, ^{131}I, ^{137}Cs and ^{134}Cs were detected in air outside the plant, and this, together with fears of larger explosions, led to the establishment of a 20 km exclusion zone around the plant. This was later extended to 30 km radius. As the fate of the molten fuel in the three reactors is not yet understood, it is has not been possible to fully quantify the total emissions of radioactivity to the environment, but a significant amount of contaminated seawater used for cooling flowed into groundwater and back into the ocean. About 900 PBq (900×10^{15} Bq) is thought to have been released to the atmosphere, with potential exposure within the exclusion zone exceeding 20 mSv y^{-1}. Towards the end of March 2011, elevated levels of ^{131}I, ^{134}Cs, ^{137}Cs, and ^{132}Te were detected at Public Health England air monitoring stations in the UK, but they returned to background levels within a month.

There were two immediate fatalities at the site, and a number (possibly as many as 45) related deaths arising from the evacuation of over 100 000 inhabitants from the exclusion zone. Several workers died during the immediate containment and clean-up operations but these may not have been due to radiation. Stabilisation of the damaged reactors is ongoing while clean-up of the site will take decades. There were no reports of radiation-related fatalities to the public, although epidemiological studies of the population will obviously take decades to complete.

19.7.3 Radioactive Waste Treatments and Disposal

An inevitable consequence of the human use of radioactivity is that radioactive waste material is produced which must then be disposed of. Although there is no universal scheme for the classification of such waste it is usual for it to be categorized in terms of its activity content as low, intermediate and high level. In the UK, for example, radioactive waste is managed at around 30 different sites.

19.7.3.1   Low Level Waste

Low level wastes are produced in large volumes by all the various medical, industrial, scientific, and military applications of radioactivity. They include contaminated solutions and solids, protective, cleaning and decontamination materials, laboratory ware, and other equipment. They also include gases and liquids operationally discharged from power stations and other facilities. Some of this low level waste is sufficiently low in activity to be directly discharged into the environment, either with or without prior dilution or chemical treatment. Typical maximum activity concentrations in low level waste are 4×10^9 Bq t^{-1} (alpha) and 12×10^9 Bq t^{-1} (beta and gamma). Much low level waste is currently disposed of by shallow burial in landfill sites, often with the co-disposal of other, non-radioactive, controlled wastes, or by discharge into surface waters in rivers, lakes, estuaries, or coastal seas or by discharge into the atmosphere. If the environmental biophysico-chemical behaviour of the radionuclides in question and their possible pathways back to humans are well understood, then it is possible to make reliable estimates of the likely resultant doses of radio-activity to those most exposed in the population. If these doses are suitably low then the disposal methods may be deemed to be acceptable. However at the present time there are many uncertainties in the understanding of such behaviour, pathways and doses, but nevertheless the very large volumes of low level waste being generated will, through lack of economically and environmentally viable alternatives, continue to be disposed of in these relatively uncontrolled ways. In the UK there are $\sim 35\,000$ m^3 of low level waste stocks (2010).

19.7.3.2   Intermediate Level Waste

Intermediate level wastes are sufficiently active to prevent their direct discharge into the environment, with maximum specific activities of typically 2×10^{12} Bq m^{-3} (α) and 2×10^{-14} Bq m^{-3} (β and γ). They comprise much of the solid and liquid wastes generated during fuel reprocessing, residues from power station effluent plants, and wastes produced by the decommissioning of nuclear facilities. Very large quantities (~ 2000 t y^{-1}) of intermediate and low level wastes have been disposed of by dumping in deep ocean waters in the NE Atlantic. Although some authorities still consider this method of disposal to be the best practicable environmental option for these categories of waste it is no longer practised and intermediate level waste produced in the UK is now stored on land, mainly at Sellafield in Cumbria, NW England, awaiting further policy decisions. The total stock of intermediate level waste in the UK is around 83 000 m^3 (2010).

19.7.3.3   High Level Waste

High level wastes mainly consist of spent fuel and its residues and very active liquids generated during fuel reprocessing. Typical maximum activities are 4×10^{14} Bq m^{-3} (α) and 8×10^{16} Bq m^{-3} (β and γ). At present such wastes generated in the UK are stored at Sellafield in storage ponds where it is proposed they will be vitrified prior to further storage (to allow the decay of shorter lived nuclides) and finally disposal in deep repositories. No such repository yet exists but deep mines and boreholes on land and sea as well as other more exotic solutions including extraterrestrial disposal have all been proposed. Current stocks of high level waste in the UK total 1600 m^3 (2010).

19.7.4   Fuel Reprocessing

In most nuclear reactors the economic lifetime of the fuel in the core is determined not by the depletion of fissile material but by the production and accumulation of fission products which progressively reduce the efficiency of the reactor. In a typical reactor the fuel is changed on a three-year cycle, generating large amounts of partially spent fuel contaminated with fission products.

Apart from the economic considerations, which may or may not be in favour of recovering the unreacted fissile material from the spent fuel, depending upon the relative costs of reprocessing and importing further uranium ore, there are several reasons why such reprocessing is carried out: it allows the production of plutonium for military purposes, it reduces dependence on imported ores, and it reduces the volume of high-level waste produced by a reactor. It is also a very large scale international commercial operation.

In the reprocessing method currently used in the UK, the short lived activity is allowed to decay by storage for several months, after which the spent fuel is dissolved in nitric acid and a sequential extraction procedure used successively to remove the uranium, plutonium, and fission products. The uranium is then re-enriched and fabricated into fuel rods and the plutonium used in the mixed oxide fuel or for military purposes. About 35 000 t of Magnox fuel has been reprocessed at Sellafield, yielding about 15 000 t of uranium for re-enrichment. The THORP thermal oxide reprocessing plant at Sellafield processes fuel from Advanced Gas-cooled and Pressurised Water Reactors at a projected rate of about 600 t y^{-1} and will eventually handle the 2500 t of waste currently being stored at Sellafield as well as that being produced now and in the future in the UK and by overseas customers. However, a leak of radioactive waste from the THORP plant, discovered in 2005, disrupted operations for several years. No releases of radioactivity to the environment are thought to have been caused by this leak. The current PHE estimates of the annual effective dose equivalents arising from fuel reprocessing to the UK population are ~ 1 mSv y^{-1} to the most exposed individuals, with a collective dose of 80 person Sv y^{-1}. The reprocessing of fuel requires the transport of large quantities of spent and reprocessed fuel around the globe, giving rise to the possibility of further accidental releases into the environment.

19.8   POLLUTION FROM NON-NUCLEAR PROCESSES

Two non-nuclear industrial processes, the burning of fossil fuels and the smelting of non-ferrous metals, release non-trivial quantities of radioactivity into the environment and require brief consideration here. Coal contains uranium and thorium in varying concentrations, typically 1–2 µg g^{-1} of both ^{238}U and ^{232}Th but at much higher concentrations (100–300 µg g^{-1}) in coal from some areas, *e.g.* the western USA. It also contains significant quantities of ^{14}C and ^{40}K. Similarly oil and natural gas both contain members of the naturally-occurring U and Th decay series. When the fuel is burned, as in a conventional power station, these nuclides and any daughters present are either released into the atmosphere in the flue gas and fly ash or retained in the bottom ash. One current estimate of the amount of ^{226}Ra emitted by a typical 1000 MW coal-fired station is 10^9–10^{10} Bq y^{-1} with a larger amount being retained in the ash. These emissions undoubtedly give rise to elevated environmental concentrations, with the whole body dose equivalents for those most exposed living in the vicinity of large coal-fired plants being possibly as high as 1 mSv per year.

The second non-nuclear industrial source of radioactivity, the smelting of nonferrous metals, arises because of the natural occurrence of radioactive isotopes of lead. The geochemistry of lead is intimately associated with that of uranium and thorium, there being four radioactive isotopes of lead: ^{210}Pb and ^{214}Pb in the ^{238}U decay chain, ^{211}Pb in the ^{235}U decay chain, and ^{212}Pb in the ^{232}Th decay chain. All except ^{210}Pb have half-lives of less than 12 hours, but the 22 y half-life of ^{210}Pb and its subsequent decay to form ^{210}Po, itself an alpha-emitter with a half-life of 138 days, makes it of environmental significance. During the primary and secondary smelting of lead and other non-ferrous metals and their ores some lead is released into the atmosphere. A fraction of this will be ^{210}Pb together with a similar quantity of ^{210}Po. This aerosol may then be inhaled, giving rise to an exposure of the lung to alpha particles. It is possible that those living close to large smelters and other sources of ^{210}Po may receive measurable and non-trivial doses of radioactivity.

Apart from the obvious public concerns over possible nuclear weapons proliferation, testing and use, three areas appear to be of growing interest at the present time. The first is the possibility of exposure of the public to radiation resulting from illegal trafficking of radioactive sources. The scale of this is not known. Secondly, given the increasing concern over the emissions of greenhouse gases to the atmosphere, particularly carbon dioxide, extending the lifetime of existing nuclear power plants may be the most viable option for the continued use of nuclear power, for a variety of economic and sociopolitical reasons, and this is raising public concerns about nuclear plant safety. Thirdly, the use of radioactivity in medical diagnostics, especially in CT scans, continues to attract interest.

ACKNOWLEDGEMENTS

I would like to thank Dr. M. Kelly for his comments on early editions of this chapter.

BIBLIOGRAPHY

R. S. Cambray, *et al.*, Observations on radioactivity from the Chernobyl accident, *Nuclear Energy*, 1987, **26**, 77–101.

K. D. Cliff, J. C. H. Miles and K. Brown, *Decay Products in Buildings*, NRPB-R159, HMSO, London, 1984.

J. H. Gittus, *et al.*, *The Chernobyl Accident and its Consequences*, United Kingdom Atomic Energy Authority, HMSO, London, 1988.

J. S. Hughes and G. C. Roberts, *The Radiation Exposure of the UK Population – 1984 Review*, NRPB-R173, HMSO, London, 1984.

R. Kathren, *Radioactivity in the Environment: Sources, Distribution, and Surveillance*, Harwood, Amsterdam, 1984.

National Radiological Protection Board, *Living with Radiation*, HMSO, London, 1986.

A. D. Wrixon, New ICRP recommendations, *J. Radiol. Protect.*, 2008, **28**, 161–168.

Organization for Economic Co-operation and Development Nuclear Energy Agency, *The Radiological Impact of the Chernobyl Accident in OECD Countries*, OECD, Paris, 1987.

F. Warner and R. M. Harrison, *Radioecology after Chernobyl*, SCOPE 50, John Wiley and Sons, Chichester, 1993.

Health Protection Agency Report HPA-RPD-001, *Ionising Radiation Exposure of the UK Population, 2005 Review*, HMSO, 2005 (and subsequent HPA reports); www.hpa.org.uk/Publications (accessed 05/07/2013).

Health Protection Agency Report HPA-CRCE-041, *Environmental Radioactivity Surveillance Programme: Results for 2011 Including Monitoring Following the Fukushima Dai-ichi Accident in Japan*, HMSO, 2012.

R. E. Hester and R. M. Harrison, Nuclear Power and the Environment, in *Issues in Environmental Science and Technology, vol. 32*, Royal Society of Chemistry, Cambridge, 2011.

CHAPTER 20

Health Effects of Environmental Chemicals

JUANA MARIA DELGADO-SABORIT* AND ROY M. HARRISON[†]

Division of Environmental Health and Risk Management, School of Geography, Earth and Environmental Sciences, University of Birmingham, Edgbaston, Birmingham B15 2TT, UK
*Email: j.m.delgadosaborit@bham.ac.uk

20.1 INTRODUCTION

With the established expectation of a long and healthy life, and growing public concern about the impact of man's activities upon the environment, questions on the relationship between environmental quality and human health have come increasingly to the fore. The increased incidence of childhood asthma, for example, has spawned much popular debate and new research initiatives on possible environmental causes. Similarly, the issue of a possible decline in human fertility has resulted in growing speculation about the role of endocrine disrupting chemicals. The health impact of exposure to small particles in the air is also attracting much attention, and there continues to be concern about environmental exposure to pesticides, toxic metals and asbestos fibres. Memories of severe accidents, such as those at Bhopal and Seveso, serve to highlight the potential consequences to health of chemicals inadvertently released into the environment.

The public's perception of risk is highly dependent upon the origin of the hazard. Chemicals that are unseen, such as residues remaining after pesticide treatment in food, are not perceived as having the same danger as those emanating, for example, from an industrial source or a vehicle exhaust pipe, which are clearly seen. This issue has recently become somewhat clouded, however, by the development of genetic modification technology which has the capacity to introduce 'natural' substances into 'unnatural' locations.

Industrialization has resulted in increased concentrations of chemicals in air, water and soil. However, the presence of a chemical in the environment cannot be taken as an indication *per se* that it is harmful to human health – it may have no relevant exposure pathway, or be present in too low a concentration to pose any real threat to health. However, for the sake of human health and that of the natural environment, particular attention must be paid to those substances which persist in

---

[†]This chapter is based upon an earlier contribution from Paul Harrison which is gratefully acknowledged.

Pollution: Causes, Effects and Control, 5th Edition
Edited by R M Harrison
© Delgado-Saborit and Harrison and The Royal Society of Chemistry 2014
Published by the Royal Society of Chemistry, www.rsc.org

the environment and in the tissues of plants and animals and which bioaccumulate in the food chain.

The health outcome resulting from exposure to a toxic substance (either directly or indirectly through the food chain) can range from effects on perceived well-being, through to chronic ill-health, cancer, and death. There may be reversible physiological changes or more severe pathological effects on the cardiovascular and respiratory systems. Or there may be adverse reproductive, immunological, neurological and developmental consequences.

The factors which will determine the outcome of exposure to a particular chemical include the inherent toxicity of the substance and its mode of action, the route of exposure, the concentration, timing and duration of exposure, and the inherent susceptibility of the individual exposed. Individuals can be at particular risk because of their genetic makeup, their age, exposure history, current health status, fitness or lifestyle, including diet. Indeed, nutritional status is increasingly being recognized as an important modifying factor that can influence health and affect responses to environmental stressors. For example, it is likely that intake of dietary antioxidants can afford some resistance to the effects of air pollutants.

Different types of exposure to environmental chemicals can be distinguished. There is catastrophic exposure, which results from the massive accidental release of material into the environment such as occurred at Seveso or Bhopal. Localized incidents can occur as a result of heavy contamination of the local environment or of the adulteration of food and water. The outbreaks of mercury poisoning in Iraq and elsewhere as the result of the use for flour making of seed grain treated with organic mercurials are examples of this kind of exposure. Then there is the lower-level but more chronic exposure which can occur with air pollutants and food and water contaminants. With the introduction of nanomaterials into our everyday life, such as in antibacterial clothes coatings or catalysts in vehicles, there is the need to ensure that the lifecycle of these new materials does not represent a risk for the nanotechnology workers, product consumers and general population. Sometimes these exposures are unavoidable in a society which depends upon the use of chemicals to maintain its way of life, but better understanding of the risks of such exposures can enable decisions to be made about more controlled use of the substances or justifiable replacement by proven safer alternative chemicals or processes. Availability of accurate scientific information on such issues is crucial for a proper assessment of risk and the appropriate application of the 'precautionary principle'.

20.2 CATASTROPHIC EXPOSURE

The two best known examples of catastrophic exposure to chemicals occurred at Seveso and Bhopal, following a series of events that produced runaway chemical reactions in the reactors, which the security systems could not withhold, causing the release of dangerous chemicals into the nearby local communities.

20.2.1 Seveso, Italy

On 10 July, 1976 there was a massive release of 2,3,7,8-tetrachlorodibenzo-*p*-dioxin (TCDD) from a chemical plant in Seveso, near Milan in northern Italy, which was manufacturing 2,4,5-trichlorophenol (2,4,5-TCP). A chemical runaway reaction was produced in one of the tanks of the factory and a safety disc in a reaction vessel ruptured emitting a plume of chemicals containing 2,4,5-TCP and TCDD. The released chemicals deposited over a cone shaped area, downwind from the factory, about 2 km long and 700 m wide. In all, an area of 3–4 km was contaminated and an estimated 3 to 16 kg of dioxin was released. There were almost 28 000 people living in the vicinity of the factory. Those who lived in the immediate area downwind were evacuated 14 days after the explosion and the area was closed off. About 5000 people in the most heavily contaminated area were allowed to stay in their homes but they were not allowed to cultivate or consume local vegetables or fruit nor to raise or keep poultry or other animals.[1]

TCDD is both extremely toxic and extremely stable and is known, at sufficient dose levels, to affect fetal development and to have porphyrinogenic effects (often manifested as digestive system and skin disorders), all of which are well documented. It is not used commercially but is found as a trace contaminant when 2,4,5-TCP is synthesized. 2,4,5-TCP itself is used to make some herbicides, such as 2,4,5-trichlorophenoxyacetic acid (2,4,5-T) and 2,4-dichlorophenoxyacetic acid (2,4-D). TCDD is often present in trace amounts in these herbicides. TCDD along with other dioxins are also produced as a by-product of waste incineration, and continues to be of major concern as a general environmental pollutant.

The population at Seveso was screened shortly after the accident happened and a number of positive findings were noted, mainly among individuals living in the most contaminated areas. The symptoms included the skin condition chloracne, which is a more severe form of acne known to be associated with high level exposure to chlorinated hydrocarbons. Chloracne was mainly related to younger age and light hair colour.[2] Subsequent follow-ups showed that the incidence of choracne had decreased and that the individuals already affected had improved.[3] Some neurological abnormalities were also noted, such as polyneuropathy with some symptoms that were due to effects on the central nervous system and the incidence of abnormal nerve conduction tests. Evidence was found of liver enlargement in about 8% of the population. Liver enzyme activity showed some abnormalities, but had returned to normal about a year after the explosion. It is interesting and noteworthy that there was no evidence that the immune system had been affected, and there has been no evidence of chromosomal abnormalities or of any damage to the fetus. No immediate deaths were recorded as being caused by the accident.[4,5] However, the affected zones showed increased mortality from circulatory diseases in the first years after the accident, from chronic obstructive pulmonary disease and from diabetes mellitus among females.[6] An analysis of cancer incidence in the exposed population also reported increased hepatobiliary (liver) cancer, elevated incidence of leukaemias and other haematological neoplasms in men, increased multiple myeloma and myeloid leukaemia (bone marrow cancers) in women, and evidence for higher incidences of soft tissue tumours and non-Hodgkin's lymphoma. A more recent study on cancer incidence confirmed the excess of lymphatic and hematopoietic tissue neoplasms, and found an increased risk of breast cancer, but no cases of soft tissue sarcomas. No cancer cases were reported among subjects diagnosed with chloracne.[7] Initial studies reported that breast cancer and endometrial cancer in women were reduced,[8] which was attributed to the dioxin being an antioestrogen – see section 20.4.5. However, a recent follow-up study has found an increased risk of breast cancer incidence among women exposed to TCDD in Seveso,[9] a double – but not significant – risk for endometriosis,[10] increased risk of uterine leiomyoma,[11] increased in time to pregnancy and infertility[12] and an increased risk of early menopause.[13] Exposure to TCDD has also been associated with a lowered male : female sex ratio in the offspring of exposed men,[14] with smaller birth weight, which was stronger for pregnancies within the first eight years after exposure,[15] and with increased risk of neonatal hypothyroidism.[16] Children exposed to TCDD showed also development of dental aberrations.[17]

The explosion at Seveso excited a great deal of public alarm, particularly because dioxin was involved. However, major harmful effects were directed towards the environment; many farm animals died and the site became a wasteland of dying plants and deserted homes. Huge amounts of top-soil were removed from the site. The incident prompted the European Community to adopt, in 1982, a Directive aimed at preventing such major chemical accidents (the so-called 'Seveso' Directive, now constituted as the adapted *Control of Major Accident Hazards – 'COMAH' – Directive 2012/18/*EU).[18]

20.2.2   Bhopal, India

The effects of the incident at Bhopal, India, were directed almost entirely onto the population living around the factory involved in a catastrophic release of methyl isocyanate (MIC). The accident occurred on 3 December, 1984 at the Union Carbide factory, which had been producing the insecticide carbaryl for about eighteen years. MIC was one of the main ingredients, and was produced from monomethylamine (MMA) and phosgene, the latter being produced on site by reacting chlorine and carbon monoxide. The MMA and chlorine were brought by tanker from other plants in India, stored and used when required. Chloroform was used as a solvent throughout the process. Thus there was in this plant, a variety of extremely toxic materials in use. On the night of the accident, it seems that some water inadvertently got into a tank where 41 tonnes of MIC were being stored, causing a runaway chemical reaction. The heat of the reaction, possibly augmented by reactions with other materials present in the tank as contaminants, produced vaporization of such momentum that it could not be contained by the safety systems. The safety valve on the tank blew open and remained open for about two hours, allowing MIC in liquid and vapour form to escape into the surroundings. The prevailing wind carried the cloud towards the north of the plant, and then towards the west, affecting approximately 100 000 people living in the vicinity. There were at least 2000 deaths and mortality remained elevated among the most severely exposed population. None of the workers on night duty at the plant were harmed. The most frequent symptoms in those who survived were burning of the eyes, coughing, watering of the eyes, and vomiting.[19] Very many individuals have continued to suffer physical and mental trauma as a consequence of this tragedy. Follow-up studies observed significant neurological, reproductive, neurobehavioral, and psychological effects.[20]

20.3   LOCALIZED CONTAMINATION INCIDENTS

Most incidents in which local communities have suffered overt signs of toxicity have involved exposure to food contaminants. Well-known examples covered in this section include Toxic Oil Syndrome in Spain and methylmercury poisoning in Iraq. In the UK, incidents have occurred in Epping (Greater London) and in North Cornwall, where the water supply was contaminated. There have been other examples where contamination has occurred indirectly as a result of an environmental pollutant entering the food chain, for example at Minamata and Niigata, Japan. The best known example of health effects being associated with environmental contamination around waste disposal sites is Love Canal in the USA, although there have been more recent concerns sparked by findings of increased incidences of congenital abnormalities around landfill sites in Europe.

20.3.1   Toxic Oil Syndrome

Oil fraudulently sold as olive oil was at the heart of a severe poisoning episode in Spain,[21] although the toxic agent responsible has not been identified satisfactorily to this day. The syndrome manifested itself in May 1981 in an eight year old boy who died with acute respiratory insufficiency. Very soon, the same symptoms were generalised in patients admitted to hospitals in Madrid, and fewer cases were found in provincial hospitals. The final toll was about 20 000 persons affected, nearly 300 deaths – a case fatality rate of about 2% – and many individuals developing chronic disease (*e.g.* scleroderma, neurologic changes).[22]

   The illness began with a fever, followed by severe respiratory symptoms and a variety of skin rashes, which led some of the victims to be diagnosed as having measles or German measles (rubella). Many of the patients developed signs of cerebral oedema (swelling of the brain) and many had cardiac abnormalities.

   The cause of the disease was traced to adulterated cooking oil which was fraudulently sold to the public as pure olive oil. The oil was sold by door-to-door salesmen in five-litre plastic bottles with no labels. Because olive oil is an expensive commodity in Spain, it was the poorer families in the working class suburbs of Madrid who almost entirely bore the brunt of the disease episode. The composition of the oil varied but rapeseed oil accounted for up to 90%. There were varying amounts of soya oil, castor oil, olive oil, and animal fats. The oil also contained between 1 and 50 ppm of aniline and between 1500 and 2000 ppm of acetanilide.

It seems that those who perpetrated the fraud tried to refine out the aniline and in doing so produced a number of other chemical species. One of these was acetanilide, which then reacted with fatty acids in the oil to produce oleic acid anilide, which was originally presumed to be the toxic agent.[23] Later work, however, showed the presence of a number of other anilides. The fatty acid esters of 3-(*N*-phenylamino)-1,2-propanediol, such as 1,2-dioleoyl ester of 3-(*N*-phenylamino)-1,2-propanediol, stand out as new markers of the toxicity. The other aniline compounds were considered to be hydrolysis products of diacyl propane-1,2-diol-3-aminophenyl or positional isomers. It has been suggested that these compounds may have been responsible for the symptoms produced by the toxic oil,[24] but despite much effort the precise toxicant or combination of toxicants has not definitively been identified.[25]

20.3.2 Rice Oil Contamination by Polychlorinated Biphenyls (PCBs)

Episodes of human poisoning with PCBs have occurred in Japan and in Taiwan. The disease first made its appearance in 1968 in the western part of Japan, when a number of families were noted to have developed chloracne. Epidemiological studies brought other cases to light and it was found that the factor which the cases had in common was exposure to a particular batch of one brand of rice oil. Chemical analysis showed that the oil was contaminated with PCBs.

The PCBs, commonly used then as an industrial heat-transfer fluid, were shown to have leaked into the oil from equipment which had been used to process the oil. By the end of 1977, 1665 individuals were considered to have met the diagnostic criteria for what has come to be known as 'Yusho disease'.[26]

In addition to chloracne, the patients with Yusho disease had a number of systemic complaints, including loss of appetite, lassitude, nausea and vomiting, weakness, and loss of sensation in the extremities. Some also had hyperpigmentation of the face and nails.

The patients were followed up from 1969 to 1975, and in 64% of cases the skin lesions improved. A number of non-specific symptoms persisted, however, including a feeling of fatigue, headache, abdominal pain, cough with sputum, numbness and pain in the extremities and changes in menstruation in women. Objective findings included a sensory neuropathy, retarded growth in children, and abnormal development of the teeth. Children who had been exposed *in utero* had lower birth weight and were hyperpigmented.

Some patients were found to be anaemic and some had other abnormalities, but the most striking observation was a marked increase in serum triglyceride levels. The mean value in the patients was 134 ± 60 mg per 100 ml compared with a mean of 74 ± 29 mg per 100 ml in normal controls. When the PCB concentrations in the serum were measured by gas chromatography it was found that the patients with Yusho disease had an isomeric pattern which was different from that seen in controls whose only exposure had been from the general environment.[27]

Yusho disease appeared in Taiwan in the spring and summer of 1979 in two prefectures in the middle part of the country. The signs and symptoms were indistinguishable from those seen in the Japanese outbreak and by the end of 1980 more than 1800 people had been affected. The source of the PCBs was again contaminated rice oil. The blood levels of those with the disease ranged from 54 to 135 ppb. Later studies showed that polychlorinated dibenzofurans (PCDFs) and polychlorinated quaterphenyls (PCQs) were also present in the blood.[28]

The contribution of PCBs to the development of the Yusho disease symptoms it is not known, since subjects were also exposed to PCDFs and to PCQs formed when PCBs are heated. In animal models, PCDFs and PCQs are more toxic than PCBs and it seems that there may have been some synergistic effects between the different compounds in the oil. The fact that the PCB isomers in the oil were different from those in the general environment may also be of importance since they may have been more toxic than those to which the population at large is ubiquitously exposed.

20.3.3 Polybrominated Biphenyls (PBBs) in Cattle Feed

PBBs are used mainly in plastics as a fire retardant. In May and June of 1973, some ten to twenty bags of PBB were sent in error (instead of a livestock food additive) to a grain elevator in the state of Michigan, USA. The chemical company which made the PBB normally supplied magnesium oxide to go into the cattle feed but both products were packed in the same colour bag and although the PBB was labelled 'Firemaster' rather than 'Nutrimaster', and although this difference was actually noted by the staff at the grain elevator, it was nevertheless incorporated into the feed and distributed throughout the state to be fed to the cows.

Reports of sick cows began to surface in August, 1973 and towards the end of the year it was realized that the feed was to blame. Despite this the contamination continued, both because there was cross-contamination of otherwise normal feed from the grain elevator and because the tainted feed was resold at a discount after it had been returned to the suppliers. Not until PBB was formally identified in the feed in May 1974 was any attempt made to limit the contamination.

From the time that the feed had become contaminated, dairy products containing PBB had been sold throughout Michigan and cows and other livestock which had been given it had been slaughtered for meat. A representative sample of 2000 people was surveyed and more than half had a concentration of PBB in their fatty tissues exceeding 10 ppb.[29] Farmers and others who consumed produce directly from contaminated farms had the highest levels of PBB.

A study carried out in which over 1000 farmers were compared with unexposed farmers from Wisconsin found some adverse effects in the exposed group such as acne, dry skin, hyperpigmentation, and discoloration of the nails. The exposed group also complained more of headaches, nausea, depression, and a number of other non-specific symptoms. Serum levels of hepatic enzymes were higher in the Michigan farmers than their neighbouring controls. Individuals with symptoms were also shown to be more likely to have elevated enzyme levels. Changes in the immune system were also found [30,31] and some individuals had enlargement of the liver and a sensory neuropathy.[32] In follow-up studies, the PBB levels in the serum were found to have decreased, but it was interesting that elevated PCB levels were seen and that these were actually higher than the PBB levels. While there was no relationship between abnormal liver function tests and serum PBB concentrations, there was a slight (statistically insignificant) negative correlation with serum PBB levels and some tests of thyroid function.[33]

It is noteworthy that in none of these studies did the subjective or the objective health effects findings correlate with serum or fat PBB concentrations. This may have been because there was another toxic contaminant present which was acting independently of the PBB, or that the levels of PBB had fallen in the interval between ingestion and the beginning of the studies. It may also be the case that levels of PBB in blood and fat are not good indicators of levels in target organs.

Two other scares involving contaminated animal feed – and the subsequent transfer of pollutants into the food chain – have occurred more recently. Both these incidents occurred in Belgium, first in May 1999, when dioxins from contaminated fats were involved, and then again in May 2000 after the discovery of high levels of PCBs in animal feed. The dioxin scare resulted in the widespread recall of products, with countries around the world banning the import of Belgian food products.

20.3.4 Mercury Poisoning in Minamata and Niigata

Mercury in its organic form has accounted for many of the episodes of endemic disease resulting from environmental exposure to this metal. Probably the best known of these is Minamata Bay disease.[34] This disease was first noted at the end of 1953 when an unusual neurological disorder began to affect the villagers who lived on Minamata Bay. About 700 people from both sexes and all ages were affected and presented with a mixture of signs relating to the peripheral and central nervous systems. The prognosis of the condition was poor; many patients became disabled and bedridden and about 40% died. The disorder was associated with the consumption of fish and shellfish caught in the bay and contaminated by mercury. The source of the mercury was related to the effluent released into the bay from a factory which was manufacturing vinyl chloride using mercuric chloride as a catalyst. Although it was claimed that the factory released *inorganic* mercury, which was methylated by microorganisms living the in sediments in the bay, this reaction is very slow. The likely source was that the mercury was actually released in the organic form, as at the time there were no regulations forbidding this in Japan.

A second outbreak of methylmercury poisoning occurred in Japan in 1965 in Niigata, affecting a further 500 or so individuals. This outbreak was sourced to the consumption by the local population of fish in which mercury had bioconcentrated after contamination of the Agano River by industrial effluent.[35]

20.3.5 Methylmercury Poisoning in Iraq

A major poisoning episode occurred in 1971–72 when the Iraqi government imported a large consignment of seed grain treated with an alkylmercury fungicide, which was then distributed to the largely illiterate rural population. The distribution was accompanied by warnings that the seed was for sowing not for eating, and the sacks were marked with warning labels in English and Spanish, but not in the local language. The seed had been treated with a red dye to distinguish it from edible grain, but the farmers found that they could remove the dye by washing and they equated this with the removal of the poison. The grain began to be used to make bread in November 1971 and the first cases of poisoning appeared in December. By the end of March 1972 there had been 6530 admissions to hospital and 459 (7%) of these had died.[36] These represented only the most severe cases and the true extent of this outbreak will probably never be known, although it has been suggested that the incidence of the disease may have been as high as 73 per 10 000.[37]

20.3.6 Aluminium Contamination of Drinking Water in North Cornwall

In July 1988, about 20 tonnes of aluminium sulfate were accidentally deposited into the treated water reservoir at Lowermoor Treatment Works, resulting in major contamination of the drinking water supply to Camelford in Cornwall and surrounding district.[35] Aluminium levels increased to over 10 mg L^{-1}, well above the 0.2 mg L^{-1} limit set by the EC on palatability grounds, and the pH of the water dropped below 5.0.

Despite reassuring messages from local sources, residents and holiday makers reported a large number of acute symptoms, and speculation about longer term effects was rife. The expert assessment of the situation, made by the Lowermoor Incident Health Advisory Group,[38] was that the early reported symptoms of gastrointestinal disturbances, rashes and mouth ulcers were indeed probably due to the incident, but that these effects were short-lived. Persistent toxic effects were thought unlikely because of the transient nature of the exposure and because all the known toxic effects of aluminium in man are associated with prolonged exposure.

However, a significant time after the incident, hundreds of people resident in the Lowermoor area continued to attribute health complaints to the contamination event. Such complaints included joint and muscle pains, malaise, fatigue and memory problems. Also, various scientific studies reported unexpected symptoms and clinical findings. The Advisory Group reconvened and concluded that, based on all the evidence available to it, some of the continuing symptoms experienced by the residents were probably induced by the sustained anxiety naturally felt by many people, and that others were wrongly attributed to the incident as a result of heightened awareness provoked by the incident and subsequent events. It was, however, recommended that a developmental follow-up of children exposed *in utero* during the incident should be undertaken, together with formal testing of the possibility of particular individual 'sensitivity' to aluminium.[38]

A complication of this incident was that the low pH of the drinking water resulted in other metals – copper, zinc, and lead – being dissolved from domestic plumbing, and flushing of the mains distribution system to remove the contaminated water also resulted in the disturbance of old sediments, mainly deposits of iron and manganese oxides.[39] Therefore various additive or synergistic effects may have occurred.

A subsequent revisit of the Camelford incident[40] in 1993 confirmed that a number of individuals showed consistent evidence of impaired information processing and memory, with no obvious relationship with measurement of anxiety and depression. Although the abnormal neuropsychological findings indicated cognitive impairment, it was uncertain whether this was caused by an acute episode of brain damage or other causes of stress resulting from the accident. A more recent study, completed some 10 years after the event,[41] investigated 55 people affected by the incident and found objective evidence of damage to cerebral function which was not related to anxiety. The authors suggested that the aluminium exposure was indeed the cause of the observed changes. However, the validity of this study has not gone unchallenged, with major criticisms relating to the bias that is inherent in the self-selection of cases. A recent report of the Committee on Toxicity of Chemicals in Food, Consumer Products and the Environment on the Lowermoor water pollution incident concluded that a causal relationship could not be found for neurotoxic effects in adults or those who were children at the time of the incident; that there was no indication of a causal effect on joint pains and/or swelling problems; that there was no evidence of an increased overall cancer risk; nor an adverse effect on the thyroid gland. A dermatologist consultant who examined individuals suffering from nail and skin problems reported that the symptoms reported were common and no further investigation was required. The investigation of the committee also concluded that no conclusions could be drawn from the higher proportion of children with Special Educational Needs or of the long-term impact of the incident on health. Nonetheless, the Committee recommended continuing with follow-up neuropsychological studies, with investigations of the cognitive, behavioural and educational development of children and that routine monitoring of the health of population exposed during the incident should continue.[39]

20.3.7 'Epping Jaundice' – Chemical Contamination of Food during Storage

The ingestion of contaminated wholemeal bread resulted, in February 1965, in an unusual outbreak of jaundice in Epping, UK.[35,42] The outbreak, which affected at least 84 people, was traced to the contamination of flour by 4,4-diaminophenylmethane. This had spilled from a container onto the floor of a van transporting both the flour and chemicals. The chemicals were absorbed by the flour through the sacking, and the flour was subsequently used to make the bread.

Jaundice and liver enlargement were preceded, in most cases, by severe pains in the upper abdomen and chest. The pains were mostly of acute onset, but in some patients the onset was insidious. Raised levels of serum bilirubin (a blood breakdown product) and the enzymes alkaline phosphatase and aspartate aminotransferase were recorded in most of the 57 patients further investigated. Needle biopsies of the liver showed considerable evidence of inflammation and cholestasis and damage to liver cells. Experimental studies subsequently showed the liver lesions to be reproducible in mice following administrations of 4,4-diaminophenylmethane.[43] All the patients eventually recovered.

20.3.8 Love Canal, USA

The reporting of chemical odours in the basements of homes in the Love Canal district, USA, led to a toxicological investigation which made this area famous in the history of waste disposal and resulted in much regulatory activity in the USA (and subsequently in other countries of the world).

Love Canal was a waste disposal site containing municipal and chemical waste disposed of over a 30 year period up to 1953.[35] Homes were then built on the site during the 1960s, and leachates began to be detected in the late 1960s. Dibenzofurans and dioxins were among the chemicals detected in the organic phase of the leachates. Animal studies indicated possible risks of immunotoxic, carcinogenic and teratogenic effects.[44,45] The episode resulted in significant fears of ill-health and much psychological stress. Limited follow-up of residents identified low birthweights in the offspring of Love Canal residents,[46] but no causal link has been established for cancer incidence in the area.

Love Canal provides an instance of considerable public anxiety and stress resulting from the identification of a potential environmental toxic hazard. There is evidence that the psychological and other consequences of such incidents might easily outweigh the actual toxic effects of chemical exposure.[47,48] Indeed, having reviewed the literature on hazardous waste disposal sites, Grishan[49] concluded that "there are few published scientific reports of health effects clearly attributable to chemicals from uncontrolled disposal sites".

Concern about possible health impacts of landfill sites has resurfaced in recent years with the publication of studies conducted in the UK and elsewhere in Europe[50,51] that suggest an association between congenital malformations and residence close to landfill sites. However, these studies have produced various anomalous findings and provided no evidence for a causal relationship. It is clear that further work is needed to investigate the question of causality.

20.3.9 Exxon Valdez, MV Braer, Prestige and other Major Oil Spill Accidents

In the last five decades there have been more than 38 accidents involving supertankers in different countries. The first heavily studied major spill accident occurred in Alaska, when the *Exxon Valdez* spilled 37 000 tons of crude oil on the 24th of March 1989. However, the major spills have occurred in UK, where the *MV Braer* spilt 85 000 tons of crude oil in the Southwest Shetland Islands on the 5th of January 1993; and the *Sea Empress* spilt 72 000 tons in front of the cost of Milford Haven. The Bay of Biscay had also its toll, with *Erika* spilling 20 000 tons in South Penmarch (Britanny, France) on the 12th of December 1999 and the *Prestige* on the 19th of November 2002, spilling 63 000 tons of crude oil in front of the Galician coasts (Spain). Other important spillages occurred in Pakistan (Tasman Spirit; 37 000 tons on July 2003)[52] and more recently the accident occurring on the Deepwater Horizon oil rig in April 2010 in the Gulf of Mexico with 627 000 tons, which is considered the largest accidental marine oil spill in the history of the petroleum industry.

The oil spillages have affected the local flora and fauna of the nearby coasts, with the potential to access the food chain. Nevertheless, the local volunteer inhabitants who had mobilised to clean-up the spillage from the coast are considered the population most exposed, whose health may potentially be affected by the toxic properties of the oil.[52] The crude oil contains polycyclic aromatic hydrocarbons (PAH), volatile organic compounds (VOC), and some heavy metals (*e.g.* Zn, Ni, V).[53] Acute exposure to PAHs is known to cause respiratory symptoms, damage the skin and mucous membranes and have been associated with skin cancers. Certain VOCs, in particular benzene, have been associated with blood cancers. The most prevalent health effects related to oil spillage exposure are nervous symptoms, skin and mucous irritations and also some psychological effects. Evidence from clean-up volunteers and fishermen exposed to the *Prestige* spillage shows genotoxic damage and endocrine alterations.[52]

A follow-up report on the health effects of clean-up workers from the *Exxon-Valdez* showed that they were at higher risk of having anxiety disorder, post-traumatic stress disorder, depression,[53] respiratory problems and dermatitis than the non-exposed population. A survey conducted 14 years after the clean-up showed greater prevalence of symptoms of chronic airway disease, as well as self-reported neurological impairment and multiple chemical sensitivity.[54] A survey on 3669 clean-up workers from the *Erika* spill found that 53% of the exposed population reported cases of headache, rash, eye redness, respiratory problems, nausea and abdominal pain. The risk of skin cancer was also increased, especially in those volunteers that cleaned up birds, as they did not use gloves to avoid damaging the feathers. Similar symptoms were associated with the *Prestige*[53] and the *Deepwater Horizon*[54] spillage clean-up volunteers, which included conjunctivitis, headache, sore throat, breathing difficulty, vomiting, skin rash and abdominal pain. A detailed study on fishermen involved in the clean-up of the *Prestige* two years after the accident showed a persistent airway injury with increased respiratory symptoms such as bronchial hyperresponsiveness, increased oxidative stress and inflammation of the airways – shown by higher 8-isoprostane levels in exhaled breath condensate (EBC) – and increase growth factors in EBC. This was associated with the fact that the constituents of the oil act as a respiratory irritant, which might have contributed to respiratory oxidative changes, which elicit a process of airway-wall remodelling. This population also showed an increase risk for structural chromosomal alterations in circulating lymphocytes, which is an early genotoxic marker associated with an increase risk for cancer.[55]

20.4   GENERALIZED ENVIRONMENTAL POLLUTION

While local disasters and incidents such as those described above are often significant and can tell us much about the consequences of high level, acute or sub-chronic exposure to environmental contaminants, of greater significance to public health are the lower level chronic exposures to air pollutants and other substances that are more commonly encountered or more widely dispersed in the environment. Human exposure to these pollutants can occur through inhalation, ingestion (from contaminated food and drink) or absorption through the skin and mucous membranes (from soil or dust, for example). Dioxins, PCBs and DDT are well known examples of chemicals that are now ubiquitous in the environment but whose impact on human health remains uncertain. This section details the potential impact on health of indoor air pollutants, toxic metals, asbestos and other fibrous materials in the environment, pesticides, an ill-defined group of substances commonly referred to as 'endocrine disrupters', and finally the novel engineered nanomaterials, for which little information is yet available.

20.4.1   Indoor Air Pollution

The presence of noxious substances in the outside air, coming from factories, domestic fuel combustion and vehicle exhausts, continues to be a major cause of concern. In terms of overall morbidity and mortality, probably nothing can surpass the harm which has been caused historically by exposure to the by-products of coal burning, and industrial emissions trapped in stagnant air by temperature inversions. There have been episodes in which sudden increases in the number of recorded deaths have followed exceptional air pollution events. For example, episodes in the Meuse Valley in 1930 and in Donora in1948, which were the consequence of industrial pollution trapped by thermal inversions, caused 60 and 20 deaths attributed to air pollution, respectively. Nonetheless, the Great Smog of London in 1952 has been the most disastrous air pollution incident. The smog originated from coal combustion emissions trapped by a thermal inversion and the estimated toll of the London smog was an overall excess in daily deaths of 3500 to 4000 (see Chapter 11).[56] Even though the levels of sulfur dioxide and smoke seen during these episodes are no longer experienced (at least in the developed world), there remains considerable concern that present day air pollution, most notably by airborne particulate matter – commonly measured as PM_{10}[‡] or $PM_{2.5}$ – continues to have a real and measurable impact on human health. These concerns about outdoor air quality are reflected in continuing discussions of the health effects of vehicle emissions, calls for tighter emission controls and the setting of standards for the common air pollutants. In the UK, there has been also an extensive programme of government funded research committed to understanding better the impacts and mechanisms of effect of air pollution.[57]

Chapter 11 of this book is dedicated to the health effects of outdoor air pollution, but there is also a need to concentrate on a compartment of the environment where people may spend up to 90% of their time – the indoor environment. Although outdoor air quality has a considerable influence on indoor air, for many pollutants the greatest proportion of total exposure is determined by exposures indoors. Indoor pollution sources, such as gas cookers, can result in pollutant levels that are much higher indoors than out, and some pollutants are found only indoors. The impetus over recent years to conserve energy has resulted in 'tighter' buildings with much reduced air exchange and therefore a greater propensity for indoor pollutants to build up.

People are exposed to a wide variety of indoor air pollutants both in their workplace and at home. Concern is directed especially to domestic air quality since the home is where many non-healthy individuals, and the very young and old, spend much of their time. Pollutants indoors arise from a variety of sources, most notably from the combustion of fuel for cooking and heating, emissions from building products and furnishings, and from the use of DIY and consumer products.[58]

This section briefly reviews the main chemical indoor air pollutants found in homes. Radon (a natural radioactive gas) and house dust mites, bacteria and fungi (biological agents), whilst important, are not included here. The health effects related to exposure to organohalogen and organophosphorus organic compounds used as flame retardants indoors is covered in Chapter 18.

20.4.1.1 Carbon Monoxide

Carbon monoxide is one of the most important indoor air pollutants[59] and continues to kill as many as 50 people a year in the UK through accidental poisoning, and 200 are left seriously ill.[60] It is especially dangerous since it has no colour, smell or taste. Carbon monoxide is produced when fuel burns with an inadequate supply of oxygen. High concentrations can be produced by faulty, incorrectly installed, inadequately maintained or poorly ventilated gas and solid fuel appliances (cooking or heating),[61] paraffin (kerosene) heaters, *etc.*, or by fumes leaking from a flue into a poorly ventilated room. Its toxic action is primarily through hypoxia, *i.e.* the displacement of oxygen in haemoglobin in the blood to form carboxyhaemoglobin, thus depriving the tissues of the body of their oxygen supply. Early symptoms of exposure include tiredness, drowsiness, headaches, dizziness, pains in the chest and stomach pains. There can be severe and permanent damage to the central nervous system. Excessive exposure can lead to loss of consciousness, coma and death. Exposure of pregnant women can also result in adverse effects on the foetus. Long term effects of chronic exposure in adults have been linked with physical symptoms, sensory-motor changes, cognitive memory deficits, emotional-psychiatric alterations, cardiac events and low birth weight.[61] People thought to be most at risk are those with existing cardiovascular disease or individuals with compromised blood oxygen levels, including pregnant women and the elderly. Epidemiological studies have demonstrated increased incidences of low birth weight, congenital defects, infant and adult mortality, cardiovascular admissions, congestive heart failure, stroke, asthma, tuberculosis, pneumonia among others.[61] Surprisingly, many details of carbon monoxide toxicity remain unresolved, especially those relating to chronic low level exposure.

[‡]Particulate matter of aerodynamic diameter less than 10 μm.

Signs and symptoms of acute carbon monoxide poisoning are often confused with those of food poisoning and this can result in misdiagnosis, with sometimes tragic consequences. Many fatal cases of carbon monoxide poisoning result from the accidental blockage of flues, but leakage of combustion products into the room air and misuse of fuel burning appliances are also important causes.

20.4.1.2   *Environmental Tobacco Smoke (ETS)*

Where it occurs, this is another very significant indoor air pollutant. It contains tar droplets and a cocktail of various other toxic chemicals including carbon monoxide, nitric oxide, ammonia, hydrogen cyanide and acrolein, together with proven animal carcinogens such as *N*-nitrosamines, polycyclic aromatic hydrocarbons and benzene. Environmental tobacco smoke can irritate the eyes, nose and throat, and exposed babies and children are more prone to chest, ear, nose and throat infections. Women exposed during pregnancy tend to have lower birthweight babies, and asthmatics may be adversely affected by acute exposures. The Californian EPA concluded that causal links have been established between ETS exposure and heart disease, lung cancer and nasal sinus cancers in adults, and sudden infant death syndrome (SIDS), asthma and middle ear disease in children.[62]

20.4.1.3   *Nitrogen Dioxide (NO$_2$)*

NO$_2$ and other oxides of nitrogen are formed when fuel is burned in air. Thus NO$_2$ is generated indoors by gas, oil and solid fuel appliances. The main sources are unflued appliances such as gas cookers, gas wall heaters and kerosene (paraffin) heaters. Indoor levels are significantly influenced by outdoor levels, but where an indoor source is present this tends to dominate. Exposure to high levels typically occurs in the kitchen during gas cooking. Nitrogen dioxide can irritate the lungs, but the mechanisms of toxic action at lower levels remain to be fully elucidated, and there is continued uncertainty and debate about the actual impact on the health of occupants of NO$_2$ levels as typically found indoors. The main health effects associated with indoor exposure to NO$_2$ are respiratory symptoms, bronchoconstriction, increased bronchial reactivity, airway inflammation, and decreases in immune defence, which produces increase sensitivity to airway infections. Overall, the weight of published evidence points to a possible hazard of respiratory illness in children, perhaps resulting from increased susceptibility to infections.[63] There is suggestive evidence of synergistic effects of NO$_2$ with allergens, such as house dust mites,[64] and increase asthma severity following respiratory viral infection. Children and asthmatics are suggested to be the most sensitive populations affected by levels of indoor NO$_2$.[61]

20.4.1.4   *Formaldehyde*

Formaldehyde[63] is a colourless gas with a pungent odour that is given off from various furnishings and fittings found in the home. One of the most important sources is pressed wood ('chipboard'), made using bonding materials containing urea–formaldehyde resin, which has become increasingly used in furniture items over the last few decades. Another major source is urea–formaldehyde foam insulation (UFFI) installed in wall cavities. Formaldehyde is also generated during the combustion of fuel and is a component of cigarette smoke. Formaldehyde gas can irritate the mucous membranes of those exposed. The odour threshold is in the region of 0.05–1.00 ppm and levels found in some homes may reach the threshold, for some individuals, for transient eye, nose and throat irritation. Although formaldehyde is a sensitizing agent, no studies to date have demonstrated an increased sensitivity to sensory irritation to formaldehyde among people generally considered susceptible, such as asthmatics, children or older people.[65] Certain individuals do appear peculiarly sensitive to and intolerant of formaldehyde exposure.

20.4.1.5   Volatile Organic Compounds (VOCs)

VOCs[§] in the indoor environment originate from a number of sources including smoking, furnishings, furniture and carpet adhesives, building materials, cosmetics, cleaning agents and DIY materials. VOCs also originate from fungi, tobacco smoke and fuel combustion. By far the greatest peak exposure to VOCs occurs during home decorating using solvent-based paints. Glues are another important source of high peak levels. It has been estimated that between 50 and 300 different compounds may occur in a typical non-industrial indoor environment,[58] including aliphatic and aromatic hydrocarbons, halogenated compounds and aldehydes. Because of the diverse range of chemical substances defined as VOCs, determination of health effects is problematic. However, it is known that at levels typically found indoors the major effects are likely to be sensory. Short exposures to high levels of solvent vapours can cause temporary dizziness; lengthy or repeated exposure can irritate the eyes and lungs and may affect the nervous system.[63] Some VOCs, such as benzene, are carcinogens causing blood dyscrasias and leukaemia.[66]

VOCs have been associated by some with the syndromes known as 'sick building syndrome' and 'multiple chemical sensitivity'[67] and it has been postulated that they may react with ozone and other substances to produce more toxic compounds.[68]

20.4.1.6   Polycyclic Aromatic Hydrocarbons (PAH)

PAHs[**] sources in the indoor environment include smoking, cooking, domestic heating with fuel stoves and open fireplaces, as well as from incense and candle emissions. There is evidence that exposure to PAHs leads to DNA damage, cytotoxic and oxidative stress, increased PAH–DNA adducts, DNA strand-breaks and impaired DNA repair capacity. Children exposed *in utero* to PAH have been associated with a lower birthweith. Benzo[a]pyrene (BaP) is one PAH known to be a carcinogen. Exposure to PAHs has been associated with lung cancer, whilst there is some evidence relating PAH exposure to bladder and breast cancer. There is limited evidence that exposure to PAH, including benzo[a]pyrene increases the risk of ischemic heart disease.[69]

20.4.2   Metals

Episodes or incidents involving exposure to mercury and aluminium, mostly through ingestion, have already been described. Added to this is the evidence for more widespread environmental contamination by metals such as mercury, cadmium and lead. For example, the lead concentration in polar ice has increased over 20-fold since 1800, and tissue levels of lead and mercury in Greenland Inuit Eskimos are 4–8 times higher than in preserved ancestors from five centuries ago. Industrial sources of metals include refineries, chemical plants, cement manufacturing, power plants, smelters and incinerators. Until recently, vehicles were a major source of environmental lead through the use of tetraethyllead as an anti-knock additive in petrol.

Metals have a variety of effects upon the human body, mostly at the cellular level. Some metals disrupt biochemical reactions while others block essential biological processes, including the absorption of nutrients. Some accumulate in the body giving rise to toxic concentrations after many years of exposure, and yet others (including arsenic, beryllium, cadmium and chromium) are carcinogens. Exposure to methylmercury and high levels of lead can cause gross developmental deformities.[62]

The health effects of lead have been studied extensively and are well documented.[70] Environmental lead exposure has been linked to reduced IQ in children and to elevated blood pressure in

---

[§]A wide range of compounds with boiling point between approximately 50 °C and 250 °C and which, at room temperature, produce vapours.
[**]PAH are a broad group of organic compounds that contain two or more fused aromatic rings.

adults, although the cause–effect relationship for the latter is not especially robust.[71] Airborne lead exposure has been much reduced in recent years by the introduction in many countries of unleaded gasoline. However, other routes of exposure (most notably ingestion) remain significant. The outbreaks of endemic lead poisoning in classical and historical times are well known: suffice it to say that the adulteration of food and drink with lead has been a significant contributor to morbidity and perhaps to mortality in the past. The most serious form of endemic lead poisoning arose from the habit of adulterating wine with lead to improve a poor vintage and make it more saleable. During the eighteenth century there were a number of famous outbreaks of lead poisoning, the 'Devonshire colic' being perhaps the most well known. In this case, the adulteration of cider with lead arose accidentally due to the presence of lead in the pounds and presses used to make the cider. The widespread use of lead in cooking utensils, in glazes, and in pewter added to the burden of lead exposure during the eighteenth and nineteenth centuries.[72] Most human exposure to lead in recent times has arisen from the use of lead in domestic water pipes.

Environmental exposure to cadmium was considered to be the cause of a disease which was first reported in 1955 as occurring in a localized area downstream from a mine on the Juntsu River in the Toyama Prefecture in Japan. The condition was almost entirely confined to elderly women who had borne several children. It was characterized by severe bone pain, waddling gait, severe osteomalacia (bone softening), pathological fractures, and some signs of renal impairment. The water which was used to irrigate crops was frequently contaminated by outpourings from the mine which contained zinc, lead, and cadmium. Levels of cadmium in rice samples were shown to be about ten times the amount normally present and the view was gradually formed that it was the cadmium which was responsible for the disease.[73] However, it seems likely that deficiencies of calcium and vitamin D were also at least partially to blame and that cadmium may have been acting only as one factor in what was a multi-factorial aetiology for this disease. Concern is occasionally raised about high levels of cadmium in soils (due either to naturally high levels or historical industrial or mining activity) and the effects this might have on the health of the local population.[62] An example of this has occurred in Shipham, UK.[74]

People who depend upon fish as the staple part of their diet may still be at risk of excessive exposure to mercury, even if not on the scale experienced at Minamata and Niigata. For example, blood methylmercury levels have been found to be almost ten times higher in a Peruvian population who ate on average 10.1 kg of fish per family of (on average) 6.2 persons, compared with a control population whose fish intake was considerably more modest. (The mean in the high fish eating population was 82 ng ml^{-1} and in the control population, 9.9 ng ml^{-1}.) Moreover, 29.5% of the heavily exposed population had signs of a sensory neuropathy.[75] In a fish-eating population in New Guinea, hair mercury concentrations were between two and three times that of a control group (6.4 µg g^{-1} compared with 2.4 µg g^{-1}); there were, however, no demonstrable ill effects in this group.[76]

20.4.3   Asbestos and Man-made Mineral Fibres (MMMF)

The industrial exploitation of asbestos began about a hundred years ago, its use probably peaking in the 1960s. Because the health effects of asbestos were unsuspected, workers in the industry were often exposed to massive levels of airborne fibres, resulting in serious, and often fatal, pulmonary interstitial fibrosis (termed 'asbestosis'), lung cancer and mesothelioma (cancer of the epithelium lining the chest cavity). First reports of these diseases started to appear in the 1900s, 1930s and 1960, respectively.[77,78] The work of Wagner in South Africa in the 1950s and 1960s was especially significant in assessing and understanding the health impacts of asbestos fibres. These health concerns resulted in a reduction, at least in the developed world, of asbestos use. The commercial materials marketed as 'asbestos' were from both the amphibole and serpentine groups of silicate minerals. Those from the amphibole group were identified as being particularly hazardous, and crocidolite (blue asbestos) use was discontinued in the UK following the voluntary ban on its import in the 1970s. Import of amosite ceased in 1980. Use of chrysotile (white asbestos; serpentine group) has also significantly declined but until recently was still allowed in many products, most notably asbestos cement and sheeting, and in friction products such as brake linings.[62] In Europe, at least, there are now formal moves to remove chrysotile from the market because of it has also been related with mesothelioma and lung cancer,[79] and the availability of less hazardous substitutes.[80]

Because asbestos is a cheap and effective insulator, fire retardant and reinforcing material, its use became widespread and it was incorporated into a wide range of products. The demonstrated health risks of asbestos have resulted in considerable concern about environmental exposure to asbestos fibres as well as worries about the legacy of historical occupational exposures. Exposure to asbestos fibres in smokers is especially relevant because of the synergistic interaction that has been demonstrated between asbestos and smoking in the induction of lung cancer.[81] Of particular interest is the evidence that, contrary to expectations, the age-specific incidence of mesothelioma in men in the UK, instead of falling in parallel with the decrease in asbestos use, is actually increasing.[82] If the projection by Peto *et al.*[82] is accurate, British mesothelioma deaths will approach 3000 per year by 2020. This is in addition to the several hundred deaths per year ascribed to asbestosis and a similar number from lung cancer. Although most of the reported deaths are a result of previous occupational exposure, the level of mortality and morbidity has given rise to concern about environmental exposures, not only to asbestos but to other fibrous materials with similar characteristics.

As the consumption of asbestos has fallen, so the production and use of manmade mineral fibres has steadily increased.[83] While some of these materials have specifically been developed as asbestos substitutes, many of the uses of these fibres are new and unrelated to the historic use of asbestos. The man-made fibres being used in the largest amounts are the insulation wools (glass or fused rock products), which are used for both thermal and acoustic insulation. Energy conservation drives in the last decade or so have resulted in very widespread use of these materials in homes as loft and cavity wall insulation. They also have important applications in horticultural products, and smaller amounts of slag wool and ceramic fibre are used in more specialized applications.

In considering the health hazards posed by asbestos and man-made mineral fibres, the prime features of importance are the physical diameter and shape of the liberated fibres, and their persistence in the lung; the chemical composition of fibres, except insofar as it influences these properties, is generally considered less important, although the crystalline habit seems to be relevant, and some scientists believe that the propensity to give rise to oxygen free radicals is a major factor. Size and shape are important because, to have effect, fibres must be fine enough to penetrate the deep lung; and the fibrous shape affects deposition and inhibits removal by the lung's natural defence systems. Many man-made fibres (notably those used domestically) pose little or no risk because they are too coarse to gain access to the lower respiratory tract, or do not liberate fibrous dust during normal handling and use, or are highly soluble and therefore do not persist in the lung. All these factors have to be taken into account when considering the risk to health of fibrous materials, especially of the newer materials, encountered in the environment.[83]

20.4.4 Pesticides

Notwithstanding the obvious beneficial effects of their use, pesticides do present recognized hazards to human health,[84] mostly through high level occupational exposure and accidental poisoning incidents, although low level exposure remains a cause of public concern.

There are a number of ways in which humans can be exposed to pesticides through the environmental route. Pesticides used domestically in wood preservation or as household insecticides may be a particularly important source of exposure for the general public. Possible effects of pesticide residues in food and water probably cause the greatest public anxiety, although reports of clinical poisoning by residues seem to be extremely rare. Analysis of reported consumer poisonings by pesticides show that most arise from spillage of pesticides onto food during storage or transport, eating a food article not intended for human consumption (*e.g.* treated grain or seed potatoes; see section 20.3.5), and improper application of pesticides.[85] Another route of exposure is through inhalation of air contaminated with pesticides, which might be of importance for local populations living in proximity to areas where pesticides are used. Concentrations of pesticides in air are very seasonal and are correlated to local agricultural use patterns. The highest concentrations tend to occur during the spring and summer, coinciding with application times and warmer temperatures. More than 40% of pesticides detected in air are herbicides, 33% are insecticides and 26% fungicides. Some of the pesticides detected in air in Europe nowadays are forbidden according to EU regulations (*e.g.* the fungicides vinclozolin, trycyclazole).[86] The detection of these pesticides in air after being banned might be a consequence of their long-term accumulation in soils. Indeed, much of the concern about pesticide use, however, revolves around long-term accumulation in the environment and the low level but chronic human exposure to these compounds which occurs either directly through the food chain or indirectly through the air (in local agricultural communities). This has almost certainly been one of the main reasons for the increased popularity of 'organic' foods.

Pesticides include insecticides, fungicides and herbicides. The main classes of insecticide are organochlorines (*e.g.* dichlorodiphenyltrichloroethane (DDT), lindane, dieldrin), which are very persistent in the environment,[87] anticholesterinases (the organophosphates, *e.g.* parathion, and the carbamates), pyrethroids and other botanicals, and fumigants (*e.g.* ethylene dibromide). Fungicides include organometals, phenols and carbamates. Most herbicides are bipyridinium compounds (*e.g.* Paraquat), phenoxy compounds, organophosphates and substituted anilines. Various other compounds (not detailed here) are used to control rodents, mites and ticks, molluscs, bacteria, birds and algae.

The best known pesticides are probably the organochlorines and organophosphates. The organochlorines act as neurotoxins to the target organisms. Although, as a class, the organochlorine pesticides are less acutely toxic to humans than some other insecticide classes, they have greater potential for chronic toxicity. Many are now banned or restricted because of their persistence in the environment and their propensity to accumulate in the tissues of living organisms.[88] They are probably the compounds of most concern with respect to chronic environmental exposure, and recently have also been implicated as possible environmental endocrine disrupters (see section 20.4.5). Organophosphates, on the other hand, which function by blocking the activity of acetylcholinesterase, break down rapidly and do not accumulate in tissues. However, they are often extremely toxic and non-selective, and more instances of acute poisoning have occurred with organophosphates than with any other insecticide class.[89]

Information on the effects of occupational exposure through their use in sheep dips[90] has given rise to concern in some quarters about possible low level environmental exposure to this group of pesticides. Children and foeti seem to be particularly sensitive to exposure to pesticides. *In utero* exposures to the organophosphates have been associated with decreased birth weight and birth length,[91] and decreases in gestational duration.[92] Children exposed to pesticides seem to have increased risk of malignancies, including leukaemia, neuroblastoma, Wilm's tumor, soft-tissue sarcoma, Ewing's sarcoma, non-Hodgkin's lymphoma and cancers of the brain, colorectum and testes.[93] Environmental exposure to the pesticides DDT, dichlorodiphenyldichloroethylene and dieldrin has been inconclusively related to increased risk of breast cancer.[94] Long-term exposure to pesticides has also been associated with the development of Parkinson's disease.[95] Nonetheless, further research is required to assess the uncertainties regarding low dose extrapolation to many pesticide compounds, and to investigate chronic health effects to the general public associated with environmental low-concentration exposure to pesticides.

20.4.5 Endocrine Disrupters (see also Chapters 1 and 3)

There is continuing public interest in the possible adverse consequences arising from the release into the environment of chemicals with the propensity to disrupt hormone function in humans and in wildlife. Originally described as 'environmental oestrogens', such agents are now known by the broader term 'endocrine disrupters' or 'endocrine disrupting chemicals'.[96] In addition to the original understanding that such substances mimicked, inhibited or otherwise interfered with the action of the sex hormones oestradiol and the androgens, the new term incorporates possible effects on other endocrine systems and organs, including the thyroid, pituitary and adrenals. In this way, growth, development, behaviour, immune response, *etc.* could all conceivably be affected by endocrine disrupters.

In humans there is an increasing body of evidence for changing trends in reproductive health.[97] In particular, elevated incidences of testicular cancer in men and breast cancer in women have been demonstrated.[98] There is evidence for reduced sperm counts and sperm quality, although there are marked regional differences.[99–101] Other effects of concern include cryptorchidism (undescended testes), hypospadias (a congenital malformation of the penis) and prostate cancer.[101] There seems to be an agreement of the common origin of these health effects in foetal life or childhood.[101] A wide range of environmental pollutants have been implicated as possible endocrine disrupters. Of these, synthetic hormones, organochlorine pesticides, polychlorinated biphenyls, phthalates, alkylphenols and bisphenol A are perhaps the most widely studied.[102] Considerable attention is also being given to the role of natural phytoestrogens in the human diet. It is clear that, as more and more chemicals are investigated, the list of those with endocrine disruptive potential will continue to expand.

The organochlorine pesticides, because of their propensity to build up in the environment and their known effects on wildlife, are a particular focus of attention. The o',p'-isomers of DDT have been demonstrated to have oestrogenic activity and to bind to the oestrogen receptor, but these isomers are relatively unstable and are rarely found in the environment. The major and persistent DDT metabolite, $p'p'$-DDE, has no significant interaction with the oestrogen receptor and possesses no inherent oestrogenic activity. However, it does bind strongly to the androgen receptor and inhibits androgen receptor action.[103,104]

There are many factors to consider when assessing the possible impact of environmental endocrine disrupters on human health. Firstly, the potency of many of the relevant environmental contaminants is low or very low when compared with endogenous oestrogens such as oestradiol, although differential protein binding complicates the picture. As with all toxicological effects, there may be additive or synergistic effects between endocrine disrupting chemicals, though the results of tests for such interactive effects have been variable. Humans have always been exposed to natural phytoestrogens present in foodstuffs and over the last 50 years or so have been exposed to high concentrations of hormonally active substances in the form of medical treatments and oral contraceptives. Against this background, and considering the likely importance of the stage of life at which exposure occurs, it is currently very difficult to assess whether or not environmental endocrine disrupters play a significant part in reproductive disorders in the general population.[102] Further research is needed before a firm assessment can be made.[105]

20.4.6 Nanoparticles

In recent years, nanotechnology research and product development has increased exponentially due to new and beneficial properties of nanomaterials.[106] Nowadays engineered nanomaterials are found in multiple applications, from electronics (*e.g.* silica and alumina nanoparticles, carbon nanotubes),[107] catalysts (*e.g.* CeO_2 nanoparticles),[108] power generation (TiO_2 nano-structures),[109] construction and composite materials (*e.g.* carbon nanotubes),[107] to cosmetics (TiO_2 nanoparticles),[110] antibacterial coatings (*e.g.* Ag nanoparticles),[111] and will also be increasingly used in medicine for purposes of diagnosis, imaging and drug delivery.[106,112] Therefore, nanomaterials are increasingly present in some workplaces and consumer products.[113] The ISO Technical Report on ultrafine, nanoparticle and nano-structured aerosols defines nanoparticles (NPs) as "a particle with a nominal diameter smaller than about 100 nm", whilst ultrafine particles (UFPs) are defined similarly, but are often used as a term for nanoparticles incidentally released to the atmosphere as a by-product process (*e.g.* combustion, welding).[114]

Nowadays, exposure to NPs is liable to occur in occupational environments, with researchers developing new nanomaterials and applications as well as workers manufacturing, handling, bagging and shipping, and using NPs (*e.g.* mixing, drying, and spraying) in factories being most at risk of exposure.[106,113] Consumers may be exposed to NPs if products that contain them (*e.g.* surface coatings) are worked or degraded, the NPs becoming freely available for exposures.[113] On the other hand, whilst exposure to NPs occurs mainly in occupational settings, exposure to UFPs is widespread in the general population since they are incidentally released to or produced in the atmosphere. There are three main sources of UFPs in the air. Ultrafine particles in the 30–100 nm size range are directly emitted from road traffic exhaust, which is the main contributor, or by industrial, commercial or residential combustion processes. UFPs can be formed in the air from the condensation of low volatility vapours formed from the oxidation of atmospheric gases, and are typically in the size range 1–10 nm. The third source of UFPs occurs during dilution of diesel exhaust in the atmosphere. Diesel exhaust is comprised largely (in mass) by particles with a core of graphitic carbon formed within the combustion chamber of the engine; and (in number) by newly formed particles that condense during dilution of the exhaust from semi-volatile vapours originated from engine oil vaporised in the combustion process. The graphitic UFPs are typically in the size range 30–100 nm, whilst the condensed UFPs are generally 10 to 30 nm in size when freshly formed.[115] Ultrafine particles are mainly composed of organic compounds (32–67%), followed by graphitic carbon (4–18%), metal oxides (3–26%) arising from fuel impurities and engine wear, sulfate (3–18%) and nitrate (1.5–18%) compounds.[116]

The three main routes for exposure to nanoparticles are inhalation, ingestion and dermal penetration. In the general population the main exposure to nanoparticles is likely to arise from breathing UFPs in air. After inhalation, nanoparticles (and UFPs) will be deposited in the respiratory tract, with 50% of 20 nm nanoparticles in the alveolar region, and 25% in the naso-pharyngeal and tracheobronchial regions. Ingestion of nanoparticles may occur when swallowing the mucous that traps the nanoparticles deposited in the airways or by sucking or licking a contaminated surface. Dermal penetration can occur in the workplace when handling nanomaterials, or in the general population when using consumer products that contain nanoparticles (*e.g.* creams). Current research is investigating the potential of nanoparticles to penetrate the epidermis.[113]

In the case of inhaled nanoparticles, once deposited in the respiratory system, they largely escape alveolar macrophage surveillance and gain access to the pulmonary interstitium,[117,118] and consequently may appear in many compartments of the body, including liver, heart and nervous system,[119] after penetrating the cell epithelium and entering the blood[120] or lymph systems.[113] NPs that have accessed the interstitium may interact with interstitial macrophages and other sensitive cells eliciting inflammation that could lead to subsequent disease.[121] Evidence also shows that NPs can access the brain *via* the olfactory epithelium[122] and through the blood–brain barrier.[123] Current research is investigating the potential of nanoparticles to penetrate the intestinal epithelium[124] and skin epidermis,[125] as well as the capability of NPs to redistribute within cells,[126,127,128] and enter organelles.[129]

The physico-chemical characteristics that confer the novel properties of engineered nanomaterials are attributable to their small size which gives them large surface area per unit mass; chemical composition, such as purity, crystallinity and electronic properties; surface structure, *e.g.* surface reactivity, surface groups, inorganic or organic coatings; solubility; shape; and aggregation.[112] However, some of these characteristics also confer upon nanoparticles their attributed toxicity. The dose of NPs in contact with the biological system is a function of their specific surface area, which depends on the NP diameter and surface reactivity which defines the capacity of the NP to generate free radicals and depends on the NP chemical composition, surface groups, specific coating and surface charge. Both specific surface area and reactivity control the ability of NPs to produce inflammation and toxicity.[130] If the NPs are soluble in biological fluid and contain reactive materials such as metals or organics on the surface, there is the potential for additional toxicity due to the soluble material. On the other hand, insoluble NP might have increased biodurability in the biological system. In any case, the stability of the NP in the biological system not only depends on the intrinsic physico-chemical NP properties, but also on the host susceptibility such as target organ and extra- or intra-cellular location, which defines the local hydro-biochemical conditions (*e.g.* pH and oxidative potential).[130]

There is as yet no epidemiological evidence deriving from long-term exposure to nanomaterials in occupational exposures since their use and manufacture is relatively new. Similarly, only a few epidemiological studies have reported health effects related to environmental exposure to ultrafine particles. Nonetheless, epidemiological studies of occupational and environmental exposures to airborne particles and fibrous materials, alongside recent toxicological research on nanoparticles, provide information on potential health effects related to nanoparticles exposure.[131]

Research on the health effects related to occupational exposures to carbon black show some adverse effects on the respiratory system, such as decreased lung function, and dust retention in the lungs associated with recent, but not cumulative exposure.[132,133] Epidemiological studies on welders, who are generally exposed to metallic UFP, show a prevalence of lung cancer[134,135] and neurodegenerative diseases,[136] although these outcomes are largely related to the type of metal welded.[137,138] There is also extensive evidence linking exposure to asbestos and other manmade fibres (see section 20.4.3) with mortality (lung cancer and mesothelioma) and respiratory morbidity.[130] The similarity between asbestos and fibrous-like nanomaterials (*e.g.* carbon nanotubes) is a cause for concern.

Environmental exposures to UFP in six panel studies with patients suffering from chronic pulmonary disease in Germany,[139] Finland[140] and UK[141] were associated with a decrease of peak expiratory flow and an increase of daily symptoms and medication.[142] A European multicentre cohort study correlated short-term estimated particle number concentration, *i.e.* a surrogate for UFP concentration, with increased cardiac readmission to hospital in survivors of a first myocardial infarction. Yet, the effect of other co-pollutants was similar and hence the effect cannot be ascribed solely to UFP.[143] A large German study of the effects of UFP number concentrations on daily mortality found a similar effect for fine (*i.e.* aerodynamic diameter <2.5 µm) and UFP, although UFP showed a more delayed effect on mortality with a lag of four days. The same study also showed a more immediate effect on mortality associated with the respiratory system, whilst a more delayed effect was associated with the cardiovascular system.[144] The larger effect of UFP upon mortality associated with the cardiovascular system than with the respiratory system is consistent with results from a later study in London, which showed an association of UFP number concentrations with daily mortality and admissions, particularly for cardiovascular diseases lagged 1 day.[145] A study in Beijing showed an association between the UFP number concentration and daily cardiovascular mortality and ischemic heart disease mortality for a 2-day delay.[146] These effects on the cardiovascular system might be associated with changes in heart rate variability, cardiac repolarization, increased blood pressure, systemic inflammation, endothelial dysfunction and blood coagulation.[147–149] This is consistent with toxicological studies showing that exposure to engineered NPs can lead to oxidative stress,[150,151] inflammation[118,152] and blood translocation[120,153] of UFP.

All the available evidence on UFPs suggests that some engineered nanomaterials could lead to oxidative stress and inflammatory responses, which can be linked to adverse health effects.

However, the total toxicity of the NP is the complex interaction between the surface reactivity, the soluble chemical composition, the surface charge, shape and biopersistence.[154]

Opinion is widespread that the fast development of the area of nanotechnology requires attention to safety issues.[112] Calls have been made to develop and implement stricter and thorough regulations. The Royal Society and Royal Academy of Engineering (2004) issued a report on nanoscience and nanotechnology that calls for precaution in the use of engineered NPs and that mechanisms should be in place to identify, reduce or remove potential waste products containing engineered NPs.[155] Similarly, the European Union promotes active development of risk assessment to generate adequate toxicological data, and to evaluate potential human and environmental exposures to nanomaterials throughout their lifecycle.[156,157] The European Union also has legislation that applies to nanomaterials, such as REACH, which provides an overarching framework of hazard assessment applicable to the manufacture, marketing and use of chemicals. Although this legislation does not make explicit reference to nanomaterials, these are included in the definition of 'substance'. Therefore, those nanomaterials that fulfil the criteria for classification as 'hazardous' under *European Regulation (EC) No. 1277/2008 on Classification, Labelling and Packaging of Substances and Mixtures* must be classified, labelled and notified to the European Chemicals Agency. The European Commission has also provided some guidance on how to do chemical safety assessment of substances at nanoscale; to evaluate materials and manage nanomaterials safely;[158] and how to classify, label and package nanomaterials[159] in accordance with REACH. The implementation of REACH and other legislation aims to ensure that appropriate testing procedures take place before the widespread use and environmental release of engineered nanomaterials.[130]

20.5   CONCLUSIONS

Following the industrial revolution, and especially since the Second World War and the ensuing rapid growth of the chemical industry, an increasing number of foreign substances have been synthesized and either deliberately, accidentally or incidentally released into the environment. In addition, man's activities in utilising the earth's natural resources and modifying the environment for benefit or gain have resulted in the local and global release of pollutants, especially combustion products. The increasing human population and continuing rapid strides in the industrial development of many countries, linked with the inherent desire for improved quality of life, have exacerbated the problem. Against this, however, is the developing trend to consider ever more seriously the impacts of man's activities on the environment. This, coupled with increasing concern about how environmental quality can affect human health, has led for example to the development of such concepts as 'sustainable development' and the 'precautionary principle', and of 'cradle-to-grave' lifecycle environmental assessments, 'eco-labelling' and 'product stewardship'.

The link between environmental chemicals and health is therefore a topic receiving increasing attention. This chapter has looked at some of the ways that sudden (catastrophic) or more incidental local exposures, as well as generalised chronic exposures to environmental pollutants can result in adverse effects on human health. The causes leading to the catastrophic and incidental exposures are associated with human actions. The accidents in Seveso and Bhopal were related to poor process management and inadequate maintenance programmes which led to a runaway chain reaction that could not be contained by the safety design measures. Most of the localised incidents resulted in contamination of the food chain as a consequence of poorly managed chemical discharge into rivers or coastal waters; wrong labelling of toxic materials; fraudulent practices; or poor handling practice, such as PCB contamination of products or concurrent transport of toxic materials with foodstuffs. The generalised chronic exposure to traditional air pollutants, pesticides and endocrine disrupters is linked to our current standard of living, since they are associated with road traffic, industrial processes, and increased production of foodstuffs.

In many cases the impacts of low level exposures on human health are suspected rather than proven. This is because of the problems associated with establishing causal links when effects and doses are small, and many other factors that confound the situation in human populations. Exposure to contaminants can occur through three major pathways; ingestion, inhalation and dermal contact. The former is of prime concern, as demonstrated by the cases reviewed here and by the obvious public interest revolving around the presence of pesticides and other chemical residues in food and water. In particular the halogenated hydrocarbons, including dioxins and organo-chlorine pesticides, continue to attract public, political and scientific attention. We are living now with the legacy of the historic use of chemicals such as DDT and PCBs that persist and circulate within the environment. Understandably, this has led to the current special concern afforded to any persistent organic chemical with the propensity to bioaccumulate.

With respect to exposure by inhalation, pollution from road vehicles and the quality of air in the indoor environment are seen to be significant issues. New concerns have arisen over exposures to novel nanomaterials, due to the rapid growth in their production and use, and the currently inadequate base of toxicological and (especially) epidemiological data on their safety. Exposure through dermal contact (*e.g.* with sunscreens) is also a cause for concern.

Clearly, total personal exposure (by whatever route) is fundamentally important, as is individual susceptibility to the effects of toxins determined, for example, by genetic makeup, age, current health status and exposure history. Outputs from the human genome mapping project and gene-environment interaction studies could well have important impacts in the area of individual susceptibility. Developments in toxicology, including, for example, the application of biomarkers and advances in modelling and risk assessment for low dose exposures, are already enabling more accurate estimates to be made of population exposure profiles and health consequences. Increased understanding of total exposure, the apportionment of exposure, and genetic susceptibilities will facilitate the adoption of appropriate control measures.

One of the biggest issues of the moment relates to the ability of some ubiquitous environmental chemicals to act as hormone mimics (or antagonists) or otherwise to interfere with the balance of sex and other hormones and therefore possibly affect human reproductive health and fertility as well as behaviour, growth and development. This question of the endocrine disruptive activity of chemicals in the environment, to which we are all commonly exposed, is likely to continue to be an issue of considerable scientific and public interest.

Some generic issues are also certain to continue to provide challenges, for example, the somewhat conflicting expectations of a totally 'safe' environment, good health and a high standard of living. The discrepancies between the public perception of risk and scientific risk assessment, the problem of discriminating physical from psychological health effects, and the general issue of assessing exposure to chemicals in the environment and estimating health risks in the face of many confounding factors are all likely to engage the minds of scientists, regulators and policy makers well into the 21st century.

REFERENCES

1. A. Giovanardi, *Proceedings of the Expert Meeting on the Problems Raised by TCDD Pollution*, 1976.
2. A. Baccarelli, A. C. Pesatori, D. Consonni, P. Mocarelli, D. G. Patterson, N. E. Caporaso, P. A. Bertazzi and M. T. Landi, *Br. J. Dermatol.*, 2005, **152**, 459–465.
3. G. Filippini, B. Bordo, P. Crenna, N. Massetto, M. Musicco and R. Boeri, *Scand. J. Work Environ. Health*, 1981, **7**, 257–262.
4. F. Pocchiari, V. Silano and A. Zampieri, *Ann. N. Y. Acad. Sci.*, 1979, **320**, 311–320.
5. G. Reggiani, *J. Toxicol. Environ. Health*, 1980, **6**, 27.

6. D. Consonni, A. C. Pesatori, C. Zocchetti, R. Sindaco, L. C. D'Oro, M. Rubagotti and P. A. Bertazzi, *Am. J. Epidemiol.*, 2008, **167**, 847–858.

7. A. C. Pesatori, D. Consonni, M. Rubagotti, P. Grillo and P. A. Bertazzi, *Environ. Health*, 2009, 8.

8. P. A. Bertazzi, A. C. Pesatori, D. Consonni, A. Tironi, M. T. Landi and C. Zocchetti, *Epidemiology*, 1993, **4**, 398–406.

9. M. Warner, B. Eskenazi, P. Mocarelli, P. M. Gerthoux, S. Samuels, L. Needham, D. Patterson and P. Brambilla, *Environ. Health Perspect.*, 2002, **110**, 625–628.

10. B. Eskenazi, P. Mocarelli, M. Warner, S. Samuels, P. Vercellini, D. Olive, L. L. Needham, D. G. Patterson, P. Brambilla, N. Gavoni, S. Casalini, S. Panazza, W. Turner and P. M. Gerthoux, *Environ. Health Perspect.*, 2002, **110**, 629–634.

11. B. Eskenazi, M. Warner, S. Samuels, J. Young, P. M. Gerthoux, L. Needham, D. Patterson, D. Olive, N. Gavoni, P. Vercellini and P. Mocarelli, *Am. J. Epidemiol.*, 2007, **166**, 79–87.

12. B. Eskenazi, M. Warner, A. R. Marks, S. Samuels, L. Needham, P. Brambilla and P. Mocarelli, *Epidemiology*, 2010, **21**, 224–231.

13. B. Eskenazi, M. Warner, A. R. Marks, S. Samuels, P. M. Gerthoux, P. Vercellini, D. L. Olive, L. Needham, D. G. Patterson and P. Mocarelli, *Environ. Health Perspect.*, 2005, **113**, 858–862.

14. P. Mocarelli, P. M. Gerthoux, E. Ferrari, D. G. Patterson, S. M. Kieszak, P. Brambilla, N. Vincoli, S. Signorini, P. Tramacere, V. Carreri, E. J. Sampson, W. E. Turner and L. L. Needham, *Lancet*, 2000, **355**, 1858–1863.

15. B. Eskenazi, P. Mocarelli, M. Warner, W. Y. Chee, P. M. Gerthoux, S. Samuels, L. L. Needham and D. G. Patterson, *Environ. Health Perspect.*, 2003, **111**, 947–953.

16. A. Baccarelli, S. M. Giacomini, C. Corbetta, M. T. Landi, M. Bonzini, D. Consonni, P. Grillo, D. G. Patterson, Jr., A. C. Pesatori and P. A. Bertazzi, *PLoS Medicine*, 2008, **5**, 1133–1142.

17. S. Alaluusua, P. Calderara, P. M. Gerthoux, P. L. Lukinmaa, O. Kovero, L. Needham, D. G. Patterson, J. Tuomisto and P. Mocareili, *Environ. Health Perspect.*, 2004, **112**, 1313–1318.

18. *Directive 2012/18/EU of the European Parliament and of the Council of 4 July 2012 on the Control Of Major-Accident Hazards involving Dangerous Substances*, amending and subsequently repealing *Council Directive 96/82/EC*.

19. N. Andersson, M. K. Muir, V. Mehra and A. G. Salmon, *Br. J. Ind. Med.*, 1988, **45**, 469–475.

20. V. R. Dhara and R. Dhara, *Arch. Environ. Health*, 2002, **57**, 391–404.

21. WHO, *Toxic Oil Syndrome*, Report of a WHO meeting, Madrid, 21–25 March 1983, Copenhagen, 1984.

22. E. Gelpi, M. P. de la Paz, B. Terracini, I. Abaitua, A. G. de la Camara, E. M. Kilbourne, C. Lahoz, B. Nemery, R. M. Philen, L. Soldevilla and S. Tarkowski, WHO/CISAT, *Environ. Health Perspect.*, 2002, **110**, 457–464.

23. J. M. Tabuenca, *Lancet*, 1981, **2**, 567–568.

24. A. V. Roncero, C. J. del Valle, R. M. Duran and E. G. Constante, *Lancet*, 1983, **2**, 1024–1025.

25. WHO, *Toxic Oil Syndrome. Ten Years of Progress*, 2004.

26. H. Urabe, H. Koda and M. Asahi, *Ann. N. Y. Acad. Sci.*, 1979, **320**, 273–276.

27. P. H. Chen, J. M. Gaw, C. K. Wong and C. J. Chen, *Bull. Environ. Contaminat. Toxicol.*, 1980, **25**, 325–329.

28. T. Kashimoto, K. Takayama, M. Mimura, S. Ohta and H. Miyata, *Fukuoka Acta Medica*, 1989, **80**, 210–220.

29. I. J. Selikoff, *A Survey of the General Population of Michigan for Health Effects of PBB Exposure*, Final Report submitted to the Michigan Department of Public Healthand, Michigan, 1979.

30. P. J. Landrigan, K. R. Wilcox, Jr., J. Silva, Jr., H. E. Humphrey, C. Kauffman and C. W. Heath, Jr., *Ann. N. Y. Acad. Sci.*, 1979, **320**, 284–294.

31. J. G. Bekesi, J. F. Holland, H. A. Anderson, A. S. Fischbein, W. Rom, M. S. Wolff and I. J. Selikoff, *Science*, 1978, **199**, 1207–1209.
32. J. K. Stross, R. K. Nixon and M. D. Anderson, *Ann. N. Y. Acad. Sci.*, 1979, **320**, 368–372.
33. K. Kreiss, C. Roberts and H. E. B. Humphrey, *Arch. Environ. Health*, 1982, **37**, 141–147.
34. M. Katsuna, *Minimata Disease*, Kumamoto University, 1968.
35. H. P. Illing, *General and Applied Toxicology*, Macmillan, 1995.
36. F. Bakir, S. F. Damluji, L. Aminzaki, M. Murtadha, A. Khalidi, N. Y. Alrawi, S. Tikriti, H. I. Dhahir, T. W. Clarkson, J. C. Smith and R. A. Doherty, *Science*, 1973, **181**, 230–241.
37. G. Kazantzis, A. W. al-Mufti, A. al-Jawad, Y. al-Shawani, M. A. Majid, R. M. Mahmoud, M. Soufi, K. Tawfiq, M. A. Ibrahim and H. Dabagh, *Bull. WHO*, 1976, 53.
38. Lowermoor Incident Health Advisory Group, *Water Pollution at Lowermoor, North Cornwall*, 2nd Report, HMSO, London, 1991.
39. Committee on Toxicity of Chemicasl in Food, Consumer Products and the Environment, *Subgroup Report on the Lowermoor Water Pollution Incident*, 2005.
40. T. M. McMillan, A. J. Freemont, A. Herxheimer, J. Denton, A. P. Taylor, M. Pazianas, A. R. C. Cummin and J. B. Eastwood, *Hum. Exp. Toxicol.*, 1993, **12**, 37–42.
41. P. Altmann, J. Cunningham, U. Dhanesha, M. Ballard, J. Thompson and F. Marsh, *Br. Med. J.*, 1999, **319**, 807–811.
42. H. Kopelman, M. H. Robertso, P. G. Sanders and I. Ash, *Br. Med. J*, 1966, **1**, 514–516.
43. R. Schoenta, *Nature*, 1968, **219**, 1162–1163.
44. J. B. Silkworth, D. S. Cutler and G. Sack, *Fundam. Appl. Toxicol.*, 1989, **12**, 303–312.
45. J. B. Silkworth, D. S. Cutler, L. Antrim, D. Houston, C. Tumasonis and L. S. Kaminsky, *Fundam. Appl. Toxicol.*, 1989, **13**, 1–15.
46. N. J. Vianna and A. K. Polan, *Science*, 1984, **226**, 1217–1219.
47. L. H. Roht, S. W. Vernon, F. W. Weir, S. M. Pier, P. Sullivan and L. J. Reed, *Am. J. Epidemiol.*, 1985, **122**, 418–433.
48. T. F. Jones, A. S. Craig, D. Hoy, E. W. Gunter, D. L. Ashley, D. B. Barr, J. W. Brock and W. Schaffner, *New Engl. J. Med.*, 2000, **342**, 96–100.
49. J. W. Grisham, *Health Aspects of the Disposal of Waste Chemicals*, Pergamon, Oxford, 1986.
50. H. M. P. Fielder, C. M. Poon-King, S. R. Palmer, N. Moss and G. Coleman, *Br. Med. J.*, 2000, **320**, 19–22.
51. H. Dolk, M. Vrijheid, B. Armstrong, L. Abramsky, F. Bianchi, E. Garne, V. Nelen, E. Robert, J. E. S. Scott, D. Stone and R. Tenconi, *Lancet*, 1998, **352**, 423–427.
52. F. Aguilera, J. Mendez, E. Pasaro and B. Laffon, *J. Appl. Toxicol.*, 2010, **30**, 291–301.
53. X. Bosch, *Lancet*, 2003, **361**, 147–147.
54. G. M. Solomon and S. Janssen, *JAMA, J. Am. Med. Assoc.*, 2010, **304**, 1118–1119.
55. G. Rodriguez-Trigo, J.-P. Zock, F. Pozo-Rodriguez, F. P. Gomez, G. Monyarch, L. Bouso, M. Dolors Coll, H. Verea, J. M. Anto, C. Fuster and J. Albert. Barbera, *Ann. Internal Med.*, 2010, **153**, 489–498.
56. W. P. D. Logan, *Lancet*, 1953, **1**, 336–338.
57. Institute for Environment and Health, *Research on Air Pollution 1999*, IEH, Leicester, 2000.
58. P. Wolkoff, Indoor Air, *Int. J. Indoor Air Qual. Climate*, 1995, Supplement No 3/95.
59. Institute for Environment and Health, *Indoor Air Quality in the Home (2): Carbon Monoxide*, IEH, Leicester, 1998.
60. NHS, *Carbon Monoxide Poisoning*, 2013; http://www.nhs.uk/conditions/Carbon-monoxide-poisoning/Pages/Introduction.aspx (acessed 06/07/2013).
61. D. Jarvis, G. Adamkiewicz, M. E. Heroux, R. Rapp and F. Kelly, *WHO Guidelines for Indoor Air Quality: Selected Pollutants*, WHO, Copenhagen, 2010.
62. P. T. C. Harrison, in *Air Pollution and Health*, ed. R. E. Hester and R. M. Harrison, Royal Society of Chemistry, Cambridge, 1998.

63. Institute for Environment and Health, *Indoor Air Quality in the Home: Nitrogen Dioxide, Formaldehyde, Volatile Organic Compounds, House Dust Mites, Fungi and Bacteria*, IEH, Leicester, 1998.

64. W. S. Tunnicliffe, P. S. Burge and J. G. Ayres, *Lancet*, 1994, **344**, 1733–1736.

65. D. Kaden, C. Mandin, G. Nielsen and P. Wolkoff, in *WHO Guidelines for Indoor Air Quality: Selected Pollutants*, WHO, Copenhagen, 2010.

66. R. Harrison, J. M. Delgado-Saborit, F. Dor and R. Henderson, in *WHO Guidelines for Indoor Air Quality: Selected Pollutants*, WHO, Copenhagen, 2010.

67. L. Molhave, in *Environmental Toxicants - Human Exposures and their Health Effects*, ed. M. Lippmann, Wiley, New York, 2000, p. 889.

68. P. Wolkoff, P. A. Clausen, B. Jensen, G. D. Nielsen and C. K. Wilkins, *Indoor Air, Int. J. Indoor Air Qual. Climate*, 1997, **7**, 92–106.

69. H. Choi, R. Harrison, H. Komulainen and J. M. Delgado-Saborit, in *WHO Guidelines for Indoor Air Quality: Selected Pollutants*, WHO, Copenhagen, 2010.

70. D. Krewski, A. Oxman and G. W. Torrance, *The Risk Assessment of Environmental Hazards*, Wiley, New York, 1989.

71. F. K. Hare, *Lead in the Canadian Environment: Science and Regulation. Final Report of the Commission on Lead in the Environment*, Royal Society of Canada, Ottawa, 1986.

72. T. Waldron, *Diet and Crafts in Towns*, BAR, Oxford, 1989.

73. L. Friberg, T. Kjellstrom, G. Nordberg and M. Piscator, *Cadmium in the Environment, III*, USEPA, Washington, DC, 1975.

74. D. Barltrop and C. D. Strehlow, *Lancet*, 1982, **2**, 1394–1395.

75. M. D. Turner, D. O. Marsh, J. C. Smith, J. B. Inglis, T. W. Clarkson, C. E. Rubio, J. Chiriboga and C. C. Chiriboga, *Arch. Environ. Health*, 1980, **35**, 367–378.

76. J. H. Kyle and N. Ghani, *Arch. Environ. Health*, 1982, **37**, 266–271.

77. D. H. K. Lee and I. J. Selikoff, *Environ. Res.*, 1979, **18**, 300–314.

78. J. C. Wagner, C. A. Sleggs and P. Marchand, *Br. J. Ind. Med.*, 1960, **17**, 260–271.

79. C. Ramazzini, *Am. J. Ind. Med.*, 2011, **54**, 168–173.

80. P. T. C. Harrison, L. S. Levy, G. Patrick, G. H. Pigott and L. L. Smith, *Environ. Health Perspect.*, 1999, **107**, 607–611.

81. P. T. C. Harrison and J. C. Heath, *Mechanisms in Fibre Carcinogenesis*, Plenum Press, New York, 1991.

82. J. Peto, F. E. Matthews, J. T. Hodgson and J. R. Jones, *Lancet*, 1995, **345**, 535–539.

83. Institute for Environment and Health, *Fibrous Materials in the Environment*, IEH, Leicester, 1997.

84. R. Levine, in *Handbook of Pesticide Toxicology*, Vol. 1 Academic Press, New York, 1991, p. 275.

85. T. C. Marrs, *General and Applied Toxicology*, Macmillan, London, 1995.

86. V. Yusa, C. Coscolla, W. Mellouki, A. Pastor and M. de la Guardia, *J. Chromatogr. A*, 2009, **1216**, 2972–2983.

87. M. Alexander, *Environ. Sci. Technol.*, 2000, **34**, 4259–4265.

88. V. T. Covello and M. W. Merkhofer, *Risk Assessment Methods*, Plenum Press, New York, 1993.

89. W. Stopford, *Industrial Toxicology*, van Nostrand Reinhold, New York, 1985.

90. R. Stephens, A. Spurgeon, I. A. Calvert, J. Beach, L. S. Levy, H. Berry and J. M. Harrington, *Lancet*, 1995, **345**, 1135–1139.

91. F. P. Perera, V. Rauh, W. Y. Tsai, P. Kinney, D. Camann, D. Barr, T. Bernert, R. Garfinkel, Y. H. Tu, D. Diaz, J. Dietrich and R. M. Whyatt, *Environ. Health Perspect.*, 2003, **111**, 201–205.

92. B. Eskenazi, K. Harley, A. Bradman, E. Weltzien, N. A. Jewell, D. B. Barr, C. E. Furlong and N. T. Holland, *Environ. Health Perspect.*, 2004, **112**, 1116–1124.

93. S. H. Zahm and M. H. Ward, *Environ. Health Perspect.*, 1998, **106**, 893–908.
94. S. M. Snedeker, *Environ. Health Perspecti.*, 2001, **109**, 35–47.
95. F. D. Dick, *Br. Med. Bull.*, 2006, **79–80**, 219–231.
96. R. E. Hester and R. M. Harrison, *Endocrine Disrupting Chemicals,* Royal Society of Chemistry, *Cambridge*, 1999.
97. Institute for Environment and Health, *Environmental Oestrogens: Consequences to Human Health and Wildlife*, IEH, Leicester, 1995.
98. S. H. Safe, *Environ. Health Perspect.*, 2000, **108**, 487–493.
99. A. Giwercman, L. Rylander and Y. L. Giwercman, *Reproduct. Biomed. Online*, 2007, **15**, 633–642.
100. C. Sonnenschein and A. M. Soto, *J. Steroid Biochem. Mol. Biol.*, 1998, **65**, 143–150.
101. J. Toppari, J. C. Larsen, P. Christiansen, A. Giwercman, P. Grandjean, L. J. Guillette, B. Jegou, T. K. Jensen, P. Jouannet, N. Keiding, H. Leffers, J. A. McLachlan, O. Meyer, J. Muller, E. Rajpert-DeMeyts, T. Scheike, R. Sharpe, J. Sumpter and N. E. Skakkebaek, *Environ. Health Perspect.*, 1996, **104**, 741–803.
102. P. T. C. Harrison, P. Holmes and C. D. N. Humfrey, *Sci. Total Environ.*, 1997, **205**, 97–106.
103. S. C. Maness, D. P. McDonnell and K. W. Gaido, *Toxicol. Appl. Pharmacol.*, 1998, **151**, 135–142.
104. B. J. Danzo, *Environ. Health Perspect.*, 1997, **105**, 294–301.
105. European Commission, *European Workshop on the Impact of Endocrine Disrupters on Human Health and Wildlife*, Report EUR 17549, European Commission, Brussels, 1997.
106. T. A. J. Kuhlbusch, C. Asbach, H. Fissan, D. Goehler and M. Stintz, *Particle Fibre Toxicol*, 2011, 8.
107. J. Wang, C. Asbach, H. Fissan, T. Huelser, T. A. J. Kuhlbusch, D. Thompson and D. Y. H. Pui, *J. Nanoparticle Res.*, 2011, **13**, 1373–1387.
108. B. Park, K. Donaldson, R. Duffin, L. Tran, F. Kelly, I. Mudway, J. P. Morin, R. Guest, P. Jenkinson, Z. Samaras, M. Giannouli, H. Kouridis and P. Martin, *Inhalation Toxicol.*, 2008, **20**, 547–566.
109. A. J. Frank, N. Kopidakis and J. van de Lagemaat, *Coordinat. Chem. Rev.*, 2004, **248**, 1165–1179.
110. A. P. Popov, A. V. Priezzhev, J. Lademann and R. Myllyla, *J. Phys. D: Appl. Phys.*, 2005, **38**, 2564–2570.
111. J. S. Kim, E. Kuk, K. N. Yu, J. H. Kim, S. J. Park, H. J. Lee, S. H. Kim, Y. K. Park, Y. H. Park, C. Y. Hwang, Y. K. Kim, Y. S. Lee, D. H. Jeong and M. H. Cho, *Nanomed. Nanotechnol. Biol. Med.*, 2007, **3**, 95–101.
112. A. Nel, T. Xia, L. Madler and N. Li, *Science*, 2006, **311**, 622–627.
113. D. Mark, in *Nanotechnology – Consequences for Human Health and the Environment*, ed. R. E. Hester and R. M. Harrison, The Royal Society of Chemistry, Cambridge, 2007, pp. 50–80.
114. ISO, *Workplace Atmospheres – Ultrafine, Nanoparticle and Nano-structured Aerosols – Inhalation Exposure Characterization and Assessment*, ISO, 2007.
115. R. M. Harrison, in *Nanotechnology – Consequences for Human Health and the Environment*, ed. R. E. Hester and R. M. Harrison, The Royal Society of Chemistry, Cambridge, 2007, pp. 35–49.
116. G. R. Cass, L. A. Hughes, P. Bhave, M. J. Kleeman, J. O. Allen and L. G. Salmon, *Philos. Trans. R. Soc. London, Ser. A*, 2000, **358**, 2581–2592.
117. M. Semmler-Behnke, S. Takenaka, S. Fertsch, A. Wenk, J. Seitz, P. Mayer, G. Oberdorster and W. G. Kreyling, *Environ. Health Perspect.*, 2007, **115**, 728–733.
118. G. Oberdorster, *Int. Arch. Occupat. Environ. Health*, 2001, **74**, 1–8.
119. M. Geiser, B. Rothen-Rutishauser, N. Kapp, S. Schurch, W. Kreyling, H. Schulz, M. Semmler, V. I. Hof, J. Heyder and P. Gehr, *Environ. Health Perspect.*, 2005, **113**, 1555–1560.

120. A. Nemmar, P. H. M. Hoet, B. Vanquickenborne, D. Dinsdale, M. Thomeer, M. F. Hoylaerts, H. Vanbilloen, L. Mortelmans and B. Nemery, *Circulation*, 2002, **105**, 411–414.

121. K. Donaldson, X. Y. Li and W. MacNee, *J. Aerosol Sci.*, 1998, **29**, 553–560.

122. G. Oberdorster, Z. Sharp, V. Atudorei, A. Elder, R. Gelein, W. Kreyling and C. Cox, *Inhalation Toxicol.*, 2004, **16**, 437–445.

123. P. R. Lockman, R. J. Mumper, M. A. Khan and D. D. Allen, *Drug Dev. Ind. Pharmacy*, 2002, **28**, 1–13.

124. A. des Rieux, E. G. E. Ragnarsson, E. Gullberg, V. Preat, Y. J. Schneider and P. Artursson, *Eur. J. Pharm. Sci.*, 2005, **25**, 455–465.

125. B. Baroli, M. G. Ennas, F. Loffredo, M. Isola, R. Pinna and M. A. Lopez-Quintela, *J. Investigat. Dermatol.*, 2007, **127**, 1701–1712.

126. B. D. Chithrani, A. A. Ghazani and W. C. W. Chan, *Nano Lett.*, 2006, **6**, 662–668.

127. P. J. Smith, M. Giroud, H. L. Wiggins, F. Gower, J. A. Thorley, B. Stolpe, J. Mazzolini, R. J. Dyson and J. Z. Rappoport, *Int. J. Nanomed.*, 2012, **7**, 2045–2055.

128. R. C. Stearns, J. D. Paulauskis and J. J. Godleski, *Am. J. Respiratory Cell Mol. Biol.*, 2001, **24**, 108–115.

129. A. E. Nel, L. Madler, D. Velegol, T. Xia, E. M. V. Hoek, P. Somasundaran, F. Klaessig, V. Castranova and M. Thompson, *Nature Mater.*, 2009, **8**, 543–557.

130. L. Tran, R. Aitken, J. Ayres, K. Donaldson and F. Hurley, in *Nanotechnology – Consequences for Human Health and the Environment*, ed. R. E. Hester and R. M. Harrison, The Royal Society of Chemistry, Cambridge, 2007, pp. 102–117.

131. A. D. Maynard and E. D. Kuempel, *J. Nanoparticle Res.*, 2005, **7**, 587–614.

132. K. Gardiner, M. van Tongeren and M. Harrington, *Occupat. Environ. Med.*, 2001, **58**, 496–503.

133. T. Sorahan, L. Hamilton, M. van Tongeren, K. Gardiner and J. M. Harrington, *Am. J. Ind. Med.*, 2001, **39**, 158–170.

134. J. J. Beaumont and N. S. Weiss, *J. Occupat. Environ. Med.*, 1981, **23**, 839–844.

135. N. Becker, J. Claude and R. Frentzelbeyme, *Scand. J. Work Environ. Health*, 1985, **11**, 75–82.

136. B. A. Racette, S. D. Tabbal, D. Jennings, L. Good, J. S. Perlmutter and B. Evanoff, *Neurology*, 2005, **64**, 230–235.

137. J. M. Antonini, M. D. Taylor, A. T. Zimmer and J. R. Roberts, *J. Toxicol. Environ. Health, Part A*, 2004, **67**, 233–249.

138. L. Simonato, A. C. Fletcher, A. Andersen, K. Anderson, N. Becker, J. Changclaude, G. Ferro, M. Gerin, C. N. Gray, K. S. Hansen, P. L. Kalliomaki, K. Kurppa, S. Langard, F. Merlo, J. J. Moulin, M. L. Newhouse, J. Peto, E. Pukkala, B. Sjogren, P. Wild, R. Winkelmann and R. Saracci, *Br. J. Ind. Med.*, 1991, **48**, 145–154.

139. A. Peters, H. E. Wichmann, T. Tuch, J. Heinrich and J. Heyder, *Am. J. Respirat. Crit. Care Med.*, 1997, **155**, 1376–1383.

140. P. Penttinen, K. L. Timonen, P. Tiittanen, A. Mirme, J. Ruuskanen and J. Pekkanen, *Environ. Health Perspect.*, 2001, **109**, 319–323.

141. T. Osunsanya, G. Prescott and A. Seaton, *Occup. Environ. Med.*, 2001, **58**, 154–159.

142. A. Ibald-Mulli, H. E. Wichmann, W. Kreyling and A. Peters, *J. Aerosol Med.*, 2002, **15**, 189–201.

143. H. E. Wichmann and A. Peters, *Philos. Trans. R. Soc. London, Ser. A*, 2000, **358**, 2751–2768.

144. H. E. Wichmann, C. Spix, T. Tuch, K. Wittmaack, J. Cyrys, G. Wolke, A. Peters, J. Heinrich, W. G. Kreyling and J. Heyder, *Daily Mortality and Fine and Ultrafine Particles in Erfurt, Germany, Part B: Role of Sources, Elemental Composition and other Pollutants*, Health Effects Institute, 2000.

145. R. W. Atkinson, G. W. Fuller, H. R. Anderson, R. M. Harrison and B. Armstrong, *Epidemiology*, 2010, **21**, 501–511.

146. S. Breitner, L. Liu, J. Cyrys, I. Brueske, U. Franck, U. Schlink, A. M. Leitte, O. Herbarth, A. Wiedensohler, B. Wehner, M. Hu, X.-C. Pan, H. E. Wichmann and A. Peters, *Sci. Total Environ.*, 2011, **409**, 5196–5204.

147. R. Ruckerl, A. Ibald-Mulli, W. Koenig, A. Schneider, G. Woelke, J. Cyrys, J. Heinrich, V. Marder, M. Frampton, H. E. Wichmann and A. Peters, *Am. J. Respirat. Crit. Care Med.*, 2006, **173**, 432–441.

148. M. W. Frampton, J. Bausch, D. Chalupa, P. K. Hopke, E. L. Little, D. Oakes, J. C. Stewart and M. J. Utell, *Inhalation Toxicol.*, 2012, **24**, 831–838.

149. M. W. Frampton, M. J. Utell, W. Zareba, G. Oberdorster, C. Cox, L.-S. Huang, P. E. Morrow, F. E.-H. Lee, D. Chalupa, L. M. Frasier, D. M. Speers and J. Stewart, *Research Report (Health Effects Institute)*, 2004, 1–47; discussion 49–63.

150. N. Li, C. Sioutas, A. Cho, D. Schmitz, C. Misra, J. Sempf, M. Y. Wang, T. Oberley, J. Froines and A. Nel, *Environ. Health Perspect.*, 2003, **111**, 455–460.

151. J. R. Gurr, A. S. S. Wang, C. H. Chen and K. Y. Jan, *Toxicology*, 2005, **213**, 66–73.

152. E. Bermudez, J. B. Mangum, B. A. Wong, B. Asgharian, P. M. Hext, D. B. Warheit and J. I. Everitt, *Toxicol. Sci.*, 2004, **77**, 347–357.

153. A. Nemmar, M. F. Hoylaerts, P. H. M. Hoet and B. Nemery, *Toxicol. Lett.*, 2004, **149**, 243–253.

154. K. Donaldson and V. Stone, in *Nanotechnology – Consequences for Human Health and the Environment*, ed. R. E. Hester and R. M. Harrison, The Royal Society of Chemistry, Cambridge, 2007, pp. 81–101.

155. The Royal Society and The Royal Academy of Engineering, Nanoscience and Nanotechnologies: Opportunites and Uncertainties, 2004; http://royalsociety.org/uploadedFiles/Royal_Society_Content/policy/publications/2004/9693.pdf (accessed06/07/2013).

156. European Commision, *Towards a European Strategy for Nanotechnology*, 2004.

157. European Commision, *Nanosciences and Nanotechnologies: An Action Plan for Europe 2005–2009*, 2005.

158. European Commission, *Follow-up of the 6th Meeting of REACH Competent Authorities for the Implementaion of Regulation (EC) 1907/2006 (REACH), Nanomaterials in REACH*, 2008.

159. European Commission, *Annex II: Final Version of Classification, Labelling and Packaging of Nanomaterials in REACH and CLP*, 2008.

CHAPTER 21

The Legal Control of Pollution

RICHARD MACRORY[a] AND WILLIAM HOWARTH*[b]

[a] Centre for Law and the Environment, Faculty of Laws, University College London, Endsleigh Gardens, London WC1H OEG, Email: r.macrory@ucl.ac.uk; [b] Kent Law School, University of Kent, Canterbury, Kent, CT2 7NS, UK
*Email: w.howarth@kent.ac.uk

21.1 INTRODUCTION

Inevitably, laws on pollution control differ between jurisdictions and each set of national laws, viewed comparatively, might be seen as unique. In the concise discussion which follows, it is impossible to explore national variations in any depth, and the legislation of the United Kingdom, and more particularly of England, provides the main focus of discussion. It is recognised that the national law of England has some distinctive characteristics, as a common law jurisdiction in which important principles are established by judges deciding cases rather than through legislatures seeking comprehensively to codify the law, as happens in many legal systems. The national legislation to be discussed might also be seen as atypical in reflecting a lengthy historical progression of attempts to address pollution problems through legislation that has taken place in the United Kingdom. The continuing influence of this legal legacy might be seen as a point of contrast to those countries where pollution control laws have been adopted only relatively recently. Nonetheless, more recent 'national' laws on pollution control are frequently adopted in response to broader global or regional international agreements and this is noted in the discussion that follows. Particularly, as a Member State of the European Union, the United Kingdom is bound to adopt legislation to transpose Union environmental legislation into national law, and may be seen as sharing this characteristic with other Member States even though the national laws that are adopted may not be identical. Recognising the inevitable differences between national laws on pollution control, it is suggested that the laws to be discussed provide an illuminating national 'case study'.

In a modern industrial society the control and management of pollution needs to be underpinned by law. The realisation of this came at a relatively early stage in the industrialization of the United Kingdom, where a series of legal enactments dating from the early nineteenth century first sought to criminalise the most serious kinds of pollution of water and air, and introduced public health measures for the prevention of 'nuisances'. By international comparisons, the early origin of

Pollution: Causes, Effects and Control, 5th Edition
Edited by R M Harrison

national environmental law is remarkable. Setting aside questions of practical effectiveness, the early attempts to legislate for environmental protection may be seen as providing a bedrock of key ideas as to how the law should be applied in this context and, in some instances, establishing legal mechanisms that have proved to be remarkably durable over time.

However, from its precocious origins, UK pollution law has undergone momentous change and developed increasing sophistication as the crudity of some the early attempts to 'legislate away' pollution came to be appreciated. The importance of allocation of enforcement responsibilities to an appropriately empowered and resourced regulatory authority came to be recognised. The use of licensing powers, to authorise emissions and activities that are not seriously harmful, became widely used as dispensation from broadly defined legislative bans on polluting activities. Not least significant, was the progressive recognition that sectoral concerns, about pollution of water, air and land, should be superseded by the need to protect the environment, *as a whole*, and that that law should facilitate this integrated approach. Moreover, from the traditionally reactive role of environmental law, in merely providing for sanctions against wrongdoers, the law has developed an important forward-looking or strategic dimension. This is seen in use of the law in requiring measures to prevent pollution and to realise a satisfactory state of environmental quality, contributing to the long-term goal of sustainable development.

Although the early development of pollution law in the UK may, in part, have been attributable to public concern and national perceptions of what kinds of conduct are environmentally unacceptable, substantial changes have been brought about as a result of the need to adapt legislation to meet more broadly formulated international obligations. This has involved giving effect to obligations agreed in regional or global international treaties. However, the major source of non-national environmental law which needs to be implemented in the UK derives from the European Union. From its origins as a regional economic grouping, the EU has developed into the most advanced supra-national source of environmental law and policy in the world, having adopted several hundred environmental legislative measures which need to be given effect to within the 28 Member States. Not only has the EU been tremendously prolific in its environmental legislative output, it has embraced a range of approaches to environmental regulation that are without counterparts in UK national law. Certainly, the increasing need to ensure that UK law and policy reflects EU obligations has had a far-reaching effect upon national thinking about how the law may best be used to protect the environment. This may be no bad thing since, as is widely recognised, 'pollution does not respect national boundaries'.

Alongside the appreciation that many environmental problems need to be addressed through coordinated supra-national initiatives, is the challenging task of law keeping abreast of changes in technical development and scientific thinking about the environment. Greater scientific knowledge of environmental impacts necessitates a rethinking of the appropriate regulatory approach towards both new processes and products, and also existing activities and their impacts. A good example of this is to be seen in the longstanding concern about air pollution, where over the years the nature of the environmental problem has been subject to fundamental re-characterisation. Unsatisfactory air quality was originally seen as a problem of local amenity and public health (addressed through national legislation) but came to be regarded as a matter having transboundary and regional impacts (addressed under EU law and some specific international treaties) and has eventually been recognised as a global environmental problem (needing to be addressed through a common approach involving all nations). Law frequently lags behind scientific knowledge. It also encounters greater difficulties in addressing more diffuse kinds of pollution, where the origins and impacts are more remote or uncertain, and particularly where international agreement between nations is needed as a response.

Given the diverse and dynamic character of environmental law, and the extremely technical nature of many individual environmental laws, the aim of this chapter is to provide a selective overview of role of the law in the control and management of pollution, with observations on some of the key themes. Although the main focus is upon the UK, and particularly England, in many respects the discussion involves scanning a range of legal approaches, many of which have their origins in EU environmental law and policy or in global international environmental treaties. Following an introductory account of the purposes and mechanisms of law, the body of the chapter considers environmental laws arising from different sources and with different institutional responsibilities, national, EU and international, and the particular importance attached to private rights in respect of protection from pollution. A sample of the possibilities drawn from these different legal sources are pursued through four case studies: looking at contrasting uses of the law in relation to statutory nuisances; water pollution and water quality law; integration of pollution control and environmental permitting; and procedural law. The concluding observations offer some reflections upon the overall contribution of environmental law.

21.2   THE PURPOSES AND MECHANISMS OF ENVIRONMENTAL LAW

Environmental law serves a variety of purposes and does so through a range of regulatory mechanisms. At the most mandatory extreme, it seeks to define the kinds of activity that society has determined are so unacceptable that they should constitute criminal offences and be subject to sanctions for violation. Drafting laws of this kind involves precise wording, to ensure precision and clarity as to what acts are being made unlawful. Every citizen has the right to know what actions constitute crimes. This is particularly important in this area where many crimes have been deliberately drafted in the legislation so as to avoid the need for the prosecution to show intention or recklessness on the part of the wrongdoer (known as 'strict liability'). Criminal prohibitions will almost invariably need to provide for various exceptions and defences, for example, in relation to actions undertaken in an emergency or actions specifically authorised for various purposes. Not least important is the allocation of enforcement responsibility and the need to specify who has power to bring legal proceedings. In some instances, in England and Wales, an environmental criminal prosecution may be instigated by anyone, including an environmental non-governmental organisation (NGO), but commonly enforcement responsibility is entrusted to a particular regulatory authority. Where enforcement responsibility is identified, a typical environmental offence will deal in considerable detail with enforcement powers and procedures, for example, the powers available to a regulatory authority to undertake monitoring and to enter premises to ascertain whether an offence has been committed.

Where a criminal environmental offence is provided for, it is often made subject to a defence where an activity is allowed under an environmental permit granted to the operator of a particular installation. The process of authorising environmentally benign activities in this way has been a longstanding part of UK environmental regulatory practice, though it has been extensively harmonised by the recent introduction of an environmental permitting, discussed below. The permitting of activities under this scheme involves the imposition of a range of administrative requirements to enable public consultation on permit applications, modification or withdrawal of permits for specified reasons, and for offences to be provided for in relation to contravention of various permitting requirements, with appropriate provision for appeals on matters of this kind. In essence, the permitting system is designed to legitimate a degree of pollution considered acceptable by society.

Prohibitions though, essentially reflect a reactive approach, even where coupled with a permitting system. Of course, they should have a deterrent effect, but there are other respects in which the law adopts a more explicit preventative approach, in seeking to impose requirements that are aimed at avoiding pollution incidents arising, rather than threatening punishment after the event. Examples of preventative uses of the law are manifold, but illustrations may arise in situations where persons have custody of polluting substances and are made subject to specific legal requirements in respect of the storage, use or disposal of the material concerned. The need for an anticipatory approach towards pollution control is also particularly strong where the location is one of special sensitivity for environmental quality or ecological reasons. For that reason, more stringent legal rules may apply where land is specially designated, for example, as a 'nitrate vulnerable zone' in respect of agricultural nitrate contamination or as an area of particular ecological importance under national or EU law.

The prohibitive approach to environmental wrongdoing may also be contrasted with instances where environment law makes provision for liability after a pollution event. This arises, for example, by the imposition of liability upon a person who has caused pollution to meet the cost of clean up operations or making good consequent ecological damage. Liability rules will also define when the polluter must compensate others for damage caused by their activities. Although primarily part of the judge-made 'common law' rather than provided for under legislation, the private rights of individuals to seek redress for environmental harm remain an important legal aspect of the subject. In principle, this branch of the law enables victims of pollution to gain compensation for the harm they have suffered and, in exceptional circumstances, a court order requiring the polluting activity to cease. An essential part of environmental law, therefore, is about the use of the law to protect the rights of individuals. Liability law may also help in encouraging conduct to reduce liability in that most businesses (and certainly their insurers) do not wish to be exposed to liability, but this is not normally the primary aim of such laws.

Alongside prohibiting and preventing polluting activities, environmental law also focuses upon the environmental implications of decision-making and policy implementation by central government and public bodies. In a society where extensive executive power is exercised by the state across a spectrum of environmental concerns, a significant part of environmental law relates to the scrutiny of public decision-making and the basis for legal challenge. Notably, in this context, the principle of the 'rule of law' requires that government or a public body has an explicit legal basis for its actions. Hence, where a public body purports to authorise an activity which may be damaging to the environment, it must be able to show that it has the legal power to grant the particular authorisation and that the procedure that led to the decision satisfies general requirements as to fairness. Decisions that are lacking in either of these respects are subject to challenge by way of 'judicial review', which has frequently allowed opponents to overturn decisions by public bodies which were to seen to have adverse environmental consequences.

The basis for an environmental challenge against a national government or executive body is not limited to matters of arising from national decision-making. Membership of the EU imports the obligation that the UK will give full effect to EU obligations and will do nothing to obstruct implementation of those obligations. Whilst provision is made for central enforcement of EU obligations, by the European Commission against Member States who fail to comply with their EU obligations, this does not preclude national challenges to government or public bodies where EU environmental legislation is not fully and correctly implemented. Hence, failure to transpose an EU environmental measure into national law in a timely and comprehensive way is a breach of law of a kind which may give rise to rights of action by an individual or environmental NGO. Beyond that, where an EU environmental measure gives rise to ongoing obligations in respect of implementation, monitoring, reporting, formulating action plans, enforcement and a diverse range of other matters, and these are not met by the UK, a right of challenge may arise. Although deriving from EU law, these are rights which are enforceable in national courts, potentially leading to compensation being payable to individuals where a sufficient interest can be shown to have been infringed.

Alongside the 'regulatory' aspects of environmental law, where legislation provides for criminalisation of polluting activities or where executive bodies are constrained to exercise legal powers lawfully, another branch of environmental law concerns liability for harms suffered by individuals.

Given this extensive range of legal purposes and mechanisms, it is evident that environmental law comprises an eclectic range of approaches, perhaps unified only by the fact that they are the kinds of law that are most likely to be relevant to addressing an 'environmental problem' (and the characterisation of this may itself sometimes be debatable). The following section seeks to structure the different kinds of environmental law according to their sources and institutional responsibilities.

21.3   SOURCES OF LAW AND INSTITUTIONAL RESPONSIBILITIES

21.3.1   National Law

Despite the early origins of environmental law, the national approach, until fairly recent times, has been reactive and piecemeal. Diverse legal controls, relating to pollution of water and air, and nuisances, were conceived of by Parliament as meeting essentially different concerns, needing to be addressed by separate legislation and with regulatory responsibility entrusted to bodies with distinct remits. The appreciation that the environment should be considered holistically and that regulatory roles should be unified, or at least co-ordinated to recognise this, has gained ground only recently. In part, the legacy of environmental sectoralism is to be seen the lack of a codified body of statutory law concerning the environment of the kind adopted in other jurisdictions. *The Environmental Protection Act 1990*, the *Water Resources Act 1991*, the *Clean Air Act 1993* and the *Environment Act 1995* are key enactments in consolidating and updating previous measures regarding the pollution of land, water and air, and regulatory responsibilities for these. Nonetheless, they need to be read alongside other legislation governing land use planning and nature conservation to gain a fuller picture of the spectrum of environmental controls.

As regards the form of national legislation on pollution control, the historic preference has been for laws to be set out in primary acts of Parliament, generally drafted and promoted by the Government. Although often lengthy and detailed, primary legislation often comprises only a skeleton framework for regulation, with powers commonly given to Ministers to make regulations or other forms of subsidiary or secondary legislation at a later date. Regulations have administrative attractions since they are much simpler to promote, change and update than primary acts of Parliament, and may contain a wealth of detail which would otherwise overload the parliamentary process and become inflexible. Notwithstanding the detail and technicality of primary and secondary environmental legislation, the final word on the meaning of this legislation, and how it should be applied to particular situations, lies with the courts. Judicial decisions on the interpretation of the legislation are an essential source of law, capable of placing an elucidation upon legislation which might not have been apparent from its literal reading.

Although primary and secondary legislation form the main sources of substantive UK law regarding pollution control, they will often be supplemented by various forms of government guidance in the form of published circulars, codes of practice or similar measures. The use of supplementary guidance or 'soft law' has become increasingly common and influential in the implementation of national law, with a similar trend in the frequent use of guidance in respect of EU legislative measures and other international environmental legislation. Guidance, as a supplement to substantive law, is attractive to policy makers because it allows implementation approaches to be expressed in flexible, non-legalistic language which may be useful particularly when dealing with a new area of policy or where formalistic rules would be inappropriate. The precise legal status of guidance, however, is variable depending upon whether it is explicitly required or merely facilitated by legislation. In some instances statutes require or allow the relevant Minister to adopt guidance, in which case this would be accorded higher status as 'statutory guidance'. So, for example, primary legislation enables the relevant Minister to approve a *Code of Good Agricultural Practice*, to give practical guidance and to promote desirable practices for minimising water pollution. Although it is stated that contravention of the Code will not of itself give rise to any criminal or civil liability the regulatory authority is bound to take this into account when deciding how it should exercise certain regulatory powers.[1] In other instances, no formal provision is made for the issue of guidance and any guidance that is provided will be accorded a lesser status, though may still be relevant, for example, when a court is seeking to ascertain what constitutes an appropriate standard of environmental management in a particular activity.

Primary legislation also has an important role in establishing regulatory responsibilities in respect of environmental offences and related requirements. *The Environment Act 1995* establishes the Environment Agency for England, setting out legal requirements for its composition, aims and objectives, functions, advisory committees and a diverse range of legal powers and duties in respect of environmental quality matters. The establishment of the Agency constituted a major step in the integration of administrative responsibility for pollution control, in bringing together regulatory functions that were previously exercised by different public bodies, particularly in relation to water, waste management and certain industrial processes. Also, the *1995 Act* was innovative in setting out the way in which environmental regulatory functions should be exercised. Specifically, the Agency's functions in respect of pollution control are to be exercised to protect and enhance the environment "taken as a whole" and to make a contribution towards attaining the objective of achieving sustainable development indicated in Ministerial guidance.[2] General duties are imposed upon the Agency in respect of the environment, requiring it to "further" or "have regard to" the conservation of flora, fauna and other natural features[3] and to have regard to the costs and benefits of action or inaction in relation to the exercise of certain legal powers.[4]

Since its establishment, the Environment Agency has been active in seeking to meet the high expectations engendered by the *1995 Act*, particularly in relation to a holistic, integrated and risk-related approach to environmental law enforcement. Modernisation of the approach towards environmental law enforcement has proved challenging for the Agency, at least in the early period of its establishment, because of the diversity of environmental laws for which it had regulatory responsibility. As will be seen, the process of securing a harmonised approach to enforcement across different sectors of activity has been greatly enhanced by the recent introduction of a unified environmental permitting regime.

As regards the actual enforcement practice of the Agency, it is notable that it has sought to instill greater transparency and consistency in its use of regulatory powers by the adoption of an explicit statement on the uses that it envisages making of its powers of enforcement. The Agency's *Enforcement and Sanctions Statement* sets out guiding principles, though this needs to be read alongside its *Guidance and Offence Response Options* document which gives numerous examples of environmental infringements of different magnitude and seeks to relate the gravity of the incidents to the proportionate legal response.[5] The intention is that it should be possible to ascertain, so far as reasonably practical given the inevitable uniqueness of every particular infringement, what regulatory or enforcement consequences will follow from any particular category of environmental non-compliance.

21.3.2 European Union Law

Over recent years, the greater part of 'national' environmental lawmaking has been undertaken for the purpose of giving effect to obligations that are entered into by the UK as a Member State of the European Union. Historically, this is remarkable because the Treaty of Rome (1957) which established the original European Economic Community, sought to establish a common market and secure harmonious development of economic activities, and contained no explicit provisions for the adoption of environmental legislation. Although legislation with an environmental dimension was adopted at a relatively early stage, ostensibly for the purpose of attaining the objectives of the common market, explicit powers in relation to the environment were not provided for until the *Single European Act (1987)*. Most recently, the adoption of the Treaty of Lisbon in 2009 has involved the replacement of the original European Community Treaty by the Treaty on the Functioning of the European Union (TFEU) and an amended version of the Treaty on the European Union (TEU), which now contain the key provisions of EU law relating to the environment.

EU Union environmental law contrasts markedly with national legislative approaches in respect of its foundations in an environmental policy based upon stated objectives and principles. At the broadest level, the environmental objectives of the EU include working for 'sustainable development' and 'a high level of protection and improvement of the quality of the environment' (Art.3(3) TEU). Not only does this involve, adopting specific environmental legislation, but environmental protection requirements must be integrated into the definition and implementation of other Union policies and activities, in particular with a view to promoting sustainable development (Art.11 TFEU). EU environment policy is to be based on the precautionary principle, and on the principles that preventive action should be taken, that environmental damage should, as a priority, be rectified at source and that the polluter should pay (Art.191(1) TFEU). Although the ideas of 'precaution', 'prevention' and 'making polluters pay' are widely endorsed in international environmental instruments, it is notable that these terms are not defined in EU treaties, and their precise meaning and application in specific contexts may be contentious. Nonetheless, the idea that particular legislation should be based on explicit, if general, environmental policy principles contrasts with the absence of explicit principles of this kind in national environmental law-making. Application of the EU environment policy in specific areas is indicated by successive Environment Action Programmes which identify priorities for legislation and other kinds of action over specified periods.[6]

In respect of the application of the environmental policy principles in particular legislative acts, EU foundational treaties provide powers to adopt particular secondary legislative measures in areas within the Union's competence. Most EU environmental legislation has been adopted in the form of 'directives'. The status of a directive is that it is binding upon Member States as to the result to be achieved, but largely leaves to the national authorities the choice of form and methods by which that result is to be achieved (Art.288 TFEU). When adopted, a directive usually allows a period for transposition into national law (usually two or three years) and, within that period a Member State is bound to give the directive effect in its national laws. This may involve putting in place administrative and enforcement arrangements to ensure that its objectives are fully met within the national jurisdiction. Hence, the EU's 'sincere cooperation' obligation requires Member States to take appropriate measures, of a general or particular kind, to ensure fulfilment of the obligations arising out of the EU treaties or resulting from particular acts of the institutions of the Union (Art.4(3) TEU).

Beyond formal legal transposition requirements, particular environmental directives may impose a range of obligations of an administrative or environmentally substantive kind. Commonly, an environmental directive requires the identification of a competent authority to administer the directive by formulating plans or programmes, determining licence applications, designating areas or installations that are subject to control, and bringing enforcement proceedings where necessary. The failure to identify or empower an appropriate public body for these purposes would be a breach of the directive for which the Member State would be liable.

By contrast to the possibilities for administrative failure to implement an environmental directive, substantive failure to implement a directive arises where a directive specifies that a particular level of environmental quality must be achieved and where a Member State fails to meet that requirement. For example, where an environmental quality directive sets out precise parameters, the failure of a Member State to meet a parameter would be a breach of the directive, irrespective of its compliance in respect of formal transposition and its administrative compliance. Hence, in relatively early cases before the Court of Justice, the UK was found guilty of failing to meet substantive environmental requirements in respect of the quality of drinking water and the quality of bathing waters.[7] Significantly, in these cases it was established, as a principle of EU law, that it is not open to a Member State to plead circumstances within its own control as a reason for failing to implement a directive. In effect, implementation of EU legislation involves a strict obligation upon a Member State to ensure the timely and complete fulfilment of both formal and substantive obligations under an environmental directive.

The need to give national effect to an EU directive should mean that each Member State enacts national transposing legislation which creates legally binding requirements to secure its timely and complete implementation. In theory, this might have the consequence that any individual who has suffered harm as a result of the inadequate implementation should be able to seek a remedy in a national court for harm caused by the Member State's failure to fulfil the implementation requirements of the directive. To some extent, the possibility of state liability for failure to implement European Union legislation is recognised under principles established by the European Court of Justice.[8] However, in environmental contexts, state liability for failure fully to implement a directive is dependent upon whether the claimant is within the range of persons that a particular directive was intended to protect and whether the loss suffered is sufficiently serious and directly caused by the breach Because most environmental directives are intended to protect the general public interest in environmental quality, rather than the specific interests of individual claimants, the possibilities for a successful claim of this kind are limited.

National courts are subject to various requirements in respect of the application of EU law within national jurisdictions which might provide a basis for challenging the adequacy of implementation of a Union measure. In circumstances, where a Member State has failed to transpose a directive into national law, the principle of 'direct effect' allows rights to be claimed under the directive providing that its requirements are so clear, precise and unconditional that no discretion is allowed to a Member State as to the manner of its implementation.[9] However, direct effect is limited to providing remedies against the Member State or public bodies and does not allow rights to be enforced against private individuals or corporate bodies.[10] Alternatively, where an EU directive has been transposed into national law, but the national law is at variance with the requirements of the directive, national courts are bound to apply a principle of 'sympathetic interpretation' which requires the national law to be interpreted consistently with the directive, so far as this is possible.[11] National courts are also obliged to adopt an active role in satisfying themselves that European Union law is being fully complied with, whether or not parties in legal proceedings specifically raise this as an issue.[12]

Most EU environmental law is in the form of Directives, allowing a large degree of discretion to Member States as to how they achieve the obligations contained in the Directive. This allows, for example, Member States with a heavily devolved system of government to designate local bodies as the key implementing authorities, whereas others with a more centralized tradition may appoint national bodies.

The other key form of EU law is a Regulation. This takes immediate effect within a Member State, and under principles of EU law will trump any conflicting national laws. No national transposition measures are normally needed. Deciding whether to choose a directive or regulation is essentially a matter of political choice rather than legal principle, and regulations are best suited where it is very important to have precisely the same legislation in all Member States. In the environmental field, regulations are very much in the minority, and have been used in areas such as product standards (where any variation in national laws between Member States will inhibit trade), transboundary movements of waste, or where the EU is implementing an international treaty to which it is party.

Although various possibilities exist for individuals to bring actions based upon EU environmental law in national courts, the alternative to this is the central enforcement of Union law under powers provided to the European Commission. As the 'guardian of the Treaties', the Commission is entitled to instigate proceedings before the Court of Justice of the European Union in circumstances where a Member State is in breach of a directive. Actions of this kind may be prompted either by a non-compliance finding arising from the Commission's monitoring of implementation of EU law by Member States or by a complaint being raised by an individual, an environmental NGO or a member of the European Parliament.

The Court of Justice of the European Union has powers to consider cases brought by the Commission where it is alleged that a Member State has failed to implement EU law. Where the Court finds that a Member State has failed to fulfil an obligation under a Union Treaty, the state is required to take the necessary measures to comply with the judgment of the Court (Art.260 TFEU). Where the Commission considers that a Member State has not taken the necessary measures to transpose a directive into national law or failed to comply with a previous judgment of the Court, it may bring the case before the Court and specify the amount of a lump sum or penalty payment to be paid by the Member State. The Court may then impose a lump sum or penalty payment on it (Art.260(2) and (3) TFEU). The Court is the only international court in the world which can directly impose financial penalties on countries, and although the power is used sparingly, it is notable that a significant number of the early cases which were subject to this procedure concerned continuing non-compliance with environmental directives.[13]

A case may also reach the Court of Justice of the European Union where a national court seeks authoritative guidance on the meaning of EU law raised in a particular case. The Court will provide the guidance, but leave it to the national court to decide its case in the light of its ruling. In this sense the Treaty deliberately encourages a partnership between national courts and the Court of Justice of the European Union rather than imposing a more familiar hierarchy of courts

21.3.3   International Law

'International law' is generally understood to mean either the customary rules which are generally accepted as governing the relationships between nations and those rules which have been mutually accepted as governing these relations and set out explicitly in bilateral, multilateral or global conventions, treaties or agreements. Documents of these kinds are couched in the language of obligations, though often comprise 'soft law' duties of the most general or unspecific kind. Such commitments are made voluntarily and with the appreciation that violation is not normally intended to result in legal proceedings or punitive sanctions. Although some modern environmental treaties incorporate fairly sophisticated internal mechanisms to monitor and review measures taken by signatories, there is no systematic institutional means for ensuring compliance with international law. In respect of environmental matters, the International Court of Justice has only rarely considered cases in this field. Possibly this is because, for diplomatic reasons, states do not seem to regard the Court as a suitable forum in which to resolve these kinds of dispute. This is not to say that, diplomatic pressure, economic interests, public opinion and a range of other matters are not good reasons why nations should feel obliged to honour international commitments, but the model of mandatory rules being enforced by a regulatory authority is generally misplaced in the international environmental law context.

Beyond the essentially consensual character of international law, it must be recognised that states are the primary actors and convention obligations usually only apply between states. They have no direct application to individuals or non-state bodies, except where certain non-state bodies such as the EU, have been given the status of parties. The inter-state character of international law is particularly evident in countries like the UK where a 'dualist' approach to the status of international conventions is adopted. The effect of this is that, even where an international convention has been ratified by the UK, it cannot be pleaded directly before a national court in the absence of national transposing legislation.

Despite the consensual and inter-state characteristics of international law, international agreements to address diverse environmental concerns have proliferated over recent years and have been of major importance in the development of regional and national environmental law. Recognising that many environmental issues involve impacts across national frontiers and require international cooperation to address them, international environment agreements have the capacity to secure improvements which cannot be accomplished by individual nations acting alone. Broadly, a distinction may be drawn between international agreements which seek to establish general objectives for the environment and those which seek to address more specific kinds of environmental problem.

At the general level, the United Nations Conference on the Human Environment, held in Stockholm in 1972, produced a Declaration of principles which may be seen as providing a foundation for international legislation on the environment. Principle 1 of this Declaration states, "Man has the fundamental right to freedom, equality and adequate conditions of life, in an environment of a quality that permits a life of dignity and well-being and he bears a solemn responsibility to protect and improve the environment for present and future generations". Principle 21 provides that, "States have, in accordance with the Charter of the United Nations and the principles of environmental law, the sovereign right to exploit their own resources pursuant to their own environmental policies, and the responsibility to ensure that activities within their jurisdictions or control do not cause damage to the environment of other states or of areas beyond the limits of national jurisdiction".[14]

The foundational ideas from the Stockholm Declaration were further developed at the 1992 United Nations Conference on Environment and Development held at Rio de Janeiro. The Rio Declaration established the concept of 'sustainable development' as the global imperative for the environment and development. Principle 1 of the Declaration states that human beings are at the centre of concerns for sustainable development, and that they are entitled to a healthy and productive life in harmony with nature. Principle 3 requires that the right to development must be fulfilled so as equitably to meet developmental and environmental needs of present and future generations. Principle 4 asserts that in order to achieve sustainable development, environmental protection must constitute an integral part of the development process and cannot be considered in isolation from it.[15] Despite the important global consensus on the need for 'sustainable development' established under the Declaration, no definition of its meaning was provided. A frequently cited definition is that it comprises "development which meets the needs of the present without compromising the ability of future generations to meet their own needs".[16] However, the absence of any elaboration of an authoritative and internationally agreed definition of the key concept in the Rio Declaration may be seen as indicative of the difficulties in reaching a consensus on more specific environmental commitments. The more specific the commitment, the more difficult it becomes to secure a widely endorsed international agreement, and the rather nebulous idea of 'sustainable development' may be seen as representing the best or most feasible overall compromise in this respect.

The general international environmental obligations set out in the Stockholm and Rio Declarations have been applied in a wide range of international instruments, bilateral and multilateral, and regional and global, which apply to specific kinds of activity which adversely impact upon the environment and ecosystems. It is beyond the scope of this chapter to provide extensive coverage of this extensive body of law, but some of the international agreements which have greatest general relevance to pollution control may be briefly noted:

- 1979 Geneva Convention on Long-range Transboundary Air Pollution and protocols on emission of sulphur dioxide (1994) nitrous oxides (1988) volatile organic compounds (1991) persistent organic pollutants (1998) and heavy metals (1998).[17]
- 1985 Vienna Convention on the Protection of the Ozone Layer (and 1990 Montreal Protocol).[18]
- 1989 Basel Convention on the Control of Transboundary Movements of Hazardous Wastes and their Disposal.[19]

- 2001 Stockholm Convention on Persistent Organic Pollutants.[20]
- 1982 United Nations Convention on the Law of the Sea.[21]
- 1992 OSPAR Convention for the Protection of the Marine Environment of the North-East Atlantic.[22]
- 1973 MARPOL International Convention on the Prevention of Pollution from Ships.[23]

The practical legal impact of conventions of these kinds is almost entirely dependent upon the implementing actions undertaken by signatories. A good illustration of this is to be seen in the 1992 United Nations Framework Convention on Climate Change.[24] The Climate Change Convention recognised that increasing concentrations of greenhouse gases are responsible for global warming and established the objective of stabilising atmospheric concentrations at a level that will prevent dangerous anthropogenic interference with climate systems. For that purpose, the Framework Convention set out general principles and objectives that ratifying countries should adhere to in reducing emissions of greenhouse gases. The Kyoto Protocol to Climate Change Convention of 1997 went significantly beyond the 1992 Convention in placing binding limits on emissions of greenhouse gases for major industrial countries, with a target for carbon dioxide of at least 5.2% below 1990 levels, needing to be achieved between 2008 and 2012. The 2009 Climate Conference ended with the 'Copenhagen Accord' which set the global objective of keeping warming to less than 2 degrees centigrade above the pre-industrial temperature. However, this outcome fell short of the establishing a legally binding global climate treaty to succeed the Kyoto Protocol from 2013. Securing agreement on post-Kyoto legally binding greenhouse gas emissions objectives remains the objective for future negotiations.

At the EU level, the primary response to international climate change agreements is the adoption of an *Emissions Trading Directive (2003/87/EC)* establishing, from 2005, a scheme for greenhouse gas emission allowance trading, which requires Member States to ensure that specified installations hold a permit for greenhouse gas emissions and to formulate national plan indicating the total quantity of allowances and how these are to be allocated between installations. Alongside, the emissions trading scheme, the EU has also adopted the *Linking Directive (Directive 2004/101/EC)* which gives effect to mechanisms under the Kyoto Protocol for measures which allow for emission reductions in other developed (*Annex 1*) countries through 'joint implementation' or in developing countries through the 'clean development mechanism'. The *Linking Directive* allows for conversion of credits obtained through these mechanisms to be set off against national reductions in emissions up to upper limits set out in national allocation plans. In addition, the EU has adopted a range of more specific measures related to renewable energy which seek further energy sources which reduce carbon emissions.

At a national level, alongside the measures that have been necessary to implement EU climate change legislation, the UK has adopted a range of specific legal instruments concerning the reduction of emissions and providing incentives for conversion to renewable energy sources. Most notable amongst these measures is the *Climate Change Act 2008*, which sets 'legally binding' targets for greenhouse gas emissions, provides powers to enable those targets to enable those targets to be met and strengthens the institutional framework needed to make a transition to a low carbon economy. Specifically, a target is set of at least an 80% cut in greenhouse gas emissions by 2050 and reductions of 34% by 2020. The constitutional power of a government to bind subsequent governments is legally questionable and the practical enforceability of the obligation to meet such targets in legal proceedings is uncertain.[25] Nonetheless, the *2008 Act* may be seen as a formal statement of the commitment of Parliament that national measures should achieve or exceed internationally agreed commitments.

This brief discussion of climate change law serves to illustrate themes which might be seen to apply in relation to any of the international environmental conventions that have been listed above. International environmental law serves to register an international consensus of concern about an environmental issue and to indicate how this should be addressed by parties to an agreement. The specification as to what national response is needed tends to be pitched at a general level (and the relatively specific targets for emission reductions under the Kyoto Protocol may be atypical in this respect). The overriding point is that 'obligations' set out in international conventions only gain a mandatory force where they are made the subject of EU or national laws for which there is a regulatory authority with an enforcement responsibility.

21.4 PRIVATE RIGHTS AND CIVIL REMEDIES

Since the introduction of an extensive body of public 'regulatory' environmental law there has been a tendency to see pollution issues as being entirely governed by provisions of this kind. However, this perception neglects the continuing importance of private rights, particularly of those individuals who suffer harm as a result of pollution. Since the earliest times, judges in common law courts have provided remedies to persons in this position and judge-made rules, alternatively referred to as 'civil law' rights or remedies under the 'law of tort'. These continue to be of importance where pollution involves an intrusion upon private rights. The contrast between public and private law approaches should be stressed: public law environmental regulation is largely concerned with the criminalisation of unacceptable behaviour and ultimately its punitive consequences; whereas private law is primarily aimed at providing redress for the victim of harm.

Typically, situations which raise the potential for an environmental civil action, involve a landholder using land in way which adversely impacts upon neighbouring land use, usually through the transmission of some kind of pollution to the neighbouring land. Potentially, situations of this kind could be legally categorised as 'trespass', where there is a sufficiently direct interference with the neigbouring land, or 'negligence' where there has been a lack of reasonable care towards neighbours, particularly where some kind of personal injury is involved. However, the most commonly used kind of civil action in respect of environmental complaints is the tort of 'private nuisance' which arises where the claimant (victim) is able to show that the defendant's activities amount to an unreasonable interference with the claimant's use or 'enjoyment' of land. The possibilities for a civil action in private nuisance are open-ended, but it is apparent from the historic origins of this kind of action that its purpose is to protect interests in land, rather than the personal interests of occupiers. Nonetheless, a successful action of this kind will allow a court to award compensation, termed 'damages', to the claimant for the harm that has been suffered in respect of land use. In exceptional cases, where damages are an inadequate remedy, to grant a court order, termed an 'injunction', to require the defendant to discontinue the activity that has given rise to the harm.

Focusing upon a civil action in private nuisance, where the claimant has suffered unreasonable interference with use of land due to the activities of the defendant (polluter) the theoretical open-endedness of the action is greatly curtailed by the factors which need to be shown for an action to succeed.

These are:

  (i) An interest in land.
 (ii) Causality.
(iii) Locality.
(iv) 'Reasonable foreseeability' of harm.

The need for a claimant to show an interest in land which has been harmed means that a person without such an interest will be unable to bring an action in private nuisance. Hence, in a leading House of Lords decision, it was found that occupants of houses claiming in a nuisance action were unable to succeed where they lacked an interest in the properties, suggesting that either a freehold or leasehold interest in land will be needed.[26] The property qualification for a nuisance action will be problematic for many prospective environmental litigants, for example, where an amenity or environmental campaigning group wishing to challenge a polluting activity, but lacks any legal interest in the affected land. Moreover, the need to show harm to interests in land is problematic in respect of natural resources that are not the subject of private ownership, such as air, flowing water and wild fauna. If these things are not the subject of ownership, the law of nuisance to land will not be directly unavailable as a means of providing redress for environmental interferences affecting them. The nuisance action only becomes available at the point where, say, pollution of air impacts upon the use of land or pollution of water impacts upon private fishery rights.

The second element, of causality, involves the claimant establishing the origin of pollution, showing that harm has been suffered and that the link between these is sufficiently direct to satisfy the civil burden of proof, requiring this to be shown 'on the balance of probabilities'. Each aspect of causality is capable of being problematic in practice. In a highly polluted industrial area the claimant may be unable to show that the harmful emission or discharge originated from the defendant's premises rather than from other polluting installations in the area. Given that civil litigation is an expensive, lengthy and inherently uncertain activity, an unsupported claimant will have to be willing to undertake to pay substantial costs both in legal fees and for expert witnesses to assist in establishing the all-important findings on matters of fact. Because of the general rule on the allocation of costs in civil proceedings, summed up as 'the loser pays all', the prospect of having to pay the costs of both sides means that there is a significant economic deterrent to civil environmental litigation for all but the wealthiest claimants or those who are poor enough to be granted legal aid. The potential difficulties for a claimant are graphically illustrated by a decided case were a farmer, unsuccessfully, brought civil proceedings claiming that the ill health of his cattle was the result of toxic emissions from a nearby high-temperature incinerator. The litigation occupied 198 days of court time, involved 21 expert witnesses and generated costs of £1.5 million for the claimant (who was legally aided) and £4.5 for the defendant.[27]

The third element of 'locality' relates to the area or neighbourhood in which an alleged nuisance takes place and the recognition that what constitutes a nuisance in one area might not be in another. As it was put in a nineteenth century case, "what would be a nuisance in Belgrave Square would not necessarily be so in Bermondsey".[28] Clearly, it would be unreasonable to expect rural tranquility and pristine environmental purity in a heavily populated city or industrial area and the concept of a 'nuisance' should reflect this relativity. To a degree the law accommodates this, but within important limits. The problem of what kinds of polluting interference a landholder in an industrial area must accept were first addressed by nineteenth century courts dealing with the worst excesses of manufacturing impacts upon the environment. A key decision drew a distinction between 'material injury to property' and mere 'sensible personal discomfort' and found that nuisances of the first kind would be actionable irrespective of locality whereas nuisances of the second kind would be dependent upon locality.[29] Hence, for nuisances involving matters such as noise, smell or dust, which do not involve physical harm to property, or the fumes generated by the operation of a copper smelting works (as in the case referred to) a court will consider the location. A complaint about an industrial activity taking place in an industrial area is, therefore, unlikely to be actionable as a nuisance unless it involves physical damage to property. Again, the implications of this for claimants are unhelpful: those living in the most polluted areas are least likely to have their complaints fall within the scope of a civilly actionable private nuisance.

The final element in a private nuisance action is foreseeability of harm. This aspect was decisive in a House of Lords decision where groundwater, abstracted for supply by a water company, had been contaminated by spillages which, on the balance of probabilities, were found to have originated from a leather tannery. In an action brought by the water company to recover the cost of establishing an alternative water supply source, it was found that recovery of damages would only be possible where the defendant could have reasonably foreseen at the time of the relevant pollution incident that the activity would cause harm to the claimant. On the facts, the court found that at the time of the spillages one might have predicted damages from fumes, but no one would have reasonably predicted the way the spillages seeped through the ground and contaminated a water supply some miles away. This outcome is difficult to reconcile with the polluter pays principle, but it should be recognised that 'foreseeability' is a standard that rises over time, with greater scientific knowledge of adverse environmental impacts. Arguably, were the same facts to arise today, a court might be willing to find that the harm was of a reasonably foreseeable kind.

Given the cumulative obstacles confronting a potential environment claimant in a private nuisance action, it might be concluded that the prospects of success are remote and that the complaint might be better addressed by public regulatory authorities being urged to exercise their prosecution and other enforcement powers. However, because of the inappropriateness of regulatory enforcement and/or the unwillingness of authorities to take action in some instances, civil law retains a remarkable durability as a means of addressing certain kinds of environmental complaint.

A recent example of this durability is to be seen in litigation brought by householders living in the proximity of waste landfill site who maintained that the odour nuisance from the site was so serious that they were unable to open their windows or venture into their gardens. The contention of the waste management company was that the operation of the site was fully in accordance with the statutory licensing requirements for the site and was, therefore, a reasonable use of the land which could not be an actionable private nuisance in civil law. In effect, the conformity with operating permit should provide a defence to what might otherwise be a nuisance. Although this argument was accepted by a lower court, the Court of Appeal firmly rejected it. As it was put,

"The common law of nuisance has co-existed with statutory controls, albeit less sophisticated, since the 19th century. There is no principle that the common law should "march with" a statutory scheme covering similar subject-matter. Short of express or implied statutory authority to commit a nuisance . . . there is no basis, in principle or authority, for using such a statutory scheme to cut down private law rights."[30]

This statement categorically affirms the continuing importance of nuisance and private rights, irrespective of public regulation of the activity giving rise to a complaint. It must be recognised that there are some instances where statutory law explicitly removes the right to a common law action for nuisance but, outside those instances, the right of action in private nuisance remains.

Although private nuisance is the most likely common law action to address an environmental complaint, another significant civil law possibility has been illustrated by recent litigation. This is the possibility of an action in *public* nuisance. A civil action in public nuisance may be brought where there is a nuisance which affects the public generally, but an individual suffers damage going beyond the inconvenience suffered to the public in general. Whilst, public nuisance is a longstanding common law remedy, its application in environmental contexts has been limited until recently.

The application of public nuisance in an environmental context is to be seen in legal proceedings brought by the 'Corby Group'. The central issue was about the liability of a local authority arising from land reclamation work on a former steelworks site that was heavily contaminated by heavy metals and various 'teratogenic' substances. It was argued by the parents of the claimants that exposure of pregnant women to these substances had caused children to be born with deformities to the upper limbs. Alongside negligence, the case was argued on the basis that liability arose in public nuisance and through breach of a statutory duty in respect of waste. In the main proceedings, concerned to establish whether there was a legal basis for liability, it was held that the local authority had been in breach of a duty care by allowing mud and dust from the reclamation sites to be spread, either by the wind or by vehicles, and deposited on public roads in the area. As a result, the claimant's mothers had been exposed to relevant substances. It was reasonably foreseeable that exposure could arise and that pregnant women could inhale or ingest the relevant contaminants and this would lead to birth defects of the type complained of. Alongside other bases for liability, the local authority was found liable for public nuisance in causing, allowing or permitting the dispersal of dangerous or noxious contaminants. Notably, establishing a basis for liability left the question of whether the condition of individuals had actually resulted from the defaults of the local authority. This was a matter which left to be determined by further proceedings, though those proceedings never took place following a substantial, but undisclosed, settlement being made between the parties.[31]

The Corby Group litigation is notable, first, because it illustrates that no interest in land is needed to succeed in action in public nuisance, in marked contrast to the requirements for land ownership to be shown in private nuisance. Second, it shows that recovery for personal injury is possible in public nuisance, whereas this is not allowed in private nuisance which serves only to protect interests in land. Third, it shows a remarkable willingness of a court, in principle at least, to accept that the extremely complex medical evidence that was brought before the Court provided a sufficient basis for the causative mechanism for the deformities that were at issue. Possibly the ruling may open up new avenues for environmental civil litigation which have been barred by the difficulties noted in relation to private nuisance. However, the impact of the case may be narrowed by the single, critically important, point of fact: the 'public' character of the nuisance in the Corby litigation followed from the contaminated mud or dust having been deposited on a *public* road which therefore, gave rise to a nuisance to the public in general. It is far from clear that many environmental complaints share this characteristic.

Although civil environmental liability arises primarily from the judge-made common law, there are some instances where it is provided for under legislation. Examples of this kind of statutory environmental liability arise in relation to liability for oil pollution, radioactive material and waste. It should also be noted that the UK government has enacted a regime for remediation of contaminated land. This allows for remediation notices to be served upon those responsible for land contamination and, in some cases, simply the present owners of land contaminated by previous uses, requiring action to be taken to remove contamination or otherwise to address the continuing polluting impacts of past land use activities.[32] In addition, the issue of liability for environmental harm is importantly addressed by EU law under the *Environmental Liability Directive (2004/35/EC)*. This Directive seeks to establish a framework for the prevention and restoration of environmental damage based on the 'polluter-pays' principle, to prevent and remedy environmental damage. Liability is imposed in respect of negligent damage to biodiversity, and damage to water and land, particularly where these involve a breach of other EU environmental legislation. In the first instance, the liability, either to prevent damage or to take remedial action, falls upon the operator of the polluting installation. However, competent national authorities are empowered to take preventative or remedial action, though the cost of this must be recovered from the responsible operator. Powers are given to persons 'having a sufficient interest', including organisations promoting environmental protection, to request competent authorities to take action and, where action is declined, there is a right to have the failure to act to be subject to a review by a court or other impartial body. The kind of 'administrative' environmental liability arising under the Directive contrasts with the common law approach insofar as a public body is made ultimately responsible for taking action, whereas under the common law it is left to individuals to instigate legal proceedings and, if necessary, to bear the costs. The EU approach is also distinct from the common law in that it envisages the operator meeting actual prevention and remedy costs, rather than making any compensatory award to an individual who suffers environmental harm. It is evident that implementing the polluter pays principle under EU law and imposing liability under the civil law are directed towards distinct, though complementary, objectives.

A final point to note in respect of private environmental rights concerns the extent to which the right to a satisfactory environmental quality can be conceived of as a fundamental human right. The legal protection of human rights has advanced significantly over recent years, most notably in the UK by the enactment of the *Human Rights Act 1998*, which provides for national implementation of the European Convention for the Protection of Human Rights and Fundamental Freedoms. The European Convention is primarily concerned with civil and political liberties, but in some respects provides for rights capable of having an environmental dimension, particularly, the right to life (Art.2), the right to respect for private and family life (Art.8) and the right to property (Art.1 of the First Protocol to the Convention). The central issue concerns the circumstances in which an environmental intrusion is capable of constituting a violation of these rights. This is legally important because, fundamental rights of these kinds are afforded a status above other kinds of legal rights to the extent that the *1998 Act* allows a court to declare national legislation incompatible with Convention rights, and victims of contravention of these rights must be allowed remedies against public bodies.

In addressing the interpretation of the rights noted above, national courts are bound to have regard to the decisions of the European Court of Human Rights concerning the interpretation of the Convention. Within the case law of this Court there are a number of decisions which indicate that environmental intrusions are capable of constituting a violation of Convention rights.[33] However, the Court has also recognised that no special status is to be given to environmental human rights, as such. It has also accepted, in the context of a UK case involving sleep disturbance due to night-time aircraft noise, that a 'fair balance' needs to be drawn between the rights of individuals to protection of their homes from noise and the public interest in maintaining aviation facilities.[34]

Given the ambivalent status of environmental rights under the Convention, there was much speculation following the enactment of the *UK Human Rights Act 1998*, as to the extent to which this would allow a prioritisation of environmental complaints insofar as they could be characterised as infringements of Convention rights. To a great extent this has not happened. Although environmental infringement of Convention rights has been argued in several instances before national courts, there are only a limited number of cases where this has been accepted and these have usually concerned provisions that have been found to be unfair on procedural grounds. The leading national decision concerned a claimant who alleged that flooding of his garden from a sewer following periods of heavy rainfall constituted a violation of Convention rights and should be compensated as such. However, the House of Lords took the view that national legislatures have a broad discretion in addressing matters of social and economic policy and courts should be reluctant to intervene on matters within a government's margin of appreciation. The fact that a statutory compensation scheme had been put in place in respect of sewer flooding indicated that the Government had sought to achieve a fair balance between private and public interests and had acted within the discretion available to it.[35] This ruling does not preclude the possibility that a polluting intrusion could be an infringement of a Convention right in an extreme case, but makes that possibility less likely than had been previously thought.

21.5   LEGAL MODELS OF POLLUTION REGULATION

21.5.1   The Law on Statutory Nuisances

The diverse body of regulatory environmental law comprises measures adopted at national, EU and international levels over quite a long period of time, and the UK provisions that have been made for 'statutory nuisances' may be placed at the early end of that chronology. As has been noted, common law courts have provided civil law remedies for private nuisance which, in theory, should allow individuals to take legal action to protect themselves from pollution and other environmental interferences. The practical constraints upon of civil litigation, however, have long been recognised, and Parliament intervened to provide that certain kinds of urban disamenity should be addressed by legislation. Under the *Nuisances Removal Act 1846* and the *Public Health*

Act 1848 powers were given to local bodies to take action in respect of a range of matters giving rise to environmental health concerns. In effect, matters which were previously only actionable in civil became subject to enforcement powers, ultimately, involving the criminalisation of certain kinds of activity. This legislative translation of civil wrongs into criminal offences proved remarkably effective and durable, and modern provisions on statutory nuisances comprise one of the most frequently used areas of environmental law in practice.

Statutory nuisances are now provided for in Part III of the *Environmental Protection Act 1990* and consist of a list of particular kinds of nuisance which are subject to enforcement responsibilities entrusted to local authorities. Each local authority is bound to inspect its area to detect statutory nuisances and to take reasonable steps to investigate any complaint of a nuisance. Where a nuisance is found to exist, the local authority is bound to serve an 'abatement notice' on the person responsible for the nuisance, requiring it to be abated and/or requiring particular works to be done to secure its abatement within a specified time. Subject to various qualifications, a person upon whom a notice is served who, without reasonable excuse, fails to comply with the notice will commit an offence. The attraction of this enforcement procedure, from the perspective of an environmental complainant, is that it is only necessary to notify a local authority environmental health department of a complaint. Thereafter, it is the duty of the local authority to investigate the complaint and pursue enforcement proceedings where necessary, and to bear the costs of enforcement action. Nevertheless, where a local authority declines to take action, perhaps because it does not accept that a statutory nuisance exists, it is open to a private individual to pursue statutory nuisances proceedings as a 'person aggrieved', who is prejudicially affected by the nuisance.[36]

The criminalisation of nuisances, subject to an explicit enforcement procedure, is a clear point of contrast with the civil law. A second point of contrast is that, whist a civil law private nuisance can comprise anything interfering with the 'enjoyment' of land, statutory nuisance powers only arise in relation to a defined range of matters that fall under a list provided for by legislation. The present list of statutory nuisances contains 11 kinds of nuisance relevant for these purposes. Examples are: any premises in such a state as to be prejudicial to health or a nuisance; fumes or gases emitted from premises so as to be prejudicial to health or a nuisance; artificial light emitted from premises so as to be prejudicial to health or a nuisance; and noises emitted from premises so as to be prejudicial to health or a nuisance. It is notable that two elements are provided for in this listing: first, nature of the nuisance, and second, the requirement that it should be a nuisance (within the common law) or that it should be prejudicial to health. Whilst matters such as insanitary housing have always been categorised as statutory nuisances, other concerns, such as security lighting, have been added relatively recently. However, by far the most practically important kind of statutory nuisance arises in relation to noise. Complaints about this problem received by local authority environmental health departments greatly outnumber complaints about all the other kinds of statutory nuisances.

As environmental law has developed increasing sophistication and specialization, the rather miscellaneous collection of urban disamenities which are listed as statutory nuisances, and the lack of a specialist environmental regulatory body to enforce them, have come to look rather dated, if not anachronistic. Certainly, by comparison with the allocation of responsibilities for the enforcement of laws relating to pollution to the Environment Agency, local authorities, as multi-purpose regulators, seem unspecialised in this respect. On the other hand, local authorities do exercise related environmental powers, particularly in relation to air pollution and land use planning, and statutory nuisances may have characteristics which place 'local knowledge' at a premium in respect of enforcement. The antiquity of many of the categories of statutory nuisance and the reluctance of the courts to interpret these categories in a way which reflects modern appreciations of what constitutes an environmental health concern is also notable.[37] However, the counter to this is to note that the list of nuisances has been extended over time to encompass modern kinds of nuisance where these are not otherwise provided for.

Another distinctive characteristic of the law on statutory nuisances, as compared with other areas of law, lies in the relatively strict enforcement duty that is provided for. Specifically, where a statutory nuisance is identified, the local authority 'shall' (meaning 'must') serve an abatement notice. The need for a regulatory authority to take a specific enforcement action is not found in other areas of environmental law. Normally, the regulator is allowed discretion as to enforcement action, usually subject to an explicit general statement as to how enforcement powers are to be exercised. If there is a justification for this mandatory feature of statutory nuisance enforcement, it must lie in its public health origins and the perception that matters of environmental health are of imperative importance. On the other hand, as has been seen, the duty to serve an abatement notice may, from a complainant's perspective, make this a more attractive way of addressing an environmental complaint.[38] Whatever its antiquity, peculiarity or limitations, the fact remains that statutory nuisances law continues to be amongst the most frequently used legal mechanisms to address environmental complaints.

21.5.2 Water Pollution and Water Quality Law

UK laws on the pollution of water are, also, of some antiquity. *The Rivers Pollution Prevention Act 1876*, created general offences in relation to pollution by solid matter, sewage and from manufacturing and mining. The responsibilities for enforcement were inadequately provided for under the *1876 Act*, but the basic offences were appropriately formulated in terms of 'causing' or 'knowingly permitting' specified matter to enter any stream. This formulation has stood the test of time, with essentially the same words being adopted in a sequence of later enactments up to the present provisions under the *Environmental Permitting Regulations 2010*, discussed below.

The formulation of the central water pollution offence as that of 'causing' the entry of polluting matter into waters has some important legal characteristics. In the first place it is clear that the offence is complete at the point where polluting matter has entered a watercourse. This means that the extent of the environmental damage which results from the entry is irrelevant to the commission of the offence, though this will be a matter which is relevant in sentencing. Beyond that, the use of the word 'cause' in the water pollution offence incorporates 'strict liability', so that no fault on the part of the defendant needs to be shown, merely an action which resulted in the entry of polluting matter. In a leading case establishing this point, an effluent storage tank at a papermaking works overflowed into a nearby stream as a result of the malfunctioning of a pump which controlled the flow of effluent into the tank. Despite the absence of evidence of negligence, the operator was found guilty of 'causing' the entry. As it was put, if the defendant were to be found not guilty, "unless the prosecution could discharge the often impossible onus of proving that the pollution was caused intentionally or negligently, a great deal of pollution would go unpunished".[39]

The strict liability formulation of the offence of causing the entry of polluting matter into a watercourse has some potentially harsh consequences where pollution results from the action of a person other than the person prosecuted for an offence. For example, in another leading case, oil had escaped from garage premises and entered a watercourse as a consequence of an alleged act of vandalism by an unknown person. It was ruled that the operator of the premises could be guilty of the offence if they had done something which allowed the escape of polluting matter even if there was another, more immediate, cause. This meant that the operator of the premises could be held liable for the entry unless there were circumstances so 'extraordinary' as to negate the causal effect. In the circumstances, vandalism was not considered sufficiently extraordinary and the operation of the premises, without more, was found to be sufficient to justify a finding that the operator has 'caused' the entry.[40]

Although UK law had provided for the criminalisation of water polluting acts, from a relatively early stage, it was not until much later that national law started to address the kinds of actions that were needed to prevent water pollution incidents occurring or to set out strategic objectives for the quality of water that legal powers should be used to achieve. Building upon earlier provisions, the *Water Resources Act 1991* provided for requirements to take precautions against pollution, water protection zones and nitrate sensitive areas for the purpose of preventing water pollution. The *1991 Act* also established a system of water quality objectives subject to a duty upon regulatory authorities to use their powers to ensure quality objectives for particular waters were met and maintained. However, relatively limited use was made of these preventative and provisions except in circumstances where it was necessary to exercise national legal powers to fulfil duties arising under EU law.

Turning to the EU dimension of water law, this was one of the first areas of environmental legislation to be addressed, with several measures predating the granting of explicit environmental lawmaking powers under the *Single European Act 1987*. Early water quality measures adopted a diverse range of regulatory approaches. Some directives took a substance-specific approach in imposing controls upon specified substances that were seen to be environmentally problematic, either to restrict levels of emissions of these substances or to require that specified ambient concentrations of the substances in the water environment were not exceeded (such as the *Dangerous Substances Directive 76/464/EEC* and the *Groundwater Directive 80/68/EEC*). Other directives related to particular kinds of waters and required that these waters should achieve specified water quality standards (the *Bathing Water Directive 76/160/EEC* and the *Freshwater Fish Waters Directive 78/659/EEC*). Other directives were focused upon particular kinds of polluting activity and tended to place greatest emphasis upon operational or technological requirements to protect water quality (the *Urban Waste Water Treatment Directive 91/271/EEC* and the *Agricultural Nitrates Directive 91/676/EEC*). Although the original directives were amended over time, it was recognised that they were founded upon environmental principles and strategies that had changed over time. By 2000, the directives were seen to be in urgent need of modernization and consolidation so far as this was possible. Although some directives were retained as self-standing measures, the bulk of the original EU water legislation was repealed and replaced with what is now the most important measure: the *Water Framework Directive (2000/60/EC)*.

The *Water Framework Directive* seeks to incorporate previous requirements for management of water into a single system, which aims to coordinate all the different objectives for which water is protected and to achieve defined environmental quality objectives. The central administrative feature of the Directive is that management should take place at river basin level. The river basin is seen as the natural geographical and hydrological unit for water management rather than areas defined according to national administrative or political boundaries. This leaves quite a lot to turn upon such as how river basins are delineated, with big differences between the relatively small river basin areas found in the UK and the multinational river basins established for large rivers in continental Europe. Nonetheless, the system envisages management of water quality in accordance with river basin management plans, which will need to be established and updated every six years. These plans provide the context for the co-ordination of measures which must be put in place to achieve stated environmental quality objectives under the Directive.

The *Water Framework Directive* seeks to provide a regulatory regime that it is least as stringent as the directives that are replaced, but establishes environmental objectives for different categories of waters that are specified in an innovative way.

For surface waters, the objectives are:

(i) To prevent deterioration.
(ii) To protect, enhance and restore with the aim of achieving 'good status' within 15 years, or, in respect of artificial and highly modified waters, to achieve the lesser standard of 'good ecological potential' within that period.
(iii) Progressively to reduce reduce pollution from certain priority substances and phase out emissions of priority hazardous substances.

Amongst these objectives, the achievement of 'good status', for all relevant waters by the deadline, is likely to be seen as a challenging task. This is because 'good status' for surface waters requires the achievement of *both* 'good ecological status' and 'good chemical status'. 'Good chemical status' is defined in terms of compliance with all the quality standards established for chemical substances at EU level. 'Good ecological status' is defined in terms of the quality of the biological community, assessed against normative definitions of ecological status classification for different kinds of waters. The novel aspect of this is that previous EU water legislation did not address the issue of ecological quality. Ensuring 'good ecological status', therefore, involves a new kind of regulatory challenge, as where, for example, an aquatic ecosystem falls below an expected level of ecological quality because it lacks the range of aquatic species expected in a water of its kind or because of the presence of an invasive non-native species of flora or fauna.

Achievement of good status is to be brought about by the coordinated application of a range of particular regulatory mechanisms. For that purpose, the Directive sets out lists of 'basic measures' which must be established to meet the requirements of existing EU environmental directives. In addition, it may be necessary to adopt 'supplementary measures', of a legal, administrative of economic kind, where the basic measures are insufficient to meet the environmental objectives of the Directive. Broadly, the diverse control measures envisaged should be sufficient to restrict the full range of activities that are potentially damaging the quality of the aquatic environment. Beyond that, it is notable that the Directive adopts a 'combined approach' to emission limits and environmental quality standards. Previously, under the *Dangerous Substances Directive 76/464/EEC* for example, Member States were able to comply with water directive requirements by *either* controlling emissions *or* by meeting ambient environmental quality standards. Under the *Water Framework Directive*, compliance with stated emission control limits will be insufficient where the environmental objectives of the Directive are not met, particularly where waters fail to achieve good status, and stricter emission controls will need to be put in place.

Certainly, the *Water Framework Directive* is an ambitious measure, though it may be noted that the time period allowed for achievement of its central objectives is relatively lengthy, as compared with other EU environmental directives. As with any piece of environmental legislation, an evaluation of its benefit, involves looking at the things that have actually been done to implement the Directive and, perhaps most important, what improvements in environmental quality have followed. In respect of the England and Wales, transposition of the Directive was through national Regulations, made in 2003, which allocated the executive and operational responsibilities for implementation respectively to the relevant Minister in England or Wales and the Environment Agency.[41] The Regulations made provision for the establishment of river basin districts in England and Wales and for put in place a strategic planning process for managing, protecting and improving the quality of water resources in these districts. The Regulations provided for general duties requiring the Ministers and the Environment Agency to exercise relevant functions so as to secure compliance with the Directive. The Agency was made responsible for carrying out various analyses, reviews and identifications of particular kinds of waters, as required by the Directive, along with monitoring programmes. By 2009, the Agency was bound to prepare draft river basin management plans, containing details of the results of the preparatory technical and planning work, along with the environmental objectives and programmes of measures proposed for each district. In addition to the legislative provisions, extensive Ministerial guidance is addressed to the Agency and other public bodies in respect of the practical implementation of the Directive.

The practical consequences of implementing the Directive give little reason for complacency over the state of the aquatic environment. The most recent information,[42] for 2012, shows that only 27% of surface waters in England and Wales were of good ecological status in 2012. 79% of surface waters were of good chemical status. However, overall good status encompasses both chemical status and ecological status, and the lesser of the two measures is taken as the overall measure. At an EU level, the most recent European Commission report on the implementation of the *Water Framework Directive*[43] indicates that 55% of surface waters are at less than good ecological status or potential. The high proportion of waters for which the reported chemical status was listed as 'unknown', meant that it was not possible to give a figure for overall good chemical status. The European Commission's assessment is that good status will not be reached in 2015 for a significant proportion of water bodies and the reasons for this lie in hydromorphological pressures, pollution and over-abstraction.

21.5.3 Integration of Pollution Control and Environmental Permitting

As has been noted, the recent development of environmental law in the UK has involved a shift away from the traditional sectoral approach to pollution control (involving separate laws and regulatory responsibilities for the control of pollution of water, air and land) towards the adoption of an integrated approach to environmental regulation. The starting point in the progression is to be found in reports of the Royal Commission on Environmental Pollution where it was proposed that the most polluting industrial installations should be subject to a unified inspectorate. This body would be responsible for ensuring the 'best practicable environmental option' for pollution control, taking into account the entire pollution from a process and the technical possibilities for dealing with it.[44] This recommendation led, in 1987, to the establishment of Her Majesty's Inspectorate of Pollution (HMIP) with a remit to exercise a cross-media approach to pollution control. Subsequently, statutory powers were provided to HMIP in respect of Integrated Pollution Control (IPC) under Part I of *Environmental Protection Act 1990*. These regulatory powers were later passed to the Environment Agency under the *Environment Act 1995*.

The initial statutory regime for IPC, applied only to the most polluting industrial installations, involving fuel production and combustion; metal production and processing; mineral industries; chemical industries; waste disposal and recycling; and certain 'other industries'.[45] In respect of these industries, IPC authorisations were to include an implied general condition that the person carrying on the process must use "the best available techniques not entailing excessive cost" (BATNEEC), for preventing or reducing the release of prescribed substances into any environmental medium and for rendering harmless any other substances which might cause harm if released. The onus of proof in showing BATNEEC was placed upon the applicant for authorisation. The BATNEEC strategy was to be applied to the environment as a whole, having regard to "the best practicable environmental option available" (BPEO). In practice, compliance with BATNEEC was largely determined by (HMIP) Guidance Notes, identifying best practice for different industrial sectors. Although the terms of authorisations were occasionally challenged in the courts, judges seemed strongly inclined to regard issues relating to BATNEEC and BPEO as matters within the discretion of the regulatory authority.[46]

The IPC approach adopted in the UK was influential in the adoption of EU legislation of a similar kind in the Directive concerning integrated pollution prevention and control (the *IPPC Directive, 96/61/EC*). However, the *EU IPPC Directive* was more broadly formulated than the national IPC, both in the range of installations being regulated and the obligations imposed upon them. The central obligation of the Directive was that Member States should ensure that the necessary measures are taken by competent authorities to ensure that installations are operated in such a way that:

(i) All the appropriate preventative measures are taken against pollution, in particular through application of the best available techniques (BAT).
(ii) No significant pollution is caused.
(iii) Waste production is avoided.
(iv) Energy is used efficiently.
(v) Necessary measures are taken to prevent accidents and limit their consequences.
(vi) Necessary measures are taken upon definitive cessation of activities to avoid any pollution risk and to return the site of operation to a satisfactory state.

Competent authorities were bound to take account of these principles in determining the conditions of a permit.

Notably, matters such as waste minimisation, energy efficiency, accident protection and the measures needed after the closure of an installation had not featured within the national IPC regime, although it may have been possible to have provided for them by other legal means. The Directive also gave quite wide regulatory discretion to national authorities in setting emission limits based upon the 'best available techniques'. To address the possibility of extreme differences of approach being taken by different Member States, provision was made for European Commission to organise the exchange of information concerning BAT between Member States and relevant industries. A European IPPC Bureau was established and made responsible for formulating best available techniques reference documents (known as 'BREFs', the EU equivalent of the national HMIP Guidance Notes) which needed to be taken into account in determining BAT generally or in relation to specific cases.

The move towards integration, at EU level, has not stopped with the adoption of the *IPPC Directive*. The European Commission reported on the operation of the Directive, noting problems of implementation by some Member States but, most importantly, concluding that the Directive is insufficiently wide in its scope and proposing that the general approach should be applied more extensively. This was followed up by a proposal that a new directive should encompass the present *IPPC Directive (2008/1/EC)* along with six other directives: the *Large Combustion Plants Directive (2001/80/EC)*, the *Waste Incineration Directive (2000/76/EC)*, the *Solvents (Volatile Organic Compounds) Emissions Directive (1999/13/EC)* and three Directives concerning *Titanium Dioxide (78/176/EEC, 82/883/EEC and 1992/13/EC)*.[47] The outcome of this process has been the recent adoption of *Directive on Industrial Emissions (Integrated Pollution Prevention and Control) (2010/75/EU)*. The *Industrial Emissions Directive* consolidates the *IPPC Directive* with the six directives noted above and applies the IPPC approach to all activities concerned, though containing some provisions which are of specific application to particular activities. However, the new Directive retains the essential features of IPPC, so that no installation may be operated without a permit and specific permits must use BAT to achieve a high level of protection of the environment as a whole. Permits must also provide measures necessary to ensure compliance with operators' environmental requirements and environmental quality standards, and a range of related matters. Also, new provision is made in respect of environmental inspections of installations, with the frequency of inspections dependent upon the level of environmental risk involved.

National IPC and EU IPPC are generally regarded as having brought significant regulatory and environmental benefits in requiring regulators to consider the standards to which different kind of industrial installations should operate, and to assess the advantages and disadvantages of different pollution control options for the environment as a whole. However, it should also be noted that neither IPC nor IPPC were ever intended to apply comprehensively, to all activities with potential polluting impacts, but only to the most environmentally hazardous industrial processes. Until recently at least, the UK operated a substantial number of differing regimes for authorising activities falling outside IPC. The integration of arrangements for granting environmental licenses and authorisations has been radically changed by the recent introduction of a harmonised environmental permitting regime.

Environmental permitting has been introduced progressively in England and Wales to replace distinct environmental licensing systems, applicable to the different regulatory sectors in respect of impacts on water, air and land. Environmental permitting involves the replacement of previous kinds of environmental licence, by a single authorisation procedure. The purpose of this is to simplify and reduce the regulatory burden of environmental regulation upon industry by harmonising and streamlining the requirements that must be met by operators of installations with impacts on different parts of the environment. Individual environmental permits will differ according to the nature and potential environmental impacts of the activity being undertaken, and the extent to which an activity is subject to particular EU environmental directives, but the essential aim is that environmental permits should operate in a way that is as uniform as possible. This has involved establishing common requirements for obtaining permits, their content, enforcement powers, public registers of permits, and regulatory powers and functions.

Initially, environmental permitting was introduced in the *Environmental Permitting (England and Wales) Regulations 2007 (SI 2007/3538)* which converted existing waste management licences and IPC authorisations into environmental permits. The *Environmental Permitting (England and Wales) Regulations 2010 (SI 2010/675)* replaced the *2007 Regulations* and extended the environmental permitting regime to encompass other areas previously covered by water discharge consents, groundwater authorisations and radioactive substances regulation. Notably, the Regulations are arranged so that the procedural matters, relating to the granting and significance of environmental permits, are set out in the main body of the regulations. The substantive environmental requirements, including requirements arising from EU environmental directives, are dealt with under *Schedules to the Regulations*.[48]

The consolidation and harmonisation of diverse kinds of authorisation under the new environmental permitting scheme appears to have been generally well received. The facility for an operator of an industrial installation to have a single authorisation for potentially complex range of activities, with impacts on water, air and land, will simplify regulatory requirements. Similarly, the use of 'standard rules', applicable for all sites which fall within a particular category, will remove the need to negotiate the conditions for each site individually and should make the authorisation process swifter and less burdensome for both regulatory authorities and for operators of regulated installations.

Although environmental permitting has brought about a major step towards the integration of environmental law in England and Wales, it should be recognised that complete integration still lies some way off. Significant areas of environmental law fall outside the scope of the environmental permitting regime and are subject to substantial differences in regulatory approach: statutory nuisances, drinking water quality, discharges to sewers, land use planning and nature conservation and biodiversity law. Notably, responsibilities for these matters are spread across a range of different environmental regulatory bodies and enforcement responsibilities may be seen as remaining significantly dis-integrated. Whether there remain good reasons for the retention of these distinct regulatory regimes and regulatory authorities remains a topical issue.

21.5.4 Procedural Environmental Rights

Over recent years there has been marked development in what may be termed 'procedural environmental rights'. A good commencement point for appreciation of the role and nature of such rights is to be found in Principle 10 of the Rio Declaration, from the 1992 United Nations Conference on Environment and Development,

"Environmental issues are best handled with the participation of all concerned citizens, at the relevant level. At the national level, each individual shall have appropriate access to information concerning the environment that is held by public authorities, including information on hazardous materials and activities in their communities, and the opportunity to participate in decision-making processes. States shall facilitate and encourage public awareness and participation by making information widely available. Effective access to judicial and administrative proceedings, including redress and remedy, shall be provided."[49]

At a European regional level, access to the kinds of rights envisaged by Principle 10 of the Rio Declaration has been significantly advanced by the 'Aarhus Convention'.[50] The Convention is essentially concerned with three matters, which are conveniently considered in turn:

(i)   Access to environmental information.
(ii) Public participation in environmental decision-making.
(iii) Access to environmental justice.

In respect of access to information, historically, national governments have restricted access to information of all kinds, including environmental information, where it is perceived that this information might cause public alarm or political embarrassment. This denial of information has been especially problematic for environmental NGOs in the past, particularly where they sought to raise awareness about an environmental concern. However, in the UK past prohibitions on disclosure of environmental information by officials have been progressively replaced by regimes allowing positive 'rights' of access to environmental information held by public bodies. Perhaps the culmination of this process has been the enactment of the *Freedom of Information Act 2000*. This Act creates a statutory public right of access to recorded information held by public authorities and specifies the conditions which need to be fulfilled before an authority is obliged to comply with a request for access. Notably, the right of access applies to *all* kinds of public information, not specifically information relating to the environment. The exercise of the right of access to information is regulated by an Information Commissioner, who is bound to consider applications for rulings that public authorities have not acted in accordance with information access rights under the Act, and investigative and enforcement powers are provided for in relation to this.

At EU level, the idea of public access to environmental information can also be seen to have been progressively recognised. Although, explicit rights of public access were provided for in particular environmental directives, the most comprehensive provision is made under the *Environmental Information Directives (90/313/EEC and 2003/4/EC)*. The *1990 Directive* had the purpose of ensuring freedom of access to, and dissemination of, information on the environment held by public authorities and setting out the basic terms and conditions on which such information should be made available. The basic scheme required Member States to ensure that public authorities made available information relating to the environment to any natural or legal person on request and without having to prove an interest. That is, that a valid request could be made by anyone (not necessarily an EU-based citizen or company) and no explanation needed to be given as to why the information was required. However, as with any kind of freedom of information, the right of access was made subject to a range of exceptions where non-disclosure is allowed because disclosure could compromise matters such as the confidentiality of proceedings of public authorities, international relations and national defence, public security *etc*. Nonetheless, subject to the exceptions, a public authority was bound to respond to a person requesting information as soon as possible and at the latest within two months. Where the request for information was declined, the reasons had to be given and a person who considered that a request for information was unreasonably refused was allowed a judicial or administrative review of the decision.

The Aarhus Convention provided for a right of access to environmental information that was significantly more extensive than required under the *1990 EU Directive*. A wider definition of 'environmental information' was provided for along with an extended definition of 'public authorities'. The Convention also made more detailed provision concerning the form in which information is to be made available; shorter deadlines for making information available; limitations on the operation of exceptions; additional duties on national authorities for collection and dissemination of information; and stricter procedures for review of actions of public authorities. In order to secure compliance with the Convention, the EU repealed the initial *Environmental Information Directive* and adopted a revised *Directive (2003/4/EC)*, incorporating amendments that were necessary to secure conformity with the Convention requirements.

The 'second pillar' of the Aarhus Convention seeks to establish the right of individuals to participate in certain kinds of environmental decision-making, however, the meaning of this requirement is rather uncertain. 'Public participation' is an ambiguous term because it seems to encompass a range of possible kinds of 'participation'. At one extreme, it might involve 'consultation', comprising an arrangement in which individuals are allowed to make representations on environmental proposals, with a view to influencing the decision-maker's final determination. At the other extreme, 'participation' might involve individuals being given some active part in determining the outcome of an environmental decision-making process. The question of what constitutes 'the public' is also problematic. Whilst 'public' participation has a strong democratic mandate in legitimizing environmental decision-making, there is also a concern that the participatory processes are capable of being hijacked by those with particular vested interests. This might be problematic, for example, where a balance needs to be drawn between local and national environmental impacts in respect of a major infrastructure project. In such cases local opposition tends to be vociferous, but spokespersons for the wider, though more diffuse, environmental benefits of a project are less willing to express a view. Beyond that, there are questions about how participation of (the present) public can fairly represent the views of future generations, as required by the overriding need to secure sustainable development.

Despite these uncertainties, the Convention contains specific articles detailing the contexts in which participation is required (Arts. 6 to 8). These relate to public participation:

(i) On specific activities which have a significant effect on the environment.
(ii) Concerning plans, programmes and policies relating to the environment.
(iii) During the preparation of executive regulations and/or generally applicable legally binding normative instruments.

Examples of the kind of thing that might come within the scope of these requirements are:

(i) Public consultation prior to the granting of an environmental permit for an industrial installation or before granting planning permission for a development of land.
(ii) The need for strategic environmental assessment of sectoral or local plans which envisage the granting of permission for specific projects within their scope, such as industrial development or land development plans.
(iii) Any kind of environmental legislative instrument made by central government, as primary or secondary legislation, or adopted by a local authority.

The third, and potentially most problematic, pillar of the Aarhus Convention concerns access to justice in environmental matters (Article 9). Where a person maintains the impairment of an environmental right, they will be entitled to a review procedure before a court of law and/or another independent and impartial body to challenge the substantive and procedural legality of any decision, act or omission. Notably, this is available in respect of any person or body with a 'sufficient interest' in the matter and certain NGOs are deemed to have sufficient interest. Perhaps most crucially, the procedures that are envisaged are required to provide adequate and effective remedies, and be fair, equitable, timely and not prohibitively expensive. The words 'not prohibitively expensive' are likely to be the most difficult to fulfil. Environmental litigation, as has been seen, is an expensive activity. Where, for example, a regulatory decision of a public body is being challenged on technical grounds in judicial review proceedings it is difficult to see how the cost of such proceedings would not be regarded as 'prohibitively expensive' unless the person mounting the challenge is extremely wealthy.

As regards the EU, the Aarhus Convention has been implemented by various measures, alongside the revised *Environmental Information Directive*. In some cases, modifications to individual environmental directives have been made to ensure participation rights.[51] To enable the EU to ratify the Convention it was necessary to ensure compliance with the Convention by all EU institutions. The legal instrument providing for this is the Aarhus Regulation,[52] which requires EU institutions, bodies, offices or agencies to adapt their internal procedures and practice to comply with the Convention. Hence, the institutions and bodies are required to provide for public participation in the preparation, modification or review of 'plans and programmes relating to the environment'. The Regulation also enables environmental NGOs meeting certain criteria to request an internal review of acts adopted, or omissions, by EU institutions and bodies.

An important administrative feature of the Aarhus Convention, which sets it apart from most international environmental conventions, is the provision that it makes in respect of compliance. On this, Article 15 of the Convention requires the establishment of arrangements for reviewing compliance with the Convention. A Decision of the Parties to the Convention in 2002 provides for the Committee to consider submissions by the Parties, referrals by the Secretariat and communications from the public, including individual complaints of non-compliance. Reports of the Compliance Committee are considered by the Meeting of the Parties, which may decide upon appropriate measures to bring about full compliance with the Convention. Measures include: providing advice and making recommendations, and, in the case of communications from the public, issuing declarations of non-compliance and taking other 'non-confrontational, non-judicial and consultative measures as may be appropriate'.[53] The non-mandatory nature of findings on compliance complaints and the lack of sanctioning powers should be noted, but an adverse finding by this Committee is of significant importance to a non-compliant party which is almost inevitably placed under diplomatic pressure to rectify the situation that gave rise to a complaint.

Since its establishment, the Compliance Committee has reached findings in respect of compliance issues by individual parties, including several decisions concerning the UK and the EU.[54] Most significantly, in relation to a complaint against the UK, concerning judicial review proceedings to challenge a licence to allow sea disposal of contaminated dredgings from the Port of Tyne, the Committee found that costs were prohibitively expensive and there were barriers to justice including time limits. The Committee recommended the UK review its system for allocating costs in environmental cases and review time limits for applications to ensure they are fair and equitable.[55] In relation to a complaint against the EU, it was alleged that the test of 'individual concern', applied where individuals or NGOs seek to challenge the decisions of EU institutions, does not fulfill requirements of the Aarhus Convention in respect of access to justice. Although the Committee was not convinced that the EU had failed to comply with the Convention, it considered that a 'new direction' in the jurisprudence of the EU Courts should be established in order to ensure compliance with the Convention and steps should be taken to overcome shortcomings.[56]

Perhaps, the 'new direction' in jurisprudence that was called for by the Compliance Committee is reflected in recent decisions before the Court of Justice of the European Union. In one case, an environmental group requested a review a Regulation setting maximum pesticide levels for certain products, but the European Commission refused this request on the basis that a challenge would only be allowed in relation to measures of individual scope. The Court found that this interpretation of standing was not compatible with the requirements of the Convention in respect of access to justice and the environmental group had been denied access to justice.[57] In another case, concerning an environmental group seeking to review a decision of the Commission in relation to the national extension of a deadline for achieving satisfactory air quality, the Commission refused this request because the matter was not of individual concern to the environmental group. Again, the Court found that the decision of the Commission, declining the request should be annulled because of its incompatibility with Convention requirements.[58]

The continuing difficulties in implementing the Aarhus Convention, at both national and EU levels, particularly in relation to standing to bring proceedings and securing inexpensive access to justice for environmental litigants, have wide-ranging ramifications. If standing and costs rules were to be modified for environmental litigants, the use of the law for this purpose would be made significantly more accessible. For so long as the prospect of a legal challenge has been a theoretical rather than a practical possibility for environmental decision-makers, there has been little cause to worry about whether decisions would withstand judicial scrutiny. That preconception seems set to change, with major implications for the culture surrounding environmental decision-making.

21.6 CONCLUDING OBSERVATIONS

Hopefully, this discussion has achieved its aim of providing a brief and necessarily selective account of the role of law with regard to pollution control. In doing so, an eclectic range of different kinds of legal measures at national, EU and international levels has been scanned and the dynamic nature of the subject emphasised. Preconceptions that the law concerning pollution control is simply a fixed set of prohibitive rules providing for sanctions against polluters should have been challenged. Pollution control law has been seen to be drawn from many different sources, taking markedly different approaches to environmental problems.

What remains is the fundamental question as to the *effectiveness* of law as a means of controlling and managing pollution. Although many of those engaged in environmental law would staunchly maintain that law has a vital role to play in securing satisfactory levels of environmental protection, providing evidence to support this view is challenging. Despite the massive development of environmental legislation over recent decades, and the increasing sophistication of the measures that have been adopted, there is scant evidence that this has actually resulted in commensurate improvements in environmental quality. Certainly, evidence can be provided showing improvements, for example, in certain aspects of water and air quality, but attributing these to legal measures, as opposed to changing industrial, technical, economic and social circumstances, would be unconvincing. At best, law might be seen as one element, amongst several, assisting in the task of securing satisfactory environmental quality.

The reasons for the limited achievements of law in curbing pollution and environmental decline more generally are manifold. Centrally, addressing pollution through law depends upon there being a sufficient consensus that law provides an appropriate response, as opposed to educational, technical, economic and other ways that environmental problems might be, perhaps more appropriately, tackled. Beyond that, the 'backward-looking' character of legal responses limits their capacity to address contemporary or future environmental concerns.

For the future, pollution prohibition laws are still likely to form a backbone of environmental law, but increasingly environmental policy is likely to focus less on emissions (to air, water, or land) and more on how the environmental impact of substances can be reduced in general. In the field of waste, for example, waste legislation has traditionally focused on improving the standards of waste management facilities such as landfill sites or incinerators. For the future, greater emphasis will be needed on the responsibility of producers (with recent examples of this in the field of electrical goods and motor vehicles), and issues such as eco-design, life-cycle analysis, and improving environmental efficiencies, to allow progress to be made towards a 'recycling society'. The principle of prevention has long been at the heart of EU environmental policy in theory and indeed was stated in early British air pollution legislation. Turning that principle into reality will be a challenge for policy makers. The extent to which that principle can or should be reflected in law will be an equal challenge for environment lawyers.

In a post-industrial society, a focus upon industrial production activities as the subject of legislation may be seen as increasingly misplaced. However, the shift from 'production' to 'consumption', as the activity most needing to be regulated, is politically problematic. Whilst there is fairly wide support for laws which restrict the worst excesses of industrial pollution, measures which would have the effect of restricting consumer lifestyle choices invariably meet with public opposition. Environmental law develops and operates within the major constraint of political feasibility and there are limitations upon what can be accomplished within that constraint.

REFERENCES

1. S.97, *Water Resources Act 1991;* and Department for Environment, Food and Rural Affairs, *Protecting our Water, Soil and Air: A Code of Good Agricultural Practice for Farmers, Growers*

and Land Managers, 2009; http://www.defra.gov.uk/publications/files/pb13558-cogap-090202. pdf. Another example of statutory guidance is the Code of Practice in Relation to the Duty of Care on Waste Producers, S.34 *Environmental Protection Act 1990*; http://archive.defra.gov.uk/ environment/waste/controls/documents/waste-man-duty-code.pdf (accessed 01/08/2013).

2. S.4, *Environment Act 1995;* and see Department for Environment, Food and Rural Affairs guidance, *The Environment Agency's Objectives and Contributions to Sustainable Development: Statutory Guidance,* 2002; http://archive.defra.gov.uk/corporate/about/with/ea/documents/ ea-susdev-guidance.pdf (accessed 01/08/2013).

3. S7, *Environment Act 1995.*

4. S.29, *Environment Act 1995.*

5. Environment Agency, *Enforcement and Sanctions Statement (2011)*; http://cdn.environment-agency.gov.uk/geho0910bszj-e-e.pdf (accessed 01/08/2013); and *Enforcement and Sanctions – Guidance* and *Offence Response Options Document* (2011).

6. European Commission, Decision No 1600/2002/EC of the European Parliament and of the Council laying down the Sixth Community Environment Action Programme and European Commission, *A Sustainable Europe for a Better World: A European Union Strategy for Sustainable Development*, COM (2001) 264 final.

7. For early examples concerning the UK, see *Case C-337/89 Commission* v *United Kingdom [1992] ECR I-6103* (concerning failure to meet water quality parameters under the *Drinking Water Quality Directive (80/778/EEC)*) and *Case C-56/90, Commission v United Kingdom [1993] ECR I-4109* (concerning failure to meet water quality parameters under the *Bathing Water Directive (76/160/EEC)*).

8. *Cases C-6/90* and *9/90 Francovich v Italy [1991] ECR I-5357.*

9. *Case C-26/62 Van Gend en Loos [1963] ECR 1.*

10. *Case C-188/89 Foster v British Gas plc [1990] ECR I-3313* and *Case C-201/02 R (on the application of Delena Wells) v Secretary of State for Transport, Local Government and the Regions [2004] ECJ 723.*

11. *Case C-106/89 Marlesing v La Comercial Internacional de Alimentacion [1990] ECR I-4135.*

12. *Case C-72/95 Aannemersbedrijf P.K. Kraaijeveld BV and others* v *Gedeputeerde Staten van Zuid-Holland Environnement et consommateurs [1996] ECR I-5403*, para.60.

13. *Case C-387/97 Commission v Hellenic Republic, [2000] ECR I-5047, Case C-278/01 Commission v Spain [2003] ECR I-14141* and *Case C-304/02 Commission v France [2005] ECR I-6263.*

14. http://www.unep.org/Documents.Multilingual/Default.asp?documentid = 97&articleid = 1503 (accessed 01/08/2013).

15. http://www.unep.org/Documents.Multilingual/Default.asp?documentid = 78&articleid = 1163 (accessed 01/08/2013).

16. World Commission on Environment and Development, *Our Common Future*, 1987, p. 43.

17. http://www.unece.org/fileadmin/DAM/env/lrtap/full%20text/1979.CLRTAP.e.pdf (accessed 01/08/2013).

18. http://ozone.unep.org/pdfs/viennaconvention2002.pdf (accessed 01/08/2013).

19. http://www.basel.int/Portals/4/Basel%20Convention/docs/text/BaselConventionText-e.pdf (accessed 01/08/2013).

20. http://www.pops.int/documents/convtext/convtext_en.pdf (accessed 01/08/2013).

21. http://www.un.org/Depts/los/convention_agreements/convention_overview_convention.htm (accessed 01/08/2013).

22. http://www.ospar.org/content/content.asp?menu = 00340108070000_000000_000000 (accessed 01/08/2013).

23. http://www.imo.org/about/conventions/listofconventions/pages/international-convention-for-the-prevention-of-pollution-from-ships-(marpol).aspx (accessed 01/08/2013).

24. http://untreaty.un.org/cod/avl/ha/ccc/ccc.html (accessed 01/08/2013).

25. See *R (on the application of Friends of the Earth)* v *Secretary of State for Energy and Climate Change [2009] EWCA Civ 810* where it was held that the failure of the Government to meet a target for its statutory duty to eliminate fuel poverty (under the *Warm Home and Energy Conservation Act 2000*) was not legally enforceable.

26. *Hunter* v *Canary Wharf Ltd [1997] Env LR 488.*

27. *Graham v. ReChem International* Ltd. *[1996] Env. LR 158* and see Anon, Record-breaking civil hearing ends in relief for ReChem, *ENDS Report,* 245, June, 1995, 18.

28. *Sturges v Bridgman* (1879) 11 Ch. D. 852, p. 865.

29. *St Helen's Smelting Co. v Tipping* (1865) 11 HL Cas 642.

30. *Barr and Others* v *Biffa Waste Management Services [2012] EWCA Civ 312*, with quote from the judgement of Lord Justice Carnwath at para.46, and contrast the lower court decision at [2011] EWHC 1003.

31. *Corby Group Litigation* v *Corby District Council [2009] EWHC 1944 TCC* and see *Claimants Appearing on the Register of the Corby Group Litigation* v *Corby Borough Council* [2008] EWCA Civ 463 (where it was established that damages for personal injury could be recovered in public nuisance) and *Corby Group Litigation* v *Corby District Council (No.2)* [2009] EWHC 2109 TCC (concerning costs).

32. *Part IIA Environmental Protection Act 1990*, as amended.

33. For examples: *Guerra and Others* v *Italy (1998) 26 EHRR 357* (where it was suggested acute arsenic poisoning for a chemical works might be contrary to the right to life); *Lopez Ostra* v *Spain (1995) 20 EHRR 277* (where the operation a waste treatment plant, located just 12 m from the applicant's home, was found to have interfered with enjoyment of private and family life); *S. v France (1990) 65 D&R 250* (where nuisances arising from a nuclear power station near the claimant's home were found to constitute an infringement of rights to the home and property); and *Fadeyeva v Russia (Application No. 55723/00) ECHR 9 June 2005* (where the operation of steel manufacturing plant, 450 m from the applicant's home, was found to have endangered her health, contrary to the right to respect for the home).

34. *Hatton and Others* v *United Kingdom* (8 July 2003) Grand Chamber ECHR (Appl. No. 36022/97).

35. *Peter Marcic* v *Thames Water Utilities Ltd [2003] UKHL 66.*

36. *Roper* v *Tussauds Theme Parks Ltd [2007] EWHC 624 (Admin)* where a 'person aggrieved' sought to challenge the terms of an abatement order in relation to levels of noise from a theme park.

37. *Birmingham City Council* v *Oakley [2001] Env LR 648*, where it was held that a toilet without washbasin did not cause residential premises to be in such a *state* as to be a statutory nuisance.

38. *R. v. Carrick District Council ex. p. Shelly [1996] Env LR 273*, and see Anon, Statutory nuisances and sewage discharges, *ENDS Report*, 255, April 1996, 48.

39. *Alphacell Ltd. v Woodward [1972] 2 All ER 475*, per Lord Salmon at p. 491.

40. *Environment Agency v Empress Cars (Abertillery) Ltd.* [1998] 1 All ER 481.

41. *Water Environment (Water Framework Directive) (England and Wales) Regulations*, 2003 SI 2003/3242.

42. Environment Agency, *2012 Check on Progress,* 2012; http://a0768b4a8a31e106d8b0-50dc802554eb38a24458b98ff72d550b.r19.cf3.rackcdn.com/LIT_7500_c23719.pdf (accessed 01/08/2013); and see Anon, River quality targets not met, *ENDS Report*, 455, December 2012, 22.

43. European Commission, *Report on the Implementation of the Water Framework Directive* COM, 2012, **670**, 6; http://ec.europa.eu/environment/water/water-framework/pdf/COM-2012-670_EN.pdf (accessed 01/08/2013); and European Commission, *Staff Working Document on the Implementation of the Water Framework Directive,* 2012, **379**, 20; and see SWD (20120) 379 29/30 on the United Kingdom; http://ec.europa.eu/environment/water/water-framework/pdf/CWD-2012-379_EN-Vol3_UK.pdf (accessed 01/08/2013).

44. Royal Commission on Environmental Pollution, *Fifth Report. Air Pollution Control: An Integrated Approach*, 1976 Cmnd.6371; and *Twelfth Report. Best Practicable Environmental Option*, 1988 Cm.310.

45. *Environmental Protection (Prescribed Processes and Substances) Regulations 1991 (SI 1991/ 472)*.

46. *R. v. Environment Agency and Redland Aggregates Ltd ex parte Gibson [1999] Env LR 73*; *R (on the application of Thornby Farms Ltd) v Daventry District Council [2002] Env LR 687*; and *Levy v Environment Agency [2002] EWHC 1663 (Admin)*.

47. European Commission, *Towards an Improved Policy on Industrial Emissions* COM (2007) 843 and European Commission, *Proposal for a Directive on Industrial Emissions (Integrated Pollution Prevention and Control)* COM (2007) 844.

48. Generally, on environmental permitting see Department for Environment, Food and Rural Affairs, *Environmental Permitting Guidance: Core Guidance,* 2010 updated; www.defra.uk.gov; and information on environmental permitting practice on the Environment Agency website: www.environment-agency.gov.uk.

49. http://www.unesco.org/pv_obj_cache/pv_obj_id_5153A80E5D000D9118833F33BE125378F10 50100/filename/RIO_E.PDF (accessed 01/08/2013).

50. *The 1998 United Nations Economic Commission for Europe Convention on Access to Information, Public Participation and Decision Making and Access to Justice in Environmental Matters adopted by European Environment Ministers meeting in Aarhus (Denmark);* http://www.unece. org/env/europe/ppconven.htm for text of the Convention (accessed 01/08/2013); and see *The Aarhus Convention: An Implementation Guide,* 2000.

51. *Directive 2003/35/EC* of the European Parliament and the Council providing for public participation in respect of the drawing up of certain plans and programmes relating to the environment and amending with the regard to access to justice *Council Directives 85/337/EEC and 96/61/EC* (OJ L 156/17 25 June 2003).

52. *Regulation (EC) No.1367/2006* on the application of the provisions of the Aarhus Convention to Community institutions and bodies and see also *Decision 2008/50/EC* laying down detailed rules for the application of *Regulation (EC) No.1367/2006* on the Aarhus Convention as regards requests for the internal review of administrative acts.

53. On the procedure of the Compliance Committee, see United Nations Economic Commission for Europe, *Guidance Document on the Aarhus Convention Compliance Mechanism* (not dated) at http://www.unece.org/env/pp/compliance/CC_GuidanceDocument.pdf (accessed 01/08/ 2013).

54. For reports of decisions: http://www.unece.org/env/pp/pubcom.htm (accessed 01/08/2013).

55. *ACCC/C/2008/33 complaint from ClientEarth and others about non-compliance* (under Art.9).

56. *Communication ACCC/C/2008/32 submitted by Client Earth against the European Community*.

57. *Case T-338/08, Stichting Natuur en Milieu v European Commission*, 14 June, 2012.

58. *Case T-396/09, Vereniging Milieudefensie v European Commission*, 14 June, 2012.

CHAPTER 22

The Regulation of Industrial Pollution

MARTIN G. BIGG[†]

Professor of Environmental Technology and Director, Environmental Technologies iNet, University of the West of England, Frenchay Campus, Coldharbour Lane, Bristol BS16 1QY, UK
Email: martingbigg@gmail.com

22.1 INTRODUCTION

The regulatory control of industrial emissions has undergone substantial change and development in the past 150 years. The most significant drivers have been the needs to protect local people and the environment. Control has also been in response to technological developments, incidents, public concerns and national and international drivers. The approaches to securing environmental protection have also changed significantly from the imposition of site-specific numerical emission limits to the development and implementation of national and international controls.[‡]

Science and international collaboration, particularly at European Union (EU) level, have had a major effect on industrial regulation. Most regulatory controls and limits, until around 1990, were based on private discussions between regulatory inspectors and industry. The advent of substantial European Community Directives and Regulations on air, waste, water pollution control and subsequently on integrated pollution controls on industry have achieved greater consistency and transparency in industrial pollution regulation across Europe.

The enforcement of control has varied between tight interpretation of individual limits to a wider focus on environmental objectives and outcomes. The many pieces of UK and European environmental control legislation have taken different routes to the setting of standards and their interpretation, and been enforced in a range of different ways by local and national regulatory authorities. Regulations and requirements have grown significantly with various efforts at consolidation and integration of legislation and the championing of "modern" and "better" regulation. Controls and guidance have developed from the core legislation sometimes resulting in complexity and confusion as to their legal basis.

The relationship between industry and regulator has also varied from "regulatory police" pursuing "industrial poachers" to "self-regulation" by industry. Many industry sectors have developed

---

[†]This chapter is based upon an earlier contribution from David Slater and Caroline John which is gratefully acknowledged.
[‡]*No man is an Island, entire of itself; every man is a piece of the Continent, a part of the main.* John Donne (1572–1631).

Pollution: Causes, Effects and Control, 5th Edition
Edited by R M Harrison
© M G Bigg and The Royal Society of Chemistry 2014
Published by the Royal Society of Chemistry, www.rsc.org

a far greater understanding of their environmental footprint and taken more responsibility for their own actions. This has resulted in stronger self-discipline of the sector and the better use of management controls to achieve environmental protection and outcomes. Unfortunately some parts of other sectors remain in environmental denial with corresponding continuing poor performance.

A significant development in pollution control has been the effect of the public awareness of and interest in environmental issues and the environmental impact of industry. This has been reflected in and sometimes driven by political interests. The perceptions and implications have not, however, always been driven by sound science or reasoned argument, leading to inconsistencies, distorted priorities and, in some cases, negative environmental implications. Overall this greater understanding has resulted in industrial sectors and individual operators engaging far more with the public and politicians and those informing them, including public interest groups and non-government organisations.

The most important change over the past 150 years is that the environment is far better than it might otherwise have been or actually was. We have cleaner air and water, less waste and less contamination of land. Generally we are making better use of natural resources and improving the efficiency of our industrial processes. We are more precautionary about the environmental effects and benefits of new products and processes. We are also more aware of climate change and its causes, although our willingness and ability to take the necessary measures to mitigate and adapt to climate change are lagging behind our awareness. This is significant as many measures to control pollution have increased energy demand and required the use of additional resources.

Finally, it must be stressed that industrial pollution control does not operate and cannot be seen in isolation. Economic and social considerations have to be reconciled with environmental effects and outcomes. Industry pollution regulation has imposed costs as well as produced opportunities for businesses-businesses which provide products, services and employment.[1] We now have separate environmental regulators for Northern Ireland, Scotland, Wales and England who must work together to deliver clarity and consistency. Effective regulation has also created a more consistency and clarity for industry across Europe and influenced global development and production. The control of industrial pollution is about achieving the right balance.

22.2 BACKGROUND

22.2.1 Alkali Act

The control of industrial pollution could be said to have started in 1273 when the use of coal was prohibited in London as being "prejudicial to health".[2] In 1306 a Royal Proclamation prohibited artificers (craftsmen) from using sea-coal (a soft coal) in their furnaces.[3] In 1648 Londoners petitioned Parliament to stop coal from Newcastle being imported due to the adverse health effects of burning it. In 1661 John Evelyn, a London diary writer, published an essay on air pollution in London that he sent to King Charles II. He was enraged by the heavy smog that often fouled the air and documented the impacts of the smog on people's health and the environment. He also put forward some solutions to help resolve the problem, which were ignored, setting many comparisons with the position today.[4] The *1845 Railway Clauses Consolidated Act* required railway engines to consume their own smoke and the *1847 Towns Improvement Clauses Act* contained a section dealing with factory smoke.[5]

The principles found in today's modern regulation can be traced back to the *1863 Alkali Act* which required that 95% of the offensive emissions (hydrogen chloride) should be arrested.[6] It introduced the precautionary principle to UK law and was the first example of the application of a numerical limit on emissions. 1864 saw the appointment of a Chief Inspector, with statutory responsibilities for setting standards and reporting progress to the Government. A further *Alkali Act* in 1874 introduced the requirement to use the "best practicable means" to control all other offensive or noxious gases from the works and in 1881 extended the range of industrial processes covered by the legislation.[7] A *Royal Commission on Noxious Vapours* concluded in 1878 that measures were only practicable if they did not involve "ruinous expenditure".

The *Alkali Act 1906* consolidated previous *Alkali Acts*.[8] It

(i) Reiterated the numerical controls on emissions to the air from alkali works (95% reduction and no more "than one-fifth part of a grain of muriatic acid").

(ii) Set a fine for first and subsequent offences.

(iii) In addition to the specified requirements to control the emission, required the use of the best practicable means (BPM) for preventing the escape of noxious or offensive gases by the exit flue of the processes, and for preventing the discharge directly or indirectly of such gases into the atmosphere, and for rendering such gases where discharged harmless and inoffensive subject to meeting the emission limit.

(iv) Prescribed gases being released and set requirements on discharges to water courses and waste disposal.

(v) Applied similar controls to specified works other than alkali works.

(vi) Required a scheduled works to be registered and the registration to be renewed each year. Stamp duty was charged on each certificate of registration.

(vii) Set out the requirements to be an inspector and their powers and responsibilities.

(viii) Included details on inspecting, examining and testing the activities at the works and the recording of results.

The Act defined "best practicable means" (BPM):

The expression "best practicable means", where used with respect to the prevention of the escape of noxious and offensive gases, has reference not only to the provision and the efficient maintenance of appliances adequate for preventing such escape, but also to the manner in which such appliances are used and to the proper supervision, by the owner, of any operation in which such gases are evolved.[9]

Through the *Alkali Act 1906* many of the principles and structures for the current regulation of industrial pollution were established. The *Alkali Act 1906* was finally repealed under the *Environmental Protection Act 1990*.[10]

22.2.2 Regulators

Between 1906 and 1990 responsibility for the delivery of industrial regulation in England and Wales by the Alkali Inspectorate was shifted between local authorities and government ministries. The new Department of the Environment was its home from 1970 to 1975 before it became the Industrial Air Pollution Inspectorate, part of the Health and Safety Executive until 1987. In Scotland and Northern Ireland similar shifts occurred. During this time the number of prescribed works increased and clear segregation emerged between industrial processes regulated by local authority environmental health officers under clean air and public health legislation and larger more technically complex and potentially more polluting industrial processes. The Alkali Inspectorate and the Industrial Air Pollution Inspectorate were seen as secretive organisations, not sharing environmental information but rather hiding behind the *Official Secrets Act*.[11]

Inspectors worked closely with industry representatives to agree control technologies and release limits. Data was collated nationally to ensure a level of consistency. Guidance notes were prepared on the interpretation of best practicable means (BPM) for each category of "Works" and individual notes on the interpretation of BPM for specific sites were drafted by the relevant inspecting officer, usually in collaboration with the site operator. BPM was what the inspector considered reasonable in the light of national guidelines but was not subject to external challenge or enforcement; rather the inspector preferred to secure change and improvements by working in partnership, cooperation and persuasion. In some cases "presumptive limits" were set which could be achieved using the BPM. If the presumptive limit was achieved then it was assumed that BPM was being used. As both the site-specific and the generic BPM notes were not linked to legislation they could be changed at any time, particularly when new technology was introduced or new scientific evidence on pollution became available.

Enforcement action was viewed as failure. The site documents were not shared more widely or made public. Emission monitoring was undertaken by dedicated teams and national reports prepared on the performance of each industrial sector. Record keeping by individual site inspectors was very variable and inspection frequency and focus were driven by national requirements and local issues. Publically available information on environmental performance by each site was minimal.

On 1 April 1987 the formation of Her Majesty's Inspectorate of Pollution (HMIP) and Her Majesty's Industrial Pollution Inspectorate (HMIPI) in Scotland saw the start of a significant change in industrial pollution regulation. HMIPI was given responsibility for all industrial pollution regulation while HMIP continued to regulate larger, technically more complex activities in parallel with local authority regulation. HMIP brought together industrial, hazardous waste and radiochemical regulation as well as government water regulation responsibilities. Initially industrial regulation continued along similar lines to the Alkali Inspectorate, with a Chief Inspector having responsibility for industry regulation. In England and Wales there were also separate Chief Inspectors covering radio-chemicals and hazardous waste and all reported to a senior civil servant in the Department of the Environment. The loss of the direct link to a government minister and the different cultures between the different inspectorates and core civil servants led to tensions. This separation was not resolved until the appointment of a single Chief Inspector in 1991.

The formation of the Pollution Inspectorates was in part due to a recommendation in the Royal Commission on Environmental Pollution *5^(th) Report on Air Pollution Control: an Integrated Approach* in January 1976.[12] The Commission objecting to consolidation into the Health and Safety Executive observed that the Alkali Inspectorate's proper concern was with the environment as a whole:

Their policies should be evolved not as an adjunct to industrial safety but as a part of an integrated approach to the control of environmental pollution.

22.2.3 Integrated Pollution Control

The *12^(th) Report on Best Practicable Environmental Option*, published in February 1988, urged the Government and industry to develop pollution control measures that took account of risks to the whole environment (Integrated Pollution Control).[13] It set out a procedure for finding the best practicable environmental option (BPEO) for processes affecting the environment and emphasised the need for transparent decision making.

Integrated Pollution Control (IPC) was introduced across England, Wales and Scotland by the *Environmental Protection Act 1990* in addition to separate air pollution control by local authorities in England and Wales. The phrase IPC was not specifically mentioned in the Act. The *Environmental Protection Act* was required to implement the European *Air Pollution from Industrial Plants Directive 84/360/EEC*, a framework directive which identified a wide range of industrial processes and releases to be controlled. Many were the subject of the previous alkali acts and their successors.

IPC introduced a more modern and prescriptive approach with clear statutory objectives and timescales. It required operators to have prior permission to operate. In order to gain an authorisation, prospective operators had to provide a detailed justification for the process selected and how they would operate it to meet the requirements of the Act. These were:

(i) Adoption of the best available techniques not entailing excessive costs (BATNEEC) to prevent, minimise or render harmless substances released into any medium.
(ii) Fulfilment of any EC or international obligations.
(iii) Minimising the pollution of the whole environment, having regard to the BPEO available as respects the substances which may be released.

The conditions attached to authorisations by the inspector had to be sufficient to ensure compliance with these objectives.

Guidance on IPC was issued by the Department of the Environment and the Welsh Office.[14] Section 7 of the Act explains that:

References to the best available techniques not entailing excessive costs, in relation to a process, include (in addition to references to any technical means and technology) references to the number, qualifications, training and supervision of persons employed in the process and the design, construction, lay-out and maintenance of the buildings in which it is carried on.

"Best" should be taken to mean the most effective in preventing, minimising or rendering harmless polluting releases. There may be more than one set of techniques that achieve comparable effectiveness and thus there may be more than one set of "best" techniques.

"Available" should be taken to mean economically and industrially feasible in the relevant industrial context. Economic feasibility is achieved where a technique is not excessively costly (see below). Industrial feasibility requires general accessibility, without implying that the technique has to be in general use. It includes techniques which have been implemented in the relevant industrial context with adequate guarantees to ensure the necessary business confidence. Sources outside the UK should not be considered to be "unavailable", nor should a technique which is subject to monopoly supply, provided that the operator can procure it.

"Techniques" includes the plant, the way it is operated, and the numbers, qualifications and supervision of the personnel who operate it.

"Not entailing excessive cost", in the context of a technique, applies if the benefits of employing the technique outweigh the costs. The benefits in this context relate primarily to the prevention of environmental damage, and the costs relate primarily to the costs of applying BAT. The act of balancing the benefits against the costs will involve judgement. In general, the greater the environmental damage, the greater the costs of BAT that can be required before costs are considered excessive.

BPEO was considered as:

the option which in the context of releases from a prescribed process, provides the most benefit or least damage to the environment as a whole, at acceptable cost, in the long term as well as the short term.

In 1997 the Environment Agency produced detailed guidance on how to assess BPEO within the context of IPC.[15] It assessed the contribution of process emissions to overall pollutant levels and compared these with identified Environmental Assessment Levels (EALs). These were summed to produce an index for different media and an overall Integrated Environmental Index (IEI). Weighting factors were not applied to the methodology itself, although individual EALs had varying safety factors built into them. Different options were then identified and an assessment of the costs associated with each option compared with the IEI was made. It was assumed that costs which were disproportionate to the gain in environmental performance were not "practicable".

The requirement to use BATNEEC was an evolution from and a clarification of BPM. It enabled IPC to begin to address resource efficiency and whether wastes should be produced, but was constrained on how they could be disposed of. A residual condition requiring the use of BATNEEC could be applied to all other possible releases from the processes. One exception was carbon dioxide, which was deliberately omitted to avoid overlap with other legislation tackling energy efficiency or the causes of climate change.

The implementation of the *Environmental Protection Act 1990* started the UK down a route by which it could readily implement European environmental legislation and consolidate existing UK regulations:

   (i) The creation and subsequent increase in size and power of a national regulator.
   (ii) Centralised and integrated working.
   (iii) Annual reporting of industrial releases.
   (iv) National standards and guidance.
   (v) Greater transparency of operation and public registers.
   (vi) Improved environmental awareness and performance.
   (vii) Greater recognition and use of management standards and self regulation.
  (viii) More aggressive enforcement.
   (ix) Cost recovery charging.

The successful UK implementation of IPC was also a strong influence on the European Commission and its environment policies. The European approach was more about rules and standards rather than objectives and private negotiation. The UK approach was of prevention and precaution balanced against requirements for remediation. As a result subsequent EU environmental regulations have reflected more of a compromise between working with industry and the enforcement of strict and tightening standards. The need to avoid constraining trade across Europe and beyond was balanced with the need for constancy for European businesses.

The complete re-permitting of industrial activities under IPC was a substantial exercise for the stretched resources of regulators across England, Wales and Scotland, coinciding with the privatisation of the power industry and the Government de-regulation drive in 1993.[16] The speed and extent of roll out of flue gas desulfurisation across the UK power industry was determined by political as well as environmental need. Regulation moved from "arms length" to a more consensual approach. Preparation for and the securing of an environmental permit required a high level of collaboration between industry and regulators. Consultants played a major role on both sides. Routine site inspections were replaced by application or enforcement based visits and greater emphasis was placed on environmental auditing.

IPC was intended to deliver the cross-media approach of the Royal Commission. The concept of truly integrated regulation to encapsulate the causes of pollution or the range of mitigation measures required, however, was not delivered. Planning, especially strategic environmental assessment, as well as transport and energy, were not included. Rather it was a joining of regulatory functions encompassing releases to air, land and water from specified industrial activities. Even here it was limited as most water regulation remained with the new National Rivers Authority. As a result it failed to tackle the inconsistencies and differences between national and local regulation. It failed to address the sources of many environmental issues, the proximity of industrial processes to housing and other intensely occupied areas, usually due to the encroachment of housing towards industrial areas. The BPEO analysis developed for IPC was rarely used as the system was too complex and confusing. Instances were very rare of where assessments were made and a change in process operation between different media resulted.

The *Environment Act 1995*[64] created the Environment Agency (EA) in England and Wales, and the Scottish Environment Protection Agency (SEPA) in 1996. The Northern Ireland Environment Agency (NIEA) was formed in 2008 from the Northern Ireland Environment and Heritage Service. The formation of these Agencies and subsequently the separate agency for Wales in 2013 brought together most environmental protection functions in one place with the necessary support services. The consolidation of the environmental protection legislation took longer.

22.2.4   Integrated Pollution Prevention and Control

The *EC Integrated Pollution Prevention and Control (IPPC) Directive* required a broader assessment of the environmental activities of prescribed processes.[17] Issues covered included:

   (i) Emission of pollutants to all media.
  (ii) Prevention, reduction, recovery, disposal of waste.
 (iii) Efficient use of energy.
 (iv) Use of raw materials.
  (v) Prevention of accidents.
 (vi) Remediation of sites after cessation of activities.
(vii) Noise.

The Directive objectives were to "achieve a high level of protection for the environment as a whole" and needed an integrated approach to environmental protection parallel to the UK approach of BPEO assessment within IPC.

The Directive was transposed and implemented by separate regulation in England and Wales, Scotland and Northern Ireland, reflecting the devolved administrations. It was rolled out in England and Wales through the *Pollution Prevention and Control (England and Wales) Regulations 2000* over seven years from December 2000 to January 2007.[18] The regulations maintained the separation between the IPC (part A) regime generally delivered by the national regulator and the continuation of the air pollution control (part B) Local Authority Pollution Prevention and Control (LAPC) regime. In addition there were some (part A2) installations which were primarily regulated by local authorities but required integrated control as they were included in the Directive. Guidance was provided to local authorities by national regulators. Industrial sectors new to integrated permitting included landfill, intensive farming and food and drink sectors. For the first time a consistent integrated regulatory approach could be achieved to controlling pollution from much of the waste industry, manufacturing industry and power industry.

The Regulations aimed to achieve "a high level of protection of the environment taken as a whole by, in particular, preventing or, where that is not practicable, reducing emissions into the air, water and land".[19] The main way of doing that was by determining and enforcing permit conditions based on BAT.

The *PPC Regulations 2000* define "best available techniques" (BAT):[20]

the most effective and advanced stage in the development of activities and their methods of operation which indicates the practical suitability of particular techniques for providing in principle the basis for emission limit values designed to prevent and, where that is not practicable, generally to reduce emissions and the impact on the environment as a whole;

and for the purpose of this definition:

   (i) *"available techniques" means those techniques which have been developed on a scale which allows implementation in the relevant industrial sector, under economically and technically viable conditions, taking into consideration the cost and advantages, whether or not the* techniques are used or produced inside the United Kingdom, as long as they are reasonably accessible to the operator;

(ii) *"best"* means, in relation to techniques, the most effective in achieving a high general level of protection of the environment as a whole;

(iii) *"techniques"* includes both the technology used and the way in which the installation is designed, built, maintained, operated and decommissioned.

The Regulations also stated:

in determining best available techniques special consideration shall be given to the following matters, bearing in mind the likely costs and benefits of a measure and the principles of precaution and prevention:

  (i) *the use of low-waste technology;*

  (ii) *the use of less hazardous substances;*

  (iii) *the furthering of recovery and recycling of substances generated and used in the process and of waste, where appropriate;*

  (iv) *comparable processes, facilities or methods of operation which have been tried with success on an industrial scale;*

  (v) *technological advances and changes in scientific knowledge and understanding;*

  (vi) *the nature, effects and volume of the emissions concerned;*

  (vii) *the commissioning dates for new or existing installations or mobile plant;*

(viii) *the length of time needed to introduce the best available technique;*

  (ix) *the consumption and nature of raw materials (including water) used in the process and the energy efficiency of the process;*

  (x) *the need to prevent or reduce to a minimum the overall impact of the emissions on the environment and the risks to it;*

  (xi) *the need to prevent accidents and to minimise the consequences for the environment;*

  (xii) *the information published by the Commission pursuant to Article 16(2) of the Directive or by international organisations.*

The Regulations established a clear and comprehensive framework for the assessment of most industrial activities.

In order to ensure that a high level of consistency could be achieved across Europe, *Article 16(2) of the IPPC Directive* required the European Commission to organise

an exchange of information between Member States and the industries concerned on best available techniques, associated monitoring and developments in them

It also required it to publish the results of the exchange. The Directive stated that:

development and exchange of information at Community level about best available techniques will help to redress the technological imbalances in the Community, will promote the world-wide dissemination of limit values and techniques used in the Community and will help the Member States in the efficient implementation of this Directive.

The European Integrated Pollution Prevention and Control Bureau (EIPPCB) was set up in 1997 in Seville, Spain to organise the exchange of information between Member States and industry on Best Available Techniques (BAT). The BAT reference documents (BREFs) are now used both with the *IPPC Directive* and the *Industrial Emissions Directive (IED)*.[32]

Both the EA and SEPA required all the existing prescribed installations to apply for new permits under the new regulations over a seven year programme. This was to secure consistency across the permitting of installations previously regulated under different regimes and a culture change within their respective organisations. While a level of integration and consistency was achieved between waste, manufacturing and power industry regulation, the regulation of the water industry and water aspects of the other industries tended to be kept separate. Any installation which came into operation on or after 31 October 1999 had to have a permit before it could operate. External contractors played a significant role in permit determination, albeit not the final decision. It again demonstrated the value of assessment and auditing skills, as well as technical knowledge in the determination of applications from a very wide range of industries.

In making an application the operator had to address various environmental issues which extended beyond the requirements of IPC. These included:

(i) Satisfactory environmental management of the installation.
(ii) Adequate compliance monitoring.
(iii) Assessment of polluting releases and the identification of the best available technique (BAT).
(iv) Compliance with other EU Directives, Community and national environmental quality standards (EQSs) and domestic regulations.
(v) Energy efficiency, waste minimisation and management.
(vi) Prevention of accidents.
(vii) For landfills, alternative requirements as specified by the *Landfill Regulations*.

The operator also had to consider the condition of the site at the time of the original application. This was then to be used in assessing the need for restoration when the installation closed.

The move to regulation of the installation meant that greater clarity was required of what precisely what the installation was. Not everything on a site warranted IPPC control, and how far an installation extended, say though a connecting pipe bridge or inclusion of an effluent treatment plant serving several different manufacturing activities, was subject to extensive debate. While guidance was developed, many cases were determined on an individual basis.[21] The anomalous exclusion of some waste activities on a site remained.

22.3 ENVIRONMENTAL PERMITTING REGULATIONS

22.3.1 Introduction

The eventual successful implementation of IPPC and the connections with wider economic sustainable development resulted in a shift in industry regulation from a technical assessment of a process and consideration of local impacts to the achievement of wider policy and environmental outcomes. These included energy, resource and waste management. There was a shift to a top down approach and the delivery of strategic targets and international commitments. The local application of BATNEEC and BPEO was superseded by the requirements of national and EU targets and obligations. The *Large Combustion Plant Directive*[22] and the *Waste Incineration Directive*[23] set international standards and timetables for change. The regulators identified the most significant pollutants and the industries with the greatest releases for regulatory attention and inspection. This was greatly aided by the obligation from 1998 in England and Wales and subsequently in Scotland and Northern Ireland to report releases of key pollutants on Pollution Inventories.[24] They in turn provided data to the *European Pollutant Release and Transfer Register*.[25] Reporting on the Resource Efficiency Physical Index (REPI) has enabled the performance of individual sites within a sector to be compared and best practice identified.[26]

The impact of the targeting of controls on the sectors which make the greatest contribution to the principal air pollutants can be readily seen in Figures 22.1, 22.2 and 22.3 which show the releases from Environment Agency regulated industries. Note that, as other industries have been brought under environmental regulation, their releases have been included, resulting in some increases.

The *Environmental Permitting (England and Wales) 2007 Regulations* implemented in 2008 and similar regulations in Scotland and Northern Ireland provided a single platform for permitting and compliance with a risk-based approach.[27] They finally combined industrial pollution regulation and waste management licensing into a simplified and standardised approach applicable to other industries. Subsequently they were used to provide the permitting requirements of the *Batteries and Mining Waste Directives*.[28,29] The regulations allowed businesses that would otherwise have required several permits for different activities falling under the regulations to have just one permit. Significantly no re-permitting of activities was undertaken with the new regulations. Instead existing permits were transmuted into the new regulatory regime. Together the changes met the

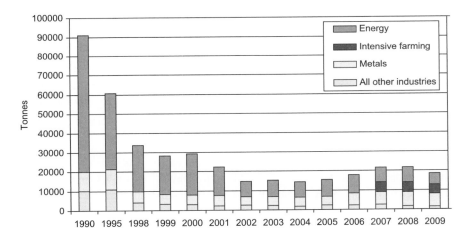

Figure 22.1   PM$_{10}$ releases from industry.

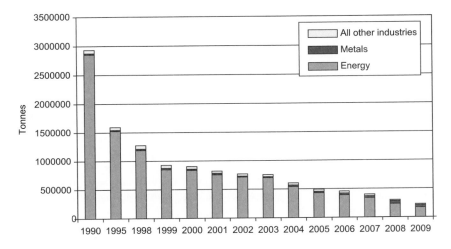

Figure 22.2   SO$_x$ releases from industry.

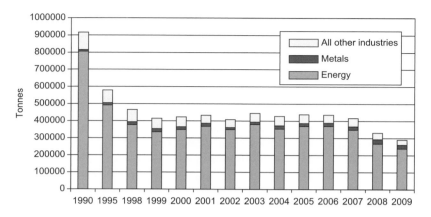

Figure 22.3 NO$_x$ releases from industry.

Government objective of reducing red tape without affecting environmental standards. Permitting was proportionate to the risks:

(i) Bespoke permits for high risk activities – usually larger, complex developments that required site-specific assessment of risks, or for activities where standard rules had not yet been developed. The application charge was usually related to environmental risk and complexity.

(ii) Standard rules permits – sets of rules for easily defined, simpler, smaller scale activities, with a fixed charge and on-line application.

(iii) Exempt activities which met specific descriptions - most were free of charge.

(iv) Deregulated activities covered by a consenting regime but such low impact there was no benefit in regulation, for example low risk waste activities.

This was followed by consolidation of:

(i) Water discharge consenting.
(ii) Groundwater authorisations.
(iii) Radioactive substances regulation.

The *Environmental Permitting Regulations 2010* covered facilities previously regulated under the *Pollution Prevention and Control Regulations 2000*, and Waste Management Licensing and exemptions schemes (as superseded by the *Environmental Permitting (England and Wales) Regulations 2007*), some parts of the *Water Resources Act 1991*, the *Radioactive Substances Act 1993* and the *Groundwater Regulations 2009*.[30] The *Environmental Permitting Regulations 2013* amended the 2010 Regulations to implement the *Directive on Industrial Emissions 2010/75/EU (IED)*.[31,32] The Directive consolidated seven existing Directives related to industrial emissions into a single clear and coherent legislative instrument. The Directive combined *the IPPC Directive*, the *Large Combustion Plants Directive*, the *Waste Incineration Directive*, the *Solvents Emissions Directive* and three *Directives on Titanium Dioxide*. As a result, at both a European level and a national level, at last a truly integrated and unified approach to the regulation of industrial pollution had been achieved.

The advent of consolidated, clear and consistent regulations facilitated the achievement of a clearer and consistent approach to the regulation of industry. At last regulators have become less concerned about the exact wording of which piece of legislation should be applied and more concerned about the achievement of environmental performance and outcomes. Industry has a clearer and simpler framework in which to operate, albeit backed up by environmental and performance standards. There is less debate about techniques and processes and more about outcomes and time scales. While the need for a high level of technical competency by the regulators may have reduced, the requirement for a good understanding of business needs and impacts has increased. In many industrial sectors the competency of staff in environmental management has significantly increased facilitating the development of a stronger risk-based approach.

The implementation of the *Environmental Permitting Regulations* was also the culmination of a steady shift towards "modern" or "better" regulation. This change started in the early twenty-first century with the publication of the National Society for Clean Air and Environmental Protection report on *Smarter Regulation*.[33] It showed how industrial pollution regulation and environmental protection should evolve to continue to deliver effective pollution control as well as sustainable development. In particular it considered the interface with health and planning, voluntary initiatives, the need for national outcome-focused strategies and the links to economic and social drivers. It also put forward the development of more negotiated agreements with industry as had been developed with the power sector under the *Large Combustion Plant Directive*.

Delivering for the Environment: A 21st Century Approach to Regulation, produced by the Environment Agency, identified the range of tools available to the regulator and heralded a more generic approach to future regulation.[34] The model of regulation is described in Figure 22.4.

Following the Hampton and Macrory reviews of compliance and enforcement, the UK regulators continued the development of a more risk-based and outcome focused approach.[35,36] The subsequent audit of the Environment Agency identified a strong commitment to better regulation and a risk-based approach.[37] It acknowledged the reduction of the regulatory burden on industry and the improvement in the quality of its written advice, forms and publications. It challenged the Environment Agency to:

   (i) Focus on environmental priorities.
  (ii) Maximise the impact of regulation on outcomes.
 (iii) Ensure enforcement action is proportionate.
 (iv) Recognise good performance.
  (v) Work with the Government on regulations.
 (vi) Improve access to and the quality of advice and guidance.

These challenges were met in the implementation of the *Environmental Permitting Regulations*.

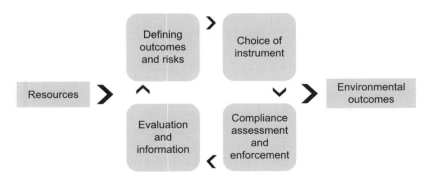

Figure 22.4 Model of Regulation.

22.3.2 Permitting

The *Environmental Permitting Regulations* require operators to obtain permits for some facilities, to register others as exempt and provides for ongoing supervision by regulators. The stated aim of the regulatory regime is to:

(i) Protect the environment so that statutory and government policy environmental targets and outcomes are achieved.
(ii) Deliver permitting and compliance with permits and certain environmental targets effectively and efficiently in a way that provides increased clarity and minimises the administrative burden on both the regulator and the operators.
(iii) Encourage regulators to promote best practice in the operation of facilities.
(iv) Continue to fully implement European legislation.

The Regulations set out the following:

(i) Descriptions of facilities that need environmental permits or need to be registered as exempt.
(ii) Process for registering exempt facilities.
(iii) How to apply for and determine permit applications.
(iv) Requirements that environmental permits contain conditions to protect the environment as required by directives and, where applicable, national policy.
(v) How environmental permits can be changed and ultimately be surrendered.
(vi) A simplified permitting system using standard rules.
(vii) Compliance obligations backed up by enforcement powers and offences.
(viii) Provisions for public participation in the permitting process.
(ix) Powers and functions of regulators and ministers.
(x) Provisions for appeals against permitting decisions.

Government guidance[38] explains how the range of regulated activities which may be undertaken on a site could be regulated. In many cases where the operator is the same for all the facilities on the same site it is in the interests of the operator and the regulator to have a single permit. Factors to be considered include:

(i) Proximity: how close facilities are together.
(ii) Common services: for example a single fenced area, supply of steam, water, waste services.
(iii) Common management systems.

A ministerial direction is required if a single regulator is desired for a site which is currently regulated by both local authority and national regulators. This could then be used to simplify regulation and permitting.

22.3.3 Planning

The town and country planning system remains completely separate from the environmental permitting regime. If a facility requires planning permission, it is recommended that applications are made in parallel for planning permission and an environmental permit. For some waste and mining facilities, however, planning permission is required before an environmental permit can be granted. This can result in anomalies. For instance, applications for municipal waste incinerators raise many environmental issues arise during the planning application process and subsequent frequent appeals. They then end up being addressed several times or the process can lead to significant frustration on behalf of the objectors. In the interests of an integrated and more efficient permitting process, it has been argued that land use and environmental planning processes should be combined. This has received a mixed reception as environmental aspects could be down-played and there is no appeal against a successful planning application except through judicial review. Operators of permitted facilities often appreciate the benefits of a multi-stage process.

Permit applicants are encouraged to hold pre-application discussions with the regulators as well engage with interested parties including local authorities and communities. These can reduce the risk of significant costs and delays further along the process due to fundamental issues around the design, materials, location or perception of the facility.[39]

The clear lesson from the officers involved in permitting is that the better the quality of a permit application and the supporting documentation, the more engagement with interested parties, especially local authorities, the quicker the determination and the greater the chance of securing a permit.

22.3.4   Risk-based Regulation

The consolidation of the regulation of the disparate range of activities and sectors within a single regulatory framework has thrown into sharp relief the different legacies, attitudes and performance of the different sectors and sub-sets within each sector. Historically the waste sector was severely tarnished by the performance of part of the sector resulting in a high number of enforcement actions and prosecutions. The water industry reported a high number of non-compliances due in part to the high amount of monitoring and reporting that it undertook. The cement and incineration sectors attracted attention more because of their feedstock than their environmental impacts.

The regulators use similar compliance – enforcement/engagement models to differentiate between the performance and attitudes of industry and the response of the regulator. Figure 22.5 describes the model used by the Environment Agency and Figure 22.6 the model used by SEPA.

The Environmental Permitting Regulations Operations Risk Appraisal Scheme (Opra for EPR) has evolved as an effective tool for assessing the risk and performance of an individual activity or site.[40] It is used by the Environment Agency for the regulation of the most complicated activities. Similar tools are used by other regulators. The Opra assessment is based on five attributes:

   (i) Complexity – the type of activities covered by the permit.
  (ii) Emissions and inputs – the amounts put into and released from an activity.
 (iii) Location – the state of the environment around the site.
 (iv) Operator performance – management systems and enforcement history.
  (v) Compliance rating – how well the permit conditions have been complied with.

The numerical result for each attribute is banded to give a final Opra profile. As a result the regulatory approach should be proportionate to the level of environmental risk and how well it is managed ("net" risk). By applying the scheme the regulator should then be able to recognise and reward good performance. This should result in a different relationship with the operator, fewer and better joined-up inspections/audits and a reduction in fees and charges.

22.3.5   Advice and Guidance

The environmental permit is now central to the delivery of environmental regulation. It contains the conditions necessary to meet the objectives of the relevant UK and European legislation. Conditions

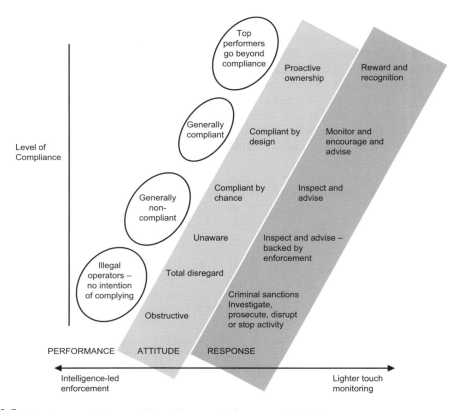

Figure 22.5   Environment Agency Compliance – Enforcement Model.

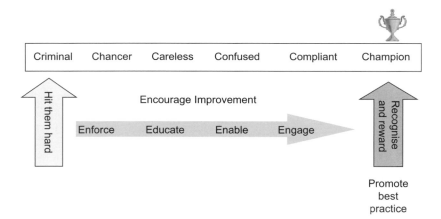

Figure 22.6   SEPA Compliance – Engagement Model.

include requirements on outcomes, limits on releases to the environment and process controls. They do not normally specify the techniques to be used. The operator is required to take responsibility for the activity and the achievement of the outcomes as well as follow relevant guidance. The regulators are expected to provide good quality and timely advice and guidance to help operators understand the options and actions needed to comply with the legislation and improve their overall environmental performance. The regulators provide a wide range of advice and guidance to those they regulate through their web sites, campaigns, face to face meetings and inspections.[41]

Environment Agency and government guidance for England consists of the following:

(i) Environmental Permitting Regulations – primary legislation.
(ii) Government Guidance – policy document explains how the regulations work.
(iii) Regulatory Guidance Notes – explains how the regulations are implemented and the meaning of terms in the regulation.
(iv) Technical Guidance – explains the basic requirements for operators.
(v) Sector Technical Guidance – explains the requirements for separate industrial sectors.
(vi) Horizontal Guidance – explains issues generic to all sectors. Notes include environmental risk assessments, energy efficiency, noise, odour, site condition report guidance, management systems.
(vii) Local Authority Manual.
(viii) Local Authority Technical Guidance.

In their engagement with business the UK regulators use similar approaches to a hierarchy of action:

(i) **Engage** with businesses by understanding them and listening to their concerns.
(ii) **Educate** businesses to help them understand how to comply with the law through providing advice and guidance as the preferred first approach.
(iii) **Enable** businesses to operate effectively through proactively providing advice and guidance and ensuring businesses can seek advice and guidance without triggering enforcement action.
(iv) **Enforce** when there are persistent breaches or significant harm has or may occur.

Interventions are determined by the level of environmental risk posed by a business, its performance in managing those risks and other public interest factors.

23.3.6 Environmental Management Systems

Ever since the development of the UK Environmental Management Standard BS7750 and the international standard ISO14001, the value of a certifiable environmental management standard as part of the tools to deliver environmental performance has been recognised by regulators and industry. The debate has been around the degree to which formal recognition of the existence of and compliance with the requirements of the standard should or could be taken into account when assessing the compliance of an operator with environmental regulations. The reactions of regulators have varied between completely ignoring the existence of the environmental management standard, to using the audit report as the starting point for compliance assessment. The standards, certification and accreditation infrastructure of the standards must complement regulatory permitting and compliance/enforcement.

Several studies have looked at the benefits of the use of an externally certified environmental management system (EMS) to the regulator.[42,43,44] The Northern Ireland Environment Agency study found:

(i) *There is a strong correlation between formal EMS and improved environmental performance.*

(ii) *There is inconclusive evidence of a correlation between formal EMS and compliance with environmental legislation, however there is strong evidence that formal EMS provides greater understanding of environmental legislation, a framework for recording the status of compliance and more stringent monitoring of emissions against permit or consent limits.*

Evidence from the research indicated that:

Organisations with a formal EMS will be more suited to a 'lighter touch' from the Regulator as they are more aware of legislation, more able to manage the risks associated with the impact of their activities on the environment and more able to provide evidence of their compliance status together with any associated monitoring programme data.

The REMAS study found:

(i) *There is strong evidence that the adoption of an accredited certified EMS improves site environmental management activities. Overall environmental management is better under ISO14001 than under an informal system; which in turn is better than under no system at all.*

(ii) *There is evidence that overall site environmental management is better under the EU Eco-management and Audit Scheme (EMAS) than under ISO14001; driven largely by better performance monitoring, documentation control and (self) reporting of environmental performance.*

(iii) *There is some evidence that improved site environmental management leads to lower average emission levels. However, the strength of the evidence differs significantly between receiving media, regions of Europe and sectors.*

(iv) *There is strong evidence that improved environmental management has an impact on the number of self recorded permit or licence breaches. The impact may be observed both positively (i.e. because it is reducing the number of breaches), or negatively (i.e. its increasing the number), and varies between regions and sectors.*

As a result there is now recognition of the benefits of an EMS and a clear hierarchy of benefits has been established:

(i) Self-certified management system – the Responsible Care scheme developed by the chemical industry provided an early example.[45]
(ii) Third party certified management system – BS7750 and ISO14001.
(iii) Third party certified management system with public reporting of performance – EU Eco Management and Audit Scheme.[46]

The benefits to operators and those affected by the activity of an effective EMS are now widely recognised:

(i) A level and consistent playing field for regulated business.
(ii) Structured and auditable method for managing environmental obligations.
(iii) Complementary and/or integrated systems to manage product/service quality, health and safety of employees and environmental risks.
(iv) Resources targeted to highest risk activities.
(v) Improved information and assurance on compliance with the law.
(vi) Earned autonomy and recognition, resulting in reduced regulatory fees and charges and reduced operating costs and improved resource efficiency.

It is important that the requirements of an EMS are integrated into the overall regulatory approach adopted by the regulator. Key elements include:

(i) Standardised, industry sector based permitting and compliance driven by specific industry facing regulatory groups.
(ii) Outcome and performance based compliance assessment.
(iii) More joined-up regulation including with other certification based approaches including standards for monitoring (Environment Agency Monitoring Certification Scheme – MCERTS) and other regulation *e.g.* EU Emission Trading Schemes, Carbon Reduction Commitment (CRC) Energy Efficiency Scheme.[47]

To achieve a high level of environmental protection, operators must have an effective management system in place. The complexity of the regulated activity will determine the nature of the management system required. Anyone holding a permit under the *Environmental Permitting Regulations 2010* is required to have an appropriate EMS in place. It must set out in detail how all the activities will be managed in accordance with the permit.

Complex facilities are encouraged to have a formal EMS externally certified to ISO 14001 by a United Kingdom Accreditation Service (UKAS) accredited certification body, or European equivalent, and to register for the EU Eco Management and Audit Scheme (EMAS). These standards include requirements for legal compliance and a commitment to continuous improvement of the environmental management system or environmental performance. EMAS requires organisations to produce an independently validated public report about their environmental performance and progress against targets and objectives.

Simpler facilities are encouraged to consider externally certified schemes but simpler schemes may be more appropriate such as the approach in the standard BS8555 (guide to the phased implementation of an environmental management system).

22.3.7 Competency

The successes of the industry regulation permitting regime depend on the operator's ability to operate the facility and meet their obligations. Operators must be technically competent to operate their facility. The operator's wider management system should contain mechanisms for assessing and maintaining technical competence. The competence of individuals should form part of those management systems.

The regulator takes account of the competency of the operator as part of the determination of a permit application and during the life of the permit. The competency of regulatory staff and operators has been an issue which used to reappear at regular intervals, especially with respect to the waste industry. As a result of collaboration by the waste industry and the regulators, several technical competency schemes have been approved. All technically competent waste operations staff are required to demonstrate continuing competence. There is a minimum site attendance requirement for operational sites which must be documented. Requirements are less strict for non-waste sites where the expectation is that staff have clearly defined roles and responsibilities. Guidance is given on demonstrating competence. Staff are expected to keep their skills up to date.

Industry is encouraged to develop and maintain sector-led competence schemes through the relevant Sector Skills Council. All schemes should be based predominantly on qualifications accredited by the Office of Qualifications and Examinations Regulation (OFQUAL), based on vocational qualifications where these exist, and agreed with the regulator and government.

The Environmental Regulators have been developing minimum levels of competence for staff regulating activities covered by an EMS. These have been established in cooperation with, and assessed by, the environment professional institutions, including the Institute for Environmental Management and Assessment (IEMA), the Chartered Institute for Waste Management (CIWM) and the Chartered Institute for Water and Environmental Management (CIWEM). The professional standing and integrity of these organisations is enhanced by strong requirements for continuing professional development. There has been significant uptake by environmental managers of membership of environment professional organisations, boosting the credibility of corporate and public commitments to environmental performance. The ambition must be that regulators and operators work to the same environmental management standards with staff of comparable competency. The development of the standing of the "Chartered Environmentalist" by the professional bodies under the auspices of the Society for the Environment is seen as the gold standard to be aimed at. As a minimum, regulators and those with environmental responsibilities in industry should achieve and maintain the standards necessary for membership of the relevant professional body.

The effective regulation of industry requires:

(i) Sufficient staff in industry with the required level of competence to manage environmental risks and compliance.
(ii) Regulators with the knowledge and expertise to enforce and advise, and the same professional standards as required from business.
(iii) Use of other experts, auditing and monitoring to provide evidence and information on company performance.
(iv) Development and adoption of professional standards for environmental management.

In order to work better together, industry and regulators also need to collaborate with:

(i) Government and its agencies to establish and maintain the "policy" framework and keep it as simple as possible.
(ii) Business and sector representative bodies to recognise and address environmental issues, identify standards and best practice, and secure compliance.
(iii) Communities and business stakeholders (*e.g.* insurers, bankers, investors) to improve corporate environmental (and social) responsibility.

Regulators and regulated organisations are now developing approaches to enable operators with good compliance records, management systems and competent staff to earn autonomy and receive more arms-length regulation. The Environment Agency has tested the approach with a cross-section of operators in England.[48] It requires commitment from operators and regulators.

The regulated organisation will:

(i) Manage their environmental risks effectively and comply with the law.
(ii) Commit at company board level to effective environmental management.
(iii) Have staff who are competent to manage environmental risks and compliance.
(iv) Provide evidence and information on company performance.
(v) Demonstrate site level performance and attitudes.

The regulators will:

(i) Be expert advisors and critical friends.
(ii) Explain and support the benefits of effective environmental management to businesses and their stakeholders.
(iii) Work with operators to maintain compliance.

Where the trials are successful, the resources required to regulate the site will be reduced, with a corresponding reduction in charges.

It is important that the national regulator takes lead responsibility for permitting as this is the single point in the regulatory cycle where standards are defined and where a step change in operation can be implemented. This principle applies to large or complex as well as simple activities. The national regulator is able to take an overview of regulatory activities and take account of the effectiveness of different forms of regulatory intervention, state of the environment and value for money. Its expertise in understanding the environmental impact of pollutants and working with others to understand possible health impacts places it in a unique position to set and monitor emission standards for sites where emissions may be significant.

Local authority regulators have a wealth of experience in dealing with nuisance sites resulting from noise and odour, balancing the needs of communities with budgets and political interest. Usually, they are the planning authority for the facility in question.

The relative allocations of responsibilities between the national and local regulators are described in Figure 22.7 and should be kept under review to ensure the delivery of value for money.

The differences within and between different industry sectors which must then be considered are summarised in Figure 22.8. Different expertise is needed by both industry and regulators to address the specific environmental risks and issues of each sector. Regulators and operators in the sectors with the most public interest, concerns and complaints need to be particularly expert in addressing these issues.

22.3.8   Consultation and Public Engagement

The Regulations require the regulators to consult with the public on environmental permit applications, but do not prescribe the methods to be used. This allows the regulators to develop proportionate and flexible approaches to public participation. They must take into consideration any representations made by consultees during the allowed time periods. The presumption is that all details of permit applications, consultations and all responses are placed on a locally accessible public register. The public can request and view the information about environmental permitting on the public registers and under the *Freedom of Information Act (FOIA)* and the *Environmental*

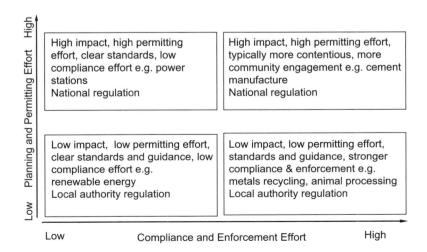

Figure 22.7   Different regulatory approaches to different industries.

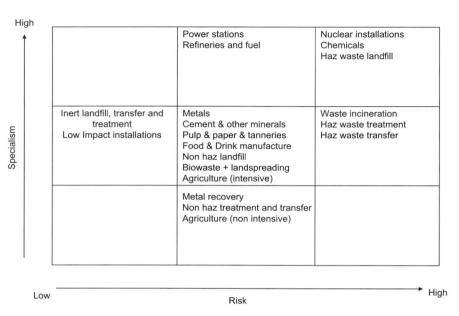

Figure 22.8   Industries, risks and expertise.

Information Regulations (EIR).[49] These latter tools are increasingly being used to gain information on controversial and high profile sites and activities.

The public register does not contain information that is excluded in the interests of national security or because it is commercially or industrially confidential. Such information is also excluded from any public consultation.

22.3.9   Enforcement

The Regulations place a duty on regulators to undertake appropriate periodic inspections of regulated facilities. While there are moves to set minimum inspection frequencies at a European level this has not been formally adopted in the UK in preference to a more risk-based approach.[50] Inspections can include reviewing information from the operator as well as carrying out independent monitoring, site inspections, in-depth audits and other compliance-related work. Permit conditions may require operators to monitor their activity, submit the results of monitoring and report both compliance and non-compliance with permit conditions.

The regulator can take the following actions:

(i) **Enforcement Notice** – where an operator has contravened, is contravening, or is likely to contravene any permit conditions. The enforcement notice specifies the steps required to remedy the problem and the timescale in which they must be taken. The enforcement notice may include steps to remedy the effects of any harm and to bring a regulated facility back into compliance.

(ii) **Suspension Notice** – where there is a risk of serious pollution. This applies whether or not the operator has breached a permit condition. The suspension notice will describe the nature of the risk of pollution, the actions necessary to remove that risk and the deadline for taking actions. When the regulator serves a suspension notice, the permit ceases to authorise the operation of the specified activities. When the operator has taken the remedial steps required by the notice, the regulator must withdraw the notice.

(iii) **Prosecution** – if an operator has committed a criminal offence, regulators should consider a prosecution. Conviction in a magistrates' court carries a fine of up to £50 000 and up to twelve months imprisonment for the most serious offences under the Regulations. Conviction in the Crown court for those offences may lead to an unlimited fine and imprisonment for up to five years. The Regulations contain an emergency defence safeguard for operators, where the operator shows that the acts were undertaken in an emergency to avoid danger to human health, all reasonable steps were taken to minimise pollution and the regulator was informed promptly.

(iv) **Revocation Notice** – the regulator can revoke a permit, in whole or in part. Revocation may be appropriate where exhaustive use of other enforcement tools has failed to protect the environment properly, where the permit holder is no longer the operator or where the operator is considered not to be competent. Unlike other types of notice, if a revocation notice is appealed, the revocation does not take effect until the appeal is determined or withdrawn. The permit ceases to authorise the operation of a regulated facility, facilities or part of a facility to the extent specified in the revocation notice. Regulators may enforce the restoration requirements by issuing enforcement notices and if necessary they can use their powers to remedy harm and recover costs.

(v) **Remediation** – if a regulated facility gives rise to a risk of serious pollution, a regulator may arrange for the risk to be removed. If an operator commits an offence that causes pollution, the regulator may arrange for steps to be taken to remedy pollution at the operator's expense. Where an incident such as a spillage occurs, the regulator should be notified and the operator should take all practical steps to address any contamination at the time of the incident. A record of the steps taken to return the site to a satisfactory state should be made available to the regulator.

22.3.10 Civil Sanctions

There has been a growing need for a more proportionate approach to environmental compliance and enforcement which avoids the need for criminal sanctions where they are not appropriate. The *Regulatory Enforcement and Sanctions Act 2008* sets a higher burden of proof, namely that the regulator must be satisfied beyond reasonable doubt that an offence has been committed, before it can issue a sanction.[51] The regulator must have reasonable grounds to believe that an offence has been committed before it can accept an offer to rectify the situation. In such cases it will still want to be in a position to establish that an offence has been committed in case further enforcement action is necessary where an offender fails to comply with the undertaking.

Civil Sanctions are additions to the regulators' enforcement toolkit and should not replace any of the traditional sanctions such as notices and prosecutions. The key features of Civil Sanctions are:

(i) Allow the regulator to apply sanctions quickly in a more proportionate way.
(ii) Focus on restoring damaged environments, stopping illegal activity quickly and removing financial gain.
(iii) Embed the "polluter pays" principle.
(iv) Benefit people and the environment.
(v) Provide flexible enforcement options.
(vi) Fill gaps in the regulator's powers.

Civil Sanctions require the same degree of engagement and work by the regulator as the traditional sanctions but have been welcomed by many businesses and communities. Businesses have seen them as a lighter touch, a more collaborative means of working, and communities have benefited from the quicker restoration and financial contributions from business. Unfortunately, some have seen them as additional regulation and particularly resisted their application to smaller businesses to which they give the greatest benefit. The enforcement undertakings have been well used but the procedures are onerous and the system is under review.

22.4 FUTURE REGULATION OF INDUSTRIAL POLLUTION

The regulation of industrial pollution continues to be important for the sake of health, the environment and the future of the planet. It has always been beyond the simple delivery of regulations:

  (i) **Regulation**
    a. Protection of health and the environment.
    b. Achievement of a level playing field.
    c. Use of the most effective and efficient tools – permitting, pricing, persuasive advice, passive guidance, relying on third parties.
  (ii) **Wider Objectives**
    a. Tackle environmental scarcity (*e.g.* water).
    b. Minimise the impact of climate change (mitigation and adaptation).
    c. Promote energy and resource efficiency.
    d. Minimise waste.

Over the past 10 years there have been increasing challenges to the Government and regulators on the need for and costs of regulation. In practice the direct charges for regulation, other than for the smallest businesses, have not been significant when compared with other costs such as energy. The costs associated with regulation, however, can be significant, such as data collection, the time and resources associated with preparing for an inspection, the inspection itself and any follow-up activity. As a result, regulators have increasingly challenged themselves in terms of the burden and benefit of their regulatory activities. Similarly, governments have become more challenging towards the work of the regulators: "regulators do not do policy or principles", they must "stick to the knitting".[52]

Soon after the election in 2010 the Prime Minister announced that he wanted the new coalition administration to be "the greenest government ever".[53] This was expected to result in the delivery of commitments on climate change and environmental improvement. The Government also wanted to move away from changing peoples' behaviour through rules and regulations by finding intelligent ways to encourage, support and enable people to make better choices for themselves. The approach was informed by the "nudge" concept set out by Richard Thaler and Cass Sunstein in *Nudge: Improving Decisions about Health, Wealth, and Happiness.*[54] The Behavioural Insights Team or "Nudge Unit", was set up in July 2010 in the Cabinet Office to make services easier to use, encourage and support people to make better choices and save the taxpayer money. In 2011 it applied its thinking to energy use.[55] The same approach can be applied to other areas of environmental regulation. For example, by providing more and better waste collection arrangements and treatment facilities, it is easier to recover and recycle. Most industries want to do the right thing for the environment as well as prosper. Reducing, recovering and recycling many waste streams make financial sense which is as good an incentive as regulation.

In April 2011 the Prime Minister launched "The Red Tape Challenge" and invited people to say which regulations should stay, go or be changed. Ministers reviewed the regulations with the presumption that burdensome regulations would go. Government Departments and regulators had to justify why they are needed. As a result there have been major challenges to some regulations on industry especially associated with waste activities and for smaller scale installations.[65]

In his 2011 Autumn Statement, the Chancellor of the Exchequer signalled a significant shift in approach to the regulation of industry:

If we burden [British businesses] with endless social and environmental goals – however worthy in their own right – then not only will we not achieve those goals, but the businesses will fail, jobs will be lost, and our country will be poorer.

With the economic downturn and the focus on growth, the Chancellor was signalling that deregulation was firmly in the ascendancy.

In 2011 the EEF reported from a survey of their members that: [56]

The cost of environmental policies is escalating, pushed up by unnecessarily burdensome regulations and an inefficient raft of climate change policies.

There is, however, continuing recognition of the contribution to the economy of environmental goods and services and commitment to environmental protection from the CBI and other organisations representing industry. Dr Neil Bentley, CBI Deputy Director General, said in 2011:[57]

Environmental regulation doesn't have to be a burden for business. Framed correctly, environmental goals can help our economic goals – help start new companies and generate new jobs and enrich all of us.

The commitment to and recognition of the value of regulation is exemplified by the approach of the UK chemical industry which has worked with the environmental regulators to ensure a robust regulatory regime. The industry is able to point to its own Responsible Care programme and effective environmental regulation in reassuring the public and governments about its environmental commitments and performance. Steve Elliott, Chief Executive of the Chemical Industries Association, said in response to the publication by SEPA of the environmental performance of industry in Scotland: [58]

We will continue to work closely with SEPA in the development of environmental policy across Scotland, to look at how we can maintain and enhance our environmental performance which at a time of economic uncertainty still remains very important to our companies and the work we do.

The Environmental Audit Committee in their twelfth report in 2012 stated:[59]

Regulations have an important role in protecting the environment. The impending consolidation of environmental regulations must not be a smokescreen for lowering the protections they afford in a short-term pursuit of growth. The Government should consult and engage fully not just with business but also with environmental stakeholders.

In March 2012 the Prime Minister commissioned a report from Lord Heseltine on how to create wealth in the UK more effectively. Lord Heseltine looked in particular at the burden and benefits of regulation.[60] He challenged:

Ask people how the Government could spur growth in the economy and most will have regulation somewhere towards the very top of the list. That is to say – regulate less. Many firms cite over regulation as acting like a tax and ask for greater freedom from central control. Yet we need look no further than the economic catastrophe of 2008 to see the problems of under regulation, where excessive risk taking by many banks caused a meltdown so massive that just about every major economy in the world has still to recover.

He also concluded:

the Government should impose an obligation on regulators to take proper account of the economic consequences of their actions.

The benefits of regulation have to be justified by government departments and regulators more than before. Regulators are challenged as to the economic impact of the regulations and their actions. In 2012 changes were proposed in Scotland for the integration of permitting across regulatory regimes and for civil sanctions.[61] Changes were also proposed to the functions of SEPA. In addition to protecting and improving the environment, including the sustainable management of natural resources, it was proposed that SEPA should contribute to improving the health and well-being of the people of Scotland and the achievement of sustainable economic growth.[62] The *Environment Act 1995* already requires SEPA to contribute to achieving "sustainable development" and it must have regard of "social and economic needs". The balance between economic, social and environmental sustainable development is shifting towards economic considerations.

In Northern Ireland the Department of the Environment is reducing the regulatory burden on business:[63]

   (i) Integrated environmental permitting.
  (ii) Simplified and harmonised inspections.
 (iii) Administrative sanctions and penalties to give a more proportionate approach to regulatory breaches.
 (iv) Criminal penalties and sanctions to give a more flexible approach in dealing with frequent offenders and serious offences.

In 2012 the Department for Environment, Food and Rural Affairs undertook a "Smarter Environmental Regulation Review" to investigate opportunities to reduce regulatory burdens while increasing environmental benefits. The first report in 2013 provided evidence to reform environmental guidance and information. It is already resulting in a consolidation of environmental regulation guidance within the government department.[66]

Environmental regulators are now far more careful to meet both industry and government expectations. They are less likely to speak out on environmental issues or environmental performance. Environmental professional bodies and environment interest groups will therefore need to maintain and improve their monitoring of the environment and the performance of those who impact on the environment. Environmental groups will need to challenge governments to ensure the application of sound science and the identification and adoption of best practice. The market on its own will not normally deliver effective environmental protection, what so often prevails is the lowest cost option, not the most sustainable. Industry will need to address the emerging environmental issues with competence and confidence. Increasingly the responsibility for delivering the environmental regulation of industry will fall to industry itself. Will industry rise to the challenge and deliver long-term sustainable solutions?

Based on recent experience of the regulation of industrial pollution, I predict that in the next five years:

   (i) There will be greater emphasis on economic growth balanced with environmental and social sustainability.
  (ii) Regulation will focus on clear environmental benefits and outcomes.
 (iii) Pressure to de-regulate/reduce regulation will continue and there will be less regulation of small scale or low impact activities.

37. *Effective Inspection and Enforcement: Implementing the Hampton Vision in the Environment Agency*, Better Regulation Executive, Department for Business, Enterprise and Regulatory Reform, London, 2008.

38. *Environmental Permitting Guidance*, Core Guidance, for the *Environmental Permitting (England and Wales) Regulations 2010*, revised March 2012, version 4.0, Defra, London, 2012.

39. *Guidance for Developments requiring Planning Permission and Environmental Permits*, Environment Agency, Bristol, 2012.

40. *Environmental Permitting Regulations, Operational Risk Appraisal Scheme*, (Opra for EPR), version 3.7, Environment Agency, Bristol, 2012.

41. *How to Comply with your Environmental Permit*, version 5, Environment Agency, Bristol, 2012.

42. R. Salmons, *REMAS Analysis of Initial Sample Data for the United Kingdom*, Environment Group, Policy Studies Institute, London, 2004. REMAS–IMPEL Requirements of Remas Criteria, European Union Network for the Implementation and Enforcement of Environmental Law, 2004.

43. New opportunities to improve environmental compliance outcomes using certified EMSs, Sniffer, Edinburgh, April 2013.

44. *Measuring the Effectiveness of Environmental Management Systems, Phase 2: Final Report Including Data & Statistical Analysis*, Northern Ireland Environment Agency, 2009.

45. *Responsible Care Global Charter*, International Council of Chemical Associations, Brussels, 2006.

46. *Regulation (EC) No 1221/2009* on the Voluntary Participation by Organisation in a Community Eco-Management and Audit Scheme (EMAS). European Commission.

47. *Directive 2003/87/EC establishing a Scheme for Greenhouse Gas Emission Allowance Trading within the Community*. European Commission. The CRC Energy Efficiency Scheme Order 2013. UK Parliament.

48. *Sustainable Business Report 2011*, Environment Agency, Bristol, 2012, p. 7.

49. Freedom of Information Act (FOIA) 2000. UK Parliament. Environmental Information Regulations (EIR) 2004. UK Parliament.

50. *Doing The Right Things III, Implementation of the Step-by-step Guidance Book on Planning of Environmental Inspections*, European Union Network for the Implementation and Enforcement of Environmental Law (IMPEL), Brussels, final report, 2008.

51. *The Regulatory Enforcement and Sanctions Act 2008 (RES Act)*. European Commission.

52. *Private comm.* with Defra officials.

53. Department of Energy and Climate Change, *press release* 10/059, 14 May 2010.

54. R. H. Thaler and C. R. Sunstein, *Nudge: Improving Decisions about Health, Wealth, and Happiness*, Penguin Books, London, 2009.

55. *Behaviour Change and Energy Use*, Cabinet Office Behavioural Insights Team, London, 2011.

56. *Manufacturing Green and Growth: Attitudes, Ambitions and Challenges*, An EEF Executive Survey, EEF, London, 2011, p. 3.

57. *CBI – Green Alliance Conference*, 12 December, 2011.

58. Chemical Industries Association, *Chemical and Pharmaceutical Companies at the top of Scotland's Environmental Performance*, press release, 4 September, 2012.

59. Environmental Audit Committee, *Twelfth Report: A Green Economy*, 2012.

60. Lord Heseltine of Thenford, *No Stone Unturned: In Pursuit of Growth*, Department for Business, Innovation and Skills, London, 2012.

61. *Better Regulation: Consultation on Proposals for a Better Regulation Bill*, The Scottish Government, Edinburgh, 2012.

62. *Consultation on Proposals for Future Funding Arrangements for the Scottish Environment Protection Agency*, The Scottish Government, Edinburgh, 2012.

63. *Northern Ireland Better Regulation Strategy*: *Annual Report 2011–2012*, Economic Policy Advisory Unit, Department of Enterprise, Trade and Investment, Belfast, 2012, p. 15.
64. The Environment Act 1995, Part 1. UK parliament.
65. Department for Environment, Food and Rural Affairs, Red Tape Challenge - Environmental Theme Implementation Plan, September 2012.
66. Department for Environment, Food and Rural Affairs, Smarter Environmental Regulation Review, Phase 1 report: guidance and information obligations, 16 May 2013.

Subject Index